Advanced Lectures in Mathematics (ALM)

www.intlpress.com/alm

AF817239

ALM 1: Superstring Theory

ALM 2: Asymptotic Theory in Probability and Statistics with Applications

ALM 3: Computational Conformal Geometry

ALM 4: Variational Principles for Discrete Surfaces

ALM 6: Geometry, Analysis and Topology of Discrete Groups

ALM 7: Handbook of Geometric Analysis, No. 1

ALM 8: Recent Developments in Algebra and Related Areas

ALM 9: Automorphic Forms and the Langlands Program

ALM 10: Trends in Partial Differential Equations

ALM 11: Recent Advances in Geometric Analysis

ALM 12: Cohomology of Groups and Algebraic K-theory

ALM 13: Handbook of Geometric Analysis, No. 2

ALM 14: Handbook of Geometric Analysis, No. 3

ALM 15: An Introduction to Groups and Lattices: Finite Groups and Positive Definite Rational Lattices

ALM 16: Transformation Groups and Moduli Spaces of Curves

ALM 17: Geometry and Analysis, No. 1

ALM 18: Geometry and Analysis, No. 2

ALM 19: Arithmetic Geometry and Automorphic Forms

ALM 20: Surveys in Geometric Analysis and Relativity

ALM 21: Advances in Geometric Analysis

ALM 22: Differential Geometry: Under the Influence of S.-S. Chern

ALM 23: Recent Developments in Geometry and Analysis

ALM 24: Handbook of Moduli, Volume I

ALM 25: Handbook of Moduli, Volume II

ALM 26: Handbook of Moduli, Volume III

ALM 27: Number Theory and Related Areas

ALM 28: Selected Expository Works of Shing-Tung Yau with Commentary, Volume I

ALM 29: Selected Expository Works of Shing-Tung Yau with Commentary, Volume II

ALM 30: Automorphic Forms and L-functions

ALM 31: Handbook of Group Actions, Volume I

ALM 32: Handbook of Group Actions, Volume II

ALM 33: Introduction to Modern Mathematics

ALM 34: Some Topics in Harmonic Analysis and Applications

ALM 35: The Legacy of Bernhard Riemann After One Hundred and Fifty Years, Volumes I and II

ALM 36: Proceedings of the Sixth International Congress of Chinese Mathematicians, Volume I

ALM 37: Proceedings of the Sixth International Congress of Chinese Mathematicians, Volume II

Advanced Lectures in Mathematics
Volume 36

Proceedings of the Sixth International Congress of Chinese Mathematicians

Volume I

Companion to the volume
Proceedings of the Sixth International Congress of Chinese Mathematicians, *Volume II*

edited by

Chang-Shou Lin · Lo Yang
Shing-Tung Yau · Jing Yu

International Press
www.intlpress.com

高等教育出版社
HIGHER EDUCATION PRESS

Advanced Lectures in Mathematics, Volume 36
Proceedings of the Sixth International Congress of Chinese Mathematicians, Volume I

Companion to the volume
Proceedings of the Sixth International Congress of Chinese Mathematicians, Volume II

Editors:
Chang-Shou Lin (Department of Mathematics, National Taiwan University)
Lo Yang (Institute of Mathematics, Chinese Academy of Sciences)
Shing-Tung Yau (Department of Mathematics, Harvard University)
Jing Yu (Department of Mathematics, National Taiwan University)

ISBN: 978-1-57146-348-7

This volume is part of a two-volume set: *Proceedings of the Sixth International Congress of Chinese Mathematicians*. ISBN 978-1-57146-3500.

Printed in the United States of America.

Contents

Plenary Lectures

Invited Lectures

Number Theory, Automorphic Forms and Arithmetic Geometry

Discrete Mathematics, Combinatorics

Algebraic Geometry, Complex Geometry

President's Speech to ICCM 2013

Shing-Tung Yau

Today is a festival celebrating the achievements of Chinese mathematicians. Fifteen years have passed by between the first congress in Beijing up until today in Taipei. Professor Lo Yang has a thorough understanding of the many hardships throughout these fifteen years, and yet, in spite of the difficulties, both academic and blood ties between scholars from Mainland, Taiwan, Hong Kong of China, and overseas have fueled enthusiastic participation.

Through the five congresses held so far, scholars have not only received academic benefits, but they have also made many new friends. Chinese mathematicians have become a strong force in the worldwide mathematical community. Based on both the list of winners, as well as the content of the speeches, we can proudly say that our congress has truly reached the world-wide first-class level. There are, of course, still some less studied mathematical disciplines which require much more work, but our young scholars have certainly made outstanding achievements in them.

For example, in probability theory, Horng-Tzer Yau solved the Dyson conjecture. In number theory, we live in an era in which we no longer hesitate. We are no longer staying in the era of Loo-keng Hua, Yuan Wang, Chengdong Pan, and Jingrun Chen. With the help of numerous overseas scholars such as John Coates, Shouwu Zhang, Ching-Li Chai, Jing Yu, Ju-Kang Yu, as well as many other Chinese and non-Chinese scholars, our young scholars have obtained important results over and over again in this important and classical field. This is definitely worthy of encouragement. Indeed, recently Yitang Zhang and Ye Tian have made enormous breakthroughs in classical problems in number theory. Their achievement dated back to the instruction and guidance of their tutors in Beijing: Professor Chengbiao Pan and Professor Chunlai Zhao.

Important achievements by Chinese mathematicians can also be found in algebraic geometry, algebra, and group representation theory, in which Chinese scholars such as Jun Li, Bong Lian, Chin-Lung Wang, Zhiwei Yun, Xuhua He, and Nanhua Xi have made many breakthroughs. They broaden the scope of mathematics disciplines in China beyond traditional strong fields such as geometry and analysis. The helps from many visitors abroad are fundamental to this development. This includes Viehweg, Esnault, Looijenga, and many others.

In geometric analysis, Yum-Tong Siu, Peter Li, Chang-Shou Lin, Kefeng Liu, Xujia Wang, Mu-Tao Wang, Conan Leung, Chiu-Chu Liu, Lizhen Ji, Jixiang Fu, Xiping Zhu, Juncheng Wei and numerous other scholars still perform first-class research. This can clearly be seen from the award winners this year, as well as their speeches. We are grateful to the constant visits of Richard Schoen, Richard Hamilton, Neil Trudinger and others.

In applied mathematics, Zhouping Xin, Sijue Wu, Chongqing Cheng, Chi-Wang Shu, Yizhao Hou, Raymond Chan, and Wen-Wei Lin have contributed excellent work in fluid theory, dynamical systems, computational fluid theory, and numerical linear algebra. In addition to these contributions, Xianfeng Gu has made outstanding advancements in image processing, as has Stephen Yau and Lei Guo on control theory. Chinese applied mathematicians are no longer just focused on classical finite element methods. The visits of Osher, Engquist, Majda and many others have been very useful for our developments.

Under the leadership of Tze-Leung Lai, Wing Hung Wong, Chien-Fu Jeff Wu, Ziliang Ying, Jianqing Fan, and Jun Liu, Chinese statisticians have also attained an important position in the world.

We would like to show our appreciation of the generous support of numerous government leaders and the Morningside Group over the years. We are especially grateful for the strong support of the Chan brothers.

In Mainland, eighteen years ago, Ronnie Chan, Yongxiang Lu (the former president of Chinese Academy of Sciences), Lo Yang, and I founded the Morningside Center of Mathematics at the Chinese Academy of Sciences in Beijing. There we trained a large class of Chinese scholars, coming from all corners of the country. In fact, the talents leading in the new directions in number theory, which I just mentioned, mostly came to the Morningside Center to learn and to perform research. They were not only from Chinese Academy of Sciences, but also from Peking University, Tsinghua University, Sun Yat-sen University, the University of Science and Technology of China, Fudan University, Nankai University, Zhejiang University, as well as many other universities. I am very glad that most of them have gone on to become very good mathematicians, and many of them will deliver speeches in the following five days. After the formation of Morningside Center of Mathematics, several centers of mathematics were formed within the mainland, this includes the center in Zhejiang University, and the Tsinghua Center of Mathematics, where many outstanding mathematicians from Europe and American are joining.

In Taiwan, Professor Zhaoxuan Liu, the former dean of "National Science Council", founded "the National Center for Theoretical Sciences", at my suggestion. This center is located at the Hsinchu Tsing Hua University, and it is split into two divisions: physics and mathematics. The first director was Chang-Shou Lin, who was very successful in fulfilling his duties. Following Chang-Shou Lin, Jing Yu and Wen-Ching Li became the directors of NCTS, and they both did a tremendous job training many talented young scholars in Taiwan, as well as increasing the rate of communication with overseas scholars. Later on, at Taiwan University, Chiao Tung University, and other mathematics research institutes were founded, which also made great contributions to bettering the academic atmosphere.

At the Chinese University of Hong Kong, at my suggestion and with President Charles Kuen Kaos help, the Institute of Mathematics was founded in 1992. Later, it was financed by my fundraising effors through my friends: Robert Kuok Hock Nien, Ka-Shing Li, William Mong Man Wai, Thomas T.T. Chen, and the Hysan Foundation. The Institute hired a group of outstanding mathematicians, and also trained a large number of talented young people. Some of the outstanding students are leading researchers including some Ph.D students who became Benjamin Peirce Fellows at Harvard University.

The first congress was held in Beijing by the Morningside Center of Mathematics, where the leaders of the academy of science and a group of distinguished government officer came to support the activities. Many prominent mathematicians from abroad representing different countries gave speeches.

The second congress was held in Taipei when, at that moment, there were not many Mainland scholars coming to Taiwan due to the problem of visas. In spite of this, all scholars were very enthusiastic about participating in the congress. We celebrated the ninetieth birthday of Professor Shiing-Shen Chern, who was unfortunately not in very good health, and so his daughter Pu Chern and his son-in-law Paul Ching-Wu Chu attended the congress on his behalf.

The third congress was held in Hong Kong by the Chinese University of Hong Kong. President Yongxiang Lu of the academy of science came to give the opening speech. State councilor Zhi Li Chen sent a letter to congratulate the success of the congress. The congress also decided to memorialize the honorable chairman of the congress, Professor Chern, who had just passed away. Professor Chern has been planning to attend the congress, and had donated one-hundred thousand yuan to the congress.

The fourth congress was held in Hangzhou, and was endorsed by the former party secretaries of the Zhejiang Province, Jinping Xi and Hongzhu Zhao. Yuanchao Li, minister of the Organization Department of the Central Committee and NPC Vice Chairman Yongxiang Lu both sent congratulatory letters. Governor Zushan Lv and other leaders attended the opening ceremony, as well as the welcoming dinner.

The fifth congress was held in Beijing, and was co-organized by Tsinghua University and the Morningside Center of the Chinese Academy of Sciences. This congress was dedicated to the hundredth anniversaries of the births of Professor Shiing-Shen Chern and of Professor Loo-keng Hua. It was held in the Great Hall of the People. There were more than 1600 participants! Its opening ceremony was truly a grand occasion.

Today the sixth congress is held in Taipei, at a time when mathematical achievements are more prosperous, which is incredibly important. We hope that, when we hold the next congress at the Chinese Academy of Sciences, our native speakers will reach the level of the Fields Medal.

We are pleased to say that fifteen years ago Chang-Shou Lin was the first Chinese mathematician residing in Taiwan to receive the Morningside Gold Award in Beijing. He did most of his work in Taiwan. After waiting fifteen years, we are pleased to see that Ye Tian and Xuhua He, who reside in Beijing and Hong Kong, respectively, are receiving the Morningside Gold Award in this congress.

It is our honor to see the participation of President Si-Chen Lee, former president of Taiwan University. He received the Shiing-Shen Chern award, due to his contributions to the mathematical society.

I am sure that many well-to-do Chinese will continue to support this congress, whether through academic works or financially. Chinese mathematics will continue to flourish. I am grateful to the support from all of you.

Department of Mathematics, Harvard University
One Oxford Street, Cambridge MA 02138, USA
E-mail address: yau@math.harvard.edu

Report of ICCM 2013 at Taipei

Jing Yu

The Sixth International Congress of Chinese Mathematicians (ICCM) was held from July 14 to July 19, 2013, in Taipei, China. The opening ceremony took place at the Grand Hotel of Taipei on July 14 Sunday, with about one thousand participants from all over the world. This monumental activity of the whole Chinese mathematical community started with welcome greetings by the ICCM honorary chair Professor Shing-Tung Yau from Harvard University, President Pan-Chyr Yang of Taiwan University, Mr. Ronnie C. Chan, co-founder of the Morningside Foundation and Chairman of Hang Lung Properties, and Professor Jing Yu from the Taida Institute of Mathematical Sciences who also served as chairman of the local organizing committee. This was the second time that ICCM held in Taiwan, the first time was 12 years ago. Despite an unexpected Typhoon invasion to the Taiwan Island on July 11/12, we are very happy to see that almost all the participants this year who scheduled to join managed to do so eventually.

As in each ICCM, people are able to witness the exciting progresses of mathematics made by Chinese mathematicians just recently.

Following now the tradition of ICCM, the big event in the opening ceremony was the Awards presentation, in particular the Morningside Medals. This year, beside the Morningside Gold and Silver Medal, a special Morningside Achievement Award was awarded to Yitang Zhang from the New Hampshire University, for his monumental breakthrough in Number Theory, proving the abundance of bounded gaps between consecutive primes. This brilliant success throws new lights on the entire development of analytic number theory. Another medal first awarded this year is the Morningside Mentor Award in Mathematics, given to Chengbiao Pan, a retired professor from China Agricultural University who held adjoint position at Peking University for many years.

There are two 2013 Morningside Gold medals. One was awarded to Xuhua He of HKUST for his fundamental contributions to arithmetic geometry, algebraic groups, and representation theory. The other Gold Medal awarded to Ye Tian, AMSS of the Chinese Academy of Sciences, for his deep and original work on the congruence number problem. The 2013 Morningside medal of applied mathematics was awarded to Xianfeng David Gu, for his contributions to computational geometry, imaging, and graphics.

There are four Morningside Silver Medals awarded this year. The recipients are Chieh-Yu Chang of Tsing Hua University, Xiaoqing Li of State University of New York at Buffalo, Hao Xu of Harvard University, and Tai-Peng Tsai of the University of British Columbia.

Professor Jean-Pierre Serre at the College de France, Fields Medalist, also recipient of both the Abel Prize, and the Wolf prize, was awarded the 2013 ICCM International Cooperation Award. Professor Serre has been visiting Taiwan biannually since 2009, and greatly impacted and encouraged many young mathematicians.

The two 2013 Chern Prize recipients are former President Si-Chen Li of the Taiwan University, for public service in support of mathematics, and Professor Bong Lian of Brandeis University, for fundamental contributions to mathematical physics.

The July 14 afternoon activity at the Grand hotel started with a Symposium, Unlocking our future: how government policies can foster fundamental science development. This symposium has Dr. Gerald L. Chan, co-founder of Morningside, as moderator, with panelists: Chi-Huey Wong, the President of "the Academia Sinica", Yan-Hwa Wu, President of Chiao Tung University, and Way Kuo, President, President of City University of Hong Kong. The second part of this Grand Hotel Symposium consist of two fascinating talks aiming at general audience, "the art of bridge building" delivered by Professor C.L. Liu of the Tsing Hua University, and "juggling mathematics and magic" delivered by Professor Ronald Graham of UC San Diego.

Honorary guest Jin-Pyng Wang, President of "the Legislative Yuan", joined the ICCM opening activity at the Grand Hotel in the evening of the opening day. Besides greeting the morning ICCM awardee, he also took part in the 2013 New World Mathematics Awards presentation ceremony. The New World Awards presentation is another tradition of ICCM, sponsored by the New World Development Company Limited, awarding prizes to the best recent Ph.D. Thesis, Master Thesis, and Bachelor Thesis in mathematics. July 14 evening, this award was presented to 27 students, among them are 7 gold Prizes for Doctor Thesis.

Venue of 2013 ICCM after the first day is sited on the Campus of Taiwan University, in a set of comfortable well-equipped lecture rooms that is situated in the beautiful drunken moon lakeside. Every morning of July 15–19, ICCM began with a distinguished one hour lecture by an international honorary guest: Jean-Pierre Serre on historical remarks of cohomology, John H. Coates on the Conjecture of Birch and Sinnerton-Dyer, Dorian Goldfeld on the relative trace formula for unipotent groups, Björn Engquist on multiscale simulations of dynamical systems with oscillatory solutions, and Stanley Osher on what sparsity and l1 optimization can do for you. These talks are followed by the one hour plenary lectures, there are six such lectures every day, covering a broad landscape of both pure and applied mathematics, and addressing to the most important recent progresses. The abstracts of all these inspiring lectures are appended with this report. We are editing a more complete account of these lectures into Proceedings of the Six ICCM to be published in near future.

There were 152 invited 45-minute talks, dividing into 10 sections:

1. Number Theory, Automorphic Forms, and Arithmetic Geometry, 25 talks.
2. Algebra, History of Mathematics, 9 talks.
3. Discrete Mathematics, Combinatorics, 8 talks.
4. Algebraic Geometry, Complex Geometry, 18 talks.
5. Geometric Analysis, Differential Geometry, Symplectic Geometry, 17 talks.
6. Topology, Geometry and Mathematical Physics, 17 talks.
7. Statistical Theory, Methods, Applications, Probability, Financial Mathematics, 11 talks.

8. Dynamical Systems, Control Theory, Optimization, Functional Analysis, Fractals, 18 talks.
9. Computational Mathematics, Imaging, Mathematical Biology, 17 talks.
10. Partial Differential Equations, 12 talks.

There were also 96 contributed talks, each given 20 minutes.

Among the participants of the Six ICCM, there are over 400 coming from mainland, it took lots efforts to process their travel documents to Taiwan. Although there are now direct flights from all major cities of mainland to Taiwan, the travel documents still have to be approved by authorities from both Mainland and Taiwan which is certainly time consuming. ICCM is by far the largest scale academic interaction across the strait. We are able to manage it this time without genuine obstacles.

The sponsoring organizations of the 2013 ICCM Conference include: Taiwan University (Taida Institute for Mathematical Sciences, Center of Advanced Studies for Theoretical Sciences, Department of Mathematics), "National Science Council" ("National Center for Theoretical Sciences", Department of International Cooperation), Ministry of Economic Affairs, the Yau Center of Chiao Tung University, and the Taipei City Government.

Abstracts of the Special Lecture, Morningside Lectures, and Plenary Lectures

Special Lecture

Historical remarks on cohomology (by *Jean-Pierre Serre*, Collège de France)

Abstract. The introduction of cohomological methods in algebra, complex variables, algebraic geometry and number theory is one of the characteristics of last century mathematics. The lecture will describe how some of these methods came into being, with special emphasis on étale cohomology and Weil conjectures.

Morningside Lectures

On the conjecture of Birch and Swinnerton-Dyer (by *John H. Coates*, University of Cambridge)

Abstract. The lecture will discuss the important progress on the conjecture of Birch and Swinnerton-Dyer made by Tian in his recent work on the classical congruent number problem, and its possible generalization to other elliptic curves.

The relative trace formula for unipotent subgroups (by *Dorian Goldfeld*, Columbia University)

Abstract. The relative trace formula of Jacquet has played a fundamental role in recent advances in number theory. The goal of my lecture will be to explain and introduce a very explicit version of the relative trace formula for the quotient $U\backslash G/U$ when $G = GL(n)$ and U is a unipotent subgroup. In the second part of the lecture, I will show how to obtain applications of the relative trace formula to some well known problems in analytic number theory.

Multiscale simulations of dynamical systems with oscillatory solutions (by *Björn Engquist*, The University of Texas at Austin)

Abstract. The heterogeneous multiscale method is a framework that can be used to design and analyze numerical methods for stiff ordinary differential equations with oscillatory solutions. Microscale simulations are applied in subsets of the computational domain in order to approximate the averaged dynamical system. We will discuss the difficulty of finding macroscale variables and present new techniques for avoiding this difficulty. We will compare with related methods and give numerical examples.

What sparsity and $l1$ optimization can do for you (by *Stanley Osher*, University of California, Los Angeles)

Abstract. Sparsity and compressive sensing have had a tremendous impact in science, technology, medicine, imaging, machine learning and now, in solving multiscale problems in applied partial differential equations, developing sparse bases for Elliptic eigenspaces and connections with viscosity solutions to Hamilton-Jacobi equations.

$l1$ and related optimization solvers are a key tool in this area. The special nature of this functional allows for very fast solvers: $l1$ actually forgives and forgets errors in Bregman iterative methods. I will describe simple, fast algorithms and new applications ranging from sparse dynamics for PDE, new regularization paths for logistic regression and support vector machine to optimal data collection and hyperspectral image processing. This is joint work with many people.

Plenary Lectures

Leaves and local Hecke symmetry (by *Ching-Li Chai*, Institute of Mathematics, "Academia Sinica"/University of Pennsylvania)

Abstract. Let M be a PEL type moduli space over an algebraic closure of the prime field with p elements; points of such a moduli space corresponds to isomorphism classes of abelian varieties with prescribed symmetries of a fixed type. A leaf in M is the locally closed subvariety obtained by fixing all p-adic invariants of these PEL type abelian varieties. A conjecture of Oort predicts that every prime-to-p Hecke orbit in M is Zariski dense in the leaf containing this Hecke orbit. In this talk we will present an overview of questions and results related to this conjecture, as well as recent progress in two aspects.

The first is a scheme-theoretic definition of the notion of leaves, more than a dozen years after its birth.

The new definition allows a natural construction of group-like structures on the formal completion at a point of a leaf, generalizing the Serre-Tate canonical coordinates for ordinary abelian varieties.

The second is a method for computing a weak asymptotic expansion of the action of the local stabilizer subgroup on the formal completion of a point of M.

This method will be illustrated in the case of the Lubin-Tate action, which occurs in moduli spaces of PEL type $U(n, 1)$.

Parameter estimation in systems and control (by *Han-Fu Chen*, Academy of Mathematics and Systems Science, Chinese Academy of Sciences)

Abstract. Many problems from systems and control such as system identification for linear systems and for a large class of nonlinear systems as well, adaptive regulation and iterative learning control (ILC) for stochastic systems, principal component analysis, adaptive filtering, blind identification in communication, and a certain kind of problems from networked systems and signal processing and many others can be reduced to problems of parameter estimation. The unknown parameters may be viewed as roots of functions, and parameter estimation can be transformed to a root-seeking problem based on observations on an unknown function called as the regression function. However, the resulting root-seeking problem can hardly be solved by the classical Robbins-Monro (RM) algorithm, because for its convergence a set of restrictive conditions are required.

The stochastic approximation (SA) algorithm with expanding truncations (SAAWET) is then introduced, and for its convergence a general convergence theorem (GCT) is presented. SAAWET is successfully applied to solving problems from systems and control. As a matter of fact, all problems listed above have been solved under reasonable and checkable conditions. Moreover, all solutions are provided in a recursive way, which is convenient for applications. This approach suggests solving the problems by two steps: first, transform problems to estimation of parameters or to root-seeking problems with appropriately selected regression function, then apply SAAWET with emphasis on proving satisfaction of the conditions required by GCT.

To demonstrate such an approach to solving problems from systems and control, the identification of the multivariate Hammerstein system is analyzed, and the recursive identification algorithms for both the linear and the nonlinear parts of the system are presented. All estimates given by the recursive algorithms are strongly consistent under reasonable conditions. Finally, the proposed algorithms are used to numerically identify a two-dimensional Hammerstein system. It is shown that the simulation results are consistent with the theoretical analysis.

Sixty years of compactifications of $\mathbb{C}^n$ (by *Baohua Fu*, Academy of Mathematics and Systems Science, Chinese Academy of Sciences)

Abstract. I shall give an overview of the study of compactifications of $\mathbb{C}^n$ as projective algebraic manifolds with $b_2 = 1$, a problem asked by Hirzebruch in 1954. Recent progress on equivariant compactifications will also be discussed. This is based on joint works with Jun-Muk Hwang.

Sliced inverse regression with interaction detection (by *Jun Liu*, Harvard University)

Abstract. Previously we have proposed a Bayesian partition model for detecting interactive variables in a classification setting with discrete covariates. This framework takes advantage of the structure of the naive Bayes classifier and introduces latent indicator variables for selecting variables and interactions. In our effort to extend the methods to continuous covariates, we found interesting connections with semi-parametric index models and the Sliced Inverse Regression method. In index models, the response is influenced by the covariates through an unknown function of several linear combinations of the predictors. Our finding of the Bayesian formulation of such models enabled us to propose a set of new models and methods that can effectively discover second-order effects and interactions among the covariates. A two-stage stepwise procedure based on likelihood ratio test is developed to select relevant predictors and a Bayesian model with dynamic slicing scheme is derived. The performance of the proposed procedure in comparison with some existing method is demonstrated through simulation studies. This is based on the joint work with Bo Jiang. .

A Langlands program for covering groups (by *Wee Teck Gan*, National University of Singapore)

Abstract. In this talk, I will survey some recent results (by many people) about a possible extension of the Langlands program to a class of covering groups studied by Beylinski and Deligne. These include the definition of the L-group of a covering group, the classification of local representations, the theory of the trace formula and endoscopy and the study of the automorphic discrete spectrum.

Ricci flow and classification of certain four-manifolds (by *Bing-Long Chen*, Sun Yat-sen University)

Abstract. It has been well known that Ricci flow has played a vital role in 3-dimensional topology. The Ricci flow with surgery in dimension 4 has already been developed in a seminal paper by R. Hamilton in 1997. It was expected that Ricci flow may also exert a great role in 4-dimensional geometry and topology. I hope this talk will serve as a step toward this subject. The main result of this talk, obtained from the study of the Ricci flow, is to show a complete classification of compact 4-manifolds whose certain generalized Yamabe invariants are positive. This classification is an analogue of a classification theorem of R. Schoen and S.T. Yau in 1970's on 3-manifolds with positive ordinary Yamabe invariants. This lecture is based on author's two joint papers with X.P. Zhu, one joint paper with S.H. Tang and X.P. Zhu.

Shock reflection-diffraction, von Neumann's conjectures (by *Gui-Qiang Chen*, University of Oxford)

Abstract. Shock waves are fundamental in nature. When a shock impinges an obstacle (steady or flying), shock reflection-diffraction phenomena occur. One of the most fundamental problems in mathematical fluid mechanics is shock reflection-diffraction by wedges. The complexity of reflection-diffraction configurations was first reported by Ernst Mach in 1878, who first observed two patterns of configurations. The problems remained dormant until the 1940s when John von Neumann, as well as other mathematical/experimental scientists, began extensive research into all aspects of shock reflection-diffraction phenomena. Moreover, the reflection-diffraction configurations are the core configurations in the structure of multidimensional Riemann solutions, which are building blocks and local structure of general entropy solutions and determine global attractors and asymptotic states of the solutions for multidimensional hyperbolic systems of conservation laws.

The shock reflection-diffraction problems involve several core mathematical difficulties we have to face in solving nonlinear partial differential equations in mechanics and geometry. These include nonlinear equations of mixed hyperbolic-elliptic type, nonlinear degenerate elliptic equations, nonlinear degenerate hyperbolic equations, free boundary problems for nonlinear degenerate equations, corner singularity/regularity especially when free boundaries meet degenerate curves, a priori estimate techniques, among others.

In this talk we will start with various shock reflection-diffraction phenomena and historic perspectives, their fundamental scientific issues, and their theoretical roles in the mathematical theory of multidimensional hyperbolic systems of conservation laws. Then we will describe how the global shock reflection-diffraction problems can be formulate as free boundary problems for nonlinear conservation laws of mixed hyperbolic-elliptic type. Finally we will discuss some recent developments in solving von Neumann's conjectures and establishing a mathematical theory of shock reflection-diffraction, including the existence, stability, and regularity of global regular configurations of shock reflection-diffraction by wedges. Further connections, trends, and open problems on the topics will also be addressed.

Bounded gaps between primes (by *Yitang Zhang*, The University of New Hampshire)

Abstract. Let p_n denote the n-th prime. It is proved that

$$\lim_{n\to\infty} \inf(p_{n+1} - p_n) < 7 \times 10^7.$$

The first step of the proof is to reduce the problem to evaluating and comparing certain arithmetic sums, following the recent work of Goldston, Pintz and Yildirim. Then the major problem we encounter is to bound the error terms efficiently. To this end we introduce a stronger version of the Bombieri-Vinogradov theorem that is motivated by the work of Bombieri, Friedlander and Iwaniec. Some important results in algebraic geometry are needed to complete the proof.

Dynamics of rational maps (by *Guizhen Cui*, Academy of Mathematics and Systems Science, Chinese Academy of Sciences)

Abstract. In 1980s, Thurston gave a topological characterization for post-critically finite rational maps. It is a fundamental result in complex dynamics with many important applications. In this talk, we will introduce our recent progress about Thurston's theory. We will present the generalization of Thurston theorem to geometrically finite rational maps, and some applications to the structure of Julia sets.

Period integral calculus (by *Bong H. Lian*, Brandeis University)

Abstract. We will discuss an interplay between special function theory and complex geometry. We begin with an example that goes back to Euler, Gauss and Legendre who studied elliptic integrals as solutions to differential equations that bears their names. We will then consider the modern incarnation and generalizations of their constructions—known as period integrals—but from a more geometrical viewpoint. We will describe recent progress in understanding these period integrals and some of their remarkable properties.

Monge-Ampère equation and optimal transportation (by *Xu-Jia Wang*, Australian National University)

Abstract. The optimal transportation is to find an optimal mapping of transferring a mass density to another one such that the total cost is minimized. This problem was first introduced by Monge in 1781. Monge's cost function is propositional to the distance the mass is transferred, namely $c(x, y) = |x-y|$, but more general costs are allowed. The optimal transportation has found a variety of applications and has been extensively studied since then. In 1940s Kantorovich introduced a dual functional, and by which, for a large class of cost functions, the optimal mapping can be determined through the potential function satisfying a Monge-Ampère type equation.

The Monge-Ampère equation also arises in many geometric problems and has been studied by Aleksandrov, Calabi, Pogorelov, Cheng-Yau, and Caffarelli, among many others. In this talk we will first introduce the optimal transportation and review the existence of optimal mappings. We then focus on the regularity of the optimal mappings. By studying the Monge-Ampère equation of the potential function, sharp conditions on the cost function have been found by the speaker and his collaborators. For Monge's cost function $|x-y|$, which does not satisfy the sharp conditions, we have also obtained the existence of optimal mappings, and established interesting regularity and singularity results of the mapping.

A Survey of local models of Shimura varieties (by *Xinwen Zhu*, Northwestern University)

Abstract. A local model is projective scheme over the ring of integers of a p-adic field which is supposed to étale locally model the integral structure of the Shimura variety of parahoric level structure. Classically, local models were constructed as moduli spaces of linear algebra structures and were studied mostly on a case by case basis. We will report the recent progress of this subject, including the group theoretical definition of local models, their algebro-geometric properties, and the applications.

Recent results on rigidity of CR morphisms between compact strongly pseudo convex CR manifolds (by *Stephen Yau*, Tsinghua University)

Abstract. Strongly pseudoconvex CR manifolds are boundaries of Stein varieties with isolated normal singularities. We prove that any non-constant CR morphism between two $(2n-1)$-dimensional strongly pseudoconvex CR manifolds lying in an n-dimensional Stein variety with isolated singularities are necessarily a CR biholomorphism. As a corollary, we prove that any non-constant self map of $(2n-1)$-dimensional strongly pseudoconvex CR manifold is a CR automorphism. We also prove that a finite étale covering map between two resolutions of isolated normal singularities must be an isomorphism. This is a joint work with Yu-Chao Tu and Huaiqing Zuo.

The interplay between computation and analysis in the study of Clay Millennium problem on Navier-Stokes equations (by *Thomas Yizhao Hou*, California Institute of Technology)

Abstract. Whether the 3D incompressible Navier-Stokes equations can develop a finite time singularity from smooth initial data with finite energy is one of the Seven Millennium Problems posted by the Clay Mathematical Institute. We review some recent theoretical and computational studies of the 3D Euler equations which show that there is a subtle dynamic depletion of nonlinear vortex stretching due to local geometric regularity of vortex filaments. Our study reveals a surprising nonlinear stabilizing effect that the convection term plays in stabilizing the solution. This is demonstrated through two reduced models of the 3D incompressible Navier-Stokes equations which shows that local flattening of the vortex structure and the effect of convection could lead to dynamic depletion of the vortex stretching term. Finally we present a new class of solutions to the 3D Euler and Navier-Stokes equations which could lead to a strong nonlinear alignment in the vortex stretching term and have the potential to develop a finite time singularity.

Congruent number problem (by *Ye Tian*, Morningside Center of Mathematics/Academy of Mathematics and Systems Science, Chinese Academy of Sciences)

Abstract. Recall that a positive integer is call a congruent number if it is the area of a right triangle with rational sides. The congruent number problem is to determine if a given positive integer is congruent and if so, find corresponding right triangles. It is known by Heegner that any prime or twice prime congruent to 5, 6, 7 modulo 8 is a congruent number. In this lecture, we construct congruent numbers with many prime factors, which extends Heegner's work.

Seiberg-Witten theory and Heegaard Floer theory (by *Yi-Jen Lee*, The Chinese University of Hong Kong)

Abstract. I will outline an equivalence proof of the Seiberg-Witten-Floer homology and the Heegaard Floer homology, and its expected extensions.

Automorphic period, L-function and relative trace formula (by *Wei Zhang*, Columbia University)

Abstract. We will present some recent progress on the study of period integrals of automorphic forms on classical groups, with an emphasis on their relation to L-functions. There are various approaches; but we will mainly discuss the approach of relative trace formula, started by Jacquet.

Compact embedded rotational hypersurfaces with constant mean curvature in a sphere (by *Haizhong Li*, Tsinghua University)

Abstract. The study of constant mean curvature (CMC) surfaces in spaces of constant curvature is one of the classical subjects in differential geometry. There are many beautiful results on this topic. By constructing a holomorphic quadratic differential for CMC surfaces, H. Hopf showed that any CMC two-sphere in Euclidean space is a round sphere. Then S.S. Chern extended Hopf's result to CMC two-spheres in 3-dimensional space forms. The famous Hopf conjecture asks that whether a compact CMC surface in Euclidean space is necessarily a round sphere? In 1980s, H.C. Wente gave a counter example of this conjecture by constructing a compact immersed CMC torus in Euclidean space. In 1950s, A.D. Alexandrov showed that if a compact CMC surface is embedded in Euclidean space, hyperbolic space or a hemisphere, then it must be totally umbilical. The minimal surface is the surface with constant mean curvature zero. It was conjectured by H.B. Lawson in 1970s that the only embedded minimal torus in three-sphere is the Clifford torus.

In 1980s, U. Pinkall and I. Sterling conjectured that embedded tori with CMC in three-sphere are surfaces of revolution. In March of 2012, Simon Brendle solved the Lawson conjecture. After that, in April of 2012, Ben Andrews and Haizhong Li gave a complete classification of CMC embedded tori in the three-sphere. When the constant mean curvature is equal to zero or $\pm 1/\sqrt{3}$, the only embedded torus is the Clifford torus. For other values of the mean curvature, there exists embedded torus which is not the Clifford torus, Ben Andrews and Haizhong Li gave a complete description of such surfaces. As a Corollary, Andrews-Li's Theorem solved the famous Pinkall-Sterling conjecture. In this lecture, we will talk about some basic facts of CMC surfaces and give explanations of key ideas of proofs of Lawson conjecture and Pinkall-Sterling conjecture.

Some mathematical problems in design of experiments (by *Ching-Shui Cheng*, Institute of Statistical Science, "Academia Sinica")

Abstract. I will discuss some mathematical problems that arise in statistical design of experiments. Several examples will be used to illustrate connections between design of experiments and finite projective geometry, coding theory and graph theory. Key results will be reviewed and some open questions will be presented.

Geometric transitions and quantum invariance (by *Chin-Lung Wang*, Taida Institute for Mathematical Sciences)

Abstract. A basic question in the study of higher dimensional geometry is to understand the relations of various invariants under geometric transitions among them. For example, the analytic continuation of quantum cohomology under K equivalent maps is extensively studied in recent years. In dimension 3, the famous Reid fantasy asks if various Calabi-Yau 3-folds are indeed connected via certain extremal transitions. It is even expected that conifold transitions may form the building blocks. This talk surveys some of the developments on the above aspects with emphasize on the relation of quantum $A+B$ models under transitions.

Global solutions of the Euler-Maxwell two-fluid system in 3D (by *Yan Guo*, Brown University)

Abstract. The fundamental "two-fluid" model for describing plasma dynamics is given by the Euler-Maxwell system, in which compressible ion and electron fluids interact with their own self-consistent electromagnetic field. We prove irrotational, smooth and localized perturbations of a constant background with small amplitude lead to global smooth solutions in three space dimensions for the Euler-Maxwell system.

Reflection groups, non-negative curvature and Tits geometry (by *Fuquan Fang*, Capital Normal University)

Abstract. A reflection in a euclidean space (sphere) is one of the fundamental notions of symmetry of geometric figures. It plays a central role in Killing and Cartan's work on Lie algebra in 19th century. Reflections groups on a hyperbolic space is important in hyperbolic geometry, and the first example goes back to F. Klein and Poincaré. In this talk I will present

(a) A complete classification of reflection groups and the equivariant structures of complete non negatively curved manifolds; and

(b) A complete classification of positively curved polar manifolds of cohomogeneity at least 2, which is achieved partially based on Tits geometry.

This is a joint work with Karsten Grove and G. Thorbergsson.

A brief survey on high order numerical methods for convection dominated problems (by *Chi-Wang Shu*, Brown University)

Abstract. Convection dominated partial differential equations are used extensively in applications including fluid dynamics, astrophysics, electro-magnetism, semiconductor devices, and biological sciences. High order accurate numerical methods are efficient for solving such partial differential equations, however they are difficult to design because solutions may contain discontinuities and other singularities or sharp gradient regions. In this talk we will survey several types of high order numerical methods for such problems, including weighted essentially nonoscillatory (WENO) finite difference methods, WENO finite volume methods, discontinuous Galerkin finite element methods, and spectral methods. We will discuss essential ingredients, properties and relative advantages of each method, and comparisons among these methods. Recent development and applications of these methods will also be discussed.

Enumerative geometry: from classical to modern (by *Yu-Jong Tzeng*, Harvard University)

Abstract. The subject of enumerative geometry goes back at least to the middle of the 19th century. It studies the number of geometric objects in a given class that satisfy a number of incidence or tangency conditions. For example:

(a) How many lines pass through distinct 2 points?
(b) How many lines pass through 4 general lines in 3-space?
(c) How many degree d rational curves on a general quintic 3-fold? (finite?)

The field has been very active in the last twenty years due to its interaction with physics, which provides motivation for many theories and conjectural formulas. One of the most famous problem in enumerative geometry is computing Severi degrees, which are the numbers of degree d plane curves with a given number of nodes and pass through an appropriate number of points in general position. The computation of Severi degrees, as well as counting nodal curves on other surfaces are great examples of classical problems which were open for long time and can finally be solved by modern techniques now. Furthermore, a recent conjecture of Gottsche states that there should be a universal formula that computes the number of nodal curves in a sufficiently ample linear system on any surface. In this talk I will first give an indication of what enumerative geometry is about, then give a survey about different curve-counting theories. Finally I will focus on the Gottsche's conjecture and its generalizations.

Rigidity in the Langlands correspondence and applications (by *Zhiwei Yun*, Stanford University)

Abstract. In the theory of Langlands correspondence for function fields, rigid automorphic representations and rigid local systems provide explicit and workable examples. We give two applications of such rigid objects. One is the construction of motives over number fields with exceptional Lie groups as Galois groups, answering a question of Serre. The other is the solution of the inverse Galois problem for certain finite groups of Lie type.

Finite and infinite soliton and kink-soliton trains of nonlinear Schrödinger equations (by *Tai-Peng Tsai*, University of British Columbia)

Abstract. I will first review known results on multi-solitons of dispersive partial differential equations, which are special solutions behaving like the sum of many weakly-interacting solitary waves. I will then describe my recent joint work with Dong Li and Stefan Le Coz: Assuming the composing solitons have sufficiently large relative speeds, we prove the existence and uniqueness of a soliton train which is a multi-soliton composed of infinitely many solitons. We also give a new construction of multi-solitons and prove uniqueness in an exponentially small neighborhood. In the 1D case, we can add to the infinite train an additional half-kink, which is a solution with a non-zero background at minus infinity.

Hall algebras and quantum groups arising from period 2 derived categories (by *Jie Xiao*, Tsinghua University)

Abstract. Inspired by the recent work of Bridgeland, we extend the derived Hall algebras to that from a period 2 derived category. This gives a successful model to realize the quantum groups globally. The talk is based on a joint work with X. Chen and F. Xu.

Universality of random matrices, Dyson's Brownian and De Giorgi-Nash-Moser theory of parabolic regularity (by *Horng-Tzer Yau*, Harvard University)

Abstract. Eugene Wigner's revolutionary vision predicted that the energy levels of large complex quantum systems exhibit a universal behavior: the statistics of energy gaps depend only on the basic symmetry type of the model. These universal statistics show strong correlations in the form of level repulsion and they represent a new paradigm of point processes that are characteristically different from the Poisson statistics of independent points.

Simplified models of Wigner's thesis have recently become mathematically accessible. For mean field models represented by large random matrices with independent entries, the celebrated Wigner-Dyson-Gaudin-Mehta (WDGM) conjecture asserts that the local eigenvalue statistics are universal. For invariant matrix models, the eigenvalue distributions are given by a log-gas with potential V and inverse temperature $\beta = 1, 2, 4$. For $\beta \notin \{1, 2, 4\}$, there is no natural random matrix ensemble behind this model, but the analogue of the WDGM conjecture asserts that the local statistics are independent of V.

In this lecture, we review the recent solution to these conjectures for both invariant and non-invariant ensembles.

We will demonstrate that the local ergodicity of the Dyson Brownian motion is the intrinsic mechanism behind the universality. Furthermore, we will show that the universality of gap distribution requires to prove a Holder regularity of a discrete parabolic equation with random coefficients. For this purpose, we incorporate the ideas of parabolic regularity via a De Giorgi-Nash-Moser approach.

Representation theories of Lie algebras and Lie superalgebras (by *Shun-Jen Cheng*, Institute of Mathematics, "Academia Sinica")

Abstract. We shall discuss the representation theories of Lie algebras and Lie superalgebras of classical types, and explain how these two apparently different theories are connected.

Structure-preserving algorithms for Riccati-type equations: rediscovery, redevelopment and applications (by *Wen-Wei Lin*, Chiao Tung University)

Abstract. In this talk, we attempt to tell the interesting story about the recent rediscovery and revival of the once faded and almost forgotten doubling algorithm for the discrete-time algebraic Riccati equation. Armed with some new insight and theoretical redevelopment, the method, now called the structure-preserving doubling algorithm (SDA), is first linked to palindromic eigenvalue problems. Generalizations were then developed for continuous-time, nonsymmetric and PCP algebraic Riccati equations and nonlinear matrix equations, for a vast array of applications in vibration analysis, surface acoustic wave simulation, optimal control, stochastic systems, transport theory, time-delayed systems, nano research, etc. We shall present some key theoretical results associated with selected important applications, and a summary of the recent discovery that the SDA can be adapted to solve large-scale problems of dimension n, with an computational complexity and memory requirement.

DEPARTMENT OF MATHEMATICS, TAIWAN UNIVERSITY
TAIPEI 10617, TAIWAN, CHINA
E-mail address: yu@math.ntu.edu.tw

Awards of ICCM 2013

Morningside Medal of Mathematics Selection Committee

The Morningside Medal of Mathematics Selection Committee comprises a panel of world renowned mathematicians and is chaired by Professor Shing-Tung Yau. A nomination committee of around 50 mathematicians from around the world nominates candidates based on their research, qualifications, and curriculum vitae. The Selection Committee reviews these nominations and recommends up to two recipients for the Morningside Gold Medal of Mathematics, up to two recipients for the Morningside Gold Medal of Applied Mathematics, and up to four recipients for the Morningside Silver Medal of Mathematics. The Selection Committee members, with the exception of the committee chair, are all non-Chinese to ensure the independence, impartiality and integrity of the awards decision.

Members of the 2013 Morningside Medal of Mathematics Selection Committee are:

Richard E. Borcherds

Professor Borcherds is Professor of Mathematics at the University of California at Berkeley. His research interests include Lie algebras, vertex algebras, and automorphic forms. He is best known for his work connecting the theory of finite groups with other areas in mathematics. Professor Borcherds was awarded the Fields Medal in 1998 for his work in algebra and geometry, in particular for his proof of the Moonshine conjecture.

John H. Coates

Professor Coates is Emeritus Sadleirian Professor of Pure Mathematics at the University of Cambridge. He is a number theorist, with particular research interests in Iwasawa theory and the arithmetic of elliptic curves. He frequently visits Asia, and is currently a Distinguished Professor at the Pohang University of Science and Technology, South Korea. Professor Coates is the first recipient of the ICCM International Cooperation Award in 2004.

Simon Donaldson

Professor Donaldson is Royal Society Research Professor at Imperial College. His research interests include differential geometry and its connections with differential topology and algebraic geometry. Professor Donaldson is a Fellow of the Royal Society, and a foreign member of the academies of France, Sweden and the United States. He is a recipient of the Fields Medal (1986), the Crafoord Prize in Mathematics (1994), the King Faisal International Prize for Science (2006), and the Shaw Prize in Mathematical Sciences (2009).

Björn Engquist

Professor Engquist is the Computational and Applied Mathematics Chair Professor at the University of Texas at Austin. His recent work includes homogenization theory, multi-scale methods, and fast algorithms for wave propagation. He is a member of the Royal Swedish Academy of Sciences, the Royal Swedish Academy of Engineering Sciences, and the Norwegian Academy of Science and Letters. He was a Guggenheim Fellow (1991), and received the first James H. Wilkinson Prize in Numerical Analysis and Scientific Computing (1982), the Peter Henrici Prize (2011), and the Birkhoff Prize in Applied Mathematics (2012).

Gerd Faltings

Professor Faltings is a Director of the Max Planck Institute for Mathematics in Germany. His research interests include number theory and algebraic and arithmetic geometry. Professor Faltings taught at Princeton University and the University of Wuppertal before joining the Max Planck Institute for Mathematics in 1994. Past honors include the Fields Medal (1986), a Guggenheim Fellowship (1988), the Gottfried Wilhelm Leibniz Prize (1996), the Karl Georg Christian von Staudt Prize (2008), and the Heinz Gumin Prize (2010).

James G. Glimm

Professor Glimm is a Distinguished Professor of Applied Mathematics at the State University of New York at Stony Brook. He has made original contributions in a variety of areas in pure and applied mathematics, including operator algebras, quantum field theory, statistical physics, partial differential equations and scientific computing. His past honors include a Guggenheim Fellowship (1963), the Dannie Heineman Prize for Mathematical Physics (1980), the Leroy P. Steele Prize (1993), and the U.S. National Medal of Science (2002).

Dorian M. Goldfeld

Professor Goldfeld is a Professor of Mathematics at Columbia University. His primary research interests include number theory and cryptography. Past honors include a Sloan Research Fellowship (1977), the Vaughan Prize (1985), and the Frank Nelson Cole Prize in Number Theory (1987). He is a Fellow of the American Academy of Arts and Sciences, the Editor of Acta Arithmetica and The Ramanujan Journal, and co-founder and board member of SecureRF.

Benedict H. Gross

Professor Gross is the George Vasmer Leverett Professor of Mathematics at Harvard University and former Dean of Harvard College. He has made several fundamental contributions to mathematics, namely number theory and algebraic geometry. Past honors include a Marshall Scholarship (1972–1974), a MacArthur Fellowship (1986), and the Frank Nelson Cole Prize in Number Theory (1987).

Victor W. Guillemin

Professor Guillemin is Professor of Mathematics at the Massachusetts Institute of Technology. His research interests are differential geometry and symplectic geometry. He is a Fellow of the American Academy of Arts and Sciences and the U.S. National Academy of Sciences. Past honors include a Guggenheim Fellowship (1988), a Humboldt Research Fellowship (1988), and the Leroy P. Steele Prize for Lifetime Achievement (2003). He is the Editor-in-Chief of the Journal of Symplectic Geometry.

Yuri I. Manin

Professor Manin is a Professor Emeritus at the Max Planck Institute for Mathematics and Northwestern University. Professor Manin's work includes algebraic geometry, number theory, differential equations, and mathematical physics, and his main research preoccupation of the last decade was quantum cohomology. Past honors include the Moscow Mathematical Society Award (1963), the Lenin Prize (1967), the Dutch Mathematical Society Brouwer Medal (1987), the Frederic Esser Nemmers Prize in Mathematics (1994), the Rolf Schock Prize in Mathematics (1999), the King Faisal International Prize for Science (2002), the Cantor Medal by the German Mathematical Society (2002), and the János Bolyai International Prize by the Hungarian Academy of Sciences (2010).

Stanley J. Osher

Professor Osher is a Professor of Mathematics at the University of California at Los Angeles, and has joint faculty appointments at the Electrical Engineering and Computer Science Departments. He is also the Director of Special Projects at the University's Institute for Pure and Applied Mathematics. He is the co-inventor and developer of many numerical methods for computational physics and image processing. Past honors include a Sloan Research Fellowship (1972), the NASA Public Service Group Achievement Award (1992), the Computational Mechanics Award by the Japan Society of Mechanical Engineering (2002), the Ralph E. Kleinman Prize (2005), and the Computational and Applied Science Award by the U.S. Association for Computational Mechanics (2007).

Shing-Tung Yau

Professor Yau is the William Caspar Graustein Professor of Mathematics at Harvard University, Director of The Institute of Mathematical Sciences at The Chinese University of Hong Kong, and Director of the Morningside Center of Mathematics at the Chinese Academy of Sciences. His research interests include differential geometry, differential equations, and general relativity. He is a member of several academies including "Academia Sinica", the Chinese Academy of Sciences, the Russian Academy of Sciences, and the U.S. National Academy of Sciences. Professor Yau has been one of the key individuals behind the founding of the Morningside Center of Mathematics, the Morningside Medal of Mathematics, and the International Congress of Chinese Mathematicians. Past honors include the Fields Medal (1982), the Crafoord Prize in Mathematics (1994), the U.S. National Medal of Science (1997), and the Wolf Prize (2010).

Morningside Special Achievement Award in Mathematics

Yitang Zhang

Dr. Zhang is awarded the Morningside Special Achievement Award in Mathematics for his proof of the restricted twin prime conjecture. For many years, number theorists have speculated that there are infinitely many twin primes, or pairs of primes with gaps equal to 2, and this has remained one of the oldest open problems in mathematics. In 2013, Dr. Zhang astounded the mathematical world by showing there exists infinitely many pairs of prime numbers with gaps less than 70000000. His paper "Bounded Gaps between Primes" was accepted for publication by the Annals of Mathematics in record time.

Dr. Zhang received his B.S. and M.S. degrees from Peking University, and later his doctorate degree from Purdue University in 1991. Before securing an academic position, he had worked at various jobs before being hired in 1999 by the University of New Hampshire as a substitute lecturer, and a full-time lecturer since 2005. Although Dr. Zhang never held a regular research position and had some hardships in his career, his passion and devotion to mathematics never waned, and in 2013 he marveled the mathematical world with his revelations. Dr. Zhang's perseverance against the odds, and his steadfast pursuit for mathematical truth, is an inspiration to us all.

Citation for Yitang Zhang (By John H. Coates, Emeritus Sadleirian Professor of Pure Mathematics, University of Cambridge)

The work of Yitang Zhang on the twin prime conjecture breaks entirely new ground, and is unquestionably one of the most surprising and important results in modern number theory.

Prime numbers are the building blocks of arithmetic, and the study of their distribution has always been one of the major themes of research in number theory. The particular question of proving that there are infinitely many primes p such that $p+2$ is also a prime is probably very ancient, although its written history only seems to have started in the 19th century. As our theoretical understanding of some aspects of the distribution of prime numbers grew, great efforts have been made by number theorists from the early 20th century until the present day to attack this so called twin prime conjecture. The best earlier result in its direction was established by the Chinese mathematician Jing-Run Chen in 1966, when he proved that there are infinitely many primes p such that $p+2$ is either a prime or the product of two primes. It is remarkable that the decisive progress on this problem has now been made by another Chinese mathematician Yitang Zhang. In a manuscript which has greatly surprised the mathematical world, Zhang has proven that there is some fixed integer c, less than 70000000, such that there are infinitely many primes p with the property that $p+c$ is also a prime. No result of this kind, establishing the existence of infinitely many prime pairs with a constant gap between them, has ever been proven before.

While it is too technical to enter into the detailed mathematical aspects of Zhang's work here, his starting point was the work of Goldston, Pintz, and Yildirim, published in 2009 and 2010, which, very roughly speaking, proved a slightly weaker form of the assertion that there are infinitely many prime pairs which differ by less than the square root of the logarithm of the smaller prime in the pair. This result created very considerable interest, but the leading experts in the field failed to make the all important improvement to get a constant gap between the pairs of primes. Now Zhang has brilliantly succeeded in doing this, by combining these methods of Goldston, Pintz, and Yildirim, with his own original variant of some deep earlier work by Bombieri, Fouvry, Friedlander, and Iwaniec from the 1980's. This is a mathematical achievement of the highest order. Yitang Zhang richly deserves a Morningside Special Achievement Award in Mathematics for this work.

Morningside Medals of Mathematics

2013 Morningside Gold Medal of Mathematics

Xuhua He

Professor He is awarded the 2013 Morningside Gold Medal of Mathematics for his contributions to several fundamental problems in arithmetic algebraic geometry, algebraic groups, and representation theory. He joined The Hong Kong University of Science and Technology (HKUST) in 2008, and is now an Associate Professor of Mathematics. After receiving his Ph.D. in mathematics from the Massachusetts Institute of Technology, he had worked at the Institute for Advanced Study and the State University of New York at Stony Brook before joining HKUST. Professor He's research interests include algebraic group theory, representation theory, combinatorics, algebraic geometry, and arithmetic geometry. Together with his collaborators, Professor He has solved long-standing open problems on Deligne-Lusztig theory and on Galois cohomology of algebraic groups. His work was reported by Professor Tonny Albert Springer at the International Congress of Mathematicians in 2006, and Professor Claire Voisin at the Bourbaki Seminar in 2011.

Ye Tian

Professor Tian is awarded the 2013 Morningside Gold Medal of Mathematics for his in-depth and original work in arithmetic algebraic geometry, especially on the congruent number problem. He is Professor at the Morningside Institute of Mathematics and the AMSS at the Chinese Academy of Sciences. He received his B.S. and M.S. degrees in pure mathematics from Sichuan University, and his Ph.D. in mathematics from Columbia University. He then worked at the Institute for Advanced Study and McGill University before joining the Morningside Institute and the AMSS in 2006. Professor Tian's research interests include arithmetic geometry and number theory. Together with his collaborators, Professor Tian solved some long-standing open problems on Birch and Swinnerton-Dyer conjecture and Diophantine equations. He is one of the recipients of the 2007 Morningside Silver Medal of Mathematics.

2013 Morningside Gold Medal of Applied Mathematics

Xianfeng David Gu

Professor Gu is awarded the 2013 Morningside Gold Medal of Applied Mathematics for his fundamental contributions to computing conformal structures, computer graphics, medical imaging, and 3-dimensional imaging. He is currently an Associate Professor of Computer Science at the State University of New York at Stony Brook. He received his B.S. in mathematics from Tsinghua University, and his M.S. and Ph.D. degrees from Harvard University under the supervision of Professor Shing-Tung Yau. Professor Gu solved with Professor Yau the fundamental problem of computing conformal structures of surfaces with arbitrary topologies based on Hodge theory. This breakthrough lays the foundation of computational conformal geometry, which Professor Gu applies to many fields in engineering and medicine. He is the winner of the U.S. National Science Foundation CAREER Award (2005), the National Natural Science Foundation of China Overseas Chinese Young Scholars Award (2006), and the recipient of several U.S. National Science Foundation and U.S. National Institute of Health awards and grants.

2013 Morningside Silver Medal of Mathematics

Chieh-Yu Chang

Professor Chang is awarded the 2013 Morningside Silver Medal of Mathematics for his important contributions to the transcendental number theory in function field. He is an Assistant Professor of Mathematics at Tsing Hua University in Hsinchu. He received his Ph.D. in mathematics from Tsing Hua University at Hsinchu, and completed his postdoctoral research at "the National Center for Theoretical Sciences" and Central University. Professor Chang's research interest revolves around the transcendence theory over function fields in positive characteristic, which originates from the foundations that Professor Jing Yu established in the 1980s and 1990s. In recent years, Professor Chang's work is focused on characteristic p multiple zeta values, their transcendence theory and associated arithmetic/geometric properties.

Xiaoqing Li

Professor Li is awarded the 2013 Morningside Silver Medal of Mathematics for her important research on the analytic properties of automorphic L-functions and their applications. She is an Associate Professor of Mathematics at the State University of New York at Buffalo. She received her Ph.D. in mathematics from Rutgers University under the supervision of Professor Henryk Iwaniec, and received her postdoctoral training at Columbia University for two years. Professor Li's research interests include analytic number theory, automorphic forms and L-functions. Her work has contributed to one of the most important problems in number theory, the Lindeloff hypothesis, which is to prove the sharp upper bound of L-functions on the one half line.

Hao Xu

Dr. Xu is awarded the 2013 Morningside Silver Medal of Mathematics for his significant contributions to the intersection theory on the moduli of curves and other related moduli spaces. He received his Ph.D. in mathematics from Zhejiang University under the supervision of Professor Kefeng Liu. He is currently a Research Fellow at Harvard University. Dr. Xu works on intersection theory on moduli spaces of curves and mathematical problems arising in deformation quantization, by using tools from integrable systems, Kähler geometry, Bergman kernels, Feynman diagrams and spectral graph theory. In 2007, he received the Zhong Jiaqing Mathematics Award from the Chinese Mathematical Society, and the New World Mathematics Award Ph.D. Thesis Gold Award at the International Congress of Chinese Mathematicians.

Tai-Peng Tsai

Professor Tsai is awarded the 2013 Morningside Silver Medal of Mathematics for his contributions to the dynamics of nonlinear Schrödinger equations and the regularity and singularity of the incompressible Navier-Stokes equations. He is Professor of Mathematics at the University of British Columbia, and had been a Courant Instructor at New York University from 1998 to 2001, and worked at the Institute for Advanced Study from 2001 to 2002. Professor Tsai received his B.S. in mathematics from Taiwan University and his Ph.D. from the University of Minnesota. His primary research interests include asymptotics and singularities of solutions of partial differential equations of fluid and quantum mechanics. In 2006, he received the André Aisenstadt Prize from the Centre de Recherches Mathématiques in Montreal.

ICCM International Cooperation Awards

The ICCM International Cooperation Award is awarded to an individual who has made significant contributions to the development of mathematics in mainland, Hong Kong, and Taiwan of China through collaboration, teaching, and supporting Chinese mathematicians. The award was established in 2004 and is presented every three years at the International Congress of Chinese Mathematicians (ICCM).

This year, the Selection Committee of the ICCM International Cooperation Award consists of six members, including Professor Shing-Tung Yau, President of ICCM (chair), Professor Shiu-Yuen Cheng of The Hong Kong University of Science and Technology, Professor Chang-Shou Lin of Taiwan University, Professor Kefeng Liu of the University of California at Los Angeles and Zhejiang University, Professor Lo Yang of the Chinese Academy of Sciences, and Professor Horng-Tzer Yau of Harvard University.

2013 ICCM International Cooperation Award

Jean-Pierre Serre

Professor Serre is awarded the 2013 ICCM International Cooperation Award for his great achievements and foundational contributions in number theory, algebraic geometry, and algebraic topology, and also for his efforts and support to the Chinese mathematical community. He received his Ph.D. from the Sorbonne in 1951, and from 1956 onwards he was the Chair in Algebra and Geometry at the Collège de France until he retired in 1994 and became an Honorary Professor. As an octogenarian, Professor Serre has been visiting Taiwan biannually since 2009, offering a series of inspiring lectures at "the National Center for Theoretical Sciences" on his cutting edge research in number theory and arithmetic geometry, and has greatly impacted the Taiwan mathematical community.

Professor Serre has made foundational contributions in algebraic topology, algebraic geometry, and number theory. In his 1951 thesis, he applied the now-called Leray-Serre spectral sequence to establish connections between the homology and homotopy groups of a space, and computed the homotopy groups of spheres. In 1955 he introduced algebraic coherent sheaves on varieties, extending cohomology theory to algebraic geometry. One year later, he introduced the GAGA theory, connecting analytic geometry and algebraic geometry. Together with John Tate, Professor Serre shaped the landscape of modern algebraic number theory, and introduced and developed algebraic K-theory, Galois cohomology and p-adic Galois representations. Professor Serre has been elected to many national academies, in particular the academies of France, Sweden, United States, and the Netherlands. He has received numerous awards, including the Fields Medal in 1954 (the youngest recipient ever), the Prix Gaston Julia in 1970, the Balzan Prize in 1985, the Leroy P. Steele Prize for Mathematical Exposition in 1995, the Wolf Prize in 2000, and the first Abel Prize in 2003. Professor Serre has been awarded honorary degrees from 13 universities, and has been made a Commander Legion d'Honneur and High Officer Ordre National du Merite by the French government.

Chern Prizes

The Chern Prize was established in 2001 in honor of Professor Shiing-Shen Chern, a mathematician, an educator, and one of the greatest geometers of the 20 century. Presented every three years at the International Congress of Chinese Mathematicians (ICCM), the Chern Prize is awarded to mathematicians of Chinese decent who have made exceptional contributions to mathematical research or to public service activities in support of mathematics.

This year, the Selection Committee of the Chern Prize consists of six members, including Professor Shing-Tung Yau, President of ICCM (chair), Professor Shiu-Yuen Cheng of The Hong Kong University of Science and Technology, Professor Chang-Shou Lin of Taiwan University, Professor Kefeng Liu of the University of California at Los Angeles and Zhejiang University, Professor Lo Yang of the Chinese Academy of Sciences, and Professor Horng-Tzer Yau of Harvard University.

2013 Chern Prize

Si-Chen Lee

Professor Lee is awarded the 2013 Chern Prize in recognition of his efforts and contribution to upgrade the research of mathematics in Taiwan, and with his vision and determination, fully endorsing the establishment of a first-class mathematical community within Taiwan University. He received his Ph.D. in electrical engineering from Stanford University in 1981. His research has led to the innovative development of ledge-type heterojunction bipolar transistor (HBT) in 1985, which has become the norm in manufacturing HBT power amplifiers, dominating the cell phone production in billions of units every year. Professor Lee is an Institute of Electrical and Electronics Engineers (IEEE) Fellow. He received many accolades, including the Dr. Sun Yat-Sen Academic Award (1987), five consecutive Outstanding Research Awards by "the National Science Council" (1986–1996), the IEEE Third Millennium Medal (2000), the Medal of Electrical Engineering from the Association of Chinese Electrical Engineers (2002), and the 47th Academic Award (Engineering and Applied Science) by "the Ministry of Education" (2003).

Bong Lian

Professor Lian is awarded the 2013 Chern Prize for his fundamental contributions to the investigation of Calabi-Yau mirror symmetry and representation theory of vertex operator algebras. He is a Professor of Mathematics at Brandeis University. He received his B.A. from the University of Toronto, and his Ph.D. from Yale University. His research areas are representation theory, Calabi-Yau geometry, and string theory. Among his achievements include his thesis on semi-infinite cohomology theory, which has become a primary tool in the representation theory of the Virasoro algebra. In 1996, he worked with Kefeng Liu and Shing-Tung Yau to solve the outstanding mirror conjecture that is fundamental in enumerative geometry and string theory. This work was also done independently by A. Givental. In 2010, Professor Lian worked with Ruifang Song and Shing-Tung Yau, and jointly developed a new approach, called tautological systems, to study the Riemann-Hilbert problem for period integrals. Recently, on a joint research work with Spencer Bloch, An Huang, Vasudevan Srinivas, Shing-Tung Yau and Xinwen Zhu, Professor Lian solved the completeness problem for tautological systems of period integrals in a number of important cases. He was a Sloan Dissertation Fellow (1990) and a Guggenheim Fellow (2003).

Morningside Mentor Award in Mathematics

Chengbiao Pan

Professor Pan is awarded the Morningside Mentor Award in Mathematics for his tireless efforts in training first class mathematicians. The Morningside Mentor Award in Mathematics is created to recognize the impact of mathematicians who invest in students' well-being through teaching, and in so doing, they pass onto their students a lifelong desire to learn, and inspire them to pursue their goals in mathematics and research work. Professor Pan is not only an inspiration to fellow mathematicians, academic professionals and students, but also a catalyst for positive change, growth, and innovation in the field of mathematics education.

Professor Pan was a Professor at the College of Science at China Agricultural University since 1986, and had held an adjunct position at the Department of Mathematics at Peking University since 1977, until his retirement in 2008. Professor Pan has devoted his work to the field of mathematics, particularly the teaching and research of analytic number theory, and has co-authored numerous academic works in Chinese including Goldbach Conjecture, Algebraic Number Theory, Elementary Number Theory, Condensed Number Theory, and Guide to Modular Forms. Besides his research work, Professor Pan was the former editor of Acta Mathematica Sinica and Advances in Mathematics, and a former committee member of the Chinese Mathematical Olympiad. He is a dedicated teacher and mentor toany aspiring mathematicians. Dr. Yitang Zhang of the University of New Hampshire and recipient of the 2013 Morningside Special Achievement Award in Mathematics, Professor Chaohua Jia of the Chinese Academy of Sciences, and Professor Wenzhi Luo of The Ohio State University, to name a few, have all benefited from his tutelage.

Plenary Lectures

Proceedings of the 6th International Congress of Chinese Mathematicians
ALM 36, pp. 3–10

Ricci Flow and Classification of Certain Four-manifolds

Bing-Long Chen

Abstract

It is well-known that Ricci flow has played a vital role in 3-dimensional topology. The Ricci flow with surgery in dimension 4 has been developed in a seminal paper by R. Hamilton in 1997. In this talk, I will show that the recent results of Ricci flow can give a complete classification of compact 4-manifolds whose certain generalized Yamabe invariant are positive. This classification is an analogue of a classification theorem of R. Schoen and S.T.Yau on 3-manifolds with positive Yamabe invariant. This talk is based on author's two joint papers with Zhu [3, 5], one joint paper with Tang and Zhu[4].

2010 Mathematics Subject Classification: Primary 53C21; Secondary 35J96.

Keywords and Phrases: Ricci flow, classification theorem.

1

One of main themes in geometry is to classify manifolds in terms of certain topological or differentiable invariants. The purpose of this talk is to discuss a classification of manifolds with certain positive naturally defined differentiable invariant: a generalized Yamabe invariant on four-manifolds.

The content of this talk is the following:

- Yamabe invariant and classification.
- Generalized Yamabe invariant.
- Conformal sphere theorem.
- Conformally invariant classification theorem.
- Ricci flow on 4-manifolds.
- Manifolds with positive isotropic curvature.

Let us start with two dimensional case. When Dimension=2, a complete classification theorem implies that $M^2 \cong \Sigma_g$ or $\Sigma_g \# RP^2$, where Σ_g is an orientable surface of genus g. All compact orientable surface M^2 may be classified topologically by a single topological invariant $\chi(M^2)$. From this, we know

$$\chi(M^2) > 0 \Rightarrow S^2, RP^2.$$

Recall Gauss-Bonnet theorem tells us

$$\frac{1}{2\pi}\int_{M^2} K d\sigma = \chi(M^2).$$

When dimension is greater than two, we are interested in the so-called Yamabe invariant. First, we review the definition of the Yamabe invariant.

Let M^n be a compact n-dimesnional manifold, g a Riemannian metric. Let $\mathcal{C}_g = \{\rho g \mid \rho > 0\}$ be the class of metrics conformal to g. We define

$$\mathcal{Y}(M^n, \mathcal{C}_g) = \inf_{g' \in \mathcal{C}_g} \frac{\int_{M^n} R_{g'} dv_{g'}}{\left(\int_{M^n} dv_{g'}\right)^{\frac{n-2}{n}}}$$

to be the Yamabe invariant of the class $\mathcal{C}_g$. Then define the (global) Yamabe invariant of M^n by

$$\mathcal{Y}(M^n) = \sup_{\mathcal{C}} \mathcal{Y}(M^n, \mathcal{C}).$$

It is clear that $\mathcal{Y}(M^n)$ is a differential invariant. When $n = 2$, $\mathcal{Y}(M^n)$ is just a multiple of the Euler characteristic number by Gauss-Bonnet formula.

When dimension=3, we have an excellent theorem originally due to Schoen and Yau. The elimination of possible homotopic three-spheres is due to the resolution of the Poincare conjecture.

Theorem 1.1. *(Schoen-Yau*[11]*, Perelman* [10]*) Let M^3 be 3-dimensional compact, orientable manifold with $\mathcal{Y}(M^3) > 0$. Then*

$$M^3 \cong S^3/\Gamma_1 \# \cdots S^3/\Gamma_k \# (S^2 \times S^1) \# \cdots (S^2 \times S^1).$$

When dimension ≥ 5, there are deep works done by Schoen-Yau, Gromov-Lawson, Rosenberg, Stolz etc. We have to say that for the purpose of a complete classification theorem, the Yamabe invariant is too weak to capture enough topological information when dimension ≥ 5. When dimension=4, manifolds with $\mathcal{Y}(M^4) > 0$ are still very far from being completely classified. This is the motivation for our study on four-manifolds.

Before introducing the new invariant, we recall in dimension 4 case, an orientation induces a decomposition $\Lambda^2 = \Lambda^2_+ \oplus \Lambda^2_-$ of two forms into self-dual and anti-self-dual two forms. Since curvature operator acts on Λ^2, this induces a block decomposition of the curvature matrix

$$Rm = \begin{pmatrix} A & B \\ {}^tB & C \end{pmatrix}$$

where $A = \frac{R}{12} + W_+$, $C = \frac{R}{12} + W_-$, $B = \overset{\circ}{Ric}$. Denote $\lambda_{\max}(W_+)$ and $\lambda_{\max}(W_-)$ largest eigenvalues of W_+ and W_-. For a conformal class $\mathcal{C}$ of metrics,we define a conformal invariant

$$\mathcal{GY}(M^4,\mathcal{C}) = \inf_{g\in\mathcal{C}} \frac{\int_{M^4}(R_g - 6\max\{\lambda_{\max}(W_+), \lambda_{\max}(W_-)\})dv_g}{(\int_{M^4} dv_g)^{\frac{1}{2}}}.$$

Then we can also define a global invariant

$$\mathcal{GY}(M^4) = \sup_{\mathcal{C}} \mathcal{GY}(M^4,\mathcal{C}).$$

It is clear that $\mathcal{Y}(M^n)$ is a differentiable invariant. We call such invariant $\mathcal{GY}(M^4)$ a generalized Yamabe invariant.

Theorem 1.2. (*Chen-Zhu* [5], *preprint* 2012) *Let M^4 be a compact four-manifold with $\mathcal{GY}(M^4) > 0$. Then M^4 is diffeomorphic to*

$$S^4\#mRP^4\#(R\times S^3)/\Gamma_1\#\cdots(R\times S^3)/\Gamma_k,$$

where these Γ_i are cocompact fixed point free subgroups of standard isometries of $R\times S^3$.

This theorem is a four-dimensional analogue of Schoen-Yau-Perelman's Theorem.

2

In this section, we give an interesting application of Theorem 1.2. We introduce the conformal sphere theorem of Chang-Gursky-Yang:

Theorem 2.1. (*Chang-Gursky-Yang* [2]) *Let (M^4, g) be a compact 4-dim Riemannian manifold satisfying*

(i) $\mathcal{Y}(M^4,\mathcal{C}_g) > 0$,

(ii) $\int_{M^4} |W_+|^2 + |W_-|^2 dvol_g < 4\pi^2\chi(M^4)$.

Then $M^4 \cong S^4$ or RP^4.

Clearly, condition (ii) is conformally invariant. Chang-Gursky-Yang's theorem also has a rigidity part: if

(a) $\mathcal{Y}(M^4,\mathcal{C}_g) > 0$,

(b) $\int_{M^4} |W_+|^2 + |W_-|^2 dvol_g = 4\pi^2\chi(M^4)$,

(c) M^4 is not diffeomorphic to S^4, RP^4.

Then Chang, Gursky and Yang proved that (M^4, g) is conformal to CP^2, or a finite cover of (M^4, g) is conformal to $S^3\times S^1$.

We remark that the Euler characteristic number $\chi(M^4) > 0$ is required in Theorem 2.1, and $\chi(M^4) \geq 0$ in their rigidity theorem. A interesting question is to generalize Theorem 2.1 to manifolds with $\chi(M^4) < 0$. i.e. establish a conformal invariant classification theorem including broader class of manifolds with arbitrary Euler characteristic.

As an easy application of Theorem 1.2, we have

Theorem 2.2. (*Chen-Zhu*[5], *preprint* 2012)
Let (M^4, g) *be a compact manifold satisfying*

(i) $\mathcal{Y}(M^4, \mathcal{C}_g) > 0$,

(ii) $\displaystyle\int_{M^4} [\max\{\lambda_{\max}(W_+), \lambda_{\max}(W_-)\}]^2 dvol_g < \frac{1}{36}\mathcal{Y}(M^4, \mathcal{C}_g)^2.$

Then

$$M^4 \cong S^4 \# mRP^4 \# (R \times S^3)/\Gamma_1 \# \cdots (R \times S^3)/\Gamma_k.$$

We remark that the conditions (i) and (ii) in our theorem are also conformally invariant. Under assumptions of our theorem:

- $\chi(M^4) > 0 \Longrightarrow S^4, RP^4$.
- $\chi(M^4) = 0 \Longrightarrow$ a finite cover $\cong S^3 \times S^1$,.
- $\chi(M^4) < 0 \Longrightarrow$ a finite cover $\cong \#_k S^3 \times S^1$.

In the following, we show Chang-Gursky-Yang's theorem can be deduced from Theorem 2.2. First of all, note that from Gauss-Bonnet-Chern formula

$$\frac{1}{8\pi^2}\int_{M^4} |W_+|^2 + |W_-|^2 + \frac{R^2}{24} - \frac{|\overset{\circ}{Ric}|^2}{2} dvol_g = \chi(M^4),$$

we know $\int_{M^4} \frac{R^2}{24} - \frac{|\overset{\circ}{Ric}|^2}{2} dvol_g$ is conformally invariant.

Suppose (M, g) satisfies the assumptions in Chang-Gursky-Yang's Theorem. Let $\tilde{g}$ be a Yamabe metric in $\mathcal{C}_g$. It is easy to see:

$$\begin{aligned}\int_{M^4} |W_+|^2 + |W_-|^2 dvol_g &< \int_{M^4} \frac{R^2}{24} - \frac{|\overset{\circ}{Ric}|^2}{2} dvol_g \\ &= \int_{M^4} \frac{R_{\tilde{g}}{}^2}{24} - \frac{|\overset{\circ}{Ric}_{\tilde{g}}|^2_{\tilde{g}}}{2} dvol_{\tilde{g}} \\ &\leq \int_{M^4} \frac{R_{\tilde{g}}{}^2}{24} = \frac{\mathcal{Y}(M^4, \mathcal{C}_g)^2}{24},\end{aligned}$$

$$|W_+|^2 = \lambda_1^2 + \lambda_2^2 + \lambda_3^2 \geq \lambda_1^2 + \frac{1}{2}(\lambda_2 + \lambda_3)^2 = \frac{3}{2}|\lambda_{\max}(W_+)|^2,$$

$$\int_{M^4} |\lambda_{\max}(W_+)|^2 + |\lambda_{\max}(W_-)|^2 dvol_g < \frac{\mathcal{Y}(M^4, \mathcal{C}_g)^2}{36}.$$

So the condition (ii) in Theorem 2.1 implies condition (ii) in Theorem 2.2. The result follows from the conclusions of Theorem 2.2. A Rigidity Theorem can also be obtained:

Theorem 2.3. (*Chen-Zhu*[5], *preprint* 2012) *Let M^4 be compact and satisfy*

(i) $\mathcal{Y}(M^4, \mathcal{C}_g) > 0$,

(ii) $\int_{M^4} [\max\{\lambda_{\max}(W_+), \lambda_{\max}(W_-)\}]^2 dvol_g = \frac{1}{36}\mathcal{Y}(M^4, \mathcal{C}_g)^2$.

Suppose M^4 is not diffeomorphic to

$$S^4 \# mRP^4 \# (R \times S^3)/\Gamma_1 \# \cdots (R \times S^3)/\Gamma_k.$$

Then it is either conformal to CP^2, or a finite cover is conformal to $S^2 \times S^2$ or $S^2 \times T^2$.

The proof of CGY's Theorem 2.1 depends on a result of Margerin (AMS Proc. of Symp. in Pure Math., Arcata, 84 (1986), 307–328): let (M^4, g) be compact and satisfy

$$R > 0, \frac{R^2}{24} > \frac{|\overset{\circ}{Ric}|}{2} + |W_+|^2 + |W_-|^2.$$

Then normalized Ricci flow converges to constant curvature metric. Condition (ii) in CGY's Theorem simply means

$$\int_{M^4} \frac{R^2}{24} - \frac{|\overset{\circ}{Ric}|^2}{2} - (|W_+|^2 + |W_-|^2) dvol_g > 0.$$

The integrand is just the quantity in Margerin' Theorem. The idea becomes very naive: find $\tilde{g} = \rho g$ in $\mathcal{C}_g$ s.t. $\frac{R_{\tilde{g}}^2}{24} > \frac{|\overset{\circ}{Ric}_{\tilde{g}}|^2}{2} + |W_+^{\tilde{g}}|^2 + |W_-^{\tilde{g}}|^2$. This naive idea is actually very hard to work out. It amounts to solve a fully nonlinear equation (Chang-Gursky-Yang: Ann math.02' 711–789):

$$\sigma_2(A_{\tilde{g}}) - \frac{\alpha}{4}|W_{\tilde{g}}|^2 \equiv \lambda > 0$$

where $A_{\tilde{g}} = Ric_{\tilde{g}} - \frac{\tilde{R}}{6}\tilde{g}$, $\tilde{g} = e^{2u}g$, $A_{\tilde{g}} = A_g - \nabla_g^2 u + 2du \otimes du + |du|^2 g$.

Now we discuss the idea of proof of our Theorems 2.2 and 1.2. It consists of the following steps:

Step 1: Consider a variant of the scalar curvature

$$\sigma_g = R_g - 6\max\{\lambda_{\max}(W_+), \lambda_{\max}(W_-)\}.$$

Solve a generalized Yamabe problem:

$$\inf_{\tilde{g} \in \mathcal{C}_g} \frac{\int_{M^n} \sigma_{\tilde{g}} dv_{\tilde{g}}}{(\int_{M^n} dv_{\tilde{g}})^{\frac{1}{2}}}.$$

Step 2: Let $\tilde{g}$ be the solution for GYP, $\tilde{g} \in C^{2,\alpha}$. Then

$$R_{\tilde{g}} - 6\max\{\lambda_{\max}(W_+^{\tilde{g}}), \lambda_{\max}(W_-^{\tilde{g}})\} \equiv c.$$

Assume $\int_{M^4} dv_{\tilde{g}} = 1$.

$$\begin{aligned} c &= \int_{M^4} R_{\tilde{g}} dv_{\tilde{g}} - 6\int_{M^4} \max\{\lambda_{\max}(W_+^{\tilde{g}}), \lambda_{\max}(W_-^{\tilde{g}})\} dv_{\tilde{g}} \\ &\geq \mathcal{Y}(M^4, \mathcal{C}_g) - 6\Big(\int_{M^4} [\max\{\lambda_{\max}(W_+^{\tilde{g}}), \lambda_{\max}(W_-^{\tilde{g}})\}]^2 dv_{\tilde{g}}\Big)^{\frac{1}{2}} \\ &= \mathcal{Y}(M^4, \mathcal{C}_g) - 6\Big(\int_{M^4} [\max\{\lambda_{\max}(W_+), \lambda_{\max}(W_-)\}]^2 dv_g\Big)^{\frac{1}{2}} \\ &> 0. \end{aligned}$$

This means that the metric $\tilde{g}$ has positive isotropic curvature.

Step 3: Classify 4-manifolds with PIC. This part is much more difficult.

Theorem 2.4. (*Chen-Tang-Zhu*[4]) *Let M^4 be a compact manifold with PIC. Then*

$$M^4 \cong S^4 \# mRP^4 \# (R \times S^3)/\Gamma_1 \# \cdots (R \times S^3)/\Gamma_k.$$

When $\pi_1(M^4)$ contains no essential incompressible space-forms, the result is due to Hamilton [7] and Chen-Zhu [3]. The proof for the above general case is discussed in the following section.

3

The approach is using Ricci flow. New features comparing with dimension three arise.

• Surgeries on $R \times (S^3 \setminus \Gamma)$ may cause orbifold singularities, so we have to deal with Ricci flow with surgeries on orbifolds.

• Since Γ may be very large a prior, the model for canonical neighborhood structure may possibly be seriously collapsed.

• Overcoming the noncollapsing issue, one can show that after performing finite number of surgeries, the Ricci flow terminates at finite time. We know each pieces left are spherical orbifold S^4/Γ_i, or an orbifold connected sum of these spherical orbifolds. This implies that M is diffeomorphic to "orbifold connected sum" $\# S^4/\Gamma_i$.

• Prove that spherical orbifolds S^4/Γ with isolated singularities have at most 2 singularities. The idea is to prove Γ satisfies so called P^2 condition and then apply Burnside theorem in finite group theory. The result is summarized into three cases:

Case 1: 0-singularity corresponds to S^4 or RP^4.

Case 2: 1-singularity corresponds to a cap with $(S^3/\Gamma') \times R$ end (after removing the singularity).

Case 3: 2-singularities corresponds to a neck $\cong (S^3/\Gamma') \times R$ (by removing the singularities).

• To glue these orbifolds together, we have to invoke Smale's conjecture[8] on groups of diffeomorphisms of $S^3 \setminus \Gamma$. Resolving the singularities pairwisely and gluing these orbifolds give standard $(R\times S^3)\backslash\Gamma$ or S^4 or RP^4. Finally, we know the original manifold is diffeomorphic to the usual connected sum of these manifolds (not orbifolds): S^4 or RP^4, $S^3 \times R/\Gamma$.

Gromov conjecture. *Let M^n be a compact manifold with positive isotropic curvature. Then $\pi_1(M)$ is virtually free, i.e. there is a free subgroup of finite index $G \subset \pi_1(M)$.*

A byproduct of Theorem 2.4 is that the above conjecture is true when $n = 4$.

Remark 3.1. In the statement of Theorem 1.1, the manifold is required to be orientable. We remark that the non-orientable case can also be handled simultaneously by using the same argument of Theorem 2.4. The result can be stated in the following general form: compact M^3 has $\mathcal{Y}(M^3) > 0$ if and only if

$$M^3 \cong S^3/\Gamma_1\#\cdots S^3/\Gamma_k\#(S^2\times R)/G_1\#\cdots(S^2\times R)/G_k,$$

where Γ_i, G_j are just standard discrete isometric group actions on S^3 and $S^2 \times R$ respectively.

From the Ricci flow point of view, if the 3-manifold is not orientable, surgery procedure may produce orbifold singularities. For instance, if you perform surgery on $\mathbb{RP}^2 \times \mathbb{R}$, then an orbifold with topology of cone over RP^2 will occur. So the argument in [10] can not apply directly. However, 3-dim case is much easier than Theorem 2.4 of 4-dim case, and the same argument of Theorem 2.4 can apply.

Acknowledgements. This work is partially supported by NSFC 11025107.

References

[1] Brendle, S. and Schoen, R., *Classification of manifolds with weakly* 1/4-*pinched curvature*, Acta Math. 200 (2008), no. 1, 1–13.

[2] Chang, S. Y. A., Gursky, M., and Yang, P., *A conformally invariant sphere theorem in four dimensions,* Publ. Math. Inst. Hautes Etudes Sci. No. **98**, (2003), 105–143.

[3] Chen, B. L. and Zhu, X. P., *Ricci flow with surgery on four-manifolds with positive isotropic curvature*, J. Diff. Geom. **74** (2006), 177–264.

[4] Chen,B. L., Tang, S. H., and Zhu, X. P., *Complete classification of compact four manifolds with positive isotropic curvature*, J. Diff. Geom., **91** (2012), 41–80.

[5] Chen, B. L. and Zhu, X. P., *A conformally invariant classification theorem in four dimensions*, http://arxiv.org/abs/1206.5051.

[6] Gromov, M. and Lawson, Jr. H. B., *The classification of simply connected manifolds of positive scalar curvature*, Ann. of Math. (2) **111** (1980), no.3, 423–434.

[7] Hamilton, R. S., *Four manifolds with positive isotropic curvature*, Commu. in Analysis and Geometry, **5** (1997), 1–92. (or see, *Collected Papers on Ricci Flow*, Edited by H.D.Cao, B.Chow, S.C.Chu, and S.T.Yau, International Press, 2002).

[8] Mccullough, D., *Isometries of elliptic 3-manifolds,* J. London Math. Soc. (2) **65** (2002), no 1, 167–182.

[9] Micallef, M. and Moore, J. D., *Minimal two-spheres and the topology of manifolds with positive curvature on totally isotropic two-planes*, Ann. of Math.(2) **127** (1988), 199–227.

[10] Perelman, G., *Ricci flow with surgery on three manifolds,* arXiv: math. DG/0303109.

[11] Schoen, R. and Yau, S. T., *On the structure of manifolds with positive scalar curvature,* Manu. Math. **28** (1979), 159–184.

DEPARTMENT OF MATHEMATICS, SUN YAT-SEN UNIVERSITY
GUANGZHOU 510275, CHINA
E-mail address: mcscbl@mail.sysu.edu.cn

Proceedings of the 6th International
Congress of Chinese Mathematicians
ALM 36, pp. 11–23

Shock Reflection-Diffraction, von Neumann's Conjectures, and Nonlinear Equations of Mixed Type

Gui-Qiang Chen and Mikhail Feldman

Abstract

Shock waves are fundamental in nature. One of the most fundamental problems in fluid mechanics is shock reflection-diffraction by wedges. The complexity of reflection-diffraction configurations was first reported by Ernst Mach in 1878. The problems remained dormant until the 1940s when John von Neumann, as well as other mathematical/experimental scientists, began extensive research into all aspects of shock reflection-diffraction phenomena. In this paper we start with shock reflection-diffraction phenomena and historic perspectives, their fundamental scientific issues and theoretical roles in the mathematical theory of hyperbolic systems of conservation laws. Then we present how the global shock reflection-diffraction problem can be formulated as a boundary value problem in an unbounded domain for nonlinear conservation laws of mixed hyperbolic-elliptic type, and describe the von Neumann conjectures: the sonic conjecture and the detachment conjecture. Finally we discuss some recent developments in solving the von Neumann conjectures and establishing a mathematical theory of shock reflection-diffraction, including the existence, regularity, and stability of global regular configurations of shock reflection-diffraction by wedges.

2010 Mathematics Subject Classification: Primary: 35M10, 35M12, 35B65, 35L65, 35L70, 35J70, 76H05, 35L67, 35R35; Secondary: 35L15, 35L20, 35J67, 76N10, 76L05.

Keywords and Phrases: Shock waves, reflection, diffraction, conservation laws, mixed elliptic-hyperbolic type, equations of mixed type, von Neumann's conjectures, free boundary problem, potential flow, Riemann problem.

1 Introduction

Shock waves are steep fronts that propagate in the compressible fluids when the convective motion dominates the diffusion. They are fundamental in nature. Examples include shock waves formed by solar winds (bow shocks), supersonic or near sonic aircrafts (transonic shocks around the body or supersonic bubbles with transonic shocks), explosions (blast waves), and various other natural processes. When a shock hits an obstacle (steady or flying), shock reflection-diffraction phenomena occur. One of the most fundamental problems in fluid mechanics is the problem of shock reflection-diffraction by wedges; see Ben-Dor [4], Courant-Friedrichs [12], von Neumann [30, 31, 32], and the references cited therein. When a plane shock hits a wedge head on, it experiences a reflection-diffraction process, and then a fundamental question is what types of wave patterns of reflection-diffraction configurations may be formed around the wedge.

The complexity of reflection-diffraction configurations was first reported by Ernst Mach [26] in 1878, who first observed two patterns of reflection-diffraction configurations: regular reflection (two-shock configuration; see Fig. 1 (left)) and Mach reflection (three-shock/one-vortex-sheet configuration; see Fig. 1 (center)); also see [4, 7, 12, 29]. The issues remained dormant until the 1940s when John von Neumann, as well as other mathematical/experimental scientists, began extensive research into all aspects of shock reflection-diffraction phenomena, due to its importance in applications. See von Neumann [30, 31, 32], Courant-Friedrichs [12], Glimm-Majda [16], and Ben-Dor [4]; also see [7, 29] and the references cited therein. It has been found that the situations are much more complicated than what Mach originally observed: The Mach reflection can be further divided into more specific sub-patterns, and various other patterns of shock reflection-diffraction may occur such as the double Mach reflection, von Neumann reflection, and Guderley reflection; see [4, 7, 12, 16, 29] and the references cited therein. The fundamental scientific issues include the following:

(i) Structure of the shock reflection-diffraction configurations;

(ii) Transition criteria between the different patterns of shock reflection-diffraction configurations;

(iii) Dependence of the patterns upon the physical parameters such as the wedge angle θ_w, the incident-shock-wave Mach number, and the adiabatic exponent $\gamma \geq 1$.

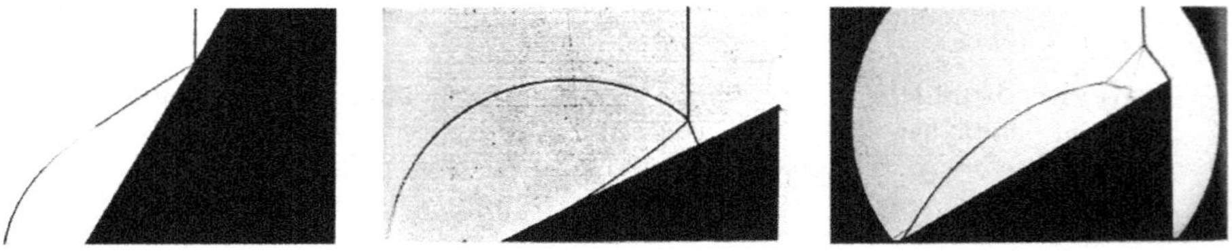

Fig. 1: Regular reflection (left); Mach reflection (center); Irregular Mach reflection (right). From Van Dyke [29], pages 142–144.

In particular, several transition criteria between the different patterns of shock reflection-diffraction configurations have been proposed, including the sonic conjecture and the detachment conjecture by von Neumann [30, 31, 32].

Careful asymptotic analysis has been made for various reflection-diffraction configurations in Lighthill [22, 23], Keller-Blank [21], Hunter-Keller [20], Harabetian [18], Morawetz [27], and the references cited therein; also see Glimm-Majda [16]. Large or small scale numerical simulations have been also made; *cf.* [4, 16, 33] and the references cited therein. However, most of the fundamental issues for shock reflection-diffraction phenomena have not been understood, especially the global structure and transition between different patterns of shock reflection-diffraction configurations. This is partially because physical and numerical experiments are hampered by various difficulties and have not yielded clear transition criteria between different patterns. In particular, numerical dissipation or physical viscosity smear the shocks and cause boundary layers that interact with the reflection-diffraction patterns and can cause spurious Mach steams; *cf.* Woodward-Colella [33]. Furthermore, some different patterns occur when the wedge angles are only fractions of a degree apart, a resolution even by sophisticated modern experiments (*cf.* [25]) has been unable to reach. For this reason, it is almost impossible to distinguish experimentally between the sonic and detachment criteria, as pointed out in [4]. In this regard, the necessary approach to understand fully the shock reflection-diffraction phenomena, especially the transition criteria, is still via rigorous mathematical analysis. To achieve this, it is essential to establish first the global existence, regularity, and structural stability of solutions of the shock reflection-diffraction problem.

Furthermore, shock reflection-diffraction configurations are the core configurations in the structure of global entropy solutions of the two-dimensional Riemann problem for hyperbolic conservation laws, while the Riemann solutions are building blocks and local structure of general solutions and determine global attractors and asymptotic states of entropy solutions, as time tends to infinity, for multidimensional hyperbolic systems of conservation laws. See [7, 16, 28, 35] and the references cited therein. In this sense, we have to understand the shock reflection-diffraction phenomena, in order to understand fully global entropy solutions to multidimensional hyperbolic systems of conservation laws.

In Section 2, we first formulate the shock reflection-diffraction problem into an initial-boundary value problem. Based on the invariance of the problem under the self-similar scaling and some basic properties of the governing equations, we reformulate the problem as a boundary value problem in an unbounded domain for nonlinear conservation laws of mixed hyperbolic-elliptic type. In Section 3, we describe the von Neumann's conjectures: the sonic conjecture and the detachment conjecture. In Section 4, we discuss some recent developments in solving the von Neumann's conjectures and establishing a mathematical theory of shock reflection-diffraction, including the existence, regularity, and stability of global regular configurations of shock reflection-diffraction by wedges. The shock reflection-diffraction problems involve several core mathematical difficulties we have to face in solving nonlinear partial differential equations in mechanics and geometry; also see [10, 11, 17, 24, 34] and the references cited therein. These include nonlinear equations of mixed hyperbolic-elliptic type, nonlinear degenerate elliptic equations, nonlinear degenerate hyperbolic equations, free boundary problems for nonlinear degenerate equations, corner singularity/regularity especially when free boundaries meet degenerate curves, and *a priori* estimate techniques.

2 Mathematical formulation and nonlinear equations of mixed type

In this section we first formulate the shock reflection-diffraction problem into an initial-boundary value problem and then reformulate the problem as a boundary value problem in an unbounded domain for a nonlinear conservation laws of mixed elliptic-hyperbolic type.

2.1 Mathematical problems

The potential flow is governed by the conservation law of mass and the Bernoulli law:

$$\partial_t \rho + \nabla_{\mathbf{x}} \cdot (\rho \nabla_{\mathbf{x}} \Phi) = 0, \tag{2.1}$$

$$\partial_t \Phi + \frac{1}{2}|\nabla_{\mathbf{x}} \Phi|^2 + h(\rho) = B \tag{2.2}$$

for the density ρ and the velocity potential Φ, where B is the Bernoulli constant determined by the incoming flow and/or boundary conditions, and

$$h'(\rho) = \frac{p'(\rho)}{\rho} = \frac{c^2(\rho)}{\rho}$$

with $c(\rho)$ being the sound speed and p the pressure. For an ideal polytropic gas, the pressure p and the sound speed c are given by

$$p(\rho) = \kappa \rho^\gamma, \quad c^2(\rho) = \kappa \gamma \rho^{\gamma-1}$$

for constants $\gamma > 1$ and $\kappa > 0$. By scaling, we may choose $\kappa = 1/\gamma$ to have

$$h(\rho) = \frac{\rho^{\gamma-1} - 1}{\gamma - 1}, \quad c^2(\rho) = \rho^{\gamma-1}. \tag{2.3}$$

From (2.2) and (2.3), we have

$$\rho(\partial_t \Phi, \nabla_{\mathbf{x}} \Phi) = h^{-1}\big(B - \partial_t \Phi - \frac{1}{2}|\nabla_{\mathbf{x}} \Phi|^2\big). \tag{2.4}$$

Then system (2.1)–(2.2) can be rewritten as

$$\partial_t \rho(\partial_t \Phi, \nabla_{\mathbf{x}} \Phi) + \nabla_{\mathbf{x}} \cdot \big(\rho(\partial_t \Phi, \nabla_{\mathbf{x}} \Phi) \nabla_{\mathbf{x}} \Phi\big) = 0 \tag{2.5}$$

with $\rho(\partial_t \Phi, \nabla_{\mathbf{x}} \Phi)$ determined by (2.4).

When a vertical planar shock perpendicular to the flow direction x_1 and separating two uniform states (0) and (1), with constant velocities $(u_0, v_0) = (0, 0)$ and $(u_1, v_1) = (u_1, 0)$, and constant densities $\rho_1 > \rho_0$ (state (0) is ahead or to the right of the shock in the figures and state (1) is behind the shock), hits a symmetric wedge:

$$W := \{|x_2| < x_1 \tan(\theta_w), x_1 > 0\}$$

head on at time $t = 0$, then a reflection-diffraction process takes place when $t > 0$. Since state (1) does not satisfy the slip boundary condition $\partial_\nu \varphi = 0$ prescribed on the wedge boundary, the solution must differ from state (1) on and near the wedge boundary. Mathematically, the shock reflection-diffraction problem is a multidimensional lateral Riemann problem in the domain $\mathbb{R}^2 \setminus \bar{W}$.

Problem 1. (Lateral Riemann Problem) *Piecewise constant initial data, consisting of state* (0) *on* $\{x_1 > 0\} \setminus \bar{W}$ *and state* (1) *on* $\{x_1 < 0\}$, *are prescribed at* $t = 0$. *Seek a solution of* (1) *for* $t \geq 0$ *subject to these initial data and the boundary condition* $\nabla\Phi \cdot \nu = 0$ *on* ∂W.

Notice that the initial-boundary value problem for (2.1)–(2.2) is invariant under the scaling:

$$(t, \mathbf{x}) \to (\alpha t, \alpha \mathbf{x}), \quad (\rho, \Phi) \to \left(\rho, \frac{\Phi}{\alpha}\right) \qquad \text{for } \alpha \neq 0.$$

Thus, we seek self-similar solutions in the form of

$$\rho(t, \mathbf{x}) = \rho(\xi, \eta), \quad \Phi(t, \mathbf{x}) = t\phi(\xi, \eta) \qquad \text{for } (\xi, \eta) = \frac{\mathbf{x}}{t}.$$

Then the pseudo-potential function $\varphi = \phi - \frac{1}{2}(\xi^2 + \eta^2)$ satisfies the following nonlinear conservation laws of mixed type:

$$\operatorname{div}\big(\rho(|D\varphi|^2, \varphi)D\varphi\big) + 2\rho(|D\varphi|^2, \varphi) = 0, \tag{2.6}$$

for

$$\rho(|D\varphi|^2, \varphi) = \left(\rho_0^{\gamma-1} - (\gamma - 1)\left(\frac{1}{2}|D\varphi|^2 + \varphi\right)\right)^{\frac{1}{\gamma-1}}, \tag{2.7}$$

where the divergence div and gradient D are with respect to (ξ, η), and we have used the initial condition to fix the Bernoulli constant B in (2.2). Then equation (2.6) is a second-order nonlinear conservation law of mixed elliptic-hyperbolic type. It is elliptic if and only if

$$|D\varphi|^2 < c^2(|D\varphi|^2, \varphi) := \rho_0^{\gamma-1} - (\gamma - 1)\left(\frac{1}{2}|D\varphi|^2 + \varphi\right), \tag{2.8}$$

which is equivalent to

$$|D\varphi| < c_*(\varphi, \rho_\infty) := \sqrt{\frac{2}{\gamma + 1}\big(\rho_0^{\gamma-1} - (\gamma - 1)\varphi\big)}. \tag{2.9}$$

The type of equation, due to its nonlinearity, depends on the solution of the equation, which makes the problem truly fundamental and challenging. Elliptic regions of the solution correspond to the subsonic flow, and hyperbolic regions to the supersonic flow.

Shocks are discontinuities in the pseudo-velocity $D\varphi$. If curve S subdivides the domain into subdomains $\Omega^{\pm}$, and $\varphi \in C^1(\Omega^{\pm} \cup S) \cap C^2(\Omega)$, then φ is a weak solution of (2.6) in Ω across S if and only if φ satisfies (2.6) in $\Omega^{\pm}$ and the Rankine-Hugoniot conditions on S:

$$[\varphi]_S = 0, \qquad [\rho(|D\varphi|^2, \varphi)D\varphi \cdot \nu]_S = 0, \tag{2.10}$$

where $[\cdot]_S$ is the jump across S, and ν is the unit normal to S.

Then Problem 1 is reformulated as the following boundary value problem in unbounded domain

$$\Lambda := \mathbb{R}^2 \setminus \{|\eta| \le \xi \tan(\theta_w), \xi > 0\}$$

in the self-similar coordinates (ξ, η).

Problem 2. (Boundary Value Problem) *Seek a solution φ of equation* (2.6) *in the self-similar domain Λ with the slip boundary condition $D\varphi \cdot \nu|_{\partial\Lambda} = 0$ on the wedge boundary $\partial\Lambda$ and the asymptotic boundary condition at infinity:*

$$\varphi \to \bar{\varphi} = \begin{cases} \varphi_0 & \text{for } \xi > \xi_0, |\eta| > \xi \tan(\theta_w), \\ \varphi_1 & \text{for } \xi < \xi_0. \end{cases} \qquad \text{when } \xi^2 + \eta^2 \to \infty.$$

By symmetry, we can restrict to the upper half-plane $\{\eta > 0\} \cap \Lambda$, with condition $\partial_\nu \varphi = 0$ on $\{\eta = 0\} \cap \Lambda$.

3 von Neumann's conjectures on shock reflection-diffraction configurations

In this section, we discuss von Neumann's conjectures, including the sonic conjecture and the detachment conjecture, on shock reflection-diffraction configurations.

If a solution has the regular reflection-diffraction configurations as shown in Fig. 2 and Fig. 3, and if φ is smooth in the subregion between the reflected shock and the wedge, then it should satisfy the boundary condition $D\varphi \cdot \nu = 0$ and the Rankine-Hugoniot conditions (2.10) at P_0 across the reflected shock separating it from state (1). We define the uniform state (2) with pseudo-potential $\varphi_2(\xi, \eta)$ such that

$$D\varphi_2(P_0) = D\varphi(P_0),$$

and the constant density ρ_2 of state (2) is equal to $\rho(|D\varphi|^2, \varphi)(P_0)$ defined by (2.7):

$$\rho_2 = \rho(|D\varphi|^2, \varphi)(P_0).$$

Then $D\varphi_2 \cdot \nu = 0$ on the wedge boundary, and the Rankine-Hugoniot conditions (2.10) hold on the flat shock $S_1 = \{\varphi_1 = \varphi_2\}$ between states (1) and (2), which passes through P_0.

State (2) can be either subsonic or supersonic at P_0. This determines the subsonic or supersonic type of regular reflection-diffraction configurations. The supersonic regular reflection-diffraction configuration as shown in Fig. 2 consists of three uniform states (0), (1), (2), and a non-uniform state in the domain Ω, where the equation is elliptic. The reflected shock $P_0P_1P_2$ has a straight part P_0P_1. The elliptic domain Ω is separated from the hyperbolic region $P_0P_1P_4$ of state (2) by a sonic arc P_1P_4. The subsonic regular reflection-diffraction configuration as shown in Fig. 3 consists of two uniform states (0) and (1), and a non-uniform state in the domain Ω, where the equation is elliptic, $\varphi_{|\Omega}(P_0) = \varphi_2(P_0)$, and $D\varphi_{|\Omega}(P_0) = D\varphi_2(P_0)$.

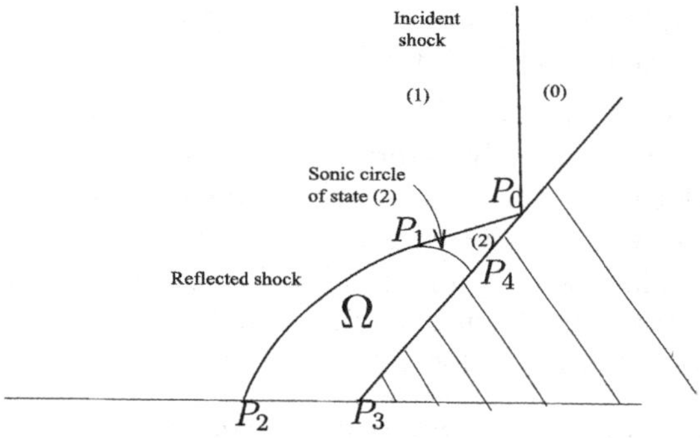

Fig. 2: Supersonic regular reflection.

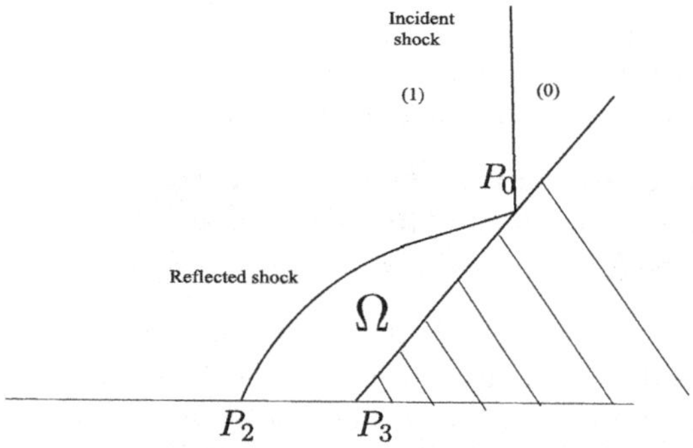

Fig. 3: Subsonic regular reflection.

Thus, a necessary condition for the existence of regular reflection solution is the existence of the uniform state (2) determined by the conditions described above. These conditions lead to an algebraic system for the constant velocity (u_2, v_2) and density ρ_2 of state (2), which has solutions for some but not all of the wedge angles. Specifically, for fixed densities $\rho_0 < \rho_1$ of states (0) and (1), there exist a sonic angle θ_w^s and a detachment angle θ_w^d satisfying

$$0 < \theta_w^d < \theta_w^s < \frac{\pi}{2}$$

such that state (2) exists for all $\theta_w \in (\theta_w^d, \frac{\pi}{2})$ and does not exist for $\theta_w \in (0, \theta_w^d)$, and the weak state (2) is

(i) supersonic at the reflection point $P_0(\theta_w)$ for $\theta_w \in (\theta_w^s, \frac{\pi}{2})$;

(ii) sonic for $\theta_w = \theta_w^s$;

(iii) subsonic for $\theta_w \in (\theta_w^d, \hat{\theta}_w^s)$

for some $\hat{\theta}_w^s \in (\theta_w^d, \theta_w^s]$. In fact, for each $\theta_w \in (\theta_w^d, \frac{\pi}{2})$, there exists also a *strong* state (2) with $\rho_2^{strong} > \rho_2^{weak}$. There had been a long debate to determine which one is physical; see Courant-Friedrichs [12], Ben-Dor [4], and the references cited therein. It is expected that strong reflection-diffraction configurations are non-physical; indeed, it is shown that the weak reflection-diffraction configuration tends to the unique normal reflection, but the strong reflection-diffraction configuration does not, when the wedge angle θ_w tends to $\frac{\pi}{2}$, as proved in Chen-Feldman [6].

If the weak state (2) is supersonic, the propagation speeds of the solution are finite and state (2) is completely determined by the local information: state (1), state (0), and the location of point P_0. That is, any information from the region of reflection-diffraction, especially the disturbance at corner P_3, cannot travel towards the reflection point P_0. However, if it is subsonic, the information can reach P_0 and interact with it, potentially altering a different reflection-diffraction configuration. This argument motivated the following conjecture von Neumann in [30, 31]:

The Sonic Conjecture. *There exists a supersonic reflection-diffraction configuration when $\theta_w \in (\theta_w^s, \frac{\pi}{2})$, i.e., the supersonicity of the weak state* (2) *implies the existence of a supersonic regular reflection-diffraction solution, as shown in Fig.* 3.

Another conjecture is that a global regular reflection-diffraction configuration is possible whenever the local regular reflection at the reflection point is possible:

The Detachment Conjecture. *There exists a regular reflection-diffraction configuration for all wedge angles $\theta_w \in (\theta_w^d, \frac{\pi}{2})$, i.e., the existence of state* (2) *implies the existence of a regular reflection-diffraction solution, as shown in Fig.* 3.

It is clear that the supersonic/subsonic regular reflection-diffraction configurations are not possible without a local two-shock configuration at the reflection point on the wedge, so this is the weakest possible criterion for the existence of (supersonic/subsonic) regular shock reflection-diffraction configurations.

4 Solution to the von Neumann conjectures

We observe that the key obstacle to the existence of regular shock reflection-diffraction configurations as conjectured by von Neumann [30, 31] is an additional possibility that, for some wedge angle $\theta_w^a \in (\theta_w^d, \frac{\pi}{2})$, shock P_0P_2 may attach to the wedge tip P_3, as observed by experimental results (*cf.* [29, Fig. 238]). To describe the conditions of such attachments, we note that

$$\rho_1 > \rho_0, \qquad u_1 = (\rho_1 - \rho_0)\sqrt{\frac{2(\rho_1^{\gamma-1} - \rho_0^{\gamma-1})}{\rho_1^2 - \rho_0^2}}.$$

Then, for each ρ_0, there exists $\rho_1^{cr} > \rho_0$ such that

$$u_1 \le c_1 \quad \text{if } \rho_1 \in (\rho_0, \rho_1^{cr}],$$
$$u_1 > c_1 \quad \text{if } \rho_1 \in (\rho_1^{cr}, \infty).$$

If $u_1 \le c_1$, we can rule out the solutions with shocks attached to the wedge tip.

If $u_1 > c_1$, this is unclear and could not be even true as experiments show (*e.g.* [29, Fig. 238]).

Thus, in [6, 7], we have obtained the following results:

(i) *If ρ_0 and ρ_1 are such that $u_1 \le c_1$, then the supersonic/subsonic regular reflection-diffraction solution exists for each wedge angle $\theta_w \in (\theta_w^d, \frac{\pi}{2})$;*

(ii) *If ρ_0, ρ_1 are such that $u_1 > c_1$, then there exists $\theta_w^a \in [\theta_w^d, \frac{\pi}{2})$ such that the regular reflection solution exists for each wedge angle $\theta_w \in (\theta_w^a, \frac{\pi}{2})$. Moreover, if $\theta_w^a > \theta_w^d$, then, for the wedge angle $\theta_w = \theta_w^a$, there exists an attached solution, i.e., a solution of Problem 2 with $P_2 = P_3$.*

The type of regular reflection (supersonic, Fig. 2, or subsonic, Fig. 3) is determined by the type of state (2) at P_0. The reflected shock P_0P_2 is $C^{2,\alpha}$-smooth. The solution φ is $C^{1,1}$ across the sonic arc for the supersonic reflection-diffraction solution, which is optimal. For the subsonic and sonic reflection-diffraction case (Fig. 3), the solution is $C^{1,\alpha}$ in Ω near P_0. Furthermore, the regular reflection-diffraction solution tends to the unique normal reflection, when the wedge angle θ_w tends to $\frac{\pi}{2}$.

These results are obtained by solving a free boundary problem for φ in Ω. The free boundary is the curved part of the reflected shock: P_1P_2 on Fig. 2, and P_0P_2 on Fig. 3. We seek a solution φ of equation (2.6) in Ω, satisfying the Rankine-Hugoniot conditions (2.10) on the free boundary, $\varphi_\nu = 0$ on the boundary of the wedge and on P_2P_3, and

$$\varphi = \varphi_2, \qquad D\varphi = D\varphi_2$$

on P_1P_4 in the supersonic case as shown in Fig. 2 and at P_0 in the subsonic case as shown in Fig. 3. The equation is expected to be elliptic in Ω for φ (unknown *a priori* before obtaining the solution), which is a part of the results.

To solve this free boundary problem, we define a class of admissible solutions of Problem 2, which are solutions φ with weak regular reflection-diffraction configuration, such that, in the supersonic reflection case, equation (2.6) is strictly elliptic for φ in $\overline{\Omega} \setminus P_1P_4$, inequalities $\varphi_2 \le \varphi \le \varphi_1$ hold in Ω, and the following monotonicity properties hold:

$$\partial_\eta(\varphi_1 - \varphi) \le 0, \quad D(\varphi_1 - \varphi) \cdot e \le 0 \qquad \text{in } \Omega,$$

where $e = \frac{P_0P_1}{|P_0P_1|}$. In the subsonic reflection case, admissible solutions are defined similarly, with changes corresponding to the structure of subsonic reflection solution. We derive uniform *a priori* estimates for admissible solutions with wedge angles $\theta_w \in [\theta_w^d + \varepsilon, \frac{\pi}{2}]$ for each $\varepsilon > 0$, and then apply the degree theory to obtain the existence for each $\theta_w \in [\theta_w^d + \varepsilon, \frac{\pi}{2}]$ in the class of admissible solutions, starting from the unique solution for $\theta_w = \frac{\pi}{2}$, called the normal reflection. To derive the *a priori* bounds, we first obtain the estimates related to the geometry of the shock: Show that the free boundary has a uniform positive distance from the sonic circle of state (1) and from the wedge boundary away from P_2 and P_0. This allows to estimate the ellipticity of (2.6) for φ in Ω (depending on the distance to the sonic arc P_1P_4 for the supersonic reflection and to P_0 for the subsonic reflection). Then we obtain the estimates near P_1P_4 (or P_0 for the subsonic reflection) in scaled and weighted $C^{2,\alpha}$ for φ and the free boundary, considering separately four cases depending on $\frac{D\varphi_2}{c_2}$ at P_0:

(i) $\frac{|D\varphi_2(P_0)|}{c_2} \geq 1 + \delta$;

(ii) Supersonic (almost sonic): $1 + \delta > \frac{|D\varphi_2(P_0)|}{c_2} > 1$;

(iii) Subsonic (almost sonic): $1 \geq \frac{|D\varphi_2(P_0)|}{c_2} \geq 1 - \delta$;

(iv) Subsonic: $\frac{|D\varphi_2(P_0)|}{c_2} \leq 1 - \delta$.

In cases (i)–(ii), equation (2.6) is degenerate elliptic in Ω near P_1P_4 on Fig. 2. In case (iii), the equation is uniformly elliptic in $\overline{\Omega}$, but the ellipticity constant is small near P_0 on Fig. 3. Thus, in cases (i)–(iii), we use the local elliptic degeneracy, which allows to find a comparison function in each case, to show the appropriately fast decay of $\varphi - \varphi_2$ near P_1P_4 in cases (i)–(ii) and near P_0 in case (iii); furthermore, combining with appropriate local non-isotropic rescaling to obtain the uniform ellipticity, we obtain the *a priori* estimates in the weighted and scaled $C^{2,\alpha}$-norms, which are different in each of cases (i)–(iii), but imply the standard $C^{1,1}$-estimates in cases (i)–(ii), and the standard $C^{2,\alpha}$-estimates in case (iii). This is an extension of methods of our earlier work [6]. In the uniformly elliptic case (iv), the solution is of subsonic reflection-diffraction configuration as shown in Fig. 3, and the estimates are more technically challenging than in cases (i)–(iii), due to the lower *a priori* regularity of the free boundary and since the uniform ellipticity does not allow a comparison function that shows the decay of $\varphi - \varphi_2$ near P_0. Thus, we prove C^{α}-estimates of $D(\varphi - \varphi_2)$ near P_0, which imply the $C^{1,\alpha}$-estimates near P_0 for φ. With all of these, we provide a solution to the von Neumann's conjectures.

In Chen-Feldman-Xiang [9], we have established the strict convexity of the curved (transonic) part of shock in the shock reflection-diffraction problem, as well as the Prandtl-Meyer reflection problem (*cf.* Bae-Chen-Feldman [2, 3]). In order to prove the convexity, we employ global properties of admissible solutions, including the existence of the cone of monotonicity discussed above.

More details can be found in Chen-Feldman [7].

Acknowledgements. The research of Gui-Qiang Chen was supported in part by the UK EPSRC Science and Innovation Award to the Oxford Centre for Nonlinear PDE (EP/E035027/1), the UK Award for the ESPRC Centre for Doctoral Training in PDE (EP/L015811/1), the NSFC under a joint project Grant 10728101, and the Royal Society–Wolfson Research Merit Award (UK). The work of Mikhail Feldman was supported in part by the National Science Foundation under Grant DMS-1101260 and by the Simons Foundation through the Simons Fellows Program.

References

[1] M. Bae, G.-Q. Chen and M. Feldman, Regularity of solutions to shock reflection problem, *Invent. Math.* **175** (2009), 505–543.

[2] M. Bae, G.-Q. Chen and M. Feldman, Prandtl-Meyer reflection for supersonic potential flow past a solid ramp, *Quart. Appl. Math.* **71** (2013), 583–600.

[3] M. Bae, G.-Q. Chen and M. Feldman, *Prandtl-Meyer Reflection Configurations, Transonic Shocks and Free Boundary Problems*, Research Monograph, Preprint, 2016.

[4] G. Ben-Dor, *Shock Wave Reflection Phenomena*, Springer-Verlag: New York, 1991.

[5] G.-Q. Chen and M. Feldman, Multidimensional transonic shocks and free boundary problems for nonlinear equations of mixed type, *J. Amer. Math. Soc.* **16** (2003), 461–494.

[6] G.-Q. Chen and M. Feldman, Global solutions to shock reflection by large-angle wedges for potential flow, *Ann. of Math.* (2) **171** (2010), 1067–1182.

[7] G.-Q. Chen and M. Feldman, *Shock Reflection-Diffraction and von Neumann's Conjectures*, Research Monograph, Annals of Mathematics Studies Series, Princeton University Press, 2016 (to appear).

[8] G.-Q. Chen and M. Feldman, Shock reflection-diffraction and nonlinear partial differential equations of mixed type. *Complex Analysis and Dynamical Systems* IV. Part 2, 55–72, Contemp. Math. **554**, AMS: Providence, 2011.

[9] G.-Q. Chen, M. Feldman and W. Xiang, Convexity of self-similar transonic shocks, Preprint, 2016.

[10] G.-Q. Chen, M. Slemrod and D. Wang, Isometric immersions and compensated compactness, *Comm. Math. Phys.* **294** (2010), 411–437.

[11] G.-Q. Chen, M. Slemrod and D. Wang, Weak continuity of the Gauss-Codazzi-Ricci system for isometric embedding, *Proc. Amer. Math. Soc.* **138** (2010), 1843–1852.

[12] R. Courant and K. O. Friedrichs, *Supersonic Flow and Shock Waves*, Springer-Verlag: New York, 1948.

[13] V. Elling, Regular reflection in self-similar potential flow and the sonic criterion, *Commun. Math. Anal.* **8** (2010), 22–69.

[14] V. Elling and T.-P. Liu, The ellipticity principle for steady and selfsimilar polytropic potential flow, *J. Hyper. Diff. Eqs.* **2** (2005), 909–917.

[15] V. Elling and T.-P. Liu, Supersonic flow onto a solid wedge. *Comm. Pure Appl. Math.* **61** (2008), 1347–1448.

[16] J. Glimm and A. Majda, *Multidimensional Hyperbolic Problems and Computations*, Springer-Verlag: New York, 1991.

[17] Q. Han and J.-X. Hong, *Isometric Embedding of Riemannian Manifolds in Euclidean Spaces*, AMS: Providence, RI, 2006.

[18] E. Harabetian, Diffraction of a weak shock by a wedge, *Comm. Pure Appl. Math.* **40** (1987), 849–863.

[19] J. K. Hunter, Transverse diffraction of nonlinear waves and singular rays, *SIAM J. Appl. Math.* **48** (1988), 1–37.

[20] J. Hunter and J. B. Keller, Weak shock diffraction, *Wave Motion* **6** (1984), 79–89.

[21] J. B. Keller and A. A. Blank, Diffraction and reflection of pulses by wedges and corners, *Comm. Pure Appl. Math.* **4** (1951), 75–94.

[22] M. J. Lighthill, The diffraction of a blast I, *Proc. Roy. Soc. London* **198A** (1949), 454–470.

[23] M. J. Lighthill, The diffraction of a blast II, *Proc. Roy. Soc. London* **200A** (1950), 554–565.

[24] C. S. Lin, The local isometric embedding in $\mathbb{R}^3$ of two-dimensional Riemannian manifolds with Gaussian curvature changing sign cleanly, *Comm. Pure Appl. Math.* **39** (1986), 867–887.

[25] G. D. Lock and J. M. Dewey, An experimental investigation of the sonic criterion for transition from regular to Mach reflection of weak shock waves, *Exp. in Fluids* **7** (1989), 289–292.

[26] E. Mach, Über den verlauf von funkenwellenin der ebene und im raume, *Sitzungsber. Akad. Wiss. Wien* **78** (1878), 819–838.

[27] C. S. Morawetz, Potential theory for regular and Mach reflection of a shock at a wedge, *Comm. Pure Appl. Math.* **47** (1994), 593–624.

[28] D. Serre, Shock reflection in gas dynamics, In: *Handbook of Mathematical Fluid Dynamics*, Vol. **4**, pp. 39–122, Elsevier: North-Holland, 2007.

[29] M. Van Dyke, *An Album of Fluid Motion*, The Parabolic Press: Stanford, 1982.

[30] J. von Neumann, Oblique reflection of shocks, *Explo. Res. Rep.* **12** (1943), Navy Department, Bureau of Ordnance, Washington, DC, USA.

[31] J. von Neumann, Refraction, intersection, and reflection of shock waves, *NAVORD Rep.* **203-45** (1945), Navy Department, Bureau of Ordnance, Washington, DC, USA.

[32] J. von Neumann, *Collect Works*, Vol. **5**, Pergamon: New York, 1963.

[33] P. Woodward and P. Colella, The numerical simulation of two-dimensional fluid flow with strong shocks, *J. Comp. Phys.* **54** (1984), 115–173.

[34] S.-T. Yau, Review of geometry and analysis, In: *Mathematics: Frontiers and Perspectives*, pp. 353–401, International Mathematics Union, Eds. V. Arnold, M. Atiyah, P. Lax, and B. Mazur, AMS: Providence, 2000.

[35] Y. Zheng, *Systems of Conservation Laws: Two-dimensional Riemann Problems*, Birkhäuser: Boston, 2001.

Mathematical Institute, University of Oxford
Oxford, OX2 6GG, England
& School of Mathematical Sciences, Fudan University
Shanghai 200433, China
E-mail address: chengq@maths.ox.ac.uk

Department of Mathematics, University of Wisconsin
Madison, WI 53706-1388, USA
E-mail address: feldman@math.wisc.edu

Proceedings of the 6th International
Congress of Chinese Mathematicians
ALM 36, pp. 25–41

Parameter Estimation in Systems and Control

Han-Fu Chen

Abstract

It is shown that a large class of problems from systems and control can be reduced to estimating unknown parameters, which can further be considered as roots of a function. The values of the function can be observed, but the observations are with errors, which may be the random noise and/or may also be caused by inaccuracy of the function itself. The classical root-seeking method known as the Robbins-Monro (RM) algorithm is aimed at estimating roots of a function on the basis of its noise-corrupted observations. However, for most of the problems arising from systems and control, the RM algorithm does not provide a convergent estimate. The stochastic approximation algorithm with expanding truncations (SAAWET) is introduced and its convergence is established. It appears that SAAWET is very successful in solving problems listed in the paper. As an example, identification of the Wiener system is carried out by the proposed approach, and a numerical example is demonstrated showing that the computation is consistent with the theoretical analysis.

Mathematics Subject Classification: 62L20, 93E.

Keywords and Phrases: System and control, parameter estimation, stochastic approximation, recursive estimation, strong consistency.

1 Introduction

Many problems from systems and control such as system identification [20], adaptive regulation and iterative learning control (ILC) for stochastic systems, principal component analysis, adaptive filtering, blind identification in communication, and a certain kind of problems from networked systems and signal processing and many others can be reduced to problems of parameter estimation.

The work is supported by NSFC (Projects Grants 61273193, 61120106011, 61134013), and by the National Center for Mathematics and Interdisciplinary Sciences, Chinese Academy of Sciences.

Let us denote the unknown parameter to be estimated by x^0. It is clear that x^0 can be considered as a root of some function $g(\cdot)$, e.g., $g(x) = A(x - x^0)$; $g(x) = \|x - x^0\|^\nu$ with $\nu > 0$; $g(x) = \sin \|x - x^0\|$ etc. Let x_k denote the kth estimate for x^0, and let O_{k+1} be the available data containing information about x^0 at time $k + 1$.

We emphasize the following point of view: O_{k+1} can always be considered as an observation on $g(x_k)$ called the regression function. This is evidenced by the following identity

$$O_{k+1} = g(x_k) + \epsilon_{k+1}, \quad \epsilon_{k+1} \triangleq O_{k+1} - g(x_k), \tag{1}$$

where ϵ_{k+1} serves as the observation noise. It is understandable that ϵ_{k+1} may contain the structural error caused by the inappropriate selection of $g(\cdot)$ and also the random noise.

Therefore, the original system and control problem has further been transformed to a root-seeking problem. To be precise, based on the available data $\{O_k\}$ given by (1), we want to estimate the root of $g(\cdot)$. This is the subject of stochastic approximation (SA) [26, 24, 2, 19]. However, such a root-seeking problem can hardly be solved by the classical Robbins-Monro (RM) algorithm [26] in SA, because for convergence of the RM algorithm a set of restrictive conditions are required.

To solve the root-seeking problems described above, we introduce a modified version of the RM algorithm called the stochastic approximation algorithm with expanding truncations (SAAWET) [4], and present a general convergence theorem (GCT) for it. SAAWET is successfully applied to solving problems from systems and control. As a matter of fact, all problems listed at beginning are solvable under reasonable and checkable conditions. Moreover, the solutions are provided in a recursive way, which is convenient for applications.

This approach suggests solving the problems by two steps: first, transform the problem to estimation of parameters or to a root-seeking problem with appropriately selected regression function, then apply SAAWET with emphasis on proving satisfaction of the conditions required by GCT.

To demonstrate such an approach, identification of the multivariate Wiener system is analyzed, and the recursive identification algorithms for both the linear and the nonlinear parts of the system are presented. All estimates given by the recursive algorithms are strongly consistent under reasonable conditions. Finally, the proposed algorithms are used to numerically identify a two-dimensional Wiener system. It is shown that the simulation results are consistent with the theoretical analysis.

The rest of the paper is organized as follows. In Section 2 various problems from diverse areas are shown to be reduced to estimating parameters and then transformed to root-seeking problems. The root-seeking algorithm SAAWET is introduced and for its convergence GCT is presented in Section 3. The proposed approach is applied to identifying the Wiener system in Section 4 to demonstrate how the method actually works. A numerical example of identification of the Wiener system is provided in Section 5, and some concluding remarks are given in Section 6.

2 Ubiquity of parameter estimation problems

The linear regression is probably the simplest parameter estimation problem. Assume that $y \in \mathbb{R}^m$ and $x \in \mathbb{R}^l$ are related in a linear way: $y = Ax$, where A is an unknown $m \times l$ matrix. Further, assume for any $x_k \in \mathbb{R}^l$ the noise-corrupted observations y_k are available:

$$y_k = Ax_k + w_k, \quad k = 1, 2, \ldots,$$

where w_k is the noise.

Based on the observations $y_1, \ldots, y_n$ we want to estimate A. For this the least squares (LS) estimate A_n is widely used, which is defined by

$$A_n \triangleq \arg\min_A \sum_{i=1}^{n} \|y_i - Ax_i\|^2.$$

The solution is $A_n \triangleq \sum_{i=1}^{n} y_i x_i^T \left(x_i x_i^T\right)^+$ and its properties have been well studied, where M^+ denotes the pseudo-inverse of a matrix M. In fact, $A_n \xrightarrow[n\to\infty]{} A$ under some conditions on $\{w_k\}$.

It is important to note that the LS estimate can recursively be derived.

Parameter estimation for linear models is the simplest estimation task. As mentioned in Introduction, parameter estimation problems may arise from various areas of systems and control. Here, we limit ourselves to show some of them.

i) Let us start with system identification for linear systems.

Assume a practical system be described by ARMAX

$$\begin{aligned}
A(z)y_k &= B(z)u_k + C(z)w_k, \\
A(z) &= I + A_1 z + \cdots + A_p z^p, \\
B(z) &= B_1 z + \cdots + B_q z^q, \\
C(z) &= I + C_1 z + \cdots + C_r z^r, \quad zy_k = y_{k-1},
\end{aligned}$$

where $y_k \in \mathbb{R}^m$ and $u_k \in \mathbb{R}^l$ are the system output and input, respectively, while $w_k \in \mathbb{R}^m$, with $Ew_k = 0$ and $E\|w_k\|^2 < \infty$ is the system noise.

The problem is to estimate unknown system orders (p, q, r) [13], the covariance matrix $R_w \triangleq Ew_k w_k^T$, and the unknown system coefficients [8, 27]

$$\begin{aligned}
&\theta_A^T \triangleq [A_1, \ldots, A_p], \qquad \theta_B^T \triangleq [B_1, \ldots, B_q], \\
&\theta_C^T \triangleq [C_1, \ldots, C_r]
\end{aligned}$$

by using the available input-output observation data $\{y_k\}$ and $\{u_k\}$.

So, identification of a linear stochastic system is, in fact, a parameter estimation problem.

ii) Let us consider the nonlinear system composed of a linear subsystem cascaded with a static nonlinearity $f(\cdot)$.

The system is called the Hammerstein system [25, 17, 1, 5, 14] when the nonlinearity is prior to the linear subsystem:

$$A(z)y_{k+1} = B(z)v_k + \xi_{k+1}, \quad v_k \triangleq f(u_k), \quad y_k = 0, \ \forall k \le 0,$$

and the cascading system is called the Wiener system [7, 16, 28, 18, 22, 23] when the nonlinearity follows the linear part:

$$y_k = f(v_k), \quad A(z)v_{k+1} = B(z)u_k, \quad v_j = 0, \ \forall j \le 0,$$

where $A(z) = I + A_1 z + \cdots + A_p z^p$ and $B(z) = B_0 + B_1 z + \cdots + B_q z^q$.

The system output y_k is observed with additive noise ε_k :

$$z_k = y_k + \varepsilon_k.$$

The task of identification is to estimate parameters

$$A_i, \ i = 1, \ldots, p, \quad B_j, \ j = 0, \ldots, q, \quad p \text{ and } q,$$

and $f(\cdot)$. Estimating $f(\cdot)$ can be carried out in two ways. We may express $f(\cdot)$ as a linear combination of basis functions with unknown coefficients. Then, estimating $f(\cdot)$ is equivalent to estimating these coefficients. We may also estimate $f(u)$ for any fixed u, and $f(u)$ may be treated as a parameter to estimate. So, in both cases identification of $f(\cdot)$ is transformed to a parameter estimation problem.

iii) Consider identification of the nonlinear ARX system [30, 3]

$$y_{k+1} = f(y_k, \ldots, y_{k+1-n_0}, u_k, \ldots, u_{k+1-n_0}) + \varepsilon_{k+1},$$

where $\{\varepsilon_k\}$ is the system noise, n_0 is the order of the system, and $f(\cdot)$ is an unknown function.

Set $\varphi_k \triangleq (y_k \ldots y_{k+1-n_0} \; u_k \ldots u_{k+1-n_0})^T$ and let $\varphi^* \in \mathbb{R}^{2n_0}$ be any fixed point. Identification of NARX consists in estimating the parameter, by which we mean the value of $f(\varphi^*)$ at any given point φ^*.

iv) Let us consider the adaptive regulation for Hammerstein and Wiener systems [7, 29].

Let y^* be the desired value of the system output.

The adaptive regulation is to design the feedback control u_k depending on the past observations $(z_0, \ldots, z_{k-1})$ to minimize

$$\limsup_{n \longrightarrow \infty} \frac{1}{n} \sum_{k=1}^{n} |y_k - y^*|^2 \quad a.s.,$$

where z_k are the noisy observations of the output of an unknown Hammerstein or Wiener system. It turns out that the optimal regulation control is a constant u^0. So, the problem is reduced to estimating "parameter" u^0.

There are many other problems that can be parameterized, for example, identification of Wiener-Hammerstein systems [21], the iterative learning control [10], some problems from networked systems such as the consensus control for multi-agent systems [15], the principal component analysis [11], blind identification in communication [9], adaptive filtering in signal processing [12], etc. Actually, parameter estimation is ubiquitous in diverse areas of systems and control.

3 Unified approach to parameter estimation

In this section we discuss how to solve the parameter estimation problems arising from systems and control.

Let x^0 denote the parameter to be estimated. As explained in Introduction, x^0 can always be considered as a root of some function $g(\cdot)$. Consequently, estimating x^0 is equivalent to approaching the root of the regression function $g(\cdot)$. It is understandable that selection of $g(\cdot)$ is important when estimating x^0.

It is clear that estimating x^0 is possible only if some data $O_k,\ k = 1, \ldots$ containing information about x^0 are available, e.g., $\{O_k\}$ may be the set of input and output data $u_k,\ y_k,\ k = 1, \ldots,$ when identifying a dynamic system.

Let x_k be the estimate for x^0 at time k. We may always treat the available data O_{k+1} as an observation on $g(x_k)$ with additive noise ϵ_{k+1} as pointed out by (1) in Introduction.

The well-known RM algorithm recursively generates the estimates x_k to approach the root x^0 of $g(\cdot)$ based on $\{O_k\}$.

Let $g(\cdot) : \mathbb{R}^m \to \mathbb{R}^m$ be with root set $J \triangleq \{x \in \mathbb{R}^m : g(x) = 0\}$.

With an arbitrary initial value x_0 the RM algorithm proposes to recursively update x_k as follows

$$x_{k+1} = x_k + a_k O_{k+1}, \quad O_{k+1} = g(x_k) + \epsilon_{k+1}, \tag{2}$$

where $\{a_k\}$ are step-sizes such that $a_k > 0,\ \ a_k \xrightarrow[k\to\infty]{} 0$, and $\sum_{k=1}^{\infty} a_k = \infty$.

In order x_k to approach to J it is required to impose a set of restrictive conditions on the regression function $g(\cdot)$ and on the noise $\{\epsilon_k\}$.

We list a set of typical conditions for convergence of the RM algorithm:

1) $\sup_{\|x-x^0\|>\epsilon} g^T(x)(x - x^0) < 0$ with $g(x^0) = 0$.

2) The growth rate of $\|g(x)\|$ as $\|x\| \to \infty$ is not faster than linearly.

3) $\{\epsilon_k\}$ is a martingale difference sequence (mds).

These conditions, however, are not easy to be met in practice. We explain this by examples.

Example 1. Let the unknown parameter $x^0 \in \mathbb{R}^m$ be considered as the root of a linear function

$$Fx^0 + f = 0, \quad F \in \mathbb{R}^{m\times m}, \quad f \in \mathbb{R}^m.$$

In other words, assume $g(x) \triangleq Fx + f$.

Assume that F and f are unknown, but at each k their approximate values F_k and f_k are available.

Then, it is natural to take $O_{k+1} \triangleq F_k x_k + f_k$, and we have

$$O_{k+1} = g(x_k) + \epsilon_{k+1}, \quad \epsilon_{k+1} \triangleq (F_k - F)x_k + f_k - f.$$

From here it is seen that the noise condition may still not be satisfied even if F_k and f_k converge to F and f, respectively, as k tends to infinity.

The following example shows that the condition 2) is also critical.

Example 2. Let the regression function be a third order polynomial $g(x) = -(x-10)^3$, and let the step-size be $a_k = \frac{1}{k+1}$.

The direct calculation shows that the RM algorithm diverges even for $\epsilon_k \equiv 0$:

$$x_0 = 0, \quad x_1 = -1000, \quad x_2 = 515149400, \dots .$$

Therefore, for convergence of the RM algorithm the linear growth rate restriction on $g(\cdot)$ is important.

Since for many practical problems the RM algorithm may not work well, we modify it by truncating it at time-varying truncation bounds. When the algorithm reaches the truncation bound we pull it back to a fixed point, and at the same time we enlarge the truncation bound for the next step. We now precisely describe it.

Let $\{M_k\}$ be a sequence of positive numbers increasingly diverging to infinity, and let $x^* \in \mathbb{R}^l$ be a fixed point. SAAWET is defined as follows:

$$x_{k+1} = (x_k + a_k O_{k+1}) I_{[\|x_k + a_k O_{k+1}\| \le M_{\sigma_k}]} + x^* I_{[\|x_k + a_k O_{k+1}\| > M_{\sigma_k}]}, \tag{3}$$

$$\sigma_k = \sum_{i=1}^{k-1} I_{[\|x_i + a_i O_{i+1}\| > M_{\sigma_i}]}, \qquad \sigma_0 = 0. \tag{4}$$

We list the conditions for convergence of SAAWET.

A1. $a_k > 0$, $a_k \xrightarrow[k\to\infty]{} 0$ and $\sum_{k=1}^{\infty} a_k = \infty$.

A2. There is a continuously differentiable function (not necessarily being non-negative) $v(\cdot) : \mathbb{R}^l \to \mathbb{R}$ such that

$$\sup_{\delta \le d(x,J) \le \Delta} g^T(x) v_x(x) < 0$$

for any $\Delta > \delta > 0$, and $v(J)$ is nowhere dense.

Further, there exists $c_0 > 0$ such that $\|x^*\| < c_0$ and $v(x^*) < \inf_{\|x\|=c_0} v(x)$ where x^* is used in the algorithm.

A3. $g(\cdot)$ is measurable and locally bounded.

A4. For the fixed sample path ω under consideration

$$\lim_{T\to 0} \limsup_{k\to\infty} \frac{1}{T} \Big\| \sum_{i=n_k}^{m(n_k,T_k)} a_i \epsilon_{i+1}(x_i(\omega), \omega) \Big\| = 0, \quad \forall T_k \in [0, T]$$

along the subscripts $\{n_k\}$ of any convergent subsequences $x_{n_k}(\omega)$, where

$$m(k,T) \triangleq \max\Big\{ m : \sum_{i=k}^{m} a_i \le T \Big\}.$$

General Convergence Theorem (GCT). *Let $\{x_k\}$ be given by SAAWET with an arbitrarily given initial value x_0. Assume A1–A3 hold. Then, $d(x_k, J) \xrightarrow[k\to\infty]{} 0$ for any sample path (ω) for which A4 holds.*

For its proof we refer to [4].

Remark 1. Condition A4 requires verification only along any convergent subsequences $\{x_{n_k}\}$ at a given trajectory of the estimates produced by the algorithm. This is a much easier task in comparison with verification along the entire sequence $\{x_k\}$, because for many problems it can be shown that $\|f(x_i) - f(x_{n_k})\| \xrightarrow[k\to\infty]{} O(T)$, $\forall i \in [n_k, n_k+1, \ldots, n_k + m(n_k, T)]$, where $f(\cdot)$ is a function. Then we have

$$\lim_{T\to 0} \limsup_{k\to\infty} \frac{1}{T}\left\| \sum_{i=n_k}^{m(n_k,T_k)} a_i(f(x_i) - f(x_{n_k}))\right\| \leq \lim_{T\to 0} O(T) = 0.$$

Remark 2. We reconsider the regression function $g(x) = -(x-10)^3$ discussed in Example 2, where it is shown that with the initial value $x_0 = 0$ and step-size $a_k = \frac{1}{k}$ the RM algorithm rapidly diverges even for the case $\epsilon_k \equiv 0$.

In contrast to this, SAAWET converges to the root 10 for a wide class of $\{\epsilon_k\}$ with an arbitrary $x^* > 0$ and any increasingly divergent $M_k > 0$.

For example, let $a_k = \frac{1}{k+1}$, $M_k = 2^{k+1}$, $x^* = 0.5$, $x_0 = 0$, $\epsilon_{k+1} - 0.9\epsilon_k = w_{k+1} + 0.5w_k$, $w_k \in \mathcal{N}(0, 0.1)$.

Then, x_k given by SAAWET converges:
$x_0 = 0$, $x_{100} = 9.26$, $x_{400} = 9.61, \ldots$.

Example 3. Assume we seek for the root x^0 of $g(\cdot)$, and let x_k be generated by (3) and (4).

The observation is $O_{k+1} = g(x_k) + \nu_{k+1}$, where the noise $\{\nu_i\}$ satisfies the following condition $\sum_{i=1}^{\infty} a_i \nu_{i+1} < \infty$.

Sometimes $g(\cdot)$ can linearly be approximated along any convergent subsequences $\{x_{n_k}\}$ such that for all sufficiently large k

$$g(x_i) - A(x^0 - x_i) = o(t), \quad \forall i \in \{n_k, \ldots, m(n_k, t)\} \quad \text{as} \quad t \to 0.$$

Then, O_{k+1} can be written as

$$O_{k+1} = A(x^0 - x_k) + \epsilon_{k+1},$$
$$\epsilon_{k+1} \triangleq g(x_k) - A(x^0 - x_k) + \nu_{k+1},$$

which clearly satisfies A4.

If we take $a_k = 1/k$, then A1 also holds.

Since $A(x^0 - x)$ is a linear function, we may take $v(x) \triangleq (x^0 - x)^T A^T A(x^0 - x)$. Then, A2 is satisfied if A is nonsingular, while A3 obviously holds.

By GCT we conclude that $x_k \xrightarrow[k\to\infty]{} x^0$. Such kind of treatment is very useful in convergence analysis.

We now consider the problem of convergence rate.

The following conditions are to be imposed.

B1. $a_k > 0$, $a_k \xrightarrow[k\to\infty]{} 0$, $\sum_{k=1}^{\infty} a_k = \infty$, and

$$\frac{a_k - a_{k+1}}{a_k a_{k+1}} \xrightarrow[k\to\infty]{} \alpha \geq 0.$$

B2. A continuously differentiable function $v(\cdot) : \mathbb{R}^l \to \mathbb{R}$ exists such that

$$\sup_{\Delta_1 \le \|x - x^0\| \le \Delta_2} g^T(x) v_x(x) < 0$$

for any $\Delta_2 > \Delta_1 > 0$, and $v(x^*) < \inf_{\|x\|=c_0} v(x)$ for some $c_0 > 0$ with $c_0 > \|x^*\|$, where x^* is used in (3).

B3. For the sample path ω under consideration the observation noise $\{\epsilon_k\}$ can be decomposed into two parts $\epsilon_k = \epsilon'_k + \epsilon''_k$ such that

$$\sum_{k=1}^{\infty} a_k^{1-\delta} \epsilon'_{k+1} < \infty, \quad \epsilon''_{k+1} = O(a_k^{\delta})$$

for some $\delta \in (0, 1]$.

B4. $g(\cdot)$ is measurable and locally bounded, and is differentiable at x^0 such that as $x \to x^0$

$$g(x) = G(x - x^0) + \delta(x), \quad \delta(x^0) = 0, \quad \delta(x) = o(\|x - x^0\|).$$

The matrix G is stable, in addition, $G + \alpha\delta I$ is also stable, where α and δ are given in B1 and B3, respectively.

Convergence Rate Theorem (CRT). *Assume B1–B4 hold. Then x_k given by* (3)–(4) *converges to x^0 with the following convergence rate:*

$$\|x_k - x^0\| = o(a_k^{\delta}),$$

where δ is the one given in B3.

For the proof we refer to [4].

4 Example of system parameter estimation

All problems listed in Section 2 have solutions with SAAWET applied, but we have no attempt to demonstrate all of them. Instead, as an example, we show how SAAWET is used to identify the Wiener system.

The Wiener system is a cascading system composed of a linear subsystem and a static nonlinearity. The control first enters the linear subsystem and then the static nonlinearity. The Wiener system is used to model diverse practical systems. For example, in a pH control problem, the linear subsystem represents the mixing dynamics of the reagent stream in a stirred vessel and the static nonlinearity describes the pH value as a function of the chemical species contained.

We adopt the nonparametric approach to identify MIMO Wiener systems by applying SAAWET incorporating with a multi-variable kernel function without requiring invertibility of the nonlinear function.

The system is expressed as follows:

$$A(z)v_k = B(z)u_k, \quad y_k = f(v_k), \quad z_k = y_k + \varepsilon_k, \tag{5}$$

where $A(z) = I + A_1 z + \cdots + A_p z^p$ and $B(z) = B_1 z + B_2 z^2 + \cdots + B_q z^q$ are the $m \times m$ and $m \times l$ matrix-valued polynomials with unknown coefficients but with known orders p and q, respectively, and z is the backward-shift operator: $zy_{k+1} = y_k$. Moreover, $u_k \in \mathbb{R}^l$, $y_k \in \mathbb{R}^m$, and $z_k \in \mathbb{R}^m$ are the system input, output, and observation, respectively; $\varepsilon_k \in \mathbb{R}^m$ is the observation noise. The unknown nonlinear function is denoted by $f(\cdot) = [f_1, f_2, \ldots, f_m]^T$, where $f_i : \mathbb{R}^m \to \mathbb{R}$. The problem is to recursively estimate A_i, $i = 1, \ldots, p$, B_j, $j = 1, \ldots, q$, and $f(v)$ at any fixed v based on the available input-output data $\{u_k, z_k\}$.

If $A(z)$ is stable, then we have

$$H(z) \triangleq A^{-1}(z)B(z) = \sum_{i=1}^{\infty} H_i z^i \quad \text{with} \quad H_1 = B_1,$$

where $\|H_i\| = O(e^{-ri})$, $\forall i > 1$ for some $r > 0$.

From here it follows that

$$\begin{aligned} & B_0 + B_1 z + \cdots + B_q z^q \\ =& (I + A_1 z + \cdots + A_p z^p)(H_0 + H_1 z + \cdots + H_i z^i + \cdots). \end{aligned} \tag{6}$$

Identifying coefficients for the same degrees of z at both sides implies

$$B_i = \sum_{j=0}^{i \wedge p} A_j H_{i-j}, \ \forall\, 1 \le i \le q, \tag{7}$$

$$H_i = -\sum_{j=1}^{i \wedge p} A_j H_{i-j}, \forall\, i \ge q+1, \tag{8}$$

where $a \wedge b$ denotes $\min\{a, b\}$.

For H_i, $q+1 \le i \le q+mp$, by (8) we obtain the following linear algebraic equation

$$[A_1, A_2, \ldots, A_p]L = -[H_{q+1}, H_{q+2}, \ldots, H_{q+mp}], \tag{9}$$

where

$$L \triangleq \begin{bmatrix} H_q & H_{q+1} & \cdots & H_{q+mp-1} \\ H_{q-1} & H_q & \cdots & H_{q+mp-2} \\ \vdots & \vdots & & \vdots \\ H_{q-p+1} & H_{q-p+2} & \cdots & H_{q+(m-1)p} \end{bmatrix}, \tag{10}$$

where $H_i \triangleq 0$ for $i < 0$.

Define

$$\theta_A^T \triangleq [A_1, \ldots, A_p], \quad G^T \triangleq -[H_{q+1}, H_{q+2}, \ldots, H_{q+mp}].$$

Then, from (9) it follows that

$$\theta_A = (LL^T)^{-1} LG, \tag{11}$$

if L is of row-full-rank.

In this case, if we can obtain estimates for $\{H_i\}$, then replacing H_i, $i = 1, 2, \dots$ in (9) with their estimates, we derive the estimate for θ_A. Finally, with the help of (7) the estimates for B_i, $\ i = 0, 1, \dots, q$ can also be obtained.

If further assume $u_k = 0, \ \forall k < 0$, then we have

$$v_k = \sum_{i=1}^{k} H_i u_{k-i}. \tag{12}$$

If $\{u_k, \ k \geq 0\}$ is a sequence of iid Gaussian random vectors $u_k \in \mathcal{N}(0, I_{l \times l})$, then $v_k \in \mathcal{N}(0, \Sigma_k)$ and $\Sigma_k = \sum_{i=1}^{k} H_i H_i^T \xrightarrow[k \to \infty]{} \Sigma \triangleq \sum_{i=1}^{\infty} H_i H_i^T$.

Before further discussing, let us introduce assumptions to be used later on.

H1. The input $\{u_k, \ k \geq 0\}$ is a sequence of iid Gaussian random vectors $u_k \in \mathcal{N}(0, I_{l \times l})$ and is independent of $\{\varepsilon_k\}$.

H2. $A(z)$ and $B(z)$ have no common left factor, $[A_p, B_q]$ is of row-full-rank, and $A(z)$ is stable: $\det A(z) \neq 0, \forall |z| \leq 1$.

H3. $f(\cdot)$ is a measurable vector-valued function satisfying the following condition:

$$|f_i(x_1, \dots, x_m)| \leq c(M + |x_1|^{\nu_1} + \cdots + |x_m|^{\nu_m}), 1 \leq i \leq m,$$

where $c > 0$, $M > 0$, and $\nu_j \geq 0, 1 \leq j \leq m$ are constants.

H4. $\{\varepsilon_k\}$ is a sequence of iid random vectors with $E\varepsilon_k = 0$ and $E[\|\varepsilon_k\|^2] < \infty$.

H5 The matrix Q is nonsingular, where $Q \triangleq E(f(v)v^T)$ with $v \in \mathcal{N}(0, \Sigma)$ and $\Sigma = \sum_{i=1}^{\infty} H_i H_i^T$.

It is important to clarify whether or not the system can uniquely be defined by the input-output $\{u_k, y_k\}$ of the system. In general, the answer is negative. To see this, let P be an $m \times m$ nonsingular matrix, and set $\tilde{A}(z) \triangleq PA(z)P^{-1}$, $\tilde{B}(z) \triangleq PB(z)$, $\tilde{f}(x) \triangleq f(P^{-1}x)$, and $\tilde{v}_k \triangleq Pv_k$. Then, the following system

$$\tilde{A}(z)\tilde{v}_k = \tilde{B}(z)u_k, \ y_k = \tilde{f}(Pv_k) = f(v_k), \ z_k = y_k + \varepsilon_k, \tag{13}$$

and the systems (5) share the same input-output $\{u_k, y_k\}$, but they have different linear and nonlinear parts. So, in order the MIMO Wiener system to be uniquely defined, the nonsingular matrix P should be fixed in advance.

Under H2, it can be shown that the matrix composed of the first m rows of L defined by (10) is of row-full-rank.

Therefore, Σ_k is nonsingular for all sufficiently large k.

By H3, $Q_{k+1} \triangleq E(f(v_{k+1})v_{k+1}^T)$ is meaningful, and $Q_k \xrightarrow[k \to \infty]{} Q$.

In order to fix coefficients of the system to be identified, let us choose P in (13) to equal $Q\Sigma^{-1}$. Then, $\tilde{P} \triangleq \tilde{Q}\tilde{\Sigma}^{-1} = I$ for system (5).

Therefore, under H5, without loss of generality, we may assume that $Q\Sigma^{-1} = I$ for system (5).

Estimation of $\{A_1, \dots, A_p, B_0, \dots, B_q\}$. It can be shown that under H1–H5 we have

$$Ez_{k+1}u_{k+1-i}^T \xrightarrow[k \to \infty]{} H_i, \ \ \forall \, i \geq 1. \tag{14}$$

We then apply SAAWET to recursively estimate H_i:

$$\begin{aligned}H_{i,k+1} =&[H_{i,k} - \frac{1}{k}(H_{i,k} - z_{k+1}u_{k+1-i}^T)]\\&\cdot I_{[\|H_{i,k}-\frac{1}{k}(H_{i,k}-z_{k+1}u_{k+1-i}^T)\|\leq M_{\delta_{i,k}}]},\\\delta_{i,k} =&\sum_{j=1}^{k-1} I_{[\|H_{i,j}-\frac{1}{j}(H_{i,j}-z_{j+1}u_{j+1-i}^T)\|>M_{\delta_{i,j}}]},\end{aligned} \tag{15}$$

where $\{M_k\}$ is an arbitrarily chosen sequence of positive real numbers increasingly diverging to infinity, $H_{i,0}$ is an arbitrary initial value.

The estimates for $\{A_1, \ldots, A_p, B_1, \ldots, B_q\}$ are naturally defined as follows:

$$[A_{1,k}, A_{2,k}, \ldots, A_{p,k}] = -[H_{q+1,k}, H_{q+2,k}, \ldots, H_{q+mp,k}]L_k^T(L_kL_k^T)^{-1}, \tag{16}$$

$$B_{i,k} = \sum_{j=0}^{i\wedge p} A_{j,k}H_{i-j,k}, \ \forall\, 1 \leq i \leq q. \tag{17}$$

Estimation of $f(\cdot)$. We now recursively estimate $f(y)$ for any fixed $y \in \mathbb{R}^m$. By using the estimates obtained for the coefficients of the linear subsystem we can estimate the internal signals v_k on the basis of the state space representations of the linear subsystem. Then, applying SAAWET incorporated with a multi-variable kernel function we obtain estimates for $f(y)$. Let us start with estimating v_k.

Define

$$C \triangleq \begin{bmatrix} -A_1 & I & & \\ \vdots & & \ddots & \\ \vdots & & & I \\ -A_s & 0 & \ldots & 0 \end{bmatrix}, \quad D \triangleq \begin{bmatrix} B_1 \\ B_2 \\ \vdots \\ B_s \end{bmatrix}, \quad H \triangleq \begin{bmatrix} I \\ 0 \\ \vdots \\ 0 \end{bmatrix},$$

and $s \triangleq \max\{p, q\}$. Then, the linear part of (5) can be presented in the state space form

$$x_{k+1} = Cx_k + Du_k, \quad v_{k+1} = H^T x_{k+1}, \tag{18}$$

where C is an $sm \times sm$ matrix, D is an $sm \times l$ matrix, and H is an $sm \times m$ matrix, $A_k = 0$ for $k > p$, and $B_k = 0$ for $k > q$.

Replacing A_i and B_j in C and D with $A_{i,k}$ and $B_{j,k}$ given by (16) and (17), respectively, $i = 1, \ldots, s$, $j = 1, \ldots, s$, we obtain the estimates C_k and D_k for C and D at time k. Thus, we define the estimate $\widehat{v}_{k+1}$ for v_{k+1} as follows:

$$\widehat{x}_{k+1} = C_{k+1}\widehat{x}_k + D_{k+1}u_k, \ \widehat{v}_{k+1} = H^T\widehat{x}_{k+1} \tag{19}$$

with an arbitrary initial value $\widehat{x}_0$.

To estimate $f(y)$, let us introduce the kernel function ω_k and its estimate $\widehat{\omega}_k$ as follows:

$$\begin{aligned}\omega_k(y) &= \frac{1}{(2\pi)^{\frac{m}{2}}(b_k)^m} e^{-\frac{(y-v_k)^T(y-v_k)}{2b_k^2}},\\ \widehat{\omega}_k(y) &= \frac{1}{(2\pi)^{\frac{m}{2}}(b_k)^m} e^{-\frac{(y-\widehat{v}_k)^T(y-\widehat{v}_k)}{2b_k^2}},\end{aligned} \tag{20}$$

where b_k is the window width of the kernel function ω_k, and $b_k = \frac{1}{k^a}$, $a > 0$ by setting.

We apply SAAWET incorporated with the kernel function to generate the estimate $\mu_k(y)$ for $f(y)$.

Take $m_k = k^b$, where $b > 0$ is such that $(m+2)a + b < \frac{1}{2}$. With $\Delta_0(y) = 0$ and arbitrary $\mu_0(y)$ we recursively estimate $f(y)$ by $\mu_k(y)$ as follows:

$$\begin{aligned}\mu_{k+1}(y) &= [\mu_k(y) - \frac{1}{k}\widehat{\omega}_k(\mu_k(y) - z_k)]\\ &\quad \cdot I_{[\|\mu_k(y)-\frac{1}{k}\widehat{\omega}_k(\mu_k(y)-z_k)\| \le m_{\Delta_k}(y)]},\\ \Delta_k(y) &= \sum_{j=1}^{k-1} I_{[\|\mu_j(y)-\frac{1}{j}\widehat{\omega}_j(\mu_j(y)-z_j)\| > m_{\Delta_j}(y)]}.\end{aligned} \tag{21}$$

Theorem 1. *Assume H1–H5 hold. Then, $H_{i,k}$, $i \ge 1$, defined by (15) converges with rate:*

$$\| H_{i,k} - H_i \| = o(k^{-\nu}) \quad a.s., \quad \forall\, \nu \in (0, 1/2). \tag{22}$$

As consequences, $A_{i,k}, i = 1, \ldots, p$ and $B_{j,k}, j = 1, \ldots, q$ defined by (16) and (17) converge and have the following convergence rate:

$$\begin{aligned}&\| A_{i,k} - A_i \| = o(k^{-\nu}) \quad a.s. \quad \forall \nu \in (0, 1/2), \quad i = 1, \ldots, p,\\ &\| B_{j,k} - B_j \| = o(k^{-\nu}) \quad a.s. \quad \forall \nu \in (0, 1/2), \quad j = 1, \ldots, q.\end{aligned}$$

Proof. We rewrite (15) as

$$\begin{aligned}H_{i,k+1} = &\left[H_{i,k} - \frac{1}{k}(H_{i,k} - H_i) - \frac{1}{k}\epsilon_{k+1}(i)\right]\\ &\cdot I_{[\|H_{i,k}-\frac{1}{k}(H_{i,k}-H_i)-\frac{1}{k}\epsilon_{k+1}(i)\| \le M_{\delta_{i,k}}]},\end{aligned}$$

where

$$\begin{aligned}\epsilon_{k+1}(i) &= H_i - z_{k+1}u_{k+1-i}^T\\ &= [H_i - Ez_{k+1}u_{k+1-i}^T] + [Ez_{k+1}u_{k+1-i}^T - z_{k+1}u_{k+1-i}^T].\end{aligned}$$

Since H_i is the single root of the linear function $-(x - H_i)$, by CRT for SAAWET it suffices to show

$$\sum_{k=1}^{\infty} \frac{1}{k^{1-\nu}}\epsilon_{k+1}(i) < \infty \quad a.s. \quad i \ge 1, \quad \forall \nu \in (0, 1/2).$$

For details we refer to [23]. □

Theorem 2. *Assume that H1–H5 hold. Then $\mu_k(y)$ defined by (21) is strongly consistent:*

$$\mu_k(y) \xrightarrow[k\to\infty]{} f(y) \quad a.s., \quad y \in \mathbb{R}^m. \tag{23}$$

Proof. The algorithm (21) can be rewritten as

$$\mu_{k+1}(y) = \left[\mu_k(y) - \frac{1}{k}\rho(y)(\mu_k(y) - f(y)) - \frac{1}{k}\bar{\epsilon}_{k+1}(y)\right] \cdot I_{[\|\mu_k(y) - \frac{1}{k}\rho(y)(\mu_k(y)-f(y)) - \frac{1}{k}\bar{\epsilon}_{k+1}(y)\| \le m_{\Delta_k}(y)]},$$

where $\bar{\epsilon}_{k+1}(y) = \widehat{\omega}_k(\mu_k(y) - z_k) - \rho(y)(\mu_k(y) - f(y))$, and

$$\rho(y) = \frac{1}{(2\pi)^{\frac{m}{2}}|\Sigma|^{\frac{1}{2}}} e^{-\frac{y^T\Sigma^{-1}y}{2}}.$$

Since $f(y)$ is the unique root of $-\rho(y)(x - f(y))$, by GCT, for (23) it suffices to prove

$$\lim_{T\to 0}\limsup_{k\to\infty} \frac{1}{T}\left\|\sum_{j=n_k}^{m(n_k,T_k)} \frac{1}{j}\bar{\epsilon}_{j+1}(y)\right\| = 0, \quad \forall T_k \in [0, T] \tag{24}$$

for any convergent subsequence $\mu_{n_k}(y)$, where

$$m(k, T) \triangleq \max\left\{m : \sum_{j=k}^{m} \frac{1}{j} \le T\right\}.$$

Write $\bar{\epsilon}_{k+1}(y)$ as $\bar{\epsilon}_{k+1}(y) = \sum_{i=1}^4 \bar{\epsilon}_{k+1}^{(i)}(y)$, where

$$\bar{\epsilon}_{k+1}^{(1)}(y) = (\widehat{\omega}_k - \omega_k)(\mu_k(y) - z_k), \quad \bar{\epsilon}_{k+1}^{(2)}(y) = (\omega_k - \rho(y))\mu_k(y),$$
$$\bar{\epsilon}_{k+1}^{(3)}(y) = \rho(y)f(y) - \omega_k f(v_k), \quad \bar{\epsilon}_{k+1}^{(4)}(y) = -\omega_k\varepsilon_k.$$

As a matter of fact, (24) can be verified with $\bar{\epsilon}_{j+1}(y)$ replaced respectively by $\bar{\epsilon}_{j+1}^{(i)}(y)$, $i = 1, 2, 3, 4$. For details we refer to [23]. □

5 Numerical example

Let the linear subsystem be given as follows

$$v_k + A_1 v_{k-1} + A_2 v_{k-2} = B_1 u_{k-1} + B_2 u_{k-2},$$

where

$$A_1 = \begin{bmatrix} 0 & 0.5 \\ -0.4 & 0 \end{bmatrix}, \quad A_2 = \begin{bmatrix} 0.3 & 0 \\ 0 & -0.5 \end{bmatrix},$$

$$B_1 = \begin{bmatrix} 0 & \sqrt{0.85} \\ -\sqrt{0.6} & 0 \end{bmatrix}, \quad B_2 = \begin{bmatrix} -0.7 & 0 \\ 0 & 0.9 \end{bmatrix},$$

and let the nonlinear function be given by $y_k = f(v_k)$ with

$$f(x,y) = \begin{bmatrix} f_1(x,y) \\ f_2(x,y) \end{bmatrix} = \begin{bmatrix} x + 0.3y^2 + 2 \\ 0.3x^2 + y + 1 \end{bmatrix}.$$

Let the observation noise $\varepsilon_k \in \mathcal{N}(0, I_2)$ be iid, and let $u_k \in \mathcal{N}(0, I_2)$ be also iid and be independent of $\{\varepsilon_k\}$.

It is noted that P for this example equals I.

The estimates of A_1, A_2, B_1, B_2 are given by Figs.1–4, respectively, the estimated curve for $f_1(\cdot)$ is given by Fig. 5, where the lower surface represents the estimation errors. The estimates for $f_1(\cdot)$ and $f_2(\cdot)$ at some particular points, namely, the estimates for $f_1(0, 0.8)$, $f_1(0.8, -0.4)$, and $f_2(-0.8, 0), f_2(-2, -1.6)$ are presented in Fig. 6.

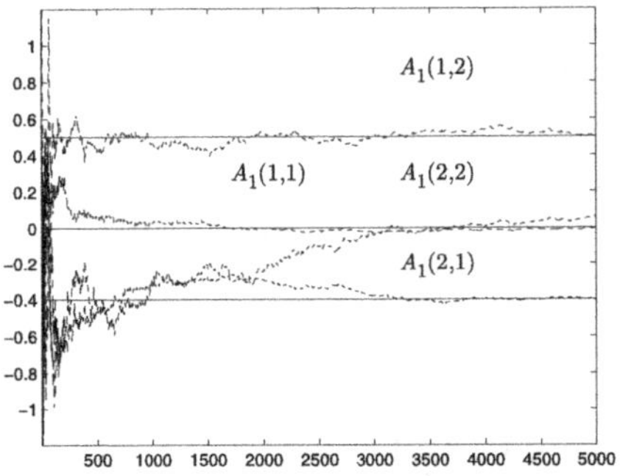

Fig. 1: Estimates for A_1.

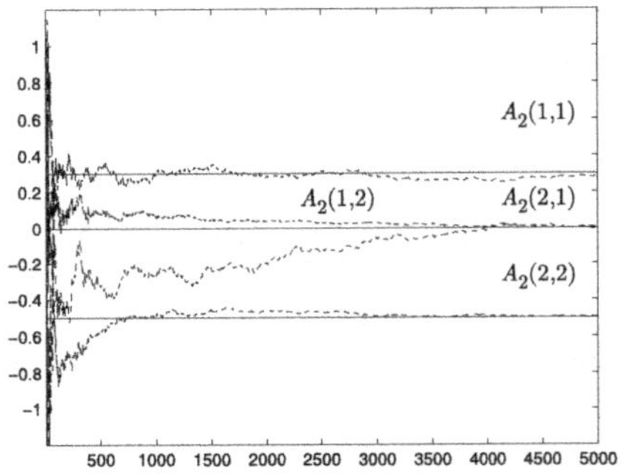

Fig. 2: Estimates for A_2.

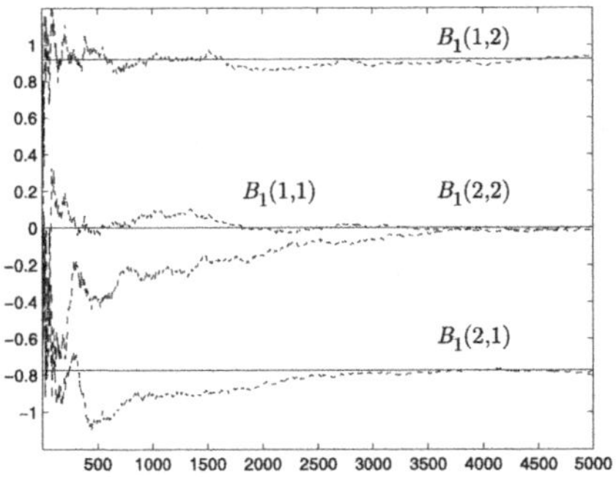

Fig. 3: Estimates for B_1.

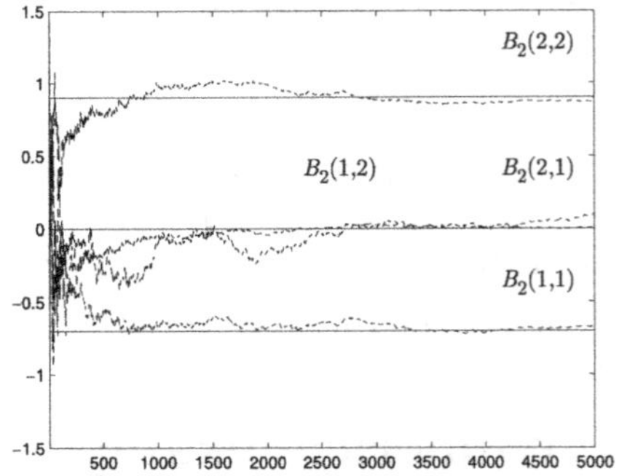

Fig. 4: Estimates for B_2.

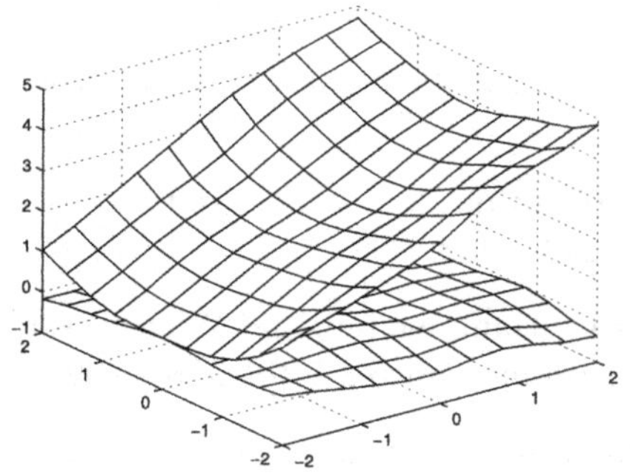

Fig. 5: Estimates for f_1 and estimation errors.

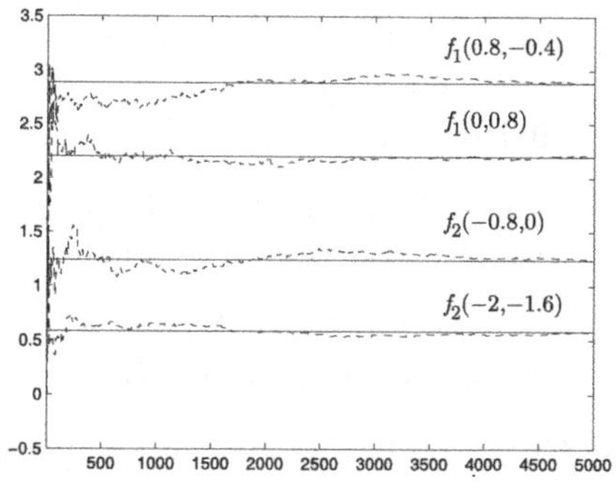

Fig. 6: Estimates for f at some particular points.

6 Concluding remarks

In the paper it is shown that many problems in systems and control can be transformed to estimating parameters, and further to root-seeking problems. The classical RM algorithm aims at solving root-seeking problems, but the conditions for its convergence are difficult to be met.

We have introduced its modified version SAAWET, which works well for the problems arising here. This provides a unified approach to solving this kind of problems.

It is natural to expect that this approach can be applied to a large class of problems in diverse areas. However, such an approach only points out the solution route, it by no means provides an automatical solution.

For a concrete realization of the solution route, the main effort is devoted to the following two things:

i) To find the appropriate observations and the regression function;

ii) To prove satisfaction of conditions required by GCT, in particular, A2 and A4.

References

[1] Bai, E. W. and D. Li (2004). Convergence of the iterative Hammerstein system identification algorithm. *IEEE Transactions on Automatic Control*, **49**, 1929–1940.

[2] Benveniste, A., M. Metivier, and P. Priouret (1990). *Adaptive Algorithms and Stochastic Approximation*, Springer-Verlag, New York.

[3] Boyd, S. and L. O. Chua (1985). Fading memory and the problem of approximating nonlinear operators with Volterra series. *IEEE Transactions on Circuits and Systems*, **32**(11), 1150–1161.

[4] Chen, H. F. (2002). *Stochastic Approximation and Its Applications*. Dordrecht, The Netherland: Kluwer.

[5] Chen, H. F. (2004). Pathwise convergence of recursive identification algorithms for Hammerstein systems. *IEEE Transactions on Automatic Control*, **49**(10), 1641–1649.

[6] Chen, H. F. (2007). Recursive identification for multivariate errors-in-variables systems. *Automatica*, **43**(7), 1234–1242.

[7] Chen, H. F. (2007). Adaptive regulator for Hammerstein and Wiener systems with noisy observations. *IEEE Trans. Autom. Control*, **52**(4), 703–709.

[8] Chen, H. F. (2010). New approach to recursive identification for ARMAX systems. *IEEE Transactions on Automatic Control*, **55**(4), 868–879.

[9] Chen H. F., X. R. Cao and J. Zhu (2002). Convergence of a stochastic approximation based algorithm for blind channel identification. *IEEE Trans. Information Theory*, **48**(5), 1214–1225.

[10] Chen, H. F. and H. T. Fang (2004). Output tracking for nonlinear stochastic systems by iterative learning control. *IEEE Trans. Autom. Control*, **49**(4), 583–588.

[11] Chen, H. F., H. T. Fang, and L. L. Zhang (2011). Recursive estimation for ordered eigenvectors of symmetric matrix with observation noise. *Journal of Mathematical Analysis and Applications*, **382**, 822–842.

[12] Chen, H.F. and G. Yin (2003). Asymptotic properties of sign algorithms for adaptive filtering. *IEEE Trans. Autom. Control*, **48**(9), 1545–1556.

[13] Chen, H. F. and W. X. Zhao (2010). New method of order estimation for ARMA/ARMAX processes. *SIAM J. Control Optim.*, **48**(6), 4157–4176.

[14] Eskinat, E., S. Johnson, and W. L. Luyben (1991). Use of Hammerstein models in identification of nonlinear systems. *AIChE Journal*, **37**(2), 255–268.

[15] Fang H. T., H. F. Chen, and L. Wen (2011). Consensus control for networked agents with noisy observations. *Journal of Systems Science and Complexity* "Online First", SpringerLink, 2011.

[16] Greblicki, W. (1997). Nonparametric approach to Wiener system identification. *IEEE Transactions on Circuits and Systems-I: Fundamental Theory and Applications*, **44**(6), 538–545.

[17] Greblicki, W. (2002). Stochastic approximation in nonparametric identification of Hammerstein systems. *IEEE Transactions on Automatic Control*, **47**(11), 1800–1810.

[18] Hu, X. L. and H. F. Chen (2005). Strong consistence of recursive identification for Wiener systems. *Automatica*, **41**, 1905–1916.

[19] Kushner, H. J. and G. Yin (2003). *Stochastic Approximation Algorithms and Applicatons*, Springer-Verlag, New York.

[20] Ljung, L and T. Söderström (1983). *Theory and Practice of Recursive Identification*, MIT Press, Cambridge, MA.

[21] Mu, B. Q. and H. F. Chen (2012). Recursive identification of Wiener-Hammerstein systems. *SIAM J. Control Optim.*, **50**(5), 2621–2658.

[22] Mu, B. Q. and H. F. Chen (2013). Recursive identification of errors-in-variables Wiener systems. *Automatica*, **49**(9), 2744–2753.

[23] Mu, B. Q. and H. F. Chen (2013-1). Recursive identification of MIMO Wiener systems. *IEEE Trans. Autom. Control*, **58**(3), 802–808.

[24] Nevelson, M. B. and R. Z. Khasminskii (1976). Stochastic approximation and recursive estimation. *Translation of Math. Monographs*, **47**, Amer. Math. Soc., Providence, RI.

[25] Pawlak, P. (1991). On the series expansion approach to the identification of Hammerstein systems. *IEEE Transactions on Automatic Control*, **36**(6), 763–767.

[26] Robbins, H. and S. Monro (1951). A Stochastic approximation method. *Ann. Math. Static.*, **22**, 400–407.

[27] Stoica, P., T. Mckelvey, and J. Mari (2000). MA estimation in polynomial time. *IEEE Trans. Signal Proceeding*, **48**(7), 1999–2012.

[28] Vörös, J. (2001). Parameter identification of Wiener systems with discontinuous nonlinearities. *Systems and Control Letters*, **44**, 363–372.

[29] Zhao, W. X. and H. F. Chen (2009). Adaptive tracking and recursive identification for Hammerstein systems. *Automatica*, **45**, 2773–2783.

[30] Zhao, W. X., H. F. Chen, and W. X. Zheng (2010). Recursive identification for nonlinear ARX systems based on stochastic approximation algorithm. *IEEE Transactions on Automatic Control*, **55**(12), 1287–1299.

KEY LABORATORY OF SYSTEMS AND CONTROL
INSTITUTE OF SYSTEMS SCIENCE, AMSS
CHINESE ACADEMY OF SCIENCES, BEIJING 100190, CHINA
E-mail address: hfchen@iss.ac.cn

Proceedings of the 6th International Congress of Chinese Mathematicians
ALM 36, pp. 43–56

Finite and Infinite Soliton and Kink-soliton Trains of Nonlinear Schrödinger Equations

Stefan Le Coz and Tai-Peng Tsai

Abstract

We will first review known results on multi-solitons of dispersive partial differential equations, which are special solutions behaving like the sum of many weakly-interacting solitary waves. We will then describe our recent joint work with Dong Li on nonlinear Schrödinger equations: Assuming the composing solitons have sufficiently large relative speeds, we prove the existence and uniqueness of a soliton train which is a multi-soliton composed of infinitely many solitons. In the 1D case, we can add to the infinite train an additional half-kink, which is a solution with a non-zero background at minus infinity.

2010 Mathematics Subject Classification: 35Q55 (35C08, 35Q51).

Keywords and Phrases: Soliton, multi-soliton, kink-soliton train, infinite soliton train, nonlinear Schrödinger equations.

1 Introduction

The nonlinear Schrödinger equation

$$\begin{cases} iu_t + \Delta u + f(u) = 0, \\ u(t=0) = u_0, \end{cases} \tag{NLS}$$

where u is a complex valued function on $\mathbb{R} \times \mathbb{R}^d$ and the nonlinearity $f : \mathbb{C} \to \mathbb{C}$ is phase covariant, i.e. there exists $g : \mathbb{R}^+ \to \mathbb{R}$ such that $f(z) = g(|z|^2)z$, appears in various physical contexts, for example in nonlinear optics or in the modelling

The research of Le Coz is supported in part by the French ANR through project ESONSE. The research of Tsai is supported in part by NSERC grant 261356-13 (Canada).

of Bose-Einstein condensates. Mathematically speaking, it is one of the model nonlinear dispersive PDE, along with the Korteweg-De Vries equation and the nonlinear wave equation.

The local Cauchy theory in the energy space $H^1(\mathbb{R}^d)$ for (NLS) is well understood (see e.g. [8] and the references cited therein). In this survey we are interested in the long time dynamics of global solutions. Essentially, two effects may be at play at large time. First of all, if the nonlinearity is not too strong, the linear part of the equation can dominate and solutions may behave as if they were solutions to the free linear Schrödinger equation. This is the *scattering effect*. On the other hand, in some cases the nonlinear term dominates and the solution tends to concentrate, with possible blow-up in finite time. This is the *focusing effect*. At the equilibrium between these two effects, one may encounter many different types of structures that neither scatter nor focus. The most common of these non-scattering global structures are the solitons, but there exist also dark solitons, kinks, etc. A generic conjecture for nonlinear dispersive PDE is the *Soliton Resolution Conjecture*. Roughly speaking, it says that, as can be observed in physical settings, any global solution will eventually decompose at large time into a scattering part and well separated non-scattering structures, usually a sum of solitons. Apart from integrable cases (see e.g. [38]), such conjecture is usually out of reach (see nevertheless the recent breakthrough [17] on energy critical wave equation). Intermediate steps toward this conjecture are existence and stability results of configurations with well separated non-scattering structure, like multi-solitons, multi-kinks, infinite soliton and kink-soliton trains, etc. Our purpose in this paper is to review some of the existing results on this topic.

2 Solitons, dark solitons and kinks

We review in this section some of the known elementary non-scattering structure solutions of (NLS).

Let us first consider special solitons, the standing waves. A *standing wave* is a solution to (NLS) of the form $u(t,x) = e^{i\omega t}\phi(x)$, where $\omega \in \mathbb{R}$ and $0 \neq \phi \in H^1(\mathbb{R}^d)$ is a localized solution to the elliptic stationary equation

$$-\Delta\phi + \omega\phi - f(\phi) = 0. \tag{1}$$

Since the ground work of Berestycki and Lions [4, 5], it is well known that (1) admits solutions in $H^1(\mathbb{R}^d)$ if $\omega > 0$ and $f(z) = g(|z|^2)z$ verify the following hypotheses.

Assumption 1. (energy-subcritical) *Let* $d \geq 1$. *Suppose* $f(z) = g(|z|^2)z$ *with* $g \in C^0([0,\infty),\mathbb{R}) \cap C^2((0,\infty),\mathbb{R})$, $g(0) = 0$,

$$|sg'(s)| + |s^2 g''(s)| \leq C \cdot (s^{\alpha_1/2} + s^{\alpha_2/2}), \qquad \forall\, s > 0,$$

where $C > 0$, $0 < \alpha_1 \leq \alpha_2 < \alpha_{\max}$, $\alpha_{\max} = +\infty$ *if* $d = 1, 2$, $\alpha_{\max} = \frac{4}{d-2}$ *if* $d \geq 3$. (focusing) *There exists* $s_0 > 0$, *such that*

$$G(s_0) := \int_0^{s_0} g(\tilde{s})d\tilde{s} > \omega s_0.$$

The profile solutions of (1) in $H^1(\mathbb{R}^d)$ are in general called *bound states*. Among bound states, it is common to distinguish between the *ground states* and the *excited states*. Recall first that three quantities are conserved along the H^1-flow of (NLS): the energy, the mass and the momentum, defined as follows.

$$E(u) := \frac{1}{2}\|\nabla u\|_2^2 - \int_{\mathbb{R}^d} F(u)dx, \qquad F(z) := \int_0^{|z|} f(s)ds,$$
$$M(u) := \frac{1}{2}\|u\|_2^2, \qquad P(u) := \Im \int_{\mathbb{R}^d} \bar{u}\nabla u dx.$$

The ground states are minimizers of the action related to (1) (defined by $S = E + \omega M$) and are in general positive, radial and unique (see [18, 23] and the recent progresses [7, 26]). For $d = 1$, there exist only ground states, whereas for $d \geq 2$ there are infinitely many excited states.

A *soliton* is a standing wave of (NLS) that has been given a speed thanks to a Galilean transform. Since (NLS) is Galilean invariant, a soliton is still a solution of (NLS). Explicitly, a soliton with frequency $\omega > 0$, speed $v \in \mathbb{R}^d$, initial phase $\gamma \in \mathbb{R}$ and position $x_0 \in \mathbb{R}^d$ has the form

$$R_{\phi,\omega,\gamma,x_0,v} := \phi(x - vt - x_0) \exp\left(i\left(\frac{1}{2}v \cdot x - \frac{1}{4}|v|^2 t + \omega t + \gamma\right)\right).$$

The dynamical properties of solitons are mostly known when the nonlinearity f is of power-type $f(z) = |z|^\alpha z$ and the profile is a ground state (we will refer to such solitons as *ground state solitons*, opposed to *excited state soliton* or in general *bound state soliton*). For L^2-subcritical α ($\alpha < \frac{4}{d}$), the ground state soliton is orbitally stable [9] (i.e. stable up to phase shifts and translations), whereas for L^2-critical and supercritical α ($\alpha = \frac{4}{d}$, $\alpha > \frac{4}{d}$) it is unstable by blow-up [3, 37] (i.e. there exists an initial data in any neighborhood of the soliton such that the corresponding solution to (NLS) blows up in finite time). In general, excited state solitons are expected to be unstable [19, 21, 31, 32, 33].

For the focusing power-type nonlinearities, solitons are the only non-scattering solutions with a fixed profile, which is always localized. However, for other nonlinearities, there exist also other types of non-scattering solutions with a fixed profile, for example the so-called dark and grey solitons or the kinks.

A *dark soliton* is a travelling wave solution of (NLS) of the form $\phi(x - ct)$, where $c \in \mathbb{R}^d$ is the velocity and ϕ is a profile which has the particularity to be non-localized but with a constant modulus at infinity. The analysis of such type of solitons is much less developed than for localized (*bright*) solitons. Most of the works deal with the case where (NLS) is the Gross-Pitaevskii equation, i.e. when $f(z) = (1 - |z|^2)z$. In this case, existence of dark solitons in dimension 1 follows from direct computations. One may refer to [10, 11] and the reference cited therein for a study of the existence and stability of dark solitons with generic nonlinearities in dimension 1. In higher dimension, existence of dark solitons has been a long time open problem. In the recent breakthrough [27], it has been proved that, for generic nonlinearities and in dimension $d \geq 3$, dark solitons exist for any speed between 0 and the speed of sound ($\sqrt{2}$ in the case of the Gross-Pitaevskii nonlinearity).

A *kink* is a soliton-type solution to (NLS) when $d = 1$,

$$K_{\phi_K,\omega,c,\gamma,x_0,v} := \phi_K(x - (c+v)t - x_0)\exp\left(i\Big(\frac{1}{2}v\cdot x - \frac{1}{4}|v|^2 t + \omega t + \gamma\Big)\right),$$

but with a profile ϕ_K which has different limits at $-\infty$ and $+\infty$. Here c is the intrinsic velocity associated to ϕ_K. When f is the Gross-Pitaevskii nonlinearity, i.e. $f(z) = (1-|z|^2)z$, there exists an explicit family of kink solutions (which are particular cases of dark solitons) given by $K(t,x) = \phi_K(x - ct)$,

$$\phi_K(x) = \sqrt{\frac{2-c^2}{2}}\tanh\left(\frac{x\sqrt{2-c^2}}{2}\right) + i\frac{c}{\sqrt{2}}, \quad |c| < \sqrt{2}. \tag{2}$$

In this paper, we will be particularly interested in cases where the kink profile ϕ has different limits *with different modulus* at $-\infty$ and $+\infty$. This will be the case if the nonlinearity f verifies the following assumption.

Assumption 2. *For some $\omega_0 > 0$, there is a first $b > 0$ such that for $h(s) = \omega_0 s - f(s)$,*

$$h(b) = 0, \quad h'(b) > 0, \quad \int_0^b h(s)ds = 0.$$

Under Assumption 2, there exists a (unique up to translation) kink-profile $\phi_K \in \mathcal{C}^2(\mathbb{R})$ with zero intrinsic velocity such that

$$\begin{cases} -\phi_K'' + \omega_0\phi_K - f(\phi_K) = 0, \\ \lim_{x\to-\infty}\phi(x) = b, \quad \lim_{x\to+\infty}\phi(x) = 0. \end{cases}$$

To our knowledge, except in our works [24, 25], such kinks have never been investigated in the analysis of the long time behavior of solutions to (NLS). One reason for that is that they are genuinely infinite energy solutions and that no proper renormalization exists to make them energy finite.

3 Multi-solitons, infinite soliton trains and kink-soliton solutions

As mentioned in Section 1, it is expected that global solutions to (NLS) will eventually decompose into a scattering part and well separated non-scattering structures. In this section, we review the existence and stability results for solutions composed of several of the elementary non-scattering structures described in Section 2. The basic example is the so-called *multi-soliton*, a solution of (NLS) build upon a finite number of solitons, Fig. 1 contains a schematic representation of a multi-soliton. To be a little more precise, let

$$R(t,x) = \sum_{j=1}^{N} R_{\phi_j,\omega_j,\gamma_j,x_j,v_j}(t,x) =: \sum_{j=1}^{N} R_j(t,x), \tag{3}$$

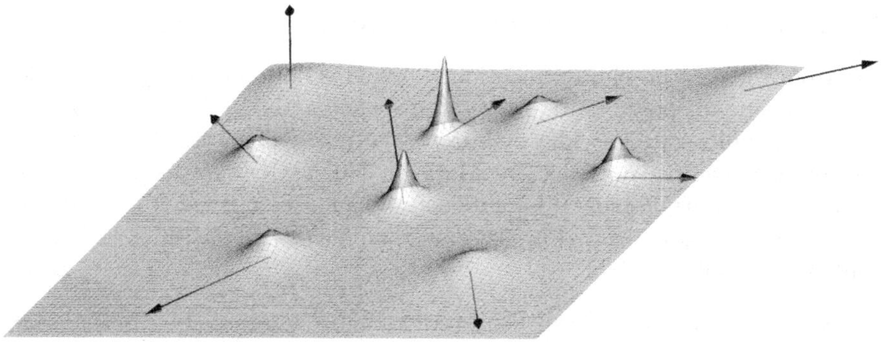

Fig. 1: Schematic representation of a multi-soliton.

where each R_j is a soliton made from some parameters $(\omega_j, \gamma_j, x_j, v_j)$ and bound state ϕ_j. As (NLS) is a nonlinear equation, R is not a solution. What we call a *multi-soliton* is a solution u of (NLS) such that

$$\lim_{t\to+\infty} \|u - R\|_{X([t,+\infty)\times\mathbb{R}^d)} = 0, \tag{4}$$

where X is some space-time function space, e.g. $L^\infty([0,+\infty), L^2(\mathbb{R}^d))$.

3.1 The integrable cases

The first result of existence of multi-solitons was obtained in Zakharov and Shabat [38] in the case of the 1-d focusing cubic (i.e. $d = 1$, $f(z) = |z|^2 z$) nonlinear Schrödinger equation via the inverse scattering method. Indeed, in this particular case the equation is completely integrable and one can obtain multi-solitons in a rather explicit manner. The companion situation to the 1-d cubic case is the Gross-Pitaevskii case, i.e. $d = 1$ and $f(z) = (1 - |z|^2)z$. In that case (NLS) is again completely integrable, however no localized solitons exist. The counter parts of the solitons in this case are the dark solitons (2), which have modulus 1 at infinity. In this situation, it is also possible to show via the inverse scattering transform the existence of a solution of (NLS) behaving at large time as decoupled well-separated dark solitons. Note that in that case, due to the non-zero condition at infinity, the profile cannot be given simply as the sum of the dark solitons.

For the 1-d cubic (NLS), Kamvissis [22] showed that it is possible to push the inverse scattering analysis forward and obtain the existence of an *infinite soliton train*, i.e. a solution u of (NLS) defined as in (4) but with $N = +\infty$ in the definition of R. In fact, the result given in [22] is sharper: it is shown that, if $d = 1$, $f(z) = |z|^2 z$ and under some technical hypotheses, any solution to (NLS) with initial data in the Schwartz class will eventually decompose at large time as an infinite soliton train and a "background radiation component". The existence of an infinite train of dark solitons has not been proved for the Gross-Pitaevskii case.

Other exotic solutions are available for 1-d cubic NLS, for example *breather* solutions which are periodic in space and whose magnitudes approach uniform values as $|t| \to \infty$, see e.g. [1].

3.2 Existence of multi-solitons, the energy method

Apart from the case previously mentioned, (NLS) is not completely integrable and it is generically not possible to apply the inverse scattering method to obtain the existence of multi-solitons. The first existence result of multi-solitons in a non-integrable setting was obtained by Merle in [30] as a by-product of the proof of existence of multiple blow-up points solutions for L^2-critical (NLS), i.e. $f(z) = |z|^{\frac{4}{d}}z$. The techniques initiated in [30] were then developed in [14, 15, 28] to obtain the following result.

Theorem 1. (Existence of multi-soliton by energy method [14, 15, 28]) *Assume $f(z) = |z|^{\alpha}z$ with $0 < \alpha < \alpha_{\max}$. Let R be the profile given in (3). Let $\omega_\star$ and $v_\star$ be given by*

$$\omega_\star = \frac{1}{2}\min\{\omega_j, j = 1, ..., N\}, \quad v_\star = \inf_{j,k=1,\dots,N, j\neq k} |v_j - v_k|.$$

There exist $\mu = \mu(d, N)$ and $v_\sharp := v_\sharp(\phi_1, ..., \phi_N) \geq 0$ such that if $v_\star > v_\sharp$ then there exist $T_0 \in \mathbb{R}$ and a solution $u \in \mathcal{C}([T_0, +\infty), H^1(\mathbb{R}^d))$ of (NLS) satisfying

$$\|u(t) - R(t)\|_{H^1} \leq e^{-\mu\sqrt{\omega_\star}v_\star t}, \quad \forall t \geq T_0.$$

If in addition all ϕ_j are ground states, then the result holds with $v_\sharp = 0$, i.e., any $v_* > 0$.

For the sake of simplicity, we have stated the result only for power-type nonlinearities, but its first part is in fact valid for any $\mathcal{C}^1$ nonlinearity verifying hypotheses a little stronger than Assumption 1. As mentioned before, the case $\alpha = \frac{4}{d}$ was treated by Merle [30], the ground state case by Martel and Merle [28] for $\alpha < \frac{4}{d}$ and Côte, Martel and Merle [15] for $\alpha > \frac{4}{d}$, and the excited state case by Côte and Le Coz [14].

The proofs in [14, 15, 28] follow a similar scheme. The idea is to choose an increasing sequence of time (T^n) with $T^n \uparrow +\infty$ and consider the solutions (u_n) to (NLS) which solve the equation backward in time with final data $u_n(T^n) = R(T^n)$. The sequence (u_n) is an approximate sequence for a multi-soliton. To show its convergence, two arguments are at play. First, one shows that there exists a time T_0 independent of n such that u_n satisfies on $[T_0, T^n]$ the *uniform estimates*

$$\|(u_n - R)(t)\|_{H^1} \leq e^{-\mu\sqrt{\omega_\star}v_\star t}.$$

Second, we have *compactness* of the sequence of initial data $u_n(T_0)$, i.e. there exists $u_0 \in H^1(\mathbb{R}^d)$ such

$$u_n \to u_0 \quad \text{strongly in } H^s(\mathbb{R}^d), \quad 0 \leq s < 1.$$

To prove the uniform estimates, one first recall that the second derivative of the action around a soliton is coercive up to some L^2 scalar products. Precisely, given a soliton R_0 with parameters $(\omega_0, \gamma_0, x_0, v_0)$, the action is given by

$$S = E + \left(\omega_0 + \frac{|v_0|^2}{4}\right)M + v_0 \cdot P$$

and there exist $K \in \mathbb{N}$ and $(\xi_k^0)_{k=1,\dots,K}$ such that we have for any $\varepsilon \in H^1(\mathbb{R}^d)$

$$\langle S''(R_0)\varepsilon, \varepsilon\rangle \gtrsim \|\varepsilon\|_{H^1}^2 - \sum_{k=1}^{K} |\left(\varepsilon, \xi_k^0\right)_2|^2.$$

Hence to control the difference $\varepsilon = u_n - R$, it is enough to construct a functional $\mathcal{S}$ similar to S but adapted to the multisoliton profile R and to get rid of the bad L^2 scalar products. The construction of the functional is done by gluing together each functional S_j suitably localized. The localization is possible since each bound state, henceforth each soliton, is exponentially decaying at infinity.

Getting rid of the bad L^2 scalar products is the trickiest part. In [28], the authors modulated the solitons in scaling, translation and phase to cancel the scalar products. This was possible only because they were in the L^2-subcritical case with ground states. In [15], the authors used a topological argument to select a final data which was not $R(T^n)$ but close enough to it, and such that for this final data the bad scalar products were vanishing. In [14], a bootstrap argument on the L^2-norm of ε was used to get an a priori control of the type $|\left(\varepsilon, \xi_k^0\right)_2| \leq \frac{1}{v_\star}\|\varepsilon\|^2$, hence allowing to control the scalar products for $v_\star$ large enough.

The compactness argument is based on the virial identity and does not present major difficulty.

The energy method is very flexible and can be adapted to many other situations, e.g. for multi-solitons of Klein-Gordon equations [2, 16] or multi-speeds solitary waves of Schrödinger systems [20]. It suits very well situations with a finite number of well localized solitons with finite energy. However, its implementation is far from being trivial when the number of solitons is infinite or when one soliton is replaced by a kink. In [24, 25], the authors have developed an approach suitable to situations where the energy technique fails to be directly applicable.

3.3 Existence of infinite trains and kink-solitons solutions, the fixed point argument

To prove the existence of solutions which have a priori infinite energy such as multi-solitons with kinks attached at both ends or infinite trains of solitons, an approach based on a fixed point argument around the desired profile has been followed in [24, 25].

We briefly recall the definition of the Strichartz space. For dimension $d \geq 1$, a pair of exponents (q, r) is said (*Schrödinger*) *admissible* if

$$\frac{2}{q} + \frac{d}{r} = \frac{d}{2}, \qquad 2 \leq q, r \leq \infty, \text{ and } (d, q, r) \neq (2, 2, \infty). \tag{5}$$

On $I \times \mathbb{R}^d$ where $I \subset \mathbb{R}$, we define the Strichartz norm

$$\|u\|_{S(I)} := \sup_{(q,\,r)\text{ admissible}} \|u\|_{L_t^q L_x^r(I\times\mathbb{R}^d)}. \tag{6}$$

For $d = 2$, we also need to impose $q > q_1$ in the above norm for some q_1 slightly larger than 2 to stay away from the forbidden endpoint. The Strichartz space $S(I)$

is the closure of all test functions in $\mathbb{R}\times\mathbb{R}^d$ under this norm. We denote by $N(I)$ the dual space of $S(I)$.

Let us first state a result on the existence of infinite soliton trains. Its conditions are far from optimal for simplicity of presentation. The same method can be applied to construct finite and infinite soliton trains for more general nonlinearities.

Theorem 2. (Existence of infinite soliton trains [24, 25]) *Let $d \geq 1$. Assume $f(z) = |z|^\alpha z + g_2(|z|^2)z$ where $g_2 \in C^0([0,\infty),\mathbb{R}) \cap C^2((0,\infty),\mathbb{R})$, $g(0) = 0$,*

$$|sg_2'(s)| + |s^2 g_2''(s)| \leq C(s^{\alpha_{1.5}/2} + s^{\alpha_2/2}), \qquad \forall\, s > 0,$$

where $C > 0$, $0 < \alpha < \alpha_{1.5} \leq \alpha_2 < \alpha_{\max}$. There exist $r_0 > \max\{1, \frac{d\alpha}{2}\}$, $c_1 > 0$ and $v_\sharp \gg 1$ such that, if an infinite soliton train profile R_∞ is given by

$$R_\infty = \sum_{j=1}^{\infty} R_j,$$

with parameters $\omega_j > 0$, $\gamma_j \in \mathbb{R}$, $x_j = 0$, $v_j \in \mathbb{R}^d$ satisfying

- *(Uniform bound for bound states) for some $0 < a < 1$ and $C > 0$,*

$$|\phi_j(x)| + \omega_j^{-1/2}|\nabla\phi_j(x)| \leq C\omega_j^{1/\alpha} e^{-a\omega_j^{1/2}|x|}, \quad \forall x \in \mathbb{R}^d, \forall j \in \mathbb{N};$$

- *(Integrability)*

$$\sum_{j=1}^{\infty} \omega_j^{\frac{1}{\alpha} - \frac{d}{2r_0}} < \infty;$$

- *(High relative speeds)*

$$v_* = \inf_{j,k\in\mathbb{N}, j\neq k} \sqrt{\omega_j}|v_k - v_j| \geq v_\sharp;$$

- *(Gradient bound)*

$$V_* = \sum_{j\in\mathbb{N}} \langle v_j\rangle \omega_j^{\frac{1}{\alpha}-\frac{d}{4}} < \infty \quad \text{if } \alpha < \frac{\alpha_2}{2+\alpha_2}.$$

Then there exists a solution u of (NLS) *satisfying, for some $T_0 \geq 0$,*

$$\|u - R_\infty\|_{S([t,\infty))} \leq e^{-c_1 v_* t}, \quad \forall t \geq T_0.$$

Moreover, such profiles R_∞ do exist for every such nonlinearity f.

Note that, in the integrable case, it was proved that an integrability condition on the parameters of the infinite trains is necessary for their existence, see [22, Remark 2].

We now switch to dimension $d = 1$ to investigate the existence of solutions of (NLS) composed of kinks and solitons as represented in Fig. 2.

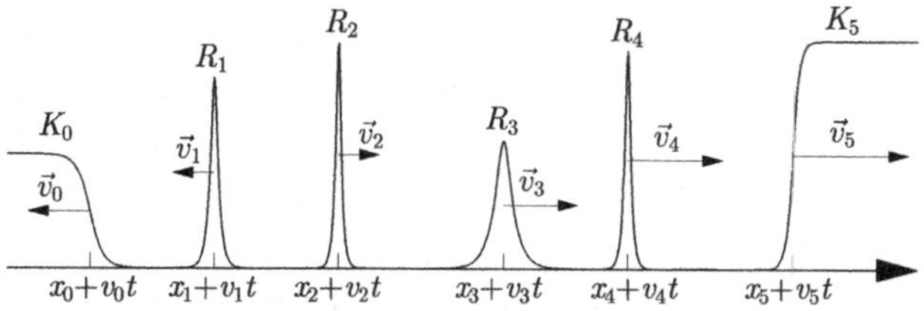

Fig. 2: Schematic representation of a kink-soliton train.

Theorem 3. (Existence of finite kink-soliton trains [24]) *Let $d = 1$. Assume that f is an energy subcritical nonlinearity such that there exists two kink profiles $\phi_{K_0}, \phi_{K_{N+1}}$ with ϕ'_{K_0} and $\phi'_{K_{N+1}}$ exponentially localized and the boundary conditions*

$$\lim_{x\to-\infty} \phi_{K_0}(x) \neq 0, \qquad \lim_{x\to+\infty} \phi_{K_0}(x) = 0,$$
$$\lim_{x\to-\infty} \phi_{K_{N+1}}(x) = 0, \qquad \lim_{x\to+\infty} \phi_{K_{N+1}}(x) \neq 0.$$

Assume also the existence of N soliton profiles $\phi_1, \dots, \phi_N$. Define the profile

$$W(t,x) := K_0(t,x) + \sum_{j=1}^{N} R_j(t,x) + K_{N+1}(t,x)$$

with parameters $(v_j, x_j, \omega_j, \gamma_j)_{j=0,\dots,N+1} \subset \mathbb{R}^4$ such that $v_0 < \dots < v_{N+1}$. Define $v_\star$ by

$$v_\star := \inf\{|v_j - v_k| : j,k = 0,\dots,N+1,\, j \neq k\}.$$

Then there exist $v_\sharp > 0$ (independent of (v_j)) large enough, $T_0 \gg 1$ and constants $\mu_1, \mu_2 > 0$ such that if $v_\star > v_\sharp$, then there exists a (unique) multi-kink solution $u \in \mathcal{C}([T_0,+\infty), H^1_{\mathrm{loc}}(\mathbb{R}))$ to (NLS) satisfying on $[T_0,+\infty)$ the estimate

$$e^{\mu_1 v_\star t}\|u - W\|_{S([t,+\infty))} + e^{\mu_2 v_\star t}\|\nabla(u - W)\|_{S([t,+\infty))} \leq 1.$$

In the above theorem, one kink can be dropped.

We may also have a kink attached to an infinite soliton train. For simplicity we choose a special nonlinearity. See [25] for general assumptions on f.

Theorem 4. (Existence of infinite kink-soliton trains [25]) *Let $d = 1$ and*

$$f(u) = |u|^\alpha u - |u|^\beta u, \quad 0 < \alpha < \beta < \infty.$$

It satisfies Assumption 2 and there is a kink profile ϕ_K. If either $0 < \alpha < 4/3$, or $4/3 \leq \alpha < \sqrt{2} < \beta = 2/\alpha$, then there exist $r_0 > 1$, $c_1 > 0$ and $v_\sharp \gg 1$ such that, if an infinite kink-soliton train profile

$$W = K + R_\infty = K + \sum_{j=1}^{\infty} R_j$$

has its parameters satisfying the same assumptions as in Theorem 2 (with index j starting from 0), then there exists a solution u of (NLS) *satisfying, for some $T_0 \geq 0$,*

$$\|u - W\|_{S([t,\infty))} \leq e^{-c_1 v_\star t}, \quad \forall t \geq T_0.$$

Moreover, such profiles W do exist.

The strategy for the proofs of Theorems 2, 3 and 4 is the following. Let W be a profile around which we want to build a solution. W can be an infinite train, a kink-soliton train, etc. Since W may be badly localized, we look for a solution of (NLS) in the form $u = W + \eta$, where η satisfies the perturbation equation

$$i\partial_t \eta + \Delta \eta + f(W + \eta) - f(W) + H = 0, \tag{7}$$

where H is a source term coming from the fact that W is not an exact solution, e.g. $H = f(R_\infty) - \sum_{j=1}^{+\infty} f(R_j)$ in the case of an infinite soliton train. In Duhamel formulation, the perturbation equation for η becomes

$$\eta(t) = -i \int_t^\infty e^{i(t-\tau)\Delta} \Big(f(W + \eta) - f(W) + H \Big) d\tau,$$

and the core of the proof is to perform a fixed point argument for this formulation. Here, two approaches are possible: One is based on a combination of the dispersive estimate

$$\|e^{it\Delta} u\|_p \lesssim |t|^{-d(\frac{1}{2} - \frac{1}{p})} \|u\|_{\frac{p}{p-1}}, \qquad \forall t \neq 0, \quad \forall p \in [2, +\infty],$$

and Strichartz estimates

$$\|u\|_{S((t_0,\infty))} \lesssim \|u_0\|_2 + \|F\|_{N((t_0,\infty))} \quad \text{for} \quad i\partial_t u + \Delta u = F, \quad u(t_0) = u_0.$$

The other uses only Strichartz estimates, but for both u and ∇u.

3.4 Open problems

We conclude this paper by reviewing some open problems related to the exotic solutions presented here.

3.4.1 Uniqueness

In Theorem 1, no uniqueness is proved for the multi-soliton. In Theorems 2, 3 and 4, the solution presented is unique *in the class of solutions satisfying a strong decay estimate towards the desired profile*; it does not preclude the possibility of existence of the same type of solution, but with a weaker decay towards the desired profile. In fact, it was proved in [14] that as soon as one of the composing soliton is linearly unstable, then there exists a one parameter family of multi-solitons converging toward the same profile. Hence in such cases uniqueness is not to be expected. However, one may hope that classification results hold, e.g. one may expect that for L^2-supercritical power-type nonlinearities, the multi-solitons converging toward a fixed N-sum of solitons form a N-parameter family. Such results were obtained for the Korteweg-de Vries equation by Combet [13]. See also [12] for partial results in that direction for (NLS). For the kink-solitons solutions or the infinite trains, no result is available yet.

3.4.2 Stability

Another natural question coming to mind when investigating the exotic solutions of nonlinear Schrödinger equation is their stability. Again, the only available results concern finite multisolitons and the problem is completely open for infinite trains or kink-soliton solutions. For power-type nonlinearities, the only case where orbital stability of a multi-soliton holds is the Gross-Pitaevskii case, as has been proved by Béthuel, Gravejat and Smets for multi-dark-solitons in [6]. Stability of multi-solitons has been proved under restrictive hypotheses in [29] for orbital stability and in [34, 35, 36] for asymptotic stability. The hypotheses on the nonlinearity (e.g. high regularity or flatness at the origin) exclude in particular the power-type nonlinearities.

3.4.3 Multikinks

One particular feature of the kinks considered in Theorem 3 is that they converge to 0 on one side. There exist however nonlinearities such that the kinks are connecting non-zero constants on both sides, for example -1 and 1 for the black solitons of Gross-Pitaevskii. There may also exist situations with kinks connecting e.g. 0 to 1 and kinks connecting 1 to 2, etc. One would expect that it is possible to construct solutions by gluing together those kinks to get a solution with modulus having increasing terraces shape. Such solutions have never been constructed for nonlinear Schrödinger equation. One reason for that is the current lack of appropriate ansatz for such terraces shape solutions.

Many other related open problems exist, e.g. constructing solutions when one of the composing element is a line-soliton in 2-d.

References

[1] N. N. Akhmediev, V. M. Eleonskiĭ, and N. E. Kulagin. First-order exact solutions of the nonlinear Schrödinger equation. *Teoret. Mat. Fiz.* **72**(2): 183–196, 1987.

[2] J. Bellazzini, M. Ghimenti, and S. Le Coz. Multi-solitary waves for the nonlinear Klein-Gordon equation. *Comm. Partial Differential Equations*, **39**(8): 1479–1522, 2014.

[3] H. Berestycki and T. Cazenave. Instabilité des états stationnaires dans les équations de Schrödinger et de Klein-Gordon non linéaires. *C. R. Acad. Sci. Paris Sér. I Math.* **293**(9): 489–492, 1981.

[4] H. Berestycki and P.-L. Lions. Nonlinear scalar field equations I. *Arch. Ration. Mech. Anal.* **82**: 313–346, 1983.

[5] H. Berestycki and P.-L. Lions. Nonlinear scalar field equations II. *Arch. Ration. Mech. Anal.* **82**(4): 347–375, 1983.

[6] F. Béthuel, P. Gravejat, and D. Smets. Stability in the energy space for chains of solitons of the one-dimensional Gross-Pitaevskii equation. *Ann. Inst. Fourier (Grenoble)*, **64**(1): 19–70, 2014.

[7] J. Byeon, L. Jeanjean, and M. Mariş. Symmetry and monotonicity of least energy solutions. *Calc. Var. Partial Differential Equations* **36**(4): 481–492, 2009.

[8] T. Cazenave. *Semilinear Schrödinger equations.* New York University – Courant Institute, New York, 2003.

[9] T. Cazenave and P.-L. Lions. Orbital stability of standing waves for some nonlinear Schrödinger equations. *Comm. Math. Phys.* **85**(4): 549–561, 1982.

[10] D. Chiron. Travelling waves for the nonlinear Schrödinger equation with general nonlinearity in dimension one. *Nonlinearity* **25**(3): 813–850, 2012.

[11] D. Chiron. Stability and instability for subsonic travelling waves of the nonlinear Schrödinger equation in dimension one. *Anal. PDE*, **6**(6): 1327–1420, 2013.

[12] V. Combet. Multi-existence of multi-solitons for the supercritical nonlinear Schrödinger equation in one dimension. *Discrete Contin. Dyn. Syst.*, **34**(5): 1961–1993, 2014.

[13] V. Combet. Multi-soliton solutions for the supercritical gKdV equations. *Comm. Partial Differential Equations* **36**(3): 380–419, 2011.

[14] R. Côte and S. Le Coz. High-speed excited multi-solitons in nonlinear Schrödinger equations. *J. Math. Pures Appl.* (9), **96**(2): 135–166, 2011.

[15] R. Côte, Y. Martel, and F. Merle. Construction of multi-soliton solutions for the L^2-supercritical gKdV and NLS equations. *Rev. Mat. Iberoam.* **27**(1): 273–302, 2011.

[16] R. Côte and C. Muñoz. Multi-solitons for nonlinear Klein-Gordon equations. *Forum Math. Sigma*, **2**: e15, 38, 2014.

[17] T. Duyckaerts, C. Kenig, and F. Merle. Classification of radial solutions of the focusing, energy-critical wave equation. *Camb. J. Math.*, **1**(1): 75–144, 2013.

[18] B. Gidas, W. M. Ni, and L. Nirenberg. Symmetry of positive solutions of nonlinear elliptic equations in $\mathbb{R}^n$. In *Mathematical Analysis and Applications, Part A*, Volume 7 of *Adv. in Math. Suppl. Stud.*, pages 369–402. Academic Press, New York, 1981.

[19] M. Grillakis. Linearized instability for nonlinear Schrödinger and Klein-Gordon equations. *Comm. Pure Appl. Math.* **41**(6): 747–774, 1988.

[20] I. Ianni and S. Le Coz. Multi-speed solitary wave solutions for nonlinear Schrödinger systems. *J. Lond. Math. Soc. (2)*, **89**(2): 623–639, 2014.

[21] C. K. R. T. Jones. An instability mechanism for radially symmetric standing waves of a nonlinear Schrödinger equation. *J. Differential Equations* **71**(1): 34–62, 1988.

[22] S. Kamvissis. Focusing nonlinear Schrödinger equation with infinitely many solitons. *J. Math. Phys.* **36**(8): 4175–4180, 1995.

[23] M. K. Kwong. Uniqueness of positive solutions of $\Delta u - u + u^p = 0$ in $\mathbb{R}^n$. *Arch. Rational Mech. Anal.* **105**(3): 243–266, 1989.

[24] S. Le Coz, D. Li, and T.-P. Tsai. Fast-moving finite and infinite trains of solitons for nonlinear Schrödinger equations. *Proc. Roy. Soc. Edinburgh Sect. A*, **145**(6): 1251–1282, 2015.

[25] S. Le Coz and T.-P. Tsai. Infinite soliton and kink-soliton trains for nonlinear Schrödinger equations. *Nonlinearity*, **27**(11): 2689–2709, 2014.

[26] M. Mariş. On the symmetry of minimizers. *Arch. Ration. Mech. Anal.* **192**(2): 311–330, 2009.

[27] M. Mariş. Traveling waves for nonlinear Schrödinger equations with nonzero conditions at infinity. *Ann. of Math.* (2), **178**(1): 107–182, 2013.

[28] Y. Martel and F. Merle. Multi solitary waves for nonlinear Schrödinger equations. *Ann. Inst. H. Poincaré Anal. Non Linéaire* **23**(6): 849–864, 2006.

[29] Y. Martel, F. Merle, and T.-P. Tsai. Stability in H^1 of the sum of K solitary waves for some nonlinear Schrödinger equations. *Duke Math. J.* **133**(3): 405–466, 2006.

[30] F. Merle. Construction of solutions with exactly k blow-up points for the Schrödinger equation with critical nonlinearity. *Comm. Math. Phys.* **129**(2): 223–240, 1990.

[31] T. Mizumachi. Instability of bound states for 2D nonlinear Schrödinger equations. *Discrete Contin. Dyn. Syst.* **13**(2): 413–428, 2005.

[32] T. Mizumachi. Vortex solitons for 2D focusing nonlinear Schrödinger equation. *Differential Integral Equations* **18**(4): 431–450, 2005.

[33] T. Mizumachi. Instability of vortex solitons for 2D focusing NLS. *Adv. Differential Equations* **12**(3): 241–264, 2007.

[34] G. Perelman. Some results on the scattering of weakly interacting solitons for nonlinear Schrödinger equations. In *Spectral Theory, Microlocal Analysis, Singular Manifolds*, Volume 14 of *Math. Top.*, pages 78–137. Akademie Verlag, Berlin, 1997.

[35] G. Perelman. Asymptotic stability of multi-soliton solutions for nonlinear Schrödinger equations. *Comm. Partial Differential Equations* **29**(7–8): 1051–1095, 2004.

[36] I. Rodnianski, W. Schlag, and A. Soffer. Asymptotic stability of N-soliton states of nonlinear Schrödinger equations, 2003. arXiv:math/0309114.

[37] M. I. Weinstein. Nonlinear Schrödinger equations and sharp interpolation estimates. *Comm. Math. Phys.* **87**(4): 567–576, 1982/83.

[38] V. E. Zakharov and A. B. Shabat. Exact theory of two-dimensional self-focusing and one-dimensional self-modulation of waves in nonlinear media. *Šoviet Physics JETP* **34**(1): 62–69, 1972.

Institut de Mathématiques de Toulouse, Université Paul Sabatier
118 route de Narbonne, 31062 Toulouse Cedex 9, France
E-mail address: slecoz@math.univ-toulouse.fr

Department of Mathematics, University of British Columbia
Vancouver BC, Canada V6T 1Z2
E-mail address: ttsai@math.ubc.ca

Proceedings of the 6th International Congress of Chinese Mathematicians
ALM 36, pp. 57–78

A Langlands Program for Covering Groups?

Wee Teck Gan

Abstract

We describe a still largely speculative and conjectural extension of the Langlands program to a class of nonlinear covering groups studied by Brylinski and Deligne. This is an exposition of the work of many people, especially the recent work of Martin Weissman, Peter McNamara and Wen-Wei Li.

Mathematics Subject Classification: Primary 54C40, 14E20; Secondary 46E25, 20C20.

Keywords and Phrases: Covering groups, Brylinski-Deligne extensions, Langlands program.

1 Covering groups

The classical Langlands program concerns representations and automorphic forms of a connected reductive linear algebraic group G defined over a local or global field F, The class of linear algebraic groups enjoy several good properties:

- Naturality: these groups first arise as connected compact Lie groups which intervene in many walks of mathematical life.
- Algebraicity: even though they started life as topological groups, their theory is in fact totally algebraic in the sense that they are the F-points of an algebraic variety over F.
- Functoriality: Being algebraic in nature, they have good functorial properties, such as with respect to base change.
- Inductivity: Levi factors of parabolic subgroups of reductive groups are reductive. This provides a framework to treat their representation theory inductively.

- Classification: they admit a simple classification by combinatorial data, namely root systems and Dynkin diagrams.

More precisely, pick a maximal torus $T \subset G$ defined over F. To (G, T), we can attach a quadruple:

$$(X(T), \Phi, Y(T), \Phi^\vee)$$

called the root datum of G:

- $X(T) = \mathrm{Hom}(T, \mathbb{G}_m)$ is the character group of T;
- $Y(T) = \mathrm{Hom}(\mathbb{G}_m, T)$ is the cocharacter group of T;
- $\Phi = \{\text{roots } \alpha \text{ of } (G, T)\} \subset X(T)$;
- $\Phi^\vee = \{\text{coroots } \alpha^\vee \text{ of } (G, T)\} \subset Y(T)$.

Such root data classify the connected reductive linear algebraic groups up to isomorphism over algebraically closed fields.

In this expository paper, we are interested in a larger class of groups called *nonlinear covering groups*. More precisely, a central extension of $G(F)$ by a discrete (finite) group μ is a short exact sequence:

$$1 \longrightarrow \mu \longrightarrow \tilde{G} \xrightarrow{\ p\ } G(F) \longrightarrow 1$$

with μ in the centre of $\tilde{G}$. One may consider such extensions in the category of abstract groups, or in the category of topological groups, in which case we require p to be open and continuous. Thus, such groups are in general not algebraic in nature: they are not the F-points of an algebraic group.

1.1 2-cocycles

Such covering groups are often described using 2-cocycles. Indeed, central extensions of $G(F)$ by μ are classified by the cohomology group:

$$H^2(G(F), \mu) \quad \text{for abstract extensions}$$

or

$$H^2_m(G(F), \mu) \quad \text{for topological extensions}$$

where one considers measurable cochains. Given

$$1 \longrightarrow \mu \longrightarrow \tilde{G} \xrightarrow{\ p\ } G(F) \longrightarrow 1,$$

one picks a (measurable) section $j : G(F) \longrightarrow \tilde{G}$ of p, and set

$$c(g, h) = j(g) \cdot j(h) \cdot j(gh)^{-1}.$$

Then c is a 2-cocyle and its class in $H^2_m(G(F), \mu)$ is well-defined.

Conversely, given a 2-cocycle c, $\tilde{G}$ is given as a set by $G(F) \times \mu$, with group law

$$(g, \alpha) \cdot (h, \beta) = (g \cdot h, \alpha \cdot \beta \cdot c(g, h)).$$

1.2 Some examples

Let us look at some examples of such covering groups. Let G be a simple split simply-connected linear algebraic group over $\mathbb{R}$. Then $G(\mathbb{R})$ is a connected real Lie group, but it is not necessarily simply-connected as a real Lie group. If $K \subset G(\mathbb{R})$ is a maximal compact subgroup, then

$$\pi_1(G(\mathbb{R})) = \pi_1(K).$$

Here are some examples:

G	$\mathrm{SL}_n(\mathbb{R})(n > 2)$	$\mathrm{Sp}_{2n}(\mathbb{R})$	$G_2(\mathbb{R})$	$F_4(\mathbb{R})$
K	$\mathrm{SO}_n(\mathbb{R})$	$\mathrm{U}_n(\mathbb{C})$	$(\mathrm{SU}_2 \times \mathrm{SU}_2)/\Delta\mu_2$	$(\mathrm{SU}_2 \times \mathrm{Sp}_6^c)/\Delta\mu_2$
$\pi_1(G)$	$\mathbb{Z}/2\mathbb{Z}$	$\mathbb{Z}$	$\mathbb{Z}/2\mathbb{Z}$	$\mathbb{Z}/2\mathbb{Z}$

When $\pi_1(K)$ is nontrivial, the universal cover $\tilde{G}$ of $G(\mathbb{R})$ is then a nonlinear covering group. It is interesting to note that Harish-Chandra's work on non-abelian harmonic analysis includes such groups.

1.3 Metaplectic group $\mathrm{Mp}_{2n}(F)$

Another classic example of a covering group is the metaplectic group $\mathrm{Mp}_{2n}(F)$ defined by Weil:

$$1 \longrightarrow \mu_2 \longrightarrow \mathrm{Mp}_{2n}(F) \xrightarrow{p} \mathrm{Sp}_{2n}(F) \longrightarrow 1.$$

It is a 2-fold cover of the symplectic group $\mathrm{Sp}_{2n}(F)$.

This group has a family of distinguished representations: the Weil representations or oscillator representations which intervene in the theory of the quantum harmonic oscillator and provides a representation theoretic framework for the theory of theta functions. It is one of the best understood covering group, but even so, it is not easy to write down a 2-cocycle.

1.4 Steinberg's group

Yet another example is a group constructed by Steinberg [St]. Let

- $G \neq \mathrm{SL}_2$ be a split, absolutely simple and simply connected linear algebraic group over F;
- T be a maximal split torus of G;
- $\Phi = \Phi(G, T)$ the associated root system, assumed to be simply laced for simplicity.
- $\{x_\alpha : \mathbb{G}_a \to U_\alpha : \alpha \in \Phi\}$ a Chevalley pinning (or épinglage).

Then one has:

$$G(F) = \langle x_\alpha(t),\ \alpha \in \Phi,\ t \in F\rangle / (\text{relations (A) and (B)}).$$

where the relations (A) and (B) are:

(A) For $\alpha + \beta \neq 0$,

$$[x_\alpha(t), x_\beta(u)] = \begin{cases} x_{\alpha+\beta}(\pm tu), & \text{if } \alpha + \beta \in \Phi; \\ 0, & \text{else.} \end{cases}$$

(B) If we set

$$w_\alpha(t) = x_{-\alpha}(t) \cdot x_\alpha(-1/t) \cdot x_{-\alpha}(t),$$
$$h_\alpha(t) = w_\alpha(t) \cdot w_\alpha(-1),$$

then

$$h_\alpha(s \cdot t) = h_\alpha(s) \cdot h_\alpha(t) = h_\alpha(t) \cdot h_\alpha(s).$$

Let E be the abstract group generated by the $x_\alpha(t)$'s subject to (A) only, so that there is a short exact sequence:

$$1 \longrightarrow C \longrightarrow E \longrightarrow G(F) \longrightarrow 1.$$

The group E is the Steinberg group and it is a nonlinear covering group of $G(F)$.

1.5 Universal abstract extension

The following is a combination of results of Steinberg [St], Moore [M] and Matsumoto [Ma]:

Theorem 1.1. (i) *The extension E is the universal abstract central extension, in the sense that for any abelian group A,*

$$H^2(G, A) = \mathrm{Hom}(C, A).$$

(ii) *When $G \neq \mathrm{Sp}_{2n}$, C is generated by the elements*

$$(s, t) = h_\alpha(st) \cdot h_\alpha(s)^{-1} \cdot h_\alpha(t)^{-1},$$

which are independent of $\alpha \in \Phi$, subject to the relations:

(i) $(s, t) \cdot (t, s) = 1$;

(ii) (s, t) *is bimultiplicative;*

(iii) $(s, t) = 1$ *if* $s + t = 1$.

1.6 Milnor-Quillen K_2

The above theorem is subsequently extended to all isotropic simply-connected G by Deodhar [D] and Prasad-Raghunathan [PR1, PR2], and E is called the universal abstract covering group. When $G \neq \mathrm{Sp}_{2n}$, its kernel C is the Milnor-Quillen K_2-group of F:

$$K_2(F) = F^\times \otimes_{\mathbb{Z}} F^\times / (\text{relations (i) – (iii)}).$$

A map

$$F^\times \times F^\times \longrightarrow A$$

satisfying the relations (i) – (iii) is called an A-valued symbol, and the map

$$(-,-) : F^\times \times F^\times \longrightarrow K_2(F)$$

is the universal symbol. Thus, any A-valued symbol arises from a group homomorphism $K_2(F) \longrightarrow A$.

1.7 Universal topological extension and Hilbert symbol

Topological coverings of $G(F)$ are obtained from the universal abstract extension by applying locally constant symbols [M, Ma].

Theorem 1.2. *The universal locally constant symbol is the Hilbert symbol*

$$(-,-)_F : F^\times \times F^\times \to K_2(F) \to \mu(F)$$

where

$$\mu(F) = \{\textit{roots of unity in } F\}.$$

For every simply-connected isotropic simple algebraic group over F, we have thus seen that there is a canonical abstract extension

$$1 \longrightarrow K_2(F) \longrightarrow E \longrightarrow G(F) \longrightarrow 1$$

and a canonical topological extension:

$$1 \longrightarrow \mu(F) \longrightarrow E \longrightarrow G(F) \longrightarrow 1.$$

However, this nice picture becomes a bit murky once G is not simply-connected or isotropic:

- there is no universal cover, because of presence of automorphisms;
- there are too many coverings, with no reasonable classification;
- it is hard to describe any given cover in terms of 2-cocycles;
- there are no good functorial properties.

Thus it is reasonable to ask if there is a class of covering groups which possesses better functorial properties, admits a sensible classification and yet is sufficient large to include all interesting examples. We will discuss such a class of covering groups singled out by Brylinski-Deligne [BD] in the next section.

2 Brylinski-Deligne extensions

Let us return to the natural extension constructed by Steinberg when G is simply-connected:

$$1 \longrightarrow K_2(F) \longrightarrow E \longrightarrow G(F) \longrightarrow 1.$$

Brylinski-Deligne's idea is to remove the F and consider the extension

$$1 \longrightarrow K_2 \longrightarrow E \longrightarrow G \longrightarrow 1$$

in an appropriate category. More precisely, they considered the problem:
Problem. Classify all central extensions of a connected reductive linear algebraic group G over F by Quillen's K_2-sheaf in the category of sheaves of groups on the big Zariski site of F-schemes of finite type.
Such extensions are also called *multiplicative K_2-torsors on G*, and we shall refer to them as BD-extensions.

2.1 Classification of BD extensions

Brylinski-Deligne [BD] gave a classification of the category (a groupoid actually) of multiplicative K_2-torsors on G. We shall describe their answer in two cases:

- $G = T$ is a torus;
- G is simply-connected absolutely simple.

The classification will be in terms of root theoretic data, with some extra enhancements.

2.2 Simply-connected G

When G is simply connected, one has the following theorem.

Theorem 2.1. *The category of multiplicative K_2-torsors on G is equivalent to the rigid groupoid with objects*

$$Q : Y(T) \to \mathbb{Z},$$

which are $W(G,T) \rtimes \mathrm{Gal}(\overline{F}/F)$-invariant $\mathbb{Z}$-valued quadratic forms. Here $W(G,T)$ is the Weyl group of T in G.

For example, if G is split and absolutely simple, then such a Q is determined by $Q(\alpha^\vee)$ where $\alpha^\vee$ is a short coroot. So the set of isomorphism classes of objects is just $\mathbb{Z}$.

2.3 Tori

Let T be a torus defined over F, with cocharacter group $Y(T)$, which is a $\mathrm{Gal}(\overline{F}/F)$-module.

Theorem 2.2. *The category of multiplicative K_2-torsors on T is equivalent to the category with*

(i) *Objects: pairs $(Q, \mathcal{E})$ where*

  - $Q : Y(T) \to \mathbb{Z}$ *is a* $\mathrm{Gal}(\overline{F}/F)$*-invariant $\mathbb{Z}$-valued quadratic form;*
  - $\mathcal{E}$ *is a* $\mathrm{Gal}(\overline{F}/F)$*-equivariant extension of abstract groups*

$$1 \longrightarrow F^\times \longrightarrow \tilde{Y} \longrightarrow Y(T) \longrightarrow 1$$

whose commutator is given by

$$[\tilde{y_1}, \tilde{y_2}] = (-1)^{b_Q(y_1,y_2)},$$

with $b_Q(y_1, y_2) = Q(y_1 + y_2) - Q(y_1) - Q(y_2)$.

(ii) *Morphisms:* $\mathrm{Hom}((Q,\mathcal{E}),(Q',\mathcal{E}'))$ *is nonempty only if* $Q = Q'$, *in which case a morphism is a* $\mathrm{Gal}(\overline{F}/F)$*-equivariant isomorphism of extensions:*

$$\begin{array}{ccccccccc} 1 & \longrightarrow & F^\times & \longrightarrow & \tilde{Y} & \longrightarrow & Y(T) & \longrightarrow & 1 \\ & & \| & & \downarrow & & \| & & \\ 1 & \longrightarrow & F^\times & \longrightarrow & \tilde{Y}' & \longrightarrow & Y(T) & \longrightarrow & 1 \end{array}$$

For general reductive G, the classification is an amalgam of the above two cases and is slightly more complicated to describe. However, the point to note here is that the classification of BD-extensions is largely root-theoretic in nature. As such, it is not too far from the familiar classification of linear algebraic groups by their root data.

2.4 BD covering groups

Starting with a multipicative K_2-torsor

$$1 \longrightarrow K_2 \longrightarrow E \longrightarrow G \longrightarrow 1,$$

one may take F-points to get an extension of abstract groups:

$$1 \longrightarrow K_2(F) \longrightarrow E(F) \longrightarrow G(F) \longrightarrow H^1(F, K_2) = 1.$$

Pushing out by the Hilbert symbol $K_2(F) \longrightarrow \mu(F)$, one obtains:

$$1 \longrightarrow \mu(F) \longrightarrow \tilde{G} \longrightarrow G(F) \longrightarrow 1.$$

If n is such that $\mu_n(\overline{F}) \subset \mu(F)$, then one has a natural map $\mu(F) \longrightarrow \mu_n(F)$ and one can push out further to get

$$1 \longrightarrow \mu_n(F) \longrightarrow \tilde{G} \longrightarrow G(F) \longrightarrow 1.$$

This gives a class of (topological) covering groups, associated to (E, n), which we will call the *BD*-covering groups.

Though the set of all BD-extensions does not exhaust all covering groups, there are some good reasons to believe that they are a good class of covering groups to consider:

- Algebraicity: they begin life in the world of algebraic geometry;
- Functoriality: BD data behaves reasonably with respect to base change, pullbacks and other natural operations in algebraic geometry;
- Exhaustive: BD coverings capture a large class of covering groups, e.g. all covers for simply-connected $G \neq \mathrm{Sp}_{2n}$;
- Classification: they admit an essentially root theoretic classification.

Given these advantages, we would like to ask:

Question. Do BD covering groups give a good framework for extending the Langlands program?

3 Local Langlands correspondence (LLC)

If F is a local field and G is a linear algebraic group over F, the LLC gives a classification

$$\mathrm{Irr}(G(F)) = \{\text{irreducible representations of } G(F)\}$$

$$\downarrow \quad \text{finite-to-one}$$

$$\Phi(G(F)) = \{L\text{-parameters for } G \text{ over } F\}.$$

We describe some of the objects appearing in this statement more precisely.

3.1 L-parameter

An L-parameter is an equivalence class of homomorphisms

$$\phi : W_F \times \mathrm{SL}_2(\mathbb{C}) \longrightarrow {}^L G(\mathbb{C})$$

(with some properties), where

- W_F = the Weil group of F; it is almost the same as the absolute Galois group $\mathrm{Gal}(\overline{F}/F)$.
- ${}^L G(\mathbb{C})$ = the *Langlands L-group of G over F*:

  $${}^L G(\mathbb{C}) = G^\vee \rtimes W_F \xrightarrow{pr} W_F$$

  where its identity component $G^\vee$ is the so-called Langlands dual group of G.

One requirement for ϕ is that

$$pr \circ \phi|_{W_F} = \text{the identity map on } W_F$$

and equivalence is defined by $G^\vee$-conjugacy. So an L-parameter is more or less a local Galois representation valued in the connected complex Lie group $G^\vee(\mathbb{C})$.

Here is another view on L-parameters. One has a (split) short exact sequence

$$1 \longrightarrow G^\vee \longrightarrow {}^L G \times \mathrm{SL}_2(\mathbb{C}) \longrightarrow W_F \times \mathrm{SL}_2(\mathbb{C}) \longrightarrow 1.$$

Then an L-parameter is simply a $G^\vee$-orbit of splittings of this short exact sequence.

3.2 The dual group $G^\vee$

Recall that for a connected reductive G, with maximal torus T, one can associate its root datum:

$$(X(T), \Phi, Y(T), \Phi^\vee).$$

The quadruple

$$(Y(T), \Phi^\vee, X(T), \Phi)$$

can be shown to be a root datum as well. Since root data classify the connected reductive linear algebraic groups over algebraically closed fields, this root datum corresponds to a linear algebraic group $G^\vee$ over $\mathbb{C}$. This complex Lie group $G^\vee$ is the Langlands dual group of G.

Here is a table of examples of $G^\vee(\mathbb{C})$:

G	GL_n	Sp_{2n}	SO_{2n+1}	SO_{2n}	E_8
$G^\vee$	$\mathrm{GL}_n(\mathbb{C})$	$\mathrm{SO}_{2n+1}(\mathbb{C})$	$\mathrm{Sp}_{2n}(\mathbb{C})$	$\mathrm{SO}_{2n}(\mathbb{C})$	$E_8(\mathbb{C})$

4 Genuine representations

For a covering group

$$1 \longrightarrow \mu \longrightarrow \tilde{G} \longrightarrow G(F) \longrightarrow 1,$$

fix an embedding

$$\epsilon : \mu \hookrightarrow \mathbb{C}^\times.$$

We are interested in the set of ϵ-genuine representations of $\tilde{G}$, namely representations π such that

$$\pi|_\mu = \epsilon.$$

The question we want to consider is:

Question. Is there a classification of $\mathrm{Irr}_\epsilon(\tilde{G})$ in the style of the LLC for linear algebraic groups?

This leads naturally to:

Question. What should be the dual group or the L-group of $\tilde{G}$?

In this section, we shall discuss some results which shed some light on these two questions.

4.1 Unramified Langlands correspondence

One of the initial evidence for the Langlands program is Langlands' re-interpretation of Satake's work on spherical functions and unramified representations, which leads to the creation of the Langlands L-group. This re-interpretation gives a classification of unramified representations, namely a bijection:

$$\{\text{Unramified representations of } G(F)\} \longleftrightarrow \{\text{Unramifed } L\text{-parameters for } G\},$$

which is a partial realization of the LLC.

Assume that G is split henceforth for simplicity. Let $K \subset G(F)$ be a hyperspecial maximal compact subgroup. We define the notions of unramified representations and L-parameters:

- An irreducible representation (π, V) of $G(F)$ is unramified if

$$V^K \neq 0.$$

  In this case, one knows that $\dim V^K = 1$.

- Let $I_F \subset W_F$ be the inertia group. An L-parameter
$$\phi : W_F \times \mathrm{SL}_2(\mathbb{C}) \to {}^L G = G^\vee \times W_F$$
is unramified if
$$\phi(w, g) = (1, w)$$
for $(w, g) \in I_F \times \mathrm{SL}_2(\mathbb{C})$.

4.2 Satake isomorphism

It is not hard to see that the map $V \mapsto V^K$ defines a bijection of the set of irreducible unramified representations of $G(F)$ with the irreducible characters of the spherical Hecke algebra:
$$H(G(F), K) = C_c^\infty(K \backslash G(F)/K).$$

The following is Langlands' re-interpretation of Satake's results:

Theorem 4.1. *There is a natural algebra isomorphism (the Satake isomorphism)*
$$H(G(F), K) \longrightarrow H(T(F), K_T)^W,$$
where T is a maximal split torus with Weyl group W and $K_T = K \cap T(F)$.

Corollary 4.2.
$$H(G(F), K) \cong \mathbb{C}[Y]^W \cong R[G^\vee],$$
where $R[G^\vee]$ denotes the Grothendieck ring of $G^\vee$. In particular, there is a bijection
$$\{\textit{irreducible unramified representations of } G(F)\}$$
$$\longleftrightarrow \{\textit{semisimple conjugacy classes in } G^\vee\}.$$

4.3 Unramified representations for BD coverings

Suppose now that $\tilde{G}$ is a BD-covering of $G(F)$ which is split over K. Fix such a splitting
$$s : K \hookrightarrow \tilde{G}.$$
Then a genuine representation π of $\tilde{G}$ is unramified (with respect to s) if
$$\pi^{s(K)} \neq 0.$$

As before, the set of such irreducible unramified representations is in natural bijection with the irreducible characters of a genuine spherical Hecke algebra:
$$H_\epsilon(\tilde{G}, K, s) = C_{c,\epsilon}^\infty(s(K) \backslash \tilde{G}/s(K)).$$

The first systematic study of such genuine unramified representations was undertaken by Savin in [S2].

4.4 Satake isomorphism for BD coverings

Recently, the Satake isomorphism has been established in the setting of BD coverings in works of McNamara [Mc], Li [L3] and Weissman [W4]:

Theorem 4.3. *One has a natural algebra isomorphism*

$$H_\epsilon(\tilde{G}, K, s) \to H_\epsilon(\tilde{T}, s(K_T))^W.$$

Could we interpret the RHS using root theoretic data or as the Grothendieck ring of some complex Lie group?

4.5 Metaplectic modification of root datum

Recall that G has a root datum $(X, \Phi, Y, \Phi^\vee)$. Suppose $\tilde{G}$ is a BD-covering with data (E, n), with $n = |\mu|$. In the Brylinski-Deligne classification of the multiplicative K_2-torsor E, one of the ingredients is a quadratic form Q. We are going to modify the root datum of G using (Q, n).

- Set
$$Y_{Q,n} = \{y \in Y : b_Q(y, z) \in n\mathbb{Z} \text{ for all } z \in Y\}.$$

- For $\alpha^\vee \in \Phi^\vee$, set
$$n_{\alpha^\vee} = \frac{n}{(n, Q(\alpha^\vee))}$$
and
$$\tilde{\alpha}^\vee = n_{\alpha^\vee} \cdot \alpha^\vee$$
and
$$\tilde{\Phi}^\vee = \{\tilde{\alpha}^\vee : \alpha^\vee \in \Phi^\vee\}.$$

- Likewise, set
$$X_{Q,n} = \mathrm{Hom}(Y_{Q,n}, \mathbb{Z})$$
and
$$\tilde{\Phi} = \{n_{\alpha^\vee}^{-1} \cdot \alpha : \alpha \in \Phi\}.$$

Then we have:

Theorem 4.4. *The quadruple*

$$(Y_{Q,n}, \tilde{\Phi}^\vee, X_{Q,n}, \tilde{\Phi}\}$$

is a root datum. If $\tilde{G}^\vee$ is the associated complex Lie group, then the Satake isomorphism givens an isomorphism

$$H_\epsilon(\tilde{G}, K, s) \cong \mathbb{C}_\epsilon[\tilde{Y}_{Q,n}]^W,$$

where we have an extension induced by $\tilde{T}$ and s:

$$1 \longrightarrow \mu \longrightarrow \tilde{Y}_{Q,n} \longrightarrow Y_{Q,n} \longrightarrow 1.$$

The latter algebra $\mathbb{C}_\epsilon[\tilde{Y}_{Q,n}]^W$ is non-canonically isomorphic to $R[\tilde{G}^\vee]$.

4.6 Dual group of a BD covering

The group $\tilde{G}^\vee$ has also arised in the geometric Langlands program, through the work of Finkelberg-Lysenko [FL] and Riech [Ri]. In any case, the Satake isomorphism suggests that the dual group of a BD covering $\tilde{G}$ should be the group $\tilde{G}^\vee$ defined above.

Here are some examples:

- if $\tilde{G} = \mathrm{Mp}_{2n}$, then
$$\tilde{G}^\vee = \mathrm{Sp}_{2n}(\mathbb{C}).$$
- if G is split simple simply connected and simply-laced, let $Z(G)_n$ be the n-torsion in the centre of G. Then
$$\tilde{G}^\vee = (G/Z(G)_n)^\vee.$$

However, because of the non-canonicity at the end of the last theorem, one cannot quite associate a semisimple conjugacy class of $\tilde{G}^\vee$ to an unramified representation of $\tilde{G}$. We shall see how this can be resolved next.

4.7 L-group of a BD covering

What about the L-group of $\tilde{G}$? If G is split, then it is natural to want to define the L-group as a direct product:
$${}^L\tilde{G} = \tilde{G}^\vee \times W_F.$$

But experience with tori, Mp_{2n} and the Satake isomorphism given in the previous theorem suggests that it is better to consider the L-group as an extension
$$1 \longrightarrow \tilde{G}^\vee \longrightarrow {}^L\tilde{G} \longrightarrow W_F \longrightarrow 1 \tag{4.1}$$
which is split, but with no canonical splitting. In other words, ${}^L\tilde{G}$ is isomorphic to $\tilde{G}^\vee \times W_F$, but not canonically so. This subtle shift in the point of view is very important, as we shall see in a moment.

In a recent paper of Weissman [W3], such a construction of ${}^L\tilde{G}$ has been given, but it seems overly complicated. Recently, Weissman has given a much simplified version of his construction, following suggestions of Deligne; see [W4].

4.8 Hope of an LLC

Having
$$1 \longrightarrow \tilde{G}^\vee \longrightarrow {}^L\tilde{G} \times \mathrm{SL}_2(\mathbb{C}) \longrightarrow W_F \times \mathrm{SL}_2(\mathbb{C}) \longrightarrow 1,$$
one can define an L-parameter for $\tilde{G}$ to be a $\tilde{G}^\vee$-orbit of splittings of the above short exact sequence. It is then natural to hope for a map:
$$\mathrm{Irr}_\epsilon(\tilde{G}) \xrightarrow{\text{finite-to-1}} \{L\text{-parameters for } \tilde{G}\}.$$

This would be an LLC for $\tilde{G}$, which I learn from M. Weissman.

Note that there is no canonical splitting of the above short exact sequence and hence no distinguished L-parameter. This reflects the fact that there is no distinguished irreducible genuine representation. This is unlike the case of linear group $G(F)$, where one has the trivial representation of $G(F)$ with corresponding distinguished L-parameter given by the "cocharacter" ρ (half sum of positive roots), which is the ρ-shift of the trivial L-parameter.

We shall now discuss some supporting evidence for the above hope.

4.9 Covering tori

The following is a theorem of Weissman [W1]:

Theorem 4.5. *If T is a split torus, and $\tilde{T}$ a BD-covering, then there is a natural map*

$$\mathrm{Irr}_\epsilon(\tilde{T}) \longrightarrow \{L\text{-parameters for } \tilde{T}\}$$

whose fiber has size 0 *or* 1*. Moreover, one can describe the image of the above map precisely.*

Weissman has results for BD-covers of more general tori, for example unramified tori, but we will not discuss these here.

4.10 Unramified representations

As shown in [W4], the splitting $s : K \to \tilde{G}$ gives rise to a splitting of the short exact sequence (4.1) over the inertia group I_F, and one says that a splitting of (4.1) is an unramified L-parameter (relative to s). Then the Satake isomorphism gives a canonical bijection

$$\{\text{irreducible unramified genuine representations of } \tilde{G}\}$$
$$\updownarrow$$
$$\{\text{unramified } L\text{-parameters for } \tilde{G}\}.$$

4.11 Iwahori-Hecke algebras

Suppose G split simply-connected, simply-laced, and $\tilde{G}$ is its degree n cover. Let

$$I \subset K \subset G(F)$$

be an Iwahori subgroup. If $(n, p) = 1$, then the covering splits uniquely over K:

$$s : K \hookrightarrow \tilde{G}.$$

Savin [S1] has shown:

Theorem 4.6. *There s a natural algebra isomorphism*

$$H_\epsilon(\tilde{G}, I, s) \cong H((G/Z(G)_n)(F), I').$$

This relates the representations of $\tilde{G}$ with $s(I)$-fixed vectors with the representations of $(G/Z(G)_n)(F)$ with I'-fixed vectors, and is in accordance with the fact that

$$\tilde{G}^\vee = (G/Z(G)_n)^\vee.$$

4.12 Mp_{2n}

The following was show by Savin and myself [GS]:

Theorem 4.7. *Fix a nontrivial additive character ψ of F. There is a bijection, depending on ψ,*

$$\mathrm{Irr}_\epsilon(\mathrm{Mp}_{2n}(F)) \longleftrightarrow \mathrm{Irr}(\mathrm{SO}_{2n+1}(F)) \sqcup \mathrm{Irr}(\mathrm{SO}'_{2n+1}(F)).$$

Assuming the LLC for SO_{2n+1}, one has a map

$$\mathrm{Irr}_\epsilon(\mathrm{Mp}_{2n}(F)) \longrightarrow \{\phi : W_F \times \mathrm{SL}_2(\mathbb{C}) \to \mathrm{Sp}_{2n}(\mathbb{C})\}$$

which depends on ψ.

If one works with the L-group of Mp_{2n} rather than of SO_{2n+1}, then one should be able to refine this to a map independent of ψ:

$$\mathrm{Irr}_\epsilon(\mathrm{Mp}_{2n}(F)) \longrightarrow \{L\text{-parameters for } \mathrm{Mp}_{2n}(F)\}.$$

5 Global Langlands correspondence

Now suppose F is a global field with ring of adeles $\mathbb{A}$. An automorphic form on G is a function

$$f : G(F)\backslash G(\mathbb{A}) \longrightarrow \mathbb{C}$$

with some smoothness and finiteness properties. One considers the $G(\mathbb{A})$-representation

$$L^2(G(F)\backslash G(\mathbb{A})) = L^2_{disc}[G] \oplus L^2_{cont}[G].$$

Then

$$L^2_{disc}[G] = \bigoplus_\pi m(\pi) \cdot \pi.$$

The global Langlands correspondence classifies the set

$$\{\text{irreducible representations } \pi \text{ of } G(\mathbb{A}) \text{ with } m(\pi) > 0\}$$

of square-integrable automorphic representations.

For BD-coverings, we would like to consider the analogous problem.

5.1 Global BD coverings

Given a multiplicative K_2-torsor

$$1 \longrightarrow K_2 \longrightarrow E \longrightarrow G \longrightarrow 1$$

one inherits:

$$1 \longrightarrow \mu(F) \longrightarrow \tilde{G}_{\mathbb{A}} \longrightarrow G(\mathbb{A}) \longrightarrow 1$$

equipped with a natural splitting

$$i : G(F) \hookrightarrow \tilde{G}_{\mathbb{A}}.$$

Using this, one may identify $G(F)$ with a subgroup of $\tilde{G}_{\mathbb{A}}$.

5.2 Genuine automorphic forms

For a fixed

$$\epsilon : \mu \longrightarrow \mathbb{C}^{\times}$$

one can consider the unitary representation

$$L^2_{\epsilon}(G(F)\backslash \tilde{G}\mathbb{A})$$

on the space of ϵ-genuine automorphic forms. We are interested in the decomposition of this unitary representation.

5.3 Harmonic analysis and trace formula

In fact, many familiar harmonic analytic results for linear groups extend to the case of covering groups with little modification.

- Locally, many results of Harish-Chandra and others on local harmonic analysis has been extended with great efficiency to $\tilde{G}$ by Li [L4]: Plancherel theorem, local integrality of character distributions, local character expansion, local trace formula, ...

- Globally, the theory of Eisenstein series and spectral decomposition of $L^2_{\epsilon}(G(F)\backslash \tilde{G})$ has been established by Moeglin-Waldspurger [MW]; this reduces one to the study of the automorphic discrete spectrum.

- In a series of recent papers [L3, L4, L5], Li has established the Invariant Trace Formula for $\tilde{G}$.

5.4 A theory of endoscopy?

The natural next question: is there a theory of endoscopy for covering groups?

- for Mp_{2n}, Li [L1] has developed a theory of endoscopy, so that the elliptic endoscopic groups of Mp_{2n} are:

$$\mathrm{SO}_{2a+1} \times \mathrm{SO}_{2b+1}$$

  where one considers all unordered pairs of non-negative integers (a, b) with $a+b=n$. There are thus $n+1$ elliptic endoscopic groups in all. This extends earlier work of Renard [R1, R2] in the real case.

- for arbitrary cover of SL_2, Hiraga and Ikeda [HI] have developed such a theory of endoscopy.

5.5 Work of Wenwei Li

We describe Li's results [L1, L2] for Mp_{2n} in greater detail here:

- For each pair (a, b) with $a+b=n$, Li defined transfer factors

$$\Delta_{a,b} : \mathrm{Mp}_{2n} \times (\mathrm{SO}_{2a+1} \times \mathrm{SO}_{2b+1}) \to \mathbb{C}.$$

- He showed existence of transfer mappings

$$\begin{array}{c} C_c^\infty(\mathrm{Mp}_{2n}) \\ {\scriptstyle t_{a,b}}\downarrow \\ C_c^\infty(\mathrm{SO}_{2a+1} \times \mathrm{SO}_{2b+1}), \end{array}$$

  giving rise to a transfer

$$\begin{array}{c} \mathcal{D}_{st}(\mathrm{SO}_{2a+1} \times \mathrm{SO}_{2b+1}) \\ {\scriptstyle t^*_{a,b}}\downarrow \\ \mathcal{D}_{gen}(\mathrm{Mp}_{2n}), \end{array}$$

  between spaces of distributions.

- Finally, he established the ordinary and weighted fundamental lemmas for the transfer $t_{a,b}$.

Now one is in a position to consider the stabilisation of the invariant trace formula for Mp_{2n}. In a recent preprint [L6], Li has completed the stabilisation of the elliptic part of the trace formula for Mp_{2n}.

5.6 Arthur's conjecture for Mp_{2n}

We describe the analog of Arthur's conjecture for Mp_{2n}. For a fixed additive automorphic character ψ, one expects that

$$L^2_{disc} = \bigoplus_{\Psi} L^2_{\Psi,\psi}$$

where

$$\Psi = \bigoplus_i \Psi_i = \bigoplus_i \Pi_i \boxtimes S_{r_i}$$

is a global discrete A-parameter for Mp_{2n}; it is also an A-parameter for SO_{2n+1}. Here, S_{r_i} is the r_i-dimensional representation of $\mathrm{SL}_2(\mathbb{C})$ and Π_i is a cuspidal representation of GL_{n_i} such that

$$\begin{cases} L(s, \Pi_i, \wedge^2) \text{ has a pole at } s = 1, \text{ if } r_i \text{ is odd;} \\ L(s, \Pi_i, \mathrm{Sym}^2) \text{ has a pole at } s = 1, \text{ if } r_i \text{ is even.} \end{cases}$$

Moreover, we have $\sum_i n_i r_i = 2n$ and the summands Ψ_i are mutually distinct.

For a given Ψ, one obtains the following extra data:

- for each v, one inherits a local A-parameter

$$\Psi_v = \bigoplus_i \Psi_{i,v} = \bigoplus_i \Pi_{i,v} \boxtimes S_{r_i}.$$

  By the local Langlands correspondence for GL_N, we may regard each $\Pi_{i,v}$ as an n_i-dimensional representation of the Weil-Deligne group WD_{F_v}. Hence, we may regard Ψ_v as a $2n$-dimensional representation of $WD_F \times \mathrm{SL}_2(\mathbb{C})$.

- one has a global component group

$$A_\Psi = \bigoplus \mathbb{Z}/2\mathbb{Z} \cdot a_i$$

  which is a $\mathbb{Z}/2\mathbb{Z}$-vector space equipped with a distinguished basis indexed by Ψ_i. Similarly, for each v, we have the local component group A_{Ψ_v} which is defined as the component group of the centralizer of the image of Ψ_v, thought of as a representation of $WD_F \times \mathrm{SL}_2(\mathbb{C})$.

- For each v, there is a natural map

$$\Delta_v : A_\Psi \longrightarrow A_{\Psi_v}$$

  by

$$\Delta_v(a_i) = \text{the image of } -Id_{\Psi_{i,v}} \text{ in } A_{\Psi_v}.$$

  Thus, there is a natural diagonal map

$$\Delta = \prod_v \Delta_v : A_\Psi \longrightarrow \prod_v A_{\Psi_v}.$$

- For each v, one has a local A-packet associated to Ψ_v and the additive character ψ_v:

$$\Pi_{\Psi_v,\psi_v} = \{\sigma_{\eta_v} : \eta_v \in \mathrm{Irr}(A_{\Psi_v})\},$$

consisting of unitary representations (possibly zero, possibly reducible) of $\mathrm{Mp}_{2n}(F_v)$ indexed by the set of irreducible characters of A_{Ψ_v}. On taking tensor products of these local A-packets, we obtain a global A-packet

$$A_{\Psi,\psi} = \{\sigma_\eta : \eta = \otimes_v \eta_v \in \mathrm{Irr}(\prod_v A_{\Psi_v})\}$$

consisting of abstract unitary representations

$$\sigma_\eta = \otimes_v \sigma_{\eta_v}$$

of $\mathrm{Mp}_{2n}(\mathbb{A})$ indexed by the irreducible characters $\eta = \otimes_v \eta_v$ of $\prod_v A_{\Psi_v}$.

- Arthur has attached to Ψ a quadratic character (possibly trivial) ϵ_Ψ of A_Ψ, This character plays an important role in the multiplicity formula for the automorphic discrete spectrum of SO_{2n+1}. For Mp_{2n}, we need to define a modification of ϵ_Ψ.

  More precisely, consider the L-parameter Φ_Ψ associated to Ψ:

$$\Phi_\Psi = \bigoplus_i \Phi_{\Psi_i}$$

  with

$$\Phi_{\Psi_i} = \bigoplus_{k=0}^{r_i-1} \Pi_i \cdot |-|^{(r_i-1-2k)/2}.$$

  Then define $\eta_\Psi \in \mathrm{Irr} A_\Psi$ by

$$\eta_\Psi(a_i) = \epsilon(1/2, \Phi_{\Psi_i}) = \begin{cases} \epsilon(1/2, \Pi_i), & \text{if } L(s, \Pi_i, \wedge^2) \text{ has a pole at } s = 1; \\ 1, & \text{if } L(s, \Pi_i, \mathrm{Sym}^2) \text{ has a pole at } s = 1. \end{cases}$$

  The modified quadratic character of A_Ψ in the metaplectic case is

$$\tilde{\epsilon}_\Psi = \epsilon_\Psi \cdot \eta_\Psi.$$

With the above data introduced, we can now state the conjecture.

Arthur Conjecture. The automorphic discrete spectrum $\mathcal{A}_2(\mathrm{Mp}_{2n})$ has a decomposition

$$L^2_{disc}(\mathrm{Mp}_{2n}) = \bigoplus_\Psi L^2_{\Psi,\psi}$$

where the sum runs over equivalence classes of discrete A-parameters of Mp_{2n} and for each such Ψ,

$$L^2_{\Psi,\psi} \cong \bigoplus_{\eta \in \mathrm{Irr}(\prod_v A_{\Psi_v}) : \Delta^*(\eta) = \tilde{\epsilon}_\Psi} \sigma_\eta.$$

This conjecture amounts to a description of $L^2_{disc}(\mathrm{Mp}_{2n})$ in terms of the discrete spectrum of SO_{2n+1}. When $n = 1$, this is the main result of Waldspurger's work on the Shimura correspondence. In ongoing work with A. Ichino, we are working towards establishing this Arthur conjecture for Mp_{2n}, at least for the tempered part of the spectrum. The stable trace formula for Mp_{2n} should establish this as well.

6 A sample of problems

We end this paper with a list of sample problems which one may want to pursue further:

- Structure theory: much remains to be understood about the map

$$\{\text{multipicative } K_2\text{-torsors}\} \longrightarrow \{\text{topological covering groups}\}.$$

  Namely, one would like to know more about its image and its fibers.

- Bruhat-Tits theory: the paper [W2] studies the problem of understanding the covering of a parahoric subgroup. It gives some partial results about determining the "special fiber" of such a covering.

- Real groups: for coverings of real groups, Harish-Chandra's classification of discrete series representations has been known for decades. Could one formulate this classification result in the spirit of the LLC? J. Adams and his collaborators, and others, have thought much about this question: see [A1, A2, ABPTV, AH, R1, R2] among others.

- p-adic groups: could one extend the Moy-Prasad theory of depth of a representation to the case of covering groups? The construction of depth zero supercuspidal representations was carried out in [HW], but could one group these into "L-packets"? Does Yu's construction of supercuspidal representations by compact induction [Y] extend readily to covering groups? Does Kim's result [K] on the exhaustion of Yu's construction (when the residue characteristic p is large) extend readily to covering groups? Together with Kim, we are working towards understanding some of these questions.

- A theory of endoscopy: as we mentioned earlier, Li's work on the invariant trace formula for covering groups has led directly to the question of whether the theory of endoscopy has a natural extension to covering groups.

- Eisenstein series and L-functions: Automorphic L-functions arises naturally when one considers the constant term of Eisenstein series. For covering groups, can one have a similar interpretation of the constant terms of Eisenstein series? What L-functions arise in this way? Addressing this question is part of the thesis work of my student Fan Gao.

- Discrete spectrum and Arthur's multiplicity formula: Can Arthur's conjecture be extended to the case of covering groups? What would the multiplicity formula look like? The case of Mp_{2n} offers some clues, but is perhaps too special a case. The ongoing work of Hiraga-Ikeda on the discrete spectrum of covering tori should shed some light on this question.

Acknowledgements. We thank Marty Weissman for his comments on a first draft of this paper and for sharing with us his paper under preparation [W4].

References

[A1] J. Adams, *Lifting of characters on orthogonal and metaplectic groups*, Duke Math. J. **92** (1998), no. 1, 129–178.

[A2] J. Adams, *Characters of covering groups of $SL(n)$*, J. Inst. Math. Jussieu **2** (2003), no. 1, 1–21.

[ABPTV] J. Adams, D. Barbasch, A. Paul, P. Trapa, and David Vogan, *Shimura correspondences for split real groups*, Journal of the American Math Society **20** (2007), 701–751.

[AH] J. Adams and R. Herb, *Lifting of characters for nonlinear simply laced groups*, Represent. Theory **14** (2010), 70–147.

[BD] J. L. Brylinski and P. Deligne, *Central extensions of reductive groups by $K2$*, Publ. Math. Inst. Hautes tudes Sci. No. **94** (2001), 5–85.

[D] V. Deodhar, *On central extensions of rational points of algebraic groups*, Amer. J. Math. **100** (1978), no. 2, 303–386.

[FL] M. Finkelberg and S. Lysenko, *Twisted geometric Satake equivalence*, J. Inst. Math. Jussieu, **9** (2010), no. 4, 719–739.

[GS] W. T. Gan and G. Savin, *Representations of metaplectic groups I: epsilon dichotomy and local Langlands correspondence*, Compos. Math. **148** (2012), no. 6, 1655–1694.

[HI] K. Hiraga and T. Ikeda, in preparation.

[HW] T. Howard and M. Weissman, *Depth-zero representations of nonlinear covers of p-adic groups*, Int. Math. Res. Not. IMRN (2009), no. 21, 3979–3995.

[K] J. L. Kim, *Supercuspidal representations: an exhaustion theorem*, J. Amer. Math. Soc. **20** (2007), no. 2, 273–320.

[L1] W. W. Li, *Transfert d'intégrales orbitales pour le groupe métaplectique*, Compos. Math. **147** (2011), no. 2, 524–590.

[L2] W. W. Li, *Le lemme fondamental pondéré pour le groupe métaplectique*, Canad. J. Math. **64** (2012), no. 3, 497–543.

[L3] W. W. Li, *La formule des traces pour les revêtements de groupes réductifs connexes. I. Le développement géométrique fin*, to appear in J.Reine Angew. Math.

[L4] W. W. Li, *La formule des traces pour les revetements de groupes réductifs connexes. II. Analyse harmonique locale*, Ann. Sci. Éc. Norm. Supér. (4) **45** (2012), no. 5, 787–859.

[L5] W. W. Li, *La formule des traces pour les revetements de groupes réductifs connexes. III. Le développement spectral fin*, Math. Ann. **356** (2013), no. 3, 1029–1064.

[L6] W. W. Li, *La formule des traces stable pour le groupe métaplectique: les termes elliptiques*, preprint, arXiv:1307.1032.

[M] C. Moore, *Group extensions of p-adic and adelic linear groups*, Inst. Hautes tudes Sci. Publ. Math. No. **35** (1968), 157–222.

[Ma] H. Matsumoto, *Sur les sous-groupes arithmétiques des groupes semi-simples déployés*, Ann. Sci. cole Norm. Sup. (4) **2** (1969), 1–62.

[Mc] P. McNamara, *Principal series representations of metaplectic groups over local fields*, in *Multiple Dirichlet series, L-functions and automorphic forms*, 299–327, Progr. Math. **300**, Birkhuser/Springer, New York, 2012.

[MW] C. Moeglin and J.-L. Waldspurger, *Spectral decomposition and Eisenstein series. Une paraphrase de l'Écriture*, Cambridge Tracts in Mathematics **113**. Cambridge University Press, Cambridge, 1995.

[PR1] G. Prasad and M. S. Raghunathan, *Topological central extensions of semisimple groups over local fields*, Ann. of Math. (2) **119** (1984), no. 1, 143–201.

[PR2] G. Prasad and M. S. Raghunathan, *Topological central extensions of semisimple groups over local fields. II*, Ann. of Math. (2) **119** (1984), no. 2, 203–268.

[R1] D. Renard, *Transfert d'intégrales orbitales entre* $\mathrm{Mp}(2n,\mathbb{R})$ *et* $\mathrm{SO}(n+1,n)$, Duke Math. J. **95** (1998), no. 2, 425–450.

[R2] D. Renard, *Endoscopy for* $\mathrm{Mp}(2n,\mathbb{R})$, Amer. J. Math. **121** (1999), no. 6, 1215–1243.

[Ri] R. Riech, *Twisted geometric Satake equivalence via gerbes on the factorizable Grassmannian*, Represent. Theory **16** (2012), 345–449.

[S1] G. Savin, *Local Shimura correspondence*, Math. Ann. **280** (1988), no. 2, 185–190.

[S2] G. Savin, *On unramified representations of covering groups*, J. Reine Angew. Math. **566** (2004), 111–134.

[St] R. Steinberg, *Générateurs, relations et revetements de groupes algébriques*, in *Colloq. Théorie des Groupes Algébriques* (*Bruxelles*, 1962), 113–127, Librairie Universitaire, Louvain; Gauthier-Villars, Paris.

[W1] M. H. Weissman, *Metaplectic tori over local fields*, Pacific J. Math. **241** (2009), no. 1, 169–200.

[W2] M. Weissman, *Managing metaplectiphobia: covering p-adic groups*, Harmonic analysis on reductive, p-adic groups, 237–277, Contemp. Math. **543**, Amer. Math. Soc., Providence, RI, 2011.

[W3] M. Weissman, *Split metaplectic groups and their L-groups*, to appear in J. Reine Angew. Math.

[W4] M. Weissman, *Covering the Langlands program*, in preparation.

[Y] J. K. Yu, *Construction of tame supercuspidal representations*, J. Amer. Math. Soc. **14** (2001), no. 3, 579–622.

Department of Mathematics, National University of Singapore
Block S17, 10 Lower Kent Ridge Road, Singapore 119076
E-mail address: matgwt@nus.edu.sg

Proceedings of the 6th International Congress of Chinese Mathematicians
ALM 36, pp. 79–93

Global Solutions of the Euler–Maxwell Two-fluid System in 3D

Yan Guo, Alexandru D. Ionescu, and Benoit Pausader

Abstract

The fundamental "two-fluid" model for describing plasma dynamics is given by the Euler–Maxwell system, in which compressible ion and electron fluids interact with their own self-consistent electromagnetic field. We prove irrotational, smooth and localized perturbations of a constant background with small amplitude lead to global smooth solutions in three space dimensions for the Euler–Maxwell system.

Mathematics Subject Classification: Partial Differential Equations, 35 and 35Q.

Keywords and Phrases: Two-fluid models, dispersion and smoothness, absence of shocks.

A plasma is a collection of fast-moving charged particles. It is believed that more than 90% of the matter in the universe is in the form of plasma, from sparse intergalactic plasma, to the interior of stars to neon signs. In addition, understanding of the instability formation in plasma is one of the main challenges for nuclear fusion, in which charged particles are accelerated at high speed to create energy. We refer to [2, 4, 9, 44, 45] for physics references in book form.

At high temperature and velocity, ions and electrons in a plasma tend to become two separate fluids due to their different physical properties (inertia, charge). One of the basic fluid models for describing plasma dynamics is the so-called "two-fluid" model, in which two compressible ion and electron fluids interact with their own self-consistent electromagnetic field. Such an Euler–Maxwell system describes the dynamical evolution of the functions $n_e, n_i : \mathbb{R}^3 \to \mathbb{R}$, $v_e, v_i, E, B : \mathbb{R}^3 \to \mathbb{R}^3$,

The first author is supported in part by NSF grant #1209437 and a CNSF grant #10828103. The second author is supported in part by a Packard Fellowship and NSF grant DMS-1065710. The third author is supported in part by NSF grant DMS-1 142293. Both the first and the third authors thank the support of Beijing International Mathematical Research Center.

which evolve according to the quasi-linear coupled system,

$$\begin{aligned}
&\partial_t n_e + \operatorname{div}(n_e v_e) = 0,\\
&n_e m_e \left[\partial_t v_e + v_e \cdot \nabla v_e\right] + \nabla p_e = -n_e e \left[E + \frac{v_e}{c} \times B\right],\\
&\partial_t n_i + \operatorname{div}(n_i v_i) = 0,\\
&n_i M_i \left[\partial_t v_i + v_i \cdot \nabla v_i\right] + \nabla p_i = Z n_i e \left[E + \frac{v_i}{c} \times B\right],\\
&\partial_t B + c\nabla \times E = 0,\\
&\partial_t E - c\nabla \times B = 4\pi e \left[n_e v_e - Z n_i v_i\right],
\end{aligned} \tag{1}$$

together with the elliptic equations

$$\operatorname{div}(B) = 0, \quad \operatorname{div}(E) = 4\pi e(Z n_i - n_e) \tag{2}$$

and two equations of state expressing p_e and p_i in terms of n_e and n_i. These equations describe a plasma composed of electrons and one species of ions. The electrons have charge $-e$, density n_e, mass m_e, velocity v_e, and pressure p_e, and the ions have charge Ze, density n_i, mass M_i, velocity v_i, and pressure p_i. In addition, c denotes the speed of light and E and B denote the electric and magnetic field. The two equations (2) are propagated by the dynamic flow, provided that we assume that they are satisfied at the initial time.

The full Euler–Maxwell system (1) with constraint (2) forms the foundation of the "two-fluid" model in the plasma theory, which captures the complex dynamics of a plasma due to electromagnetic interactions present in the model. Even at the linear level, there are new ion-acoustic waves, Langmuir waves, as well as light waves etc. At the nonlinear level, the Euler–Maxwell system is the origin of many well-known dispersive PDE, such as KdV [25], KP [37, 40], Zakharov [46], Zakharov–Kuznetsov [37, 40] and NLS, which can be derived from (1) and (2) via different scaling and asymptotic expansions. We also refer to [3, 12, 13] for derivation of the cold-ion and quasi-neutral equations and to [5] for a study of a similar model for semiconductors.

From a PDE viewpoint, the full Euler–Maxwell system (1) with constraint (2) can be classified as a system of nonlinear hyperbolic conservation laws with *no dissipation and no relaxation effects*[1]. Despite major advances in the mathematical study of hyperbolic conservation laws in one space dimension over the years, no general mathematical theory exists for the compressible Euler equations in higher space dimension. One of the reasons is that, for these equations, shock waves (i.e., discontinuities) will generically develop even from smooth initial data (see e.g. John [33]). Even worse, a classical result of Sideris [43] demonstrates that, for the compressible Euler equation for a neutral gas, shock waves will develop even for smooth irrotational initial data with small amplitude. This shock formation was recently further described in [7, 8] (see also [1]).

[1]When dissipation or relaxation is present, one expects stronger decay, even at the level of the L^2-norm, see e.g. [5, 39] and the references therein. In our case however, the evolution is time-reversible and we need a different mechanism of decay based on dispersion.

In this paper we consider perturbations of the flat neutral equilibrium, namely $(n_e^0, v_e^0, n_i^0, v_i^0, E^0, B^0) = (Zn_0, 0, n_0, 0, 0, 0)$, for constant $n_0 > 0$ to the Euler–Maxwell system (1) and (2). In order to state our main result, we normalize the Euler–Maxwell system in the following way. Assume the pressures are given by the formulas[2]:

$$p_e = P_e \frac{n_e^2}{2}, \quad p_i = P_i Z^2 \frac{n_i^2}{2} \tag{3}$$

with constants P_e and P_i. The physical parameters are then the effective ion and electron temperatures

$$k_B T_e = n_0 P_e, \quad k_B T_i = n_0 Z P_i,$$

where k_B denotes the Boltzmann constant, with corresponding electron and ion thermal speeds[3]

$$V_e = \sqrt{\frac{n_0 P_e}{m_e}} = \sqrt{\frac{k_B T_e}{m_e}}, \quad V_i = \sqrt{\frac{n_0 P_i Z}{M_i}} = \sqrt{\frac{k_B T_i}{M_i}}.$$

We also have the Debye length

$$\frac{1}{\lambda_D^2} = 4\pi e^2 \left[\frac{n_0}{k_B T_e} + \frac{Z n_0}{k_B T_i}\right] = 4\pi e^2 \left[\frac{1}{P_e} + \frac{1}{P_i}\right].$$

The Euler–Maxwell system can be adimensionalized to depend only on three parameters: the ratio of the electron to ion masses (per charge)

$$\varepsilon := Z m_e / M_i, \tag{4}$$

the ratio of the temperatures

$$T := P_e / P_i = Z T_e / T_i, \tag{5}$$

and the (normalized) ratio of the speed of light to the ion velocity

$$C_b := \varepsilon \frac{c^2}{V_i^2} = \frac{c^2}{V_e V_i} \sqrt{T\varepsilon} = \frac{c^2 m_e}{n_0 P_i}. \tag{6}$$

More precisely, let

$$\lambda := \sqrt{\frac{4\pi e^2}{P_i}}, \qquad \beta := \sqrt{\frac{4\pi n_0 Z e^2}{M_i}},$$

[2]In fact, our approach allows to treat any sufficiently smooth *barotropic* pressure law, in particular the typical power law $p_e \sim n_e^{\gamma_e}$ for some $\gamma_e > 0$ and similarly for p_i. We refer to Appendix [GIP] for more precise statements. We use the particular quadratic laws for the pressure here only for the sake of concreteness and since it minimizes the nonlinear terms we have to consider.

[3]These correspond to the speed of inertial (linearized) waves if one neglects the electromagnetic field.

and

$$\begin{aligned} n_e(x,t) &= n_0\big[n(\lambda x,\beta t)+1\big], & n_i(x,t) &= (n_0/Z)\big[\rho(\lambda x,\beta t)+1\big],\\ v_e(x,t) &= (\beta/\lambda)v(\lambda x,\beta t), & v_i(x,t) &= (\beta/\lambda)u(\lambda x,\beta t),\\ E(x,t) &= (4\pi e n_0/\lambda)\tilde{E}(\lambda x,\beta t), & B(x,t) &= (cM_i\beta/(Ze))\tilde{B}(\lambda x,\beta t). \end{aligned} \tag{7}$$

The parameter β is the *ion plasma frequency* and $\beta/\lambda = V_i$ is the ion thermal velocity. In terms of $n, v, \rho, u, \tilde{E}, \tilde{B}$ the system (1)–(2) becomes

$$\begin{aligned} &\partial_t n + \mathrm{div}((n+1)v) = 0,\\ &\varepsilon\left(\partial_t v + v\cdot\nabla v\right) + T\nabla n + \tilde{E} + v\times\tilde{B} = 0,\\ &\partial_t \rho + \mathrm{div}((\rho+1)u) = 0,\\ &(\partial_t u + u\cdot\nabla u) + \nabla\rho - \tilde{E} - u\times\tilde{B} = 0,\\ &\partial_t\tilde{B} + \nabla\times\tilde{E} = 0,\\ &\partial_t\tilde{E} - \frac{C_b}{\varepsilon}\nabla\times\tilde{B} = \left[(n+1)v - (\rho+1)u\right],\\ &\mathrm{div}(\tilde{B}) = 0, \quad \mathrm{div}(\tilde{E}) = \rho - n, \end{aligned} \tag{8}$$

where ε, T and C_b have been defined above. We will assume throughout the paper that

$$\varepsilon \leq 10^{-3}, \qquad T\in[1,100], \qquad C_b \geq 6T. \tag{9}$$

We will make two additional simplifications. Using the system (8) it is easy to see that

$$\begin{aligned} \partial_t\big[\tilde{B} - \varepsilon\nabla\times v\big] &= \nabla\times\big[v\times(\tilde{B} - \varepsilon\nabla\times v)\big],\\ \partial_t\big[\tilde{B} + \nabla\times u\big] &= \nabla\times\big[u\times(\tilde{B} + \nabla\times u)\big], \end{aligned}$$

Therefore "*generalized irrotational flows*" with the property that

$$\tilde{B} = \varepsilon\nabla\times v = -\nabla\times u \tag{10}$$

are naturally preserved for all time.

Our main theorem is as follows [GIP]:

Theorem 1. *Assume* (9)*. Let* $N_0 = 10^4$ *and assume that*

$$\begin{aligned} &\|(n^0,v^0,\rho^0,u^0,\tilde{E}^0,\tilde{B}^0)\|_{H^{N_0}} + \|(n^0,v^0,\rho^0,u^0,\tilde{E}^0,\tilde{B}^0)\|_Z = \delta_0 \leq \overline{\delta},\\ &\mathrm{div}(\tilde{E}^0) + n^0 - \rho^0 = 0, \qquad \tilde{B}^0 = \varepsilon\nabla\times v^0 = -\nabla\times u^0, \end{aligned} \tag{11}$$

where $\overline{\delta} = \overline{\delta}(C_b, T, \varepsilon) > 0$ *is sufficiently small, and the* Z *norm is defined in* [GIP]*. Then there exists a unique global solution* $(n,v,\rho,u,\tilde{E},\tilde{B}) \in C([0,\infty) : H^{N_0})$ *of the system* (8) *with initial data* $(n(0),v(0),\rho(0),u(0),\tilde{E}(0),\tilde{B}(0)) = (n^0,v^0,\rho^0,u^0,\tilde{E}^0,$ $\tilde{B}^0)$*. Moreover, for any* $t\in[0,\infty)$,

$$\mathrm{div}(\tilde{E})(t) + n(t) - \rho(t) = 0, \qquad \tilde{B}(t) = \varepsilon\nabla\times v(t) = -\nabla\times u(t), \tag{12}$$
(generalized irrotationality)

and, with $\beta := 1/100$,

$$\begin{aligned}&\|(n(t), v(t), \rho(t), u(t), \tilde{E}(t), \tilde{B}(t))\|_{H^{N_0}} \\ &+ \sup_{|\alpha|\le 4}(1+t)^{1+\beta/2}\|(D_x^\alpha n(t), D_x^\alpha v(t), D_x^\alpha \rho(t), D_x^\alpha u(t), D_x^\alpha \tilde{E}(t), D_x^\alpha \tilde{B}(t)\|_{L^\infty} \lesssim \delta_0.\end{aligned} \tag{13}$$

Our main result demonstrates that even though the Euler–Maxwell system (1) and (2) is much more complicated than the pure Euler system for a neutral gas, it is in fact *more stable* in the sense that global smooth solutions can persist globally without any shock formations. This is a stark and surprising contrast to Sideris's result for the pure Euler equations [43].

Remark 2. *We make a few remarks about the assumptions in Theorem* 1.

- *Condition* (9) *is needed for our careful analysis of the dispersion relations that appear in the study of the linearized system. It is consistent with the relevant physical ranges of the parameters.*
- *Our hypothesis imply in particular that the perturbation is electrically neutral, i.e.*

$$\int_{\mathbb{R}^3}\left[Zn_0(1+\rho^0(x)) - n_0(1+n^0(x))\right]dx = 0.$$

This is however forced by Maxwell's relation (2) *if we assume that the electric perturbation is integrable.*

- *The smallness assumption is needed: large deviations from an equilibrium do create shocks* [26].

1 Important simplified models

The result of blow-up of Sideris for the pure compressible Euler equations [43] can be understood from the fact that small and irrotational perturbations of a constant background for the pure compressible Euler equations obey a *quasilinear wave equation without null-structure* of the form

$$(\partial_{tt} - \Delta)\,\alpha = \mathcal{Q}(\alpha, \nabla\alpha, \nabla^2\alpha) \tag{14}$$

where α is related to the unknown and the right-hand side denotes a quadratic nonlinearity in up to two derivatives of α. This type of equation has slow decay of linear waves (decay like $1/t$) and strong resonances and therefore blow-up or formation of shocks is expected.

The Euler–Maxwell system (8) contains a nonlinearity $\mathcal{Q}$ essentially similar to the pure compressible Euler case. However, due to self-consistent electromagnetic interaction, the linearized Euler–Maxwell system exhibits much more complex and subtle linear and bilinear dispersive effects than that from the wave equation. The main task in the present work is to systematically track down and exploit such dispersive effects mathematically to preserve smoothness globally in time and prevent shock formation.

In order to put our result in the right context as well as to understand the wealth of dynamics involved in small perturbations of (1)–(2), we need to introduce some intermediate models. The Euler–Maxwell system (1) and (2) is such a "master equations" describing very rich and complex plasma dynamics, that it contains several well-known simplified models in plasma physics. For instance, in all physical situations[4], $m_e \ll M_i$. It is then natural to formally set $\varepsilon = 0$ in (8), which leads to simplified *one fluid* models for either ions ($M_i = 1$, $m_e = 0$) or electrons ($M_i = \infty$, $m_e = 1$). Moreover, if all the velocities are much smaller than the speed of light, then $C_b \gg 1$. Formally setting[5] $C_b = \infty$ and $B \equiv 0$ replaces the Maxwell equations by the much simpler *Poisson equation.* We refer to [3, 12] for other examples.

In the following, we will consider the simplified models in a form which is consistent with the reformulation (8) given appropriate approximations. This might look somewhat different from the classical form of these models. However, after an appropriate rescaling the equations should be the same up to cubic and higher-order terms which can be ignored in our situation.

1.1 Single-fluid models

The simplest model we can derive is the *Euler–Poisson model for the electrons*

$$\begin{aligned} \partial_t n + \operatorname{div}((1+n)v) &= 0, \\ \partial_t v + v \cdot \nabla v + \nabla n &= \nabla\phi, \\ \Delta\phi &= n. \end{aligned} \tag{15}$$

Here the magnetic field vanishes $B \equiv 0$, and the ions are treated as motionless with a constant density and only form a fixed charged background. Such a simplified system is used for describing Langmuir waves in the two-fluid theory. After suitable change of unknown, (15) can be reformulated as

$$(\partial_{tt} - \Delta + 1)\,\alpha = \mathcal{Q}(\alpha, \nabla\alpha, \nabla^2\alpha). \tag{16}$$

The linearized Euler–Poisson system for irrotational flows is no longer the acoustic (wave) equation as in the pure Euler system (14), but the Klein–Gordon system with "mass term" created by the plasma frequency due to the electrostatic interaction. Taking advantage of the much better properties of Klein–Gordon equations (faster time decay of linear waves like $t^{-3/2}$, absence of quadratic resonances), global smooth irrotational flows were constructed in [23] via the normal form method of Shatah [42]:

Theorem 3. (Stability of a neutral equilibrium [23]) *Solutions of* (15) *with initial data* (n^0, v^0) *small, smooth, neutral and irrotational in the sense that*

$$\int_{\mathbb{R}^3} n^0(x)dx = 0, \qquad \nabla \times v^0 \equiv 0$$

[4]Indeed, the ratio m_e/M_i is no bigger than the ratio of the electron mass to the proton mass which equals 1/1836.

[5]This is called the *electrostatic approximation.*

remain globally smooth and decay to 0 in L^∞ as $t \to +\infty$.

The neutral assumption was later removed in [16] and this result was extended to two spatial dimensions independently in [29, 38] (see also [31, 32]). Theorem 3 was the first positive result indicating that the dispersive effect alone in the two-fluid theory may prevent shock formation[6] and it started an investigation to understand to which extent the introduction of electromagnetic forces could stabilize the full Euler–Maxwell system.

Recently, further progress was made in this direction in the study of another simplified model: the *Euler–Poisson equation for the ions*[7]:

$$\begin{aligned} \partial_t \rho + \operatorname{div}((1+\rho)u) &= 0, \\ \partial_t u + u \cdot \nabla u + \nabla \rho &= -\nabla \phi, \\ -\Delta \phi &= \rho - \phi. \end{aligned} \tag{17}$$

Here the electron dynamics with constant temperature is decoupled from the ion dynamics via the Boltzmann relation. The model equation then becomes

$$\left(\partial_{tt} - \Delta + (-\Delta)(1-\Delta)^{-1}\right)\alpha = |\nabla|\mathcal{Q}(\alpha, \nabla\alpha). \tag{18}$$

This system has intermediate behavior between (14) and (16). The linearized solutions decay slowly (like $t^{-4/3}$) and create many strong degeneracies near the zero frequency, where the dispersion relation is similar to the wave dispersion up to third order. Nevertheless, the first and third authors were able to obtain an analogue of Theorem 3 for perturbations of a neutral equilibrium by using a variation on the normal form method, controlling bilinear multipliers with rough coefficients using arguments inspired by [27]. Here, a crucial property is the fact that the nonlinearity is an exact derivative, which helps compensate for the degeneracy at the 0 frequency.

1.2 Two-fluid models with different speeds

Both systems (15) and (17) can be reduced (under the irrotational assumption) to a (complex) scalar quasilinear equation with one speed. This is no longer the case for more complicated two-fluid models which yield quasilinear system with different speeds. Bilinear interactions in quasilinear systems generically create resonant sets of 2D spheres in the phase space, which are very challenging to control analytically. This was first studied in [14] for the case of semilinear systems of Klein–Gordon equations with different speeds (see also [11] for a study of a system with different

[6] Another way to prevent shock formation is to introduce exponential damping of the perturbation via dissipation or relaxation (see e.g. [39]). We will not discuss this at all in this paper.

[7] In many works (including [24] and [13, 25, 37]), the Poisson relation in (17) is replaced by

$$-\Delta\phi = 1 + \rho - e^{\phi},$$

but, for small perturbations, this agrees with (17) up to nonlinear corrections which can be easily handled.

masses) and led in [15] to the first construction of global smooth solutions for the Euler–Maxwell equation for electrons,

$$\begin{aligned} \partial_t n + \operatorname{div}((1+n)v) &= 0, \\ \partial_t v + v\cdot\nabla v + \nabla n &= -\left[E + v\times B\right], \\ \partial_t B + \nabla\times E &= 0, \\ \partial_t E - C\nabla\times B &= (1+n)v \end{aligned} \tag{19}$$

with constraints $\operatorname{div}(B) = 0$ and $\operatorname{div}(E) = n$:

Theorem 4. (Stability in the Euler–Maxwell system for electrons [15, 30]) *A solution of* (19) *with initial data* (n^0, v^0, E^0, B^0) *small, smooth, compactly supported, neutral and irrotational in the sense that*

$$\int_{\mathbb{R}^3} n^0(x)dx = 0, \quad \int_{\mathbb{R}^3} B^0(x)dx = 0, \quad \nabla\times v^0 + CB^0 \equiv 0$$

remains global and smooth and decays to 0 *in* L^∞.

This was first shown in [15] under additional generic conditions on the parameters. Later in [30], the generic condition was removed and a stronger (integrable) decay was obtained, providing a robust approach even in the quasilinear case. The model system is

$$\begin{aligned} (\partial_{tt} - \Delta + 1)\,\alpha &= \mathcal{Q}_1(\alpha, \beta, \nabla\alpha, \nabla\beta, \nabla^2\alpha, \nabla^2\beta), \\ (\partial_{tt} - C\Delta + 1)\,\beta &= \mathcal{Q}_2(\alpha, \beta, \nabla\alpha, \nabla\beta, \nabla^2\alpha, \nabla^2\beta). \end{aligned} \tag{20}$$

It is important to note that the speed of the electron fluids is different from the speed of the magnetic field, so that new analytical tools are needed to estimate the 2D resonant sphere in the phase space. The main result of [30] is the natural analogue of Theorem 3 and it is the foundation of the approach we use in this work. Note that in this case, we also need to introduce a decay condition on the initial data in order to be able to perform a more refined analysis of the solutions.

2 Description of the method

We use a combination of dispersive analysis and energy estimate, relying heavily on the Fourier transform (see [6, 17, 18, 27, 35, 36, 42] for previous seminal works). To overcome the quasilinear nature of the nonlinearity and ensure global existence, we use classical high order energy estimates to make up for the loss of derivatives in the nonlinearity. Global existence follows if a certain norm of lower regularity remains bounded and decays faster than $1/t$.

Such a crucial decay property is established by a semilinear analysis of systems of dispersive equations. Expecting some form of scattering, we express the solution as a free evolution[8] from a profile which varies more slowly in time. After suitable

[8] I.e. a solution to the linearized equation.

algebraic manipulations, and appropriate use of the Fourier transform, we need to study bilinear operators T of the form

$$\widehat{T[f,g]}(\xi) = \int_{\mathbb{R}} \int_{\mathbb{R}^3} e^{it\Phi(\xi,\eta)} m(\xi,\eta) \widehat{f}(\xi-\eta,t)\widehat{g}(\eta,t) d\eta dt \tag{21}$$

with a phase Φ which is specific to each interaction and which is of the form

$$\Phi(\xi,\eta) = \Lambda_0(\xi) \pm \Lambda_1(\xi-\eta) \pm \Lambda_2(\eta), \quad \Lambda_j \in \{\Lambda_i, \Lambda_e, \Lambda_b\}, \tag{22}$$

where the functions Λ_j are defined from the linearized system. As a first approximation, one may think of f, g as being smooth bump functions and m being essentially a smooth cut-off, and the main challenge is to estimate efficiently the infinite time integral. It then becomes clear that a key role is played by the properties of the function Φ and in particular by the points where it is stationary,

$$\nabla_{(t,\eta)}[t\Phi(\xi,\eta)] = 0, \quad \text{i.e.} \quad \Phi(\xi,\eta) = 0 \text{ and } \nabla_\eta \Phi(\xi,\eta) = 0. \tag{23}$$

The collection of such points form the *space-time resonant set.* This was already highlighted in [17] and forms the basis of the space-time resonance method. In some situations, one has no or few fully stationary points and the task is mainly to propagate enough smoothness of $\hat{f}$, $\hat{g}$ to exploit (non)-stationary phase arguments.

However, as in the case of the Euler–Maxwell equation for electrons (19), the space-time resonance set can be a 2D sphere. In addition, the linearized system for the full two fluid model is now coupled, which makes the derivation of the dispersive system more involved and requires a careful study of the dispersion relations Λ_j appearing in (22) (see [GIP]). Moreover, the appearance of an "ion-like" dispersion relation Λ_i, similar to that in (17), leads to additional mathematical difficulties of slower time decay of linear solutions, rough bilinear multipliers and rough phase around the zero frequency.

To overcome these new difficulties, we employ and extend the method developed in [29, 30]. We seek an appropriate space B satisfying two requirements: first, the bilinear operator T in (21) needs to be bounded

$$T : B \times B \to B, \tag{24}$$

second, the free flow of the linearized Euler–Maxwell system (8) with initial data bounded in the space B should belong to a space like $L^1_t L^\infty_x$, which has sufficiently strong time decay to close the energy estimate.

This strategy, initiated in the previous works [29, 30] shares similarities with *the space-time resonance method* as developed in [14, 15, 17, 18, 19, 20] but with new types of function space localized in both space and frequency, which are naturally compatible with the introduction of fractional powers of the weights (like $x^{-1-\varepsilon}L^2$ for the B^1-norm below), and with new bilinear estimates. We find this approach more precise and flexible analytically, which is crucial to analyze the complicated phase function (22) arising in the Euler–Maxwell system. Together with orthogonality arguments and localized decay estimates, this allows to overcome the central difficulty of controlling delicate space-time resonant points in (23).

We also mention the works in [10, 41] which consider global existence in dispersive equations or systems with nonlinearity with small power without assuming any weights on the initial data (see also [22] and the references therein). It appears, however that such approach would be very difficult to carry on in the context of a system like (1) due to the loss of derivative in the nonlinearity. In addition, even in a purely semilinear setting, the analysis of the most delicate space-time resonances seems presently out of reach when the functions have too rough Fourier transforms.

2.1 Choice of the norms

In order to define such a space B, we measure localization both in space and in frequency. We quantify all these "coordinates" all the way to the uncertainty principle and decompose an arbitrary function as a sum of "atoms":

$$f = \sum_{X \cdot N \geq 1} Q_X P_N f, \quad (Q_X f)(x) \simeq \mathbb{1}_{X \leq |x| \leq 2X}(x) f(x),$$
$$\widehat{(P_N f)}(\xi) \simeq \mathbb{1}_{N \leq |\xi| \leq 2N}(\xi) \hat{f}(\xi).$$

We can then define the norms for the space B on each atom. The simplest norm giving the appropriate decay would be a weighted space $x^{-1-\varepsilon} L^2$ and this is the main motivation for our "strong" norm B^1. Unfortunately, some interactions seem to produce outputs which are not bounded in this norm around a 2D resonant sphere. To account for this, we also introduce another kind of atoms, the "weak" atoms, bounded only in B^2 which barely fail to be in $x^{-1} L^2$, but are essentially concentrated on the $2D$-resonant spheres. Finally, each atom is allowed to be a combination of the two above types:

$$\|f\| = \sup_{X \cdot N \geq 1} \|Q_X P_N f\|_{B_{X,N}},$$
$$\|g\|_{B_{X,N}} = \|g\|_{B^1_{X,N} + B^2_{X,N}} = \inf_{g = g_1 + g_2} \{\|g_1\|_{B^1_{X,N}} + \|g_2\|_{B^2_{X,N}}\}.$$

We refer to [GIP] for the precise definition of the Z norm that we use and for the proof that these norms yield the desired integrability upon application of the linear flow.

2.2 Analysis of the bilinear operators

Once the form of the norm has been assessed, the main difficulty is to understand the bilinear interactions in (21) and to fine-tune the norms to ensure that they are appropriately bounded as in (24). After quantifying all the information, one needs to treat a huge sum of elementary interactions (even for a single atom as an output). However, appropriate use of energy-estimate, simple orthogonality arguments and finite speed of propagation quickly limit the cases to only a few possibilities, see [GIP]. Then non-stationary phase analysis allows to focus on the *space-time resonant sets* as in (23). This is where the bulk of the work is done, following a previous work [30] where such an analysis was performed on the simplified Euler–Maxwell equation for the electrons.

After a careful analysis of the interactions done in [GIP], we isolate three different problematic space-time resonant sets $\mathcal{S}$.

- Case A: we have the case of a "classical" $2D$ sphere

$$\mathcal{S}_A = \{(\xi,\eta) = (R\omega, r\omega),\ \omega \in \mathbb{S}^2\}, \quad R \neq 0,\ r \neq 0$$

  which already appears in the analysis of (20) and which is responsible for the introduction of the "weak" atoms. Fortunately, here the phase is nondegenerate and we can perform an efficient stationary phase analysis and use additional refined orthogonality arguments as in [30].

- Case B: we have a first degenerate sphere

$$\mathcal{S}_B = \{(\xi,\eta) = (R'\omega, 0),\ \omega \in \mathbb{S}^2\}, \quad R' \neq 0$$

  where in addition, the phase is not smooth in η.

  In this case, we use the fact that the speed of propagation of the singular perturbation is slower than expected, the fact that the phase is weakly elliptic and a careful adaptation of the refined orthogonality analysis of Case A, keeping track of how the bounds deteriorate as $\eta \to 0$.

- Case C: the presence of an "ion-like" dispersion relation brings in a strong degenerate set at 0

$$\mathcal{S}_D = \{(\xi,\eta) = (0, r'\omega),\ \omega \in \mathbb{S}^2\}, \quad r' \neq 0 \text{ or } r' = 0.$$

  Here the problem comes from the strong degeneracy of the phase. Similar problems already appeared for the Euler–Poisson equation for the ions (17), but for (1) we need more refined multiplier estimate and orthogonality arguments, combined with additional finite speed of propagation estimates and use of the null-form structure coming from the presence of a derivative in front of the nonlinearity as in (18) in order to overcome the loss coming from the roughness of the multiplier after application of a normal form transformation.

References

[1] S. Alinhac, Temps de vie des solutions régulières des équations d'Euler compressibles axisymétriques en dimension deux. (French) [Life spans of the classical solutions of two-dimensional axisymmetric compressible Euler equations] Invent. Math. **111** (1993), no. 3, 627–670.

[2] J. A. Bittencourt, Fundamentals of plasma physics, 3rd edition, 2004, Springer.

[3] S. Cordier and E. Grenier. Quasineutral limit of an Euler–Poisson system arising from plasma physics. Comm. Partial Differential Equations **25** (2000), no. 5–6, 1099–1113.

[4] F. Chen, introduction to plasma physics, 1995, Springer.

[5] G.-Q. Chen, J. Jerome, and D. Wang, Compressible Euler–Maxwell equations. Proceedings of the Fifth International Workshop on Mathematical Aspects of Fluid and Plasma Dynamics (Maui, HI, 1998). Transport Theory Statist. Phys. **29** (2000), no. 3–5, 311–331.

[6] D. Christodoulou, Global solutions of nonlinear hyperbolic equations for small initial data, Comm. Pure Appl. Math. **39** (1986), 267–282.

[7] D. Christodoulou, The Formation of Shocks in 3-Dimensional Fluids, EMS Monographs in Mathematics, EMS Publishing House, 2007.

[8] D. Christodoulou and S. Miao, Compressible flow and Euler's equations, arXiv:1212.2867.

[9] J.-L. Delcroix and A. Bers, Physique des plasmas, InterEditions/ CNRS Editions, Paris, 1994.

[10] J.-M. Delort and D. Fang, Almost global existence for solutions of semilinear Klein–Gordon equations with small weakly decaying Cauchy data. Comm. Partial Differential Equations **25** (2000), no. 11–12, 2119–2169.

[11] J.-M. Delort, D. Fang, and R. Xue, Global existence of small solutions for quadratic quasilinear Klein–Gordon systems in two space dimensions. J. Funct. Anal. **211** (2004), 288–323.

[12] P. Degond, F. Deluzet, and D. Savelief, Numerical approximation of the Euler–Maxwell model in the quasineutral limit, J. Comput. Phys. **231** (2012), no. 4, 1917–1946.

[13] D. Gérard-Varet, D. Han–Kwan, and F. Rousset, Quasineutral limit of the Euler–Poisson system for ions in a domain with boundaries, Indiana Univ. Math. J., to appear.

[14] P. Germain, Global existence for coupled Klein–Gordon equations with different speeds. Ann. Inst. Fourier (Grenoble) **61** (2011), no. 6, 2463–2506.

[15] P. Germain and N. Masmoudi, Global existence for the Euler–Maxwell system, arXiv:1107.1595.

[16] P. Germain, N. Masmoudi, and B. Pausader, Non-neutral global solutions for the electron Euler–Poisson system in 3D, Siam. J. Math. Anal., to appear.

[17] P. Germain, N. Masmoudi, and J. Shatah, Global solutions for 3D quadratic Schrödinger equations, Int. Math. Res. Not. (2009), 414–432.

[18] P. Germain, N. Masmoudi, and J. Shatah, Global solutions for the gravity water waves equation in dimension 3, Ann. of Math., to appear.

[19] P. Germain, N. Masmoudi, and J. Shatah, Global solutions for 2D quadratic Schrödinger equations, J. Math. Pures Appl. (9) **97** (2012), no. 5, 505–543.

[20] P. Germain, N. Masmoudi, and J. Shatah, Global existence for capillary water-waves, arXiv:1210.1601.

[21] E. Gourgoulhon, An introduction to relativistic hydrodynamics, EAS Publications Series **21** (2006), 43–79.

[22] S. Guo, On the 1D cubic NLS in an almost critical space, master thesis, U. Bonn.

[23] Y. Guo, Smooth irrotational flows in the large to the Euler–Poisson system in $\mathbb{R}^{3+1}$, Comm. Math. Phys. **195** (1998), no. 2, 249–265.

[GIP] Y. Guo, A. D. Ionescu, and B. Pausader, Global solutions of the Euler–Maxwell two-fluid system in 3D, arXiv:1303.1060.

[24] Y. Guo and B. Pausader, Global smooth ion dynamics in the Euler–Poisson system, Comm. Math. Phys. **303** (2011), no. 1, 89–125.

[25] Y. Guo and X. Pu, KdV limit of the Euler–Poisson system, arXiv:1202.1830.

[26] Y. Guo and S. Tahvildar-Zadeh, Formation of singularities in relativistic fluid dynamics and in spherically symmetric plasma Dynamics, Contemporary Mathematics **238** (1999), 151–161.

[27] S. Gustafson, K. Nakanishi, and T.-P. Tsai, Scattering theory for the Gross–Pitaevskii equation in three dimensions, Commun. Contemp. Math. **11** (2009), 657–707.

[28] Z. Hani, F. Pusateri, and J. Shatah, Scattering for the Zakharov system in 3 dimensions, Comm. Math. Phys., arXiv: 1206.3473.

[29] A. D. Ionescu and B. Pausader, The Euler–Poisson system in 2D: global stability of the constant equilibrium solution, Int. Math. Res. Not. (2013), 761–826.

[30] A. D. Ionescu and B. Pausader, Global solutions of quasilinear systems of Klein–Gordon equations in 3D, J. Eur. Math. Soc., to appear.

[31] J. Jang, The two-dimensional Euler–Poisson system with spherical symmetry. J. Math. Phys. **53** (2012), no. 2, 023701, 4 pp.

[32] J. Jang, D. Li, and X. Zhang, Smooth global solutions for the two dimensional Euler Poisson system, Forum Mathematicum, to appear.

[33] F. John, Blow-up of solutions of nonlinear wave equations in three space dimensions, Manuscripta Math. **28** (1979), 235–268.

[34] T. Kato, The Cauchy problem for quasi-linear symmetric hyperbolic systems, Arch. Rational Mech. Anal. **58** (1975), 181–205.

[35] S. Klainerman, Global existence of small amplitude solutions to nonlinear Klein–Gordon equations in four space-time dimensions, Comm. Pure Appl. Math. **38** (1985), 631–641.

[36] S. Klainerman, The null condition and global existence to nonlinear wave equations. Nonlinear systems of partial differential equations in applied mathematics, Part 1 (Santa Fe, N.M., 1984), 293–326, Lectures in Appl. Math. **23**, Amer. Math. Soc., Providence, RI, 1986.

[37] D. Lannes, F. Linares, and J.-C. Saut, The Cauchy problem for the Euler–Poisson system and derivation of the Zakharov–Kuznetsov equation, "Perspectives in Phase Space Analysis of PDE's" to appear in Birkhäuser series "Progress in Nonlinear Differential Equations and Their Applications".

[38] D. Li and Y. Wu, The Cauchy problem for the two dimensional Euler–Poisson system, arXiv:1109.5980.

[39] Y. J. Peng, Global existence and long-time behavior of smooth solutions of two-fluid Euler–Maxwell equations, Ann. Inst. H. Poincaré Anal. Non Linéaire **29** (2012), no. 5, 737–759.

[40] X. Pu, Dispersive limit of the Euler–Poisson system in higher dimensions, arXiv:1204.5435.

[41] T. Schottdorf, Global existence without decay for quadratic Klein–Gordon equations, preprint, arXiv:1209.1518.

[42] J. Shatah, Normal forms and quadratic nonlinear Klein–Gordon equations, Comm. Pure Appl. Math. **38** (1985), 685–696.

[43] T. Sideris, Formation of singularities in three-dimensional compressible fluids, Comm. Math. Phys. **101** (1985), 475–485.

[44] P.A. Sturrock, Plasma Physics, Cambridge University Press, 1994.

[45] D. G. Swanson, Plasma Waves, Academic Press, Boston, 1989.

[46] B. Texier, Derivation of the Zakharov equations, Arch. Ration. Mech. Anal. **184** (2007), no. 1, 121–183.

Division of Applied Mathematics, Brown University
Providence, RI 02912, USA
E-mail address: guoy@cfm.brown.edu

Department of Mathematics, Princeton University
Princeton, NJ 08544, USA

E-mail address: aionescu@math.princeton.edu

LAGA, Université Paris 13 (UMR 7539)
93430 Villetaneuse, France
E-mail address: pausader@math.univ-paris13.fr

Proceedings of the 6th International
Congress of Chinese Mathematicians
ALM 36, pp. 95–117

Compact Embedded Rotational Hypersurfaces with Constant Mean Curvature in a Sphere

Zhijie Huang and Haizhong Li

Abstract

In the first part of this paper, we give a survey of complete classification of embedded CMC tori in a 3-dimensional sphere, which was proved by Ben Andrews and Haizhong Li in 2012. In the second part of this paper, by studying the monotonicity of the period function $K(H, n, C)$, we give a complete classification of n-dimensional compact embedded rotational CMC hypersurfaces in S^{n+1} for $n = 3, 4$. In last part of this paper, we give a survey of compact embedded rotational hypersurfaces with constant m^{th} mean curvature in a sphere.

Mathematics Subject Classification: Primary 53C42; Secondary 53A10.

Keywords and Phrases: Lawson conjecture, Pinkall-Sterling conjecture, classification of embedded CMC tori, rotational CMC hypersurfaces, monotonicity of period function.

1 Lawson conjecture and Pinkall–Sterling conjecture

Minimal surfaces or constant mean curvature surfaces in 3-dimensional space form $R^3(c)$ are one of the most important and most fruitful topics in differential geometry, when $c = 0, R^3(c) = \mathbb{R}^3$ is Euclidean space, when $c = 1, R^3(c) = S^3$ is a unit sphere, when $c = -1, R^3(c) = H^3$ is a hyperbolic space. Let $x : \Sigma \to R^3(c)$ be a surface, with two principal curvatures k_1 and k_2. Then the mean curvature function is defined by

$$H = \frac{1}{2}(k_1 + k_2).$$

Supported by grants of NSFC No. 11271214.

When H is a constant, we call Σ to be a constant mean curvature (CMC) surface. When $H = 0$, we call Σ to be a minimal surface. We know that surfaces with constant mean curvature $H \neq 0$ in $R^3(c)$ is a critical point of area functional which preserves the volume it enclosed (see Barbosa–do Carmo [BD]) and a minimal surface in $R^3(c)$ is a critical point of the area functional. H. Hopf proved the following well-known global result by constructing a holomorphic quadratic differential for CMC surfaces in R^3 in 1951:

Theorem 1.1. (Hopf theorem [Ho1]) *Let M be a compact surface in $\mathbb{R}^3$ with $H =$ constant and the genus $g(M) = 0$, then M is a round sphere.*

Remark 1.2. Chern [Ch] extended Hopf's result to compact surfaces in 3-dimensional space form $R^3(c)$ with $H =$ constant and $g(M) = 0$.

H. Hopf proposed the following famous conjecture:

Conjecture 1. (Hopf conjecture [Ho1, Ho2]) Any compact surfaces with $H =$ constant in $\mathbb{R}^3$ must be a round sphere.

In 1956, Alexandrov [Al1] checked Hopf's conjecture under extra condition "embeddedness".

Theorem 1.3. (Alexandrov's uniqueness theorem [Al1]) *If a compact CMC surface is embedded in $\mathbb{R}^3$, a hemisphere S^3_+, or hyperbolic space H^3, then it must be totally umbilical.*

In 1984, Wente [We] constructed counterexamples (non-trivial CMC tori) for Hopf's conjecture by use of integrable systems. Wente's paper was followed by a series of papers by Abresch [Ab], Bobenko [Bo], Pinkall–Sterling [PS] and many others. In particular, they constructed many CMC tori in $\mathbb{R}^3$, S^3 and H^3.

Except totally geodesic and Clifford torus (or totally umbilical surface and $S^1(r) \times S^1(\sqrt{1-r^2})$, $0 < r < 1$), it is a difficult problem to construct embedded minimal surfaces (or CMC surfaces). In 1970, H. B. Lawson discovered an infinite family of embedded minimal surfaces of higher genus:

Theorem 1.4. (Lawson [La2]) *Given any pair of positive integers m and k, there exists an embedded minimal surfaces of any given genus g, and there are at least two such surfaces unless g is a prime number.*

In 1988, Karcher–Pinkall–Sterling [KPS] constructed more embedded minimal surfaces:

Theorem 1.5. (Karcher–Pinkall–Sterling [KPS]) *There exist additional examples of embedded minimal surfaces in S^3, which are not part of the family obtained by Lawson in [La2]. These surfaces have genus $3, 5, 6, 7, 11, 19, 73$ and 601.*

In 1966, Almgren proved the following uniqueness result for minimal surface with genus 0:

Theorem 1.6. (Almgren [Alm]) *Any compact minimal surface with genus 0 in S^3 is the equator.*

In 1970, for minimal tori Lawson [La1] conjectured that

Conjecture 2. (Lawson conjecture [La1]) The only embedded minimal torus in S^3 is the Clifford torus $S^1(\frac{\sqrt{2}}{2}) \times S^1(\frac{\sqrt{2}}{2})$.

In 2012, Brendle [Br1] solved Lawson conjecture by use of his ingenious ideas and "non-collapsing" technique. About recent progress and new results of minimal surfaces, readers can consult Brendle's excellent survey paper [Br3].

In 1989, for embedded tori with constant mean curvature H in S^3, Pinkall and Sterling [PS] conjectured that

Conjecture 3. (Pinkall-Sterling conjecture [PS]) All embedded tori with constant mean curvature in S^3 are surfaces of revolution.

In 2012, Andrews and Li [AL] solved Pinkall–Sterling conjecture by use of "non-collapsing" technique and adaption of Brendle's arguments. Moreover, Andrews and Li [AL] gave a complete classification of all embedded CMC tori in S^3.

2 Classification of embedded constant mean curvature tori in S^3

Combining Lawson conjecture proved by Brendle, we can state the result of Andrews and Li as followings

Theorem 2.1. (Andrews–Li [AL])

1. *Every embedded CMC torus Σ in S^3 is a rotational surface.*

2. *If $H \in \{0, \frac{1}{\sqrt{3}}, -\frac{1}{\sqrt{3}}\}$ then every embedded torus with mean curvature H is congruent to the Clifford torus $S^1(\frac{\sqrt{2}}{2}) \times S^1(\frac{\sqrt{2}}{2})$ and torus $S^1(\frac{1}{2}) \times S^1(\frac{\sqrt{3}}{2})$.*

3. *If Σ is an embedded CMC torus which is not congruent to an isoparametric torus $S^1(r) \times S^1(\sqrt{1-r^2}), 0 < r < 1$ then there exists a maximal integer $m \geq 2$ such that Σ has m-fold symmetry.*

4. *For given $m \geq 2$, there exists at most one such CMC torus (up to congruence).*

5. *For given $m \geq 2$, there exists an embedded CMC torus with mean curvature H and maximal symmetry $S^1 \times Z_m$ if $|H|$ lies strictly between $\cot \frac{\pi}{m}$ and $\frac{m^2-2}{2\sqrt{m^2-1}}$.*

Remark 2.2.

1. The case $H = 0$ is the Lawson conjecture which was proved by Brendle [Br1]. The rigidity appearing for $H = \pm\frac{1}{\sqrt{3}}$ is unexpected.

2. For $H \neq \{0, \frac{1}{\sqrt{3}}, -\frac{1}{\sqrt{3}}\}$ and Σ is not $S^1(r) \times S^1(\sqrt{1-r^2})$, $0 < r < 1$, CMC embedded tori are the analogues of Delaunay Surface in R^3. Number of these CMC embedded tori depends on the value of H.

3. The embeddedness assumption in Theorem 2.1 is crucial: There exists an infinite family of non-rotationally symmetric immersed CMC tori in S^3 (see [Bo, Per]).

In order to prove Andrews–Li's Theorem 2.1, they used the non-collapsing argument originating from [An], together with the modifications of techniques introduced by Brendle [Br1] to conclude that all embedded CMC tori are rotational surfaces. If $H = 0$, they can get that Σ must be Clifford torus $S^1(\frac{\sqrt{2}}{2}) \times S^1(\frac{\sqrt{2}}{2})$ by use of Otsuki's uniqueness result of embedded rotational surfaces [Ot1, Ot2]. But for $H \neq 0$, they need to prove the function $K(H, C)$ has monotonicity, which is key step in classification of embedded rotational surfaces with constant mean curvature H in S^3.

Now we will give some ideas of the proof of Theorem 2.1.

In 2011, Andrews [An] gave a direct proof of the non-collapsing result (due to Sheng and Wang [SW]) for mean-convex hypersurface in $\mathbb{R}^{n+1}$ moving under the mean curvature flow.

Definition 2.3. (δ-non-collapsing) For $\delta > 0$, an embedded compact mean-convex hypersurface $M \subset \mathbb{R}^{n+1}$ is called δ-non-collapsing, if for every point x of M there is a sphere of radius $\delta/H(x)$ enclosed by M which touches M at x.

Theorem 2.4. (Non-collapsing result) *For any embedded compact mean-convex hypersurface $M \subset \mathbb{R}^{n+1}$ moving under the mean curvature flow, δ-non-collapsing is preserved if the initial embedded compact mean-convex hypersurface is δ-non-collapsing.*

Andrews [An] observed that the noncollapsing condition is equivalent to that $Z : M \times M \to \mathbb{R}$ satisfies

$$Z(x, y) = \frac{H(x)}{\delta}\|F(y) - F(x)\|^2 + \langle F(y) - F(x), \nu(x)\rangle \geq 0 \tag{2.1}$$

for $(x, y) \in M \times M$ and $\nu(x)$ is an unit outward normal vector of $F(x)$. This function was shown to admit a maximum principle argument to preserve initial non-negativity.

The idea of working with functions of pairs of points was in turn inspired by earlier work of Huisken and Hamilton for the curve shortening flow and for Ricci flow on surfaces. See [Ha1, Ha2, Hu].

The key geometric idea in the non-collapsing argument is to compare the curvature of enclosed balls touching the surface to a suitable function at the touching point.

Let $M^n = F(\Sigma^n)$ be an embedded hypersurface in $S^{n+1} \subset \mathbb{R}^{n+2}$ given by an embedding F, and bounding a region $\Omega \subset S^{n+1}$. The ball in S^{n+1} with boundary curvature Φ which is tangent to $F(\Sigma)$ at the point $F(x)$ is $B = B_{\Phi^{-1}}(p)$, where $p = F(x) - \Phi^{-1}\nu(x)$, and ν is the unit normal to $F(\Sigma)$ at $F(x)$ in S^{n+1} which points out of Ω. The statement that this ball lies entirely in Ω is equivalent to the statement that for any $y \in \Sigma$, $\|F(y) - p\|^2 \geq \Phi^{-2}$, which can be written as follows:

$$\|F(y) - (F(x) - \Phi^{-1}\nu(x))\|^2 - \Phi^{-2} \geq 0.$$

This is equivalent to

$$Z(\Phi, x, y) := \frac{\Phi(x)}{2}\|F(y) - F(x)\|^2 + \langle F(y) - F(x), \nu(x)\rangle \geq 0. \tag{2.2}$$

Since $F(x), F(y) \in S^{n+1}$ we have $\|F(x)\|^2 = \|F(y)\|^2 = 1$ and $\langle F(x), \nu(x)\rangle = 0$, so that

$$Z(\Phi, x, y) = \Phi(x)(1 - F(x) \cdot F(y)) + \langle F(y), \nu(x)\rangle. \tag{2.3}$$

The case $H = 0$ was proved by Brendle [Br1], so we assume that $H > 0$. We denote by $\lambda(x) = \lambda_1(x)$ the largest principal curvature at x, and by $\mu(x) = \lambda(x) - H$.

The first step of the proof of Theorem 2.1 is to prove that for any $x \in \Sigma$ and $y \in \Sigma$, we have

$$Z(\lambda, x, y) = \lambda(x)(1 - F(x) \cdot F(y)) + \langle F(y), \nu(x)\rangle \geq 0. \tag{2.4}$$

Now we choose

$$\Phi(x) = \kappa\mu(x) + H$$

where κ is a positive constant. We need the following Simons' identity (for example, see [Si, NS, Li]):

Proposition 2.5. *Suppose that $F : \Sigma \to S^3$ is an embedded CMC torus in S^3. Then the function μ is strictly positive and satisfies*

$$\Delta\mu - \frac{|\nabla\mu|^2}{\mu} + 2(\mu^2 - 1 - H^2)\mu = 0.$$

Since Σ is compact and embedded, for sufficiently large κ, then $Z(\kappa\mu + H, x, y)$ is nonnegative.

Along any geodesic $\gamma(s)$ in Σ through x we have

$$Z(\kappa\mu + H, x, \gamma(s)) = \frac{1}{2}\left(\kappa\mu + H - h_x(\gamma', \gamma')\right)s^2 + O(s^3).$$

Choose $\gamma'(0)$ to be in the direction of the largest principal curvature, so that $h_x(\gamma', \gamma') = \lambda = H + \mu$. Then

$$Z(\kappa\mu + H, x, \gamma(s)) = \frac{1}{2}(\kappa - 1)\mu s^2 + O(s^3).$$

If $\kappa < 1$, Z takes negative values for small s.

If $\kappa > 1$, Z is positive in a neighborhood of the diagonal $\{(x, x) : x \in \Sigma\}$ in $\Sigma \times \Sigma$.

We choose

$$\bar{\kappa} = \inf\{\kappa > 0 : \ Z(\kappa\mu + H, x, y) \geq 0 \text{ for all } x, y \in \Sigma\}.$$

Then $1 \leq \bar{\kappa} < \infty$. If $\bar{\kappa} > 1$, then there must exist $(\bar{x}, \bar{y})$ in $\Sigma \times \Sigma$ with $\bar{x} \neq \bar{y}$ such that $Z(\bar{x}, \bar{y}) = 0$, while $Z(x, y) \geq 0$ for every $(x, y) \in \Sigma \times \Sigma$. Here we write $Z(x, y) = Z(\bar{\kappa}\mu + H, x, y)$.

Then the second derivatives of Z are nonnegative at $(\bar{x}, \bar{y})$. However, by a *delicate calculations* and using Proposition 2.5, Andrews and Li can have in normal coordinates of $\bar{x}$ and $\bar{y}$ (see [AL])

$$\sum_{i=1}^{2}\left(\frac{\partial}{\partial x_i}+\frac{\partial}{\partial y^i}\right)^2\Big|_{(\bar{x},\bar{y})} Z \le -(\bar{\kappa}^2-1)d^2\mu^2 H < 0, \tag{2.5}$$

which is a contradiction. So we conclude that $\bar{\kappa} = 1$, that is, for any $x \in \Sigma$ and $y \in \Sigma$

$$Z(\lambda, x, y) = \lambda(x)(1 - F(x)\cdot F(y)) + \langle F(y), \nu(x)\rangle \ge 0. \tag{2.6}$$

The second step of the proof of Theorem 2.1 is to prove the following Proposition:

Proposition 2.6. *Let $F : \Sigma \to S^3$ be a CMC embedded torus for which $Z(\lambda(x), x, y) \ge 0$ for every $x, y \in \Sigma$. Then Σ is rotationally symmetric.*

Since Σ is a CMC torus and therefore has no umbilical points, we have global smooth eigenvector fields e_1 and e_2 such that $h(e_1, e_1) = \lambda_1 = \lambda$ and $h(e_2, e_2) = \lambda_2 = 2H - \lambda$, and $h(e_1, e_2) = 0$. We can deduce that $(\nabla_{e_1} h)(e_1, e_1) = 0$, and consequently also $(\nabla_{e_1} h)(e_2, e_2) = 0$ everywhere on Σ. So that $e_1\lambda = 0$ and then $e_1\mu = 0$.

Remark 2.7. If $H = 0$, we also have $(\nabla_{e_2} h)(e_2, e_2) = 0$ by changing e_1 with e_2, $\nu(x)$ into $-\nu(x)$, we have that h is parallel, so $M = S^1(\frac{\sqrt{2}}{2}) \times S^1(\frac{\sqrt{2}}{2})$.

By use of Codazzi equations and through delicate analysis, Andrews and Li can solve

$$F(u, v) = (r(u)\cos v, r(u)\sin v, \sqrt{1 - r^2(u)}\cos\theta(u), \sqrt{1 - r^2(u)}\sin\theta(u)) \tag{2.7}$$

where $0 \le v < 2\pi$, $0 \le u < \frac{m\pi}{\sqrt{1+H^2}}$ and m is some positive integer, where

$$\theta(u) = \int_0^u \frac{r(\tau)(\mu(\tau) + H)}{1 - r^2(u)} d\tau,$$

$$r(u) = \frac{\mu^{-1/2}}{\sqrt{C}} = \sqrt{\frac{C - 2H + \sqrt{C^2 - 4 - 4HC}\sin(2\sqrt{1 + H^2}u)}{2(1 + H^2)C}},$$

where C is a constant which is not less than $2(H + \sqrt{1 + H^2})$. This gives a outline of proof of Proposition 2.6.

Define

$$K(H, C) = \int_{\frac{t_1^2}{C}}^{\frac{t_2^2}{C}} \frac{(Hu + C^{-1})}{\sqrt{u}(1 - u)\sqrt{-u^2(1 + H^2) + (1 - 2HC^{-1})u - C^{-2}}} du,$$

where t_1 and t_2 are defined by

$$\begin{aligned} t_1 &= \sqrt{\frac{C-2H-\sqrt{C^2-4HC-4}}{2(1+H^2)}}, \\ t_2 &= \sqrt{\frac{C-2H+\sqrt{C^2-4HC-4}}{2(1+H^2)}}. \end{aligned} \tag{2.8}$$

The third step of the proof of Theorem 2.1 is to prove the following key result:

Proposition 2.8. *For any nonnegative real number H, $K(H,C)$ is monotone decreasing in $2(H+\sqrt{1+H^2}) \leq C < \infty$.*

Remark 2.9. When $H=0$, Proposition 2.8 was proved by Otsuki [Ot2, Ot3] and Furuya [Fu].

In the proof of Theorem 2.1, the following two propositions are needed:

Proposition 2.10. (Otsuki [Ot1], Perdomo [Per]) *Suppose that $F : \Sigma \to S^3$ is a rotational torus in S^3, which is not a Clifford torus and is given by (2.7). Then $F(\Sigma)$ is an embedded torus if and only if*

$$K(H,C) = \frac{2\pi}{m} \tag{2.9}$$

for some positive integer m.

Proposition 2.11. (Perdomo [Per]) *For any integer $m \geq 2$, if H satisfies*

$$\cot\frac{\pi}{m} < H < \frac{m^2-2}{2\sqrt{m^2-1}}, \tag{2.10}$$

then there exists a compact embedded rotational torus in S^3 with constant mean curvature H whose isometry group contains $O(2) \times Z_m$ which is not torus $S^1(r) \times S^1(\sqrt{1-r^2})$, $0 < r < 1$.

This gives a outline of the proof of Andrews–Li's Theorem 2.1, readers can see Andrews–Li's paper [AL] for details.

At end of this section, we would like to propose the following problems:

Problem 2.12. *Classify some class of embedded Weingarten surfaces in S^3.*

Problem 2.13. *Classify all embedded minimal Lagrangian tori in CP^2 and CH^2, respectively.*

Problem 2.14. *Classify all embedded minimal tori (or CMC tori) in $S^2 \times \mathbb{R}$ and $H^2 \times \mathbb{R}$.*

Problem 2.15. *Classify all embedded CMC tori in 3-dimensional Berger sphere S^3_b.*

Problem 2.16. *Classify all embedded CMC tori in 3-dimensional metric Lie groups.*

Some Related backgrounds about above problems can be found in the literatures, Brendle's recent paper [Br2], Castro and Urbano [CU1, CU2], Abresch and Rosenberg [AR], Fernandez and Mira [FM], Torralbo and Urbano [TU], Torralbo [To], Meeks III and Perez [MP].

3 n-dimensional rotational CMC hypersurfaces in a sphere

In 1970, Otsuki [Ot1] studied n-dimensional embedded minimal hypersurfaces with two distinct principal curvatures in a unit sphere S^{n+1} and proved that these kind of hypersurfaces must be congruent to isoparametric hypersurfaces. The study of hypersurfaces with two principal curvatures has attracted the attention of many mathematician, for example, do Carmo–Dajczer [DD], Otsuki [Ot2, Ot3], Ripoll [Ri], Brito–Leite [BL], Leite [Le]. Li–Wei [LW1, LW2], Wei–Cheng–Li [WCL], Perdomo [Per], Andrews–Huang–Li [AHL], and many others.

Let Σ be a hypersurface in a unit sphere $S^{n+1} \subset \mathbb{R}^{n+2}$, ν be a unit normal vector field on Σ, then $\nu : \Sigma \to S^{n+1}$ is also called Gauss map. Let $S_p : T_p\Sigma \to T_p\Sigma$ be the shape operator, which is defined by

$$S_p(X) = -\bar{\nabla}_X \nu, \quad \forall X \in T_p\Sigma,$$

where $\bar{\nabla}$ is the Levi–Civita connection of $\mathbb{R}^{n+2}$. Assume $\lambda_1, \dots, \lambda_n$ are principal curvatures of Σ, which are defined as the eigenvalues of the shape operator S_p, $H = \frac{\lambda_1 + \cdots + \lambda_n}{n}$ is the mean curvature of Σ. We will assume that Σ is a rotational hypersurface with constant mean curvature H, that is,

$$\lambda_1 = \lambda, \quad \lambda_2 = \cdots = \lambda_n = \mu, \quad \lambda + (n-1)\mu = nH = \text{const}.$$

Without loss of generality, we can assume that $\mu - \lambda > 0$ by choosing suitable direction of unit normal vector ν. Choose a local orthonormal frame $\{e_1, e_2, \dots, e_n\}$ such that

$$S(e_1) = \lambda e_1, \quad S(e_i) = \mu e_i, \; i = 2, \dots, n. \tag{3.1}$$

Then we have

Theorem 3.1. (Perdomo [Per]) *Assume $\Sigma \subset S^{n+1}$ is an n-dimensional rotational hypersurface with constant mean curvature H and $n \geq 2$. If the local orthonormal frame is chosen as above, then we have*

1. $X(\lambda) = X(\mu) = 0$, *for any* $X \in \text{span}\{e_2, \dots, e_n\}$;
2. $\nabla_X e_1 = \frac{e_1(\mu)}{\lambda - \mu} X$, *for any* $X \in \text{span}\{e_2, \dots, e_n\}$;
3. $\nabla_{e_1} e_1 = 0$;

4. $1+\lambda\mu = e_1\left(\frac{e_1(\mu)}{\mu-\lambda}\right) - \left(\frac{e_1(\mu)}{\mu-\lambda}\right)^2$.

Since $\lambda + (n-1)\mu = nH$, we have

$$\mu - \lambda = \mu - (nH - (n-1)\mu) = n(\mu - H) = nw^{-n},$$

where $w = (\mu - H)^{-\frac{1}{n}}$. Since $\mu - \lambda > 0$, it implies that w is a smooth function, moreover,

$$\mu = H + w^{-n}, \quad \lambda = H - (n-1)w^{-n}. \tag{3.2}$$

The conclusion (4) in Theorem 3.1 can be written as an equation of w,

$$\frac{e_1(e_1(w))}{w} - \frac{n-1}{w^{2n}} - \frac{(n-2)H}{w^n} + H^2 + 1 = 0. \tag{3.3}$$

Multiplying the both sides of (3.3) by $2we_1(w)$, we can get

$$(e_1(w))^2 + w^{2-2n} + (1+H^2)w^2 + 2Hw^{2-n} = C. \tag{3.4}$$

Notice that $\nabla_{e_1} e_1 = 0$, therefore the integral curve of e_1 is a geodesic. For any $p_0 \in \Sigma$, let $\gamma(u)$ be the geodesic starting from p_0 with $\gamma'(0) = e_1(p_0)$. Then by the uniqueness of geodesic, we know that $\gamma'(u) = e_1(\gamma(u))$. Let $g(u) = w(\gamma(u))$, then equation (3.4) reduces to an ordinary differential equation of g:

$$(g')^2 + g^{2-2n} + (1+H^2)g^2 + 2Hg^{2-n} = C, \tag{3.5}$$

where C is a constant with

$$C \geq C_0 = n(2+2H^2)^{\frac{n-2}{n}} \frac{2 + nH^2 + H\sqrt{n^2H^2 + 4(n-1)}}{\left((n-2)H + \sqrt{n^2H^2+4(n-1)}\right)^{\frac{2n-2}{n}}}$$

and $C = C_0$ if and only if g is a constant g_0,

$$g_0 = \left(\frac{(n-2)H + \sqrt{n^2H^2 + 4(n-1)}}{2(1+H^2)}\right)^{\frac{1}{n}}$$

and M is an isoparametric hypersurface. Thus from now on, we can assume $C > C_0$.

Write (see Perdomo [Per])

$$T = 2\int_{t_1}^{t_2} \frac{t^{n-1}}{\sqrt{Ct^{2n-2} - 1 - (1+H^2)t^{2n} - 2Ht^n}}dt,$$

where $t_1 < t_2$ are positive solutions of the following equation:

$$Cs^{2n-2} - (1+H^2)s^{2n} - 1 - 2Hs^n = 0.$$

Let $r(u) = \frac{g(u)}{\sqrt{C}}$, it is easy to check that $r(u)$ satisfies (see Perdomo [Per])

$$\frac{r''}{r} + 1 + \lambda\mu = 0, \quad (r')^2 + r^2(1+\mu^2) = 1, \quad \mu' = -(\mu - \lambda)\frac{r'}{r}. \tag{3.6}$$

We also have the following result (see Perdomo [Per])

Theorem 3.2. (Perdomo [Per]) *Assume $n \geq 2$, $H > 0$ is a positive real number, then*

1. *Assume $g_C : \mathbb{R} \to \mathbb{R}$ is a solution of equation (3.5) with period T, $\lambda, r, \theta : \mathbb{R} \to \mathbb{R}$ are defined as follows:*

$$r = \frac{g_C}{\sqrt{C}}, \quad \mu = H + g_C^{-n}, \quad \theta(u) = \int_0^u \frac{r(s)\mu(s)}{1 - r^2(s)} ds$$

then the immersed hypersurface $\phi : S^{n-1} \times \mathbb{R} \to S^{n+1}$ given by

$$\phi(y, u) = (\, r(u)\, y, \sqrt{1 - r(u)^2}\, \cos(\theta(u)), \sqrt{1 - r(u)^2}\, \sin(\theta(u))\,) \tag{3.7}$$

has constant mean curvature H.

2. *Such ϕ gives a compact embedded rotational hypersurface with constant mean curvature H in S^{n+1} if and only if there exists a positive integer m satisfying*

$$K(H, n, C) := 2 \int_0^{\frac{T}{2}} \frac{r(u)\mu(u)}{1 - r^2(u)} du = \frac{2\pi}{m}. \tag{3.8}$$

In general, if there exists a pair of positive integer (k, m) such that $K(H, n, C) = \frac{2k\pi}{m}$, the image of ϕ is an n-dimensional compact immersed hypersurface in S^{n+1}.

We can check (see Perdomo [Per])

$$\lim_{C \to C_0^+} K(H, n, C) = \pi \sqrt{2 - \frac{2nH}{\sqrt{4(n-1) + H^2 n^2}}} =: f(n, H); \tag{3.9}$$

$$\lim_{C \to +\infty} K(H, n, C) = 2 \int_0^{\frac{1}{\sqrt{1+H^2}}} \frac{Ht}{(1 - t^2)\sqrt{1 - (1 + H^2)t^2}} dt = 2\text{arccot}(H). \tag{3.10}$$

Remark 3.3. If we denote $t = \frac{1}{H}$, and denote $h(t) = 2\text{arccot}H = 2\arctan t$ and $g(n, t) = f(n, H) = \pi \sqrt{2 - \frac{2n}{\sqrt{4(n-1)t^2 + n^2}}}$, then we have $h'(0) = 2$ and

$$g'(n, t) = \frac{4(n-1)n\pi t}{(n^2 + 4(n-1)t^2)^{3/2} \sqrt{2 - \frac{2n}{\sqrt{4(n-1)t^2 + n^2}}}}.$$

Notice that

$$2 - \frac{2n}{\sqrt{4(n-1)t^2 + n^2}} = \frac{2\sqrt{4(n-1)t^2 + n^2} - 2n}{\sqrt{4(n-1)t^2 + n^2}}$$
$$= \frac{16(n-1)t^2}{\left(2\sqrt{4(n-1)t^2 + n^2} + 2n\right)\sqrt{4(n-1)t^2 + n^2}},$$

therefore we have

$$\lim_{t\to 0} g'(n,t) = \frac{4(n-1)n\pi}{n^3 \cdot \frac{4\sqrt{n-1}}{2n}} = \frac{2\sqrt{n-1}}{n}\pi.$$

When $n \geq 9$, we have $g'(n,0) = \frac{2\sqrt{n-1}}{n}\pi < 2 = h'(0)$ and $g(n,0) = h(0) = 0$, therefore for small t we have $g(n,t) < h(t)$. It means that when $n \geq 9$, for large H, $K(H,n,C_0^+) = f(n,H)$ is less than $K(H,n,\infty) = 2\text{arccot}H$, therefore $K(n,H,C)$ can not be monotone decreasing.

4 Classification of 3-dimensional and 4-dimensional compact embedded rotational hypersurfaces in a sphere

In 2010, Perdomo [Per] proved that

Theorem 4.1. (Perdomo [Per]) *For any positive integer $m \geq 2$ and $n \geq 2$, if*

$$a_{n,m} = \cot\frac{\pi}{m} < H < \frac{(m^2-2)\sqrt{n-1}}{n\sqrt{m^2-1}} = b_{n,m}, \tag{4.1}$$

then there exists an n-dimensional compact embedded non-isoparametric rotational hypersurface Σ in a unit sphere S^{n+1} with constant mean curvature H and m-fold symmetry.

Recall that Σ is a rotational hypersurface in S^{n+1} if and only if there exists a n-dimensional subspace Π such that Σ is invariant under the action of orthogonal transformation group of Π. Moreover, we say that Σ has m-fold symmetry if and only if Σ is also invariant under the action of Z_m, where Z_m is the group generated by the rotation which rotates $\Pi^\perp$ by angle $\frac{2\pi}{m}$ while keeping Π invariant.

A natural question comes up: for given m and H satisfying equation (4.1), is such embedded rotational CMC hypersurface unique or not? For $n = 2$, Andrews and Li gave a positive answer to this question by proving that the function $K(2,H,C)$ is monotone decreasing with respect to C (see Proposition 2.8). In this paper, we will extend Andrews–Li's result to $n = 3,4$. Combining Perdomo's Theorem 4.1, we get the following theorem:

Theorem 4.2. *Assume $n = 3,4$, for any positive integer m, then there exists a unique n-dimensional compact embedded non-isoparametric rotational hypersurface Σ in a unit sphere S^{n+1} with constant mean curvature H and m-fold symmetry if and only if*

$$a_{n,m} = \cot\frac{\pi}{m} < H < \frac{(m^2-2)\sqrt{n-1}}{n\sqrt{m^2-1}} = b_{n,m}.$$

It's also very interesting to see whether the intervals $(a_{n,m}, b_{n,m})$ will cover the half real axis or not. For $n = 2$, Andrews and Li showed that the union of the intervals $(a_{2,m}, b_{2,m})$ covers $(0, +\infty)$ except a simple point $\frac{1}{\sqrt{3}}$. Thus if $H = \pm\frac{1}{\sqrt{3}}$, the only embedded torus with constant mean curvature H must be $S^1(\frac{1}{2}) \times S^1(\frac{\sqrt{3}}{2})$. For $n = 3, 4$, we have the following result

Corollary 4.3. 1. *For $n = 3$, if $H = 0$ or $\frac{2\sqrt{2}}{3\sqrt{3}} \leq H \leq \frac{1}{\sqrt{3}}$, then there exists no compact embedded rotational hypersurface with constant mean curvature H in S^4 other than product of S^1 and S^2. For $n = 3$, if $H \neq 0$ and $H \notin [\frac{2\sqrt{2}}{3\sqrt{3}}, \frac{1}{\sqrt{3}}]$, then exist compact embedded non-isoparametric rotational hypersurfaces with constant mean curvature $H \geq 0$.*

2. *For $n = 4$, if $H = 0$ or $\frac{1}{2} \leq H \leq \frac{1}{\sqrt{3}}$, then there exists no embedded rotational hypersurface with constant mean curvature H in S^5 other than product of S^1 and S^3. For $n = 4$, if $H \neq 0$ and $H \notin [\frac{1}{2}, \frac{1}{\sqrt{3}}]$, then exist compact embedded non-isoparametric rotational hypersurfaces with constant mean curvature $H \geq 0$.*

Using Mathematica to make numerical analysis of the figure of the function $K(n, H, C)$, we find that when $n \geq 8$, there exists H_0, such that $K(n, H_0, C)$ is not monotone decreasing with respect to C, in fact, for $n \geq 9$, we proved this conclusion in Remark 3.3. Thus the following problem is interesting.

Problem 4.4. *Can we prove that when $n \leq 7$, $K(n, H, C)$ is monotone decreasing with respect to C?*

Just as proof of Andrews–Li's Theorem 2.1, in order to prove Theorem 4.2, we only need to prove that for $n = 3, 4$, the function $K(H, n, C)$ defined by equation (3.8) is monotone decreasing with respect to C. We state it as following:

Theorem 4.5. *Assume $n = 3, 4$, then $K(H, n, C)$ is monotone decreasing with respect to C in $(C_0, +\infty)$.*

The ideas of proof of Theorem 4.5 come from Otsuki [Ot2, Ot3], and Andrews–Li [AL]. The proof uses the method of complex analysis, for basic theory of complex analysis, readers can refer Ahlfors [Ah].

By changing variables, we can have (see Perdomo [Per])

$$K(H, n, C) = 2 \int_{\frac{t_1(C)}{\sqrt{C}}}^{\frac{t_2(C)}{\sqrt{C}}} \frac{t(H + C^{-\frac{n}{2}} t^{-n})}{(1 - t^2)\sqrt{1 - t^2(1 + (H + C^{-\frac{n}{2}} t^{-n})^2)}} dt, \tag{4.2}$$

where $\frac{t_1(C)}{\sqrt{C}}$ and $\frac{t_2(C)}{\sqrt{C}}$ are the only two positive roots of the following equation

$$1 - (1 + H^2)t^2 - C^{-n} t^{2-2n} - 2HC^{-\frac{n}{2}} t^{2-n} = 0.$$

Let $a = C^{-\frac{n}{2}}$, denote $T(H, n, a) = K(H, n, C)$, and then we have

$$T(H, n, a) = 2 \int_{x_1}^{x_2} \frac{Ht^n + a}{(1 - t^2)\sqrt{t^{2n-2} - (t^{2n} + (Ht^n + a)^2)}} dt, \tag{4.3}$$

where $0 < a < a_0 = C_0^{-\frac{n}{2}}$, $x_{1,2} = \frac{t_{1,2}(C)}{\sqrt{C}}$.

Let $f(z) = z^{2n-2} - (z^{2n} + (Hz^n + a)^2)$. Define a Riemann surface S by

$$S = \{(z, w) \in \mathbb{C}^2 : w^2 = f(z)\}. \tag{4.4}$$

Now we consider a simple closed curve γ_1 on S, such that γ_1 enclose only two singular points $(x_1, 0), (x_2, 0)$ in its interior, and the projection of γ_1 in z-plane consists of the following four parts: two circles σ_1, σ_3 centered at $(x_1, 0)$ and $(x_2, 0)$ with radius ε, two lines σ_2, σ_4 coming from moving the interval $[x_1 + \varepsilon, x_2 - \varepsilon]$, $\mp\varepsilon'$-distance in imaginary axis direction, here ε' is a small positive number. And the corresponding part of σ_2 on w-plane lies around negative part of real axis, the corresponding part of σ_4 on w-plane lies around positive part of real axis. (See Figs. 1 and 2.)

Lemma 4.6. *Under the assumptions above, we have*

$$T(H, n, a) = -\oint_{\gamma_1} \frac{Hz^n + a}{(1 - z^2)\, w(z)} dz. \tag{4.5}$$

Proof. First, we claim that

$$\lim_{\varepsilon \to 0} \int_{\sigma_1} \frac{Hz^n + a}{(1 - z^2)\, w(z)} dz = \lim_{\varepsilon \to 0} \int_{\sigma_3} \frac{Hz^n + a}{(1 - z^2)\, w(z)} dz = 0. \tag{4.6}$$

Since the integral on σ_1 and σ_3 are similar, we only prove the claim on σ_1. On σ_1, $z - x_1 = \varepsilon e^{i\theta}, 0 < \theta < 2\pi$, $\frac{Hz^n + a}{(1 - z^2) w(z)} = \frac{O(1)}{\sqrt{\varepsilon}}$, thus we have

$$\begin{aligned} &\lim_{\varepsilon \to 0} \int_{\sigma_1} \frac{Hz^n + a}{(1 - z^2)\, w(z)} dz \\ =& \lim_{\varepsilon \to 0} O(1)\sqrt{\varepsilon} = 0. \end{aligned}$$

Next we shall calculate the integral of $\frac{Hz^n + a}{(1 - z^2)\, w(z)}$ on $\sigma_2 = \{x - \varepsilon' i : x_1 + \varepsilon \leq x \leq x_2 - \varepsilon, \varepsilon' > 0\}$. When $\varepsilon' \to 0$, $w(z) \to -\sqrt{f(x)}$, therefore

$$\begin{aligned} &\lim_{\varepsilon' \to 0, \varepsilon \to 0} \int_{\sigma_2} \frac{Hz^n + a}{(1 - z^2) w(z)} dz \\ =& \int_{x_1}^{x_2} \left[\lim_{\varepsilon' \to 0} \frac{Hz^n + a}{(1 - z^2) w(z)} \right] dx \\ =& \int_{x_1}^{x_2} -\frac{Hx^n + a}{(1 - x^2)\sqrt{f(x)}} dx \\ =& -\frac{1}{2} T(H, n, a). \end{aligned} \tag{4.7}$$

Similarly, on σ_4, we have

$$\lim_{\varepsilon' \to 0, \varepsilon \to 0} \int_{\sigma_4} \frac{Hz^n + a}{(1 - z^2) w(z)} dz = -\frac{1}{2} T(H, n, a). \tag{4.8}$$

Combining equations (4.6), (4.7) and (4.8), we complete the proof of Lemma 4.6. □

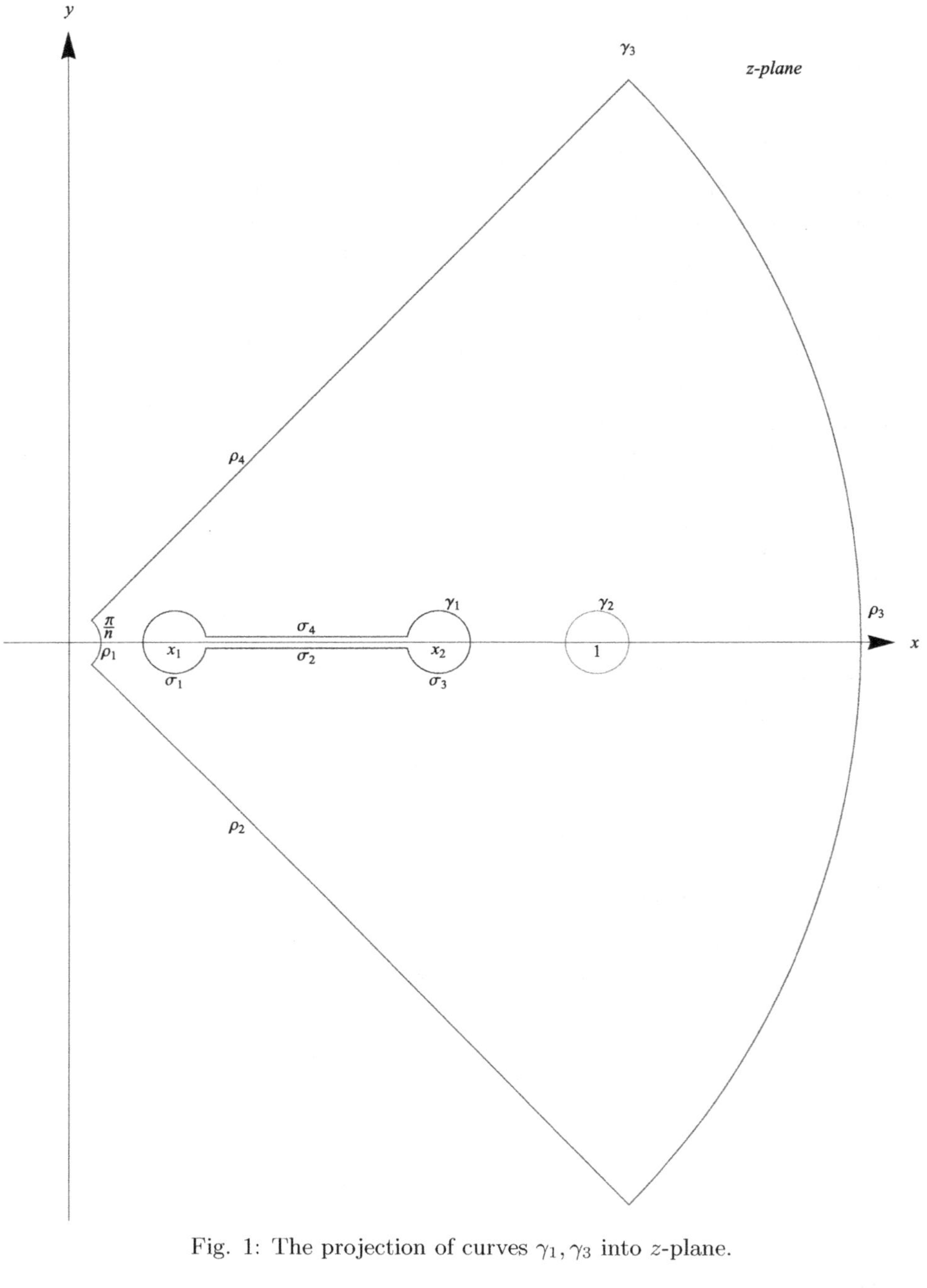

Fig. 1: The projection of curves γ_1, γ_3 into z-plane.

Lemma 4.7. *$T(H, n, a)$ is differentiable with respect to a, and*

$$T'(H,n,a) = \frac{dT(H,n,a)}{da} = -\oint_{\gamma_1} \frac{z^{2n-2}}{w(z)^3}dz. \tag{4.9}$$

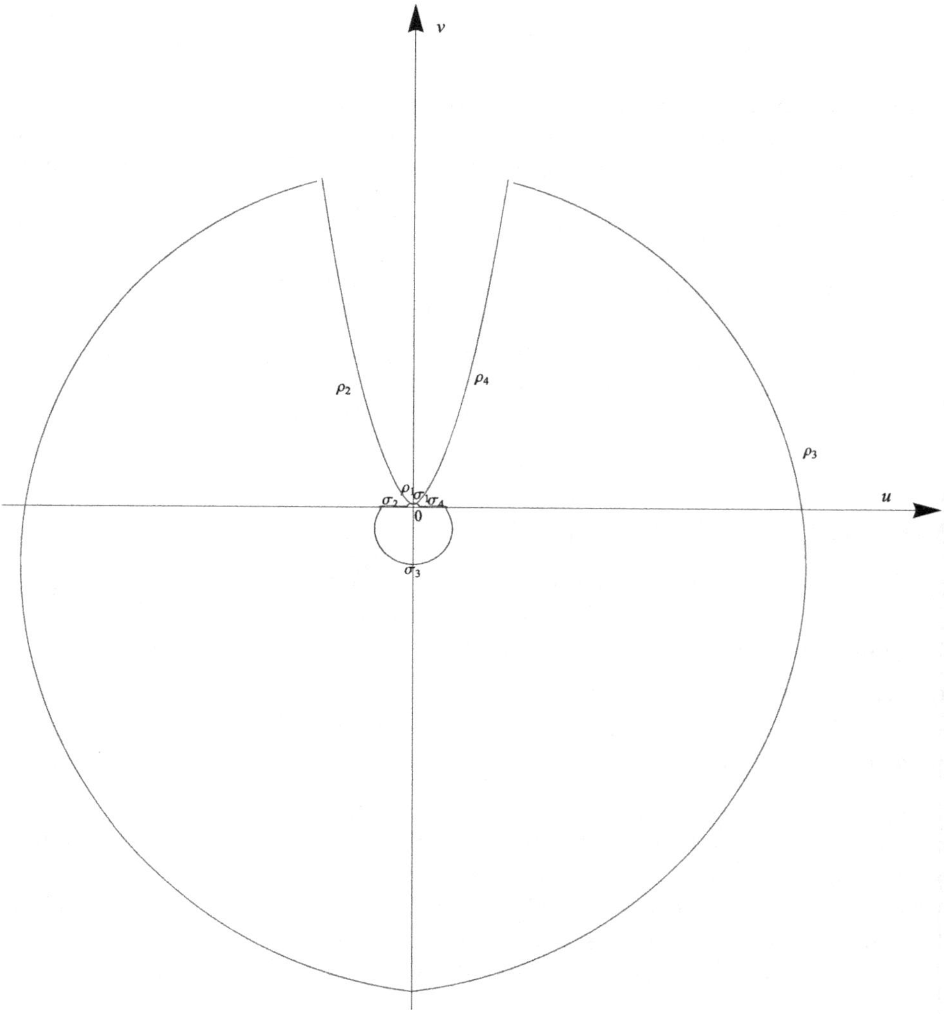

Fig. 2: The projection of curves γ_1, γ_3 into w-plane.

Proof. Since $w(z)^2 = f(z) = z^{2n-2} - z^{2n} - (Hz^n + a)^2$, differentiating the both sides with respect to a, we have

$$2w(z)\frac{dw(z)}{da} = -2(Hz^n + a). \tag{4.10}$$

Therefore,

$$\begin{aligned}\frac{d}{da}\left(\frac{Hz^n + a}{w(z)}\right) &= \frac{1}{w(z)} - \frac{Hz^n + a}{w(z)^2}\frac{dw(z)}{da} \\ &= \frac{w(z)^2 + (Hz^n + a)(Hz^n + a)}{w(z)^3}\end{aligned}$$

$$= \frac{f(z) + (Hz^n + a)^2}{w(z)^3} = \frac{z^{2n-2}(1-z^2)}{w(z)^3}. \qquad (4.11)$$

For any given $a \in (0, a_0)$, since the positive roots of $f(z)$, x_1, x_2, are both continuous in a, we can choose γ_1 sufficiently large that for a neighborhood of a, the roots $x_1(a), x_2(a)$ always lie in the interior of γ_1. Then by Cauchy integral formula, we have

$$T(H, n, a) = -\oint_{\gamma_1} \frac{Hz^n + a}{(1-z^2)w(z)} dz.$$

Putting equation (4.11) into the equation above, we have

$$\begin{aligned} T'(H, n, a) &= -\oint_{\gamma_1} \frac{d}{da}\left(\frac{Hz^n + a}{(1-z^2)w(z)}\right) dz \\ &= -\oint_{\gamma_1} \frac{z^{2n-2}}{w(z)^3} dz. \end{aligned}$$

We complete the proof of Lemma 4.7. □

Since $a = C^{-\frac{n}{2}}$, by the chain rule of differential, we have $\frac{dK}{dC} = \frac{dT}{dC} = \frac{dT}{da}\frac{da}{dC} = -\frac{n}{2}T'C^{-\frac{n}{2}-1}$. Thus in order to prove Theorem 4.5, we only need to prove the following theorem:

Theorem 4.8. *Assume $n = 3, 4$, for any positive real number $H > 0$, $a \in (0, a_0)$, we have $T'(H, n, a) > 0$. That is for any $H > 0$, $T(H, n, a)$ is monotone increasing with respect to a in $(0, a_0)$.*

Proof. We shall deform γ_1 into a simple closed curve γ_3, the projection of γ_3 on z-plane consists: two half-lines ρ_2, ρ_4 with argument $-\frac{\pi}{n}$ and $\frac{\pi}{n}$, two parts of circles ρ_1, ρ_3, lying in the interior of the two half-line, centered at origin with radius ε, R, where $0 < \varepsilon \ll 1 \ll R$. (See the above Figs. 1 and 2.) During the deformation, the curve will pass through a singular point $z = 1$. When $z \to 0$, $w(z) \to ai$.

Assume γ_2 is a small circle centered at $z = 1$, we have

$$-\oint_{\gamma_3} \frac{Hz^n + a}{(1-z^2)w(z)} dz = -\oint_{\gamma_1} \frac{Hz^n + a}{(1-z^2)w(z)} dz - \oint_{\gamma_2} \frac{Hz^2 + a}{(1-z^2)w(z)} dz. \qquad (4.12)$$

Noting

$$-\oint_{\gamma_2} \frac{Hz^n + a}{(1-z^2)w(z)} dz = \frac{H+a}{2w(1)} \oint_{\gamma_2} \frac{1}{z-1} dz = \frac{H+a}{-2(H+a)i} \cdot 2\pi i = -\pi, \qquad (4.13)$$

thus we have

$$-\oint_{\gamma_3} \frac{Hz^n + a}{(1-z^2)w(z)} dz = -\oint_{\gamma_1} \frac{Hz^n + a}{(1-z^2)w(z)} dz - \pi. \qquad (4.14)$$

Take derivative with respect to a on both side, we get

$$T'(H, a) = -\oint_{\gamma_1} \frac{z^{2n-2}}{w(z)^3} dz = -\oint_{\gamma_3} \frac{z^{2n-2}}{w(z)^3} dz. \qquad (4.15)$$

We shall compute the integral of $\frac{z^{2n-2}}{w(z)^3}$ on those 4 parts of γ_3.

On ρ_1, $\frac{z^{2n-2}}{w(z)^3} = o(\varepsilon)$, and on ρ_3, $\frac{z^{2n-2}}{w(z)^3} = O(R^{-n-2})$, thus it is easy to check that

$$\lim_{\varepsilon\to 0}\int_{\rho_1}\frac{z^{2n-2}}{w(z)^3}dz = \lim_{R\to\infty}\int_{\rho_3}\frac{z^{2n-2}}{w(z)^3}dz = 0. \tag{4.16}$$

Denote $\omega_n = e^{\frac{\pi}{n}i}$, and then on ρ_2, we have $z = y\bar{\omega}_n, y \in (\varepsilon, R)$, therefore

$$\int_{\rho_2}\frac{z^{2n-2}}{w(z)^3}dz = \int_\varepsilon^R \frac{(y\bar{\omega}_n)^{2n-2}}{[w(y\bar{\omega}_n)]^3}d(y\bar{\omega}_n) = \int_\varepsilon^R \frac{y^{2n-2}\bar{\omega}_n^{2n-1}}{[w(y\bar{\omega}_n)]^3}dy. \tag{4.17}$$

Similarly, on ρ_4, we have $z = y\omega_n, y \in (\varepsilon, R)$, thus

$$\int_{\rho_4}\frac{z^{2n-2}}{w(z)^3}dz = -\int_\varepsilon^R \frac{(y\omega_n)^{2n-2}}{[w(y\omega_n)]^3}d(y\omega_n) = -\int_\varepsilon^R \frac{y^{2n-2}{\omega_n}^{2n-1}}{[w(y\omega_n)]^3}dy. \tag{4.18}$$

Since $w(0) = i\sqrt{-f(0)} = ai$, by continuity of w, and $\overline{w(y\bar{\omega}_n)^2} = w(y\omega_n)^2$, we know that $w(y\bar{\omega}_n) = -\overline{w(y\omega_n)}$. Thus we have

$$\int_{\rho_2}\frac{z^{2n-2}}{w(z)^3}dz = -\int_\varepsilon^R \frac{y^{2n-2}\bar{\omega}_n^{2n-1}}{\overline{w(y\omega_n)}^3}dy = \overline{\int_{\rho_4}\frac{z^{2n-2}}{w(z)^3}dz}. \tag{4.19}$$

It implies that

$$\int_{\rho_2}\frac{z^{2n-2}}{w(z)^3}dz + \int_{\rho_4}\frac{z^{2n-2}}{w(z)^3}dz = 2\mathrm{Re}\left(\int_{\rho_4}\frac{z^{2n-2}}{w(z)^3}dz\right). \tag{4.20}$$

From equation (4.18), and $\omega_n = e^{\frac{\pi}{n}i}$, $\omega_n^{2n} = 1$, we conclude that

$$2\mathrm{Re}\left(\int_{\rho_4}\frac{z^{2n-2}}{w(z)^3}dz\right) = -\int_\varepsilon^R 2\mathrm{Re}\left(\frac{y^{2n-2}}{\omega_n w(y\omega_n)^3}\right)dy. \tag{4.21}$$

If the real part of the integral term doesn't change sign along ρ_4, we will get that the integral above have sign. And when $n = 3, 4$, we can achieve this by studying the argument of $w(y\omega_n)$, more precisely, the argument of $\omega_n w(y\omega_n)^3, y \in (\varepsilon, R)$. Firstly, $\arg\omega_n = \frac{\pi}{n}$. By the definition of $w(z)$, we also have

$$w(y\omega_n)^2 = f(y\omega_n) = y^{2n-2}\omega_n^{2n-2} - (Hy^n - a)^2 - y^{2n},$$

therefore $\arg[w(y\omega_n)^2] \in (\pi, \frac{2n-2}{n}\pi)$. Since $w(0) = i\sqrt{-f(0)} = ai$, by the continuity of w, we know that $\arg w(y\omega_n) \in (\frac{\pi}{2}, \frac{n-1}{n}\pi)$. This implies $\arg\left(\omega_n(y\omega_n)^3\right) \in (\frac{3\pi}{2} + \frac{\pi}{n}, \frac{(3n-2)\pi}{n})$.

When $n = 3$ or 4, it's easy to get that

$$n = 3, \quad \left(\frac{3\pi}{2} + \frac{\pi}{n}, \frac{(3n-2)\pi}{n}\right) = \left(-\frac{\pi}{6}, \frac{\pi}{3}\right);$$

$$n = 4, \quad \left(\frac{3\pi}{2} + \frac{\pi}{n}, \frac{(3n-2)\pi}{n}\right) = \left(-\frac{\pi}{4}, \frac{\pi}{2}\right).$$

In both cases, we have $\arg[\omega_n w(y\omega_n)^3] \in (-\frac{\pi}{2}+\frac{\pi}{n}, \frac{(n-2)\pi}{n}) \subset (-\frac{\pi}{2}, \frac{\pi}{2})$, therefore $\mathrm{Re}\left(w_n w(y\omega_n)^3\right) > 0$. It means that as long as $y \in (\varepsilon, R)$, we have

$$\mathrm{Re}\left(\frac{y^{2n-2}}{\omega_n w(y\omega_n)^3}\right) = \frac{y^{2n-2}\mathrm{Re}\left(\omega_n w(y\omega_n)^3\right)}{|w(y\omega_n)|^6} > 0. \tag{4.22}$$

We conclude that

$$\int_{\rho_2} \frac{z^{2n-2}}{w(z)^3} dz + \int_{\rho_4} \frac{z^{2n-2}}{w(z)^3} dz = -\int_{\varepsilon}^{R} 2\mathrm{Re}\left(\frac{y^{2n-2}}{\omega_n w(y\omega_n)^3}\right) dy < 0. \tag{4.23}$$

Combining the arguments above, we prove that when $n = 3, 4$,

$$T'(H, n, a) = -\oint_{\gamma_3} \frac{z^{2n-2}}{w(z)^3} dz = -\left(\int_{\rho_1} + \int_{\rho_2} + \int_{\rho_3} + \int_{\rho_4}\right) \frac{z^{2n-2}}{w(z)^3} dz > 0. \tag{4.24}$$

This completes the proof of Theorem 4.8. □

5 Compact embedded rotational hypersurfaces with constant m^{th} mean curvature in a sphere

Let Σ be a compact hypersurface in a space form $R^{n+1}(c)$, when $c = 1, R^{n+1}(c)$ is a unit sphere S^{n+1}; when $c = 0, R^{n+1}(c)$ is a Euclidean space $\mathbb{R}^{n+1}$; when $c = -1, R^{n+1}(c)$ is a hyperbolic space H^{n+1}. Then m^{th} mean curvature H_m of Σ are defined by

$$C_n^m H_m = \sum_{1 \le i_1 < i_2 < \cdots < i_m \le n} \lambda_{i_1} \cdots \lambda_{i_m},$$

where $C_n^m = \frac{n!}{m!(n-m)!}$ and λ_i are principal curvatures of hypersurface Σ. We note that H_1 is the mean curvature H.

It is well known that Alexandrov [Al2] and Montiel–Ros [MR] proved that totally umbilical hypersurfaces are the only compact embedded hypersurfaces with constant m^{th} mean curvature H_m in a Euclidean space $\mathbb{R}^{n+1}$, or a hemisphere S_+^{n+1}, or hyperbolic space H^{n+1}, for $1 \le m \le n$. On the other hand, one knows that standard round spheres and isoparametric hypersurfaces $S^l(r) \times S^{n-l}(\sqrt{1-r^2})$, $1 \le l \le n-1$ are compact embedded hypersurfaces with constant m^{th} mean curvature H_m in a unit sphere $S^{n+1}(1)$. Hence, it is natural to ask the following problem:

Problem 5.1. *Do there exist compact embedded rotational hypersurfaces with constant m^{th} mean curvature H_m in $S^{n+1}(1)$ other than the standard round spheres and isoparametric hypersurfaces $S^l(r) \times S^{n-l}(\sqrt{1-r^2})$, $1 \le l \le n-1$?*

When $m = 1$, Perdomo [Per] proved

Theorem 5.2. ([Per]) *For any $n \geq 2$ and any integer $m \geq 2$, if constant mean curvature H satisfies*

$$\cot\frac{\pi}{m} < H < \frac{(m^2-2)\sqrt{n-1}}{n\sqrt{m^2-1}},$$

then there exists an n-dimensional compact nontrivial embedded rotational hypersurface with constant mean curvature H in $S^{n+1}(1)$.

When $m = 2$, Wei, Cheng and Li [WCL] proved

Theorem 5.3. ([WCL]) *For any $n \geq 3$ and any integer $m \geq 2$, if constant 2th mean curvature H_2 satisfies*

$$\frac{1}{(\tan\frac{\pi}{m})^2} < H_2 < \frac{m^2-2}{n},$$

then there exists an n-dimensional compact non-isoparametric embedded rotational hypersurface M with constant 2th mean curvature H_2 in $S^{n+1}(1)$.

Theorem 5.4. ([WCL]) *For any $n \geq 5$ and any integer $m \geq 3$, if constant 4th mean curvature H_4 satisfies*

$$\frac{1}{(\tan\frac{\pi}{m})^4} < H_4 < \frac{m^4-4}{n(n-4)},$$

then there exists an n-dimensional compact non-isoparametric embedded rotational hypersurface with constant H_4 in $S^{n+1}(1)$.

For constant $H_m = 0$, Li and Wei proved

Theorem 5.5. ([LW2]) *There are no compact minimal embedded rotational hypersurfaces with $H_m = 0$ $(1 \leq m \leq n-1)$ other than the Riemannian product $S^{n-1}(\sqrt{\frac{n-m}{n}}) \times S^1(\sqrt{\frac{m}{n}})$ and round geodesic spheres.*

Remark 5.6. When $m = 1$, Theorem 5.5 was proved by Otsuki [Ot1].

For constant $H_m > 0$, we have

Theorem 5.7. ([WCL]) *For any integer $1 \leq m \leq n-1$, there exist many nontrivial embedded rotational hypersurfaces with constant $H_m > 0$ in $S^{n+1}(1)$.*

Remark 5.8. When $m = 1$, Theorem 5.7 was proved by Brito–Leite [BL].

Finally, Andrews, Huang and Li [AHL] proved

Theorem 5.9. (Andrews–Huang–Li [AHL]) *Suppose M is a compact embedded hypersurface in S^{n+1} with two distinct principal curvatures λ and μ, whose multiplicities are m and $n-m$ respectively. If for some positive number α,*

$$\lambda + \alpha\mu = 0,$$

then λ and μ are both constant and M is congruent to tori

$$S^m\left(\sqrt{\frac{1}{\alpha+1}}\right) \times S^{n-m}\left(\sqrt{\frac{\alpha}{\alpha+1}}\right), \quad 1 \leq m \leq n-1.$$

References

[Ab] U. Abresch, *Constant mean curvature tori in terms of elliptic functions,* J. Reine Angew. Math. **374** (1987), 169–192.

[AR] U. Abresch and H. Rosenberg, *A Hopf differential for constant mean curvature surfaces in $S^2 \times R$ and $H^2 \times R$*, Acta Math. **193** (2004), no. 2, 141–174.

[Ah] L. V. Ahlfors, *Complex analysis*, 3rd edition, McGraw-Hill, Inc., 1979.

[Al1] A. D. Alexandrov, *Uniqueness theorems for surfaces in the large. I*, Vestnik Leningrad. Univ. **11** (1956), no. 19, 5–17.

[Al2] A. D. Alexandrov, *Uniqueness theorems for surfaces in the large*, V.Vestnik, Leningrad Univ. **13** (1958), 5–8.

[Alm] F. J. Almgren, *Some interior regularity theorems for minimal surfaces and an extension of Bernstein's theorem*, Ann. Math. **84** (1966), 277–292.

[An] B. Andrews, *Non-collapsing in mean-convex mean curvature flow*, Geom. Topol. **16** (2012), 1413–1418.

[AHL] B. Andrews, Z. Huang, and H. Li, *Uniqueness for a class of embedded Weingarten hypersurfaces in S^{n+1}*, ALM 33, Introduction to Modern Mathematics, 95–107, Higher Education Press and International Press, 2015.

[AL] B. Andrews and H. Li, *Embedded constant mean curvature tori in the three-sphere*, arXiv:1204.5007, 2012, J. Differential Geom. **99** (2015), 169–189.

[BD] J. L. Barbosa and M. do Carmo, *Stability of hypersurfaces with constant mean curvature*, Math. Z. **185** (1984), 339–353.

[Bo] A. I. Bobenko, *All constant mean curvature tori in $\mathbb{R}^3$, S^3, H^3 in terms of theta-functions*, Math. Ann. **290** (1991), 209–245.

[Br1] S. Brendle, *Embedded minimal tori in S^3 and the Lawson conjecture*, Acta Math. **211** (2013), no.2, 177–190.

[Br2] S. Brendle, *Embedded Weingarten tori in S^3*, Adv. Math. **257** (2014), 462–475.

[Br3] S. Brendle, *Minimal surfaces in S^3: a survey of recent results*, Bull. Math. Sci. **3**(2013), 133–171.

[BL] F. Brito and M. L. Leite, *A remark on rotational hypersurfaces of S^m*, Bull. Soc. Math. Belg. -Tijdschr. Belg. Wisk. Gen. **42** (1990), 303–318.

[DD] M. do Carmo and M. Dajczer, *Rotational hypersurfaces in spaces of constant curvature*, Trans. Amer. Math. Soc. **277** (1983), 685–709.

[CU1] I. Castro and F. Urbano, *New examples of minimal Lagrangian tori in the complex projective plane*, Manuscripta Math. **85** (1994), 265–281.

[CU2] I. Castro and F. Urbano, *Twistor holomorphic Lagrangian surfaces in the complex projective plane and hyperbolic planes*, Annals of Global Analysis and Geometry **13** (1995), 59–67.

[Ch] S.-S. Chern, *On surfaces of constant mean curvature in a three-dimensional space of constant curvature*, Lecture Notes in Math. **1007** (1983), 104–108.

[FM] I. Fernandez and P. Mira, *Constant mean curvature surfaces in 3-dimensional Thurston geometries*, Proceedings of ICCM in 2010, Vol. **2**, 830–861.

[Fu] S. Furuya, *On periods of periodic solutions of a certain nonlinear differential equation*, Lecture Notes in Math. **243** (1971), 320–323.

[Ha1] R. Hamilton, *An isoperimetric estimate for the Ricci flow on the two-sphere*, Ann. of Math. Stud. **137** (1995), 191–200.

[Ha2] R. Hamilton, *Isoperimetric estimates for the curve shrinking flow in the plane*, Modern methods in complex analysis (Princeton, NJ, 1992), Ann. of Math. Stud., Vol. **137**, Princeton Univ. Press, Princeton, NJ, 1995, 201–222.

[Ho1] H. Hopf, *Über Flächen mit einer Relation zwischen den Hauptkrümmungen*, Math. Nachr. **4** (1951), 232–249.

[Ho2] H. Hopf, *Differential Geometry in the Large*, Lecture Notes in Mathematics **1000** (1983), Springer-Verlag, Berlin.

[Hu] G. Huisken, *A distance comparison principle for evolving curves*, Asian J. Math. **2** (1998), 127–133.

[KPS] H. Karcher, U. Pinkall, and I. Sterling, *New minimal surfaces in S^3*, J. Differ. Geom. **28** (1988), 169–185.

[La1] H. B. Lawson, Jr., *The unknottedness of minimal embeddings*, Invent. Math. **11** (1970), 183–187.

[La2] H. B. Lawson, Jr., *Complete minimal surfaces in S^3*, Ann. of Math. **92** (1970), 335–374.

[Le] M. L. Leite, *Rotational hypersurfaces of space forms with constant scalar curvature*, Manuscripta Math. **67** (1990), 285–304.

[Li] H. Li, *Stability of surfaces with constant mean curvature*, Proc. Amer. Math. Soc. **105** (1989), 992–997.

[LW1] H. Li and G. Wei, *Embedded rotational hypersurfaces with constant scalar curvature in S^n: a correction to a statement in M. L. Leite, Manuscripta Math.* 67 (1990), 285–304; Manuscripta Math. **120** (2006), 319–323, 81–99.

[LW2] H. Li and G. Wei, *Compact embedded rotation hypersurfaces of S^{n+1}*, Bull. Braz. Math. Sco. New Series **38** (2007), 81–99.

[MP] W. H. Meeks III and J. Perez, *Constant mean curvature surfaces in metric Lie groups*, Geometric Analysis: Partial Differential Equations and Surfaces, Contemp. Math., Vol. **570**, Amer. Math. Soc., Providence, RI, 2012, 25–110.

[MR] S. Montiel and A. Ros, *Compact hypersurfaces: the Alexandrov theorem for higher order mean curvatures*, Differential Geometry, 279–296, Pitman Monogr. Surveys Pure Appl. Math. **52**, Longman Sci. Tech., Harlow, 1991.

[NS] K. Nomizu and B. Smyth, *A formula of Simons' type and hypersurfaces with constant mean curvature*, J. Differential Geometry **3** (1969), 367–377.

[Ot1] T. Otsuki, *Minimal hypersurfaces in a Riemannian manifold of constant curvature*, Amer. J. Math. **92** (1970), 145–173.

[Ot2] T. Otsuki, *Geodesics of* O_n^2 *and an analysis on a related Riemann surface*, Tôhoku Math. J. **28** (1976), 411–427.

[Ot3] T. Otsuki, *On a differential equation related with differential geometry*, Mem. Fac. Sci. Kyushu Univ. Ser. A **47** (1993), 245–281.

[Per] O. Perdomo, *Embedded constant mean curvature hypersurfaces on spheres*, Asian J. Math. **14** (2010), 73–108.

[PS] U. Pinkall and I. Sterling, *On the classification of constant mean curvature tori*, Ann. of Math. (2) **130** (1989), 407–451.

[Ri] J. B. Ripoll, *Superficies invariantes de curvatura memia constanto*, IMPA Ph. D. thesis, 1986.

[SW] W. Sheng and X. -J. Wang, *Singularity profile in the mean curvature flow*, Methods Appl. Anal. **16** (2009), 139–155.

[Si] J. Simons, *Minimal varieties in Riemannian manifolds*, Ann. of Math. (2) **88** (1968), 62–105.

[To] F. Torralbo, *Compact minimal surfaces in the Berger spheres*, Annals of Global Analysis and Geometry **41** (2012), 391–405.

[TU] F. Torralbo and F. Urbano, *Compact stable constant mean curvature surfaces in homogeneous 3-manifolds*, Indiana University Math. J. **61** (2012), 1129–1156.

[WCL] G. Wei, Q. -M. Cheng, and H. Li, *Embedded hypersurfaces with constant mth mean curvature in a unit sphere*, Commun. Contemp. Math. **12** (2010), 997–1013.

[We] H. C. Wente, *Counterexample to a conjecture of H. Hopf*, Pacific J. Math. **121** (1986), 193–243.

Department of Mathematical Sciences, Tsinghua University
Beijing 100084, China
E-mail address: huangzj07@mails.tsinghua.edu.cn

Department of Mathematical Sciences & Mathematical Sciences Center, Tsinghua University
Beijing 100084, China
E-mail address: hli@math.tsinghua.edu.cn

Proceedings of the 6th International Congress of Chinese Mathematicians
ALM 36, pp. 119–133

A Brief Survey on High Order Numerical Methods for Convection Dominated Problems

Chi-Wang Shu

Abstract

Convection dominated partial differential equations are used extensively in applications including fluid dynamics, astrophysics, electro-magnetism, semi-conductor devices, and biological sciences. High order accurate numerical methods are efficient for solving such partial differential equations, however they are difficult to design because solutions may contain discontinuities and other singularities or sharp gradient regions. In this presentation we will survey several types of high order numerical methods for such problems, including weighted essentially non-oscillatory (WENO) finite difference methods, WENO finite volume methods, discontinuous Galerkin finite element methods, and spectral methods. We will discuss essential ingredients, properties and relative advantages of each method, and comparisons among these methods. Examples of recent development and applications of these methods will also be discussed.

Mathematics Subject Classification: 65M06, 65M60, 65M70.

Keywords and Phrases: Convection dominated PDE, high order accurate numerical methods, weighted essentially non-oscillatory, discontinuous Galerkin methods, spectral methods.

1 Introduction

In this presentation, we are interested in numerically solving hyperbolic conservation laws

$$u_t + f(u)_x + g(u)_y + h(u)_z = 0 \tag{1}$$

where $\xi_1 f'(u) + \xi_2 g'(u) + \xi_3 h'(u)$ is diagonalizable with real eigenvalues for any real $\xi = (\xi_1, \xi_2, \xi_3)$. The equation (1) may have discontinuous solutions (containing shocks, contact discontinuities, etc.) even if the initial and boundary conditions are smooth. For properties of equation (1) we refer to, e.g., [33].

Hyperbolic conservation laws (1) appear often in applications. A sample and clearly incomplete list of applications include aeroacoustics, astrophysics, certain problems in computational biology, computational fluid dynamics, especially high speed flows, electro-magnetic waves, semiconductor device simulations, and traffic flows.

Also, in many applications, we often encounter the so-called *convection dominated* problems. i.e. the partial differential equations (PDEs) (1) with additional terms rather than zero on the right hand side, however these additional terms have a small coefficient. For example, a one dimensional convection dominated convection-diffusion equation is given by

$$u_t + f(u)_x = \varepsilon (a(u) u_x)_x$$

where $a(u) \geq 0$ and $0 < \varepsilon \ll 1$. A one dimensional convection dominated convection-dispersion equation is given by

$$u_t + f(u)_x = \varepsilon a(u)_{xxx}$$

where $a(u)$ is an arbitrary function and $0 < \varepsilon \ll 1$. For $f(u) = \alpha u + \beta u^2$ and $a(u) = u$ this is the famous Korteweg-de Vries (KdV) equation. Another good example is the compressible Navier-Stokes equation with high Reynolds numbers, which is the Euler equation plus a diffusion right hand side with a small coefficient $\frac{1}{Re}$ where Re is the Reynolds number.

For convection dominated problems, the main difficulty for the numerical solutions is the treatment of the convection terms. Therefore, we will mainly discuss the conservation laws (1) in this presentation.

In the following we will summarize the importance of using high order methods for solving convection dominated problems.

For smooth solutions, especially for time dependent problems with large time simulation, high order methods are much more efficient than lower order ones to reach the same error tolerance. The smaller this tolerance, the longer the simulation time, the more efficient high order methods will be. See, for example, [12] for a detailed discussion.

For non-smooth or even discontinuous solutions of *linear* PDEs, such as the Maxwell equations in computational electro-magnetism and linearized Euler equations in aeroacoustics, it can be proved that a high order scheme is still high order accurate, measured in a suitable *negative norm*. Then a suitable post-processor can recover high order accuracy in a strong norm such as the L^2 norm. We refer to, e.g. Lax and Mock [20] for finite difference methods, Majda, McDonough and Osher [23] for spectral methods, and Cockburn, Luskin, Shu and Süli [5] and Yang and Shu [43] for discontinuous Galerkin finite element methods.

For non-smooth or even discontinuous solutions of *nonlinear* PDEs, such as the Euler equations of compressible gas dynamics, a rigorous mathematical proof of the high order convergence in a suitable negative norm of high order schemes is not known. It is also true that many of the unmodulated high order schemes such as the spectral method and high order compact schemes may become nonlinearly unstable when computing solutions containing strong discontinuities, rendering the computer codes to crash. However, there are still reasons to believe that high order accurate schemes, especially those modern high order accurate, "high resolution" schemes which have nonlinear mechanisms to control spurious oscillations, such as the weighted essentially non-oscillatory (WENO) schemes [22, 18, 31], are suitable choices for solving nonlinear PDEs with discontinuous solutions, especially when the solutions contain both discontinuities and complicated smooth region structures. Lax [19] argued that, for a nonlinear system, high order information is retained by a high order scheme and may be extracted by postprocessing. In fact, Lax's argument indicates that more high order information is retained in high order solutions of nonlinear systems than of linear ones, since in the nonlinear case the solution operator is contractive.

2 Brief comparison of several high order methods

In this section we give a brief comparison of several commonly used high order methods.

If the solution of the PDE is *smooth*, the spectral method is the most efficient numerical method [14]. Using N grid points or expansion terms, the error is smaller than $O(N^{-k})$ for *any* finite k, if the solution is C^{k+1}. If the solution is analytic, then the error is exponential, that is, the error decays as $O(e^{-\alpha N})$ for some constant $\alpha > 0$. As an example: a Fourier spectral method applied to linear convection with a smooth initial condition with $N = 16$ already produces an error of 10^{-11}, while a second order accurate finite difference scheme with the same number of unknowns produces an error of 10^{-1} and a fourth order finite difference scheme is barely better at the error level of 10^{-2}.

High order accurate compact schemes, which can give "spectral-like" resolution for many wave problems, are also popular for solving PDEs with smooth solutions [21]. They are more suitable for non-periodic problems, as they can be used on uniform meshes. In comparison, spectral methods for non-periodic problems would require special non-uniform grids to achieve both stability and accuracy.

If the computational domain is not a box, then the spectral element method, which combines the advantages of the spectral method in its high order accuracy and the finite element method in its flexibility for arbitrary geometry, would be a good choice [24]. The discontinuous Galerkin method, because of its flexibility in meshes, adaptivity, and parallel implementation, is also a good choice [8].

If the solution of the PDE has discontinuities or sharp gradient regions, we could use high order methods (spectral, compact scheme, etc.) with filtering, e.g. [38, 45]. The filtering can be either linear or nonlinear. The advantage of using filtering is that it is simple to code, as the base algorithm and the filtering can be coded separately and applied sequentially. A disadvantage is that, usually there are parameters in the filtering to be tuned. If these parameters are not tuned well, the result could be either too dissipative or too oscillatory.

We could also use high order WENO finite difference schemes [18, 31]. Such schemes approximate the point values $u_j \approx u(x_j,t)$ by a conservative difference scheme

$$\frac{d}{dt}u_j + \frac{1}{\Delta x}\left(\hat{f}_{j+1/2} - \hat{f}_{j-1/2}\right) = 0.$$

The numerical flux

$$\hat{f}_{j+1/2} = \hat{f}\left(u_{j-p}, \dots, u_{j+q}\right)$$

is obtained by a WENO approximation providing both high order accuracy in smooth regions and essentially non-oscillatory shock transitions. Advantages include the robustness of schemes with no user-tuned parameters. The WENO finite difference schemes are stable for strong shocks. The code for multi-dimensions is essentially the same as the one for one dimension. The main disadvantage is that it can only be used on rectangular uniform meshes or smooth curvilinear meshes, if both conservation and at least third order accuracy are desired. Also, such schemes need local characteristic decomposition to avoid oscillations during wave interactions. The CPU cost is thus 4 to 10 times more than that of a second order scheme depending on the specific equation and implementation.

A good example where the solution contains both discontinuities and rich structures in smooth regions is the problem of a strong shock interacting with a pair of vortices [49], simulated by the fifth order finite difference WENO scheme on the compressible Navier-Stokes equations. A three-dimensional problem, involving shock interaction with a longitudinal vortex, is studied in [48] also by the fifth order finite difference WENO scheme on the compressible Navier-Stokes equations.

Alternatively, high order WENO finite volume schemes [22, 31] can be used. Such schemes approximate the cell averages $\bar{u}_j \approx \frac{1}{\Delta x}\int_{x_{j-1/2}}^{x_{j+1/2}} u(x,t)dx$ by a conservative scheme

$$\frac{d}{dt}\bar{u}_j + \frac{1}{\Delta x}\left(\hat{f}_{j+1/2} - \hat{f}_{j-1/2}\right) = 0.$$

The numerical flux is given by

$$\hat{f}_{j+1/2} = h\left(u^-_{j+1/2}, u^+_{j+1/2}\right)$$

where $h(u^-, u^+)$ is a monotone flux, that is, h is increasing in its first argument and decreasing in its second argument, symbolically $h(\uparrow, \downarrow)$, and $u^{\pm}_{j+1/2}$ are approximations to $u(x_{j+1/2}, t)$ obtained from the neighboring cell averages $\{\bar{u}_{j-p}, \dots, \bar{u}_{j+q}\}$ by a WENO reconstruction providing both high order accuracy in smooth regions and essentially non-oscillatory shock transitions. Advantages again include a robustness of these schemes without user-tuned parameters. Similar to finite difference WENO schemes, finite volume WENO schemes are stable for strong shocks. An additional advantage over finite difference schemes is that finite volume schemes can be applied to non-smooth meshes and arbitrary triangulations, e.g. [16, 59]. The code for multi-dimensional tensor product meshes (the meshes do not need to be smooth) is similar as the one for one dimension. However, high order finite volume schemes are very expensive for multi-dimensional calculation. It is up to 4 times more expensive than a finite difference WENO scheme on the same mesh with the same order of accuracy for 2D, and up to 9 times for 3D. The algorithm formulation and coding are both very complicated for arbitrary (not tensor product) triangulation. We refer to, e.g. [2, 46] for a detailed comparison of finite difference and finite volume schemes.

Finally, we mention the discontinuous Galerkin (DG) finite element methods [6, 7, 8]. These methods approximate the whole k-th degree polynomial $u_j(x,t) \approx u(x,t)$ in the cell I_j by requiring

$$\frac{d}{dt}\int_{I_j} u_j(x,t)\varphi dx - \int_{I_j} f(u_j(x,t))\varphi_x dx + \hat{f}_{j+1/2}\varphi(x_{j+1/2}) - \hat{f}_{j-1/2}\varphi(x_{j-1/2}) = 0$$

to hold for all k-th degree polynomials φ on I_j. The numerical flux

$$\hat{f}_{j+1/2} = h\left(u_j(x_{j+1/2},t), u_{j+1}(x_{j+1/2},t)\right)$$

is the same as that for the finite volume scheme. A major advantage of the DG schemes is that they can be applied to very general geometry, triangulation and boundary conditions. They are also easy for adaptive computation and parallel implementation. DG schemes have very good wave resolution capability [61]. More theoretical stability results are available than finite difference and finite volume WENO schemes. We refer to, e.g. [8, 32, 15] for more details. Disadvantages of DG schemes include the fact that coding is more complicated than finite difference schemes. Also, limiters to control spurious oscillations for strong shocks are not as robust as the WENO methodology. This problem has been intensively studied in recent literature, for example WENO limiters have been designed by Qiu and Shu [27], Zhong and Shu [62] and Zhu, Zhong, Shu and Qiu [63]. Another disadvantage of DG methods is that computer memory requirement is very large, especially for 3D computation. Recent study on hybridizing finite volume and DG methods, for example the Hermite WENO schemes of Qiu and Shu [26, 28] or the P^nP^m method of Dumbser and Zanotti [9], attempts to address this memory problem.

3 Examples of recent developments on high order methods

In this section we give two examples of recent developments on high order methods.

3.1 Bound preserving high order schemes

A general framework is recently established to preserve strict bounds (maximum principle for scalar problems and positivity of relevant quantities for scalar problems or systems), while maintaining provable high order accuracy of the original schemes. These techniques apply to multi-dimensions in general unstructured triangulations as well. See [51, 52, 57].

First order monotone schemes can maintain the maximum principle. For the one-dimensional conservation law

$$u_t + f(u)_x = 0,$$

the first order monotone scheme

$$u_j^{n+1} = H_\lambda(u_{j-1}^n, u_j^n, u_{j+1}^n) = u_j^n - \lambda[h(u_j^n, u_{j+1}^n) - h(u_{j-1}^n, u_j^n)]$$

where $\lambda = \frac{\Delta t}{\Delta x}$ and $h(u^-, u^+)$ is a monotone flux, satisfies monotonicity of H (H is an increasing function of all three arguments, symbolically $H_\lambda(\uparrow, \uparrow, \uparrow)$) under a suitable CFL condition

$$\lambda \leq \lambda_0.$$

Therefore, if

$$m \leq u_{j-1}^n, u_j^n, u_{j+1}^n \leq M$$

then

$$u_j^{n+1} = H_\lambda(u_{j-1}^n, u_j^n, u_{j+1}^n) \geq H_\lambda(m, m, m) = m,$$

and

$$u_j^{n+1} = H_\lambda(u_{j-1}^n, u_j^n, u_{j+1}^n) \leq H_\lambda(M, M, M) = M.$$

This verifies the maximum principle for first order monotone schemes.

The evolution of the cell average for a higher order finite volume or DG scheme satisfies

$$\bar{u}_j^{n+1} = G(\bar{u}_j^n, u_{j-\frac{1}{2}}^-, u_{j-\frac{1}{2}}^+, u_{j+\frac{1}{2}}^-, u_{j+\frac{1}{2}}^+) = \bar{u}_j^n - \lambda[h(u_{j+\frac{1}{2}}^-, u_{j+\frac{1}{2}}^+) - h(u_{j-\frac{1}{2}}^-, u_{j-\frac{1}{2}}^+)],$$

where G is increasing of its first, second and fifth arguments but is decreasing in the other two arguments, symbolically $G(\uparrow, \uparrow, \downarrow, \downarrow, \uparrow)$, therefore there is no maximum principle. The problem is with the two arguments $u_{j-\frac{1}{2}}^+$ and $u_{j+\frac{1}{2}}^-$ which are values at points *inside* the cell I_j.

The polynomial $p_j(x)$ (either reconstructed in a finite volume method or evolved in a DG method) is of degree k, defined on I_j such that $\bar{u}_j^n$ is its cell average on I_j, $u_{j-\frac{1}{2}}^+ = p_j(x_{j-\frac{1}{2}})$ and $u_{j+\frac{1}{2}}^- = p_j(x_{j+\frac{1}{2}})$. We can therefore take a Legendre Gauss-Lobatto quadrature rule which is exact for polynomials of degree k, then

$$\bar{u}_j^n = \sum_{\ell=0}^{m} \omega_\ell p_j(y_\ell)$$

with $y_0 = x_{j-\frac{1}{2}}$, $y_m = x_{j+\frac{1}{2}}$. The scheme for the cell average can therefore be rewritten as

$$\begin{aligned}
\bar{u}_j^{n+1} &= \omega_m \left[u_{j+\frac{1}{2}}^- - \frac{\lambda}{\omega_m} \left(h(u_{j+\frac{1}{2}}^-, u_{j+\frac{1}{2}}^+) - h(u_{j-\frac{1}{2}}^+, u_{j+\frac{1}{2}}^-) \right) \right] \\
&\quad + \omega_0 \left[u_{j-\frac{1}{2}}^+ - \frac{\lambda}{\omega_0} \left(h(u_{j-\frac{1}{2}}^+, u_{j+\frac{1}{2}}^-) - h(u_{j-\frac{1}{2}}^-, u_{j-\frac{1}{2}}^+) \right) \right] + \sum_{\ell=1}^{m-1} \omega_\ell p_j(y_\ell) \\
&= \omega_m H_{\lambda/\omega_m}(u_{j-\frac{1}{2}}^+, u_{j+\frac{1}{2}}^-, u_{j+\frac{1}{2}}^+) + \omega_0 H_{\lambda/\omega_0}(u_{j-\frac{1}{2}}^-, u_{j-\frac{1}{2}}^+, u_{j+\frac{1}{2}}^-) \\
&\quad + \sum_{\ell=1}^{m-1} \omega_\ell p_j(y_\ell).
\end{aligned}$$

Therefore, if

$$m \le p_j(y_\ell) \le M$$

at all Legendre Gauss-Lobatto quadrature points and a reduced CFL condition

$$\lambda/\omega_m = \lambda/\omega_0 \le \lambda_0$$

is satisfied, then

$$m \le \bar{u}_j^{n+1} \le M.$$

We replace $p_j(x)$ by the limited polynomial $\tilde{p}_j(x)$ defined by

$$\tilde{p}_j(x) = \theta_j(p_j(x) - \bar{u}_j^n) + \bar{u}_j^n$$

where

$$\theta_j = \min\left\{ \left|\frac{M - \bar{u}_j^n}{M_j - \bar{u}_j^n}\right|, \left|\frac{m - \bar{u}_j^n}{m_j - \bar{u}_j^n}\right|, 1 \right\},$$

with

$$M_j = \max_{x \in S_j} p_j(x), \qquad m_j = \min_{x \in S_j} p_j(x)$$

where S_j is the set of Legendre Gauss-Lobatto quadrature points of cell I_j.

Clearly, this limiter is just a simple scaling of the original polynomial around its average. The following lemma, guaranteeing the maintenance of accuracy of this simple limiter, is proved in Zhang and Shu [52]:

Lemma. *Assume $\bar{u}_j^n \in [m, M]$ and $p_j(x)$ is an $O(\Delta x^p)$ approximation, then $\tilde{p}_j(x)$ is also an $O(\Delta x^p)$ approximation.*

For scalar nonlinear conservation laws, passive convection in a divergence-free velocity field, and 2D incompressible Euler equations in the vorticity-streamfunction formulation, high order finite volume WENO and DG schemes maintaining maximum principle have been designed in Zhang and Shu [51] and in Zhang, Xia and Shu [57].

For scalar nonlinear convection diffusion equations, high order finite volume WENO schemes (on structured meshes) and second order DG schemes on unstructured triangulations maintaining maximum principle have been designed in Zhang, Liu and Shu [50] and in Zhang, Zhang and Shu [60] respectively.

For Euler equations of gas dynamics, high order finite volume WENO, DG schemes and finite difference WENO maintaining positivity of density and pressure (or internal energy) have been designed in Zhang and Shu [52, 53, 54, 55, 56] and in Zhang, Xia and Shu [57].

For shallow water equations, high order finite volume WENO and DG schemes maintaining non-negativity of water height and well-balancedness for problems with dry areas have been designed in Xing, Zhang and Shu [41] and in Xing and Shu [40].

For PDEs involving global integral terms including a hierarchical size-structured population model and Vlasov-Boltzmann transport equations, high order finite volume WENO and DG schemes maintaining non-negativity of density have been designed in Zhang, Zhang and Shu [47] and in Cheng, Gamba and Proft [4], respectively.

Positivity-preserving semi-Lagrangian schemes have been designed in Qiu and Shu [25] and in Rossmanith and Seal [29]. Positivity-preserving finite volume Lagrangian schemes for multi-material flows have been designed in Cheng and Shu [3].

Bound-preserving limiters for problems containing δ-function singularities are designed in [43, 44]. It is even more crucial for such limiters to deal with the extreme situation of δ-function singularities in a stable fashion while maintaining accuracy and resolution.

In [52], a Mach 80 (i.e. the Mach number of the jet inflow is 80 with respect to the soundspeed in the jet gas) jet flow problem with radiative cooling is computed using the third order DG scheme. Without the positivity-preserving limiter the code blows up. In [39], we study a shock diffraction problem in which the shock passes a backward facing corner of 135°. It is easy to get negative density and/or pressure below the corner. This problem also involves mixed triangular / rectangular meshes for the DG method.

3.2 Inverse Lax-Wendroff procedure for boundary conditions

For finite difference schemes approximating PDEs, there are two major difficulties associated with numerical boundary conditions. The first difficulty is that high order finite difference schemes involve a wide stencil, hence there are several points near the boundary (either as ghost points outside the computational domain or as the first few points inside the computational domain near the boundary) which need different treatment.

For example, if we have the following scheme

$$u_j^{n+1} = au_{j-2}^n + bu_{j-1}^n + cu_j^n + du_{j+1}^n$$

with suitably chosen constants a, b, c and d (which depend on $\lambda = \frac{\Delta t}{\Delta x}$), approximating the PDE

$$u_t + u_x = 0,$$

$$u(x,0) = f(x), \qquad u(0,t) = g(t)$$

to third order accuracy, then either a ghost point u_{-1}^n is needed, or the scheme cannot be used to compute u_1^{n+1}.

The second difficulty is that the boundary of the computational domain may not coincide with grid points. For example, in 1D, we may have the physical boundary $x = 0$ located anywhere between two grid points. While this seems artificial, it is unavoidable for a moving boundary computed on a fixed grid. This difficulty is more profound in 2D (complicated geometry computed on Cartesian meshes).

One of the major difficulties is the small cell near the boundary and the resulting small time step required for stability. Special treatments are designed in the literature to address this difficulty [1, 13].

Let us first review on the traditional Lax-Wendroff procedure for solving, e.g.

$$u_t + u_x = 0.$$

We first perform Taylor expansion in time

$$u_j^{n+1} = u_j + (u_t)_j \Delta t + \frac{1}{2}(u_{tt})_j \Delta t^2 + \cdots .$$

Then we replace the time derivatives by spatial derivatives by repeatedly using the PDE:

$$(u_t)_j = -(u_x)_j;$$
$$(u_{tt})_j = -((u_x)_t)_j = -((u_t)_x)_j = (u_{xx})_j;$$

etc., and approximate the spatial derivatives by finite differences of suitable order of accuracy.

We now look at the basic idea of the inverse Lax-Wendroff procedure, by switching the roles of x and t in the traditional Lax-Wendroff procedure. Suppose we are solving

$$u_t + u_x = 0, \qquad u(0,t) = g(t)$$

and suppose the boundary $x = 0$ is of distance $a\Delta x$ from x_1 (with a constant a), the inverse Lax-Wendroff procedure to determine u_1 is as follows:

- Taylor expansion in space

$$u_1 = u(0,t) + u_x(0,t)a\Delta x + \frac{1}{2}u_{xx}(0,t)(a\Delta x)^2 + \cdots .$$

- Replace the spatial derivatives by time derivatives by repeatedly using the PDE:

$$u_x = -u_t; \qquad u_x(0,t) = -u_t(0,t) = -g'(t);$$
$$u_{xx} = (-u_t)_x = -(u_x)_t = u_{tt};$$
$$u_{xx}(0,t) = u_{tt}(0,t) = g''(t);$$

  etc.

- Compute $g'(t)$, $g''(t)$, etc. either analytically or by finite difference.

For earlier work in this direction see [10, 11]. For steady state Hamilton-Jacobi equations, this technique is used in the fast sweeping framework for high order WENO schemes in Huang, Shu and Zhang [17] and in Xiong, Zhang, Zhang and Shu [42], and for discontinuous Galerkin methods in Zhang, Chen, Li, Zhao and Shu [58].

For scalar and systems of time-dependent conservation laws, this technique is used to compute solution in arbitrary domains using rectangular meshes which are not aligned with the physical boundary in Tan and Shu [34] and in Tan, Wang, Shu and Ning [37].

For compressible inviscid flows involving complex moving geometries, this technique is used to compute solution on a fixed rectangular mesh in Tan and Shu [35, 36].

We are most interested in applying our method to the solid wall boundary conditions $(u, v) \cdot \mathbf{n} = 0$, when the wall is not aligned with the grid and can be curved. A typical example of this kind is the double Mach reflection problem. This problem is initialized by sending a horizontally moving shock into a wedge inclined by a $30°$ angle. In order to impose the solid wall condition by the reflection technique, people usually solve an equivalent problem that puts the solid wall horizontal and puts the shock $60°$ angle inclined to the wall. Another way to avoid the trouble of imposing boundary conditions is to use a multidomain WENO method [30]. With the usage of our inverse Lax-Wendroff procedure, we are able to solve the original problem with a uniform mesh in a single domain, obtaining similarly good results as other more complicated approaches, see [34].

Another typical example involves a curved wall which is a circular cylinder of unit radius positioned at the origin on a x-y plane. The problem is initialized by a Mach 3 flow moving toward the cylinder from the left. In order to impose the solid wall boundary condition at the surface of the cylinder by the reflection technique, a particular mapping from the unit square to the physical domain is usually used in traditional finite difference methods, e.g. in [18]. Using our inverse Lax-Wendroff procedure, we are able to solve this problem directly in the physical domain, obtaining similarly good results, see [34].

The last example shows that our high order method can also treat a rigid body whose motion is induced by the fluid. We test the so-called cylinder lift-off problem. In this problem, a rigid cylinder initially resting on the floor of a 2D channel is driven and lifted by a strong shock. The computational domain is $[0, 1] \times [0, 0.2]$. A rigid cylinder with radius 0.05 and density 10.77 is initially centered at $(0.15, 0.05)$. A Mach 3 shock starts at $x = 0.08$ moving towards the cylinder. The density and pressure of the resting gas are $\rho = 1.4$ and $p = 1.0$ respectively. The top and bottom of the domain are rigid walls. The left boundary is set to the post-shock state and the right boundary is supersonic outflow. The good results, comparable to those in the literature obtained from more complicated algorithms, can be seen in [35, 36].

4 Concluding remarks

In this presentation, we have argued that high order methods have a good potential for large scale simulation of complex applications.

One aspect in algorithm design that we have emphasized is that numerical methods should obey physical properties of the PDE without compromising accuracy (e.g. bound-preserving, positivity-preserving).

Another aspect in algorithm design that we have emphasized is that numerical boundary conditions should be designed to maintain accuracy and stability, and to allow computation of complex geometry problem on simple meshes.

We hope this presentation will motivate more research on the design and application of high order schemes to convection dominated problems.

References

[1] M.J. Berger, C. Helzel, and R.J. LeVeque, *h-box methods for the approximation of hyperbolic conservation laws on irregular grids*, SIAM Journal on Numerical Analysis **41** (2003), 893–918.

[2] J. Casper, C.-W. Shu, and H.L. Atkins, *Comparison of two formulations for high-order accurate essentially nonoscillatory schemes*, AIAA Journal **32** (1994), 1970–1977.

[3] J. Cheng and C.-W. Shu, *Positivity-preserving Lagrangian scheme for multi-material compressible flow*, Journal of Computational Physics **257** (2014), 143–168.

[4] Y. Cheng, I.M. Gamba, and J. Proft, *Positivity-preserving discontinuous Galerkin schemes for linear Vlasov-Boltzmann transport equations*, Mathematics of Computation **81** (2012), 153–190.

[5] B. Cockburn, M. Luskin, C.-W. Shu, and E. Süli, *Enhanced accuracy by post-processing for finite element methods for hyperbolic equations*, Mathematics of Computation **72** (2003), 577–606.

[6] B. Cockburn and C.-W. Shu, *TVB Runge-Kutta local projection discontinuous Galerkin finite element method for conservation laws II: general framework*, Mathematics of Computation **52** (1989), 411–435.

[7] B. Cockburn and C.-W. Shu, *The Runge-Kutta discontinuous Galerkin method for conservation laws V: multidimensional systems*, Journal of Computational Physics **141** (1998), 199–224.

[8] B. Cockburn and C.-W. Shu, *Runge-Kutta discontinuous Galerkin methods for convection-dominated problems*, Journal of Scientific Computing **16** (2001), 173–261.

[9] M. Dumbser and O. Zanotti, *Very high order PNPM schemes on unstructured meshes for the resistive relativistic MHD equations*, Journal of Computational Physics **228** (2009), 6991–7006.

[10] M. Goldberg and E. Tadmor, *Scheme-independent stability criteria for difference approximations of hyperbolic initial-boundary value problems. I*, Mathematics of Computation **32** (1978), 1097–1107.

[11] M. Goldberg and E. Tadmor, *Scheme-independent stability criteria for difference approximations of hyperbolic initial-boundary value problems. II*, Mathematics of Computation **36** (1981), 603–626.

[12] B. Gustafsson, H.-O. Kreiss, and J. Oliger, *Time Dependent Problems and Difference Methods*, John Wiley & Sons, New York, 1995.

[13] C. Helzel, M.J. Berger, and R.J. LeVeque, *A high-resolution rotated grid method for conservation laws with embedded geometries*, SIAM Journal on Scientific Computing **26** (2005), 785–809.

[14] J. Hesthaven, S. Gottlieb, and D. Gottlieb, *Spectral Methods for Time-dependent Problems*, Cambridge University Press, Cambridge, 2007.

[15] J. Hesthaven and T. Warburton, *Nodal Discontinuous Galerkin Methods*, Springer, New York, 2008.

[16] C. Hu and C.-W. Shu, *Weighted essentially non-oscillatory schemes on triangular meshes*, Journal of Computational Physics **150** (1999), 97–127.

[17] L. Huang, C.-W. Shu, and M. Zhang, *Numerical boundary conditions for the fast sweeping high order WENO methods for solving the Eikonal equation*, Journal of Computational Mathematics **26** (2008), 336–346.

[18] G.-S. Jiang and C.-W. Shu, *Efficient implementation of weighted ENO schemes*, Journal of Computational Physics **126** (1996), 202–228.

[19] P.D. Lax, *Accuracy and resolution in the computation of solutions of linear and nonlinear equations*, in *Recent Advances in Numerical Analysis*, edited by C. De Boor and G. Golub, Academic Press, 1978.

[20] P.D. Lax and M. Mock, *The computation of discontinuous solutions of linear hyperbolic equations*, Communications on Pure and Applied Mathematics **31** (1978), 423–430.

[21] S.K. Lele, *Compact finite difference schemes with spectral-like resolution*, Journal of Computational Physics **103** (1992), 16–42.

[22] X.-D. Liu, S. Osher, and T. Chan, *Weighted essentially non-oscillatory schemes*, Journal of Computational Physics **115** (1994), 200–212.

[23] A. Majda, J. McDonough, and S. Osher, *The Fourier method for nonsmooth initial data*, Mathematics of Computation **32** (1978), 1041–1081.

[24] A.T. Patera, *A spectral element method for fluid dynamics: laminar flow in a channel expansion*, Journal of Computational Physics **54** (1984), 468–488.

[25] J.-M. Qiu and C.-W. Shu, *Positivity preserving semi-Lagrangian discontinuous Galerkin formulation: theoretical analysis and application to the Vlasov-Poisson system*, Journal of Computational Physics **230** (2011), 8386–8409.

[26] J.-X. Qiu and C.-W. Shu, *Hermite WENO schemes and their application as limiters for Runge-Kutta discontinuous Galerkin method: one-dimensional case*, Journal of Computational Physics **193** (2003), 115–135.

[27] J.-X. Qiu and C.-W. Shu, *Runge-Kutta discontinuous Galerkin method using WENO limiters*, SIAM Journal on Scientific Computing **26** (2005), 907–929.

[28] J.-X. Qiu and C.-W. Shu, *Hermite WENO schemes and their application as limiters for Runge-Kutta discontinuous Galerkin method II: two dimensional case*, Computers and Fluids **34** (2005), 642–663.

[29] J.A. Rossmanith and D.C. Seal, *A positivity-preserving high-order semi-Lagrangian discontinuous Galerkin scheme for the Vlasov-Poisson equations*, Journal of Computational Physics **230** (2011), 6203–6232.

[30] K. Sebastian and C.-W. Shu, *Multi domain WENO finite difference method with interpolation at subdomain interfaces*, Journal of Scientific Computing **19** (2003), 405–438.

[31] C.-W. Shu, *High order weighted essentially non-oscillatory schemes for convection dominated problems*, SIAM Review **51** (2009), 82–126.

[32] C.-W. Shu, *Discontinuous Galerkin methods: general approach and stability*, Numerical Solutions of Partial Differential Equations, S. Bertoluzza, S. Falletta, G. Russo and C.-W. Shu, Advanced Courses in Mathematics CRM Barcelona, Birkhäuser, Basel, 2009, 149–201.

[33] J. Smoller, *Shock Waves and Reaction-diffusion Equations*, second edition, Springer-Verlag, New York, 1994.

[34] S. Tan and C.-W. Shu, *Inverse Lax-Wendroff procedure for numerical boundary conditions of conservation laws*, Journal of Computational Physics **229** (2010), 8144–8166.

[35] S. Tan and C.-W. Shu, *A high order moving boundary treatment for compressible inviscid flows*, Journal of Computational Physics **230** (2011), 6023–6036.

[36] S. Tan and C.-W. Shu, *Inverse Lax-Wendroff procedure for numerical boundary conditions of hyperbolic equations: survey and new developments*, in "Advances in Applied Mathematics, Modeling and Computational Science", R. Melnik and I. Kotsireas, Editors, Fields Institute Communications **66**, Springer, New York, 2013, 41–63.

[37] S. Tan, C. Wang, C.-W. Shu, and J. Ning, *Efficient implementation of high order inverse Lax-Wendroff boundary treatment for conservation laws*, Journal of Computational Physics **231** (2012), 2510–2527.

[38] H. Vandeven, *Family of spectral filters for discontinuous problems*, Journal of Scientific Computing **6** (1991), 159–192.

[39] C. Wang, X. Zhang, C.-W. Shu, and J. Ning, *Robust high order discontinuous Galerkin schemes for two-dimensional gaseous detonations*, Journal of Computational Physics **231** (2012), 653–665.

[40] Y. Xing and C.-W. Shu, *High-order finite volume WENO schemes for the shallow water equations with dry states*, Advances in Water Resources **34** (2011), 1026–1038.

[41] Y. Xing, X. Zhang, and C.-W. Shu, *Positivity preserving high order well balanced discontinuous Galerkin methods for the shallow water equations*, Advances in Water Resources **33** (2010), 1476–1493.

[42] T. Xiong, M. Zhang, Y.-T. Zhang, and C.-W. Shu, *Fifth order fast sweeping WENO scheme for static Hamilton-Jacobi equations with accurate boundary treatment*, Journal of Scientific Computing **45** (2010), 514–536.

[43] Y. Yang and C.-W. Shu, *Discontinuous Galerkin method for hyperbolic equations involving δ-singularities: negative-order norm error estimates and applications*, Numerische Mathematik **124** (2013), 753–781.

[44] Y. Yang, D. Wei, and C.-W. Shu, *Discontinuous Galerkin method for Krause's consensus models and pressureless Euler equations*, Journal of Computational Physics **252** (2013), 109–127.

[45] H.C. Yee and B. Sjögreen, *Development of low dissipative high order filter schemes for multiscale Navier-Stokes/MHD systems*, Journal of Computational Physics **68** (2007), 151–179.

[46] R. Zhang, M. Zhang, and C.-W. Shu, *On the order of accuracy and numerical performance of two classes of finite volume WENO schemes*, Communications in Computational Physics **9** (2011), 807–827.

[47] R. Zhang, M. Zhang, and C.-W. Shu, *High order positivity-preserving finite volume WENO schemes for a hierarchical size-structured population model*, Journal of Computational and Applied Mathematics **236** (2011), 937–949.

[48] S. Zhang, H. Zhang, and C.-W. Shu, *Topological structure of shock induced vortex breakdown*, Journal of Fluid Mechanics **639** (2009), 343–372.

[49] S. Zhang, Y.-T. Zhang, and C.-W. Shu, *Interaction of a shock wave with an oblique vortex pair: shock dynamics and mechanism of sound generation*, Physics of Fluids **18** (2006), 126101.

[50] X. Zhang, Y.-Y. Liu, and C.-W. Shu, *Maximum-principle-satisfying high order finite volume WENO schemes for convection-diffusion equations*, SIAM Journal on Scientific Computing **34** (2012), A627–A658.

[51] X. Zhang and C.-W. Shu, *On maximum-principle-satisfying high order schemes for scalar conservation laws*, Journal of Computational Physics **229** (2010), 3091–3120.

[52] X. Zhang and C.-W. Shu, *On positivity preserving high order discontinuous Galerkin schemes for compressible Euler equations on rectangular meshes*, Journal of Computational Physics **229** (2010), 8918–8934.

[53] X. Zhang and C.-W. Shu, *Positivity-preserving high order discontinuous Galerkin schemes for compressible Euler equations with source terms*, Journal of Computational Physics **230** (2011), 1238–1248.

[54] X. Zhang and C.-W. Shu, *Maximum-principle-satisfying and positivity-preserving high order schemes for conservation laws: Survey and new developments*, Proceedings of the Royal Society A **467** (2011), 2752–2776.

[55] X. Zhang and C.-W. Shu, *A minimum entropy principle of high order schemes for gas dynamics equations*, Numerische Mathematik **121** (2012), 545–563.

[56] X. Zhang and C.-W. Shu, *Positivity-preserving high order finite difference WENO schemes for compressible Euler equations*, Journal of Computational Physics **231** (2012), 2245–2258.

[57] X. Zhang, Y. Xia, and C.-W. Shu, *Maximum-principle-satisfying and positivity-preserving high order discontinuous Galerkin schemes for conservation laws on triangular meshes*, Journal of Scientific Computing **50** (2012), 29–62.

[58] Y.-T. Zhang, S. Chen, F. Li, H. Zhao, and C.-W. Shu, *Uniformly accurate discontinuous Galerkin fast sweeping methods for Eikonal equations*, SIAM Journal on Scientific Computing **33** (2011), 1873–1896.

[59] Y.-T. Zhang and C.-W. Shu, *Third order WENO scheme on three dimensional tetrahedral meshes*, Communications in Computational Physics **5** (2009), 836–848.

[60] Y. Zhang, X. Zhang, and C.-W. Shu, *Maximum-principle-satisfying second order discontinuous Galerkin schemes for convection-diffusion equations on triangular meshes*, Journal of Computational Physics **234** (2013), 295–316.

[61] X. Zhong and C.-W. Shu, *Numerical resolution of discontinuous Galerkin methods for time dependent wave equations*, Computer Methods in Applied Mechanics and Engineering **200** (2011), 2814–2827.

[62] X. Zhong and C.-W. Shu, *A simple weighted essentially nonoscillatory limiter for Runge-Kutta discontinuous Galerkin methods*, Journal of Computational Physics **232** (2013), 397–415.

[63] J. Zhu, X. Zhong, C.-W. Shu, and J.-X. Qiu, *Runge-Kutta discontinuous Galerkin method using a new type of WENO limiters on unstructured meshes*, Journal of Computational Physics **248** (2013), 200–220.

Division of Applied Mathematics, Brown University
Providence, RI 02912, USA
E-mail address: shu@dam.brown.edu

Proceedings of the 6th International Congress of Chinese Mathematicians
ALM 36, pp. 135–151

Congruent Number Problem

Ye Tian

Abstract

We introduce our recent progress on the congruent number problem. We also give some explicit methods to construct quadratic twists with rank one Mordell-Weil groups for a given elliptic curve over $\mathbb{Q}$. Non-triviality of Heegner points plays a central role on these results.

2002 Mathematics Subject Classification: 11G40, 11F67.

Keywords and Phrases: Congruent number, Heegner points, BSD conjecture, explicit Gross-Zagier formula.

1 Congruent numbers with many prime factors

A positive integer n is defined to be a *congruent number* if it is the area of a right-angled triangle, all of whose sides have rational length. The *congruent number problem* is the question of determining whether a given positive integer n is a congruent number or not, and also to find corresponding right-angled triangles when it is congruent.

It has long been known that $5, 6, 7$ are congruent numbers with corresponding right-angled triangles having side lengths

$$(9/6, 40/6, 41/6), \qquad (3, 4, 5), \qquad (175/60, 288/60, 337/60),$$

respectively. Fermat was the first person to prove that 1 is not a congruent number, and discovered in this way the method of infinite descent. Similar arguments show that $2, 3$ are not congruent numbers.

It is clear that a positive integer n is congruent if and only if the elliptic curve over $\mathbb{Q}$, as the intersection of two quadratic surfaces in $\mathbb{P}^3$:

$$E^{(n)}: \qquad a^2 + b^2 = c^2, \quad \frac{1}{2}ab = nd^2$$

has a rational point with $d \neq 0$. It turns out that the Mordell-Weil group $E^{(n)}(\mathbb{Q})$ of $E^{(n)}$ over $\mathbb{Q}$ has the torsion subgroup $E^{(n)}(\mathbb{Q})_{\mathrm{tor}} \cong (\mathbb{Z}/2\mathbb{Z})^2$ exactly consisting of points with $d = 0$. Thus n is a congruent number if and only if $E^{(n)}(\mathbb{Q})$ has rank larger than 0. By Euclidean parametrization of right-angled triangles:

$$a/d = 2xz/yz, \quad b/d = (x^2 - z^2)/yz, \quad c/d = (x^2 + z^2)/yz,$$

we obtain a plane model of $E^{(n)}$: $ny^2z = x^3 - xz^2$. Note that the isomorphism class of $E^{(n)}$ only depends on the class of n in $\mathbb{Q}^\times/\mathbb{Q}^{\times 2}$. *From now on, we assume that n is a square-free positive integer.* It is now easy to see that n is a congruent number if and only if there exist positive integers x, z such that $xz(x+z)(x-z)$ is the product of n and a non-zero square. For example, taking $(x, z) = (5, 4), (2, 1), (16, 9)$, we know that $5, 6, 7$ are congruent numbers. Taking $(x, z) = (x, 1)$ with integer $x \geq 2$, we see that numbers of form $(x-1)x(x+1)$ are congruent, and therefore that there are infinitely many square-free congruent numbers in each residue class of $1, 2, 3, 5, 6, 7$ modulo 8.

Let $\epsilon(n) = \pm 1$ denote the sign of $L(E^{(n)}, s)$, the L-function of $E^{(n)}$ over $\mathbb{Q}$. It is known that $\epsilon(n) = +1$ if $n \equiv 1, 2, 3 \bmod 8$ and -1 if $n \equiv 5, 6, 7 \bmod 8$. Thus the vanishing order of $L(E^{(n)}, s)$ at $s = 1$ is odd if and only if $n \equiv 5, 6, 7 \bmod 8$, and therefore the Birch and Swinnerton-Dyer conjecture predicts that

Conjecture 1.1. *Any positive integer $n \equiv 5, 6, 7 \bmod 8$ is a congruent number.*

Heegner showed this conjecture in the case n has exactly one odd prime factor.

Theorem 1.2. (Heegner, 1952) *Any positive integer $n \equiv 5, 6, 7 \bmod 8$ with exactly one odd prime factor is a congruent number.*

Actually Heegner is the first mathematician who found (in [12]) a method to construct fairly general solutions to cubic Diophantine equations. In [25] and [24], we extended Heegner's result to congruent numbers with many prime factors based on his construction, and using an induction argument and special values of L-function and its derivative of elliptic curves.

Theorem 1.3. (Tian, 2012) *Let $m \equiv 5$, 6, or 7 $\bmod 8$ be a positive integer and let n denote its odd part. Assume that*

1. *all except one prime factors of n are $\equiv 1 \bmod 8$;*
2. *$\mathbb{Q}(\sqrt{-n})$ has no ideal class of exact order 4.*

Then m is a congruent number. Moreover, for the elliptic curve $E^{(m)} : my^2 = x^3 - x$,

$$\mathrm{rank}_{\mathbb{Z}}\, E^{(m)}(\mathbb{Q}) = 1 = \mathrm{ord}_{s=1} L(E^{(m)}, s),$$

and the Shafarevich-Tate group of $E^{(m)}$ is finite and of odd cardinality.

Corollary 1.4. *For any integer $k \geq 1$, there are infinitely many square-free congruent numbers with exactly k odd prime factors in each residue class of 5, 6, 7 modulo 8.*

We now sketch of the proof of Theorem 1.3. Let $\Gamma_0(32)$ be the subgroup of $SL_2(\mathbb{Z})$ consisting of all matrices $\left(\begin{smallmatrix} a & b \\ c & d \end{smallmatrix}\right)$ with $c \equiv 0 \bmod 32$. As usual, we write $X_0(32)$ for the modular curve over $\mathbb{Q}$ of level $\Gamma_0(32)$ so that its complex points are given as

$$X_0(32)(\mathbb{C}) = \Gamma_0(32)\backslash \left(\mathcal{H} \cup \mathbb{P}^1(\mathbb{Q})\right),$$

where $\mathcal{H}$ is the upper half complex plane, and $\mathbb{P}^1(\mathbb{Q})$ is the projective line over $\mathbb{Q}$. This curve has genus 1, and the cusp at infinity, which we denote by ∞, is also rational. In fact, it is easy to see that the elliptic curve $(X_0(32), \infty)$ is isomorphic over $\mathbb{Q}$ to the elliptic curve $E' : y^2 = x^3 + 4x$. Let E be the elliptic curve $y^2 = x^3 - x$. We can therefore fix an isogeny $f : X_0(32) \to E$ of degree 2, defined over $\mathbb{Q}$ and mapping ∞ to the zero element of E.

Let $n = p_0 p_1 \cdots p_k$ be a product of distinct odd primes with $p_1, \ldots, p_k \equiv 1 \bmod 8$. Let $m = n$ or $2n$ such that $m \equiv 5, 6$ or $7 \bmod 8$. Note that the condition (2) is equivalent to that the 4-rank of ideal class group $\mathcal{A}$ of $K = \mathbb{Q}(\sqrt{-2n})$ is zero if $n \equiv \pm 3 \bmod 8$ and 1 otherwise. Let H be the Hilbert class field of the imaginary quadratic field K. Let $m^* = (-1)^{(n-1)/2} m$ so that $K(\sqrt{m^*}) \subset H$ and let $E(\mathbb{Q}(\sqrt{m^*}))^-$ be the group of points $x \in E(\mathbb{Q}(\sqrt{m^*}))$ such that $\sigma x = -x$ where σ is the non-trivial element in the Galois group $\mathrm{Gal}(\mathbb{Q}(\sqrt{m^*})/\mathbb{Q})$. Then $E(\mathbb{Q}(\sqrt{m^*}))^- \cong E^{(m)}(\mathbb{Q})$, whose torsion subgroup is $E[2]$. We now construct so called Heegner point $y_m \in E(\mathbb{Q}(\sqrt{m^*}))^- \cong E^{(m)}(\mathbb{Q})$, which will be shown to be of infinite order for m satisfying the condition (2) in Theorem 1.3.

1. If $n \equiv 5 \bmod 8$, then $m = m^* = n$. Take $P \in X_0(32)$ be the image of $i\sqrt{2n}/8$ on the upper half plane $\mathcal{H}$ via the complex uniformzation $X_0(32) = \Gamma_0(32)\backslash(\mathcal{H} \cup \mathbb{P}^1(\mathbb{Q}))$. It turns out that the point $z := f(P) + (1 + \sqrt{2}, 2 + \sqrt{2})$ on E is defined over the Hilbert class field H, and that

$$y_n := \mathrm{Tr}_{H/K(\sqrt{n})} z \in E(\mathbb{Q}(\sqrt{n})).$$

   Moreover, y_n (resp. $2y_n$) belongs to $E(\mathbb{Q}(\sqrt{n}))^-$ if $k \geq 1$ (resp. $k = 0$).

2. If $n \equiv 3, 7 \bmod 8$, then $m^* = -m$. Let $P \in X_0(32)$ be the image of $(i\sqrt{2n} + 2)/8$ via the complex uniformization and let $z := f(P) + (1 + \sqrt{2}, 2 + \sqrt{2})$ which turns out to be a point defined over $H(i)$. Identify $\mathcal{A}$ with $G = \mathrm{Gal}(H(i)/K(i))$ and let $\sigma_0 \in \mathrm{Gal}(H(i)/K(i))$ be the element corresponding to the ideal class of $(2, \sqrt{-2n})$. Let χ be the character of G factor through $\mathrm{Gal}(K(i, \sqrt{m^*})/K(i))$ which is non-trivial if $m \neq 2n$. Let $\phi \subset \mathcal{A}$ be a complete representative of $G/\langle\sigma_0\rangle$ and define

$$y_m := y_{m,\phi} = \sum_{\sigma \in \phi} \chi(\sigma) z^\sigma.$$

Our main result is the following theorem, from which Theorem 1.3 follows by the Gross-Zagier formula and a Kolyvagin's result,

Theorem 1.5. *Let $n = p_0 p_1 \cdots p_k$ be a product of distinct odd primes with $p_i \equiv 1 \bmod 8$ for $1 \leq i \leq k$. Let $m = n$ or $2n$ such that $m \equiv 5, 6$ or $7 \bmod 8$. Let y_m*

be the point defined as above. Then we have that $y_m \in 2^{k-1}E(\mathbb{Q}(\sqrt{m^*}))^- + E[2]$. *Moreover, if the condition* (2) *in Theorem* 1.3 *is satisfied, then the point* $y_m \notin 2^k E(\mathbb{Q}(\sqrt{m^*}))^- + E[2]$.

We now explain our method in the case $p_0 \equiv 5 \bmod 8$ (other cases are similar). In this case, $m^* = m = n$ in Theorem 1.5 and the condition (2) says that the ideal class group $\mathcal{A}$ of $K = \mathbb{Q}(\sqrt{-2n})$ satisfies that $2\mathcal{A}$ has odd cardinality. Here are some notations:

- fix an embedding of the algebraic closure $\bar{K}$ of K in $\mathbb{C}$;
- H: the Hilbert class field of K; $H_0 = K(\sqrt{p_0}, \ldots, \sqrt{p_k})$ its genus subfield; and $H_0^+ = H_0 \cap \mathbb{R}$;
- K^{ab} – the maximal abelian extension of K;
- $\widehat{\mathbb{Z}} = \prod_p \mathbb{Z}_p$; $\widehat{K} = K \otimes_{\mathbb{Z}} \widehat{\mathbb{Z}}$ the finite adéles of K; and $\widehat{\mathcal{O}}_K = \mathcal{O}_K \otimes \widehat{\mathbb{Z}}$;
- $\sigma: \widehat{K}^\times / K^\times \longrightarrow \mathrm{Gal}(K^{ab}/K)$, the Artin reciprocity law.
- K_2: the local field of K at the unique prime above 2, $\mathcal{O}_2$ its ring of integers, and $\varpi = \sqrt{-2n} \in K_2^\times$ its uniformizor. The unit group $\mathcal{O}_2^\times$ is generated by $\pm 1, 5, 1+\varpi$ as a $\mathbb{Z}_2$-module.

A key fact is the following result on the Galois conjugation of the point $z = f(P) + (1+\sqrt{2}, 2+\sqrt{2})$ on the elliptic curve E: $y^2 = x^3 - x$.

Theorem 1.6. *The Galois conjugations* $z_t := z^{\sigma_t}$ *of* z, $t \in \widehat{K}^\times$, *satisfy the following properties:*

1. *the point* $z_t \in E(H)$ *for any* $t \in \widehat{K}^\times$ *and only depends on the class of* t *modulo* $K^\times \widehat{\mathcal{O}}_K^\times$;
2. *the complex conjugation of* z_t, *denoted by* $\bar{z}_t$, *is equal to* $z_{t^{-1}}$; *and*
3. $z_{\varpi t} + z_t = 0$ *or* $(0,0)$ *according to* $n \equiv 1 \bmod 8$ *or* $\equiv 5 \bmod 8$.

Let $\mathrm{PGL}_2(\mathbb{Q})_+$ denote the subgroup of $\mathrm{PGL}_2(\mathbb{Q})$ of elements with positive determinants. Let N be the normalizer of $\Gamma_0(32)$ in $\mathrm{PGL}_2(\mathbb{Q})_+$, which acts naturally on $X_0(32)$. One can show that translations by $\mathbb{Q}(i)$-torsion points of $E' = (X_0(32), \infty)$ can be described explicitly in the term of elements in N. An equivalent result on $X_0(32)$ of Theorem 1.6 follows from this explicit description and Shimura's reciprocity law.

First, consider the case when $n = p_0$ is a prime congruent to 5 modulo 8. Now $H_0 = K(\sqrt{p_0})$ is the genus subfield of H and $2\mathcal{A} \cong \mathrm{Gal}(H/H_0)$. By Gauss' genus theory, the condition (2) automatically holds, i.e. $\mathcal{A}$ has no order 4 elements, or equivalently, $2\mathcal{A}$ is of odd cardinality. By Theorem 1.6,

$$y_{p_0} + y_{p_0}^{\sigma_\varpi} = \#2\mathcal{A} \cdot (0,0) = (0,0),$$

and then $2y_{p_0} \in E(\mathbb{Q}(\sqrt{p_0}))^-$ but $y_{p_0} \notin E(\mathbb{Q}(\sqrt{p_0}))^-$ since $\sigma_\varpi(\sqrt{p_0}) = -\sqrt{p_0}$. We claim that $2y_{p_0} \notin 2E(\mathbb{Q}(\sqrt{p_0}))^- + E[2]$ and thus $2y_{p_0}$ is not torsion. Otherwise, assume $2y_{p_0} = 2y' + t$ for some $y' \in E(\mathbb{Q}(\sqrt{p_0}))^-$ and $t \in E[2]$, then $y_{p_0} - y' \in E[4] \cap E(H_0)$. Considering the ramification, we know that $y_{p_0} - y' \in E[2]$ and therefore $y_{p_0} \in E(\mathbb{Q}(\sqrt{p_0}))^-$, a contradiction. We have shown Theorem 1.5 in the case $k = 0$, which is due to Heegner and Monsky.

Our method to show Theorem 1.5 is to induct on k with the above initial case $k = 0$. Now assume $k \geq 1$. Let $y_0 = \mathrm{Tr}_{H/H_0} z \in E(H_0^+)$. For any positive divisor d of n divisible by p_0, let $y_d = \mathrm{Tr}_{H/K(\sqrt{d})} z$, which actually belongs to $E(\mathbb{Q}(\sqrt{d}))^-$. These points y_d and y_0 are related by the following relation:

$$2^k y_0 - \sum_{p_0|d|n} y_d \in E[2]. \tag{1.1}$$

We claim that for each proper divisor d of n divisible by p_0, $y_d \in 2^k E(\mathbb{Q}(\sqrt{d}))^- + E[2]$ (and we will show this key result at the end using induction hypothesis). Assume this, we can write $y_d = 2^k y'_d + t'_d$ for some $y'_d \in E(\mathbb{Q}(\sqrt{d}))^-$ and $t_d \in E[2]$. Thus

$$y_n = 2^k \Big(y_0 - \sum_{p_0|d|n, d \neq n} y'_d \Big) + t$$

for some $t \in E[2]$. Consider the (injective) Kummer map

$$E(\mathbb{Q}(\sqrt{n}))/2^{k+1}E(\mathbb{Q}(\sqrt{n})) \longrightarrow H^1(\mathbb{Q}(\sqrt{n}), E[2^{k+1}]).$$

and the exact inflation-restriction sequence

$$\begin{aligned} 1 &\to H^1(\mathrm{Gal}(H_0^+/\mathbb{Q}(\sqrt{n})), E[2]) \\ &\to H^1(\mathbb{Q}(\sqrt{n}), E[2^{k+1}]) \to H^1(H_0^+, E[2^{k+1}]). \end{aligned}$$

Here we use $E[2^{k+1}](H_0^+) = E[2]$. Since $y_0 - \sum_{p_0|d|n, d\neq n} y'_d \in E(H_0^+)$, we know that the image of $2y_n$ in the Kummer map factors through $H^1(\mathrm{Gal}(H_0^+/\mathbb{Q}(\sqrt{n}))$, $E[2])$ and then is killed by 2, thus $4y_n \in 2^{k+1}E(\mathbb{Q}(\sqrt{n}))$, or $4(y_n - 2^{k-1}y'_n) = 0$ for some $y'_n \in E(\mathbb{Q}(\sqrt{n}))$. Note any non-trivial sub-extension of $H_0^+/\mathbb{Q}(\sqrt{n})$ must be ramified at some odd prime and $E[2^\infty](\mathbb{Q}(\sqrt{n})) = E[2]$, we know that $y_n = 2^{k-1}y'_n$ modulo $E[2]$. On the other hand, $y_n = 2^k \left(y_0 - \sum_{p_0|d|n, d\neq n} y'_d \right)$ mod $E[2]$, thus

$$2^{k-1}\Big(y'_n - 2y_0 + \sum_{d \neq n} 2y'_d \Big) \in E[2],$$

which implies that $y'_n = 2y_0 - \sum_{d\neq n} 2y'_d + t$ for some $t \in E[2]$. Note that for any $0 < d|n$ with $p_0|d$, we have that $\sigma_\varpi(\sqrt{d}) = -\sqrt{d}$ and therefore $y_d'^{\sigma_\varpi} = -y'_d$. Thus by Theorem 1.6 again, $y_n'^{\sigma_\varpi} + y'_n = 2(y_0^{\sigma_\varpi} + y_0) = 0$, i.e. $y'_n \in E(\mathbb{Q}(\sqrt{n}))^-$. This shows that $y_n \in 2^{k-1}E(\mathbb{Q}(\sqrt{n}))^- + E[2]$.

Now assume the condition (2) in Theorem 1.3, i.e. that the ideal class group $\mathcal{A}$ of $K = \mathbb{Q}(\sqrt{-2n})$ satisfies $2\mathcal{A}$ is of odd cardinality. It follows that $y_0 + y_0^{\sigma_\varpi} = (0,0) \in E[2]$. On the other hand, suppose that $y_n = 2^k y'_n + t_n$ for some $y'_n \in E(\mathbb{Q}(\sqrt{n}))^-$ and $t_n \in E[2]$, then by the relation (1.1) we have that

$$2^k \Big(y_0 - \sum_{p_0|d|n} y'_d \Big) \in E[2], \qquad \text{with } y_0 - \sum_{p_0|d|n} y'_d \in E(H_0^+).$$

Again, any non-trivial sub-extension of $H_0^+/\mathbb{Q}$ must be ramified at some odd prime, we have that $y_0 - \sum_{p_0|d|n} y'_d = t$ for some $t \in E[2]$. Thus

$$y_0 + y_0^{\sigma_\varpi} = \sum_{p_0|d|n} (y'_d + y_d'^{\sigma_\varpi}) + (t + t^{\sigma_\varpi}) = 0.$$

It is contradict to $y_0 + y_0^{\sigma_\varpi} = (0, 0)$.

Using induction hypothesis, we now show the above claim that for a proper divisor d of n divisible by p_0, $y_d \in 2^k E(\mathbb{Q}(\sqrt{d}))^- + E[2]$. Fix such a divisor d, let χ be the non-trivial character of $\mathrm{Gal}(H/K)$ factoring through $\mathrm{Gal}(K(\sqrt{d})/K)$ and define

$$P^\chi(f) = \sum_{\sigma \in \mathrm{Gal}(H/K)} f(P)^\sigma \chi(\sigma) \in E(\mathbb{Q}(\sqrt{d}))^-,$$

which is equal to $2y_d$ above. Here we use Theorem 1.6 to see that $P^\chi(f) \in E(\mathbb{Q}(\sqrt{d}))^-$. Now replacing K by $K_0 = \mathbb{Q}(\sqrt{-2d})$, P by $P_0 \in X_0(32)$ corresponding to $i\sqrt{2d}/8 \in \mathcal{H}$, and χ by corresponding χ_0, one can similarly constrict a point $P_0^{\chi_0}(f) \in E(\mathbb{Q}(\sqrt{d}))^-$. Thus we obtain two points $P^\chi(f)$, $P_0^{\chi_0}(f)$ in the group $E(\mathbb{Q}(\sqrt{d}))^-$. Our claim is derived immediately from the following comparison on 2-divisibilities of these two points.

Theorem 1.7. *Let $n = p_0p_1\cdots p_k$ and $d|n$ be integers as above. Assume that $d = p_0p_{i_1}\cdots p_{i_{k-s}}$ is a proper divisor of n, i.e. $s > 0$, and that $P_0^{\chi_0}(f)$ belongs to $2^tE(\mathbb{Q}(\sqrt{d}))^- + E[2]$ for some integer $t \geq 0$. Then*

$$P^\chi(f) \in 2^{t+s+1}E(\mathbb{Q}(\sqrt{d}))^- + E[2].$$

To prove Theorem 1.7, we combine the following three results:

(1) If either $P_0^{\chi_0}(f)$ or $P_\chi(f)$ is not torsion, then $E(\mathbb{Q}(\sqrt{d}))^- \otimes_\mathbb{Z} \mathbb{Q}$ is one dimensional $\mathbb{Q}$-vector space. In this case, the ratio $[P^\chi(f) : P_0^{\chi_0}(f)] \in \mathbb{Q} \cup \{\infty\}$ of these two points in the one dimensional space is given by

$$[P^\chi(f) : P_0^{\chi_0}(f)]^2 = [\widehat{h}(P^\chi(f)) : \widehat{h}(P_0^{\chi_0}(f))],$$

where $\widehat{h} : E(\bar{\mathbb{Q}}) \to \mathbb{R}$ denotes the Néron-Tate height function. This result follows from a theorem of Kolyvagin on Euler system.

(2) It is known that the algebraic part of L-values $L(E^{(m)}, 1)$, defined by

$$L^{\mathrm{alg}}(E^{(m)}, 1) := L(E^{(m)}, 1)/\Omega^{(m)},$$

is a rational number, where for any square-free positive integer m, we denote by $\Omega^{(m)} := \frac{2}{\sqrt{m}} \int_1^\infty \frac{dx}{\sqrt{x^3 - x}}$ the real period of the elliptic curve $E^{(m)}$. Note that $L^{\mathrm{alg}}(E^{(2)}, 1) \neq 0$ and in fact, its 2-adic valuation is -2. (see [2], p. 87). Then the following equality holds:

$$\widehat{h}(P^\chi(f)) = \frac{L^{\mathrm{alg}}(E^{(2n/d)}, 1)}{L^{\mathrm{alg}}(E^{(2)}, 1)} \cdot \widehat{h}(P_0^{\chi_0}(f)),$$

This result follows from Gross-Zagier formula of Yuan-Zhang-Zhang (Theorem 1.2 in [28]).

(3) Let $s \geq 1$ be an integer and let $m = 2p_1 \cdots p_s$ be a product of distinct primes $p_i \equiv 1 \bmod 8$. Then the 2-adic valuation of $L^{\mathrm{alg}}(E^{(m)}, 1)$ is not less than $2s-1$. This result is Zhao's estimation [29].

Now the proof of Theorem 1.5, in the case $n \equiv 5 \bmod 8$, is complete. We next discuss the generalizations of our method in the remain two sections of the paper.

2 General cases and explicit formulas

The induction method in last section can be applied to prove much more cases of non-triviality of Heegner points. Basically, this method works when the algebraic-part of special values of derivative of L-functions of $E^{(m)}$ has minimal 2-adic valuation. To make this generalization, we use Waldspurger's special value formula to obtained estimation on 2-adic valuation of special L-values in the step (3) in the proof of Theorem 1.7. We also obtained the explicit Gross-Zagier from the version of Yuan-Zhang-Zhang in [28] so that we may state our results on Heegner points in term of special values of derivative of L-functions. The results in this section are our joint work with Yuan and Zhang in [28].

Definition 2.1. *For any elliptic curve E over $\mathbb{Q}$, let $\epsilon(E)$ be its sign and define $\mathscr{L}(E) \in \mathbb{R}_{\geq 0}$ by*

$$\prod_{v \leq \infty} \Omega_v(E) \cdot \frac{R(E)}{|E(\mathbb{Q})_{\mathrm{tor}}|^2} \cdot \mathscr{L}(E)^2 = \begin{cases} L(E,1), & \text{if } \epsilon(E) = 1, \\ L'(E,1), & \text{if } \epsilon(E) = -1, \end{cases}$$

where $\Omega_v(E)$ is the local Tamagawa number of E at v and $R(E)$ is the regulator of E. We write $\mathscr{L}(n)$ for $\mathscr{L}(E^{(n)})$, and $\epsilon(n)$ for $\epsilon(E^{(n)})$, respectively, for the elliptic curve $E^{(n)} : ny^2 = x^3 - x$.

By a theorem of Coates-Wiles, if $n \geq 1$ is an integer with $\epsilon(n) = 1$, then n is a non-congruent number provided $\mathscr{L}(n) \neq 0$; by Gross-Zagier formula, if $n \geq 1$ is an integer with $\epsilon(n) = -1$, then n is a congruent number provided $\mathscr{L}(n) \neq 0$. We know that $\mathscr{L}(n)$ is always an integer and we will discuss its parity in term of genus class groups of imaginary quadratic fields.

Definition 2.2. *For integer $n \geq 1$, define $g(n)$ to be an element in $\mathbb{F}_2$ by that $g(n) = 1$ if and only if $\mathbb{Q}(\sqrt{-n})$ has no order 4 ideal class.*

Gauss' genus theory implies that $g(n)$ is easily determined by Legendre symbols among prime factors of n, -1, and 2. For example, any positive integer $n \equiv 1 \bmod 8$ has $g(n) = 0$ if either all of its prime factors $\equiv 1 \bmod 4$ or all of its prime factors $\equiv \pm 1 \bmod 8$.

Theorem 2.3. *Let $n > 1$ be square-free with $\epsilon(n) = 1$. Then $\mathscr{L}(n) \in \mathbb{Z}$, and moreover, $\mathscr{L}(n)$ is odd if and only if*

$$\sum_{\substack{n = d_1 \cdots d_r \\ d_i \equiv 1(8),\ i > 1}} \prod_i g(d_i) = 1,$$

where the sum runs over non-ordered decompositions $n = d_1 \cdots d_r$ with all $d_i > 1$.

Remark. 1. If a positive integer $n \equiv 1 \bmod 8$ with all prime factors $\equiv 1 \bmod 4$ or all prime factors $\equiv \pm 1 \bmod 8$, then $\mathscr{L}(n)$ must be even.

2. It is expected that among all positive square-free odd integers n with $\epsilon(n) = 1$, the density of ones with $2 \nmid \mathscr{L}(n)$ is

$$\prod_{k=1}^{\infty}(1+2^{-k})^{-1} = 0.4194\cdots,$$

in sprit of Heath-Brown's result and the following special case of BSD conjecture.

Conjecture 2.4. *Let n be a positive square-free integer with $\epsilon(n) = 1$, then the following are equivalent:*

1. *$\mathscr{L}(n)$ is odd, i.e.* $\displaystyle\sum_{\substack{n=d_1\cdots d_r,\\ d_i\equiv 1(8),\ i>1}} \prod_i g(d_i) = 1.$

2. $\dim_{\mathbb{F}_2} \mathrm{Sel}_2(E^{(n)})/E^{(n)}[2] = 0.$

We now discuss the case with $\epsilon(n) = -1$. Let $A^{(n)}$ denote the elliptic curve $ny^2 = x^3 + 4x$ and $f : A^{(n)} \to E^{(n)}$ a degree 2-isogeny over $\mathbb{Q}$. Let ρ be the non-negative integer such that

$$2^{\rho(n)} := [E^{(n)}(\mathbb{Q}) : f(A^{(n)}(\mathbb{Q})) + E^{(n)}[2]].$$

Theorem 2.5. *Let $n > 1$ be a square-free odd integer with $\epsilon(n) = -1$. Then $2^{-\rho(n)}\mathscr{L}(n) \in \mathbb{Z}$. Moreover $2^{-\rho(n)}\mathscr{L}(n)$ is even only if*

$$\sum_{\substack{n=d_0\cdots d_r\\ d_i\equiv 1(8),\ i>0}} \prod_i g(d_i) = \sum_{\substack{n=d_0\cdots d_r,\\ \epsilon(d_0)=-1,\epsilon(d_1)=1,\ d_i\equiv 1(8),\ i>1}} \prod_i g(d_i) = 0,$$

where the sums run over non-ordered decompositions $n = d_0 \cdots d_r$ with all $d_i > 1$.

Corollary 2.6. *Let n be a positive integer such that $\mathbb{Q}(\sqrt{-n})$ has no ideal classes of exact order* 4.

1. *If $n \equiv 5 \bmod 8$ with all prime factors $\equiv 1 \bmod 4$, then n is a congruent number.*

2. *If $n \equiv 7 \bmod 8$ with prime factors $\equiv \pm 1 \bmod 8$, then both n and $2n$ are congruent numbers.*

It is expected among all positive odd square-free n with $\epsilon(n) = -1$, the ones satisfying the genus class number condition in Theorem 2.5 has density more than 60%, in spite of the work of Heath-Brown and the following conjecture.

To obtain Theorem 2.3 and Theorem 2.5, beside a refinement of induction argument in first section, we also use the explicit forms of Waldspurger's formula and Gross-Zagier formula with respect to primitive integral Gross-Prasad test vectors.

Let F be a number field and $\mathbb{A}$ its adeles. Let $|\ | : \mathbb{A}^\times \to \mathbb{R}^\times$ be the normalized absolute valuation. Let B be a quaternion algebra over F and π an irreducible unitary representation of $B_{\mathbb{A}}^\times$. Let K/F be a quadratic extension of number fields and $\chi : K^\times \backslash \mathbb{A}_K^\times \to \mathbb{C}^\times$. Assume that

- $\omega_\pi \cdot \chi|_{\mathbb{A}^\times} = 1$, where ω_π is the central character of π;
- for each place v of F, $\epsilon(1/2, \pi_v, \chi_v) \cdot \eta_v \chi_v(-1) = \epsilon(B_v)$, where $\eta = \otimes \eta_v$ denotes the quadratic Hecke character of $\mathbb{A}^\times$ associated to the extension K/F, and $\epsilon(B_v) = 1$ or -1 if B is split or not over F_v, respectively.

Thus the root number of the Rankin-Selberg L-series $L(s, \pi, \chi)$ is 1. Fix an F-embedding K into B, via which $T = K^\times$ is a sub-tours of G. We give an explicit form of Waldspurger's period formula under the following assumptions:

1. F is totally real and K is totally imaginary quadratic over F;
2. for each place $v \nmid \infty$, either π_v or χ_v is unramified;
3. for any $v|\infty$, π_v^{JL} is a discrete series of weight k_v on $\mathrm{GL}_2(\mathbb{R})$ with $k_v \geq 2$ even integer.

Let $N, c, D, d \in \mathbb{A}_f^\times$ such that for any finite place v of F, N_v generates the conductor of π_v, $c_v \in \mathcal{O}_v$ is of minimal valuation such that χ_v is trivial on $(\mathcal{O}_v + c_v\mathcal{O}_{K_v})^\times$, D_v generates the relative discriminant of K/F, and d_v generates the different of F_v. For each $v \nmid \infty$, let R_v be an order of $B(F_v)$ with discriminant $N_v\mathcal{O}_v$ such that $R_v \cap K_v = \mathcal{O}_{c_v} := \mathcal{O}_v + c_v\mathcal{O}_{K_v}$. Such order exists and unique up to conjugacy of $K_v^\times$. Recall that a Gross-Prasad test vector is a pure vector $f = \prod_v f_v \in \pi$ for the pair π and χ is chosen as follows.

(i) if $v \nmid \infty$ with $\mathrm{ord}_v(N_v) \leq 1$ or K_v/F_v is unramified, then $\pi_v^{R_v^\times}$ is of dimension one and $f_v \in \pi_v^{R_v^\times}$ is a non-zero vector;

(ii) if $v \nmid \infty$ with $\mathrm{ord}_v(N_v) \geq 2$ and K_v/F_v is ramified, then vectors $f \in \pi_v$ satisfying that $\pi_v(t)f = \chi_v^{-1}(t)f$ for all $t \in K_v^\times$ form a one dimension space, and a Gross-Prasad test vector f is any non-zero such vector.

(iii) if $v|\infty$, let f_v be such that $\pi_v(t)f_v = \chi_v(t)^{-1}f_v$ for any $t \in K_v^\times$, which is unique up to scalar.

Theorem 2.7. (Explicit Waldspurger Formula) *Assume that the sign of $L(s, \pi, \chi)$ is 1 and the conditions (1)–(3) are satisfied. Let f' be the newform of π^{JL} and f a Gross-Prasad test form. Denote by $(f', f')_{\mathrm{Pet}}$ and $(f, f)_{\mathrm{Pet}}$ their Peterson norms with respect to Tamagawa measures on $Z\backslash\mathrm{GL}_2$ and $Z\backslash G$, respectively. Then*

$$\Big| \sum_{t \in \widehat{K}^\times / K^\times \widehat{F}^\times \widehat{\mathcal{O}}_c^\times} f(t)\chi(t) \Big|^2 \cdot (f, f)_{\mathrm{Pet}}^{-1}$$

$$= \kappa^2 \cdot [\mathcal{O}_c^\times : \mathcal{O}_F^\times]^2 \cdot 2^{[F:\mathbb{Q}]} \cdot \frac{L^{((N,D)\infty)}(1/2, \pi, \chi)}{(f', f')_{\mathrm{Pet}} |Dc^2|^{1/2}} \cdot \prod_{v|\infty} (4\pi)^{-(k_v+1)} \Gamma(k_v)$$

$$\cdot \prod_{v|N, inert} (1 - q_v^{-1})(1 + q_v^{-1})^{-1} \cdot \prod_{v||N, ramified} 2(1 + q_v^{-1})^{-1} \cdot \prod_{v^2|N, ramified} 2(1 - q_v^{-1}),$$

where $\kappa = 1$ or 2 is the order of the kernel of the natural morphism from the ideal class group of F to that of K, (N, D) is the set of places v divides both N and D.

Let F be a totally real field and $\mathbb{A}$ its adéles . Let X be a Shimura curve over F associated to a incoherent quaternion algebra $\mathbb{B}$ over $\mathbb{A}$ with ramification set Σ (containing all infinite places), ξ a Hodge bundle on X, and J its Jacobian. Let A be a simple abelian variety defined over F parameterized by X. Then

$$\pi_A := \mathrm{Hom}^0_\xi(X, A) = \mathrm{Hom}^0(J, A)$$

is a representation $\mathbb{B}^\times$ over $\mathbb{Q}$ whose infinite components are all trivial. It is known that $M := \mathrm{End}_{\mathbb{B}^\times}(\pi_A) = \mathrm{End}^0(A)$ is a number field of degree $\dim A$ over $\mathbb{Q}$. Let $(\cdot,\cdot) : \pi_A \times \pi_{A^\vee} \to M$ be the perfect $\mathbb{B}^\times$-pairing given by $(f_1, f_2) = \mathrm{Vol}(X_U)^{-1}(f_{1,U} \circ f_{2,U}^\vee)$, where $U \subset \mathbb{B}_f^\times$ is an open compact subgroup such that f_i is represented by $f_{i,U}$, $i = 1, 2$, the composition using the canonical isomorphism $J_U^\vee \cong J_U$ and the volume using the measure $dxdy/(2\pi y^2)$ on $\mathcal{H}$. Let

$$\langle \cdot,\cdot \rangle_M : A(\bar{F})_{\mathbb{Q}} \otimes_M A^\vee(\bar{F})_{\mathbb{Q}} \longrightarrow M \otimes_{\mathbb{Q}} \mathbb{R}$$

be the M-bilinear height pairing whose trace to $\mathbb{R}$ is the usual height pairing.

Let K be a totally imaginary quadratic extension with a fixed embedding $K_{\mathbb{A}} \hookrightarrow \mathbb{B}$ over $\mathbb{A}$. Let $\chi : \widehat{K}^\times/K^\times \to L^\times$ be a finite order Hecke character valued in a number field $L \supset M$, and also viewed as a Galois character via class field theory. Assume that

- $\omega_{\pi_A} \cdot \chi|_{\mathbb{A}^\times} = 1$.
- for each place v of F: $\epsilon(1/2, \pi_v, \chi_v)\chi_v\eta_v(-1) = \mathrm{inv}(\mathbb{B}_v)$.

Theorem 2.8. (Explicit Formula of Gross-Zagier) *Assume that π and χ has disjoint ramification. Let $f \in \pi_A$ be a Gross-Prasad test vector as in Theorem 2.7. With the same notations as in Theorem 2.7,*

$$\begin{aligned}&(f,f)^{-1} \cdot \widehat{h}\left(\sum_{t \in \widehat{K}^\times/\widehat{F}^\times K^\times \widehat{\mathcal{O}}_c^\times} f(P)^t \chi(t)\right)\\
&= \kappa^2 \cdot [\mathcal{O}_c^\times : \mathcal{O}_F^\times]^2 \cdot 2^{[F:\mathbb{Q}]+1} \cdot \frac{L'^{((N,D)\infty)}(1/2, \pi, \chi)}{(f', f')_{\mathrm{Pet}}|Dc^2|^{1/2}(4\pi)^{3[F:\mathbb{Q}]}}\\
&\cdot \prod_{v|N, inert} (1-q_v^{-1})(1+q_v^{-1})^{-1} \cdot \prod_{v||N, ramified} 2(1+q_v^{-1})^{-1} \cdot \prod_{v^2|N, ramified} 2(1-q_v^{-1}).\end{aligned}$$

The following theorem is in the special case where the primitive Gross-Prasad test vector is the classical modular parametrization of elliptic curves over $\mathbb{Q}$ by $\Gamma_0(N)$-level modular curve. It generalizes classical Gross-Zagier formula in [11], and will be used in Section 3.

Theorem 2.9. *Let E be an elliptic curve over $\mathbb{Q}$ of conductor N with associated newform $\phi = \sum_n a_n q^n, a_1 = 1$, and let $f : X_0(N) \to E$ be a modular parametrization. Let K be an imaginary quadratic field of discriminant D_K and χ a ring class character over K with conductor c prime to N. Assume that*

1. *the Heegner hypothesis: any prime $p|N$ is not inert over K and must be split over K if $p^2|N$.*

2. *for any prime $p|(N, D_K)$, let ϖ be a uniformizor of K_p above p, then $\chi(\varpi) \neq a_p$.*

Let x denote the CM point on $X_0(N)$ represented by the N-siogeny $(\mathbb{C}/\mathcal{O} \to \mathbb{C}/\mathfrak{n})$ of elliptic curves, where $\mathcal{O}$ is the order of K of conductor c and $\mathfrak{n}$ is a proper ideal of $\mathcal{O}$ such that $\mathcal{O}/\mathfrak{n} \cong \mathbb{Z}/N\mathbb{Z}$, and let

$$P_\chi(f) := \sum_{\sigma \in \mathrm{Gal}(H_c/K)} f(x)^\sigma \chi(\sigma) \in E(H_c)_{\mathbb{C}}.$$

Then

$$L'(1, E, \chi) = \frac{8\pi^2 (\phi, \phi)_{\Gamma_0(N)}}{u^2 \sqrt{|D_K c^2|}} \cdot \frac{\widehat{h}_K(P_\chi(f))}{\deg f} \cdot \prod_{p \,|(N, D_K)} 2^{-1},$$

where $\widehat{h}_K$ is the Néron -Tate height on E over K, $u = [\mathcal{O}_c^\times : \mathbb{Z}^\times]$, and the Peterson norm

$$(\phi, \phi)_{\Gamma_0(N)} = \iint_{\Gamma_0(N)\backslash\mathcal{H}} |\phi(z)|^2 dx dy, \qquad z = x + iy.$$

3 Quadratic twists of elliptic curves

After Heegner's celebrated work on congruent number problem, Birch was the first to recognize the significance of the divisor classes supported on certain CM points in the arithmetic of the Jacobin $J_0(N)$ of the modular curve $X_0(N)$. Using them, Birch was able to construct many quadratic twists of elliptic curves over $\mathbb{Q}$ by prime discriminant with infinite order rational points when they possesses a cuspidal group of even order, see below Lemma 3.1. Using our induction method in first section, together with technique of Kolyvagin, we may extend Birch's work to quadratic twists of elliptic curves by discriminants with many prime factors. The results in this section are our joint work with Coates, Li, and Zhai in [5].

Let E be an elliptic curve over $\mathbb{Q}$ of conductor N. Let $K \neq \mathbb{Q}(i), \mathbb{Q}(\sqrt{-3})$ be an imaginary quadratic field, which satisfies the Heegner hypothesis for E: any prime factor of N splits in K. Then there exists $\mathcal{N} \subset \mathcal{O}_K$ such that $\mathcal{O}_K/\mathcal{N} \cong \mathbb{Z}/N\mathbb{Z}$. For each positive integer m with $(m, N) = 1$, let $\mathcal{O}_m = \mathbb{Z} + m\mathcal{O}_K$ be the order of K of conductor m. Let $\mathcal{N}_m = \mathcal{N} \cap \mathcal{O}_m$, the point

$$x_m := (\mathbb{C}/\mathcal{O}_m \to \mathbb{C}/\mathcal{N}_m^{-1}) \in X_0(N)$$

is defined over the ring class field H_m, which is the abelian extension over K characterized by $\widehat{K}^\times / K^\times \widehat{\mathcal{O}}_m^\times \xrightarrow{\sim} \mathrm{Gal}(H_m/K)$ via the reciprocity law map in the class field theory. Let $f : X_0(N) \to E$ be a modular parametrization mapping $[\infty]$ to O and let $y_m = f(x_m) \in E(H_m)$. (Note that the notation y_m in this section is different from what we used in Section 1!) The Euler system of Heegner points y_m has the following properties:

(i) Let ℓ be a prime inert in K and prime to m and $(\ell m, N) = 1$, then

$$\mathrm{Tr}_{H_{m\ell}/H_m} y_{m\ell} = a_\ell y_m.$$

(ii) Let τ be the complex conjugation and w_N the Fricker involution. Then the following equality of two sets holds:

$$w_N(\mathrm{Gal}(H_m/K) \cdot y_m) = \tau(\mathrm{Gal}(H_m/K) \cdot y_m).$$

Lemma 3.1. (Birch) *Let E be an elliptic curve defined over $\mathbb{Q}$ and $f : X_0(N) \to E$ a modular parametrization mapping $[\infty]$ to $O \in E$ and let $\epsilon = \pm 1$ be the eigenvalue of Fricke involution w_N on the eigenform $\sum a_n q^n$. Assume that*

1. *E has no order 4 $\mathbb{Q}$-rational torsion point,*
2. *the cusp torsion point $T := f([0])$ has even order.*

Let $K = \mathbb{Q}(\sqrt{-p_0})$ be an imaginary quadratic field, where $p_0 \equiv 3 \bmod 4$ is a prime, satisfying the Heegner hypothesis for E. Then the Heegner point

$$y_K := \mathrm{Tr}_{H_1/K} y_1 \in E(K)$$

satisfies $2y_K \in E(K)^\epsilon$, but $y_K \notin E(K)^\epsilon + E(K)_{\mathrm{tor}}$.

Corollary 3.2. *Let E be an elliptic curve over $\mathbb{Q}$ of sign -1 satisfying the conditions (1) and (2) in Lemma 3.1. Then $\operatorname{rank} E(\mathbb{Q}) = \operatorname{ord}_{s=1} L(E, s) = 1$.*

Proof. Note that $f \circ w_N - \epsilon f$ is a constant morphism and then for all $P \in X_0(N)$,

$$f(P^{w_N}) - f(P) = f([\infty]^{w_N}) - f([\infty]) = f([0]) := T.$$

Since the ideal class number of K is odd, $\overline{y}_K - \epsilon y_K = T$. It follows that $2y_K \in E(K)^\epsilon$.

Now suppose that $y_K \in E(K)^\epsilon + E(K)_{\mathrm{tor}}$. Let C be an odd integer killing the odd part of $E(K)_{\mathrm{tor}}$, then $Cy_K = Cy + t$ for some $y \in E(K)^\epsilon$ and $t \in E(K)_{\mathrm{tor}}[2^\infty] = E(\mathbb{Q})[2]$ (Suppose there exists $S \in E(K)[2^\infty] \setminus E(\mathbb{Q})[2]$, then $S \notin E(\mathbb{Q})$ since E has no 4-torsion point, and therefore $\mathbb{Q}(P) = K$. But $\mathbb{Q}(P)$ is unramified outside $2N$, and K is ramified at p_0, a contradiction.) It follows that

$$C(\overline{y}_K - \epsilon y_K) = 0,$$

i.e. $\overline{y}_K - \epsilon y_K$ is of odd order, which is contradict to $\overline{y}_K - \epsilon y_K = T$. □

Theorem 3.3. *Let E be an elliptic curve over $\mathbb{Q}$ of sign $+1$ satisfying*

1. *E has no $\mathbb{Q}$-rational torsion point of order 4 and the torsion point $T := f([0])$ has even order,*
2. *there is a supersingular prime q_0 for E with $\left(\frac{-1}{q_0}\right) = \left(\frac{N}{q_0}\right) = 1$;*

The for any $k \geq 1$, there are infinitely many square-free integer n with exactly k prime factors, such that $E^{(n)}(\mathbb{Q})$ is of rank one.

Example. 1. $E = (X_0(14), [\infty])$ and $q_0 = 5$.

2. $E = (X_0(49), [\infty])$ and any prime $q_0 \equiv 1 \bmod 4$ which is inert in $\mathbb{Q}(\sqrt{-7})$.

Remark. 1. For some elliptic curves, such as $y^2 = x^3 - x$, there is no supersingular prime $\equiv 1 \bmod 4$ so that the above result does not contain Theorem 1.3.

2. It is clear that Theorem 3.3 follows from the Lemma 3.1 and the below theorem.

Theorem 3.4. *Let $f : X_0(N) \to E$ and supersingular prime q_0 for E be as in Theorem 3.3. Let K be an imaginary quadratic field of odd ideal class number such that Heegner hypothesis holds for E and that q_0 is inert in K. For each $r \geq 1$, let Σ_r denote the (infinite) set of primes $q \neq q_0$ which satisfy*

- $a_q \equiv 0 \bmod 2^{r+1}$, *$q$ is inert in K, and* $\left(\frac{-1}{q}\right) = \left(\frac{N}{q}\right) = 1$.

For any $q_1, \ldots, q_r \in \Sigma_r$, let $m = q_0 \cdots q_r$ and $n = p_0 m$. Let $x_m = (\mathbb{C}/\mathcal{O}_m \to \mathbb{C}/\mathcal{N}_m^{-1}) \in X_0(N)$ and $y_m = f(x_m)$. Then

$$P_m := \mathrm{Tr}_{H_m/K(\sqrt{m})} y_m \in 2^{r-1} E(\mathbb{Q}(\sqrt{-n}))^- + E(\mathbb{Q})[2],$$

but

$$P_m \notin 2^r E(\mathbb{Q}(\sqrt{-n}))^- + E(\mathbb{Q}(\sqrt{-n}))_{\mathrm{tor}}.$$

In particular, P_m is of infinite order. Here for $r = 0$, the statement is understood as $2P_m \in E(\mathbb{Q}(\sqrt{-n}))^-$ and $P_m \notin E(\mathbb{Q}(\sqrt{-n}))^- + E(\mathbb{Q}(\sqrt{-n}))_{\mathrm{tor}}$.

Proof. By Chebatorev density theorem, there are infinite many primes $q \neq q_0$, such that for the extension $K(\sqrt{N}, E[2^r])$, $\mathrm{Frob}_q = \mathrm{Frob}_{q_0}$ when restricted to $\mathbb{Q}(\sqrt{N}, E[2^r])$ and Frob_q is non-trivial when restricted to K. Σ_r contains such primes q and therefore is infinite with positive density.

Let $H_0 = K(\sqrt{q_0}, \sqrt{q_1}, \ldots, \sqrt{q_r})$, then $[H_m : H_0]$ is odd and $E(H)[2^\infty] = E(\mathbb{Q})[2]$. Let $P_0 := \mathrm{Tr}_{H/H_0} y_m$. Note the facts that $[H_m : H_0]$ is odd, $\left(\frac{N}{q}\right) = 1$ for all $q|m$, and that $f(P^{w_N}) + f(P) = T$. We have that

$$\overline{P}_0 + P_0 = T.$$

For each positive divisor $d|m$, let $P_d = \mathrm{Tr}_{H/K(\sqrt{d})} y_m$ and $P_d^0 = \mathrm{Tr}_{H_d/K(\sqrt{d})} y_d$, then $P_d = (\prod_{q|M/d} a_q) P_d^0$. We have

$$P_m + \sum_{d|m, d \neq m} P_d = 2^{r+1} P_0 + 2^r \sum_{\sigma \neq 1} P_0^\sigma = 2^r P_0 + \mathrm{Tr}_{H_m/K} y_m = 2^r P_0.$$

Here we use the property: $\mathrm{Tr}_q y_{qn} = a_q y_n$ if $(q, n) = 1$ and q is inert in K. Let C be an odd integer killing the odd part of $E(H_0)_{\mathrm{tor}}$.

We first consider the case with $r = 0$, i.e. $m = q_0$, and $P_m = P_0$. In this case, we have that $P_m^\sigma + P_m = 0$ and $\overline{P}_m + P_m = T$. It follows that $2P_m \in E(\mathbb{Q}(\sqrt{-n}))^-$.

Suppose now that $P_m = P + t$ for some $P \in E(\mathbb{Q}(\sqrt{-n}))^-$ and $t \in E(\mathbb{Q}(\sqrt{-n}))_{\text{tor}}$. Then $CP_m = CP + s$ for some $s \in E(\mathbb{Q}(\sqrt{-n}))[2^\infty] = E(\mathbb{Q})[2]$. We now have that $C\overline{P}_m = -CP + s$ and $C(\overline{P}_m + P_m) = 0$, i.e. $\overline{P}_m + P_m$ has odd order, which is contradict to $\overline{P}_m + P_m = T$ has order 2.

Now assume that $r \geq 1$. We have that

$$P_m = 2^r Q, \quad \text{where} \quad Q := \left(P_0 - \sum_{d|m, d \neq m} \frac{\prod_{q|m/d} a_q}{2^r} P_d^0 \right) \in E(H_0).$$

Considering the descent

$$E(K(\sqrt{m}))/2^r E(K(\sqrt{m})) \longrightarrow H^1(K(\sqrt{m}), E[2^r])$$

and the inflation-restriction exact sequence

$$0 \longrightarrow H^1(H_0/K(\sqrt{m}), E(H_0)[2^r]) \longrightarrow H^1(K(\sqrt{m}), E[2^r]) \longrightarrow H^1(H_0, E[2^r]),$$

we see that $2P_m \in 2^r E(K(\sqrt{m}))$ and then

$$P_m = 2^{r-1} P + t, \qquad P = 2Q + s$$

for some $P \in E(K(\sqrt{m}))$ and $t, s \in E(\mathbb{Q})[2]$. We now show that $P \in E(\mathbb{Q}(\sqrt{-n}))^-$.

Let $\sigma \in \mathrm{Gal}(H_0/K)$ maps $\sqrt{q_0} = -\sqrt{q_0}$ and fixes all $\sqrt{q_i}$ for $1 \leq i \leq r$. Note that $P_m^\sigma + P_m = 0$, that $P_d^{0\sigma} + P_d^0 = 0$ if $q_0 | d$, and that $\prod_{q|m/d} a_q = 0$ if $q_0 \nmid d$. Thus we have that $2^r(P_0^\sigma + P_0) = 0$ and it follows that

$$P^\sigma + P = 2(Q^\sigma + Q) = 2(P_0^\sigma + P_0) = 0.$$

The property that $\left(\frac{N}{q_i}\right) = 1$ implies from the relation $f(P^{w_N}) + f(P) = T$ that $\overline{P}_0 + P_0 = T$ and thus

$$\overline{P} + P = 2(\overline{Q} + Q) = 0.$$

It follows that $P \in E(\mathbb{Q}(\sqrt{-n}))^-$ and thus $P_m \in 2^{r-1} E(\mathbb{Q}(\sqrt{-n}))^- + E(\mathbb{Q})[2]$.

Suppose that $P_m = 2^r P' + t$ with $P' \in E(\mathbb{Q}(\sqrt{-n}))^-$ and $t \in E(K(\sqrt{-n}))_{\text{tor}}$. Recall that C is an odd integer killing the odd part of $E(\mathbb{Q}(\sqrt{-n}))_{\text{tor}}$, then we have that

$$C \left(P_0 - P' - \sum_{d|m} \frac{\prod_{q|m/d} a_q}{2^r} P_d^0 \right) \in E(H_0)[2^\infty] = E(\mathbb{Q})[2].$$

It follows that $C(\overline{P}_0 + P_0) = 0$, which is contradict to that $\overline{P}_0 + P_0 = T$. □

Combining the above technique and the induction process using L-values, we may obtain more quadratic twists of elliptic curves with infinite order rational points by constructing non-trivial Heegner points. The elliptic curve $E = (X_0(49), [\infty])$ gives a typical example.

Theorem 3.5. *Let $N = p_0p_1\cdots p_k$ and $M = q_1\cdots q_r$ be products of distinct primes satisfying the following conditions*

1. $p_0 \equiv 3 \bmod 4$ *inert in* $\mathbb{Q}(\sqrt{-7})$,
2. $q_1, \ldots, q_r \equiv 1 \bmod 4$ *inert in both* $\mathbb{Q}(\sqrt{-7})$ *and* $\mathbb{Q}(\sqrt{-p_0})$,
3. *for all* $1 \le i \le k$ *primes* $p_i \equiv 1 \bmod 4$ *completely splits in*
$$\mathbb{Q}(i, \sqrt[4]{-7}, \sqrt{q_1}, \ldots, \sqrt{q_r}),$$
4. *the field* $\mathbb{Q}(\sqrt{-N})$ *has no order* 4 *ideal class.*

Let E be the elliptic curve $(X_0(49), [\infty])$. Then we have that

$$\operatorname{rank} E^{(-NM)}(\mathbb{Q}) = 1 = \operatorname{ord}_{s=1} L(E^{(-NM)}, s)$$

and that $Ш(E^{(-NM)}/\mathbb{Q})$ is finite and of odd order. The 2-part of BSD conjecture holds for $E^{(-NM)}$.

Let $K = \mathbb{Q}(\sqrt{-N})$ and let H_M be the ring class field of K of conductor M. Let $H_0 = \mathbb{Q}(\sqrt{q_1}, \ldots, \sqrt{q_r}, \sqrt{-p_0}, \sqrt{p_1}, \ldots, \sqrt{p_k})$ and it is clear that $[H_M : H_0]$ is of odd cardinality if K has no order 4 ideal class. For any divisor $D \ge 1$ with $p_0|D|N$, we define a quadratic character χ_D over $K(\sqrt{M})$ defining the extension $K(\sqrt{M}, \sqrt{-D})$, and define a point

$$Y_D = \sum_{\sigma \in \mathrm{Gal}(H_M/K(\sqrt{M}))} y_M^\sigma \chi_D(\sigma).$$

We can actually prove the following theorem, from which Theorem 3.5 follows.

Theorem 3.6. *Assume that $r \ge 1$ and $k \ge 0$. Then the point*

$$Y_N \in 2^{k+r-2} E(\mathbb{Q}(\sqrt{-NM}))^- + E(\mathbb{Q}(\sqrt{-NM}))_{\mathrm{tor}}.$$

Moreover if K has no order 4 ideal class, then

$$Y_N \notin 2^{k+r-1} E(\mathbb{Q}(\sqrt{-NM}))^- + E(\mathbb{Q}(\sqrt{-NM}))_{\mathrm{tor}}.$$

Proof. We do incaution on $k \ge 0$. The initial case $k = 0$ is proved by Theorem 3.4. The induction process is similar to the one in the proof of Theorem 1.5. Here the explicit Gross-Zagier formula in Theorem 2.9 with ramified characters is needed in the induction process. □

Acknowledgements. The author thanks Li Cai, John Coates, Yifeng Liu, Xinyi Yuan, Shouwu Zhang, Wei Zhang for many useful discussions and comments. The author thanks John Coates, Benedict Gross, Dihua Jiang, Victor Kolyvagin, Yuan Wang, Tonghai Yang, Lo Yang, Shing-Tung Yau, and Shouwu Zhang for constant encouragement and support.

References

[1] B.J. Birch, *Elliptic curves and modular functions*, Symposia Mathematica, Indam Rome 1968/1969, Vol. **4**, 27–32. London: Academic Press (1970).

[2] B.J. Birch and H.P. Swinnerton-Dyer, *Notes on elliptic curves (II)*, J. Reine Angew. Math, **218** (1965), 79–108.

[3] D. Bump, *Automorphic Forms and Representations*, Cambridge Studies in Advanced Mathematics **55**, 1998.

[4] J. Coates and A. Wiles, *On the conjecture of Birch and Swinnerton-Dyer*, Invent. Math. **39** (1977), 233–251.

[5] J. Coates, Y. Li, Y. Tian, and S. Zhai, *Quadratic twists of elliptic curves*, preprint.

[6] A. Diaconu and Y. Tian, *Twisted Fermat curves over toatlly real fields*, Annals of Math. **162** (2005).

[7] L. E. Dickson, *History of the Theory of Numbers*, Volume II. Carnegie Institute, Washington DC (1920), reprinted by Chelsea (1966).

[8] K. Feng, *Non-congruent numbers, odd graphs and the Birch-Swinnerton-Dyer conjecture*, Acta Arithmetica, LXXV. **1** (1996).

[9] D. Goldfeld, *Conjectures on elliptic curves over quadratic fields*, in Number theory, Carbondale 1979 (ed. M. B. Nathanson), Lecture Notes in Math. **751**, Springer, Berlin, 1979, 108–118.

[10] B. Gross, *Kolyvagin's work on modular elliptic curves*, in L-function and Arithmetic (ed. J. Coates and M.J.Talyor), Cambridge University Press, 1991.

[11] B. Gross and D. Zagier, *Heegner points and derivatives of L-series*, Invent. Math. **84** (1986), no. 2, 225–320.

[12] K. Heegner, *Diophantische analysis und modulfunktionen*, Math. Z. **56** (1952), 227–253.

[13] N. M. Katz and P. Sarnak, *Zeros of zeta functions and symmetry*, Bull. Amer. Math. Soc. (N.S.) **36** (1999), no. 1, 1–26.

[14] V.A.Kolyvagain, *Euler System*, The Grothendieck Festschrift. Prog. in ath., Boston, Birkhauser, 1990.

[15] V.A.Kolyvagain, *Finiteness of $E(\mathbb{Q})$ and $Ш(E,\mathbb{Q})$ for a subclass of Weil curves*, Math.USSR Izvestiya **32** (1989), no. 3.

[16] S. Lang, *Elliptic Functions*. GTM 112, Springer, 1987.

[17] D. Li and Y. Tian, *On the Birch-Swinnerton-Dyer conjecture of elliptic curves $E_D : y^2 = x^3 - D^2x$*, Acta Mathematica Sinca, English Series 2000, April, Vol. **16**, no. 2, 229–236.

[18] P. Monsky, *Mock Heegner points and congruent numbers*, Math. Z. **204** (1990), 45–68.

[19] J. Nekovář, *The Euler system method for CM points on Shimura curves*, in L-functions and Galois Representations (Durham, July 2004), LMS Lecture Note Series **320**, Cambridge University Press, 2007, 471–547.

[20] B. Perrin-Riou, *Points de Heegner et derivees de fonctions L p-adiques*, Invent. Math. **89** (1987), no. 3, 455–510.

[21] G. Shimura, *Introduction to the Arithmetic Theory of Automorphic Functions*, Princeton University Press, 1971.

[22] J. Silverman, *Arithmetic of Elliptic Curves*, GTM **106**, Springer-Verlag, 1986.

[23] N.M. Stephens, *Congruence properties of congruent numbers*, Bull. Lond. Math. Soc. **7** (1975), 182–184.

[24] Y. Tian, *Congruent numbers with many prime factoirs*, PNAS **109**, no. 52, 21256–21258.

[25] Y. Tian, *Congruent numbers and Heegner points*, preprint.

[26] Y. Tian, X. Yuan, and S. Zhang, *Genius periods, genius points, and congruent number problem*, preprint.

[27] J.B. Tunnell, *A classical diophantine problem and modular forms*, Invent. Math. **72** (1983), 323–334.

[28] X. Yuan, S. Zhang, and W. Zhang, *The Gross-Zagier Formula on Shimura Curves*, Annals of Mathematics Studies Number **184**, 2012.

[29] C. Zhao, *A criterion for elliptic curves with second lowest 2-power in $L(1)$*. Math. Proc. Camb. Phil. Soc. **131** (2001), 385–404.

Morningside Center of Mathematics, Chinese Academy of Sciences
Beijing 100190, China
E-mail address: ytian@math.ac.cn

Proceedings of the 6th International Congress of Chinese Mathematicians
ALM 36, pp. 153–172

Monge-Ampère Equation and Optimal Transportation

Xu-Jia Wang

Abstract

In this paper we consider the regularity of potential functions in the optimal transportation, which satisfies a Monge-Ampère type equation of the form $\det[D^2u - A(x, Du)] = b(x, u, Du)$, subject to a second boundary condition, where A is a matrix determined by the potential function u and the cost function c. The regularity has been studied by many authors in the last two decades, and necessary and sufficient conditions have been found.

2000 Mathematics Subject Classification: 35J96, 35J20.

Keywords and Phrases: Monge-Ampère equation, optimal transportation, regularity.

1 Introduction

Let Ω, Ω^* be two bounded domains in the Euclidean space $\mathbb{R}^n$, and f, g be two nonnegative functions defined respectively on Ω and Ω^*, satisfying

$$\int_\Omega f = \int_{\Omega^*} g. \tag{1.1}$$

Let c be a smooth cost function defined on $\mathbb{R}^n \times \mathbb{R}^n$. The optimal transportation problem concerns the existence and regularity of a measure preserving mapping $T : \Omega \to \Omega^*$ which minimizes the cost functional

$$\mathcal{C}(s) = \int_\Omega f(x)c(x, s(x))\, dx \tag{1.2}$$

This work was supported by the Australian Research Council.

among the set $\mathcal{S}$ of all measure preserving mappings from Ω to Ω^*. A map s is called measure preserving if it is Borel measurable and for any Borel set $E \subset \Omega^*$,

$$\int_{s^{-1}(E)} f = \int_E g; \tag{1.3}$$

or equivalently, for any continuous function $h \in C(\Omega^*)$,

$$\int_\Omega h(s(x))f(x)\,dx = \int_{\Omega^*} h(y)g(y)\,dy. \tag{1.3a}$$

The optimal transportation was first studied by Monge [34], in which the transport cost is proportional to the distance the mass is transported, namely

$$c(x,y) = |x-y|. \tag{1.4}$$

This is probably the most natural cost function. In the last few decades, optimal transportation has found many new applications, in areas such as probability, image processing, fluid mechanics, reflector antenna design, and geometry, etc, where more general cost functions are allowed. An important one is the quadratic cost

$$c(x,y) = |x-y|^2. \tag{1.5}$$

Since the work of Monge, the optimal transportation has been studied by numerous researchers, but the progress was limited because of the lack of compactness of the set $\mathcal{S}$. A breakthrough was made by Kantorovich, who introduced a relaxed functional and the dual functional

$$I(\varphi,\psi) = \int_\Omega f(x)\varphi(x)dx + \int_{\Omega^*} g(y)\psi(y)dy, \quad (\varphi,\psi) \in K, \tag{1.6}$$

where

$$K = \{(\varphi,\psi) \in C(\Omega)\times C(\Omega^*) : \varphi(x)+\psi(y) \le c(x,y),\ \forall\, x\in\Omega, y\in\Omega^*\}.$$

The duality means the following relation

$$\inf_{s\in\mathcal{S}} \mathcal{C}(s) = \sup_{(\varphi,\psi)\in K} I(\varphi,\psi). \tag{1.7}$$

Note that the functional I is linear and the set K is convex. One can easily prove that there is a maximizer (u,v) for the supremum $\sup_K I(u,v)$. The maximizer is unique up to a constant, and is called potential functions of the optimal transportation. It was shown that the optimal mapping T can be determined by the potential function, through the formula

$$D_x c(x,T(x)) = Du(x), \tag{1.8}$$

assuming that $D_x c(x,\cdot)$ is a diffeomorphism between Ω^* and its image (which excludes the important cost function (1.4)). Therefore the optimal mapping is a function of x and Du. Moreover, the potential function u satisfies the Monge-Ampère equation

$$\det[A(x, Du) - D^2 u] = b(x, u, Du), \tag{1.9}$$

where

$$\begin{aligned} A(x, Du) &= D_x^2 c(x, T(x)), \\ b(x, u, Du) &= \left| \det \frac{\partial^2 c}{\partial x_i \partial y_j} \right| \frac{f(x)}{g(T(x))}. \end{aligned} \tag{1.10}$$

As we will see below, the quadratic cost function corresponds to the standard Monge-Ampère equation

$$\det D^2 u = b(x, u, Du). \tag{1.11}$$

This equation is particularly important in geometry and has been extensively studied in the last century.

The existence of optimal mappings is determined by (1.8). But the regularity of the optimal mappings for non-quadratic cost functions was a major open problem in optimal transportation, as was pointed out by Villani in [43] and also raised by Caffarelli [6]. An important advance was made in [32], where the regularity was proved, in addition to some natural conditions, if the cost function c satisfies a structural condition (A3) and if the target domain Ω^* is c-convex with respect to the domain Ω.

Both the c-convexity of the target domain and the structural condition (A3) are necessary for the regularity [32, 29]. Since the work [32], several other interesting results on the regularity have been obtained, which include the following

- a weak form of the structural condition (A3) is also necessary [29];
- the strict c-convexity and $C^{1,\alpha}$ regularity of the potential functions [41, 29, 26];
- the $C^{2,\alpha}$ and $W^{2,p}$ estimates [27, 28];

Though equations of the form (1.9) in dimension two has been studied before [36], our results for the optimal transportation are much more precise. Our conditions for the regularity are not only sufficient but also necessary. Our treatment uses many ideas from that for the standard Monge-Ampère equation (1.11), but also introduced several new ideas. In this note we will mainly focus on these new observations which are not seen before. The main two of them are

- to understand the notions of convexities relative to the cost functions (or c-convexity for brevity); and
- the local geometry of potential functions and cost functions in a normalization process.

For the local geometry, our assumption (A3) is essentially equivalent to the connectedness of the contact set of the potential function with its c-support functions.

Under this condition we are able to show the cost function and the potential functions converges to quadratic functions in a blow-up process. In this note we will mention only very briefly those techniques which was developed in the study of the standard Monge-Ampère equation (1.11).

Very recently the regularity of optimal mappings under a weak form of (A3), namely (A3w), was also investigated in [18, 21]. The optimal transportation on manifold has also been studied by many authors. The existence of optimal mappings can be obtained by a similar idea [33] but the regularity is very complicated. It is not smooth in general [23] but it is under some strong conditions, see [11, 12, 19, 20, 31, 45] and the references therein. We will not discuss these two topics in this note. The interested readers are referred to the above mentioned papers.

This paper is arranged as follows. In Section 2 we review briefly the existence of optimal mappings. In Section 3 we introduce the notions of c-convexity. We then recall the a priori estimates for equation (1.9) in Section 4. In Section 5 we consider the geometric implication of the assumption (A3). The final Section 6 contains some remarks on Monge's problem.

2 Existence and uniqueness of optimal mappings

First we show that there exist maximizers to the supremum in (1.7). For any $(u, v) \in K$, let

$$\begin{aligned} v^*(x) &= \inf_{y\in\Omega^*}\{c(x,y) - u(y)\}, \\ u^*(y) &= \inf_{x\in\Omega}\{c(x,y) - v^*(x)\}. \end{aligned} \tag{2.1}$$

Apparently $u^* \geq u$ and $v^* \geq v$, and $(u^*, v^*) \in K$. Hence

$$I(u,v) \leq I(u^*, v^*). \tag{2.2}$$

Choosing a sequence $(u_k, v_k) \in K$ such that

$$I(u_k, v_k) \to \sup_K I(u,v).$$

By (2.2) we may assume that

$$\begin{aligned} u_k(x) &= \inf_{y\in\Omega^*}\{c(x,y) - v_k(y)\}, \\ v_k(y) &= \inf_{x\in\Omega}\{c(x,y) - u_k(x)\}. \end{aligned} \tag{2.3}$$

The above formulas imply that (u_k, v_k) are Lipschitz continuous, with the Lipschitz constant controlled by that of the cost function c. By the mass balance condition (1.1), we have $I(u,v) = I(u - C, v + C)$. Hence we may add a constant to u_k such that $\inf u_k = 0$. Then both u_k and v_k are uniformly bounded and Lipschitz continuous. So (u_k, v_k) converges by a subsequence to a pair of Lipschitz functions $(u, v) \in K$, and

$$I(u,v) = \sup\{I(\phi, \psi) : (\phi, \psi) \in K\}. \tag{2.4}$$

The component functions u and v are called the potential functions of the optimal transportation problem. By (2.3) we have

$$\begin{aligned} u(x) &= \inf_{y\in\Omega^*}\{c(x,y)-v(y)\}, \\ v(y) &= \inf_{x\in\Omega}\{c(x,y)-u(x)\}, \end{aligned} \tag{2.5}$$

which is a duality.

For any given $x_0 \in \Omega$, let $y_0 \in \Omega^*$ such that

$$\begin{aligned} u(x_0) &= c(x_0,y_0) - v(y_0), \\ u(x) &\le c(x,y_0) - v(y_0), \quad \forall\, x\in\Omega. \end{aligned} \tag{2.6}$$

Hence u is semi-concave, namely there exists $C > 0$ such that $u - C|x|^2$ is concave. Therefore u is twice differentiable a.e.. Suppose Du exists at x_0. Then at x_0 we have

$$Du(x_0) = D_x c(x_0, y_0). \tag{2.7}$$

For the existence of optimal mappings, we make the following assumption.

(A1) $\forall\, x, z \in \mathbb{R}^n$, $\exists_1\, y \in \mathbb{R}^n$ such that $D_x c(x,y) = z$; and $\forall\, y, z \in \mathbb{R}^n$, $\exists_1\, x \in \mathbb{R}^n$ such that $D_y c(x,y) = z$.

For convenience in the above assumption we assume that the cost c is defined in the whole space $\mathbb{R}^n \times \mathbb{R}^n$. By assumption (A1), we obtain from (2.7) a mapping $T:\ \Omega \to \Omega^*$ such that

$$Du(x) = D_x c(x, T(x)). \tag{2.8}$$

Note that if we write the above equation as $p = D_x c(x,T)$, then $T = T(x,p)$, namely T depends on x and Du.

Theorem 2.1. *Under the assumption* (A1), *there is a unique optimal mapping T given by* (2.8).

This result was obtained by Brenier, Caffarelli, Gangbo and McCann. The proof is as follows. First one shows that the mapping T is measure preserving, namely (1.3a) holds. Hence for any $(\phi, \psi) \in K$,

$$\begin{aligned} \mathcal{C}(s) &= \int_\Omega c(x, s(x)) f(x) \\ &\ge \int_\Omega [\phi(x) + \psi(s(x))] f(x) \\ &= \int_\Omega \phi(x) f(x) + \int_{\Omega^*} \psi(y) g(y) = I(\phi, \psi). \end{aligned}$$

By (2.6), we see that the inequality becomes an equality when $(\phi,\psi) = (u,v)$ are the potential functions and $s = T$. Hence T is optimal.

The above inequality also implies the uniqueness of optimal mappings. Indeed, if t is another optimal mapping, then by (1.3a) we have

$$\int_\Omega f(x)u(x) + \int_{\Omega^*} g(y)v(y) = \int_\Omega f(x)u(x) + \int_\Omega f(x)v(t(x))dx$$
$$= \int_\Omega f(x)[u(x) + v(t(x))]dx.$$

Since $(u,v) \in K$, we have $u(x) + v(t(x) \le c(x,t(x))$; and the equality holds if and only if $u(x) + v(t(x)) = c(x,t(x))$ a.e.. Hence by (A1) we have $t(x) = T(x)$ a.e.

One may also consider the optimal transportation from Ω^* to Ω with the cost function c. Then similarly there is an optimal mapping T^*, determined by

$$Dv(y) = D_y c(T^*(y), y). \tag{2.9}$$

It is easy to see that $T^* = T^{-1}$ if they are 1-1.

3 Notions of convexity

To study the regularity of the optimal mappings, by (2.8) it suffices to study the regularity of the potential function. A key point is to understand the notions of convexity relative to the cost function c, or c-convexity for brevity. When $c(x,y) = x \cdot y$, the c-convexity coincides with the usual convexity. In this section we introduce some terminology.

- **supergradient** $\partial^+ u$: Given $x_0 \in \Omega$, the supergradient $\partial^+ u$ is defined by

  $$\partial^+ u(x_0) = \{p \in \mathbb{R}^n : \quad u(x) \le u(x_0) + p \cdot (x - x_0) + o(|x - x_0|) \quad x \text{ near } x_0\}.$$

- c-**supergradient** $\partial^+ u$: Given $x_0 \in \Omega$, the c-supergradient $\partial^+ u$ is defined by

  $$\partial_c^+ u(x_0) = \{y \in \mathbb{R}^n : \quad u(x) \le c(x,y) - c(x_0,y) + u(x_0) + o(|x - x_0|) \quad x \text{ near } x_0\}.$$

  For any set $E \in \Omega$, we denote $\partial^+ u(E) = \bigcup_{x\in E} \partial^+ u(x)$ and $\partial_c^+ u(E) = \bigcup_{x\in E} \partial_c^+ u(E)$. Then

  $$D_x c(x_0, \partial_c^+ u(x_0)) = \partial^+ u(x_0). \tag{3.1}$$

  Note that $\partial^+ u(x_0)$ is a closed, convex set. Hence $\partial_c^+ u(x_0)$ is closed and c-convex with respect to x_0. The definition of the c-convexity of domains will be given below.

- c-**support function**: A *local c-support* of u at $x_0 \in \Omega$ is a function of the form

  $$h = c(\cdot, y_0) + a_0,$$

where $a_0 = u(x_0) - c(x_0, y_0)$, and $y_0 \in \mathbb{R}^n$, such that

$$\begin{aligned} u(x_0) &= h(x_0), \\ u(x) &\le h(x) \quad \text{near} \quad x_0. \end{aligned} \tag{3.2}$$

If $h = c(\cdot, y_0) + a_0$ is a local c-support of u at x_0, then $y_0 \in \partial_c^+ u(x_0)$.

If $u(x) \le h(x)$ for all $x \in \Omega$, then h is a *global c-support* (or c-support for short) of u at x_0.

- **c-concavity of functions**: Assume u is semi-concave. We say
  - u is *locally c-concave* if $\forall\, x_0 \in \Omega$ and $y \in \partial_c^+ u(x_0)$, $h = c(\cdot, y) - c(x_0, y) + u(x_0)$ is a local c-support of u at x_0;
  - u is *c-concave* if $\forall$ point $x_0 \in \overline{\Omega}$, $\exists$ a global c-support at x_0 in Ω;
  - u is *fully c-concave* if it is locally c-concave and every local c-support of u is a global c-support;
  - u is *strictly c-concave* if it is fully c-concave and every c-support of u contacts its graph at one point only.

- **c-segment**: A set of points $\ell \subset \mathbb{R}^n$ is a c-segment with respect to a point $y_0 \in \mathbb{R}^n$ if $D_y c(\ell, y_0)$ is a line segment in $\mathbb{R}^n$.

- **c-convexity of domains**: We say a set U is c-convex with respect to another set V if the image $c_y(U, y)$ is convex for each $y \in V$. Namely for any two points $x_0, x_1 \in U$ and any $y \in V$, the c-segment relative to y connecting x_0 and x_1 lies in U.

- **c-Legendre transform**: Let φ be an upper semi-continuous function defined on Ω. The c-transform of φ is a function φ^* defined on $\mathbb{R}^n$, given by

$$\varphi^*(y) = \inf\{c(x, y) - \varphi(x) : x \in \Omega\}. \tag{3.3}$$

Similarly for an upper semi-continuous function ψ defined on Ω^*, the c-transform of ψ is the function

$$\psi^*(x) = \inf\{c(x, y) - \psi(y) : y \in \Omega^*\}.$$

- **c-normal mapping**: N_u is defined by

$$N_u(x_0) = \{y \in \mathbb{R}^n : \ u(x) \le c(x, y) - c(x_0, y) + u(x_0), \ \forall\, x \in \Omega\}.$$

For any subset $E \subset \Omega$, we denote $N_u(E) = \bigcup_{x \in E} N_u(x)$. This c-normal mapping is the classical normal mapping when $c(x, y) = x \cdot y$. If u is the potential function, we have $N_u(x) = T(x)$ at all points where u is differentiable, where T is the mapping determined by (2.8).

Remarks 3.1. (i) By definition, a c-concave function u can be represented as

$$u(x) = \inf\{c(x,y) - a(y) : \quad y \in N_u(\Omega)\},$$

where a is a function of y only. When $c(x,y) = x \cdot y$, the above formula means a concave function is the infimum of linear functions.

(ii) Let (u, v) be the potential function to the optimal transportation. Then by (2.5), v is the c-Legendre transform of u and vice verse.

(iii) From the definition, φ^* and ψ^* are c-concave. In particular, by (2.5), potential functions are c-concave.

Following the treatment for the standard Monge-Ampère equation, we can introduce the notion of generalized solutions to the Monge-Ampère equation (1.9). Indeed, by the c-normal mapping we can introduce a measure $\mu = \mu_{u,g}$ in Ω, where $g \in L^1(\mathbb{R}^n)$ is a nonnegative measurable function, such that for any Borel set $E \subset \Omega$,

$$\mu(E) = \int_{N_u(E)} g(x)dx. \tag{3.4}$$

Then we can prove that μ is a Radon measure, using the c-Legendre transform and almost everywhere differentiability of c-concave function. One can then prove the weak continuity of the measure with respect to uniform convergence of c-concave functions. By the weak continuity, we can prove the existence of generalized solutions to equation (1.9) under the Dirichlet or the second boundary condition. The uniqueness of generalized solutions can be proved similarly to that for the standard Monge-Ampère equation. The proof of these results are similar to those for the standard Monge-Ampère equation (1.11) and is omitted here.

4 The Monge-Ampère equation

Suppose the potential function $u \in C^2(\Omega)$. Differentiating (2.7) we have

$$D^2u(x) = D_x^2c(x, T(x)) + D_{xy}^2 c \cdot DT. \tag{4.1}$$

Note that $|\det DT| = \frac{f(x)}{g(T(x))}$. Hence we obtain equation (1.9), namely

$$\det(D_x^2 c - D^2 u) = |\det(D_{xy}^2 c)| \frac{f(x)}{g(T(x))}, \quad x \in \Omega, \tag{4.2}$$

This is an equation of Monge-Ampère type. The boundary condition is

$$T(\Omega) = \Omega^*. \tag{4.3}$$

Namely $T(x) \in \partial\Omega^*$, $\forall\, x \in \partial\Omega$.

Note that if $c(x,y) = x \cdot y$, the above equation becomes the standard Monge-Ampère equation (1.11) with $b = f(x)/g(Du)$. If $c(x,y) = \frac{1}{2}|x-y|^2$, let $u' = u - \frac{1}{2}|x|^2$, $v' = v - \frac{1}{2}|y|^2$, then u' satisfies equation (1.11).

We remark that, instead of the supremum $\sup_K I(\phi, \psi)$, one may also consider the infimum $\inf_K I(\phi, \psi)$. Then the minimizers are semi-convex and satisfy the equation

$$\det(D^2u - D_x^2c) = |\det(D_{xy}^2c)|\frac{f(x)}{g(T(x))}, \quad x \in \Omega,$$

where the matrix $(D^2u - D_x^2c) \geq 0$.

Moreover, by duality, the potential function v satisfies the equation

$$\det(D_y^2c - D^2v) = |\det(D_{xy}^2c)|\frac{g(y)}{f(T_v(y))}, \quad x \in \Omega,$$
$$T^*(\Omega^*) = \Omega.$$

Let us review very briefly the regularity of solutions to the standard Monge-Ampère equation (1.11). We consider the second boundary problem

$$\det D^2u = \frac{f(x)}{g(Du(x))} \quad \text{in } \Omega, \tag{4.4}$$
$$Du(\Omega) = \Omega^*. \tag{4.5}$$

This boundary value problem corresponds to the optimal transportation with cost $c(x, y) = x \cdot y$, or $c(x, y) = |x - y|^2$. For regularity of the Dirichlet problem and the Neumann problem of the Monge-Ampère equation, we refer the readers to [39].

For the a priori estimates, first note that (4.3) implies a uniform bound for $|Du|$. Higher regularity of the problem (4.4)–(4.5) is mainly due to Caffarelli [4, 5]. He proved the following regularity results.

(i) Assume f, g satisfy (1.1), $C^{-1} \leq f, g \leq C$, and Ω^* is convex. Then u is strictly convex in Ω and $u \in C^{1,\alpha}(\Omega)$ for some $\alpha \in (0, 1)$.

(ii) If in addition f and g are continuous, then $u \in W_{loc}^{2,p}(\Omega)$ for any $p > 1$. If $f \in C^\alpha(\Omega)$ and $g \in C^\alpha(\Omega^*)$, where $\alpha \in (0, 1)$, then $u \in C^{2,\alpha}(\Omega)$.

(iii) If furthermore $\partial\Omega$, $\partial\Omega^*$ are uniformly convex, $\in C^{2,\alpha}$ for some $\alpha \in (0, 1)$, and $f \in C^\alpha(\overline{\Omega})$, $g \in C^\alpha(\overline{\Omega}^*)$, then u satisfies the estimate

$$\|u\|_{C^{2,\alpha}(\overline{\Omega})} \leq C, \tag{4.6}$$

where C depends on n, f, g, Ω and Ω^*.

The global regularity (4.6) was also obtained by Delanoë [10] for dimension $n = 2$ and Urbas [42] for all dimensions, under the stronger conditions $f, g \in C^{1,1}$ and $\partial\Omega \in C^{3,1}$.

For the regularity of the potential function u with general cost functions, we always assume that $c \in C^\infty(\mathbb{R}^n \times \mathbb{R}^n)$. We also assume that

(A1) $\forall\, x, z \in \mathbb{R}^n$, $\exists_1\, y \in \mathbb{R}^n$ such that $D_xc(x, y) = z$; and $\forall\, y, z \in \mathbb{R}^n$, $\exists_1\, x \in \mathbb{R}^n$ such that $D_yc(x, y) = z$.

(A2) For any $x, y \in \mathbb{R}^n$,

$$\det D^2_{xy} c \neq 0.$$

(A3) $\exists$ a constant $c_0 > 0$ such that $\forall\ x \in \Omega, y \in \Omega^*$, and $\xi, \eta \in \mathbb{R}^n$ with $\xi \perp \eta$,

$$\sum_{i,j,k,l,p,q,r,s} (c^{p,q} c_{ij,p} c_{q,rs} - c_{ij,rs}) c^{r,k} c^{s,l} \xi_i \xi_j \eta_k \eta_l \geq c_0 |\xi|^2 |\eta|^2, \tag{4.7}$$

where $c_{i,j}(x, y) = \frac{\partial^2 c(x,y)}{\partial x_i \partial y_j}$, and $(c^{i,j})$ is the inverse matrix of $(c_{i,j})$.

Assumption (A1) was needed in Section 2 for the existence of optimal mappings. Assumption (A2) is such that the equation (5.1) is nondegenerate. The key condition is (A3), which is a complicated structural condition. However as we will see below, this condition is necessary for the regularity.

First we have the following regularity result [32].

Theorem 4.1. *Suppose $f \in C^{1,1}(\Omega)$, $g \in C^{1,1}(\Omega^*)$, f, g satisfy* (1.1) *and $C^{-1} \leq f, g \leq C$. Suppose the cost function c satisfies assumptions* (A1)–(A3). *Then if the domain Ω^* is c-convex relative to Ω, the potential function u is $C^{3,\alpha}$ smooth in Ω, for any $\alpha \in (0, 1)$.*

The proof is by choosing an appropriate auxiliary function to obtain the interior second derivative estimate. More precisely, denote $w_{ij} = c_{ij}(x, T(x)) - u_{ij}$, and let

$$G(x, \xi) = \eta^2(x) w_{\xi\xi}, \tag{4.8}$$

where η is a cut-off function, and $w_{\xi\xi} = \sum \xi_i \xi_j w_{ij}$ for a unit vector $\xi \in \mathbb{R}^n$. Suppose G attains a maximum at x_0 and $\xi_0 = (1, 0, \ldots, 0)$. Then at x_0 we have

$$w_{11} \sum w^{ij} (\log G)_{ij} \leq 0.$$

By a careful computation, we arrive at

$$w^{ii} [c^{k,l} c_{ii,k} c_{l,st} - c_{ii,st}] c^{s,1} c^{t,1} w_{11}^2 \leq K.$$

Hence by assumption (A3), we obtain $G \leq C$.

Once the second derivatives are bounded, equation (4.2) becomes uniformly elliptic and higher regularity follows from Evans and Krylov's regularity theory.

In Theorem 4.1 we also assume that the target domain Ω^* is c-convex with respect to Ω. This assumption is used in proving that the potential function, which is a generalized solution to (4.2), is smooth. The proof uses the uniqueness of generalized solutions to the Dirichlet problem.

Remark 4.1. In [32], it was also proved, by a contradiction argument, that if the target domain Ω^* is not c-convex, there exist smooth and positive f, g such that the optimal mapping is not continuity.

The above interior regularity result has been extended to boundary [41].

Theorem 4.2. *Suppose $f \in C^{1,1}(\overline{\Omega})$, $g \in C^{1,1}(\overline{\Omega}^*)$, f, g satisfy* (1.1) *and $C^{-1} \le f, g \le C$. Suppose the cost function c satisfies assumptions* (A1)–(A3). *Then if Ω and Ω^* are uniformly c-convex with respect to each other and both domains have $C^{3,1}$ smooth boundaries, then $u \in C^{3,\alpha}(\overline{\Omega})$ for any $\alpha \in (0,1)$.*

To prove Theorem 4.2, a key ingredient is to establish the uniform obliqueness of the boundary condition (4.3), for which an idea of Urbas was used [42].

Remark 4.2. By a direct calculation we have the formula

$$A_{ij,kl} =: \frac{\partial^2 A_{ij}}{\partial z_k \partial z_l} = (c_{ij,rs} - c^{p,q} c_{ij,p} c_{q,rs}) c^{r,k} c^{s,l}, \tag{4.9}$$

where

$$A_{ij}(x,z) = c_{ij}(x, T(x,z)), \tag{4.10}$$

and $T(x,z)$ is the mapping determined by (2.8), namely $D_x c(x, T(x,z)) = z$. Hence condition (A3) may be written in the form

$$A_{ij,kl}\xi_i\xi_j\eta_k\eta_l \le -c_0|\xi|^2|\eta|^2 \tag{4.11}$$

for all $\xi \perp \eta \in \mathbb{R}^n$.

Also by direct computation, one can verify that the following functions satisfies assumption (A3).

- The power cost $c(x,y) = -|x-y|^m, \qquad m \in (-2,1), m \ne 0$.
- $c(x,y) = -\log|x-y|$.
- $c(x,y) = \sqrt{1+|x-y|^2}$.
- $c(x,y) = \sqrt{1-|x-y|^2}$.
- $c(x,y) = -[x \cdot y + f(x)g(y)]$, where f, g are convex with $|Df|, |Dg| < 1$.

For the quadratic function $c(x,y) = |x-y|^2$, we have $A_{ij,kl} \equiv 0$, where $A_{ij,kl}$ is given in (4.9). Hence the quadratic function satisfies a weak form of (A3), namely

(A3w) For any vectors $\xi, \eta \in \mathbb{R}^n$ with $\xi \perp \eta$,

$$\sum_{i,j,k,l,p,q,r,s} (c^{p,q} c_{ij,p} c_{q,rs} - c_{ij,rs}) c^{r,k} c^{s,l} \xi_i\xi_j\eta_k\eta_l \ge 0. \tag{4.12}$$

Theorems 4.1 and 4.2 can be extended to manifolds. But the verification of assumption (A3) is very complicated. At the moment it is known that the cost function

$$c(x,y) = \text{dist}(x,y)^2$$

satisfies (A3) on the unit sphere [29], or on manifolds which are smooth perturbation of the sphere, where $\text{dist}(x,y)$ is the geodesic distance between the points x and y.

We note that if c satisfies (A3), then $-c$ does not.

5 Geometric properties of cost functions under (A3)

The condition (A3) is a strong condition. However without this condition, the optimal mapping T_u is not continuous in general. A counterexample was found by Loeper [29]. He shows that if there exist vector ξ and η with $\xi \perp \eta$ such that the quantity

$$\sum_{i,j,k,l,p,q,r,s} (c^{p,q}c_{ij,p}c_{q,rs} - c_{ij,rs})c^{r,k}c^{s,l}\xi_i\xi_j\eta_k\eta_l < 0 \tag{5.1}$$

at some point x_0 in Ω, then there exist positive and smooth f, g such that the optimal mapping is not continuous near x_0, even though all other conditions in Theorem 4.1 are satisfied. The proof is based on a geometric property of the condition A3 and a contradiction argument. This geometric property is important for further investigation of regularity problem.

5.1 A geometric property of the condition (A3)

Let y_0, y_1 be two points in $\overline{\Omega^*}$. Let $\widehat{y_0y_1}$ be the c-segment relative to a point $x_0 \in \Omega$, connecting y_0 and y_1. By definition,

$$\widehat{y_0y_1} = \{y_t : \ c_x(x_0, y_t) = p_t, \ t \in [0,1]\}, \tag{5.2}$$

where $p_t = tp_1 + (1-t)p_0$, and $p_0 = c_x(x_0, y_0)$, $p_1 = c_x(x_0, y_1)$. Let

$$\begin{aligned} h_i(x) &= c(x, y_i) - a_i, \quad i = 0, 1, \\ h_t(x) &= c(x, y_t) - a_t, \end{aligned} \tag{5.3}$$

where $t \in (0,1)$, a_0, a_1 and a_t are constants such that $h_0(x_0) = h_1(x_0) = h_t(x_0)$.

Lemma 5.1. *Suppose* (A3) *holds. Then for $x \neq x_0$, near $x_0, 0 < t < 1$, we have the inequality*

$$h_t(x) > \min\{h_0(x), h_1(x)\}, \tag{5.4}$$

This property was discovered by Loeper [29]. Inequality (5.4) follows from (4.7). Indeed, by a rotation of axes, we assume that $p_1 = p_0 + \delta e_n$, where δ is a positive constant and $e_n = (0, \ldots, 0, 1)$ is the unit vector in the x_n-axis. Let $T = T(x,p)$ be the optimal mapping, obtained from (2.8). Then by (4.7),

$$\frac{d^2}{dt^2} c_{ij}(x, T(x, p_t))\xi_i\xi_j \leq -c_0\delta^2 \tag{5.5}$$

for any unit vector ξ orthogonal to e_n. Now (5.4) follows from (5.5).

We note that if the condition (A3) is relaxed to (A3w), then correspondingly we have

$$h_t(x) \geq \min\{h_0(x), h_1(x)\}. \tag{5.4w}$$

Inequality (5.4) or (5.4w) implies a monotonicity of the set

$$\mathcal{N}_t = \{x \in \mathbb{R}^n : c(x, y_t) - c(0, y_t) - c(x, y_0) - c(0, y_0)\}.$$

That is if $t_0 < t_1 < t_2$, then $\mathcal{N}_{t_1}$ stays between $\mathcal{N}_{t_0}$ and $\mathcal{N}_{t_2}$. Moreover, from Lemma 5.1 we also have

Corollary 5.1. *Let (u, v) be the potential function to the optimal transportation. Assume that Ω is c-convex with respect to Ω^* and the cost function c satisfies* (A3w). *Then for any point x_0, the contact set $E = \{x \in \Omega : u(x) = h(x)\}$ is connected, where $h(x)$ is a c-support of u at x_0.*

We remark that if u is not C^1 at some point x_0, the set E may be disconnected.

Corollary 5.2. *Assume that Ω^* is c-convex with respect to Ω and the cost function c satisfies* (A1)–(A3). *Assume also that $C^{-1} \le f/g \le C$ for some positive constant C. Then v is strictly c-concave and u is C^1 smooth.*

Note that by the c-Legendre transform, the strict concavity of v is equivalent to the C^1 regularity of u.

Corollary 5.2 was proved in [40]. The C^1 regularity can be strengthened to $C^{1,\alpha}$ for some $\alpha > 0$ [30, 26].

Theorem 5.1. *Assume the cost function c satisfies conditions* (A1)–(A3), *Ω^* is c-convex with respect to Ω, and $f \ge 0$, $f \in L^p(\Omega)$ for some $p \in (\frac{n+1}{2}, +\infty]$. Then $u \in C^{1,\alpha}(\overline{\Omega})$, where $\alpha = \frac{\beta(n+1)}{2n^2+\beta(n-1)}$ and $\beta = 1 - \frac{n+1}{2p}$.*

An interesting phenomenon is that when $p = \infty$, the exponent $\alpha = \frac{\beta(n+1)}{2n^2+\beta(n-1)}$ is optimal [26]. For the standard Monge-Ampère equation, the exponent depends in f and g. But for general cost functions, it depends on the cost function too.

5.2 Super-level sets[1]

To study the regularity of optimal mappings, by Remark 4.1, we always assume that Ω^* is c-convex. By Corollary 5.2, we also assume in the following that u is a strictly c-concave and C^1 smooth.

For any given point $x_0 \in \Omega$ and any small positive constant $h > 0$, denote

$$S^0_{h,u}(x_0) = \{x \in \Omega : u(x) > \phi(x) - h\} \tag{5.6}$$

the *super-level set* of u, where $\phi = c(\cdot, y_0) + a_0$ is a c-support of u at x_0, $a_0 = u(x_0) - c(x_0, y_0)$, $y_0 = T_u(x_0)$. For simplicity, we write sometimes $S^0_{h,u}(x_0)$ as S^0_h when no confusion arises.

For a fixed $x_0 \in \Omega$, we make the changes

$$\begin{aligned} c(x,y) &\longrightarrow [c(x,y) - c(x,y_0)] - [c(x_0,y) - c(x_0,y_0)], \\ u(x) &\longrightarrow [u(x) - u(x_0)] - [c(x,y_0) - c(x_0,y_0)], \\ v(y) &\longrightarrow [v(y) - v(y_0)] - [c(x_0,y) - c(x_0,y_0)], \end{aligned} \tag{5.7}$$

[1]In most papers on Monge-Ampère equations, one considers sub-level sets. In this note we consider c-concave functions and accordingly the sub-level sets need to change to super-level sets.

where $y_0 = T_u(x_0)$ and v is the dual potential function. Then the cost function c satisfies

$$\begin{aligned} c(x, y_0) &\equiv 0, \quad \forall\, x \in \Omega, \\ c(x_0, y) &\equiv 0, \quad \forall\, y \in \Omega^*, \end{aligned} \tag{5.8}$$

and the potential functions u, v satisfy

$$\begin{aligned} u(x_0) = 0, \quad Du(x_0) = 0, \quad u \le 0 \quad \text{in } \Omega, \\ v(y_0) = 0, \quad Dv(y_0) = 0, \quad v \le 0 \quad \text{in } \Omega^*. \end{aligned} \tag{5.9}$$

Lemma 5.2. *Assume that $S^0_{h,u}(x_0) \Subset \Omega$. Then the super-level set $S^0_{h,u}(x_0)$ is c-convex w.r.t. y_0. Moreover, by making the coordinate transform*

$$\begin{aligned} x &\to D_y c(x, y_0), \\ y &\to D_x c(x_0, y), \end{aligned} \tag{5.10}$$

the super-level set $S^0_{h,u}(x_0)$ can be made convex in the usual sense.

It is well known that a bounded convex domain D has a unique minimum ellipsoid, which attains the minimum volume among all ellipsoids containing the set D, such that

$$\frac{1}{n} E \subset D \subset E, \tag{5.11}$$

where $\frac{1}{n}E = \{z + \frac{1}{n}(x - z) : x \in E\}$ and z is the centre of E. We say D is *normalized* if E is a unit ball.

5.3 Normalization

By the coordinate transform (5.10), the super-level set $S^0_{h,u}(x_0)$ is convex. By (5.7) we also assume that (5.8) and (5.9) hold. By a translation of coordinates we also assume $x_0 = 0, \quad y_0 = 0$. The normalization of u in S^0_h is the rescaled solution

$$\bar{u}(\bar{x}) = \frac{1}{h} u(x), \tag{5.12}$$

together with the change of the cost function

$$\bar{c}(\bar{x}, \bar{y}) = \frac{1}{h} c(x, y), \tag{5.13}$$

where

$$\bar{x} = G_h x, \qquad \bar{y} = G^*_h y, \tag{5.14}$$

and G_h is a linear transform which normalized the convex set S^0_h, G^*_h is the unique linear transform such that $G^*_h \cdot G_h = h^{-1} I$, where I is the unit matrix.

Denote $U = G(S^0_h)$. Then $\sup_U \bar{u} = 0$ and $\bar{u} = 1$ on ∂U. One deduces from (1.9) that $\bar{u}$ satisfies the equation

$$\det [\bar{A}(\cdot, D\bar{u}) - D^2 \bar{u}] = \bar{f}, \tag{5.15}$$

for an appropriate $\bar{f}$. By the normalization, one has

$$\partial_{p_k} \bar{A}_{ij}(\cdot, 0) = 0. \tag{5.16}$$

The following convergences are critical.

Theorem 5.2. *Assume that the cost function c satisfies* (A1)–(A3), Ω^* *is c-convex with respect to* Ω, f, g *satisfy* (1.1) *and* $C^{-1} \leq f, g \leq C$. *Then*

$$\bar{c}(x, y) \to x \cdot y \quad \text{as} \quad h \to 0. \tag{5.17}$$

If f, g *are continuous, then*

$$\bar{u}(x) \to |x|^2 \quad \text{as} \quad h \to 0. \tag{5.18}$$

We remark that these properties are not true if c does not satisfy (A3). From these properties we infer that

Corollary 5.3. *Letting* $h \to 0$, *we have*

$$\bar{A}(\cdot, D\bar{u}) \to 0. \tag{5.19}$$

To prove the above results, a key point is to prove that $\bar{A}$ is uniformly bounded. One then prove its derivatives converges to zero as $h \to 0$.

5.4 The $C^{2,\alpha}$ and $W^{2,p}$ estimates

From the above results, one sees that the Monge-Ampère equation (5.15) is *close* to the standard Monge-Ampère equation. Hence one expects that Caffareli's $C^{2,\alpha}$ and $W^{2,p}$ estimates hold for equation (1.9), and they are indeed obtained in [27, 28].

For simplicity let us assume that the right hand side b is a function of x. Denote

$$\omega_b(r) = \sup\{|b(x) - b(y)| : \ |x - y| < r\}. \tag{5.20}$$

Then we have the following estimate [27].

Theorem 5.3. *Assume the cost function c satisfies* (A1)–(A3) *and b satisfies*

$$C_1 \leq b \leq C_2 \tag{5.21}$$

for some positive constants $C_1, C_2 > 0$. *Let* $u \in C^2(\Omega)$ *be an elliptic solution of* (1.1). *Then for all* $x, y \in \Omega_\delta$, *we have the estimate*

$$|D^2u(x) - D^2u(y)| \leq C\left[d + \int_0^d \frac{\omega_b(r)}{r} + d\int_d^1 \frac{\omega_b(r)}{r^2}\right], \tag{5.22}$$

where $d = |x - y|$, $\Omega_\delta = \{x \in \Omega : \ \mathrm{dist}(x, \partial\Omega) > \delta\}$, $C > 0$ *depends only on* $n, \delta, C_1, C_2, A, \sup|Du|$, *and the modulus of continuity of* Du. *It follows that:*

(i) *If b is Dini continuous, then the modulus of continuity of* D^2u *can be estimated by* (5.22) *above;*

(ii) *If* $b \in C^\alpha(\Omega)$ *for some* $\alpha \in (0, 1)$, *then*

$$\|u\|_{C^{2,\alpha}(\Omega_\delta)} \leq C\left[1 + \frac{\|b\|_{C^\alpha(\Omega)}}{\alpha(1-\alpha)}\right]; \tag{5.23}$$

(iii) *If* $b \in C^{0,1}(\Omega)$, *then*

$$|D^2u(x) - D^2u(y)| \le Cd[1 + \|b\|_{C^{0,1}}|\log d|], \quad \forall x, y \in \Omega_\delta. \tag{5.24}$$

The $C^{2,\alpha}$ estimate for the Monge-Ampère equation was found by Caffarelli [3] and Heinz (for $n = 2$) [36]. A simplified proof was given in [22], using a perturbation method introduced in [48]. This method also gave more precise estimate for the modulus of continuity for the second derivatives, as shown in (5.22).

To use the perturbation argument, one needs to establish the interior second derivative estimate of Pogorelov type, for the normalized equation (5.15). Note that after the normalization in Section 5.3, the cost function satisfies (A3w) but not (A3).

The $W^{2,p}$ estimate for the Monge-Ampère equation was also discovered by Caffarelli [3]. Following Caffarelli's argument, and using Theorem 5.2 and Corollary 5.3, we have the $W^{2,p}$ estimate for the Monge-Ampère equation (1.9).

Theorem 5.4. *Assume the cost function* c *satisfies* (A1)–(A3), f, g *are continuous,* $C_1 \le f, g \le C_2$, *and* Ω^* *is* c^**-convex with respect to* Ω. *Let* u *be the potential function. Then* $D^2u \in L^p(\Omega')$, $\forall\ p \ge 1$, $\Omega' \subset\subset \Omega$, *and we have the estimate*

$$\|u\|_{W^{2,p}(\Omega')} \le C, \tag{5.25}$$

where C *depends on* n, p, C_1, C_2, $\Omega, \Omega', \Omega^*$, *and the modulus of continuity of* f.

We remark that Caffarelli's proof for the $W^{2,p}$ can also be simplified, which was given in the lecture notes by the author at Tsinghua University in 2011.

6 Remark on Monge's problem

The cost function $c(x, y) = |x - y|$, which is the most natural one in optimal transportation and first appeared in Monge's paper [34], does not satisfy any of the conditions (A1)–(A3). Since Monge's paper, this problem has been extensively in the last two centuries. We refer the readers to [13, 14] for excellent account of the history of this problem.

In this note we mention some recent works on this problem. The existence of optimal mapping was obtained by Sudakov [37, 1], using a probability approach. However his argument contains a gap, namely he needs to show that the set of end points of the transport rays has measure zero. The existence was later obtained by Evans and Gangbo [14] using approximation by the p-Laplace equation with $p \to \infty$. This approach is quite complicated and requires huge amount of analysis for the PDE, and so some regularity of the domains and mass densities are needed. A different approach, using the cost $c_p(x, y) = |x - y|^p$ to approximate the cost $|x - y|$, was found in [7, 38]. The existence has also been obtained if the standard norm $|x|$ is replaced by a more general norm in Euclidean space [8, 9].

When the cost function satisfies assumption (A1), the optimal mapping is unique. But the optimal mapping for the cost $c(x, y) = |x - y|$ is not unique.

Indeed, for any measure preserving mapping $s: (0,1) \to (0,1)$, the total cost $\mathcal{C}$, defined in (1.2), is a constant independent of the mapping s. However there is a unique mapping T among all the optimal mappings which is monotone [17], namely

$$(T(y) - T(x)) \cdot (y - x) \geq 0.$$

This unique mapping is the limit of the optimal mappings for the cost $c_p(x,y) = |x-y|^p$ with $p \searrow 1$, or for the $c_\epsilon(x,y) = \sqrt{\epsilon^2 + |x-y|^2}$ with $\epsilon \searrow 0$, if the optimal mappings are convergent. In a recent paper [25], some uniform estimates for the optimal mapping T_ϵ, corresponding to the cost function c_ϵ, were obtained. More precisely, assuming that the domains and mass densities satisfy the assumptions in Theorem 4.1, they proved that the eigenvalues of the Jacobian of T_ϵ (which are positive) are uniformly bounded for $\epsilon > 0$, but they also found an example which shows that the Jacobian matrix is not uniformly bounded.

References

[1] L. Ambrosio, Optimal transport maps in Monge-Kantorovich problem, *Proc. Inter. Cong. Math.* **III** (Beijing, 2002), 131–140.

[2] Y. Brenier, Polar factorization and monotone rearrangement of vector-valued functions, *Comm. Pure Appl. Math.* **44** (1991), 375–417.

[3] L. Caffarelli, Interior $W^{2,p}$ estimates for solutions of Monge-Ampère equations, *Ann. of Math.* **131** (1990), 135–150.

[4] L. Caffarelli, The regularity of mappings with a convex potential, *J. Amer. Math. Soc.* **5** (1992), 99–104.

[5] L. Caffarelli, Boundary regularity of maps with convex potentials II. *Ann. of Math.* **144** (1996), no. 3, 453–496.

[6] L. Caffarelli, Nonlinear elliptic theory and the Monge-Ampère equation, *Proc. Inter. Cong. Math.* **I** (Beijing, 2002), 179–187.

[7] L. Caffarelli, M. Feldman, and R. McCann, Constructing optimal maps for Monge's transport problem as a limit of strictly convex costs, *J. Amer. Math. Soc.* **15** (2002), 1–26.

[8] T. Champion and L. De Pascale, The Monge problem for strictly convex norms in $\mathbf{R}^d$, *J. Eur. Math. Soc.* **12** (2010), 1355–1369.

[9] T. Champion and L. De Pascale, The Monge problem in $\mathbf{R}^d$, *Duke Math. J.* **157** (2011), 551–572.

[10] P. Delanoë, Classical solvability in dimension two of the second boundary value problem associated with the Monge-Ampère operator, *Ann. Inst. Henri Poincaré, Analyse Non Linéaire* **8** (1991), 443–457.

[11] P. Delanoë and Y. Ge, Regularity of optimal transport on compact locally nearly spherical manifolds, *J. Reine Angew. Math.* **646** (2010), 65–115.

[12] S.Z. Du and Q.-R. Li, Positivity of Ma-Trudinger-Wang curvature on Riemannian manifolds, preprint.

[13] L.C. Evans, Partial differential equations and Monge-Kantorovich mass transfer, *Current Developments in Mathematics*, 65–126, Int. Press, Boston, MA, 1999.

[14] L.C. Evans and W. Gangbo, Differential equations methods for the Monge-Kantorovich mass transfer problem, *Mem. Amer. Math. Soc.* **653** (1999).

[15] W. Gangbo and R.J. McCann, Optimal maps in Monge's mass transport problem, *C.R. Acad. Sci. Paris* Series I, Math. **321** (1995), 1653–1658.

[16] W. Gangbo and R.J. McCann, The geometry of optimal transportation, *Acta Math.* **177** (1996), 113–161.

[17] M. Feldman and R. McCann, Uniqueness and transport density in Monge's mass transportation problem, *Cal. Var. PDE* **15** (2002), 81–113.

[18] A. Figalli, Y.-M. Kim, and R. McCann, Hölder continuity and injectivity of optimal maps, *Arch. Rational Mech. Anal.* **209** (2013), 747–795.

[19] A. Figalli, L. Rifford, and C. Villani, On the Ma-Trudinger-Wang curvature on surfaces, *Calc. Var. PDE* **39** (2010), 307–332.

[20] A. Figalli, L. Rifford, and C. Villani, Nearly round spheres look convex, *Amer. J. Math.* **134** (2012), 109–139.

[21] N. Guillen and J. Kitagawa, On the local geometry of maps with c-convex potential, arXiv:1212.4865.

[22] H.Y. Jian and X.-J. Wang, Continuity estimates for the Monge-Ampère equation, *SIAM J. Math. Anal.* **39** (2007), 608–626.

[23] Y.H. Kim, Counterexamples to continuity of optimal transport maps on positively curved Riemannian manifolds, *Int. Math. Res. Notice*, 2008, Art. ID rnn120, 15 pp.

[24] Y.H. Kim and R. McCann, Continuity, curvature, and the general covariance of optimal transportation, *J. Eur. Math. Soc.* **12** (2010), 1009–1040.

[25] Q.R. Li, F. Santambrogio, and X.-J. Wang, Regularity in Monge's mass transfer problem, preprint.

[26] L.K. Liu, Hölder regularity of optimal mappings in optimal transportation, *Calc. Var. PDE* **34** (2009), 435–451.

[27] L.K. Liu, N.S. Trudinger, and X.-J. Wang, Interior $C^{2,\alpha}$ regularity for potential functions in optimal transportation, *Comm. Partial Diff. Eqns* **35** (2010), 165–184.

[28] L.K. Liu, N.S. Trudinger, and X.-J. Wang, On asymptotic behaviour and $W^{2,p}$ regularity of potentials in optimal transportation, preprint.

[29] G. Loeper, On the regularity of solutions of optimal transportation problems, *Acta Math.* **202** (2009), 241–283.

[30] G. Loeper, Regularity of optimal maps on the sphere: the quadratic cost and the reflector antenna, *Arch. Ration. Mech. Anal.* **199** (2011), 269–289.

[31] G. Loeper and C. Villani, Regularity of optimal transport in curved geometry: the nonfocal case, *Duke Math. J.* **151** (2010), 431–485.

[32] X.N. Ma, N.S. Trudinger, and X.-J. Wang, Regularity of potential functions of the optimal transportation problem, *Arch. Rat. Mech. Anal.* **177** (2005), 151–183.

[33] R.J. McCann, Polar factorization of maps on Riemannian manifolds, *Geom. Funct. Anal.* **11** (2001), 589–608.

[34] G. Monge, Memoire sur la Theorie des Déblais et des Remblais, *Historie de l'Académie Royale des Sciences de Paris, avec les Mémoires de Mathématique et de Physique pour la Même année,* 1781, 666–704.

[35] S.T. Rachev and L. Ruschendorff, Mass Transportation Problems, Springer, Berlin, 1998.

[36] F. Schulz, Regularity theory for quasilinear elliptic systems and MongeAmpère equation in two dimension, *Lecture Notes in Mathematics* **1445** (1990), Springer.

[37] V.N. Sudakov, Geometric problems in the theory of infinite dimensional probability distributions, *Proceedings of Steklov Institute* **141** (1979), 1–178.

[38] N.S. Trudinger and X.-J. Wang, On the Monge mass transfer problem, *Calc. Var. PDE* **13** (2001), 19–31.

[39] N.S. Trudinger and X.-J. Wang, The Monge-Ampère equation and its geometric applications, *Handbook of geometric analysis*, Vol. **1**, 467–524, Int. Press, Somerville, MA, 2008.

[40] N.S. Trudinger and X.-J. Wang, On strict convexity and continuous differentiability of potential functions in optimal transportation, *Arch. Rat. Mech. Anal.* **192** (2009), 403–418.

[41] N.S. Trudinger and X.-J. Wang, On the second boundary value problem for Monge-Ampère type equations and optimal transportation, *Ann. Sc. Norm. Super. Pisa Cl. Sci.* (5) **8** (2009), 143–174.

[42] J. Urbas, On the second boundary value problem for equations of Monge-Ampère type, *J. Reine Angew. Math.* **487** (1997), 115–124.

[43] C. Villani, Topics in the optimal transportation, Graduate Studies in Mathematics **58**, *Amer. Math. Soc.*, 2003.

[44] C. Villani, Optimal Transport, Old and New, Springer-Verlag, Berlin, 2009.

[45] C. Villani, Stability of a 4th-order curvature condition arising in optimal transport theory, *J. Funct. Anal.* **255** (2008), 2683–2708.

[46] X.-J. Wang, On the design of a reflector antenna, *Inverse Problems* **12** (1996), 351–375.

[47] X.-J. Wang, On the design of a reflector antenna II, *Calc. Var. PDE* **20** (2004), 329–341.

[48] X.-J. Wang, Schauder estimates for elliptic and parabolic equations, Chinese Ann. Math. Series B, **27** (2006), 637–642.

Centre for Mathematics and Its Applications
The Australian National University
Canberra, ACT 0200, Australia
E-mail address: Xu-Jia.Wang@anu.edu.au

Proceedings of the 6th International Congress of Chinese Mathematicians
ALM 36, pp. 173–197

Recent Results on Rigidity of CR Morphisms Between Compact Strongly Pseudoconvex CR Manifolds

Stephen Yau and Huaiqing Zuo

Dedicated to Professor George Daniel Mostow on the occasion of his 90th birthday

Abstract

The purpose of this paper is to summarize the results on rigidity of CR morphisms between compact strongly pseudoconvex CR manifolds developed by the first author and his coauthors. This theory is related to the rigidity of etale covering between resolutions of isolated singularities.

2010 Mathematics Subject Classification: 32V10, 14B05, 32S05, 32S45.

Keywords and Phrases: CR morphism, CR manifold, rigidity, CR automorphism.

1 Introduction

Rigidity phenomena in complex geometry have received a lot of attention historically. Borel and Narasimhan [Bo-Na67] obtained some general results on rigidity of morphisms with domain space carrying no non-constant pseudoconvex function which is bounded above and target space being covered by analytic subset of a bounded domain in $\mathbb{C}^n$. In 1975, Kobayashi and Ochiai [Ko-Oc75] proved that there are only finitely many surjective morphisms between two fixed projective manifolds of general type. In 1981, Kalka, Shiffman and Wong [KSW81] developed a

Stephen Yau is supported by NSFC (grant no. 11531007) and the start-up fund from Tsinghua University. Huaiqing Zuo is supported by the NSFC (grant nos. 11401335, 11531007), Tsinghua University Initiative Scientific Research Program and the start-up fund from Tsinghua University.

general theory to study the finiteness and rigidity theorems for holomorphic mappings. Let $\mathrm{Hol}_k(X,Y)$ denote the complex space of holomorphic maps of rank $\geqslant k$ from the compact complex space X into the complex manifold Y. They proved that if Y satisfies certain convexity or cohomological conditions, then for suitable k, $\mathrm{Hol}_k(X,Y)$ is either discrete or finite. They also showed that if the tangent bundle of Y satisfies k-pseudoconvexity condition, then $\mathrm{Hol}_k(X,Y)$ is discrete. As a corollary they asserted that if Y is a compact Hermitian manifold with negative holomorphic sectional curvature, then the set of surjective holomorphic maps from X onto Y is finite. On the other hand, if Y is a compact Kähler manifold with $c_1(Y)$ represented by a negative semi-definite form and either $c_n(Y) \neq 0$ or $c_1^n(Y) \neq 0$, Kalka, Shiffman and Wong showed that $\mathrm{Hol}_n(X,Y)$ is discrete.

Inspiring by the Grauert-Oka principle, Mok [Mok84] formulated the following problem on n-dimensional Stein spaces.

Problem. Suppose V is an n-dimensional Stein space which has the homotopy type of a real n-dimensional CW-complex. Suppose $f : V \to V$ is a holomorphic homotopy equivalence. Does it follow that f is a biholomorphism?

Mok [Mok84] pointed out that obvious counter-example on the punctured disc show that the answer is in general negative even in the case of bounded domains and injective holomorphic self-mappings. In [Mok84] , instead of assuming that f is a homotopy equivalence, Mok assumes that V is hyperbolic in the sense of Caratheodory and f induces an isomorphism on the n-th integral homology group $H_n(V,\mathbb{Z})$, which is assumed to be nonzero and finitely generated. Then he proved that f is a biholomorphism.

Recently Huang, Kebekus and Peternell [H-K-P06] have proved the following beautiful result. Let Y be a projective n-dimensional manifold which is not uniruled. If either $\pi_1(Y)$ is finite or $c_n(Y) \neq 0$, then for each connected normal compact complex variety X, the space of surjective morphism from X to Y is discrete.

In 1977, Wong [Wo77] proved an important result in complex geometry that any strongly pseudoconvex domain with smooth boundary in $\mathbb{C}^n$ with noncompact automorphism group must be biholomorphically equivalent to the unit ball. In 1978, Yau [Ya78] proved the Calabi Conjecture. As a consequence, he [Ya77] proved that $3c_2 \geqslant c_1^2$ for any Kähler surface with ample canonical bundle, and the equality holds if and only if the surface is covered by the ball in $\mathbb{C}^2$. Using this results, he [Ya77] proved the Severi Conjecture that every complex surface which is homotopic to the complex projective plane $\mathbb{CP}^2$ is biholomorphic to $\mathbb{CP}^2$.

As a consequence of his famous strong rigidity theorem [Mo83] Mostow showed that two compact quotients of the ball of complex dimension $\geqslant 2$ with isomorphic fundamental groups are either biholomorphic or conjugate biholomorphic. S.-T. Yau conjectured that this phenomenon of strong rigidity should hold also for compact Kähler manifolds of complex dimension $\geqslant 2$ with same homotopy type and negative sectional curvature. In his paper [Siu80], Siu proved that Yau's conjecture is true when the curvature tensor of one of the two compact Kähler manifolds is strongly negative with no curvature assumption on the other manifold.

CR manifolds are abstract models of boundary of complex manifolds. Strongly pseudoconvex CR manifolds have rich geometric and analytic structures. The harmonic theory for the $\overline{\partial}_b$ complex on compact strongly pseudoconvex CR manifolds was developed by Kohn [Kohn65]. Using this theory, Boutet de Monvel [BM75] proved that if X is a compact strongly pseudoconvex CR manifold of real dimension $2n-1$, $n \geqslant 3$ then there exist C^∞ functions $f_1, \ldots, f_N$ on X such that each $\overline{\partial}_b f_j = 0$ and $f = (f_1, \ldots, f_N)$ defines an embedding of X in $\mathbb{C}^N$. Thus, any compact strongly pseudoconvex CR manifold of dimension $\geqslant 5$ can be CR embedded in some complex Euclidean space. On the other hand, 3-dimensional strongly pseudoconvex compact orientable CR manifolds are not necessarily embeddable. Throughout this paper, our strongly pseudoconvex CR manifolds are always assumed to be compact orientable and embedded in some $\mathbb{C}^N$. By a beautiful theorem of Harvey and Lawson [Ha-La00, Ha-La75], these CR manifolds are the boundaries of subvarieties with only isolated normal singularities.

Rigidity problems of CR immersions into spheres and hyperquadrics were studied by Ebenfelt, Huang and Zaitsev [EHZ04, EHZ05]. Let X be a strongly pseudoconvex CR manifold of dimension $2n-1$. For $p \in X$, let $f\colon (X, p) \longrightarrow S^{2n+2d-1}$ be a local CR immersion of X near p into unit sphere $S^{2n+2d-1}$ in $\mathbb{C}^{n+d}$. The result of [EHZ04] states that if $d < \frac{n}{2} - 1$, then f is rigid in the sense that any other immersion of (X, p) into $S^{2n+2d-1}$ is of the form $\phi \circ f$, where ϕ is biholomorphic automorphism of the unit ball $B \subseteq \mathbb{C}^{n+d}$. As a striking corollary, they show that if X and X' are two strongly pseudoconvex CR manifolds of dimension $2n-1$ in $S^{2n+2d-1}$ with $d < \frac{n}{2} - 1$ and if X and X' are locally CR equivalent at some points $p \in X$ and $p' \in X'$, then there exists a unitary linear transformation which maps X to X'.

In a remarkable paper [Pin74], Pinchuk showed that a proper holomorphic mapping between strongly pseudoconvex domains in $\mathbb{C}^n$ is locally biholomorphic. In fact he proved that proper holomorphic self-maps of strongly pseudoconvex domains are necessarily biholomorphic. It was proved in [Be-Ca82] and [Di-Fo82] that proper holomorphic maps extend smoothly to the boundaries and hence induce CR morphisms between the boundaries.

In [Yau11], the first author of this paper took another point of view. In view of Fornaess theory on strongly pseudoconvex domains [For76], he investigated the rigidity property of CR morphisms between strongly pseudoconvex CR manifolds by means of the singularities theory. Later in his joint work with Tu and Zuo [TYZ13], the rigidity of CR morphisms between CR manifolds lying in the same variety is proved. The purpose of this paper is to summarize the results on rigidity of CR morphisms between strongly pseudoconvex CR manifolds by means of the singularity theory.

In Section 2, we shall recall some basic notions and theorems of CR manifolds and the interplay between CR manifold and singularities. In Section 3, we shall discuss the rigidity of CR morphisms between any two compact strongly pseudoconvex CR manifolds. In Section 4, we will show that the rigidity of CR morphisms between two compact strongly pseudoconvex CR manifolds lying in a same variety with isolated singularities is related to the rigidity of etale covering between resolutions of isolated singularities. In Section 5, we give examples of strongly pseudoconvex 3-dimensional CR manifolds and computes their non-constant CR endomorphisms.

2 Preliminaries on CR manifolds

Definition 2.1. *Let X be a connected oriented orientable manifold of real dimension $2n-1$. A CR structure on X is an $(n-1)$-dimensional subbundle S of the complexified tangent bundle $\mathbb{C}TX$ such that*

(1) $S \cap \overline{S} = \{0\}$;

(2) *If L, L' are local sections of X, then so is $[L, L']$.*

A manifold with a CR structure is called a CR manifold. There is a unique subbundle $\mathcal{H}$ of the tangent bundle $T(X)$ such that $\mathbb{C}\mathcal{H} = S \oplus \overline{S}$. Furthermore, there is a unique homomorphism $J : \mathcal{H} \to \mathcal{H}$ such that $J^2 = -1$ and $S = \{v - iJv : v \in \mathcal{H}\}$. The pair $(\mathcal{H}, J)$ is called the real expression of the CR structure.

Definition 2.2. *Let $L_1, \ldots, L_{n-1}$ be a local frame of S. Then $\overline{L}_1, \ldots, \overline{L}_{n-1}$ is a local frame of $\overline{S}$ and one may choose a local section N of $T(X)$ which is purely imaginary, such that $L_1, \ldots, L_{n-1}, \overline{L}_1, \ldots, \overline{L}_{n-1}, N$ is a local frame of $\mathbb{C}T(X)$. The matrix (c_{ij}) defined by*

$$[L_i, \overline{L}_j] = \sum a_{ij}^k L_k + \sum b_{ij}^k \overline{L}_k + \sqrt{-1} c_{ij} N$$

is Hermitian and called the Levi form of X.

Proposition 2.1. *The number of nonzero eigenvalues and the absolute value of the signature of the Levi form (c_{ij}) at each point are independent of the choice of $L_1, \ldots, L_{n-1}, N$.*

Definition 2.3. *The CR manifold X is called strongly pseudoconvex if the Levi form is definite at each point of X.*

Theorem 2.1. (Boutet de Monvel [BM75]) *If X is a compact strongly pseudoconvex CR manifold of dimension $(2n-1)$ and $n \geq 3$, then X is CR embeddable in $\mathbb{C}^N$.*

Although there are non-embeddable compact 3-dimensional CR manifolds, in this paper all CR manifolds are assumed to be embeddable in complex Euclidean space.

Theorem 2.2. (Harvey-Lawson [Ha-La75, Ha-La00]) *For any compact connected embeddable CR manifold X, there is a unique complex variety V in $\mathbb{C}^N$ for some N such that the boundary of V is X and V has only normal isolated singularities.*

With the notation in Definition 2.1, for any C^∞ function u, there is a section $\overline{\partial}_b u \in \Gamma(\overline{S}^*)$ defined by $(\overline{\partial}_b u)(\overline{L}) = \overline{L}u$ for any $L \in \Gamma(S)$. This can be generalized as follows:

Definition 2.4. *A complex vector bundle E over X is said to be holomorphic if there is a differential operator $\overline{\partial}_E : \Gamma(E) \to \Gamma(E \otimes \overline{S}^*)$ such that if $\overline{L}u$ denotes $(\overline{\partial}_E u)(\overline{L})$ for $u \in \Gamma(E)$ and $L \in \Gamma(S)$, then for any L_1, $L_2 \in \Gamma(S)$ and any C^∞ function f on X*

(1) $\overline{L}(fu) = (\overline{L}f)u + f(\overline{L}u)$;

(2) $[\overline{L}_1, \overline{L}_2]u = \overline{L}_1\overline{L}_2 u - \overline{L}_2\overline{L}_1 u$.

A solution u of the equation $\overline{\partial}_E u = 0$ is called a holomorphic section.

The vector bundle $\widehat{T}(X) = \mathbb{C}T(X)/\overline{S}$ is holomorphic with respect to the following $\overline{\partial} = \overline{\partial}_{\widehat{T}(X)}$. Let ω be the projection from $\mathbb{C}T(X)$ to $\widehat{T}(X)$. Take any $u \in \Gamma(\widehat{T}(X))$ and express it as $u = \omega(Z)$, $Z \in \Gamma(\mathbb{C}T(X))$. For any $L \in \Gamma(S)$, define $(\overline{\partial}u)(\overline{L}) = \omega([\overline{L}, Z])$. The section $(\overline{\partial}u)(\overline{L})$ of $\widehat{T}(X)$ does not depend on the choice of Z and $\overline{\partial}u$ gives a section of $\widehat{T}(X) \otimes \overline{S}^*$. Further the operator $\overline{\partial}$ satisfies the conditions in Definition 2.4. The resulting holomorphic vector bundle $\widehat{T}(X)$ is called the holomorphic tangent bundle of X.

Lemma 2.1. *If X is a real hypersurface in a complex manifold M, then the holomorphic tangent bundle $\widehat{T}(X)$ is naturally isomorphic to the restriction to X of the bundle $T^{1,0}(M)$ of all $(1,0)$ tangent vectors to M.*

Definition 2.5. *Let X be a compact CR manifold of real dimension $2n-1$. X is said to be a Calabi-Yau CR manifold if there exists a nowhere vanishing holomorphic section in $\Gamma(\Lambda^n\widehat{T}(X)^*)$ where $\widehat{T}(X) = \mathbb{C}T(X)/S^*$ is the holomorphic tangent bundle of X.*

Remark 2.1. (a) *Let X be a compact CR manifold of real dimension $2n-1$ which is a boundary of domain in $\mathbb{C}^n$. Then X is a Calabi-Yau CR manifold.*

(b) *Let X be a strongly pseudoconvex CR manifold of real dimension $2n-1$ contained in the boundary of a bounded strongly pseudoconvex domain in $\mathbb{C}^{n+1}$. Then X is a Calabi-Yau CR manifold.*

(c) *More generally, let V be a Stein variety V with singular set S such that the canonical bundle of $V - S$ is trivial. If X is a real hypersurface in $V - S$, then X is a Calabi-Yau CR manifold.*

Definition 2.6. *Let X be a compact connected strongly pseudoconvex embeddable CR manifold of real dimension $2n-1$. Let V be the normal subvariety in $\mathbb{C}^N$ such that the boundary of V is X. Let $\pi : M \to V$ be a resolution of singularities of V. The geometric genus of X denoted by $p_g(X)$ is defined to be $\dim H^{n-1}(M, \mathcal{O})$.*

Proposition 2.2. ([Ya-Yu02]) *Let X be a connected compact strongly pseudoconvex CR manifold of real dimension $2n-1$ and $n \geq 2$. Suppose that X bounds a normal variety $V \subseteq \mathbb{C}^N$ with isolated singularities $Y = \{q_1, \dots, q_m\}$. Let $\pi : M \to V$ be a resolution of singularities of V. Then the geometric genus $p_g(X) := \dim H^{n-1}(M, \mathcal{O})$ is a CR invariant of X. In fact, let U be any small strongly pseudoconvex neighborhood of Y. Then*

$$p_g(X) = \dim H^0(U - Y, \Omega^n)/L^2(U - Y, \Omega^n)$$

where Ω^n is the sheaf of germs of holomorphic n-forms and $L^2(U - Y, \Omega^n)$ is the space of holomorphic n-forms ω on $U - Y$ which are L^2-integrable for some small neighborhood $U' \subset\subset U$, i.e., $\int_{U'-Y} \omega \wedge \overline{\omega} < \infty$.

Definition 2.7. *Let (V, q) be a complex analytic with isolated normal singularity at q of dimension n. The geometric genus $p_g(V, q)$ of the singularity (V, q) is defined to be* $\dim H^0(U - \{q\}, \Omega^n)/L^2(U - \{q\}, \Omega^n)$ *where U is any strongly pseudoconvex neighborhood of q.*

Remark 2.2. (a) *Let $\pi : (M, A) \to (U, q)$ be a resolution of singularity. By a theorem of* [Yau77], $H^{n-1}(M, \mathcal{O}) = \dim H^0(V - \{q\}, \Omega^n)/L^2(V - \{q\}, \Omega^n)$. *It follows easily that $p_g(V, q)$ is independent of U in the above definition.*

(b) *With the notation in Proposition* 2.2, *we have* $p_g(X) = \sum_{i=1}^{m} p_g(V, q_i)$.

3 General theory on rigidity of CR morphism

The starting point of our investigation on CR morphisms between two embeddable strongly pseudoconvex compact CR manifolds is the following result.

Theorem 3.1. *Let X_1 and X_2 be two compact connected strongly pseudoconvex CR manifolds of dimension $2n - 1 \geq 3$ which bound complex varieties V_1 and V_2 in $\mathbb{C}^{N_1}$ and $\mathbb{C}^{N_2}$ respectively. Suppose the singular set S_i of V_i, $i = 1, 2$ is either an empty set or a set consisting of only isolated normal singularities. If $\phi\colon X_1 \to X_2$ is a non-constant CR morphism, then ϕ is surjective and ϕ can be extended to a proper surjective holomorphic map from V_1 to V_2 such that $\phi(S_1) \subseteq S_2$, $\phi^{-1}(X_2) = X_1$ and $\phi\colon V_1 - \phi^{-1}(S_2) \to V_2 - S_2$ is a covering map. Moreover, if S_2 does not have quotient singularity, then $\phi^{-1}(S_2) = S_1$.*

Proof. Let $\phi_1, \ldots, \phi_{N_2}$ be the component functions of ϕ. Then ϕ_i as CR holomorphic function on X_1 can be extended onto a one-sided neighborhood of X_1 in V_1. By Andreotti and Grauert ([An-Gr62], Théorème 15), ϕ_i can be holomorphically extended to $V_1 - S_1$ where S_1 is the singular set of V_1. Since S_1 is either an empty set or a set consisting of only isolated normal singularities, ϕ_i can be holomorphically extended onto V_1.

We claim that $\phi(V_1) \subseteq V_2$. To see this, let $f_1, \ldots, f_k$ be the defining equations of V_2, i.e. $V_2 = \{y \in \mathbb{C}^{N_2} : f_1(y) = \cdots = f_k(y) = 0\}$. Clearly $\phi^*(f_i) = f_i \circ \phi$ is a holomorphic function on V_1 which vanishes on X_1 for $1 \leq i \leq k$. Since X_1 is of real codimension one in V_1, $\phi^*(f_i)$ is identically zero on V_1 for $1 \leq i \leq k$. This implies that $\phi(V_1) \subseteq V_2$. By the maximum principle, $\phi(X_1) \cap \phi(V_1 - X_1) = \emptyset$. It follows that $\phi^{-1}(X_2) = X_1$ and ϕ is a proper map from V_1 to V_2. By the proper mapping theorem, $\phi(V_1)$ is a complex variety.

We next claim that $\dim \phi(V_1) = n$. If $\dim \phi(V_1) < n$, then for some q in $\phi(V_1)$, $\phi^{-1}(q)$ is a compact variety of dimension at least one sitting inside V_1. This gives a contradiction since V_1 is Stein. As $\phi(V_1) \subseteq V_2$ and $\dim V_1 = n = \dim V_2$, we have $\phi(V_1) = V_2$. It follows that $\phi(X_1) = X_2$. A local computation of Fornaess ([For76], Proposition 12) would apply to show that ϕ is a local biholomorphism near X_1. (This was observed independently by Pinchuk [Pin74].) In particular, $\phi : V_1 - \phi^{-1}(S_2 \cup \phi(S_1)) \to V_2 - (S_2 \cup \phi(S_1))$ is locally biholomorphic and hence is a finite covering.

Let $p \in S_1$ and $q = \phi(p)$. We claim that $q \in S_2$. Suppose on the contrary that q is a smooth point in V_2; then ϕ maps a neighborhood U_1 of p to a neighborhood U_2 of q as a branch covering. Since p is a normal singularity, the punctured neighborhood $U_1 - \{p\}$ of p is connected. On the other hand, the punctured neighborhood $U_2 - \{q\}$ of q is simply connected because q is a smooth point. We conclude that $\phi|_{U_1} : U_1 \to U_2$ is one-to-one and onto. By the Hartogs extension theorem, the inverse map $\phi^{-1}|_{U_2-\{q\}} : U_2 - \{q\} \to U_1 - \{p\}$ can be holomorphically extended onto U_2. It follows that $\phi|_{U_1} : U_1 \to U_2$ is a biholomorphic map. This leads to a contradiction. Therefore $\phi(S_1) \subseteq S_2$ and $\phi : V_1 - \phi^{-1}(S_2) \to V_2 - S_2$ is a covering map.

Now assume that S_2 does not have a quotient singularity. Let q be any point in S_2. We need to show that $\phi^{-1}(q) \subseteq S_1$. If $\phi^{-1}(q)$ is not contained in S_1, then there exists a smooth point q' of V_1 in $\phi^{-1}(q)$. Recall that $\phi^{-1}(q)$ is a finite set. We can find an open neighborhood U of q' which is biholomorphic to a domain in $\mathbb{C}^n$ such that $\phi|_U$ from U to the germ of (V_2, q) is a branch covering with ramification locus $\{q'\}$. By Theorem 1 of [Pr67], we conclude that (V_2, q) is a quotient singularity. This is a contradiction. □

As a corollary of Theorem 3.1, we have the following super-rigidity results of CR morphisms between strongly pseudoconvex manifolds.

Corollary 3.1. *Let X_1 be a compact strongly pseudoconvex CR manifold of dimension $2n-1 \geq 3$ which bounds a complex variety V_1 in $\mathbb{C}^{N_1}$ with isolated normal singularities. Let X_2 be a compact strongly pseudoconvex CR manifold of dimension $2n-1$ which bounds a complex submanifold V_2 in $\mathbb{C}^{N_2}$. Then there is no non-constant CR morphism from X_1 to X_2.*

It is a natural question to ask what happens if we interchange the roles of X_1 and X_2 in Corollary 3.1.

Proposition 3.1. *Let X_1 be a compact strongly pseudoconvex CR manifold of dimension $2n-1 \geq 3$ which bounds a complex submanifold V_1 in $\mathbb{C}^{N_1}$. Let X_2 be a compact strongly pseudoconvex CR manifold of dimension $2n-1$ with either* (i) *geometric genus $p_g(X_2) > 0$, or* (ii) *$p_g(X_2) \geq 0$ and X_2 bounds a complex variety V_2 in $\mathbb{C}^{N_2}$ with a non-quotient singularity. Then there is no non-constant CR morphism from X_1 to X_2.*

Proof. By Theorem 3.1, if there exists a non-constant CR morphism $\phi : X_1 \to X_2$, then ϕ can be extended as a ramified covering map from V_1 to V_2 with ramification locus S_2. Since V_1 is smooth, by the proof of Theorem 3.1, S_2 consists of only quotient singularities and hence the geometric genus of these singularities are zero. It follows that $p_g(X_2) = 0$ in view of Remark 2.2(b). This leads to a contradiction. □

Remark 3.1. *Proposition* 3.1 *is false if $p_g(X_2) = 0$ and V_2 has only quotient singularities in the interior, as we can see from the following example.*

Example 3.1. *Let $B = \{(x,y) \in \mathbb{C}^2 : |x|^2+|y|^2 < 1\}$ and $S = \partial B$. In the notation of Proposition* 3.1, *let $X_1 = S$ be the standard sphere and $V_1 = B$. Let $\sigma : B \to B$*

be the map given by $\sigma(x,y)=(-x,-y)$. *Let* V_2 *be the quotient of* V_1 *by the cyclic group of order 2 generated by* σ. *Then* V_2 *is a strongly pseudoconvex variety with* A_1 *singularity* $\{(z_1,z_2,z_3)\in\mathbb{C}^3 : z_1z_2=z_3^2\}$. *The quotient map* $\phi: V_1\to V_2$ *is given by* $(z_1,z_2,z_3)=(x^2,y^2,xy)$. *Clearly* ϕ *sends* X_1 *surjectively onto* $X_2=\partial V_2$ *and* ϕ *is a non-constant CR morphism.*

Proposition 3.2. *Let* X_1 *and* X_2 *be two compact strongly pseudoconvex embeddable CR manifolds of dimension* $2n-1\geq 3$. *If there is a non-constant CR morphism from* X_1 *to* X_2, *then* $p_g(X_1)\geq p_g(X_2)$.

Proof. Let V_i be a normal variety in $\mathbb{C}^{N_i}$ with only isolated singularities such that $\partial V_i = X_i, i=1,2$. Let S_1 and S_2 be the singular set of V_1 and V_2 respectively. Let $\phi: X_1\to X_2$ be a non-constant CR morphism. In view of Theorem 3.1, ϕ can be extended to a proper surjective holomorphic map from V_1 and V_2 such that $\phi(S_1)=S_2$, and $\phi: V_1-\phi^{-1}(S_2)\to V_2-S_2$ is a covering map. There is a natural map

$$\phi^*: \frac{\Gamma(V_2-S_2,\Omega^n)}{L^2(V_2-S_2,\Omega^n)} \to \frac{\Gamma(V_1-\phi^{-1}(S_2),\Omega^n)}{L^2(V_1-\phi^{-1}(S_2),\Omega^n)}.$$

Since $\phi: V_1-\phi^{-1}(S_2)\to V_2-S_2$ is a finite covering map, a form $w\in\Gamma(V_2-S_2,\Omega^n)$ is L^2-integrable if and only if $\phi^*(w)$ is L^2-integrable. Thus ϕ^* is injective. Observe that $\phi^{-1}(S_2)-S_1$ is a discrete subset in the smooth part of V_1. By Hartog's theorem, $\Gamma(V_1-\phi^{-1}(S_2),\Omega^n)=\Gamma(V_1-S_1,\Omega^n)$ and $L^2(V_1-\phi^{-1}(S_2),\Omega^n)=L^2(V_1-S_1,\Omega^n)$. It follows that $p_g(X_2)\leq p_g(X_1)$. □

Corollary 3.2. *Let* X_1, X_2 *be two compact strongly pseudoconvex embeddable CR manifolds of dimension* $2n-1\geq 3$. *If* $p_g(X_1)<p_g(X_2)$, *then there is no non-constant CR morphism from* X_1 *to* X_2.

The following theorem says that if the codimension of X_2 is small and $\dim X_2\geq 5$, then there is no non-constant CR morphism from X_1 to X_2 except CR biholomorphic maps. This rigidity phenomenon does not require any curvature assumption on X_1 or X_2.

Theorem 3.2. *Let* X_1 *and* X_2 *be two compact strongly pseudoconvex CR manifolds of dimension* $2n-1\geq 5$ *which bound complex varieties* V_1 *and* V_2 *with only isolated normal singularities in* $\mathbb{C}^{N_1}$ *and* $\mathbb{C}^{N_2}$ *respectively. Let* S_1 *and* S_2 *be the singular sets of* V_1 *and* V_2 *respectively and* S_2 *is non-empty. Suppose* $2n-N_2-1\geq 1$. *Then there exists no non-constant CR morphism from* X_1 *to* X_2 *if* $|S_1|$ *is not divisible by* $|S_2|$. *If* $|S_1|=|S_2|$, *then* X_1 *is CR biholomorphic to* X_2.

Proof. Let $\phi: X_1\to X_2$ be a non-constant CR morphism. Theorem 3.1 says that ϕ can be extended to a proper surjective holomorphic map from V_1 to V_2 such that $\phi(S_1)\subseteq S_2$ and $\phi: V_1-\phi^{-1}(S_2)\to V_2-S_2$ is a covering map of degree d. For any $q\in S_2$, we know that the punctured neighborhood of q in V_2 is $(2n-N_2-1)$-connected in view of a theorem of Hamm [Ham81]. Since $2n-N_2-1\geq 1$ by assumption, the punctured neighborhood of q is simply connected. We claim that $\phi^{-1}(q)\subseteq S_1$. If $\phi^{-1}(q)$ is not contained in S_1, then there exists a smooth point q' of V_1 in $\phi^{-1}(q)$. Recall that $\phi^{-1}(q)$ is a finite set. We can find an open neighborhood

U of q' which is biholomorphic to a domain in $\mathbb{C}^n$ such that $\phi|_U$ from U to the germ of (V_2, q) is a branch covering with ramification locus $\{q'\}$. Since the punctured neighborhood of q in V_2 is simply connected, this implies $\phi|_U$ is injective and hence $\phi|_U$ is a biholomorphism. This leads to a contradiction because q is a singular point. We have shown that $\phi^{-1}(q) = \{q'_1, \ldots, q'_d\} \subseteq S_1$. There are exactly d points in $\phi^{-1}(q)$ because the punctured neighborhood of q is simply connected. Therefore $|S_1|$ is divisible by $|S_2|$. On the other hand if $|S_1| = |S_2|$, then $d = 1$ and hence ϕ is a biholomorphism. □

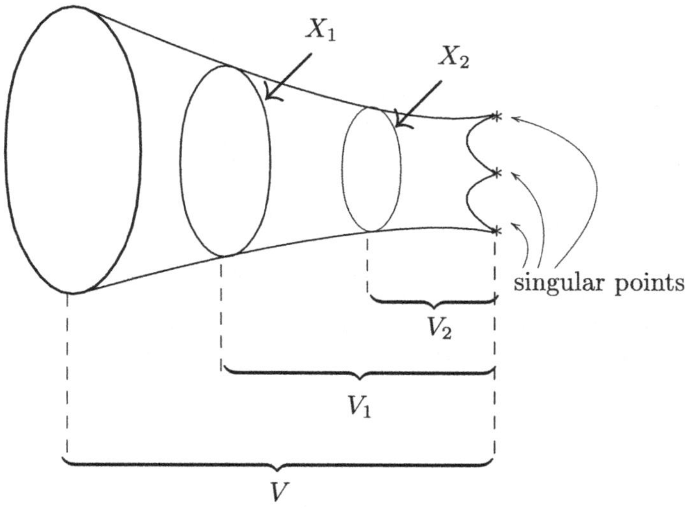

4 CR morphisms between two compact strongly pseudoconvex CR manifolds lying in a same variety

The main purpose of this section is to prove the following theorem.

Theorem 4.1. *Let X_1 and X_2 be two $(2n-1)$-dimensional compact strongly pseudoconvex CR manifolds lying in a Stein variety V of dimension n in $\mathbb{C}^N$. Let $V_1 \subseteq V$, $V_2 \subseteq V$ such that $\partial V_1 = X_1$ and $\partial V_2 = X_2$. Assume that the singular set S of V is nonempty and is equal to the singular set of V_i, $i = 1, 2$. Then nontrivial CR morphisms from X_1 to X_2 are necessarily CR biholomorphisms.*

Proof. Let $\Phi\colon X_1 \longrightarrow X_2$ be a non-constant CR morphism. In view of Theorem 3.1, Φ can be extended to a proper holomorphic map from V_1 to V_2 such that $\Phi\colon V_1 - \Phi^{-1}(S) \longrightarrow V_2 - S$ is a covering map of degree d and $\Phi(S) = S$. Let $S = \{q_1, \ldots, q_m\}$. Then $\Phi^{-1}(S) = \{q_1, \ldots, q_m, p_1, \ldots, p_k\}$. We shall prove that $\Phi^{-1}(S) = S$. Let $\pi\colon M \longrightarrow V_2$ be a resolution of singularities of V_2 such that the exceptional sets

$$E_1 = \pi^{-1}(q_1) = \bigcup_{i=1}^{\ell_1} A_i^1, \ \ldots, \ E_m = \pi^{-1}(q_m) = \bigcup_{i=1}^{\ell_m} A_i^m$$

are normal crossing divisors.

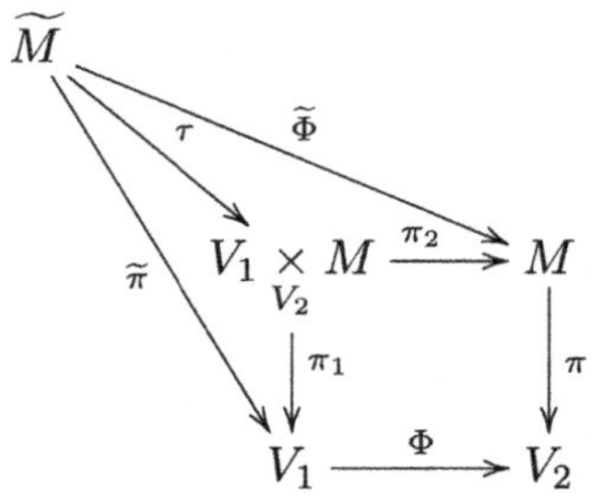

Consider the fiber product $V_1 \underset{V_2}{\times} M$ of the maps $\Phi\colon V_1 \longrightarrow V_2$ and $\pi\colon M \longrightarrow V_2$. Let $\tau\colon \widetilde{M} \longrightarrow V_1 \underset{V_2}{\times} M$ be the normalization map. Then we have the following commutative diagram where π_1 and π_2 are natural projections. Notice that $\pi_1\colon V_1 \underset{V_2}{\times} M \longrightarrow V_1$ is a biholomorphism outside $\pi_1^{-1}(\Phi^{-1}(S))$ and $\pi_2\colon V_1 \underset{V_2}{\times} M \to M$ is a covering map outside $\bigcup_{i=1}^{m} E_i$. Thus

$$\widetilde{\Phi} := \pi_2 \circ \tau\colon \widetilde{M} \longrightarrow M$$

is a d-fold branch covering. For each $A_i^j \subseteq E_j$, and any point $q_i^j \in A_i^j$ which is a smooth point in $\bigcup_{i=1}^{m} E_i$, we choose a germ of a curve Γ_i^j at the point q_i^j which intersects with $\bigcup_{i=1}^{m} E_i$ only at q_i^j and the intersection of A_i^j and Γ_i^j is transversal at q_i^j. Let $\Gamma = \bigcup \Gamma_i^j$, $1 \leqslant j \leqslant m$, $1 \leqslant i \leqslant \ell_j$. Notice that $\widetilde{\pi} := \pi_1 \circ \tau$ is a proper map which is a biholomorphism outside $\widetilde{E} := \widetilde{\pi}^{-1}(\Phi^{-1}(S)) = \widetilde{\Phi}^{-1}(E)$ where $E = E_1 \cup \cdots \cup E_m$. Observe that $\widetilde{E}$ has exactly $m + k$ connected components $\widetilde{E} = \widetilde{E}_1 \cup \cdots \cup \widetilde{E}_{m+k}$. Clearly

$$\widetilde{\Phi}_*(\widetilde{E}) = \sum_{i,j} d_i^j A_i^j, \text{ where } d_i^j \leqslant d.$$

By the projection formula (cf. p. 34 of [Ful98], or p. 426 of [Har77]),

$$\begin{aligned} \sum_{j=1}^{m} \sum_{i=1}^{\ell_j} d_i^j &= \Gamma \cdot \widetilde{\Phi}_*(\widetilde{E}) = \widetilde{\Phi}^*(\Gamma) \cdot \widetilde{E} \\ &\geqslant (\ell_1 + \cdots + \ell_m)d. \end{aligned}$$

The last inequality comes from the fact that $\widetilde{\Phi}^*(\Gamma_i^j)$ has d distinct branches because $\widetilde{\Phi}\colon \widetilde{M} - \widetilde{\Phi}^{-1}(E) \longrightarrow M - E$ is a d-fold covering space. Since $d_i^j \leqslant d$, we conclude that $d_i^j = d$ for all i, j. It follows that the branch locus of $\widetilde{\Phi}$ is contained in the singular locus of $\bigcup_{i=1}^{m} E_i$ which is of dimension $n-2$. As $\widetilde{M}$ is normal and M is smooth, $\widetilde{\Phi}\colon \widetilde{M} \longrightarrow M$ is a covering map by purity of branch locus. In particular, $\widetilde{M}$ is smooth.

Now we are ready to prove that $\Phi^{-1}(S) = S$, i.e., there are no $p_1, \ldots, p_k$ points in $\Phi^{-1}(S)$. Observe that $\widetilde{\pi}^{-1}(p_i)$, $1 \leqslant i \leqslant k$ and E_j, $1 \leqslant j \leqslant m$ are maximal compact connected analytic subsets in $\widetilde{M}$ and M respectively. Since $\widetilde{\Phi}$ is a covering map, there is a neighborhood $\widetilde{U}_i$ of $\widetilde{\pi}^{-1}(p_i)$ which maps biholomorphically to a neighborhood U_j of E_j for some j via $\widetilde{\Phi}$. As $\widetilde{\pi}\colon \widetilde{M} \longrightarrow V_1$ is a point modification in a neighborhood of p_i, there is a neighborhood D_i of p_i such that

$$\widetilde{\pi}\colon \widetilde{\pi}^{-1}(D_i) - \widetilde{\pi}^{-1}(p_i) \longrightarrow D_i - \{p_i\}$$

is a biholomorphism. Similarly, there is a neighborhood O_j of q_j such that

$$\pi\colon \pi^{-1}(O_j) - E_j \longrightarrow O_j - \{q_j\}$$

is a biholomorphism. Therefore

$$\pi \circ \widetilde{\Phi} \circ \widetilde{\pi}^{-1}\colon D_i - \{p_i\} \longrightarrow O_j - \{q_j\}$$

is a biholomorphism. Observe that p_i is a smooth point of D_i and q_j is an isolated normal singularity. It follows that $\pi \circ \widetilde{\Phi} \circ \widetilde{\pi}^{-1}$ extends to a biholomorphism from D_i to O_j. In particular q_j is not a singular point. This contradiction shows that $\Phi^{-1}(S) = S$ and hence $\widetilde{\pi}\colon \widetilde{M} \longrightarrow V_1$ is also a resolution of singularities of V_1.

If $\Phi(q_i) = q_j$, then (V, q_i) is isomorphic to (V, q_j) as germs of singularities. This is because the resolution of (V, q_j) is a resolution of (V, q_i). The proof of Theorem 4.1 is completed in view of Theorem 4.3 below. □

Let (V, x) be a germ of complex analytic space with only one isolated singularity x. By Hironaka's paper [Hir63], it is biholomorphic equivalent to a germ of a complex algebraic singularity. Now let V be a complex analytic variety with only finitely many isolated singularities. By the equivalence and resolution theorems of algebraic varieties over field of characteristic 0, we can construct a resolution $\pi : \widetilde{V} \to V$ of V such that $\widetilde{V}$ is smooth and π is a bimeromorphic proper morphism.

The key point in the proof of Theorem C of [Yau11] for surface case is applying the minimal resolution. But in higher dimensional cases, there is no minimal resolution in general. Fortunately, by [BCHM10], there is a unique partial resolution $f : V^{can} \to V$ called the relative canonical model of V such that V^{can} has canonical singularity and the canonical divisor $K_{V^{can}}$ is f-ample. For surface, the relative canonical model is obtained by contracting all (-2)-rational curves in the minimal resolution of V. In general, the relative canonical model is isomorphic to

$$Proj \bigoplus_{m \geq 0} g_* \mathcal{O}(mK_Z)$$

where $g : Z \to V$ is any resolution of V.

Definition 4.1. *Let (V, x) be a germ such that x is the only isolated singularity. Take the relative canonical model $f : V^{can} \to V$ of V and denote E to be the exceptional set. Define*

$$cv_V(x) = (K_{V^{can}})^{\dim E} \cdot E,$$

which is called the canonical volume *of x.*

If V^{can} is not isomorphic to V, we have E being nonempty and $cv_V(x) = (K_{V^{can}})^{\dim E} \cdot E > 0$ by the f-ampleness. If V^{can} is isomorphic to V and x is a singular point, we set $cv_V(x) = (K_{V^{can}})^{\dim E} \cdot E = (K_{V^{can}})^0 \cdot x = 1$. Finally, if x is a smooth point, we set $cv_V(x) = 0$.

In general, if V has finitely many isolated normal singularities $x_i, i = 1, ..., m$, then we consider the sum of canonical volume

$$\sum_{i=1}^{m} cv_V(x_i) = \sum_{i=1}^{m} (K_{V^{can}})^{\dim E_{x_i}} \cdot E_{x_i}$$

where $f : V^{can} \to V$ is the relative canonical model of V and E_{x_i} is the exceptional set over x_i. From the definition, we see that $\sum_{i=1}^{m} cv_V(x_i) > 0$ for nonempty isolated normal singularities $x_i, i = 1, ..., m$, on V.

Theorem 4.2. *Let V be an algebraic variety or a complex space with finitely many normal isolated singularities $x_i, i = 1, ..., m$, on V. Let $cv_V(x_i)$ be the canonical volume of x_i. Then $\sum_{i=1}^{m} cv_V(x_i)$ is multiplicative in étale covering maps between resolutions. That is, if W is another algebraic variety or complex space with normal isolated singularities $y_j, j = 1, ..., s$, $p_1 : \widetilde{W} \to W$ and $p_2 : \widetilde{V} \to V$ are resolutions of W and V respectively, $\Phi : \widetilde{W} \to \widetilde{V}$ is an étale covering map, then we have $\sum_{j=1}^{s} cv_W(y_j) = d \sum_{i=1}^{m} cv_V(x_i)$ where d is the degree of Φ.*

Proof. Suppose we have resolutions $p_1 : \widetilde{W} \to W$ and $p_2 : \widetilde{V} \to V$, and we consider relative canonical models $p_1' : W^{can} \to W$ and $p_2' : V^{can} \to V$. We start from the following claim:

Claim: If $\Phi : W^{can} \to V^{can}$ is an étale covering map, then $\sum_{j=1}^{s} cv_W(y_j) = d \sum_{i=1}^{m} cv_V(x_i)$ where d is the degree of Φ.

Let E_1 be the exceptional set of p_1' and E_2 the exceptional set of p_2'. $E_1 = \sum_{j=1}^{s} E_{y_j}$ where E_{y_j} is the exceptional set over y_j, and similarly $E_2 = \sum_{i=1}^{m} E_{x_i}$ where E_{x_i} is the exceptional set over x_i. Since the canonical divisors $K_{W^{can}}$ and $K_{V^{can}}$ are p_1'-ample and p_2'-ample respectively, if E_1 is not empty, we have $\sum_{j=1}^{s} cv_W(y_j) =$

$\sum_{j=1}^{s}(K_{W^{can}})^{\dim E_{y_j}} \cdot E_{y_j} > 0$. Similarly, $\sum_{i=1}^{m} cv_V(x_i) = \sum_{i=1}^{m}(K_{V^{can}})^{\dim E_{x_i}} \cdot E_{x_i} > 0$ if E_2 is not empty.

By Φ being an étale covering map, we have the pullback $\Phi^* K_{V^{can}} = K_{W^{can}}$. Also, $\Phi^* E_2 = E_1$ since E_1 and E_2 are the only proper sets in W^{can} and in V^{can} respectively if we shrink V and W. Therefore, from

$$\sum_{j=1}^{s}(K_{W^{can}})^{\dim E_{y_j}} \cdot E_{y_j} = \sum_{i=1}^{m}(\Phi^* K_{V^{can}})^{\dim E_{x_i}} \cdot \Phi^* E_{x_i} = d\sum_{i=1}^{m}(K_{V^{can}})^{\dim E_{x_i}} \cdot E_{x_i}$$

where d is the degree of Φ, we have $\sum_{j=1}^{s} cv_W(y_j) = d\sum_{i=1}^{m} cv_V(x_i)$.

If $V^{can} \cong V$, E_2 has dimension 0 and $E_2 = \sum_{i=1}^{m} x_i$. Since étale morphisms are locally isomorphisms, Φ sends singular points to singular points, and we have $E_1 = \sum_{j=1}^{s} y_j$ is also 0-dimensional and $W^{can} \cong W$. The intersection $(K_{V^{can}})^0 \cdot x_i$ is just 1 by definition. By counting the singular points, we have $\sum_{j=1}^{s} cv_W(y_j) = s = dm = d\sum_{i=1}^{m} cv_V(x_i) > 0$. The claim is proved.

Now, for two resolutions $p_1 : \widetilde{W} \to W$ and $p_2 : \widetilde{V} \to V$, we have birational map $\phi_2 : \widetilde{V} \dashrightarrow V^{can}$ over V (see the diagram below).

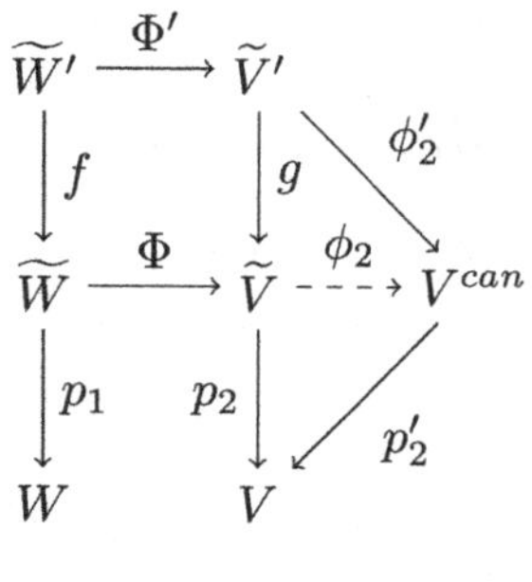

Take a common resolution $\widetilde{V}'$ of $\widetilde{V}$ and V^{can} with birational morphisms $g : \widetilde{V}' \to \widetilde{V}$ and $\phi_2' : \widetilde{V}' \to V^{can}$, $\Phi' : \widetilde{W}' \to \widetilde{V}'$ is the base-change of $\Phi : \widetilde{W} \to \widetilde{V}$. Then Φ' is also an étale covering map with the same degree of Φ. After replacing Φ and ϕ_2 by Φ' and ϕ_2' respectively, we can assume $\phi_2 : \widetilde{V} \to V^{can}$ a birational morphism.

By the property of canonical model, V^{can} and W^{can} have canonical singularities. We use a theorem in [Tak03]:

Theorem. ([Tak03], Theorem 1.1) Let V be a normal analytic space and let $f : \widetilde{V} \to V$ be a resolution of singularities. Then the induced homomorphism $f_* : \pi_1(\widetilde{V}) \to \pi_1(V)$ is an isomorphism if (V, Δ) is Kawamata log-terminal (klt) for some divisor Δ.

Definition 4.2. *A pair (X,Δ) of a normal variety and an effective $\mathbb{Q}$-divisor such that $K_X+\Delta$ is $\mathbb{Q}$-Cartier and let $\Delta=\sum d_i\Delta_i$ be the prime decomposition. We say that (X,Δ) is* (1) *Kawamata log-terminal (klt) iff $d_i<1$ for all i and there exists a projective birational morphism $\mu:Y\to X$ from a smooth variety Y with a normal crossing divisor E_i such that $K_Y\equiv\mu^*(K_X+\Delta)+\sum e_iE_i$ holds with $e_i>-1$;* (2) *canonical iff there exists a projective birational morphism $\mu:Y\to X$ from a smooth variety Y with a normal crossing divisor E_i such that $K_Y\equiv\mu^*(K_X+\Delta)+\sum e_iE_i$ holds with $e_i\geq 0$ for all i.*

Continuation of the proof of Theorem 4.2.

We see that $\pi_1(V^{can})\cong\pi_1(\widetilde{V})$ by the above theorem since V^{can} has canonical singularities, $(V^{can},0)$ is *klt*. Now we take the étale cover $\beta:W'\to V^{can}$ which gives the subgroup $\beta_*\pi_1(W')\subseteq\pi_1(V^{can})$ isomorphic to the subgroup $\Phi_*\pi_1(\widetilde{W})\subseteq\pi_1(\widetilde{V})$.

$$\begin{array}{ccc}\widetilde{W} & \xrightarrow{\ \Phi\ } & \widetilde{V}\\ {\scriptstyle h}\downarrow & & \downarrow{\scriptstyle \phi_2}\\ W' & \xrightarrow[\ \beta\ ]{} & V^{can}\end{array}$$

We see that β is an étale covering map with the same degree as Φ. Note that W' has canonical singularities since étale morphisms are locally isomorphisms. Because $\phi_{2*}\Phi_*\pi_1(\widetilde{W})=\beta_*\pi_1(W')$, there is a morphism $h:\widetilde{W}\to W'$ coming from the morphism $\phi_2:\widetilde{V}\to V^{can}$ extending to the étale covers $\widetilde{W}$ and W' of $\widetilde{V}$ and V^{can} respectively, and h is birational since ϕ_2 is. In fact, h is a resolution morphism from $\widetilde{W}$ to W'. We want to construct a morphism $q:W'\to W$ such that $p_1=q\circ h$. Let z_i be a coordinate function defined on W. Since h is proper with connected fiber, $p_1^*(z_i)$ is a function on $\widetilde{W}$ which descends to W' as a continuous function $h_*p_1^*(z_i)$ which is holomorphic outside codimension one subvariety of W'. $h_*p_1^*(z_i)$ is actually holomorphic on the smooth part of W' because it is a continuous function on W'. Recall that the singular set of W' consists of isolated normal singularities. So $h_*p_1^*(z_i)$ is actually holomorphic on W'. This gives a morphism $q:W'\to W$ such that $p_1=q\circ h$. As p_1 and q are birational, q is also birational. If $E\subseteq W'$ is an exceptional curve over W, by projection formula, we have

$$K_{W'}\cdot E=\beta^*K_{V^{can}}\cdot E=K_{V^{can}}\cdot\beta_*E>0$$

since β_*E is a sum of exceptional curves in V^{can} over V and $K_{V^{can}}$ is relative ample over V. So $K_{W'}$ is relatively ample over W. Then, by the uniqueness of relative canonical model, we have W' isomorphic to W^{can}. Replace W' by W^{can}, we have an étale covering map $\beta:W^{can}\to V^{can}$ and this is the claim above, which gives $\sum_{j=1}^{s}cv_W(y_j)=d\sum_{i=1}^{m}cv_V(x_i)$ where $d=\deg\beta=\deg\Phi$. □

Theorem 4.3. *Let V be a normal Stein space whose singular set is nonempty and finite. Let $f_1 : \widetilde{V_1} \to V$ and $f_2 : \widetilde{V_2} \to V$ be two resolutions of V. If $\widetilde{\Phi} : \widetilde{V_1} \to \widetilde{V_2}$ is a finite étale covering map, $\widetilde{\Phi}$ must be an isomorphism.*

Proof. We take $W = V$ in Theorem 4.2. Since the isolated singular points are nonempty, we have $\sum_{i=1}^{m} cv_V(x_i) > 0$. The equation $\sum_{i=1}^{m} cv_V(x_i) = d \sum_{i=1}^{m} cv_V(x_i)$, where d is the degree of $\widetilde{\Phi}$, gives $d = 1$. Hence $\widetilde{\Phi}$ is an isomorphism. □

Remark 4.1. *Note that even if V is a smooth projective variety, an étale endomorphism $\Phi : V \to V$ is not necessary an isomorphism. For example, let V be an abelian variety, $\hat{n} : V \to V$ is the morphism sending point $x \in V$ to its n times, nx, then $\hat{n}$ is an étale covering map which is not an isomorphism if $n > 1$. Therefore, we need some restraints on V to force the degree to be 1.*

In [BFF12], a notion of volume for an isolated singular point of a normal variety is defined and the volume is also multiplicative in étale covering maps. By theorem A of [BFF12], if K_V is $\mathbb{Q}$-Cartier and V is not log canonical, the volume of the singularity is nonzero and we can determine the degree of an étale morphism. Our definition of canonical volume is like another multiplicative number between resolutions, and it determines the degree of étale coverings. Our method of proving the multiplicativity of canonical volume by taking the étale cover corresponding to a subgroup of the fundamental group is like the proof in discussion of nearly étale map in [NZ09].

The following example shows that strongly pseudoconvexity plays an important role in the above theory.

Example 4.1. *Let $X_1 = \{(x, y, z) \in \mathbb{C}^3 : xy = z^2, a|x|^4 + |y|^4 + |z|^4 = \epsilon_0\}$ and $X_2 = \{(x, y, z) \in \mathbb{C}^3 : xy = z^2, a|x|^2 + |y|^2 + |z|^2 = \epsilon_0\}$ where a is a positive real number. Let $\psi : X_1 \to X_2$ be given by $\psi(x, y, z) = (x^2, y^2, z^2)$. Then ψ is a surjective CR morphism from X_1 to X_2, but ψ is not a CR biholomorphism. Note that X_2 is strongly pseudoconvex, but X_1 is only weakly pseudoconvex.*

5 Explicit computation of CR automorphisms of strongly pseudoconvex 3-dimensional CR manifolds

The purpose of this section is to prove the following results by direct computation.

Theorem 5.1. *Let a be a positive real number and $X_a = \{(x, y, z) \in \mathbb{C}^3 : xy = z^2, a|x|^2 + |y|^2 + |z|^2 = \epsilon_0\}$.*

(1) *If there exists a non-constant CR morphism from X_a to X_b, then $a = b$.*

(2) *For $a \neq \frac{1}{4}$, any non-constant CR morphism ψ from X_a to itself must be a CR biholomorphism and ψ must be one of the following forms:*

(a) $\psi(x,y,z) = (e^{i\theta_1}x, e^{i\theta_2}y, e^{i\frac{\theta_1+\theta_2}{2}}z)$, $0 \le \theta_1,\ \theta_2 \le 2\pi$.

(b) $\psi(x,y,z) = (e^{i\theta_1}y/\sqrt{a}, e^{i\theta_2}\sqrt{a}x, e^{i\frac{\theta_1+\theta_2}{2}}z)$, $0 \le \theta_1,\ \theta_2 \le 2\pi$.

(3) *For $a = \frac{1}{4}$, any non-constant CR morphism ψ from $X_{\frac{1}{4}}$ to itself must be a CR biholomorphism and ψ must be one of the following form:*

(a) $\psi(x,y,z) = (e^{i\theta_1}x, e^{i\theta_2}y, e^{i\frac{\theta_1+\theta_2}{2}}z)$, $0 \le \theta_1,\ \theta_2 \le 2\pi$.

(b) $\psi(x,y,z) = (2e^{i\theta_1}y, \frac{1}{2}e^{i\theta_2}x, e^{i\frac{\theta_1+\theta_2}{2}}z)$, $0 \le \theta_1,\ \theta_2 \le 2\pi$.

(c) $$\psi(x,y,z) = \begin{pmatrix} \frac{-2r^2}{2r^2+1}e^{i(\theta_{31}-\theta)} & \frac{2}{2r^2+1}e^{i(\theta_{32}-\theta)} & \frac{4r}{2r^2+1}e^{i\theta_{13}} \\ \frac{-1}{2(2r^2+1)}e^{i(\theta_{31}-\theta)} & \frac{2r^2}{2r^2+1}e^{i(\theta_{32}+\theta)} & \frac{-2r}{2r^2+1}e^{i(\theta_{13}+2\theta)} \\ \frac{r}{2r^2+1}e^{i\theta_{31}} & \frac{2r}{2r^2+1}e^{i\theta_{32}} & \frac{2r^2-1}{2r^2+1}e^{i(\theta_{13}+\theta)} \end{pmatrix}$$
$$\times \begin{pmatrix} x \\ y \\ z \end{pmatrix} + \text{higher order terms in } x, y \text{ and } z,$$

where $0 \le \theta,\ \theta_{31},\ \theta_{32},\ \theta_{13} \le 2\pi$, $\theta_{13} = \frac{\pi}{2} + \frac{\theta_{31}}{2} + \frac{\theta_{32}}{2} - \theta$ *and* $r > 0$.

Corollary 5.1. *The automorphism group of the compact strongly pseudoconvex CR manifold* $X_{\frac{1}{4}} = \{(x,y,z) \in \mathbb{C}^3 : xy = z^2,\ \frac{1}{4}|x|^2 + |y|^2 + |z|^2 = \epsilon_0\}$ *consists of the following mappings:*

(a) $\psi(x,y,z) = (e^{i\theta_1}x, e^{i\theta_2}y, e^{i\frac{\theta_1+\theta_2}{2}}z)$, $0 \le \theta_1,\ \theta_2 \le 2\pi$.

(b) $\psi(x,y,z) = (2e^{i\theta_2}y, \frac{1}{2}e^{i\theta_2}x, e^{i\frac{\theta_1+\theta_2}{2}}z)$, $0 \le \theta_1, \theta_2 \le 2\pi$.

(c) $$\psi(x,y,z) = \begin{pmatrix} \frac{-2r^2}{2r^2+1}e^{i(\theta_{31}-\theta)} & \frac{2e^{i(\theta_{32}-\theta)}}{2r^2+1} & \frac{4re^{i\theta_{13}}}{2r^2+1} \\ \frac{-e^{i(\theta_{31}-\theta)}}{2(2r^2+1)} & \frac{2r^2e^{i(\theta_{32}+\theta)}}{2r^2+1} & \frac{-2re^{i(\theta_{13}+2\theta)}}{2r^2+1} \\ \frac{re^{i\theta_{31}}}{2r^2+1} & \frac{2re^{i\theta_{32}}}{2r^2+1} & \frac{2r^2-1}{2r^2+1}e^{i(\theta_{13}+\theta)} \end{pmatrix} \begin{pmatrix} x \\ y \\ z \end{pmatrix}.$$

Proof of Theorem 5.1. Firstly, we prove the theorem for $b \neq \frac{1}{4}$.

Let $V_a = \{(x,y,z) \in \mathbb{C}^3 : xy = z^2$ and $a|x|^2 + |y|^2 + |z|^2 \le \epsilon_0\}$. Let $\psi : X_a \to X_b$ be a non-constant CR morphism. Then Theorem 3.1 says that ψ can be extended to a proper surjective holomorphic map $\psi : V_a \to V_b$ such that $\psi(0) = 0$, $\psi(X_a) = X_b$ and $X_a = \Psi^{-1}(X_b)$. Write ψ in the following form:

$$\begin{pmatrix} \psi_1(x,y,z) \\ \psi_2(x,y,z) \\ \psi_3(x,y,z) \end{pmatrix} = \begin{pmatrix} a_{11} & a_{12} & a_{13} \\ a_{21} & a_{22} & a_{23} \\ a_{31} & a_{32} & a_{33} \end{pmatrix} \begin{pmatrix} x \\ y \\ z \end{pmatrix} + \text{higher order terms in } x, y \text{ and } z.$$

Then the constants a_{ij} satisfy the following equations:

$$a_{11}a_{21} - a_{31}^2 = 0, \tag{5.1}$$
$$a_{12}a_{22} - a_{32}^2 = 0, \tag{5.2}$$
$$a_{13}a_{23} - a_{33}^2 + a_{11}a_{22} + a_{12}a_{21} - 2a_{31}a_{32} = 0, \tag{5.3}$$
$$a_{11}a_{23} + a_{13}a_{21} - 2a_{31}a_{33} = 0, \tag{5.4}$$
$$a_{12}a_{23} + a_{13}a_{22} - 2a_{32}a_{33} = 0. \tag{5.5}$$

For any $(x, y, z) \in X_a$, we have the following equation:

$$
\begin{aligned}
& b|\psi_1(x,y,z)|^2 + |\psi_2(x,y,z)|^2 + |\psi_3(x,y,z)|^2 = a|x|^2 + |y|^2 + |z|^2 \\
\Longrightarrow\ & (b|a_{11}|^2 + |a_{21}|^2 + |a_{31}|^2)|x|^2 + (b|a_{12}|^2 + |a_{22}|^2 + |a_{32}|^2)|y|^2 \\
& +(b|a_{13}|^2 + |a_{23}|^2 + |a_{33}|^2)|z|^2 + (ba_{11}\overline{a}_{12} + a_{21}\overline{a}_{22} + a_{31}\overline{a}_{32})x\overline{y} \\
& +(ba_{12}\overline{a}_{11} + a_{22}\overline{a}_{21} + a_{32}\overline{a}_{31})\overline{x}y + (ba_{11}\overline{a}_{13} + a_{21}\overline{a}_{23} + a_{21}\overline{a}_{33})x\overline{z} \\
& +(ba_{13}\overline{a}_{11} + a_{23}\overline{a}_{21} + a_{33}\overline{a}_{31})\overline{x}z + (ba_{12}\overline{a}_{13} + a_{22}\overline{a}_{23} + a_{32}\overline{a}_{33})y\overline{z} \\
& +(ba_{13}\overline{a}_{12} + a_{23}\overline{a}_{22} + a_{33}\overline{a}_{32})\overline{y}z = a|x|^2 + |y|^2 + |z|^2.
\end{aligned}
$$

It follows that

$$b|a_{11}|^2 + |a_{21}|^2 + |a_{31}|^2 = a, \tag{5.6}$$
$$b|a_{12}|^2 + |a_{22}|^2 + |a_{32}|^2 = 1, \tag{5.7}$$
$$b|a_{13}|^2 + |a_{23}|^2 + |a_{33}|^2 = 1, \tag{5.8}$$
$$ba_{11}\overline{a}_{12} + a_{21}\overline{a}_{22} + a_{31}\overline{a}_{32} = 0, \tag{5.9}$$
$$ba_{11}\overline{a}_{13} + a_{21}\overline{a}_{23} + a_{31}\overline{a}_{33} = 0, \tag{5.10}$$
$$ba_{12}\overline{a}_{13} + a_{22}\overline{a}_{23} + a_{32}\overline{a}_{33} = 0. \tag{5.11}$$

Case 1. $a_{31} \neq 0$ and $a_{32} \neq 0$. In view of (5.1) and (5.2), we have $a_{11} \neq 0$, $a_{21} \neq 0$, $a_{12} \neq 0$ and $a_{22} \neq 0$ in this case.

$$(5.1) \Rightarrow \frac{a_{11}}{a_{31}} = \frac{a_{31}}{a_{21}} := r_1 \neq 0$$
$$\Rightarrow a_{11} = r_1 a_{31}, a_{21} = \frac{1}{r_1} a_{31}, \tag{5.12}$$
$$(5.2) \Rightarrow \frac{a_{22}}{a_{32}} = \frac{a_{32}}{a_{12}} := r_2 \neq 0$$
$$\Rightarrow a_{22} = r_2 a_{32}, a_{12} = \frac{1}{r_2} a_{32}, \tag{5.13}$$
$$(5.4) \text{ and } (5.12) \Rightarrow r_1 a_{23} + \frac{1}{r_1} a_{13} - 2a_{33} = 0, \tag{5.14}$$
$$(5.5) \text{ and } (5.13) \Rightarrow \frac{1}{r_2} a_{23} + r_2 a_{13} - 2a_{33} = 0, \tag{5.15}$$
$$(5.14) \text{ and } (5.15) \Rightarrow \left(r_1 - \frac{1}{r_2}\right) a_{23} + \left(\frac{1}{r_1} - r_2\right) a_{13} = 0. \tag{5.16}$$

There are two cases to be considered.

Case 1 (a): $r_1 - \dfrac{1}{r_2} = 0$, i.e. $r_2 = \dfrac{1}{r_1}$.

$$(5.13) \Rightarrow a_{22} = \frac{1}{r_1} a_{32}, a_{12} = r_1 a_{32}, \tag{5.17}$$
$$(5.14) \Rightarrow a_{33} = \frac{1}{2} r_1 a_{23} + \frac{1}{2r_1} a_{13}, \tag{5.18}$$
$$(5.3), (5.12) \text{ and } (5.17) \Rightarrow a_{13}a_{23} - a_{33}^2 = 0, \tag{5.19}$$
$$(5.19) \text{ and } (5.18) \Rightarrow a_{13} = r_1^2 a_{23}, \tag{5.20}$$

$$(5.18) \text{ and } (5.20) \Rightarrow a_{33} = r_1 a_{23} = \frac{1}{r_1} a_{13},$$

$$(5.9), (5.12) \text{ and } (5.17) \Rightarrow b|r_1|^2 + \frac{1}{|r_1|^2} + 1 = 0,$$

which is a contradiction. Hence Case 1 (a) cannot happen.

Case 1 (b): $r_1 - \dfrac{1}{r_2} \neq 0.$

$$(5.16) \Rightarrow a_{23} = \frac{r_2}{r_1} a_{13}, \tag{5.21}$$

$$(5.14) \text{ and } (5.21) \Rightarrow a_{33} = \left(\frac{r_2}{2} + \frac{1}{2r_1}\right) a_{13}. \tag{5.22}$$

In view of (5.21) and (5.22), we have $a_{13} \neq 0$ because of (5.8).

$$(5.9), (5.12) \text{ and } (5.13) \Rightarrow b\frac{r_1}{\overline{r}_2} + \frac{\overline{r}_2}{r_1} + 1 = 0. \tag{5.23}$$

(5.10), (5.12), (5.21), (5.22) and the fact that $a_{31} \neq 0$, $a_{13} \neq 0$ imply

$$br_1 + \frac{\overline{r}_2}{|r_1|^2} + \frac{\overline{r}_2}{2} + \frac{1}{2\overline{r}_1} = 0. \tag{5.24}$$

(5.11), (5.13), (5.21) and (5.22) and the fact that $a_{32} \neq 0$ and $a_{13} \neq 0$ imply

$$\frac{b}{r_2} + \frac{|r_2|^2}{\overline{r}_1} + \frac{\overline{r}_2}{2} + \frac{1}{2\overline{r}_1} = 0. \tag{5.25}$$

Clearly (5.23), (5.24) and (5.25) imply

$$\det \begin{pmatrix} \dfrac{r_1}{\overline{r}_2} & \dfrac{\overline{r}_2}{r_1} & 1 \\ r_1 & \dfrac{\overline{r}_2}{|r_1|^2} & \dfrac{\overline{r}_2}{2} + \dfrac{1}{2\overline{r}_1} \\ \dfrac{1}{r_2} & \dfrac{|r_2|^2}{\overline{r}_1} & \dfrac{\overline{r}_2}{2} + \dfrac{1}{2\overline{r}_1} \end{pmatrix} = 0$$

$$\Rightarrow (\overline{r_1 r_2} + 1)(r_1 r_2 + \overline{r_1 r_2})(r_1 r_2 - 1) + 2\overline{r}_1\overline{r}_2(1 - r_1 r_2)(1 + r_1 r_2) = 0.$$

Since $r_1 - \frac{1}{r_2} \neq 0$, i.e. $r_1 r_2 - 1 \neq 0$, we have

$$\begin{aligned} &(\overline{r}_1\overline{r}_2 + 1)(r_1 r_2 + \overline{r}_1\overline{r}_2) - 2\overline{r}_1\overline{r}_2(1 + r_1 r_2) = 0 \\ &\Rightarrow (r_1 r_2 - \overline{r}_1\overline{r}_2)(1 - \overline{r}_1\overline{r}_2) = 0. \end{aligned}$$

Since $\overline{r}_1\overline{r}_2 - 1 \neq 0$, we have

$$r_1 r_2 = \overline{r}_1\overline{r}_2. \tag{5.26}$$

Let $\alpha = \frac{r_1}{\overline{r}_2}$. Then $\alpha = \overline{\alpha}$, $r_1 = \alpha \overline{r}_2$ and $\overline{r}_1 = \alpha r_2$. (5.23), (5.24) and (5.25) can be rewritten as

$$b\alpha^2 + \alpha + 1 = 0, \tag{5.27}$$

$$\frac{b}{r_2} + \frac{\overline{r}_2}{\alpha} + \frac{\overline{r}_2}{2} + \frac{1}{2\alpha r_2} = 0, \tag{5.28}$$

$$b\alpha + \frac{\alpha |r_2|^2}{2} + \frac{1}{2} + |r_2|^2 = 0. \tag{5.29}$$

(5.28) $-\alpha$ (5.29) $\Rightarrow$

$$(b\alpha^2 - 1)(\alpha |r_2|^2 - 1) = 0,$$
$$\text{i.e. } b\alpha^2 = 1 \text{ or } \alpha |r_2|^2 = 1. \tag{5.30}$$

If $\alpha |r_2|^2 = 1$, then (5.29) implies

$$\frac{b}{|r_2|^2} + 1 + |r_2|^2 = 0,$$

which is absurd since the left hand side is positive. Therefore we conclude that $\alpha^2 = \frac{1}{b}$. Then (5.27) and (5.29) imply

$$\alpha = -2 \text{ and } b = \frac{1}{4}. \tag{5.31}$$

Case 2. $a_{31} = 0$. By (5.1), we have either $a_{11} = 0$ or $a_{21} = 0$.
Case 2 (a): $a_{31} = 0$ and $a_{11} = 0$. By (5.6), we have $a_{21} \neq 0$.

$$(5.9) \Rightarrow a_{21}\overline{a}_{22} = 0 \Rightarrow a_{22} = 0, \tag{5.32}$$
$$(5.4) \Rightarrow a_{13}a_{21} = 0 \Rightarrow a_{13} = 0, \tag{5.33}$$
$$(5.2) \text{ and } (5.32) \Rightarrow a_{32} = 0. \tag{5.34}$$

By (5.7), (5.32) and (5.34), we have $a_{12} \neq 0$.

$$(5.5), (5.32) \text{ and } (5.34) \Rightarrow a_{12}a_{23} = 0 \Rightarrow a_{23} = 0, \tag{5.35}$$
$$(5.3) \text{ and } (5.35) \Rightarrow -a_{33}^2 + a_{12}a_{21} = 0, \tag{5.36}$$
$$(5.6) \Rightarrow |a_{21}|^2 = a, \tag{5.37}$$
$$(5.7), (5.32) \text{ and } (5.34) \Rightarrow |a_{12}|^2 = \frac{1}{b}, \tag{5.38}$$
$$(5.8), (5.33) \text{ and } (5.35) \Rightarrow |a_{33}|^2 = 1. \tag{5.39}$$

(5.36), (5.37), (5.38) and (5.39) imply $a = b$ and

$$\begin{pmatrix} a_{11} & a_{12} & a_{13} \\ a_{21} & a_{22} & a_{23} \\ a_{31} & a_{32} & a_{33} \end{pmatrix} = \begin{pmatrix} 0 & a_{12} & 0 \\ a_{21} & 0 & 0 \\ 0 & 0 & a_{33} \end{pmatrix} = \begin{pmatrix} 0 & e^{i\theta_1}/\sqrt{a} & 0 \\ e^{i\theta_2}\sqrt{a} & 0 & 0 \\ 0 & 0 & e^{i(\theta_1+\theta_2)} \end{pmatrix}.$$

Case 2 (b): $a_{31} = 0$ and $a_{21} = 0$. By (5.6), we have $a_{11} \neq 0$.

$$(5.9) \Rightarrow a_{12} = 0, \tag{5.40}$$

$$(5.4) \Rightarrow a_{23} = 0, \tag{5.41}$$

$$(5.2) \text{ and } (5.40) \Rightarrow a_{32} = 0, \tag{5.42}$$

$$(5.42), (5.40) \text{ and } (5.7) \Rightarrow a_{22} \neq 0,$$

$$(5.42), (5.40), (5.5) \text{ and } a_{22} \neq 0 \Rightarrow a_{13} = 0, \tag{5.43}$$

$$(5.3), (5.43) \text{ and } (5.40) \Rightarrow -a_{33}^2 + a_{11}a_{22} = 0, \tag{5.44}$$

$$(5.6) \Rightarrow b|a_{11}|^2 = a, \tag{5.45}$$

$$(5.7) \Rightarrow |a_{22}|^2 = 1, \tag{5.46}$$

$$(5.8) \Rightarrow |a_{33}|^2 = 1. \tag{5.47}$$

(5.44), (5.45), (5.46) and (5.47) $\Rightarrow a = b$ and

$$\begin{pmatrix} a_{11} & a_{12} & a_{13} \\ a_{21} & a_{22} & a_{23} \\ a_{31} & a_{32} & a_{33} \end{pmatrix} = \begin{pmatrix} a_{11} & 0 & 0 \\ 0 & a_{22} & 0 \\ 0 & 0 & a_{33} \end{pmatrix} = \begin{pmatrix} e^{i\theta_1} & 0 & 0 \\ 0 & e^{i\theta_2} & 0 \\ 0 & 0 & e^{i(\theta_1+\theta_2)} \end{pmatrix}.$$

Case 3. $a_{32} = 0$. By (5.2), we have either $a_{12} = 0$ or $a_{22} = 0$.
Case 3 (a): $a_{32} = 0$ and $a_{12} = 0$. By (5.7), we have $a_{22} \neq 0$.

$$(5.9) \Rightarrow a_{21} = 0, \tag{5.48}$$

$$(5.5) \Rightarrow a_{13} = 0, \tag{5.49}$$

$$(5.1) \text{ and } (5.48) \Rightarrow a_{31} = 0, \tag{5.50}$$

$$(5.6), (5.48) \text{ and } (5.50) \Rightarrow b|a_{11}|^2 = a \text{ and } a_{11} \neq 0, \tag{5.51}$$

$$(5.4), (5.48) \text{ and } (5.5) \Rightarrow a_{23} = 0, \tag{5.52}$$

$$(5.7) \Rightarrow |a_{22}|^2 = 1, \tag{5.53}$$

$$(5.8) \Rightarrow |a_{33}|^2 = 1, \tag{5.54}$$

$$(5.3) \Rightarrow a_{33}^2 = a_{11}a_{22}. \tag{5.55}$$

(5.51), (5.53), (5.54) and (5.55) $\Rightarrow a = b$ and

$$\begin{pmatrix} a_{11} & a_{12} & a_{13} \\ a_{21} & a_{22} & a_{23} \\ a_{31} & a_{32} & a_{33} \end{pmatrix} = \begin{pmatrix} a_{11} & 0 & 0 \\ 0 & a_{22} & 0 \\ 0 & 0 & a_{33} \end{pmatrix} = \begin{pmatrix} e^{i\theta_1} & 0 & 0 \\ 0 & e^{i\theta_2} & 0 \\ 0 & 0 & e^{i\frac{\theta_1+\theta_2}{2}} \end{pmatrix}.$$

Case 3 (b): $a_{32} = 0 = a_{22}$. By (5.7), we have $a_{12} \neq 0$.

$$(5.9) \Rightarrow a_{11} = 0, \tag{5.56}$$

$$(5.1) \text{ and } (5.50) \Rightarrow a_{31} = 0, \tag{5.57}$$

$$(5.6), (5.56) \text{ and } (5.57) \Rightarrow a_{21} \neq 0, \tag{5.58}$$

$$(5.10), (5.56), (5.58) \text{ and } (5.57) \Rightarrow a_{23} = 0, \tag{5.59}$$

$$(5.11) \Rightarrow a_{13} = 0, \tag{5.60}$$

$$(5.6), (5.56) \text{ and } (5.57) \Rightarrow |a_{21}|^2 = a, \tag{5.61}$$

$$(5.7) \Rightarrow b|a_{12}|^2 = 1, \tag{5.62}$$
$$(5.8), (5.59) \text{ and } (5.60) \Rightarrow |a_{33}|^2 = 1, \tag{5.63}$$
$$(5.3), (5.56), (5.57) \text{ and } (5.59) \Rightarrow a_{33}^2 = a_{12}a_{21}. \tag{5.64}$$

(5.61), (5.62), (5.63) and (5.64) $\Rightarrow a = b$ and

$$\begin{pmatrix} a_{11} & a_{12} & a_{13} \\ a_{21} & a_{22} & a_{33} \\ a_{31} & a_{32} & a_{33} \end{pmatrix} = \begin{pmatrix} 0 & a_{12} & 0 \\ a_{21} & 0 & 0 \\ 0 & 0 & a_{33} \end{pmatrix} = \begin{pmatrix} 0 & \frac{e^{i\theta_1}}{\sqrt{a}} & 0 \\ e^{i\theta_2}\sqrt{a} & 0 & 0 \\ 0 & 0 & e^{\frac{i(\theta_1+\theta_2)}{2}} \end{pmatrix}.$$

We have shown that if there exists a non-constant CR morphism $\psi : X_a \longrightarrow X_b$, then $a = b$ and ψ must be one of the following forms.

(i) $\psi_1 = a_{11}x + \psi_1^{(2)} + \psi_1^{(3)} + \cdots,$

$\psi_2 = a_{22}y + \psi_2^{(2)} + \psi_2^{(3)} + \cdots,$

$\psi_3 = a_{33}z + \psi_3^{(2)} + \psi_3^{(3)} + \cdots,$

where $a_{11} = e^{i\theta_1}$, $a_{22} = e^{i\theta_2}$ and $a_{33} = e^{i\frac{\theta_1+\theta_2}{2}}$, $\psi_i^{(j)}$ = homogeneous polynomial of degree j.

(ii) $\psi_1 = a_{12}y + \psi_1^{(2)} + \psi_1^{(3)} + \cdots,$

$\psi_2 = a_{21}x + \psi_2^{(2)} + \psi_1^{(3)} + \cdots,$

$\psi_3 = a_{33}z + \psi_3^{(2)} + \psi_3^{(3)} + \cdots,$

where $a_{12} = e^{i\theta_1}/\sqrt{a}$, $a_{21} = e^{i\theta_2}\sqrt{a}$ and $a_{33} = e^{i\frac{\theta_1+\theta_2}{2}}$. In both case (i) and case (ii), we have

$$a\psi_1\overline{\psi}_1 + \psi_2\overline{\psi}_2 + \psi_3\overline{\psi}_3 = a|x|^2 + |y|^2 + |z|^2. \tag{5.65}$$

By comparing the 3rd order terms in (5.65), we see easily that the 2nd order terms of (ψ_1, ψ_2, ψ_3) are zero. Repeating this argument, we see that (ψ_1, ψ_2, ψ_3) has only linear terms.

Secondly, we prove the theorem for $b = \frac{1}{4}$.

The proof is the same as those for $b \neq \frac{1}{4}$ above except in Case 1 (b). Here we shall follow our previous notations. Let us summarize what we have proved in this situation:

$$a_{31} \neq 0, a_{32} \neq 0, a_{11} \neq 0, a_{21} \neq 0, a_{12} \neq 0 \text{ and } a_{22} \neq 0, \tag{5.66}$$
$$r_1r_2 \neq 1, r_1 \neq 0, r_2 \neq 0, \tag{5.67}$$
$$r_1 = -2\overline{r}_2. \tag{5.68}$$

(5.1) implies

$$a_{11} = r_1a_{31}, \; a_{21} = \frac{1}{r_1}a_{31}, \tag{5.69}$$

(5.2) implies

$$a_{22} = r_2 a_{32},\ a_{12} = \frac{1}{r_2} a_{32},$$

(5.4) and (5.5) imply

$$a_{23} = \frac{r_2}{r_1} a_{13},\ a_{33} = \left(\frac{r_2}{2} + \frac{1}{2r_1}\right) a_{13}. \tag{5.70}$$

Notice that (5.9), (5.10) and (5.11) are equivalent to (5.68) in this situation

$$(5.6) \Rightarrow |a_{31}|^2 = \frac{4ar^2}{(2r^2+1)^2}, \tag{5.71}$$

$$(5.7) \Rightarrow |a_{32}|^2 = \frac{4r^2}{(2r^2+1)^2}, \tag{5.72}$$

$$(5.8) \Rightarrow |a_{13}|^2 = \frac{16r^2}{(2r^2+1)^2}, \tag{5.73}$$

$$(5.3) \Rightarrow \frac{a_{13}^2}{4r_1} = \frac{a_{31}a_{32}}{r_2}. \tag{5.74}$$

Let

$$r_2 = re^{i\theta}, \tag{5.75}$$

$$(5.68) \Rightarrow r_1 = -2re^{-i\theta}, \tag{5.76}$$

$$(5.74), (5.75) \text{ and } (5.76) \Rightarrow a_{13}^2 = -8e^{-2i\theta} a_{31}a_{32}, \tag{5.77}$$

$$(5.77), (5.71), (5.72) \text{ and } (5.73) \Rightarrow a = \frac{1}{4}. \tag{5.78}$$

$$a_{31} = \frac{r}{2r^2+1} e^{i\theta_{31}}, \tag{5.79}$$

$$a_{32} = \frac{2r}{2r^2+1} e^{i\theta_{32}}, \tag{5.80}$$

$$a_{13} = \frac{4r}{2r^2+1} e^{i\theta_{13}}, \tag{5.81}$$

$$\theta_{13} = \frac{\pi}{2} + \frac{\theta_{31}}{2} + \frac{\theta_{32}}{2} - \theta. \tag{5.82}$$

It follows that the automorphism group of $X_{\frac{1}{4}}$ contains a linear subgroup of dimension 4 in the following form

$$\psi(x,y,z) = \begin{pmatrix} -2re^{-i\theta}a_{31} & \frac{1}{r}e^{-i\theta}a_{32} & a_{13} \\ -\frac{1}{2r}e^{i\theta}a_{31} & re^{i\theta}a_{32} & -\frac{1}{2}e^{2i\theta}a_{13} \\ a_{31} & a_{32} & \frac{2r^2-1}{4r}e^{i\theta}a_{13} \end{pmatrix} \begin{pmatrix} x \\ y \\ z \end{pmatrix}$$

$$= \begin{pmatrix} \frac{-2r^2}{2r^2+1}e^{i(\theta_{31}-\theta)} & \frac{2}{2r^2+1}e^{i(\theta_{32}-\theta)} & \frac{4r}{2r^2+1}e^{i\theta_{13}} \\ \frac{-1}{2(2r^2+1)}e^{i(\theta_{31}+\theta)} & \frac{2r^2}{2r^2+1}e^{i(\theta_{32}+\theta)} & \frac{-2r}{2r^2+1}e^{i(\theta_{13}+2\theta)} \\ \frac{r}{2r^2+1}e^{i\theta_{31}} & \frac{2r}{2r^2+1}e^{i\theta_{32}} & \frac{2r^2-1}{2r^2+1}e^{i(\theta_{13}+\theta)} \end{pmatrix} \begin{pmatrix} x \\ y \\ z \end{pmatrix}.$$

It can be shown that

$$\det\begin{pmatrix} -2re^{i\theta}a_{31} & \frac{1}{r}e^{-i\theta}a_{32} & a_{13} \\ -\frac{1}{2r}e^{i\theta}a_{31} & re^{i\theta}a_{32} & -\frac{1}{2}e^{2i\theta}a_{13} \\ a_{31} & a_{32} & \frac{2r^2-1}{4r}e^{i\theta}a_{13} \end{pmatrix}$$

$$= a_{31}a_{32}a_{33}\left(-r^3 - \frac{3r}{2} - \frac{6r^2+1}{8r^3}\right) \neq 0.$$

References

[An-Gr62] Andreotti, A. and Grauert, H., Théorèmes de finitude pour la cohomologie des espaces complexes, Bull. Soc. Math. France **90** (1962), 193–259.

[Be-Ca82] Bell, S. and Catlin, D., Boundary regularity of proper holomorphic mappings, Duke Math. J. **49** (1982), no. 2, 385–396.

[BCHM10] Birkar, C., Cascini, P., Hacon, C., and McKernan, J., Existence of minimal models for varieties of log general type, J. Amer. Math. Soc. **23** (2010), 405–468.

[Bo-Na67] Borel, A. and Narasimhan, R., Uniqueness conditions for certain holomorphic mappings, Invent. Math. **2** (1967), 247–255.

[BFF12] Boucksoml, S., Fernex, de T., and Favre, C., The volume of an isolated singularity, Duke Math. J. **161** (2012), no. 8, 1455–1520.

[BM75] Boutet de Monvel, L., Intégration des équations de Cauchy Riemann, Séminaire Goulaouic-Lions-Schwartz, Exré IX, 1974-1975.

[Di-Fo82] Diederich, K. and Fornaess, J. E., Boundary regularity of proper holomorphic mappings, Invent. Math. **67** (1982), no. 3, 363–384.

[EHZ04] Ebenfelt, P., Huang, X., and Zaitsev, D., Rigidity of CR-immersions into spheres, Comm. Anal. Geom. **12** (2004), no. 3, 631–670.

[EHZ05] Ebenfelt, P., Huang, X., and Zaitsev, D., The equivalence problem and rigidity for hypersurfaces embedded into hyperquadrics, Amer. J. Math. **127** (2005), no. 1, 169–191.

[For76] Fornaess, J. E., Embedding strictly pseudoconvex domains in convex domains, Amer. J. Math. **98** (1976), no. 1, 529–569.

[Ful98] Fulton, W., Intersection Theory, 2nd ed., Springer, 1998.

[Ham81] Hamm, H., On the vanishing of local homotopy groups for isolated singularities of complex spaces, J. Reine Angew. Math. **323** (1981), 172–176.

[Ha-La75] Harvey, R. and Lawson, B., On boundaries of complex analytic varieties I, Ann. of Math. **102** (1975), 233–290.

[Ha-La00] Harvey, R. and Lawson, B., Addendum to Theorem 10.4 in Boundaries of complex analytic varieties, arXiv: math/0002195v1 [math CV] 23 Feb., 2000.

[Har77] Hartshorne, R., Alebraic Geometry, Graduate Texts in Mathematics, Springer-Verlag, 1977.

[Hir63] Hironaka, H., On the equivalence of singularities. i, Arithmetical Algebraic Geometry, Proc. Conf. Purdue Univ. (1963), 153-200.

[H-K-P06] Huang, J.-M., Kebekus, S., and Peternell, T., Holomorphic maps onto varieties of non-negative Kodaira dimension, Journal of Alg. Geom. **15** (2006), 551–561.

[Ko-Oc75] Kobayashi, S. and Ochiai, T., Meromorphic mappings onto compact spaces of general type, Invent. Math. **31** (1975), 7–16.

[Kohn65] Kohn, J., Boundaries of complex manifolds, Proc. Conference on Complex Manifolds (Minneapolis), Springer-Verlag, New York, 1965, 81–94.

[KSW81] Kalka, M., Shiffman, B., and Wong, B., Finiteness and rigidity theorems for holomorphic mappings, Michigan Math. J. **28** (1981), 289–295.

[Mok84] Mok, N., Rigidity of holomorphic self-mappings and the automorphism groups of hyperbolic Stein spaces, Math. Ann. **266** (1984), 433–447.

[Mo83] Mostow, G. D., Strong Rigidity of Locally Symmetric Spaces, Ann. of Math. Studies, Vol. **78**, Princeton University Press, 1983.

[NZ09] Nakayama, N. and Zhang, D.-Q., Building blocks of étale endomorphisms of complex projective manifolds, Proc. Lond. Math. Soc. **99** (2009), no. 3, 725–756.

[Pin74] Pinchuk, S., On proper holomorphic mappings of strictly pseudoconvex domains, Siberian Math. J. **15** (1974), 909–917.

[Pr67] Prill, D., Local classification of quotients of complex manifolds by discontinuous groups, Duke Math. Journal **34** (1967), no. 3, 375–386.

[Siu80] Siu, Y.-T., The complex analyticity of harmonic map and the strong rigidity of compact Kähler manifolds, Ann. of Math. **112** (1980), no. 2, 73–111.

[Tak03] Takayama, S., Local simple connectedness of resolutions of log-terminal singularities, Internat. J. Math. **14** (2003), no. 8, 825–836.

[TYZ13] Tu, Y. C., Yau, S.S.-T., and Zuo, H.Q., Nonconstant CR morphisms between compact strongly pseudoconvex CR manifolds and étale covering between resolutions of isolated singularities, J. Differential Geom. **95** (2013), no. 2, 337–354.

[Wo77] Wong, B., Characterization of the unit ball in $\mathbb{C}^n$ by its automorphism group, Invent. Math. **41** (1977), no. 3, 253–257.

[Ya77] Yau, S.-T., Calabi's conjecture and some new results in algebraic geometry, PNAS **74** (1977), no. 5, 1798–1799.

[Ya78] Yau, S.-T., On the Ricci curvature of a compact Kähler manifold and the complex Monge-Ampére equation I, Comm. Pure Appl. Math. **31** (1978), no. 3, 339–341.

[Yau77] Yau, S.S.-T., Two theorems on higher dimensional singularities, Mathematische Annalen **231** (1977), 55–59.

[Yau11] Yau, S.S.-T., Rigidity of CR morphisms between compact strongly pseudoconvex CR manifolds, Journal of European Mathematical Society **13** (2011), 175–184.

[Ya-Yu02] Yau, S.S.-T. and Yu, Y., Algebraic classification of rational CR structures on topological 5-sphere with transversal holomorphic S^1-action in $\mathbb{C}^4$, Math. Nachr. (2002), 207–233.

Department of Mathematical Sciences, Tsinghua University
Beijing 100084, China
E-mail address: yau@uic.edu

Mathematical Sciences Center, Tsinghua University
Beijing, 100084, China
E-mail address: hqzuo@math.tsinghua.edu.cn

Proceedings of the 6th International Congress of Chinese Mathematicians
ALM 36, pp. 199–234

Rigidity in the Langlands Correspondence and Applications

Zhiwei Yun

Abstract

We give a survey on some new construction of local systems on open curves, using the idea of rigid automorphic representations. These constructions have applications to classical problems in algebraic geometry and number theory.

2010 Mathematics Subject Classification: 14D24, 12F12, 20G41.

Keywords and Phrases: Langlands correspondence, rigid local systems.

1 Introduction

In this report, we give a survey on some new construction of local systems on open curves. The construction uses the idea of rigid automorphic representations which we will try to axiomatize here. We also give applications of these constructions to more classical problems in algebraic geometry and number theory. The main references are [17] and [36]. More details on rigid automorphic representations will appear in the article [38].

1.1 The goal

Let us fix an algebraic curve X over an algebraically closed field k. Let $S \subset X$ be a finite set of closed points. Let ℓ be a prime different from $\mathrm{char}(k)$. A local system $\mathcal{F}$ (with $\overline{\mathbb{Q}}_\ell$-coefficients, in the étale topology) over $U := X - S$ is *physically rigid* if it is determined up to isomorphism by its local monodromy around points $x \in S$. A local system $\mathcal{F}$ over U is *cohomologically rigid* if $\mathrm{H}^1(X, j_*\mathrm{End}^\circ(\mathcal{F})) = 0$, where $\mathrm{End}^\circ(\mathcal{F})$ is the local system of trace-free endomorphisms of $\mathcal{F}$, and j_* means the sheaf (not derived) push-forward along $j : X - S \to X$. These notions were defined

Supported by the NSF grant DMS-1302071.

and studied in depth by Katz in [20]. The main result of [20] is an algorithmic description of tame local systems.

We shall be interested in local systems in a broader sense. Let H be a connected reductive algebraic group over $\overline{\mathbb{Q}}_\ell$. An H-local system on U is a continuous homomorphism $\rho : \pi_1(U) \to H(\overline{\mathbb{Q}}_\ell)$, where $\pi_1(U)$ is the étale fundamental group of U. This specializes to the notion of rank n local systems when $H = \mathrm{GL}_n$. Both notions of rigidity can easily be extended to H-local systems.

We would like to construct many examples of rigid H-local systems, especially when H is of exceptional types. The key observation is the following: using the Langlands correspondence for function fields, it is easier to construct the automorphic counterpart of these rigid local systems than constructing the local systems themselves.

1.2 The method

Suppose for the moment that the complete algebraic curve X is defined over a finite field k. Let G be a split reductive group k. Langlands philosophy predicts that there should be a finite-to-one correspondence from automorphic representations π of $G(\mathbb{A}_F)$ (where $\mathbb{A}_F$ is the ring of adèles of F) to $\widehat{G}$-local system ρ_π over an open subset of X. Here $\widehat{G}$ is a group over $\overline{\mathbb{Q}}_\ell$ whose root system is dual to that of G. When $G = \mathrm{GL}_n$, this correspondence is proved by Lafforgue [21], and it is in fact one-to-one when restricted to cuspidal automorphic representations. For general connected reductive G, the assignment $\pi \mapsto \rho_\pi$ has recently be established by Lafforgue [22]. However, the construction of ρ_π is not easy to carry out explicitly in concrete examples.

We shall propose a notion of a rigid automorphic datum with respect to a given finite set of closed points $S \subset X$. Here is the simplest version of it. For each $x \in S$ one imposes a condition on the local component π_x of an automorphic representation π. This set of local conditions is called a rigid automorphic datum if there is up to isomorphism only one automorphic representation π which is unramified outside S and satisfies the given local conditions at each $x \in S$. In practice, it is easy to write down a Hecke eigenform in π as a function f on the double coset $G(F)\backslash G(\mathbb{A}_F)/K$ (for a suitable compact open subgroup $K \subset G(\mathbb{A}_F)$).

Once we have the Hecke eigenform f in π, via the Satake isomorphism, the Hecke eigenvalues of f at a place $x \notin S$ determine the image of the Frobenius element Frob_x under ρ_π. Although this much information sometimes determines the isomorphism type of ρ_π, we still do not have the local system ρ_π as a concrete geometric object. To get ρ_π geometrically, we appeal to the geometric Langlands correspondence, which is a sheaf-theoretic upgrading of the Langlands correspondence for function fields. As suggested by an observation of Weil, one should think of Bun_G, the moduli space of G-bundles over X (with suitable level structure), as a geometric incarnation of the double coset $G(F)\backslash G(\mathbb{A}_F)/K$. In the setting of geometric Langlands, one should try to upgrade the Hecke eigenform f to a sheaf (or complexes of sheaves) $\mathcal{F}$ on Bun_G. The precise meaning of "upgrade" is encoded in the "sheaf-to-function" correspondence of Grothendieck. Once the Hecke eigensheaf $\mathcal{F}$ is constructed, we can construct the corresponding $\widehat{G}$-local system

ρ_π in a completely explicit manner, via geometric Hecke operators. Of course in general it is not an easy task to construct Hecke eigensheaves $\mathcal{F}$. However, when π satisfies a rigid automorphic datum, there is essentially only one choice for $\mathcal{F}$, which forces it to be a Hecke eigensheaf.

To summarize, we start from a rigid automorphic datum for $G(\mathbb{A}_F)$, find the unique automorphic representation π satisfying this rigid automorphic datum and a Hecke eigenform f in it, then we use geometric Langlands to upgrade f into a sheaf $\mathcal{F}$, and finally we apply geometric Hecke operators to $\mathcal{F}$ to get a $\widehat{G}$-local system ρ_π.

1.3 Applications

We shall give three applications of the rigid objects in the Langlands correspondence.

1.3.1 Local systems with exceptional monodromy groups

Katz [19] has constructed an example of a local system over $\mathbb{P}^1_{\mathbb{F}_p} - \{0, \infty\}$ whose geometric monodromy lies in the exceptional group G_2 and is Zariski dense there. This G_2-local system comes from a rank 7 rigid local system which is an example of Katz's hypergeometric sheaves. For other exceptional groups, there weren't explicitly constructed examples of local systems with Zariski dense monodromy in them.

In joint work with Heinloth and Ngô [17], for any reductive group $\widehat{G}$ over $\overline{\mathbb{Q}}_\ell$, we constructed a $\widehat{G}$-local system $\mathrm{Kl}_{\widehat{G}}$ over $\mathbb{P}^1_{\mathbb{F}_p} - \{0, \infty\}$. The geometric monodromy of these local systems is quite large. For example, when $\widehat{G}$ is of type E_7, E_8, F_4 or G_2, the monodromy is Zariski dense in $\widehat{G}$. When $\widehat{G} = \mathrm{GL}_n$ or Sp_n, $\mathrm{Kl}_{\widehat{G}}$ is the same as the Kloosterman sheaf constructed by Deligne [5]; when $\widehat{G} = G_2$, $\mathrm{Kl}_{\widehat{G}}$ is the same as the G_2 example of Katz. These generalized Kloosterman sheaves $\mathrm{Kl}_{\widehat{G}}$ give first examples of motivic local systems with Zariski dense monodromy in exceptional groups other than G_2. The associated Frobenius traces give new exponential sums generalizing the Kloosterman sums, and they are equidistributed according to the Sato-Tate measure for compact Lie groups of type E_7, E_8, F_4 or G_2.

Our construction in [17] was inspired by a result of Gross [13] and the work of Frenkel and Gross [10] on rigid irregular connections. Gross showed in [13] that there exists a unique automorphic representation of G over the rational function field $F = k(t)$ whose local component at 0 is a Steinberg representation and at ∞ is a simple supercuspidal representation. He then conjectured that when $G = \mathrm{GL}_n$, the Satake parameters of this automorphic representation should give the classical Kloosterman sums. Our work [17] confirmed this conjecture and generalized it to other reductive groups.

1.3.2 Motives over number fields with exceptional motivic Galois groups

In early 1990s, Serre asked the following question [32]: Is there a motive over a number field whose motivic Galois group is of exceptional type such as G_2 or E_8?

A motive M over a number field K is, roughly speaking, part of the cohomology $\mathrm{H}^i(X)$ for some (smooth projective) algebraic variety X over K and some integer i, which is cut out by geometric operations (such as group actions). For each prime ℓ, the motive M has the associated ℓ-adic cohomology $\mathrm{H}_\ell(M) \subset \mathrm{H}^i(X_{\overline{K}}, \overline{\mathbb{Q}}_\ell)$, which admits a Galois action:

$$\rho_{M,\ell} : \mathrm{Gal}(\overline{K}/K) \to \mathrm{GL}(\mathrm{H}_\ell(M)).$$

The *ℓ-adic motivic Galois group* $G_{M,\ell}$ of M is the Zariski closure of the image of $\rho_{M,\ell}$. This is an algebraic group over $\overline{\mathbb{Q}}_\ell$. Since the motivic Galois groups that appear in the original question of Serre are only well-defined assuming the standard conjectures in algebraic geometry, we shall use the ℓ-adic motivic Galois group as a working substitute for the actual motivic Galois group (conjecturally they should be isomorphic to each other). Classical groups appear as motivic Galois groups of abelian varieties. This is why Serre raised the question for exceptional groups only. Until recently, the only known case of Serre's question was G_2, by the work of Dettweiler and Reiter [8].

In [36], we construct motivic local systems on $\mathbb{P}^1_{\mathbb{Q}} - \{0, 1, \infty\}$ with Zariski dense monodromy in exceptional groups E_7, E_8 and G_2 in a uniform way. As a consequence of this construction, we give an affirmative answer to the ℓ-adic version of Serre's question for E_7, E_8 and G_2: these groups can be realized as the ℓ-adic motivic Galois groups for motives over number fields (in fact the number field is either $\mathbb{Q}$ in the case of E_8 and G_2, or $\mathbb{Q}(i)$ in the case of E_7). With a bit more work, one can also realize F_4 as a motivic Galois group over $\mathbb{Q}$.

We remark that naive attempts to finding motives with motivic Galois group of type E_8 can be quite painful, if not hopeless. Usually people look for motives from Shimura varieties or abelian varieties. However, it is known that abelian varieties do not have exceptional motivic Galois groups; nor is there a Shimura variety of type E_8. To find an E_8-motive, at the very least, one needs to find an algebraic variety over a number field K whose cohomology in some degree has dimension at least 248 (because the smallest nontrivial representation of E_8 has dimension 248, the adjoint representation). Our construction gives a 248-dimensional Galois representation in a natural way from the geometry of the algebraic group E_8 itself.

1.3.3 Inverse Galois problem

The inverse Galois problem over $\mathbb{Q}$ asks whether every finite group can be realized as the Galois group of some Galois extension $K/\mathbb{Q}$. The problem is still open for many finite simple groups, especially those of Lie type. The same rigid local systems over $\mathbb{P}^1_{\mathbb{Q}} - \{0, 1, \infty\}$ constructed to answer Serre's question can be used to solve new cases of the inverse Galois problem. We show in [36] that for sufficiently large primes ℓ, the finite simple groups $G_2(\mathbb{F}_\ell)$ and $E_8(\mathbb{F}_\ell)$ can be realized as Galois groups over $\mathbb{Q}$. With a bit more work, one can show that $F_4(\mathbb{F}_\ell)$ is also a Galois group over $\mathbb{Q}$.

In inverse Galois theory, people use the "rigidity method" to prove certain finite groups H are Galois groups over $\mathbb{Q}$. In particular, the case of $G_2(\mathbb{F}_\ell)$ for all primes $\ell \geq 5$ was known by the work of Thompson [35] and Feit and Fong [9]. However, the case of $E_8(\mathbb{F}_\ell)$ was known only for primes ℓ satisfying a certain congruence condition modulo 31 (see the book of Malle and Matzat [27, II.10] for a summary of what was known before). Therefore our result for $E_8(\mathbb{F}_\ell)$ for all ℓ sufficiently large is new.

The input data of the rigid method in inverse Galois theory is a triple of conjugacy classes in the target group H. The notion of rigidity of such a triple is quite similar to the notion of physically rigidity for local systems over $\mathbb{P}^1_{\mathbb{C}} - \{0, 1, \infty\}$. However, connection between the rigidity method and automorphic representations has not been explored before. Our result shows that this connection can be useful in solving the inverse Galois problem, and it even sheds some light to the rigidity method itself. In fact, our construction of the local system over $\mathbb{P}^1 - \{0, 1, \infty\}$ suggests a triple in $E_8(\mathbb{F}_\ell)$ which should be a rigid triple (see [36, Conjecture 5.16]). This has been confirmed by Guralnick and Malle [16], and using this triple they are able to show that $E_8(\mathbb{F}_\ell)$ is a Galois group over $\mathbb{Q}$ for all primes $\ell > 7$.

1.4 Acknowledgements

The author would like to thank the organizers of the International Congress of Chinese Mathematicians for inviting him to speak at the conference.

2 Rigidity for automorphic representations

In this section we recall some basic concepts for automorphic representations over a function field. We will define the notion of a geometric automorphic datum (a collection of local conditions for automorphic representations to satisfy), and define what it means for such a datum to be rigid.

2.1 The setting

2.1.1 Function field

Let k be a finite field. Let X be a projective, smooth and geometrically connected curve X over k. Let $F = k(X)$ be the field of rational functions on X.

Let $|X|$ be the set of closed points of X (places of F). For each $x \in |X|$, let F_x denote the completion of F at the place x. The valuation ring and residue field of F_x are denoted by $\mathcal{O}_x$ and k_x. The maximal ideal of $\mathcal{O}_x$ is denoted by $\mathfrak{m}_x$. The ring of adèles is the restricted product

$$\mathbb{A}_F := \prod_{x \in |X|}{}' F_x$$

where for almost all x, the x-component of an element $a \in \mathbb{A}_F$ lies in $\mathcal{O}_x$. There is a natural topology on $\mathbb{A}_F$ making it a locally compact topological ring.

2.1.2 Groups

Although most of the results in this report are valid for connected quasisplit reductive groups over F, but for the simplicity of the presentation, we restrict ourselves to the following situation: G **is a connected, split, simply-connected and almost simple group over** k. We will fix a pinning of G, namely a split maximal torus T, a Borel subgroup $B \supset T$ and generators of simple roots spaces. It makes sense to talk about $G(R)$ whenever R is a k-algebra. For example, $G(F)$, $G(\mathcal{O}_x)$ and $G(F_x)$, etc.

Later we shall also need to view $G(F_x)$ and $G(\mathcal{O}_x)$ as infinite-dimensional groups over k. More precisely, for a k_x-algebra R we let $L_xG(R) = G(R\widehat{\otimes}_{k_x}F_x)$ and let $L_x^+G(R) = G(R\widehat{\otimes}_{k_x}\mathcal{O}_x)$. Here $R\widehat{\otimes}_{k_x}F_x$ or $R\widehat{\otimes}_{k_x}\mathcal{O}_x$ means the completion of the tensor product with respect to the $\mathfrak{m}_x$-adic topology on F_x or $\mathcal{O}_x$. The functor L_xG (resp. L_x^+G) is representable by a group indscheme (resp. group scheme of infinite type) over k_x. There is a reduction map $L_x^+G \to G \otimes_k k_x$.

We will also need the notion of *parahoric subgroups* of $G(F_x)$. Bruhat and Tits [4] constructed certain smooth models $\mathcal{P}$ of G over $\mathcal{O}_x$ whose special fibers are connected. By a parahoric subgroup $\mathbf{P}$ of L_xG we shall mean the group scheme over k_x whose R-point is $\mathcal{P}(R\widehat{\otimes}_{k_x}\mathcal{O}_x)$ where $\mathcal{P}$ is a Bruhat-Tits group scheme over $\mathcal{O}_x$. The k_x-points $\mathbf{P}(k_x)$ of $\mathbf{P}$ is a parahoric subgroup of $G(F_x)$ in the usual sense. Viewed as a proalgebraic group, the parahoric $\mathbf{P} \subset L_xG$ has a maximal reductive quotient $L_{\mathbf{P}}$ which is a connected reductive group over k_x. For example, the preimage $\mathbf{I}_x$ of $B \otimes_k k_x$ under the reduction map $L_x^+G \to G \otimes_k k_x$ is called the *standard Iwahori subgroup* of L_xG (with respect to B), and any parahoric subgroup of L_xG contains a conjugate of $\mathbf{I}_x$.

Finally, the Langlands dual group $\widehat{G}$ to G is a reductive group over $\overline{\mathbb{Q}}_\ell$, equipped with a pinning $(\widehat{T}, \widehat{B}, \ldots)$ such that the corresponding based root system $(\mathbb{X}^*(\widehat{T}), \Phi(\widehat{G}, \widehat{T}), \Delta_{\widehat{B}})$ of $\widehat{G}$ is the same as the based coroot system $(\mathbb{X}_*(T), \Phi^\vee(G,T), \Delta_B^\vee)$ of G with respect to T and B.

2.1.3 Automorphic representations

The group $G(\mathbb{A}_F)$ of $\mathbb{A}_F$-points of G can also be expressed as the restricted product

$$G(\mathbb{A}_F) = \prod_{x\in|X|}{}' G(F_x)$$

where most components lie in $G(\mathcal{O}_x)$. This is a locally compact topological group. With this topology on $G(\mathbb{A}_F)$ we may talk about locally constant functions on $G(\mathbb{A}_F)$. Denote the space of locally constant $\overline{\mathbb{Q}}_\ell$-valued functions on $G(\mathbb{A}_F)$ by $C^\infty(G(\mathbb{A}_F), \overline{\mathbb{Q}}_\ell)$. Define $C^\infty(G(F)\backslash G(\mathbb{A}_F), \overline{\mathbb{Q}}_\ell)$ as the subspace of $C^\infty(G(\mathbb{A}_F), \overline{\mathbb{Q}}_\ell)$ consisting of left-$G(F)$-invariant functions.

2.1.4 Definition (See [3, Definition 5.8])

1. A function $f \in C^\infty(G(F)\backslash G(\mathbb{A}_F), \overline{\mathbb{Q}}_\ell)$ is called an *automorphic form* if for some (equivalently any) $x \in |X|$, the $G(F_x)$-module spanned by right $G(F_x)$-

translations on f is admissible. Denote the space of automorphic forms by $\mathcal{A}_G$. This is a $G(\mathbb{A}_F)$-module under right translation.

2. An irreducible representation π of $G(\mathbb{A}_F)$ is an *automorphic representation* if it is a subquotient of $\mathcal{A}_G$.

2.1.5 Cusp forms

A function $f \in C^\infty(G(F)\backslash G(\mathbb{A}_F), \overline{\mathbb{Q}}_\ell)$ is a *cusp form* if for every parabolic subgroup $P \subset G$ defined over F with unipotent radical N_P, we have

$$\int_{N_P(F)\backslash N_P(\mathbb{A}_F)} f(ng)dn = 0$$

for all $g \in G(\mathbb{A}_F)$. All cusp forms have compact support in $G(F)\backslash G(\mathbb{A}_F)$. The space of cusp forms is a sub-$G(\mathbb{A}_F)$-module $\mathcal{A}_G^{\mathrm{cusp}}$ of all automorphic forms $\mathcal{A}_G$. It is known that $\mathcal{A}_G^{\mathrm{cusp}}$ decomposes discretely into a direct sum of irreducible $G(\mathbb{A}_F)$-modules, called *cuspidal automorphic representations*.

2.2 Weil's interpretation

We temporarily allow G to be a reductive group over k. When $G = \mathrm{GL}_n$, Weil has given a geometric interpretation of the double coset $G(F)\backslash G(\mathbb{A}_F)/\prod_{x\in|X|} G(\mathcal{O}_x)$ in terms of vector bundles over X. We recall his interpretation for general reductive G.

A (right) G-torsor over X is a scheme $Y \to X$ together with a fiberwise action of G that looks like $G \times X$ (with G acting on itself by right translation) étale locally over X. An isomorphism between G-torsors Y and Y' is a G-equivariant isomorphism $Y \cong Y'$ over X.

Let $\mathrm{Bun}_G(k)$ be the groupoid of G-torsors over X: this is a category whose objects are G-torsors over X and morphisms are isomorphisms between G-torsors. The groupoid $\mathrm{Bun}_G(k)$ is in fact the groupoid of k-points of an algebraic stack Bun_G. For a k-algebra R the groupoid $\mathrm{Bun}_G(R)$ is the groupoid of G-torsors over $X \otimes_k R$.

2.2.1 Example

When $G = \mathrm{GL}_n$, there is a natural way to assign a vector bundle $\mathcal{V}$ of rank n over X to a GL_n-torsor $\mathcal{E}$ and vice versa: $\mathcal{V} = \mathcal{E} \overset{\mathrm{GL}_n}{\times} \mathbb{A}_k^n$ and $\mathcal{E} = \underline{\mathrm{Isom}}_X(\mathcal{O}_X^n, \mathcal{V})$. Therefore $\mathrm{Bun}_{\mathrm{GL}_n}$ is isomorphic to the algebraic stack Bun_n classifying rank n vector bundles over X. In particular, for $n = 1$, $\mathrm{Bun}_{\mathrm{GL}_1} \cong \mathrm{Bun}_1 \cong \mathrm{Pic}(X)$.

Similarly, $\mathrm{Bun}_{\mathrm{SL}_n}$ classifies pairs $(\mathcal{V}, \iota)$ where $\mathcal{V}$ is a vector bundle of rank n over X and $\iota : \wedge^n\mathcal{V} \xrightarrow{\sim} \mathcal{O}_X$ is a trivialization of the determinant of $\mathcal{V}$.

When $G = \mathrm{PGL}_n$, $\mathrm{Bun}_{\mathrm{PGL}_n}$ is equivalent to the quotient stack $\mathrm{Bun}_n/\mathrm{Pic}(X)$ with $\mathrm{Pic}(X)$ acting on Bun_n via tensor product. More concretely, the objects in $\mathrm{Bun}_{\mathrm{PGL}_n}(k)$ are the same as the objects in $\mathrm{Bun}_n(k)$, and the morphism set between two vector bundles $\mathcal{V}$ and $\mathcal{V}'$ in the groupoid $\mathrm{Bun}_{\mathrm{PGL}_n}(k)$ is the set of isomorphism classes of line bundles $\mathcal{L}$ on X such that $\mathcal{V} \otimes \mathcal{L} \cong \mathcal{V}'$. Here we have used the fact that the Brauer group of X is trivial, so that all PGL_n-torsors over X lift to GL_n-torsors.

Weil observed that there is a natural bijection

$$e : G(F)\backslash G(\mathbb{A}_F)/\prod_{x\in|X|} G(\mathcal{O}_x) \xrightarrow{\sim} \mathrm{Bun}_G(k). \tag{2.1}$$

This is not just a bijection of sets, but in fact an isomorphism of groupoids. In other words, for any double coset $[g] = G(F)g\prod_{x\in|X|} G(\mathcal{O}_x)$, the automorphism group of $e([g])$ (as a G-torsor) is isomorphic to the stabilizer of the coset $g\prod_{x\in|X|} G(\mathcal{O}_x)$ under $G(F)$.

We give the definition of the map e on the level of sets. For any finite $S \subset |X|$, the double coset $G(\mathcal{O}_{X-S})\backslash\prod_{x\in S} G(F_x)/G(\mathcal{O}_x)$ is the subset of

$$G(F)\backslash G(\mathbb{A}_F)/\prod_{x\in|X|} G(\mathcal{O}_x)$$

consisting of the those classes represented by $g = (g_x) \in G(\mathbb{A}_F)$ with $g_x \in G(\mathcal{O}_x), \forall x \notin S$. Clearly,

$$G(F)\backslash G(\mathbb{A}_F)/\prod_{x\in|X|} G(\mathcal{O}_x) = \bigcup_{S\subset|X|,\text{finite}} G(\mathcal{O}_{X-S})\backslash\prod_{x\in S} G(F_x)/G(\mathcal{O}_x).$$

Thus it suffices to construct a compatible system of maps

$$e_S : G(\mathcal{O}_{X-S})\backslash\prod_{x\in S} G(F_x)/G(\mathcal{O}_x) \to \mathrm{Bun}_G(k).$$

For $g = (g_x) \in \prod_{x\in S} G(F_x)$, we take the trivial G-torsor $\mathcal{E}^{\mathrm{triv}}_{X-S} = (X - S) \times_k G$ over $X - S$ and glue it with the trivial G-torsor $\mathcal{E}^{\mathrm{triv}}_x = \mathrm{Spec}\mathcal{O}_x \times G$ over $\mathrm{Spec}\mathcal{O}_x$ for each $x \in S$ along $\mathrm{Spec}F_x$. The gluing means to give an isomorphism $\mathcal{E}^{\mathrm{triv}}_x|_{\mathrm{Spec}F_x} \xrightarrow{\sim} \mathcal{E}^{\mathrm{triv}}_{X-S}|_{\mathrm{Spec}F_x}$, which is given by g_x. Changing the trivializations of $\mathcal{E}^{\mathrm{triv}}_{X-S}$ and $\mathcal{E}^{\mathrm{triv}}_x$ amounts to right multiply g_x by an element in $G(\mathcal{O}_x)$ and left multiply g_x by an element in $G(\mathcal{O}_{X-S})$. The isomorphism type of the resulting G-torsor $\mathcal{E}_g$ after gluing only depends on the double coset $G(\mathcal{O}_{X-S})g\prod_{x\in S} G(\mathcal{O}_x)$, and we define $e_S(g)$ to be $\mathcal{E}_g$.

2.2.2 Level structures

One can generalize Bun_G to G-torsors with level structures. Fix a finite set $S \subset |X|$, and for each $x \in S$ let $\mathbf{K}_x \subset L_xG$ be a proalgebraic subgroup commensurable with L^+_xG. Then we may talk about G-torsors over X with $\mathbf{K}_S$-level structures: these are G-torsors $\mathcal{E}$ over X together with trivializations $\iota_x : \mathcal{E}|_{\mathrm{Spec}\mathcal{O}_x} \cong G|_{\mathrm{Spec}\mathcal{O}_x}$ (for each $x \in S$) up to left multiplication by $\mathbf{K}_x$ (via the intuitive action if $\mathbf{K}_x \subset L^+_xG$, but it requires some thought to define the action in general). We shall denote the corresponding moduli stack by $\mathrm{Bun}_G(\mathbf{K}_S)$. Then the isomorphism (2.1) generalizes to an isomorphism of groupoids

$$G(F)\backslash G(\mathbb{A}_F)/\prod_{x\notin S} G(\mathcal{O}_x) \times \prod_{x\in S} \mathbf{K}_x(k_x) \xrightarrow{\sim} \mathrm{Bun}_G(\mathbf{K}_S)(k). \tag{2.2}$$

2.3 Sheaf-to-function correspondence

2.3.1 The dictionary

Let X be a scheme of finite type over a finite field k and let $\mathcal{F}$ be a constructible complex of $\overline{\mathbb{Q}}_\ell$-sheaves for the étale topology of X. For each closed point $x \in X$, the geometric Frobenius element Frob_x at x acts on the geometric stalk $\mathcal{F}_{\overline{x}}$, which is a complex of $\overline{\mathbb{Q}}_\ell$-vector spaces. We consider the function

$$\begin{aligned} f_{\mathcal{F},k} : X(k) &\to \overline{\mathbb{Q}}_\ell, \\ x &\mapsto \sum_{i\in\mathbb{Z}} (-1)^i \mathrm{Tr}(\mathrm{Frob}_x, \mathrm{H}^i\mathcal{F}_{\overline{x}}). \end{aligned}$$

Similarly we can define a function $f_{\mathcal{F},k'} : X(k') \to \overline{\mathbb{Q}}_\ell$ for any finite extension k' of k. The correspondence

$$\mathcal{F} \mapsto \{f_{\mathcal{F},k'}\}_{k'/k}$$

is called the *sheaf-to-function* correspondence. This construction enjoys various functorial properties. For a morphism $\phi : X \to Y$ over k, the derived push forward $f_!$ transforms into integration of functions along the fibers (this is a consequence of the Lefschetz trace formula for the Frobenius endomorphism); the derived pullback f^* transforms into pullback of functions. It also transforms tensor product of sheaves into pointwise multiplication of functions.

2.3.2 Definition

Let L be a connected algebraic group over k with the multiplication map $m : L \times L \to L$ and the identity point $e : \mathrm{Spec} k \to L$. A *rank one character sheaf* $\mathcal{K}$ on L is a local system of rank one on L equipped with two isomorphisms

$$\begin{aligned} \mu &: m^*\mathcal{K} \xrightarrow{\sim} \mathcal{K} \boxtimes \mathcal{K}, \\ u &: \overline{\mathbb{Q}}_\ell \xrightarrow{\sim} e^*\mathcal{K}. \end{aligned}$$

These isomorphisms should be compatible in the sense that

$$\begin{aligned} \mu|_{L\times\{e\}} &= \mathrm{id}_\mathcal{K} \otimes u : \mathcal{K} = \overline{\mathbb{Q}}_\ell \otimes_{\overline{\mathbb{Q}}_\ell} \mathcal{K} \xrightarrow{\sim} e^*\mathcal{K} \otimes \mathcal{K}, \\ \mu|_{\{e\}\times L} &= u \otimes \mathrm{id}_\mathcal{K} : \mathcal{K} = \mathcal{K} \otimes_{\overline{\mathbb{Q}}_\ell} \overline{\mathbb{Q}}_\ell \xrightarrow{\sim} \mathcal{K} \otimes e^*\mathcal{K}. \end{aligned}$$

Since L is connected, the isomorphism μ in Definition 2.3.2 automatically satisfies the usual cocycle relation on L^3. A local system $\mathcal{K}$ of rank one on L being a character sheaf is a property rather than extra structure on $\mathcal{K}$. Let $\mathcal{CS}_1(L)$ be the category (groupoid) of rank one character sheaves $(\mathcal{K}, \mu, u)$ on L, then it carries a symmetric monoidal structure given by the tensor product of character sheaves with the unit object given by the constant sheaf. Let $\mathrm{CS}_1(L)$ be set of isomorphism classes of objects in $\mathcal{CS}_1(L)$, then it is an abelian group. The groupoid $\mathcal{CS}_1(L)$ has trivial automorphisms, and hence is equivalent to the set $\mathrm{CS}_1(L)$. We can similarly define the notion of rank one character sheaves over $\overline{k}$ and form the category $\mathcal{CS}_1(L/\overline{k})$ and the abelian group $\mathrm{CS}_1(L/\overline{k})$. The base change map $\mathrm{CS}_1(L) \to \mathrm{CS}_1(L/\overline{k})$ is injective, and the image consists of $\mathrm{Gal}(\overline{k}/k)$-invariants.

2.3.3 Remark

One can make a similar definition of rank one character sheaves when L is not necessarily connected, then the cocycle condition has to be imposed on μ as an extra requirement and μ itself is an extra datum. For example, if L is a discrete group scheme over $\overline{k}$ and we fix an isomorphism $\mathcal{K} \cong \overline{\mathbb{Q}}_\ell$ (as local systems) extending u, then μ (which satisfies the cocycle relation) gives a cocycle $\xi \in Z^2(L, \overline{\mathbb{Q}}_\ell^\times)$ satisfying $\xi_{1,\gamma} = \xi_{\gamma,1} = 1$ for all $\gamma \in L$.

For the rest of this subsection we assume k is finite. For each $\mathcal{K} \in \mathcal{CS}_1(L)$, the sheaf-to-function correspondence gives a function $f_\mathcal{K} : L(k) \to \overline{\mathbb{Q}}_\ell^\times$ which is in fact a group homomorphism because of the isomorphism μ. This way we obtain a homomorphism

$$f_L : \mathrm{CS}_1(L) \to \mathrm{Hom}(L(k), \overline{\mathbb{Q}}_\ell^\times). \tag{2.3}$$

One can show that f_L is always injective. The following result gives descriptions of $\mathrm{CS}_1(L)$ in various special cases.

2.3.4 Theorem

1. *Let L be a connected commutative algebraic group over k. Then f_L is an isomorphism of abelian groups*

$$f_L : \mathrm{CS}_1(L) \xrightarrow{\sim} \mathrm{Hom}(L(k), \overline{\mathbb{Q}}_\ell^\times).$$

2. *Let L be a connected reductive group over k and $L^{\mathrm{sc}} \to L$ be the simply-connected cover of its derived group. Then f_L induces an isomorphism of abelian groups*

$$\mathrm{CS}_1(L) \xrightarrow{\sim} \mathrm{Hom}(L(k)/L^{\mathrm{sc}}(k), \overline{\mathbb{Q}}_\ell^\times).$$

Let T be a maximal torus in L and $T^{\mathrm{sc}} \subset L^{\mathrm{sc}}$ be its preimage in L^{sc}. Then we also have

$$\mathrm{CS}_1(L) \xrightarrow{\sim} \mathrm{Hom}(T(k)/T^{\mathrm{sc}}(k), \overline{\mathbb{Q}}_\ell^\times).$$

The construction of an inverse to f_L uses the Lang torsor $L \to L$ sending $g \mapsto F_L(g)g^{-1}$, where F_L is the Frobenius endomorphism of L (relative to k).

2.3.5 Example

When $L = T$ is a split torus over $k = \mathbb{F}_q$, any object in $\mathcal{CS}_1(L/\overline{k})$ is obtained as a direct summand of $[n]_!\overline{\mathbb{Q}}_\ell$ where $[n] : T \to T$ is the n-th power morphism and n is prime to $\mathrm{char}(k)$. An object $\mathcal{K} \in \mathcal{CS}_1(L/\overline{k})$ belongs to $\mathcal{CS}_1(L)$ if and only if $\mathcal{K}^{\otimes(q-1)} \cong \overline{\mathbb{Q}}_\ell$, or equivalently it is a direct summand of $[q-1]_!\overline{\mathbb{Q}}_\ell$. Objects in $\mathcal{CS}_1(L)$ are also called *Kummer sheaves.*

2.3.6 Example

When $L = \mathbb{G}_a$ is the additive group over $k = \mathbb{F}_q$, let $\psi : k \to \overline{\mathbb{Q}}_\ell^\times$ be an additive character. Let $\lambda_{\mathbb{G}_a} : \mathbb{G}_a \to \mathbb{G}_a$ be the $\mathbb{G}_a(k)$-torsor given by $a \mapsto a^q - a$. Objects in $\mathcal{CS}_1(L)$ are of the form $\mathrm{AS}_\psi := (\lambda_{\mathbb{G}_a,!}\overline{\mathbb{Q}}_\ell)_\psi$, the direct summand of $\lambda_{\mathbb{G}_a,!}\overline{\mathbb{Q}}_\ell$ on which $\mathbb{G}_a(k)$ acts through ψ. Such local systems are called *Artin-Schreier sheaves.* Now fix a nontrivial additive character ψ of k. Let $L = V$ be a vector space over k viewed as an additive group. Then objects in $\mathcal{CS}_1(V)$ are of the form $\mathrm{AS}_\phi := \phi^*\mathrm{AS}_\psi$ for a unique $\phi \in V^*$ viewed as a homomorphism $\phi : V \to \mathbb{G}_a$.

2.3.7 Remark

(1) Definition 2.3.2 makes sense for any base field k. In this generality we have an alternative way to describe rank one character sheaves. Let $1 \to C \to \widetilde{L} \xrightarrow{v} L \to 1$ be a central extension of algebraic groups with $\widetilde{L}$ connected and C a *discrete* finite group scheme over k. For any character $\chi_C : C \to \overline{\mathbb{Q}}_\ell^\times$, the corresponding direct summand $(v_!\overline{\mathbb{Q}}_\ell)_{\chi_C} \in \mathcal{CS}_1(L)$. Conversely, all objects in $\mathcal{CS}_1(L)$ arise this way.

(2) One can define the category $\mathcal{CS}_1(L)$ for L proalgebraic and connected. In fact, write L as the inverse limit of finite-dimensional quotients L_i, and define $\mathcal{CS}_1(L)$ as the direct 2-limit of $\mathcal{CS}_1(L_i)$.

2.4 Geometric automorphic data

We resume with the setting in §2.1. Let $S \subset |X|$ be finite.

2.4.1 Definition

A pair $(\mathbf{K}_S, \mathcal{K}_S)$ is a *geometric automorphic datum* with respect to S if

1. $\mathbf{K}_S$ is a collection $\{\mathbf{K}_x\}_{x\in S}$, where $\mathbf{K}_x \subset L_xG$ is a connected proalgebraic subgroup which is commensurable with L_x^+G (i.e., $\mathbf{K}_x \cap L_x^+G$ has finite codimension in both $\mathbf{K}_x$ and L_x^+G).

2. $\mathcal{K}_S$ is a collection $\{\mathcal{K}_x\}_{x\in S}$ where $\mathcal{K}_x \in \mathcal{CS}_1(\mathbf{K}_x)$.

Recall from Remark 2.3.7(2) that $\mathcal{K}_x \in \mathcal{CS}_1(\mathbf{K}_x)$ means that $\mathcal{K}_x$ is the pullback of a rank one character sheaf from a finite-dimensional quotient $\mathbf{K}_x \twoheadrightarrow L_x$.

2.4.2 Remark

(1) Since k is a finite field, the character sheaf $\mathcal{K}_x$ is uniquely determined by the associated character $\chi_x : \mathbf{K}_x(k_x) \to \overline{\mathbb{Q}}_\ell^\times$ via the sheaf-to-function correspondence (see (2.3)). Therefore we also call the pair $(\mathbf{K}_S, \chi_S)$ a geometric automorphic datum, provided χ_x does arise from an object in $\mathcal{CS}_1(\mathbf{K}_x)$ for all $x \in S$.

(2) Definition 2.4.1 makes sense over any base field k, and we shall work in such generality in §4.

2.4.3 Definition

Let $(\mathbf{K}_S, \chi_S)$ be a geometric automorphic datum. An automorphic representation π of $G(\mathbb{A}_F)$ is called $(\mathbf{K}_S, \chi_S)$-*typical* if

1. For every $x \in S$, the local component π_x of π has a nonzero vector on which $\mathbf{K}_x(k_x)$ acts via the character χ_x.

2. For every $x \notin S$, π_x is spherical; i.e., $\pi_x^{G(\mathcal{O}_x)} \neq 0$.

Consider the space of functions

$$C_G(\mathbf{K}_S, \chi_S) := C(G(F)\backslash G(\mathbb{A}_F)/\prod_{x \notin S} G(\mathcal{O}_x) \times \prod_{x \in S} (\mathbf{K}_x(k_x), \chi_x), \overline{\mathbb{Q}}_\ell). \qquad (2.4)$$

which are invariant under $\prod_{x \notin S} G(\mathcal{O}_x)$ and are eigenvectors under the action of $\mathbf{K}_x(k_x)$ with eigenvalue χ_x, for all $x \in S$.

2.4.4 Lemma

Let $(\mathbf{K}_S, \chi_S)$ be a geometric automorphic datum. Then

1. *A necessary condition for $(\mathbf{K}_S, \chi_S)$-typical automorphic representations to exist is that $C_G(\mathbf{K}_S, \chi_S) \neq 0$.*

2. *If there is a compact subset $\Sigma \subset G(F)\backslash G(\mathbb{A}_F)$ such that all functions in $C_G(\mathbf{K}_S, \chi_S)$ vanish outside Σ, then any $(\mathbf{K}_S, \chi_S)$-typical automorphic representation is cuspidal. In this case, $C_G(\mathbf{K}_S, \chi_S)$ is the direct sum of*

$$(\otimes_{x \notin S} \pi_x^{G(\mathcal{O}_x)}) \otimes (\otimes_{x \in S} \pi_x^{(\mathbf{K}_x(k_x), \chi_x)})$$

over $(\mathbf{K}_S, \chi_S)$-typical automorphic representations π.

2.4.5 Base change of geometric automorphic data

Let k'/k be a finite extension. Let S' be the preimage of S in $X \otimes_k k'$ (S' may have more elements than S). Given a geometric automorphic datum $(\mathbf{K}_S, \chi_S)$, we may define a corresponding geometric automorphic datum $(\mathbf{K}_{S'}, \chi_{S'})$ for G and the function field $F' = F \otimes_k k'$. For each $y \in S'$ with image $x \in S$, let $\mathbf{K}_y = \mathbf{K}_x \otimes_{k_x} k_y$. Suppose χ_x corresponds to the character sheaf $\mathcal{K}_x$ on $\mathbf{K}_x$, χ_y is then the character of $\mathbf{K}_y(k_y)$ corresponding to the pullback of $\mathcal{K}_x$ along $\mathbf{K}_y \to \mathbf{K}_x$.

2.4.6 Definition

Let Z be the center of G (note by our assumption Z is finite over k). A central character $\omega : Z(F)\backslash Z(\mathbb{A}_F) \to \overline{\mathbb{Q}}_\ell^\times$ is said to be *compatible with the geometric automorphic datum* $(\mathbf{K}_S, \chi_S)$, if ω is unramified away rom S, and $\omega_x|_{Z(F_x) \cap \mathbf{K}_x(k_x)} = \chi_x|_{Z(F_x) \cap \mathbf{K}_x(k_x)}$ for each $x \in S$.

Clearly, the central character ω of a $(\mathbf{K}_S, \chi_S)$-typical automorphic representation is compatible with $(\mathbf{K}_S, \chi_S)$.

2.4.7 Definition

Let $(\mathbf{K}_S, \chi_S)$ be a geometric automorphic datum.

1. It is called *strongly rigid*, if for *every* finite extension k'/k, there are at most one $(\mathbf{K}_{S'}, \chi_{S'})$-typical automorphic representation of $G(\mathbb{A}_{F'})$ (where $F' = F \otimes_k k'$) with any given central character, and if for *some* finite extension k'/k, a $(\mathbf{K}_{S'}, \chi_{S'})$-typical automorphic representation exists.

2. It is called *weakly rigid*, if there is a constant N such that for *every* finite extension k'/k, there are at most N $(\mathbf{K}_{S'}, \chi_{S'})$-typical automorphic representations of $G(\mathbb{A}_{F'})$.

2.5 Automorphic sheaves

Suggested by Weil's interpretation in §2.2 and the sheaf-to-function correspondence §2.3, we shall seek to upgrade the function space (2.4) into a category of sheaves on the moduli stack of G-torsors over X with level structures. This idea was implemented by Drinfeld in the case $G = \mathrm{GL}_2$, and for general G it has been formulated as the geometric Langlands correspondence by Laumon [24] and Beilinson and Drinfeld [2].

2.5.1 The category of automorphic sheaves

Consider the situation in §2.4. Let $\mathbf{K}_x^+ \subset \mathbf{K}_x$ be a connected normal subgroup of finite codimension such that the rank one character sheaf $\mathcal{K}_x$ on $\mathbf{K}_x$ is pulled back from the quotient $L_x = \mathbf{K}_x/\mathbf{K}_x^+$. Let $\mathrm{Bun}_G(\mathbf{K}_S)$ and $\mathrm{Bun}_G(\mathbf{K}_S^+)$ be the moduli stack of G-torsors over X with the respective level structures, as defined in §2.2.2. The morphism $\mathrm{Bun}_G(\mathbf{K}_S^+) \to \mathrm{Bun}_G(\mathbf{K}_S)$ is an $L_S := \prod_{x \in S} L_x$-torsor. The tensor product $\mathcal{K}_S := \boxtimes_{x \in S} \mathcal{K}_x$ is an object in $\mathcal{CS}_1(L_S)$. It makes sense to talk about $\overline{\mathbb{Q}}_\ell$-complexes of sheaves on $\mathrm{Bun}_G(\mathbf{K}_S^+)$ which are $(L_S, \mathcal{K}_S)$-equivariant.

Without giving the detailed definition, we have the derived category

$$D_G(\mathbf{K}_S, \chi_S) := D^b_{(L_S, \mathcal{K}_S)}(\mathrm{Bun}_G(\mathbf{K}_S^+), \overline{\mathbb{Q}}_\ell)$$

consisting of bounded constructible $\overline{\mathbb{Q}}_\ell$-complexes on $\mathrm{Bun}_G(\mathbf{K}_S^+)$ equipped with $(L_S, \mathcal{K}_S)$-equivariant structures. There is subtle in defining this category because $\mathrm{Bun}_G(\mathbf{K}_S^+)$ is a stack that is not of finite type. However we will only consider sheaves that are supported on finite-type substacks of $\mathrm{Bun}_G(\mathbf{K}_S^+)$. Objects in the category $D_G(\mathbf{K}_S, \chi_S)$ are called *automorphic sheaves* with respect to the geometric automorphic datum $(\mathbf{K}_S, \chi_S)$.

2.5.2 Relevant points

Consider a point $\mathcal{E} \in \mathrm{Bun}_G(\mathbf{K}_S)(\overline{k})$, which represents a G-torsor over $X_{\overline{k}}$ with $\mathbf{K}_y$-level structures at $y \in S(\overline{k})$. The automorphism group $\mathrm{Aut}(\mathcal{E})$ of the point $\mathcal{E}$ is an affine algebraic group over $\overline{k}$. For each $y \in S(\overline{k})$ with image $x \in S$, restricting an automorphism of $\mathcal{E}$ gives an element in $\mathbf{K}_y = \mathbf{K}_x \otimes_{k_x} \overline{k}$ (this depends on the choice a trivialization of $\mathcal{E}$ around x), and thus a homomorphism

$$\mathrm{ev}_{S,\mathcal{E}} : \mathrm{Aut}(\mathcal{E}) \to \prod_{y \in S(\overline{k})} \mathbf{K}_y \to \prod_{y \in S(\overline{k})} L_y = L_S \otimes_k \overline{k} \tag{2.5}$$

which is well-defined up to conjugacy if one changes the trivialization of $\mathcal{E}$ around x.

2.5.3 Definition

Let $(\mathbf{K}_S, \chi_S)$ be a geometric automorphic datum. A point $\mathcal{E} \in \mathrm{Bun}_G(\mathbf{K}_S)(\overline{k})$ is *relevant to* $(\mathbf{K}_S, \chi_S)$ if the restriction of $\mathrm{ev}^*_{S,\mathcal{E}}\mathcal{K}_S$ to the neutral component $\mathrm{Aut}^\circ(\mathcal{E})$ of $\mathrm{Aut}(\mathcal{E})$ is isomorphic to the constant sheaf. Otherwise the point $\mathcal{E}$ is called *irrelevant.*

According to (2.2), we may identify the double coset

$$G(F)\backslash G(\mathbb{A}_F)/\prod_{x \notin S} G(\mathcal{O}_x) \times \prod_{x \in S} \mathbf{K}^+_x$$

with $\mathrm{Bun}_G(\mathbf{K}^+_S)(k)$, and view functions in $C_G(\mathbf{K}_S, \chi_S)$ as functions on the underlying set of $\mathrm{Bun}_G(\mathbf{K}^+_S)(k)$ that are (L_S, χ_S)-equivariant. The sheaf-to-function correspondence then gives an additive map

$$\mathrm{Ob}\, D_G(\mathbf{K}_S, \chi_S) \to C_G(\mathbf{K}_S, \chi_S).$$

We shall also need to consider similar objects after a base change from k to a finite extension k' or to $k' = \overline{k}$. We shall use the notation

$$D_G(k'; \mathbf{K}_S, \chi_S) \text{ and } C_G(k'; \mathbf{K}_S, \chi_S)$$

to denote the analogs of $D_G(\mathbf{K}_S, \chi_S)$ and $C_G(\mathbf{K}_S, \chi_S)$ defined for the function field $F' = F \otimes_k k'$ and the automorphic data $(\mathbf{K}_{S'}, \chi_{S'})$ obtained by base change from $(\mathbf{K}_S, \chi_S)$ as in §2.4.5.

2.5.4 Lemma

1. *Let $\mathcal{E} \in \mathrm{Bun}_G(\mathbf{K}_S)(\overline{k})$ be an irrelevant point. Then for any object $\mathcal{F} \in D_G(\overline{k}; \mathbf{K}_S, \chi_S)$, $i^*_{\mathcal{E}}\mathcal{F} = 0$ and $i^!_{\mathcal{E}}\mathcal{F} = 0$. Here $i_{\mathcal{E}}$ denotes the inclusion map of the fiber of $\mathcal{E}$ in $\mathrm{Bun}_G(\mathbf{K}^+_S)$.*

2. *Let $[g] \in G(F)\backslash G(\mathbb{A}_F)/\prod_{x \notin S} G(\mathcal{O}_x) \times \prod_{x \in S} \mathbf{K}_x(k_x) = \mathrm{Bun}_G(\mathbf{K}_S)(k)$ be an irrelevant point (when viewed as a $\overline{k}$-point). Then any $f \in C_G(\mathbf{K}_S, \chi_S)$ vanishes on the preimage of $[g]$ in $G(F)\backslash G(\mathbb{A}_F)/\prod_{x \notin S} G(\mathcal{O}_x) \times \prod_{x \in S} \mathbf{K}^+_x(k_x)$. Similar statement holds when k is replaced by a finite extension k'.*

2.5.5 Lemma

Let $(\mathbf{K}_S, \chi_S)$ be a geometric automorphic datum.

1. *If there is no relevant $\overline{k}$-point on $\mathrm{Bun}_G(\mathbf{K}_S)$, then $D_G(k'; \mathbf{K}_S, \chi_S) = 0$ and $C_G(k'; \mathbf{K}_S, \chi_S) = 0$ for any finite extension k' of k.*

2. *If $\mathrm{Bun}_G(\mathbf{K}_S)$ contains only finitely many relevant $\overline{k}$-points, then $(\mathbf{K}_S, \chi_S)$ is weakly rigid. Moreover for any finite extension k' of k, any $(\mathbf{K}'_S, \chi'_S)$-typical automorphic representation of $G(\mathbb{A}_{F'})$ is cuspidal.*

2.5.6 Example

Let $G = \mathrm{SL}_2$ and $F = k(t)$, the function field of $X = \mathbb{P}^1_k$. Let $S = \{0, 1, \infty\} \subset |\mathbb{P}^1_k| = |X|$. For each closed point $x \in |X|$, we let $\mathbf{K}_x = \mathbf{I}_x$ be the standard Iwahori subgroup of $L_x G$, i.e.,

$$\mathbf{I}_x(k) = \left\{ \begin{pmatrix} a & b \\ c & d \end{pmatrix} \in \mathrm{GL}_2(\mathcal{O}_x) : c \in \mathfrak{m}_x \right\}.$$

For each $x \in S$, we choose a character $\chi_x : k^\times \to \overline{\mathbb{Q}}_\ell^\times$ and view it as a character of $\mathbf{I}_x(k)$ by sending $\begin{pmatrix} a & b \\ c & d \end{pmatrix}$ to $\chi_x(\overline{a})$, where $\overline{a} \in k^\times$ is the image of $a \in \mathcal{O}_x^\times$. We consider the geometric automorphic datum $(\mathbf{I}_S, \chi_S)$.

A central character ω compatible with $(\mathbf{I}_S, \chi_S)$ exists if and only if

$$\prod_{x \in S} \chi_x(-1) = 1. \tag{2.6}$$

If this condition is satisfied, the central character ω compatible with $(\mathbf{I}_S, \chi_S)$ is unique. The condition (2.6) can always be satisfied by passing to a quadratic extension k'/k and using the base-changed automorphic datum $(\mathbf{I}_{S'}, \chi_{S'})$.

2.5.7 Proposition

Suppose for any map $\epsilon : S \to \{\pm 1\}$, $\prod_{x \in S} \chi_x^{\epsilon(x)}$ is not the trivial character on $k^\times$. Then the geometric automorphic datum $(\mathbf{I}_S, \chi_S)$ is strictly rigid.

The moduli stack $\mathrm{Bun}_G(\mathbf{I}_S)$ in this case classifies $(\mathcal{V}, \iota, \{\ell_x\}_{x \in S})$ where $\mathcal{V}$ is a rank two vector bundle over X, $\iota : \wedge^2(\mathcal{V}) \cong \mathcal{O}_X$ and ℓ_x is a line in the fiber $\mathcal{V}_x$. There is a unique relevant point in $\mathrm{Bun}_G(\mathbf{I}_S)$, which corresponds to the trivial bundle $\mathcal{O}_X^2$ with three lines $\{\ell_x\}_{x \in S}$ in generic position (i.e., three distinct lines in k^2).

3 Rigidity for local systems

In this section k is any field. We shall review the notion of a local system in étale topology, and introduce the notion of rigidity for them.

3.1 Local systems

3.1.1 Local systems in étale topology

Let U be a scheme of finite type over k. We recall some definitions from [18, §1.2, §1.4.2, §1.4.3]. A $\mathbb{Z}_\ell$-local system on U is a projective system $(\mathcal{F}_n)_{n\geq 1}$ of locally constant locally free $\mathbb{Z}/\ell^n\mathbb{Z}$-sheaves $\mathcal{F}_n$ of finite rank on X under the étale topology such that the natural map $\mathcal{F}_n \otimes \mathbb{Z}/\ell^{n-1}\mathbb{Z} \to \mathcal{F}_{n-1}$ is an isomorphism for all n. Denote the category of $\mathbb{Z}_\ell$-local systems on U by $\mathrm{Loc}(U, \mathbb{Z}_\ell)$, which is a $\mathbb{Z}_\ell$-linear abelian category. The category of $\mathbb{Q}_\ell$-local systems is by definition the abelian category $\mathrm{Loc}(U, \mathbb{Z}_\ell) \otimes \mathbb{Q}_\ell$ obtained by inverting ℓ in the Hom groups in $\mathrm{Loc}(U, \mathbb{Z}_\ell)$. Similar definition gives $\mathrm{Loc}(U, \mathcal{O}_E)$ and $\mathrm{Loc}(U, E)$ for any finite extension E of $\mathbb{Q}_\ell$. Finally define $\mathrm{Loc}(U) := \mathrm{Loc}(U, \overline{\mathbb{Q}}_\ell)$ to be the inductive limit $\varinjlim_E \mathrm{Loc}(U, E)$ over all finite extensions E of $\mathbb{Q}_\ell$.

In the sequel we assume U is connected. Fix a geometric point $u \in U$. Let $\mathcal{F}$ be an $\overline{\mathbb{Q}}_\ell$-local system on U of rank n. The stalk $\mathcal{F}_u$ is an $\overline{\mathbb{Q}}_\ell$-vector space of dimension n which carries the action of the étale fundamental group $\pi_1(U, u)$ defined in [12, V, §7]. Thus $\mathcal{F}$ determines a *continuous* homomorphism

$$\pi_1(U, u) \to \mathrm{Aut}_{\overline{\mathbb{Q}}_\ell}(\mathcal{F}_u) \cong \mathrm{GL}_n(\overline{\mathbb{Q}}_\ell)$$

where $\overline{\mathbb{Q}}_\ell$ is topologized as the inductive limit of finite extensions E of $\mathbb{Q}_\ell$, each with the ℓ-adic topology. The last isomorphism above depends upon a choice of a basis of $\mathcal{F}_u$. We get a functor

$$\omega_u : \mathrm{Loc}(U) \to \mathrm{Rep}_{\mathrm{cont}}(\pi_1(U, u), \overline{\mathbb{Q}}_\ell) \tag{3.1}$$

where $\mathrm{Rep}_{\mathrm{cont}}(\pi_1(U, u), \overline{\mathbb{Q}}_\ell)$ is the category of continuous representation of $\pi_1(U, u)$ on finite-dimensional $\overline{\mathbb{Q}}_\ell$-vector spaces. Both sides of (3.1) carry tensor structures and ω_u is in fact an equivalence of tensor categories ([18, Proposition 1.2.5]).

Suppose furthermore that U is normal and connected and $F = k(U)$ is the function field of U. When u is a geometric generic point of X, $\pi_1(U, u)$ is a quotient of the absolute Galois group $\mathrm{Gal}(F^s/F)$ ([12, V, Proposition 8.2]). Therefore every local system of rank n over U determines a continuous Galois representations

$$\rho : \mathrm{Gal}(F^s/F) \to \mathrm{GL}_n(\overline{\mathbb{Q}}_\ell).$$

We shall denote by $\pi_1^{\mathrm{geom}}(U, u)$ the fundamental group of $\pi_1(U \otimes_k \overline{k}, u)$, and call it the *geometric fundamental group* of U (with respect to the base point u). This is a normal subgroup of $\pi_1(U, u)$ which fits into an exact sequence ([12, V, Proposition 6.13])

$$1 \to \pi_1^{\mathrm{geom}}(U, u) \to \pi_1(U, u) \to \mathrm{Gal}(k^s/k) \to 1.$$

3.1.2 H-local systems

Let H be an affine algebraic group over $\overline{\mathbb{Q}}_\ell$ of finite type. We may also define the notion of H-local system on a connected U. There are two ways to do this.

First definition. Fix a geometric point $u \in U$. An H-local system on U is a continuous homomorphism

$$\rho : \pi_1(U, u) \to H(\overline{\mathbb{Q}}_\ell). \tag{3.2}$$

Such homomorphisms form a category $\mathrm{Loc}_H(U)$, in which isomorphisms are given by $H(\overline{\mathbb{Q}}_\ell)$-conjugacy of representations. For example, $\mathrm{Loc}_{\mathrm{GL}_n}(U)$ is the full subcategory of $\mathrm{Loc}(U)$ consisting of local systems of rank n.

Second (and more canonical) definition. Let $\mathrm{Rep}(H, \overline{\mathbb{Q}}_\ell)$ be the tensor category of algebraic representations of H on finite-dimensional $\overline{\mathbb{Q}}_\ell$-vector spaces. We define the category

$$\mathrm{Loc}_H(U) := \mathrm{Fun}^{\otimes}(\mathrm{Rep}(H, \overline{\mathbb{Q}}_\ell), \mathrm{Loc}(U)).$$

Here $\mathrm{Fun}^{\otimes}$ denotes the category of tensor functors between two tensor categories.

The two notions of H-local systems are equivalent. Given a representation ρ as in (3.2) and for $V \in \mathrm{Rep}(H, \overline{\mathbb{Q}}_\ell)$, the composition

$$\rho_V : \pi_1(U, u) \xrightarrow{\rho} H(\overline{\mathbb{Q}}_\ell) \to \mathrm{GL}(V)$$

is an object in $\mathrm{Loc}(U)$ of rank equal to $\dim V$. The assignment $V \mapsto \rho_V$ gives a tensor functor $\mathrm{Rep}(H, \overline{\mathbb{Q}}_\ell) \to \mathrm{Loc}(U)$. Conversely, given a tensor functor $\mathcal{F}$: $\mathrm{Rep}(H, \overline{\mathbb{Q}}_\ell) \to \mathrm{Loc}(U)$, using the equivalence (3.1), this can be viewed as a tensor functor $\mathrm{Rep}(H, \overline{\mathbb{Q}}_\ell) \to \mathrm{Rep}_{\mathrm{cont}}(\pi_1(U, u), \overline{\mathbb{Q}}_\ell)$. The Tannakian formalism [7] then implies that such a tensor functor comes from a group homomorphism ρ as in (3.2), well-defined up to conjugacy.

3.1.3 Definition

Let $\rho : \pi_1(U, u) \to H(\overline{\mathbb{Q}}_\ell)$ be an H-local system. The *global geometric monodromy group* H_ρ^{geom} of the H-local system ρ is the Zariski closure of $\rho(\pi_1^{\mathrm{geom}}(U, u))$ in H.

3.1.4 Local monodromy

Let X be a projective, smooth and geometrically connected curve over a perfect field k. We fix a finite set of closed points $S \subset |X|$ and let $U = X - S$. Let $j : U \hookrightarrow X$ be the open inclusion. Let $x \in S$, and let $i_x : \{x\} = \mathrm{Spec} k_x \hookrightarrow X$ be the inclusion. Fix an algebraic closure $\overline{F}_x$ of F_x. This gives a geometric generic point $\eta_x \in X$. The morphism $\mathrm{Spec} F_x \to U$ then induces an injective homomorphism of fundamental groups

$$\mathrm{Gal}(F_x^s/F_x) \hookrightarrow \pi_1(U, \eta_x) \cong \pi_1(U, u) \tag{3.3}$$

where the second map is well-defined up to conjugacy. Since F_x is a complete discrete valuation field with perfect residue field k_x, we have an exact sequence

$$1 \to \mathcal{I}_x \to \mathrm{Gal}(F_x^s/F_x) \to \mathrm{Gal}(\overline{k}/k_x) \to 1.$$

The normal subgroup $\mathcal{I}_x$ of $\mathrm{Gal}(F_x^s/F_x)$ is the *inertia group* at x. Under (3.3), $\mathcal{I}_x$ is contained in the normal subgroup $\pi_1^{\mathrm{geom}}(U, u) \lhd \pi_1(U, u)$. By the *local monodromy* of ρ at $x \in S$ we mean the restriction $\rho|_{\mathcal{I}_x}$.

When $\mathrm{char}(k) = p > 0$, We have a normal subgroup $\mathcal{I}_x^w \lhd \mathcal{I}_x$ called the *wild inertia group* such that the quotient $\mathcal{I}_x^t := \mathcal{I}_x/\mathcal{I}_x^w$ is the maximal prime-to-p quotient of $\mathcal{I}_x$, called the *tame inertia group.* We have a canonical isomorphism of $\mathrm{Gal}(\overline{k}/k_x)$-modules

$$\mathcal{I}_x^t \xrightarrow{\sim} \varprojlim_{(n,p)=1} \mu_n(\overline{k}).$$

The local system ρ is said to be *tame at $x \in S$* if $\rho|_{\mathcal{I}_x}$ factors through the tame inertia group $\mathcal{I}_x^t$.

3.2 Two notions of rigidity for local systems

Rigidity of a local system is a geometric property, therefore we assume the base field k to be algebraically closed in this subsection. Let X be a complete smooth connected algebraic curve over k. Fix an open subset $U \subset X$ with finite complement S.

3.2.1 Definition (extending Katz [20, §1.0.3])

Let H be an algebraic group over $\overline{\mathbb{Q}}_\ell$. Let ρ be an H-local system on U. Then ρ is called *physically rigid* if, for any other H-local system ρ' such that $\rho'|_{\mathcal{I}_x} \cong \rho|_{\mathcal{I}_x}$ (meaning conjugate in $H(\overline{\mathbb{Q}}_\ell)$) for each $x \in S$, we have $\rho \cong \rho'$ as objects in $\mathrm{Loc}_H(U)$.

Although the definition uses U as an input, the notion of physical rigidity is in fact independent of the open subset U: for any nonempty open subset $V \subset U$, ρ is rigid over U if and only if $\rho|_V$ is rigid over V. Therefore, physical rigidity is a property of the Galois representation $\rho_\eta : \mathrm{Gal}(F^s/F) \to H(\overline{\mathbb{Q}}_\ell)$ obtained by restricting ρ to a geometric generic point η of the X.

Next we introduce cohomological rigidity. For this we assume H is a connected semisimple group over $\overline{\mathbb{Q}}_\ell$. Let $\mathfrak{h}$ be the Lie algebra of H, and let $\mathrm{Ad}(\rho)$ be the composition $\pi_1(U) \xrightarrow{\rho} H(\overline{\mathbb{Q}}_\ell) \to \mathrm{GL}(\mathfrak{h})$, viewed as a local system on U of rank $\dim \mathfrak{h}$. Let $j : U \hookrightarrow X$ be the open embedding and $j_{!*}\mathrm{Ad}(\rho)$ be the non-derived direct image of $\mathrm{Ad}(\rho)$ along j. Concretely, the stalk of $j_{!*}\mathrm{Ad}(\rho)$ at $x \in S$ is the $\mathcal{I}_x$-invariants on $\mathfrak{h}$.

3.2.2 Definition (extending Katz [20, §5.0.1])

Let H be a connected semisimple group over $\overline{\mathbb{Q}}_\ell$. An object $\rho \in \mathrm{Loc}_H(U)$ is called *cohomogically rigid*, if

$$\mathrm{Rig}(\rho) := \mathrm{H}^1(X, j_{!*}\mathrm{Ad}(\rho)) = 0.$$

The vector space $\mathrm{Rig}(\rho)$ does not change if we shrink U to a smaller open subset. Therefore cohomological rigidity is also a property of the Galois representation $\rho_\eta : \mathrm{Gal}(F^s/F) \to H(\overline{\mathbb{Q}}_\ell)$.

3.2.3 Remark

When we work over the base field $\mathbb{C}$ and view U as a topological surface, one can define a moduli stack $\mathcal{M}$ of H-local systems over U with prescribed local monodromy around the punctures S. The same formula $\mathrm{Rig}(\rho)$ then calculates the dimension of the tangent space $T_\rho\mathcal{M}$ at ρ. The condition $\mathrm{Rig}(\rho) = 0$ in this topological setting says that ρ does not admit infinitesimal deformations with prescribed local monodromy around S, i.e., $\mathcal{F}$ is rigid. This interpretation is the motivation for Definition 3.2.2. However, defining a moduli stack of ℓ-adic local systems is much subtler, and this topological interpretation only serves as a heuristic.

Using the Grothendieck-Ogg-Shafarevich formula, it is easy to give the following numerical criterion for cohomological rigidity.

3.2.4 Proposition

Let ρ be an H-local system on $U = X - S$. Let g_X be the genus of X. Then ρ is cohomologically rigid if and only if

$$\frac{1}{2}\sum_{x\in S} a_x(\mathrm{Ad}(\rho)) = \dim \mathfrak{h}/(\mathfrak{h})^{\pi_1(U,u)} - g_X \dim \mathfrak{h}. \tag{3.4}$$

Here $a_x(\mathrm{Ad}(\rho))$ is the Artin conductor of the action of $\mathcal{I}_x$ on $\mathfrak{h}$.

From (3.4) we see that cohomologically rigid H-local systems exist only when $g_X \leq 1$. When $g_X = 1$, rigid examples are very limited (see [20, §1.4]). Most examples of rigid local systems are over open subsets of $\mathbb{P}^1$.

When $H = \mathrm{SL}_n$, the two notions of rigidity are related by the following theorem.

3.2.5 Theorem (Katz [20, Theorem 5.0.2])

For $X = \mathbb{P}^1$ and $H = \mathrm{SL}_n$, cohomological rigidity of an SL_n-local system implies its physical rigidity.

3.2.6 Remark

An alternative approach to define the notion of rigidity for a local system ρ over U over a finite field k is by requiring that the adjoint L-function of ρ to be trivial (constant function 1). This is the approach taken by Gross in [14]. When $\mathrm{H}^0(U_{\overline{k}}, \mathrm{Ad}(\rho)) = 0$, triviality of the adjoint L-function of ρ is equivalent to cohomological rigidity of ρ.

3.3 Rigidity in the inverse Galois theory

It is instructive to compare the notion of rigidity for local systems with the notion of a rigid tuple in inverse Galois theory. We give a quick review following [33, Chapter 8]. Let H be a *finite* group with trivial center.

3.3.1 Definition

A tuple of conjugacy classes $(C_1, C_2, \ldots, C_n)$ in H is called (*strictly*) *rigid*, if

- The equation
$$g_1 g_2 \cdots g_n = 1 \tag{3.5}$$
has a solution with $g_i \in C_i$, and the solution is unique up to simultaneous H-conjugacy;
- For any solutions $(g_1, \ldots, g_n)$ of (3.5), $\{g_i\}_{i=1,\ldots,n}$ generate H.

The connection between rigid tuples and local systems is given by the following theorem. Let $S = \{P_1, \ldots, P_n\} \subset \mathbb{P}^1(\mathbb{Q})$, and let $U = \mathbb{P}^1_{\mathbb{Q}} - S$.

3.3.2 Theorem (Belyi, Fried, Matzat, Shih, and Thompson)

Let $(C_1, \ldots, C_n)$ be a rigid tuple in H. Then up to isomorphism there is a unique connected unramified Galois H-cover $\pi : Y \to U \otimes_{\mathbb{Q}} \overline{\mathbb{Q}}$ such that a topological generator of the (tame) inertia group at P_i acts on Y as an element in C_i.

Furthermore, if each C_i is rational (i.e., C_i takes rational values for all irreducible characters of H), then the H-cover $Y \to U \otimes_{\mathbb{Q}} \overline{\mathbb{Q}}$ is defined over $\mathbb{Q}$.

From the above theorem we see that the notion of a rigid tuple is an analog of physical rigidity for H-local systems when the algebraic group H is a finite group.

Rigid tuples combined with the Hilbert irreducibility theorem solves the inverse Galois problem for H.

3.3.3 Corollary

Suppose there exists a rational rigid tuple in H, then H can be realized as $\mathrm{Gal}(K/\mathbb{Q})$ *for some Galois number field $K/\mathbb{Q}$.*

For a comprehensive survey of finite simple groups that are realized as Galois groups over $\mathbb{Q}$ using rigidity tuples, we refer the readers to the book [27] by Malle and Matzat. Recent work of Guralnick and Malle [16] establishes a rational rigid triple in $E_8(\mathbb{F}_\ell)$ following the suggestion of [36].

4 Calculus of geometric Hecke operators

This is the most technical part of this report. We will review the basic setting of the geometric Langlands program. The main result is Theorem 4.4.3, which roughly says that for rigid automorphic representations, their Galois representations under the Langlands correspondence can be explicitly constructed.

4.1 Working with a more general base field k

In this section, we work with geometric objects such as Bun_G and sheaves on them instead of talking about functions on $G(F)\backslash G(\mathbb{A}_F)$. Therefore k can be a more general field. In fact we assume k **is a finite extension of a prime field**, i.e., it is either a finite field or a number field. This assumption is not essential, and is to get the Satake equivalence (see §4.2) to work in the simplest way.

We continue with the situation in §2.1, with all geometric objects, e.g., the curve X and the group G, defined over the more general field k.

By Remark 2.3.7(1), the notion of rank one character sheaves makes sense over k, so is the notion of a geometric automorphic datum in Definition 2.4.1. For general k, we shall write $(\mathbf{K}_S, \mathcal{K}_S)$ for a geometric automorphic datum instead of $(\mathbf{K}_S, \chi_S)$ since the correspondence $\mathcal{K}_x \leftrightarrow \chi_x$ only works for finite fields. As a result, we shall also change a few notations. For example, $D_G(\mathbf{K}_S, \mathcal{K}_S)$ shall replace $D_G(\mathbf{K}_S, \chi_S)$.

We will also be working with derived categories of $\overline{\mathbb{Q}}_\ell$-sheaves and perverse sheaves on algebraic stacks. For foundational material we refer to the articles of Laszlo and Olsson [23] and Liu and Zheng [25]. **Note: All sheaf-theoretic functors are derived.**

4.2 The Satake category

The Satake category is an upgraded version of the spherical Hecke algebra

$$C_c(G(\mathcal{O}_x)\backslash G(F_x)/G(\mathcal{O}_x))$$

under the sheaf-to-function correspondence.

Let LG and L^+G be the group objects over k defined similarly as L_xG and L_x^+G, using the "standard" local field $k((t))$ in place of F_x. The fppf quotient $\mathrm{Gr} = LG/L^+G$ is called the *affine Grassmannian* of G. Then L^+G acts on Gr via left translation. The L^+G-orbits on Gr are indexed by dominant coweights $\lambda \in \mathbb{X}_*(T)^+$. The orbit containing the element $t^\lambda \in T(k((t)))$ is denoted by Gr_λ and its closure is denoted by $\mathrm{Gr}_{\leq\lambda}$. We have $\dim \mathrm{Gr}_\lambda = \langle 2\rho, \lambda\rangle$, where 2ρ is the sum of positive roots in G. Each $\mathrm{Gr}_{\leq\lambda}$ is a projective but usually singular variety over k. We denote the intersection complex of $\mathrm{Gr}_{\leq\lambda}$ by IC_λ: this is the middle extension of the shifted and twisted constant sheaf $\overline{\mathbb{Q}}_\ell[\langle 2\rho, \lambda\rangle](\langle \rho, \lambda\rangle)$ on Gr_λ (note that $\langle \rho, \lambda\rangle \in \mathbb{Z}$ since G is simply-connected).

Let $\widetilde{\mathrm{Sat}} = \mathrm{Perv}_{L^+G}(\mathrm{Gr})$ be the category of L^+G-equivariant perverse sheaves on Gr (with $\overline{\mathbb{Q}}_\ell$-coefficients) with finite type supports. The *Satake category* $\mathrm{Sat} \subset \widetilde{\mathrm{Sat}}$ is the full subcategory consisting of direct sums of IC_λ's ($\lambda \in \mathbb{X}_*(T)^+$). In [26], [11] and [28], it was shown that when k is algebraically closed, Sat carries a natural tensor structure, such that the global cohomology functor $h = H^*(\mathrm{Gr}, -) : \mathrm{Sat} \to \mathrm{Vec}$ is a fiber functor. It is also shown that the Tannakian group of the tensor category Sat is the Langlands dual group $\widehat{G}$. When k is a finite field or a number field, these results still hold, see the discussion in [36, §4.1.1, towards the end]. The Tannakian formalism gives the *geometric Satake equivalence* of tensor categories

$$\mathrm{Sat} \cong \mathrm{Rep}(\widehat{G}, \overline{\mathbb{Q}}_\ell).$$

For $V \in \mathrm{Rep}(\widehat{G}, \overline{\mathbb{Q}}_\ell)$, we denote the corresponding object in Sat by IC_V.

4.3 Geometric Hecke operators

We consider the situation of §2.5.1. In particular, we have a geometric automorphic datum $(\mathbf{K}_S, \mathcal{K}_S)$, and moduli stacks $\mathrm{Bun}_G(\mathbf{K}_S)$ and $\mathrm{Bun}_G(\mathbf{K}_S^+)$.

4.3.1 Hecke correspondence

Consider the following diagram

$$\begin{array}{ccccc} & & \mathrm{Hk}(\mathbf{K}_S^+) & \xrightarrow{\pi} & U := X - S \\ & \overleftarrow{h} \swarrow & & \searrow \overrightarrow{h} & \\ \mathrm{Bun}_G(\mathbf{K}_S^+) & & & & \mathrm{Bun}_G(\mathbf{K}_S^+) \end{array} \tag{4.1}$$

Here, the stack $\mathrm{Hk}(\mathbf{K}_S^+)$ classifies the data $(x, \mathcal{E}, \mathcal{E}', \tau)$ where $x \in U := X - S$, $\mathcal{E}, \mathcal{E}' \in \mathrm{Bun}_G(\mathbf{K}_S^+)$ and $\tau : \mathcal{E}|_{X-\{x\}} \xrightarrow{\sim} \mathcal{E}'|_{X-\{x\}}$ is an isomorphism of G-torsors over $X - \{x\}$ preserving the $\mathbf{K}_x^+$-level structures at each $x \in S$. The morphisms $\overleftarrow{h}$, $\overrightarrow{h}$ and π send $(x, \mathcal{E}, \mathcal{E}', \tau)$ to $\mathcal{E}, \mathcal{E}'$ and x respectively.

For $x \in U$, we denote its preimage under π by $\mathrm{Hk}_x(\mathbf{K}_S^+)$. We have an evaluation morphism

$$\mathrm{ev}_x : \mathrm{Hk}_x(\mathbf{K}_S^+) \to L_x^+G \backslash L_xG / L_x^+G. \tag{4.2}$$

In fact, for a point $(x, \mathcal{E}, \mathcal{E}', \tau) \in \mathrm{Hk}_x(\mathbf{K}_S^+)$, if we fix trivializations of $\mathcal{E}$ and $\mathcal{E}'$ over $\mathrm{Spec}\mathcal{O}_x$, the isomorphism τ restricted to $\mathrm{Spec}F_x$ is an isomorphism between the trivial G-torsors over $\mathrm{Spec}F_x$, hence given by a point $g_\tau \in L_xG$. Changing the trivializations of $\mathcal{E}|_{\mathrm{Spec}\mathcal{O}_x}$ and $\mathcal{E}'|_{\mathrm{Spec}\mathcal{O}_x}$ will result in left and right multiplication of g_τ by elements in L_x^+G. Therefore we have a well-defined morphism ev_x as above between stacks.

As x moves along U, we may identify the target of (4.2) as $L^+G\backslash LG/L^+G$ by choosing a local coordinate t at x. Modulo the ambiguity caused by the choice of the local coordinates, we obtain a well defined morphism

$$\mathrm{ev} : \mathrm{Hk}(\mathbf{K}_S^+) \to \left[\frac{L^+G\backslash LG/L^+G}{\mathrm{Aut}^+}\right] \tag{4.3}$$

where Aut^+ is the group scheme over k of continuous ring automorphisms of $k[[t]]$, and it acts on LG and L^+G via its action on $k[[t]]$.

4.3.2 Geometric Hecke operators

For each object $V \in \mathrm{Rep}(\widehat{G}, \overline{\mathbb{Q}}_\ell)$, the corresponding object $\mathrm{IC}_V \in \mathrm{Sat}$ under the geometric Satake equivalence defines a complex on the quotient stack $\left[\frac{L^+G\backslash LG/L^+G}{\mathrm{Aut}^+}\right]$. The geometric Hecke operator associated with V is the functor

$$\begin{aligned} \mathbb{T}_V : D_{(L_S,\mathcal{K}_S)}(\mathrm{Bun}_G(\mathbf{K}_S^+) \times U) &\to D_{(L_S,\mathcal{K}_S)}(\mathrm{Bun}_G(\mathbf{K}_S^+) \times U), \\ \mathcal{F} &\mapsto (\overrightarrow{h} \times \pi)_! \left((\overleftarrow{h} \times \pi)^* \mathcal{F} \otimes \mathrm{ev}^* \mathrm{IC}_V \right). \end{aligned}$$

The composition of these functors are compatible with the tensor structure of Sat: there is a natural isomorphism of functors

$$\mathbb{T}_V \circ \mathbb{T}_W \cong \mathbb{T}_{V\otimes W}, \forall V, W \in \mathrm{Rep}(\widehat{G}, \overline{\mathbb{Q}}_\ell)$$

which is compatible with the associativity of the tensor product in $\mathrm{Rep}(\widehat{G}, \overline{\mathbb{Q}}_\ell)$ and the associativity of composition of functors $\mathbb{T}_{V_1} \circ \mathbb{T}_{V_2} \circ \mathbb{T}_{V_3}$ in the obvious sense.

4.3.3 Definition

A Hecke eigensheaf in $D_G(\mathbf{K}_S, \mathcal{K}_S)$ consists of $\mathcal{F} \in D_G(\mathbf{K}_S, \mathcal{K}_S)$, $\rho \in \mathrm{Loc}_{\widehat{G}}(U)$ and for every $V \in \mathrm{Rep}(\widehat{G}, \overline{\mathbb{Q}}_\ell)$, an isomorphism

$$\varphi_V : \mathbb{T}_V(\mathcal{F} \boxtimes \overline{\mathbb{Q}}_\ell) \cong \mathcal{F} \boxtimes \rho_V$$

such that the following conditions are satisfied.

1. As V varies in $\mathrm{Rep}(\widehat{G}, \overline{\mathbb{Q}}_\ell)$, φ_V gives a natural isomorphism between the functors $\mathbb{T}_{(-)}(\mathcal{F} \boxtimes \overline{\mathbb{Q}}_\ell)$ and $\mathcal{F} \boxtimes \rho_{(-)}$.
2. For $V = \overline{\mathbb{Q}}_\ell$, we require that φ_V be the identity map for $\mathcal{F} \boxtimes \overline{\mathbb{Q}}_\ell$.
3. For $V, W \in \mathrm{Rep}(\widehat{G}, \overline{\mathbb{Q}}_\ell)$, $\varphi_{V\otimes W}$ is equal to the composition

$$\begin{aligned}\mathbb{T}_{V\otimes W}(\mathcal{F} \boxtimes \overline{\mathbb{Q}}_\ell) &\cong \mathbb{T}_V \circ \mathbb{T}_W(\mathcal{F} \boxtimes \overline{\mathbb{Q}}_\ell) \xrightarrow{\varphi_W} \mathbb{T}_V(\mathcal{F} \boxtimes \rho_W) \\ &\cong \mathbb{T}_V(\mathcal{F} \otimes \overline{\mathbb{Q}}_\ell) \otimes \rho_W \xrightarrow{\varphi_V} \mathcal{F} \boxtimes (\rho_V \otimes \rho_W) \cong \mathcal{F} \boxtimes \rho_{V\otimes W}.\end{aligned}$$

We sometime abuse the language and say that $\mathcal{F} \in D_G(\mathbf{K}_S, \mathcal{K}_S)$ is a Hecke eigensheaf, if there exists a $\widehat{G}$-local system ρ and isomorphisms φ_V as in the above definition. In this case, we call ρ the Hecke eigen $\widehat{G}$-local system associated with $\mathcal{F}$.

4.4 Rigid Hecke eigensheaves

The main purpose of this subsection is to establish the existence of Hecke eigensheaves in a special situation. We fix a geometric automorphic datum $(\mathbf{K}_S, \mathcal{K}_S)$.

4.4.1 Twisted representations

Let Γ be a group and $\xi \in Z^2(\Gamma, \overline{\mathbb{Q}}_\ell^\times)$ a cocycle such that $\xi_{1,\gamma} = \xi_{\gamma,1} = 1$ for all $\gamma \in \Gamma$. A ξ-twisted representation of Γ is a finite-dimensional $\overline{\mathbb{Q}}_\ell$-vector space V with automorphisms $T_\gamma : V \to V$, one for each $\gamma \in \Gamma$, such that $T_1 = \mathrm{id}_V$ and

$$T_{\gamma\delta} = \xi_{\gamma,\delta} T_\gamma T_\delta, \forall \gamma, \delta \in \Gamma.$$

Let $\mathrm{Rep}_\xi(\Gamma, \overline{\mathbb{Q}}_\ell)$ be the category of ξ-twisted representations of Γ. This is a $\overline{\mathbb{Q}}_\ell$-linear abelian category which, up to equivalence, only depends on the cohomology class $[\xi] \in \mathrm{H}^2(\Gamma, \overline{\mathbb{Q}}_\ell^\times)$.

4.4.2

Let $\mathcal{E} \in \mathrm{Bun}_G(\mathbf{K}_S)(\overline{k})$ be a relevant point. Let $A = \pi_0(\mathrm{Aut}(\mathcal{E}))$. Since $\mathrm{ev}^*_{S,\,\mathcal{E}}\mathcal{K}_S$ is trivial on $\mathrm{Aut}^\circ(\mathcal{E})$, it descends to A and gives a cocycle $\xi \in Z^2(A(\overline{k}), \overline{\mathbb{Q}}_\ell)$ satisfying $\xi_{1,a} = \xi_{a,1} = 1$ for all $a \in A(\overline{k})$ whose cohomology class is well-defined (see Remark 2.3.3).

Let Z_0 denote the inverse image of $\prod_{x\in S} L_x Z \cap \mathbf{K}_x$ under the diagonal embedding of Z. Then $Z_0(\overline{k}) \cap \mathbf{K}_S \subset \mathrm{Aut}(\mathcal{E})$. Let $Z_0(\overline{k})^*_\xi$ be the set of ξ-twisted 1-dimensional representations of $Z_0(\overline{k})$ (after restricting the cocycle ξ to $Z_0(\overline{k})$). For $\eta \in Z_0(\overline{k})^*_\xi$, let $\mathrm{Rep}_\xi(A(\overline{k}), \overline{\mathbb{Q}}_\ell)_\eta$ be the full subcategory of $\mathrm{Rep}_\xi(A(\overline{k}), \overline{\mathbb{Q}}_\ell)$ consisting of those ξ-twisted representations of $A(\overline{k})$ whose restriction to $Z_0(\overline{k})$ is the scalar multiplication via η. Then one can decompose $D_G(\mathbf{K}_S, \mathcal{K}_S)$ into a product of categories $D_G(\mathbf{K}_S, \mathcal{K}_S)_\eta$, one for each $\eta \in Z_0(\overline{k})^*_\xi$.

4.4.3 Theorem

We assume $\mathrm{Bun}_G(\mathbf{K}_S)$ *contains a unique relevant* $\overline{k}$*-point* $\mathcal{E}$*, and let* A, ξ, Z_0*,* $Z_0(\overline{k})^*_\xi$ *etc. be as in* §4.4.2.

1. *For each* $\eta \in Z_0(\overline{k})^*_\xi$*, the category of perverse sheaves in* $D_G(\overline{k}; \mathbf{K}_S, \mathcal{K}_S)_\eta$ *is equivalent to* $\mathrm{Rep}_\xi(A(\overline{k}), \overline{\mathbb{Q}}_\ell)_\eta$.
2. *Suppose* $\mathrm{Aut}(\mathcal{E})$ *is finite (hence equal to* A*), and the point* $\mathcal{E}$ *is open in* $\mathrm{Bun}_G(\mathbf{K}_S)$*. Suppose further that for some* $\eta \in Z_0(\overline{k})^*_\xi$*,* $\mathrm{Rep}_\xi(A(\overline{k}), \overline{\mathbb{Q}}_\ell)_\eta$ *contains only one irreducible object. Then* $D_G(\overline{k}; \mathbf{K}_S, \mathcal{K}_S)_\eta$ *contains a unique irreducible perverse sheaf* $\mathcal{F}_\eta$ *up to isomorphism, and* $\mathcal{F}_\eta$ *is a Hecke eigensheaf.*

The simplest case of the theorem is when $\mathrm{Aut}(\mathcal{E})$ is trivial. Then $D_G(\overline{k}; \mathbf{K}_S, \mathcal{K}_S)$ contains a unique irreducible perverse sheaf, which is a Hecke eigensheaf. In this case, the geometric automorphic datum $(\mathbf{K}_S, \mathcal{K}_S)$ is strongly rigid if k is a finite field.

4.4.4 Description of the Hecke eigen local system

Suppose we are in the situation of Theorem 4.4.3(2), i.e., $\mathcal{E}$ is the unique relevant $\overline{k}$-point in $\mathrm{Bun}_G(\mathbf{K}_S)$, and it is an open point with $A = \mathrm{Aut}(\mathcal{E})$ finite. Let ρ_η be the Hecke eigen $\widehat{G}$-local system associated with the Hecke eigensheaf $\mathcal{F}_\eta$. We shall give a description of $\rho_{\eta,V}$ for $V \in \mathrm{Rep}(\widehat{G}, \overline{\mathbb{Q}}_\ell)$.

For each point $x \in U$ define $\mathfrak{G}_x$ to be the group of G-automorphisms $\mathcal{E}|_{X-\{x\}}$ preserving the $\mathbf{K}_S$-level structures. This is representable by a group ind-scheme. As x varies over U, we get a group ind-schemes $\mathfrak{G}_U$ over U. The $\mathbf{K}_S$-analog of the diagram (4.1) restricted to the open relevant point becomes the diagram

$$\begin{array}{ccccc} & & A\backslash\mathfrak{G}_U/A & \xrightarrow{\ \pi\ } & U \\ & \overleftarrow{h}\swarrow & & \searrow\overrightarrow{h} & \\ \mathbb{B}A & & & & \mathbb{B}A \end{array}$$

Since $A \subset \mathfrak{G}_x$ for each x, A acts on each fiber $\mathfrak{G}_x$ by left and right translations.

By Remark 2.3.7(1), the character sheaf $\mathcal{K}_S$ on L_S can be obtained from a central extension

$$1 \to C \to \widetilde{L}_S \xrightarrow{v} L_S \to 1$$

where $\widetilde{L}_S$ is connected and C is a finite discrete (necessarily abelian) group scheme over k, and $\chi_C : C \to \overline{\mathbb{Q}}_\ell^\times$ is a character. Concretely $\mathcal{K}_S \cong (v_! \overline{\mathbb{Q}}_\ell)_{\chi_C}$. Let $\widetilde{A}$ and $\widetilde{\mathfrak{G}}_U$ be the base change of the cover $\widetilde{L}_S \to L_S$ along $\mathrm{ev}_{S,\mathcal{E}} : A \to L_S$ (see (2.5)). Then $\widetilde{A} \times \widetilde{A}$ acts on $\widetilde{\mathfrak{G}}_U$ via left and right translations

$$\widetilde{A} \times \widetilde{A} \times \widetilde{\mathfrak{G}}_U \to \widetilde{\mathfrak{G}}_U, \tag{4.4}$$

$$(a_1, a_2, g) \mapsto a_1 g a_2^{-1}. \tag{4.5}$$

Let $\widetilde{Z}_0 \subset \widetilde{A}$ be the preimage of $Z_0 \subset A$. Via the central extension $\widetilde{A}$ of A, we can interpret $\mathrm{Rep}_\xi(A(\overline{k}), \overline{\mathbb{Q}}_\ell)$ as (ordinary) representations of $\widetilde{A}(\overline{k})$ on which C acts via χ_C. Similarly, a twisted central character $\eta \in Z_0(\overline{k})^*_\xi$ can be viewed as a character of $\widetilde{Z}_0(\overline{k})$ whose restriction to C is χ_C.

The next result gives a concrete description of the Hecke eigen $\widehat{G}$-local systems ρ_η using the geometry of $\widetilde{\mathfrak{G}}_U$ and the $\widetilde{A} \times \widetilde{A}$ action on it. We only need the following maps to state the result

$$U \xleftarrow{\widetilde{\pi}} \widetilde{\mathfrak{G}}_U \xrightarrow{\widetilde{\mathrm{ev}}} \left[\frac{L^+G \backslash LG / L^+G}{\mathrm{Aut}^+}\right].$$

4.4.5 Proposition

In the situation of Theorem 4.4.3(2), let $\mathcal{F}_\eta$ be the Hecke eigensheaf corresponding to the irreducible object $W_\eta \in \mathrm{Rep}_\xi(A(\overline{k}), \overline{\mathbb{Q}}_\ell)_\eta$, viewed as an irreducible representation of $\widetilde{A}(\overline{k})$ as above. Let ρ_η be the corresponding Hecke eigen $\widehat{G}$-local system over $U \otimes_k \overline{k}$. For any $V \in \mathrm{Rep}(\widehat{G}, \overline{\mathbb{Q}}_\ell)$, we have an isomorphism of sheaves over $U \otimes_k \overline{k}$:

$$\rho_{\eta,V} \cong \mathrm{Hom}_{\widetilde{A}(\overline{k}) \times \widetilde{A}(\overline{k})}(W_\eta \boxtimes W_\eta^\vee, \widetilde{\pi}_! \widetilde{\mathrm{ev}}^* \mathrm{IC}_V). \tag{4.6}$$

Here we are using the left action of $\widetilde{A}(\overline{k}) \times \widetilde{A}(\overline{k})$ on $\widetilde{\pi}_! \widetilde{\mathrm{ev}}^* \mathrm{IC}_V$ induced from the left action defined in (4.4).

4.4.6 Rationality issue

Suppose that the unique relevant point $\mathcal{E}$ is defined over k and that the representation W_η can be extended to $\widetilde{A}(\overline{k}) \rtimes \mathrm{Gal}(\overline{k}/k)$, then the Hecke eigensheaf $\mathcal{F}_\eta$ is also defined over k (i.e., it descends to an object in $D_G(\mathbf{K}_S, \chi_S)$), and the Hecke eigen local system ρ_η is a $\widehat{G}$-local system over U. The identity (4.6) now holds over U.

5 Kloosterman sheaves as rigid objects over $\mathbb{P}^1 - \{0, \infty\}$

In this section we review the work [17], in which we used rigid automorphic representations ramified at two places to construct generalizations of Kloosterman sheaves.

5.1 Kloosterman automorphic data

Let $k = \mathbb{F}_q$ be a finite field. The curve is $X = \mathbb{P}^1$ with function field $F = k(t)$. Let $S = \{0, \infty\}$. We shall define a geometric automorphic datum for G with respect to S, to be called the *Kloosterman automorphic datum.*

Recall as part of the pinning we have fixed a pair of opposite Borels $B, B^{\text{opp}} \subset G$ with $T = B \cap B^{\text{opp}}$. They determine Iwahori subgroups $\mathbf{I}_\infty \subset L_\infty G$ (preimage of $B \subset G$ under the evaluation map $L_\infty^+ G \to G$) and $\mathbf{I}_0^{\text{opp}} \subset L_0 G$ (preimage of $B^{\text{opp}} \subset G$ under the evaluation map $L_0^+ G \to G$). Let $\mathbf{K}_0 = \mathbf{I}_0^{\text{opp}}$ and $\mathbf{K}_\infty = \mathbf{I}_\infty^+$, the pro-unipotent radical of $\mathbf{I}_\infty$.

Fix a character $\chi : T(k) \to \overline{\mathbb{Q}}_\ell^\times$, and view it as a character of $\mathbf{K}_0(k)$ via $\mathbf{K}_0(k) \to T(k)$. The Kloosterman automorphic datum at $x = 0$ is given by $(\mathbf{I}_0^{\text{opp}}, \chi)$.

The pro-unipotent group $\mathbf{I}_\infty^+$ is generated by the root groups $(L_\infty G)(\widetilde{\alpha})$ of all positive affine roots of the loop group $L_\infty G$. There is a projection

$$\mathbf{K}_\infty = \mathbf{I}_\infty^+ \twoheadrightarrow \prod_{i=0}^{r} (L_\infty G)(\alpha_i) \tag{5.1}$$

onto the product of root groups corresponding to simple affine roots. Each

$$(L_\infty G)(\alpha_i)$$

is isomorphic to $\mathbb{G}_a$ over k. A linear function $\phi : \prod_{i=0}^{r} (L_\infty G)(\alpha_i) \to k$ is said to be *generic* if it does not vanish on any of the factors $(L_\infty G)(\alpha_i)$. Fix such a generic linear function ϕ and fix a nontrivial additive character ψ of k. The composition $\psi \circ \phi$ gives a character of $\mathbf{I}_\infty^+(k)$ which factors through the quotient (5.1). The Kloosterman automorphic datum at $x = \infty$ is given by $(\mathbf{I}_\infty^+, \psi \circ \phi)$.

A central character compatible with the Kloosterman automorphic datum exists and is unique: its local component at 0 is given by $\chi|_{Z(k)}$ and at 0 is given by $\chi^{-1}|_{Z(k)}$.

5.1.1 Remark

The automorphic datum $(\mathbf{I}_\infty^+, \psi \circ \phi)$ picks up those representations of $G(F_\infty)$ that contain nonzero eigenvectors of $\mathbf{K}_\infty$ on which $\mathbf{K}_\infty$ acts through the character $\psi \circ \phi$. These representations are first discovered by Gross and Reeder [15, §9.3], and they call them *simple supercuspidal representations.* When G is simply-connected, any such representation is given by compact induction $\text{ind}_{Z(k)\mathbf{I}_\infty^+(k)}^{G(F_\infty)} (\omega_\infty \boxtimes \psi \circ \phi)$ for some central character $\omega_\infty : Z(k) \to \overline{\mathbb{Q}}_\ell^\times$.

5.1.2 Theorem (Gross [13], simple proof by Heinloth-Ngô-Yun [17])

1. *The Kloosterman automorphic datum is strongly rigid.*
2. *When* $\chi = 1$, *the local component at* 0 *of any* $((\mathbf{I}_0^{\mathrm{opp}}, 1), (\mathbf{I}_\infty^+, \psi \circ \phi))$*-typical automorphic representation is isomorphic to the Steinberg representation of* $G(F_0)$.

Part (1) above is a special case of Theorem 4.4.3, and the automorphism group of the unique relevant point is trivial. By Theorem 4.4.3, we have a Hecke eigen $\widehat{G}$-local system $\mathrm{Kl}_{\widehat{G}}(\chi, \phi)$ over $\mathbb{P}^1 - \{0, \infty\}$ associated with the unique irreducible perverse sheaf in $D_G(\mathbf{K}_S, \chi_S)$.

5.2 Kloosterman sheaves

5.2.1 The classical Kloosterman sheaf

We first recall the definition of Kloosterman sums. Let p be a prime number. Fix a nontrivial additive character $\psi : \mathbb{F}_p \to \overline{\mathbb{Q}}_\ell^\times$. Let $n \geq 2$ be an integer. Then the n-variable Kloosterman sum over $\mathbb{F}_p$ is a function on $\mathbb{F}_p^\times$ whose value at $a \in \mathbb{F}_p^\times$ is

$$\mathrm{Kl}_n(p; a) = \sum_{x_1, \ldots, x_n \in \mathbb{F}_p^\times; x_1 x_2 \cdots x_n = a} \psi(x_1 + \cdots + x_n).$$

These exponential sums arise naturally in the study of automorphic forms for GL_n.

Deligne [5] gave a geometric interpretation of the Kloosterman sum. He considered the following diagram of schemes over $\mathbb{F}_p$

$$\begin{array}{ccccc} & & \mathbb{G}_m^n & & \\ & \swarrow^{\pi} & & \searrow^{\sigma} & \\ \mathbb{G}_m & & & & \mathbb{A}^1 \end{array}$$

Here π is the morphism of taking the product and σ is the morphism of taking the sum.

5.2.2 Definition (Deligne [5])

The *Kloosterman sheaf* is

$$\mathrm{Kl}_n := \mathbf{R}^{n-1}\pi_! \sigma^* \mathrm{AS}_\psi,$$

over $\mathbb{P}^1_{\mathbb{F}_p} - \{0, \infty\}$. Here AS_ψ is the Artin-Schreier sheaf as in Example 2.3.6.

The relationship between the local system Kl_n and the Kloosterman sum $\mathrm{Kl}_n(p; a)$ is explained by the following identity

$$\mathrm{Kl}_n(p; a) = (-1)^{n-1} \mathrm{Tr}(\mathrm{Frob}_a, (\mathrm{Kl}_n)_a).$$

Here Frob_a is the geometric Frobenius operator acting on the geometric stalk $(\mathrm{Kl}_n)_a$ of Kl_n at $a \in \mathbb{G}_m(\mathbb{F}_p) = \mathbb{F}_p^\times$.

In [5, Théorème 7.4, 7.8], Deligne proved:

1. Kl_n is a local system of rank n.

2. Kl_n is tamely ramified at 0, and the monodromy is unipotent with a single Jordan block.

3. Kl_n is totally wild at ∞ (i.e., the wild inertia at ∞ has no nonzero fixed vector on the stalk of Kl_n), and the Swan conductor $\mathrm{Sw}_\infty(\mathrm{Kl}_n) = 1$.

The main results of [17] can be summarized as follows.

5.2.3 Theorem (Heinloth-Ngô-Yun [17])

The $\widehat{G}$-local system $\mathrm{Kl}_{\widehat{G}}(\chi, \phi)$ constructed as the Hecke eigen $\widehat{G}$-local system associated with the Kloosterman automorphic datum satisfies the following properties.

1. *$\mathrm{Kl}_{\widehat{G}}(\chi, \phi)$ is tame at 0. A generator of $\mathcal{I}_0^t$ maps to an element in $\widehat{G}$ with semisimple part given by χ, viewed as an element in $\widehat{T}$. When $\chi = 1$, a generator of $\mathcal{I}_0^t$ maps to a regular unipotent element in $\widehat{G}$.*

2. *The local monodromy of $\mathrm{Kl}_{\widehat{G}}(\chi, \phi)$ at ∞ is a* simple wild parameter *in the sense of Gross and Reeder [15, §5]. For more details see §5.2.5.*

3. *Assume $\chi = 1$. Then the global geometric monodromy group of $\mathrm{Kl}_{\widehat{G}}(1, \phi)$ is a connected almost simple subgroup of $\widehat{G}$ of types given by the following table*

$\widehat{G}$	$G^{\mathrm{geom}}_{\mathrm{Kl}_{\widehat{G}}(1,\phi)}$	condition on char(k)
A_{2n}	A_{2n}	$p > 2$
A_{2n-1}, C_n	C_n	$p > 2$
B_n, D_{n+1} $(n \geq 4)$	B_n	$p > 2$
E_7	E_7	$p > 2$
E_8	E_8	$p > 2$
E_6, F_4	F_4	$p > 2$
B_3, D_4, G_2	G_2	$p > 3$

5.2.4 Remark

For $\widehat{G}$ of type A_{n-1} (resp. C_n), the Kloosterman sheaf $\mathrm{Kl}_{\widehat{G}}(1, \phi)$ is essentially the same as Kl_n (resp. Kl_{2n}) of Deligne. When $\widehat{G}$ is of type B or G_2, $\mathrm{Kl}_{\widehat{G}}(1, \phi)$ were constructed by Katz in [19] by different methods (as special cases of hypergeometric sheaves). The above theorem treats all $\widehat{G}$ uniformly, and in particular gives the first construction of motivic local systems with geometric monodromy group F_4, E_7 and E_8.

5.2.5 Local monodromy at ∞

Let us explain in more detail what a *simple wild parameter* looks like, following Gross and Reeder [15, Proposition 5.6]. Assume $p = \mathrm{char}(k)$ does not divide $\#W$ (W is the Weyl group of G). Let $\rho|_{\mathcal{I}_\infty} : \mathcal{I}_\infty \to \widehat{G}(\overline{\mathbb{Q}}_\ell)$ be the local monodromy of

$\mathrm{Kl}_{\widehat{G}}(\chi, \phi)$ at ∞. Then up to conjugacy in $\widehat{G}$, the wild inertia $\mathcal{I}_\infty^w$ has image in $\widehat{T}[p]$, the p-torsion part of the dual maximal torus $\widehat{T}$. The image of $\mathcal{I}_\infty$ must normalize the image of $\mathcal{I}_\infty^w$, and in this case it must normalize the whole torus $\widehat{T}$. Therefore we have a commutative diagram

$$\begin{array}{ccccccccc} 1 & \longrightarrow & \mathcal{I}_\infty^w & \longrightarrow & \mathcal{I}_\infty & \longrightarrow & \mathcal{I}_\infty^t & \longrightarrow & 1 \\ & & \downarrow{\scriptstyle \rho|_{\mathcal{I}_\infty^w}} & & \downarrow{\scriptstyle \rho|_{\mathcal{I}_\infty}} & & \downarrow & & \\ 1 & \longrightarrow & \widehat{T} & \longrightarrow & N_{\widehat{G}}(\widehat{T}) & \longrightarrow & W & \longrightarrow & 1 \end{array}$$

The image of $\mathcal{I}_\infty^t$ in W is the cyclic group generated by a Coxeter element $\mathrm{Cox} \in W$, whose order is the Coxeter number h of G. The image $\rho(\mathcal{I}_\infty^w)$ is a $\mathbb{F}_p$-vector space equipped with the action of Cox. In fact $\rho(\mathcal{I}_\infty^w) \cong \mathbb{F}_p[\zeta_h]$, the extension of $\mathbb{F}_p$ by adjoining h^{th} roots of unity, and the Coxeter element acts by multiplication by a primitive h^{th} root of unity.

When $p \mid \#W$, a simple wild parameter can be more complicated. For example, when $\widehat{G} = \mathrm{PGL}_2$ and $p = 2$, the image $\rho(\mathcal{I}_\infty)$ is isomorphic to the alternating group A_4 embedded in $\mathrm{PGL}_2(\overline{\mathbb{Q}}_\ell) = \mathrm{SO}_3(\overline{\mathbb{Q}}_\ell)$ as the symmetry of a regular tetrahedron.

5.2.6 More general Kloosterman sheaves

In [37] we give further generalizations of Kloosterman sheaves. We replace $\mathbf{I}_0^{\mathrm{opp}}$ by a more general parahoric level structure $\mathbf{P}_0^{\mathrm{opp}}$, and accordingly $\mathbf{I}_\infty^+$ is replaced by a $\mathbf{P}_\infty^+$, the pro-unipotent radical of a parahoric of the same type. The geometric automorphic datum at 0 is again given by a multiplicative character χ of the reductive quotient of $\mathbf{I}_0^{\mathrm{opp}}$. The geometric automorphic datum at ∞ is given by a generic linear function ϕ from $\mathbf{P}_\infty^+$ (with a suitable notion of genericity). The representations of $G(F_\infty)$ picked up by $(\mathbf{P}_\infty^+, \psi \circ \phi)$ are the *epipelagic representations* discovered by Reeder and Yu [31]. One new feature of this generalization is that when ϕ varies in $V^{*,\circ}$ (the parameter space of generic linear functions on $\mathbf{P}_\infty^+$), the corresponding generalized Kloosterman sheaves "glue" together to give a $\widehat{G}$-local system over $V^{*,\circ} \times (\mathbb{P}^1 - \{0, \infty\})$.

6 Rigid objects over $\mathbb{P}^1 - \{0, 1, \infty\}$

In this section, we review the work [36], in which we use rigid automorphic representations to construct local systems on $\mathbb{P}^1_{\mathbb{Q}} - \{0, 1, \infty\}$. These local systems are the key objects that lead to the answer to Serre's question and the solution of the inverse Galois problem for certain finite simple groups of exceptional Lie type.

6.1 The automorphic data

Let k be a field with $\mathrm{char}(k) \neq 2$. Let $X = \mathbb{P}^1_k$ and $S = \{0, 1, \infty\}$. Assume that the longest element w_0 in the Weyl group W of G acts by inversion on T. Equivalently, this means that the Chevalley involution of G is inner. Recall that a Chevalley involution of G is an involution τ such that $\dim G^\tau$ has the minimal possible dimension, namely $\#\Phi^+$ (the number of positive roots of G). All Chevalley involutions are conjugate to each other under $G^{\mathrm{ad}}(\overline{k})$.

When the Chevalley involution of G is not inner, i.e., G is of type A_n $(n \geq 2)$, D_{2n+1} or E_6, one should consider a quasi-split form of G over the function field $F = k(t)$ ramified at 0 and ∞. We do not discuss it here.

6.1.1 A parahoric subgroup

Up to conjugacy, there is a unique parahoric subgroup $\mathbf{P} \subset L_0G$ such that its reductive quotient $L_{\mathbf{P}}$ is isomorphic to the fixed point subgroup G^τ of a Chevalley involution. For example, we can take $\mathbf{P}$ to be the parahoric subgroup corresponding to the facet containing the element $\rho^\vee/2$ in the T-apartment of the building of L_0G ($\rho^\vee$ is half the sum of positive coroots of G).

The Dynkin diagram of the reductive quotient $L_{\mathbf{P}} \cong G^\tau$ of $\mathbf{P}$ is obtained by removing one or two nodes from the extended Dynkin diagram of G. We tabulate the type of $L_{\mathbf{P}}$ and the nodes to be removed in each case.

G	$L_{\mathbf{P}}$	nodes to be removed
B_{2n}	$B_n \times D_n$	the $(n+1)^{\mathrm{th}}$ counting from the short node
B_{2n+1}	$B_n \times D_{n+1}$	the $(n+1)^{\mathrm{th}}$ counting from the short node
C_n	$A_{n-1} \times \mathbb{G}_m$	the two ends
D_{2n}	$D_n \times D_n$	the middle node
E_7	A_7	the end of the leg of length 1
E_8	D_8	the end of the leg of length 2
F_4	$A_1 \times C_3$	second from the long node end
G_2	$A_1 \times A_1$	middle node

6.1.2 Lemma

If G is not of type C, then $L^{\mathrm{sc}}_{\mathbf{P}} \to L_{\mathbf{P}}$ is a double cover (i.e., kernel is μ_2).

Even if G is of type C_n, $L_{\mathbf{P}} \cong \mathrm{GL}_n$ still admits a unique nontrivial double cover. Therefore, in all cases, there is a canonical nontrivial double cover $v : \widetilde{L}_{\mathbf{P}} \to L_{\mathbf{P}}$. In particular,

$$\mathcal{K}_0 := (v_!\overline{\mathbb{Q}}_\ell)_{\mathrm{sgn}}$$

is a rank one character sheaf on $L_{\mathbf{P}}$ (here sgn denotes the nontrivial character of $\ker(v)(k) = \mu_2(k)\{\pm 1\}$, since $\mathrm{char}(k) \neq 2$). When k is a finite field, we have an exact sequence

$$1 \to \mu_2(k) \to \widetilde{L}_{\mathbf{P}}(k) \to L_{\mathbf{P}}(k) \to \mathrm{H}^1(k, \mu_2) = \{\pm 1\}.$$

The character χ corresponding to $\mathcal{K}_0$ is given by the last arrow above.

6.1.3 The automorphic datum

Let $\mathbf{P}_0 \subset L_0G$ be the standard parahoric subgroup of the type defined in §6.1.1. Let $\mathbf{P}^{\mathrm{opp}}_\infty \subset L_\infty G$ be the parahoric subgroup of the same type but contains the Iwahori subgroup $\mathbf{I}_\infty^{\mathrm{opp}}$ (preimage of B^{opp} under the evaluation map $L_\infty^+G \to G$). Let $\mathbf{I}_1 \subset L_1G$ be the standard Iwahori subgroup. We consider the geometric automorphic datum given by

$$\begin{aligned}(\mathbf{K}_0, \mathcal{K}_0) &= (\mathbf{P}_0, \mathcal{K}_0);\\ (\mathbf{K}_1, \mathcal{K}_1) &= (\mathbf{I}_1, \overline{\mathbb{Q}}_\ell);\\ (\mathbf{K}_\infty, \mathcal{K}_\infty) &= (\mathbf{P}_\infty^{\mathrm{opp}}, \overline{\mathbb{Q}}_\ell).\end{aligned}$$

The central character compatible with $(\mathbf{K}_S, \chi_S)$ above is unique, and it exists if and only if $\chi|_{Z(k)} = 1$, which can always be achieved by passing to a quadratic extension k'/k.

The main technical result of [36] is the following.

6.1.4 Theorem ([36])

Let k be a field with $\mathrm{char}(k) \neq 2$. *Assume G is either simply-laced or of type G_2. Also assume that k contains $\sqrt{-1}$ when G is of type A_1, D_{4n+2} or E_7. Then there are exactly $\#Z(\overline{k})$ isomorphism classes of irreducible perverse sheaves in the category $D_G(\overline{k}; \mathbf{K}_S, \mathcal{K}_S)$, all of which can be defined over k. Each of these irreducible perverse sheaves is a Hecke eigensheaf, and all of them give the same Hecke eigen $\widehat{G}$-local system $\rho \in \mathrm{Loc}_{\widehat{G}}(U)$.*

Note that the condition put on G in the above theorem limits G to be simply-connected of type A_1, D_{2n}, E_7, E_8 and G_2. There is a version of the above theorem for groups of all types; however, the automorphic datum (6.1) will no longer be weakly rigid in general.

6.1.5

We briefly explain how Theorem 6.1.4 follows from Theorem 4.4.3(2). We first exhibit an open point in $\mathrm{Bun}_G(\mathbf{K}_S)$ which is relevant. The moduli stack $\mathrm{Bun}_G(\mathbf{P}_0, \mathbf{P}_\infty^{\mathrm{opp}})$ contains an open substack isomorphic to $\mathbb{B}L_\mathbf{P}$, the classifying stack of $L_\mathbf{P}$. The preimage of $\mathbb{B}L_\mathbf{P}$ in $\mathrm{Bun}_G(\mathbf{K}_S)$ is isomorphic to $L_\mathbf{P}\backslash G/B$. Now a general result of Springer [34, Corollary 4.3(i)] says that any symmetric subgroup of G acts on the flag variety of G with an open orbit. Let $O \subset G/B$ denote the unique open $L_\mathbf{P}$-orbit. We thus get an open point

$$j : L_\mathbf{P}\backslash O \hookrightarrow \mathrm{Bun}_G(\mathbf{K}_S).$$

When G is either simply-laced or of type G_2, this turns out to be the unique relevant point.

The stabilizer A_u of $L_\mathbf{P}$ on any point $u \in O$ is canonically isomorphic to $T[2]$. Taking the preimage of A_u in the double cover $\widetilde{L}_\mathbf{P}$ we get a central extension

$$1 \to \mu_2 \to \widetilde{A}_u \to T[2] \to 1.$$

The center of $\widetilde{A}_u(\overline{k})$ is exactly the preimage $\widetilde{Z}(\overline{k})$ in $\widetilde{A}_u(\overline{k})$ of $Z(\overline{k}) \subset T[2](\overline{k})$. By Stone-von Neumann theorem, for any central character $\eta : \widetilde{Z}(\overline{k}) \to \overline{\mathbb{Q}}_\ell^\times$ which is nontrivial on $\mu_2(\overline{k}) = \ker(\widetilde{A}_u(\overline{k}) \to T[2](\overline{k}))$, there is a unique irreducible representation W_η of $\widetilde{A}_u(\overline{k})$ with central character η. This W_η then gives an irreducible local system on the preimage of the open point $L_{\mathbf{P}}\backslash O$ in $\mathrm{Bun}_G(\mathbf{K}_S^+)$, whose extension to $\mathrm{Bun}_G(\mathbf{K}_S^+)$ by zero gives an object $\mathcal{F}_\eta \in D_G(\overline{k}; \mathbf{K}_S, \mathcal{K}_S)$. The $\mathcal{F}_\eta$ are exactly the irreducible perverse sheaves in $D_G(\overline{k}; \mathbf{K}_S, \mathcal{K}_S)$, and they turn out to be Hecke eigensheaves.

The following theorem summarizes the local and global monodromy of the $\widehat{G}$-local system ρ.

6.1.6 Theorem ([36])

Let G and ρ be as in Theorem 6.1.4.

1. *The local system ρ is tame.*
2. *A topological generator of $\mathcal{I}_1^t$ maps to a regular unipotent element in $\widehat{G}$.*
3. *A topological generator of $\mathcal{I}_\infty^t$ maps to a unipotent element in $\widehat{G}$ that is neither trivial nor regular if G is not of type A_1; in case G is of type A_1, a topological generator of $\mathcal{I}_\infty^t$ maps also to a regular unipotent element in $\widehat{G}$.*
4. *A topological generator of $\mathcal{I}_0^t$ maps to an element in $\widehat{G}$ whose semisimple part is a Chevalley involution in $\widehat{G}$.*
5. *When G is of type A_1, E_7, E_8 and G_2, the global geometric monodromy of ρ is Zariski dense in $\widehat{G}$. When G is of type D_{2n}, the Zariski closure of the global geometric monodromy of $\mathcal{E}$ contains $\mathrm{SO}_{4n-1} \subset \mathrm{PSO}_{4n} = \widehat{G}$ of $n \geq 3$, and contains $G_2 \subset \mathrm{PSO}_8$ if $n = 2$.*

6.2 Applications

By a descent argument (using rigidity), we have the following strengthening of Theorem 6.1.4.

6.2.1 Theorem ([36])

Let k be a prime field with $\mathrm{char}(k) \neq 2$ (i.e., $\mathbb{F}_p$ for p an odd prime or $\mathbb{Q}$). Then the eigen local system ρ can be defined over k. Moreover, the image of ρ can be conjugated to $\widehat{G}(\mathbb{Q}_\ell)$ inside $\widehat{G}(\overline{\mathbb{Q}}_\ell)$.

6.2.2 Application to the construction of motives

Assume G is of type A_1, E_7, E_8 or G_2. Applying the above theorem to $k = \mathbb{Q}$, we get a $\widehat{G}$-local systems $\rho : \pi_1(U_{\mathbb{Q}}) \to \widehat{G}(\mathbb{Q}_\ell)$ whose geometric monodromy is Zariski dense. For each $\mathbb{Q}$-point $a \in U(\mathbb{Q}) = \mathbb{Q} - \{0, 1\}$, restricting ρ to the point $a = \mathrm{Spec}\ \mathbb{Q}$ gives a continuous Galois representation

$$\rho_a : \mathrm{Gal}(\overline{\mathbb{Q}}/\mathbb{Q}) \to \widehat{G}(\mathbb{Q}_\ell). \tag{6.1}$$

By Proposition 4.4.5, one sees that for each $V \in \mathrm{Rep}(\widehat{G}, \mathbb{Q}_\ell)$, ρ_V is obtained as part of the middle dimensional cohomology of some family of varieties over U. Using this fact, it can be shown that each ρ_a is obtained from motives over $\mathbb{Q}$ (if G is type E_8 or G_2) or $\mathbb{Q}(i)$ (if G is of type A_1 or E_7).

6.2.3 Theorem ([36])

Assume G is of type A_1, E_7, E_8 or G_2. There are infinitely many $a \in \mathbb{Q} - \{0,1\}$ such that the ρ_a's are mutually non-isomorphic and all have Zariski dense image in $\widehat{G}$. Consequently, there are infinitely many motives over $\mathbb{Q}$ (if G is type E_8 or G_2) or $\mathbb{Q}(i)$ (if G is of type A_1 or E_7) whose ℓ-adic motivic Galois group is isomorphic to $\widehat{G}$ for any prime ℓ.

This result then gives an affirmative answer to the ℓ-adic analog of Serre question (see §1.3.2) for motivic Galois groups of type E_7, E_8 and G_2. The case of G_2 was settled earlier by Dettweiler and Reiter [8], using Katz's algorithmic construction of rigid local systems. Our local system ρ in the case $G = G_2$ is the same as Dettweiler and Reiter's.

6.2.4 Application to the inverse Galois problem

Let ℓ be a prime number. To emphasize on the dependence on ℓ, we denote the Galois representation ρ_a in (6.1) by $\rho_{a,\ell}$. To solve the inverse Galois problem for the groups $\widehat{G}(\mathbb{F}_\ell)$, we would like to choose $a \in \mathbb{Q} - \{0,1\}$ such that $\rho_{a,\ell}$ has image in $\widehat{G}(\mathbb{Z}_\ell)$ (which is always true up to conjugation), and its reduction modulo ℓ is surjective. This latter condition is hard to satisfy even if we know that the image of $\rho_{a,\ell}$ is Zariski dense in $\widehat{G}(\mathbb{Q}_\ell)$.

To proceed, let us consider the Betti version of Theorem 6.1.4 and Theorem 6.2.1. Namely we consider the base field $k = \mathbb{C}$ and talk about sheaves in $\mathbb{Q}$-vector spaces on the various complex algebraic moduli stacks. The same argument gives topological local system ρ^{top} over $U_{\mathbb{C}} = \mathbb{P}^1_{\mathbb{C}} - \{0, 1, \infty\}$

$$\rho^{\mathrm{top}} : \pi_1^{\mathrm{top}}(U_{\mathbb{C}}) \to \widehat{G}(\mathbb{Q})$$

whose image is Zariski dense. Since $\pi_1^{\mathrm{top}}(U_{\mathbb{C}})$ is finitely generated (in fact generated by two elements), the image of ρ^{top} lies in $\widehat{G}(\mathbb{Z}_\ell)$ for almost all primes ℓ. Therefore it makes sense to reduce modulo ℓ for large primes ℓ and we get

$$\overline{\rho}_\ell^{\mathrm{top}} : \pi_1^{\mathrm{top}}(U_{\mathbb{C}}) \to \widehat{G}(\mathbb{F}_\ell).$$

A deep theorem of Matthews, Vaserstein and Weisfeiler [29, Theorem in the Introduction] (see also Nori [30, Theorem 5.1]) says that $\overline{\rho}_\ell^{\mathrm{top}}$ is surjective for sufficiently large ℓ, when $\widehat{G}$ is simply-connected. This is the case when G is of type E_8 and G_2. Using the comparison between Betti cohomology and ℓ-adic cohomology, we conclude that for general $a \in \mathbb{Q} - \{0,1\}$ (general in the sense of Hilbert irreducibility), the reduction $\overline{\rho}_{a,\ell}$ of $\rho_{a,\ell}$ is also onto $\widehat{G}(\mathbb{F}_\ell)$. This solves the inverse Galois problem for $E_8(\mathbb{F}_\ell)$ and $G_2(\mathbb{F}_\ell)$ for sufficiently large primes ℓ (without an effective bound).

When G is of type A_1 or E_7, $\widehat{G}$ is the adjoint form. In this case, the result in [29] says that for sufficiently large prime ℓ, the image of $\overline{\rho}_\ell^{\mathrm{top}}$ contains the image of $\widehat{G}^{\mathrm{sc}}(\mathbb{F}_\ell) \to \widehat{G}(\mathbb{F}_\ell)$. We deduce that the same is true for $\rho_{a,\ell}$ for general $a \in \mathbb{Q} - \{0, 1\}$.

References

[1] Beilinson, A.; Bernstein, J.; Deligne, P. Faisceaux pervers. In Analysis and Topology on Singular Spaces, I, 5–171, Astérisque, **100**, Soc. Math. France, Paris, 1982.

[2] Beilinson, A.; Drinfeld, V. Quantization of Hitchin's integrable system and Hecke eigensheaves. Preprint.

[3] Borel, A.; Jacquet, H. Automorphic forms and automorphic representations. With a supplement "On the notion of an automorphic representation" by R. P. Langlands. Proc. Sympos. Pure Math., **XXXIII**, Automorphic forms, representations and L-functions, Part 1, pp. 189–207, Amer. Math. Soc., Providence, R.I., 1979.

[4] Bruhat, F.; Tits, J. Groupes réductifs sur un corps locale I, II, Publ. Math. IHES **41**, **60** (1972, 1984), 5–251, 197–376.

[5] Deligne, P. Applications de la formule des traces aux sommes trigonométriques. In Cohomologie étale. SGA $4\frac{1}{2}$. Lecture Notes in Mathematics, Vol. **569**. Springer-Verlag, Berlin-New York, 1977.

[6] Deligne, P. La conjecture de Weil. II. Inst. Hautes Études Sci. Publ. Math. No. **52** (1980), 137–252.

[7] Deligne, P.; Milne, J. S. Tannakian categories. In Hodge Cycles, Motives, and Shimura Varieties, 101–228. Lecture Notes in Mathematics, Vol. **900**. Springer-Verlag, Berlin-New York, 1982.

[8] Dettweiler, M.; Reiter, S. Rigid local systems and motives of type G_2. With an appendix by M. Dettweiler and N. M. Katz. Compositio Math. **146** (2010), no. 4, 929–963.

[9] Feit, W.; Fong, P. Rational rigidity of $G_2(p)$ for any prime $p > 5$. Proceedings of the Rutgers Group Theory Year, 1983–1984 (New Brunswick, N.J., 1983–1984), 323–326, Cambridge University Press, Cambridge, 1985.

[10] Frenkel, E.; Gross, B. A rigid irregular connection on the projective line. Ann. of Math. (2) **170** (2009), no. 3, 1469–1512.

[11] Ginzburg, V. Perverse sheaves on a Loop group and Langlands duality. arXiv:math/9511007.

[12] Grothendieck, A. Revêtements étales et groupe fondamental. Séminaire de Géométrie Algébrique, 1960/61. Institut des Hautes Études Scientifiques, Paris, 1963.

[13] Gross, B. H. Irreducible cuspidal representations with prescribed local behavior. Amer. J. Math. **133** (2011), no. 5, 1231–1258.

[14] Gross, B. H. Trivial L-functions for the rational function field. J. Number Theory **133** (2013), no. 3, 970–976.

[15] Gross, B.; Reeder, M. Arithmetic invariants of discrete Langlands parameters. Duke Math. J. **154** (2010), no. 3, 431–508.

[16] Guralnick, R.; Malle, G. Rational rigidity for $E_8(p)$. Preprint. arXiv:1207.1464.

[17] Heinloth, J.; Ngô, B-C.; Yun, Z. Kloosterman sheaves for reductive groups. Ann. of Math. (2) **177** (2013), no. 1, 241–310.

[18] Jouanolou, J.-P. Cohomologie ℓ-adique. Exposé VI in Séminaire de Géométrie Algébrique du Bois-Marie 1965–66 (SGA 5), Lecture Notes in Mathematics, Vol. **589**. Springer-Verlag, Berlin-New York, 1977.

[19] Katz, N. M. Exponential Sums and Differential Equations. Ann. of Math. Studies, **124**, Princeton University Press, Princeton, NJ, 1990.

[20] Katz, N. M. Rigid Local Systems. Annals of Mathematics Studies, **139**. Princeton University Press, Princeton, NJ, 1996.

[21] Lafforgue, L. Chtoucas de Drinfeld et correspondance de Langlands. Invent. Math. **147** (2002), no. 1, 1–241.

[22] Lafforgue, V. Chtoucas pour les groupes réductifs et paramétrisation de Langlands globale. arXiv:1209.5352.

[23] Laszlo, Y.; Olsson, M. The six operations for sheaves on Artin stacks. I. Finite coefficients. II. Adic coefficients. Publ. Math. Inst. Hautes Études Sci. No. **107** (2008), 109–168, 169–210.

[24] Laumon, G. Correspondance de Langlands géométrique pour les corps de fonctions. Duke Math. J. **54** (1987), no. 2, 309–359.

[25] Liu, Y.; Zheng, W. Enhanced six operations and base change theorem for sheaves on Artin stacks. arXiv:1211.5948.

[26] Lusztig, G. Singularities, character formulas, and a q-analog of weight multiplicities. Analysis and Topology on Singular Spaces, II, III (Luminy, 1981), 208–229, Astérisque, 101–102, Soc. Math. France, Paris, 1983.

[27] Malle, G.; Matzat, B. H. Inverse Galois Theory. Springer Monographs in Mathematics. Springer-Verlag, Berlin, 1999.

[28] Mirković, I.; Vilonen, K. Geometric Langlands duality and representations of algebraic groups over commutative rings. Ann. of Math. (2) **166** (2007), no. 1, 95–143.

[29] Matthews, C. R.; Vaserstein, L. N.; Weisfeiler, B. Congruence properties of Zariski-dense subgroups. I. Proc. London Math. Soc. (3) **48** (1984), no. 3, 514–532.

[30] Nori, M. On subgroups of $\mathrm{GL}_n(\mathbb{F}_p)$. Invent. Math. **88** (1987), no. 2, 257–275.

[31] Reeder, M.; Yu, J.-K. Epipelagic representations and invariant theory. J. Amer. Math. Soc., electronically published on August 5, 2013.

[32] Serre, J. P. Propriétés conjecturales des groupes de Galois motiviques et des représentations ℓ-adiques. In Motives (Seattle, WA, 1991), 377–400, Proc. Sympos. Pure Math., **55**, Part 1, Amer. Math. Soc., Providence, RI, 1994.

[33] Serre, J. P. Topics in Galois Theory. Second edition. With notes by Henri Darmon. Research Notes in Mathematics, **1**. A. K. Peters, Ltd., Wellesley, MA, 2008.

[34] Springer, T. A. Some results on algebraic groups with involutions. In Algebraic Groups and Related Topics (Kyoto/Nagoya, 1983), 525–543, Adv. Stud. Pure Math., **6**, North-Holland, Amsterdam, 1985.

[35] Thompson, J. G. Rational rigidity of $G_2(5)$. Proceedings of the Rutgers Group Theory Year, 1983–1984 (New Brunswick, N.J., 1983–1984), 321–322, Cambridge University Press, Cambridge, 1985.

[36] Yun, Z. Motives with exceptional Galois groups and the inverse Galois problem. Invent. Math. (e-published)

[37] Yun, Z. Generalized Kloosterman sheaves and Lusztig's correspondence. Preprint.

[38] Yun, Z. Rigidity in automorphic representations and local systems. To appear in the proceedings for Currents Development of Mathematics 2013.

Department of Mathematics, Stanford University
450 Serra Mall, Building 380, Stanford, CA 94305, USA
E-mail address: zwyun@stanford.edu

Proceedings of the 6th International Congress of Chinese Mathematicians
ALM 36, pp. 235–246

Bounded Gaps Between Primes

Yitang Zhang

Abstract

It is proved that

$$\liminf_{n\to\infty}(p_{n+1} - p_n) < 7 \times 10^7,$$

where p_n is the n-th prime.

Our method is a refinement of the recent work of Goldston, Pintz and Yildirim on the small gaps between consecutive primes. A major ingredient of the proof is a stronger version of the Bombieri-Vinogradov theorem that is applicable when the moduli are free from large prime divisors only, but it is adequate for our purpose.

2010 Mathematics Subject Classification: 11A41.

Keywords and Phrases: Birch-Bombieri result, Bombieri-Vinogradov theorem, gaps between primes, Klooserman sums, Weil's bound.

1 Introduction

Let p_n denote the n-th prime and write $d_n = p_{n+1} - p_n$. The prime number theorem implies that the average value of d_n is $\sim \log p_n$. The estimates of the upper and lower bounds for d_n is one of the central problems in analytic number theory.

Currently, the best result on the upper bound for d_n is due to Baker, Harman and Pintz [1] that gives

$$d_n \ll p_n^{21/40}.$$

The lower bound for d_n has attracted great interest from mathematicians. For 80 years, the work on this problem had concentrated on estimating the quantity

$$\Delta = \liminf_{n\to\infty} \frac{d_n}{\log p_n}.$$

The inequality $\Delta \leq 1$ is a trivial consequence of the prime number theorem. In 1926 Hardy and Littlewood [13] obtained $\Delta \leq 2/3$ on assuming GRH. In 1940 Erdös [8] obtained $\Delta < 1 - c$ unconditionally, with a small unspecified computable constant $c > 0$. In 1966 Bombieri and Davenport [2] obtained $\Delta \leq 0.4665...$. In 1988 Maier [15] obtained $\Delta \leq 0.2484...$. In 2009 Goldston, Pintz and Yildirim [11] eventually proved

$$\Delta = 0.$$

Recently Zhang [17] has proved the following

Theorem 1.1.

$$\liminf_{n\to\infty} d_n < \infty. \tag{1.1}$$

This result may be regarded as a weak form of the twin prime conjecture

$$\liminf_{n\to\infty} d_n = 2. \tag{1.2}$$

In fact, Zhang proves (1.1) with the right side replaced by 7×10^7. This specific number may be considerably reduced.

2 Admissible sets

Let

$$\mathcal{H} = \{h_1, h_2, ..., h_k\}$$

be a set composed of distinct non-negative integers. We say that $\mathcal{H}$ is admissible if $\nu_p(\mathcal{H}) < p$ for every prime p, where $\nu_p(\mathcal{H})$ denotes the number of distinct residue classes modulo p occupied by the h_j.

The reason for introducing admissible sets can be described as follows. Assume that $\mathcal{H}$ is not admissible. Then there is a prime p such that, for any positive integer n, the tuple

$$(n + h_1, n + h_2, ..., n + h_k) \tag{2.1}$$

contains at least one multiple of p. Thus we conclude

Suppose that there are infinitely many positive integers n such that each element in the tuple (2.1) *is a prime. Then $\mathcal{H}$ must be admissible.*

A conjecture of Hardy and Littlewood asserts the converse. Here we state a weak form of their conjecture only.

Conjecture 2.1. *Suppose that $\mathcal{H}$ is admissible. There are infinitely many positive integers n such that each element in the tuple* (2.1) *is a prime.*

Note that the twin prime conjecture is a special case of Conjecture 2.1 with $\mathcal{H} = \{0, 2\}$.

It is easy to see that Theorem 1.1 can be deduced from the following

Theorem 2.2. *Suppose that $\mathcal{H}$ is admissible with k sufficiently large. There are infinitely many positive integers n such that the tuple* (2.1) *contains at least two primes.*

Theorem 2.2 actually gives

$$\liminf_{n\to\infty} d_n \le \max_{i\ne j} |h_1 - h_j|. \tag{2.2}$$

The initial step in the proof of Theorem 2.2 consists in transferring the problem to evaluate and compare some arithmetic sums. Write

$$\theta(n) = \begin{cases} \log n & \text{if } n \text{ is prime,} \\ 0 & \text{otherwise.} \end{cases}$$

Let x denote a large number, and let f be defined on $(x, 2x) \cap \mathbf{Z}$ such that

$$f(n) \ge 0. \tag{2.3}$$

We introduce

$$S_1(x, f) = \sum_{x<n<2x} f(n) \tag{2.4}$$

and

$$S_2(x, f; \mathcal{H}) = \sum_{x<n<2x} \left(\sum_{j=1}^{k} \theta(n + h_j) \right) f(n). \tag{2.5}$$

The key point is to prove that, for any sufficiently large x, there is a function f satisfying (2.3), such that

$$\frac{S_2(x, f; \mathcal{H})}{S_1(x, f)} > \log 3x. \tag{2.6}$$

This implies that there is an integer $n \in (x, 2x)$ such that the tuple (2.1) contains at least two primes.

3 The choice of f by Goldston-Pintz-Yildirim

From now on, the set $\mathcal{H}$ is assumed to be admissible and fixed.

To choose f and show that (2.6) holds, the following conditions are necessary.

(1) f should be non-negative.

(2) One should be able to evaluate the sums $S_1(x, f)$ and $S_2(x, f; \mathcal{H})$ efficiently.

(3) The sum $S_2(x, f; \mathcal{H})$ should be "large" in comparison with $S_1(x, f)$.

Let $\Lambda(n)$ denote the von Mangoldt function. To satisfy the condition (3), one may choose

$$f(n) = \sum_{j=1}^{k} \Lambda(n + h_j)$$

(the difference $\theta - \Lambda$ is negligible). However, this is not valid in practice since one is unable to evaluate $S_2(x, f; \mathcal{H})$.

It is known that

$$\Lambda(n)=\sum_{d|n}\mu(d)\log\frac{n}{d}, \tag{3.1}$$

where $\mu(d)$ denotes the Möbius function. In many sieve problems, $\Lambda(n)$ is approximated (replaced) by

$$\Lambda_D(n)=\sum_{\substack{d|n\\ d<D}}\mu(d)\log\frac{D}{d} \tag{3.2}$$

for $n<2x$, where D is a parameter $<x$. In practice, D is usually of the form

$$D=x^b, \qquad 0<b<\frac{1}{2}. \tag{3.3}$$

If one chooses

$$f(n)=\sum_{j=1}^{k}\Lambda_D(n+h_j),$$

then, with D given by (3.3), $S_1(x,f)$ and $S_2(x,f;\mathcal{H})$ can be evaluated. However, the condition (1) may not hold.

In order to satisfy the condition (1), a simple idea which comes from the theory of Selberg's sieve is that f takes the form

$$f(n)=\lambda(n)^2, \qquad \lambda(n)\in\mathbf{R}. \tag{3.4}$$

From now on, we assume that f is of the form (3.4) and write $D=x^b$. The problem is reduced to choosing λ.

It was Goldston, Pintz and Yildirim who found a form of $\lambda(n)$ which proves very efficient. Let

$$P_{\mathcal{H}}(n)=\prod_{j=1}^{k}(n+h_j), \tag{3.5}$$

and let l be a fixed positive integer. They choose

$$\lambda(n)=\frac{1}{(k+l)!}\sum_{\substack{d|P_{\mathcal{H}}(n)\\ d<D}}\mu(d)\left(\log\frac{D}{d}\right)^{k+l}. \tag{3.6}$$

One may also consider the general form

$$\lambda(n)=\sum_{\substack{d|P_{\mathcal{H}}(n)\\ d<D}}\mu(d)h\left(\frac{\log(D/d)}{\log D}\right), \tag{3.7}$$

where h is a real polynomial satisfying

$$h^{(i)}(0)=0, \qquad 0\le i\le k.$$

The form (3.7) and its variants also find application to other problems in analytic number theory, which is usually called the GPY sieve.

From now on, we assume $b < 1/2$ and write ν_p, $P(n)$, S_1 and S_2 for $\nu_p(\mathcal{H})$, $P_{\mathcal{H}}(n)$, $S_1(x, f)$ and $S_2(x, f; \mathcal{H})$ respectively; similar abbreviations will apply in the sequel. Let $\lambda(n)$ be as in (3.6). Write

$$g(d) = \frac{1}{(k+l)!}\left(\log \frac{D}{d}\right)^{k+l} \quad \text{if} \quad d < D \tag{3.8}$$

and

$$g(d) = 0 \quad \text{if} \quad d \geq D, \tag{3.9}$$

so that

$$\lambda(n) = \sum_{d|P(n)} \mu(d)g(d). \tag{3.10}$$

The evaluation of S_1 is not difficult. Squaring out $\lambda(n)^2$ and changing the order of summation we obtain

$$S_1 = \sum_{d_1}\sum_{d_2} \mu(d_1)\mu(d_2)g(d_1)g(d_2) \sum_{\substack{x<n<2x \\ P(n)\equiv 0([d_1,d_2])}} 1, \tag{3.11}$$

where $[d_1, d_2]$ denotes the l.c.m. of d_1 and d_2. Note that $[d_1, d_2]$ is square-free and

$$[d_1, d_2] < D^2 = x^{2b}$$

if $\mu(d_1)\mu(d_2)g(d_1)g(d_2) \neq 0$. Let

$$\varrho_1(d) = |\mu(d)| \prod_{p|d} \nu_p. \tag{3.12}$$

For any square-free d, $d|P(n)$ if and only if n lies in exactly $\varrho_1(d)$ distinct residue classes $(\bmod\, d)$. It follows that the innermost sum in (3.11) is equal to

$$\frac{\varrho_1([d_1, d_2])}{[d_1, d_2]} x + O(\varrho_1([d_1, d_2])).$$

This yields

$$S_1 = xT_1 + O(x^{2b+\varepsilon}), \tag{3.13}$$

where

$$T_1 = \sum_{d_1}\sum_{d_2} \frac{\mu(d_1)\mu(d_2)g(d_1)g(d_2)\varrho_1([d_1, d_2])}{[d_1, d_2]}. \tag{3.14}$$

The error term in (3.13) is acceptable since $b < 1/2$.

We now turn to S_2. In a way similar to the proof of (3.11), we deduce that

$$S_2 = \sum_{d_1}\sum_{d_2} \mu(d_1)\mu(d_2)g(d_1)g(d_2) \sum_{j=1}^{k} \sum_{\substack{x<n<2x \\ P(n)\equiv 0([d_1,d_2])}} \theta(n + h_j). \tag{3.15}$$

If d is square-free, $d < D^2$, $n \in (x, 2x)$ and $n + h_j$ is a prime, then $d|P(n)$ if and only if

$$d \equiv c(\mathrm{mod}\, d) \quad \text{for some} \quad c \in \mathcal{C}_j(d),$$

where

$$\mathcal{C}_j(d) = \{c\colon 1 \le c \le d,\ (c, d) = 1,\ P(c - h_j) \equiv 0(\mathrm{mod}\, d)\}. \tag{3.16}$$

Thus the innermost sum in (3.15) is equal to

$$\sum_{c \in \mathcal{C}_j([d_1, d_2])} \sum_{\substack{x < n < 2x \\ n \equiv c([d_1, d_2])}} \theta(n + h_j). \tag{3.17}$$

Let

$$\varrho_2(d) = |\mu(d)| \prod_{p|d} (\nu_p - 1). \tag{3.18}$$

For square-free d it can be shown that

$$|\mathcal{C}_j(d)| = \varrho_2(d).$$

It follows from (3.17) and the prime number theorem that the innermost sum in (3.15) is equal to

$$\frac{\varrho_2(d)}{\varphi(d)} x + O\Big(\sum_{c \in \mathcal{C}_j(d)} \mathcal{R}(x; c, d) \Big) + O(x^{\varepsilon}), \tag{3.19}$$

where $\varphi(d)$ denotes the Euler function, and where

$$\mathcal{R}(x; d, c) = \Bigg| \sum_{\substack{x < n < 2x \\ n \equiv c(d)}} \theta(n) - \frac{x}{\varphi(d)} \Bigg|. \tag{3.20}$$

Inserting (3.19) into (3.15) we obtain

$$S_2 = xT_2 + O((\log x)^{2k+2l} E) + O(x^{2b+\varepsilon}), \tag{3.21}$$

where

$$T_2 = k \sum_{d_1} \sum_{d_2} \frac{\mu(d_1)\mu(d_2)g(d_1)g(d_2)\varrho_2([d_1, d_2])}{\varphi([d_1, d_2])} \tag{3.22}$$

and

$$E = \sum_{d < D^2} |\mu(d)| \tau_3(d) \varrho_2(d) \sum_{j=1}^{k} \sum_{c \in \mathcal{C}_j(d)} \mathcal{R}(x; c, d),$$

where $\tau_3(d)$ denotes the 3-fold divisor function.

The problem is now reduced to evaluation T_1 and T_2 and estimating E.

4 The main term

By (3.13) and (3.21), we need only prove the following which lead to (2.6).

(1) With k sufficiently large and with l appropriately chosen,

$$\frac{T_2}{T_1} > (1+c)\log x \tag{4.1}$$

for some positive constant c.

(2) The sum E can be efficiently bounded, namely,

$$E \ll x(\log x)^{-A} \tag{4.2}$$

for any large constant A.

Note that ϱ_1 is supported on square-free integers. Substituting $d_0 = (d_1, d_2)$ and rewriting d_1 and d_2 for d_1/d_0 and d_2/d_0 respectively, we deduce that

$$T_1 = \sum_{d_0}\sum_{d_1}\sum_{d_2} \frac{\mu(d_1d_2)\varrho_1(d_0d_1d_2)}{d_0d_1d_2} g(d_0d_1)g(d_0d_2). \tag{4.3}$$

Similarly,

$$T_2 = k\sum_{d_0}\sum_{d_1}\sum_{d_2} \frac{\mu(d_1d_2)\varrho_2(d_0d_1d_2)}{\varphi(d_0d_1d_2)} g(d_0d_1)g(d_0d_2). \tag{4.4}$$

The right sides of (4.3) and (4.4) can be evaluated by standard methods. It is shown that

$$T_1 \sim \frac{1}{(k+2l)!}\binom{2l}{l}\mathfrak{S}(\log D)^{k+2l}$$

and

$$T_2 \sim \frac{k}{(k+2l+1)!}\binom{2l+2}{l+1}\mathfrak{S}(\log D)^{k+2l+1},$$

where

$$\mathfrak{S} = \prod_p \left(1 - \frac{\nu_p}{p}\right)\left(1 - \frac{1}{p}\right)^{-k}.$$

(Note that $\mathfrak{S} = 0$ if $\mathcal{H}$ is not admissible.) It follows that

$$\frac{T_2}{T_1} \sim \frac{bk}{k+2l+1}\frac{(2l+2)(2l+1)}{(l+1)^2}\log x. \tag{4.5}$$

If $b > 1/4$, then, with k is sufficiently large in terms of b and with l appropriately chosen, we have

$$\frac{bk}{k+2l+1}\frac{(2l+2)(2l+1)}{(l+1)^2} > 1. \tag{4.6}$$

This leads to (4.1). On the other hand, the relation (4.6) can not be valid for $b \le 1/4$.

Thus we conclude

Theorem 2.2 will be valid if (4.2) holds for some $b > 1/4$.

5 The error term

We say that the primes have level of distribution ϑ if, for any large constant A and for any small positive constant ε,

$$\sum_{d<Q} \max_{y \le x} \max_{(c,d)=1} \left| \sum_{\substack{n \le y \\ n \equiv c(d)}} \Lambda(n) - \frac{y}{\varphi(d)} \right| \ll x(\log x)^{-A} \tag{5.1}$$

with $Q = x^{\vartheta-\varepsilon}$. Since the estimate

$$\sum_{\substack{n \le y \\ n \equiv c(d)}} \Lambda(n) - \frac{y}{\varphi(d)} \ll y^{1/2+\varepsilon}$$

is implied by the GRH, the primes have level of distribution 1/2 if GRH is true. This result was unconditionally proved by Bombieri and Vinogradov in 1965.

Theorem 5.1. *The primes have level of distribution* 1/2.

Elliott and Halberstam [7] conjectured that the primes have level of distribution 1. By Cauchy's inequality, we see that (4.2) will be valid if the primes have level of distribution $\vartheta = 2b + \varepsilon$. Thus, by the discussion in Section 4, we conclude

Theorem 2.2 *will be valid if the primes have level of distribution* ϑ *for some* $\vartheta > 1/2$.

6 A refinement of the GPY method

One is unable to prove that the primes have level of distribution ϑ for some $\vartheta > 1/2$, even if the GRH is assumed. We introduce a refinement of the GPY method which applies to the main and error terms equally.

Let $\varpi > 0$ be a small constant, and let $b = 1/4 + \varpi$. Our first observation is that, on the right sides of (4.3) and (4.4), the terms with $d_0 d_1 d_2$ having a large prime factor p, say $p \ge x^{\varpi}$, make minor contribution. Thus, if we modify the sum in (3.6) by imposing the constraint that $p|d$ implies $p < x^{\varpi}$, the resulting main terms in S_1 and S_2 will have minor changes only. This was independently observed by Motohashi and Pintz. Our second observation, which is the most novel part of the proof, is that with such a constraint imposed in (3.6), the resulting error in S_2 can be efficiently bounded.

From now on, let S_1, S_2 and $f(n)$ be as in (2.4), (2.5) and (3.4) respectively, but let $\lambda(n)$ be redefined by

$$\lambda(n) = \sum_{d|(P(n),\mathcal{P})} \mu(d) g(d)$$

with $g(d)$ given by (3.8) and (3.9), where

$$\mathcal{P} = \prod_{p<x^{\varpi}} p.$$

Repeating the arguments in Section 4 we obtain

$$S_1 = xT_1^* + O(x^{2b+\varepsilon})$$

and

$$S_2 = xT_2^* + O((\log x)^{2k+2l}E^*) + O(x^{2b+\varepsilon}),$$

where

$$T_1^* = \sum_{d_1|\mathcal{P}} \sum_{d_2|\mathcal{P}} \frac{\mu(d_1)\mu(d_2)g(d_1)g(d_2)\varrho_1([d_1,d_2])}{[d_1,d_2]},$$

$$T_2^* = k\sum_{d_1|\mathcal{P}} \sum_{d_2|\mathcal{P}} \frac{\mu(d_1)\mu(d_2)g(d_1)g(d_2)\varrho_2([d_1,d_2])}{\varphi([d_1,d_2])}$$

and

$$E^* = \sum_{\substack{d<D^2 \\ d|\mathcal{P}}} |\mu(d)|\tau_3(d)\varrho_2(d) \sum_{j=1}^{k} \sum_{c\in\mathcal{C}_j(d)} \mathcal{R}(x;c,d).$$

By routine estimations of the differences $T_1 - T_1^*$ and $T_2 - T_2^*$ and (4.5), it can be shown that

$$\frac{T_1^*}{T_2^*} > (1+c)\log x$$

for some constant $c > 0$, if k is sufficiently large and if l is appropriately chosen. The proof of Theorem 2.2 is therefore reduced to proving

Theorem 6.1. *If ϖ is sufficiently small, then*

$$E^* \ll x(\log x)^{-A}$$

for any large constant A.

7 Sketch of the proof of Theorem 6.1

Note that $D^2 = x^{1/2+2\varpi}$. By Theorem 5.1 and Cauchy's inequality, the proof of Theorem 6.1 is reduced to showing that

$$\sum_{\substack{x^{1/2-\varepsilon}<d<D^2 \\ d|\mathcal{P}}} \sum_{c\in\mathcal{C}_j(d)} \mathcal{R}(x;c,d) \ll x(\log x)^{-2A}, \qquad 1\le j\le k. \tag{7.1}$$

The constraint $d|\mathcal{P}$ is crucial in the proof of (7.1). This is originally due to the simple fact that if $1 < R < d$, $d > x^{1/2-\varepsilon}$ and $d|\mathcal{P}$, then d can be factored as

$$d = rq \qquad \text{with} \qquad R < r < x^{\varpi}R. \tag{7.2}$$

We can replace θ by Λ in the expression for $\mathcal{R}(x;c,d)$. The proof of (7.1) is described as follows. For any function γ supported on $(x,2x)\cap \mathbf{Z}$ we define

$$\Delta(\gamma;c,d) = \left|\sum_{n\equiv c(d)} \gamma(n) - \frac{1}{\varphi(d)}\sum_{(n,d)=1}\gamma(n)\right|, \qquad (c,d)=1.$$

Using a combinatorial identity for Λ due to Heath-Brown[14], the proof of (7.1) is reduced to estimating the sum of $\Delta(\gamma;c,d)$ for certain Dirichlet convolutions γ. There are three types of the convolutions involved in the proof. Write

$$\eta = 1 + (\log x)^{-2A}, \qquad x_1 = x^{3/8+8\varpi}, \qquad x_2 = x^{1/2-4\varpi}.$$

In the first two types the function γ is of the form $\gamma = \alpha * \beta$ such that

(i) $\alpha = (\alpha(m))$ is supported on $[M,\, \eta^{19}M)$,

(ii) $\beta = (\beta(n))$ is supported on $[N,\, \eta^{19}N)$,

(iii) $MN \in (x, 2x)$.

(There are some other conditions on $\alpha(m)$ and $\beta(n)$.) We say that γ is of Type I if $x_1 < N \le x_2$; we say that γ is of Type II if $x_2 < N < 2x^{1/2}$; we say that γ is of Type III if it is of the form $\gamma = \alpha * \varkappa_{N_1} * \varkappa_{N_2} * \varkappa_{N_3}$ where α is as in (i) and $\varkappa_N$ is the characteristic function of $[N,\, \eta N) \cap \mathbf{Z}$, such that

(iv) $N_3 \le N_2 \le N_1$, $\qquad MN_1 \le x_1$,

(v) $MN_1N_2N_3 \in (x, 2x)$.

It should be stressed that, without the constraint $d|\mathcal{P}$, we are unable to efficiently bound the sum

$$\sum_{x^{1/2-\varepsilon}<d<D^2} \sum_{c\in\mathcal{C}_j(d)} \Delta(\gamma;c,d)$$

if γ is of Type I or II. However, by (7.2), the Type I and II estimates are reduced to bounding sums of the type

$$\sum_{R<r<2R} \sum_{Q<q<2Q} \sum_{c\in\mathcal{C}_j(rq)} \Delta(\gamma;c,rq) \tag{7.3}$$

with R appropriately chosen such that it is close to N in the logarithmic scale. Thus, using the dispersion method due to Bombieri, Fouvry, Friedlander and Iwaniec [3–5, 9], and Weil's bound for Kloosterman sums, we can prove that the sum (7.3) is

$$\ll x(\log x)^{-43A}. \tag{7.4}$$

The Type III estimate essentially relies on the Birch-Bombieri result in the appendix to Friedlander and Iwaniec[10]. This result in turn relies on Deligne's proof of the Riemann Hypothesis for varieties over finite fields (the Weil Conjecture)[6]. Assuming γ is of Type III, we estimate each $\Delta(\gamma;c,d)$ directly. However, without the constraint $d|\mathcal{P}$, efficient bounds for $\Delta(\gamma;c,d)$ can not be obtained unless we relax the condition $MN_1 \le x_1$. Our argument is carried out by combining the method in [10] and the factorization (7.2) (here r is relatively small), the latter will allow us to save a factor $r^{1/2}$. Thus we obtain

$$\Delta(\gamma;c,d) \ll \frac{x^{1-\varepsilon}}{d} \tag{7.5}$$

for $x^{1/2-\varepsilon} < d < D^2$, $d|\mathcal{P}$ and $(c,d)=1$.

The estimate (7.1) follows from (7.4) and (7.5).

References

[1] R. C. Baker, G. Harman, and J. Pintz, The difference between consecutive primes II, *Proc. London Math. Soc.* (3) **83** (2001), 532–562.

[2] E. Bombieri and H. Davenport, Small differences between prime numbers, *Proc. Roy. Soc. Ser.* A **293** (1966), 1–18.

[3] E. Bombieri, J. B. Friedlander, and H. Iwaniec, Primes in arithmetic progressions to large moduli, *Acta Math.* **156** (1986), 203–251.

[4] E. Bombieri, J. B. Friedlander, and H. Iwaniec, Primes in arithmetic progressions to large moduli II, *Math. Ann.* **277** (1987), 361–393.

[5] E. Bombieri, J. B. Friedlander, and H. Iwaniec, Primes in arithmetic progressions to large moduli III, *J. Am. Math. Soc.* **2** (1989), 215–224.

[6] P. Deligne, La conjecture de Weil I, *Publ. Math. IHES.* **43** (1974), 273–307.

[7] P. D. T. A. Elliott and H. Halberstam, A conjecture in prime number theory, in *Symposia Mathematica,* Vol. **IV** (INDAM, *Roma,* 1968/1969), Academic Press, London, 1970, 59–72.

[8] P. Erdös, The difference of consecutive primes, *Duke Math. J.* **6** (1940), 438–441.

[9] J. B. Fouvry and H. Iwaniec, Primes in arithmetic progressions, *Acta Arith.* **42** (1983), 197–218.

[10] J. B. Friedlander and H. Iwaniec, Incomplete Kloosterman sums and a divisor problem, *Ann. Math.* **121** (1985), 319–350.

[11] D. A. Goldston, J. Pintz, and C. Y. Yildirim, Primes in Tuples I, *Ann. Math.* **170** (2009), 819–862.

[12] D. A. Goldston, J. Pintz, and C. Y. Yildirim, Primes in Tuples II, *Acta Math.* **204** (2010), 1–47.

[13] G. H. Hardy and J. E. Littlewood, *unpublished manuscript.*

[14] D. R. Heath-Brown, Prime numbers in short intervals and a generalized Vaughan identity, *Canad. J. Math.* **34** (1982), 1365–1377.

[15] H. Maier, Small differences between prime numbers, *Michigan Math. J.* **35** (1988), 323–344.

[16] K. Soundararajan, Small gaps between prime numbers: the work of Goldston-Pintz-Yildirim, *Bull. Amer. Math. Soc.* **44** (2007), 1–18.

[17] Y. Zhang, Bounded gaps between primes, *to appear in Ann. Math.*

Department of Mathematics and Statistics
University of New Hampshire
Durham, NH 03824, USA
E-mail address: yitangz@unh.edu

Proceedings of the 6th International Congress of Chinese Mathematicians
ALM 36, pp. 247–265

A Survey of Local Models of Shimura Varieties

Xinwen Zhu

Abstract

We survey some recent development of the theory of local models for Shimura varieties.

2010 Mathematics Subject Classification: Primary 14G35; Secondary 11G18, 14M15.

Keywords and Phrases: Local models, Shimura varieties, nearby cycles, affine flag varieties.

1 Introduction

Local models were systematically introduced by Rapoport and Zink in [RZ] as a tool to analyze the étale local structure of certain integral models of (PEL-type) Shimura varieties with parahoric level structures over p-adic fields (special cases were constructed earlier by Deligne and Pappas [DP] and independently by de Jong [dJ]). Unlike the Shimura varieties themselves, which are usually moduli spaces of abelian varieties, local models are defined in terms of linear algebra and therefore are much easier to study. Among others, Görtz [Go1, Go2] proved the flatness of certain PEL-type Shimura varieties associated to unramified unitary groups and symplectic groups using local models (some special cases were obtained in earlier works [CN, dJ, DP]), and Haines and Ngô [HN] calculated the sheaves of nearby cycles of these Shimura varieties under the further assumption that the groups are split.

On the other hand, a discovery of Pappas [Pa] showed that the originally defined integral models of Shimura varieties in [RZ] are usually not flat when the

The author is supported by NSF grant DMS-1313894 and DMS-1303296 and the AMS Centennial Fellowship.

groups are ramified. Therefore, nowadays the (local) models defined in [RZ] are usually called the naive models, and the search for a definition of the flat integral models of Shimura varieties (which more or less is equivalent to a definition of the flat local models) became an active research area in the past 15 years. In a series of papers [PR1, PR2, PR4], Pappas and Rapoport investigated the correct definition of local models in the ramified PEL case. These are flat by definition but they do not always have a neat moduli interpretation. In addition, the constructions in these works are somewhat ad-hoc and are mostly done case-by-case: Ultimately, they are based on representing the corresponding reductive group explicitly as the neutral component of the group of automorphisms of a suitable bilinear (symmetric, alternating, hermitian etc.) form on a vector space over a division ring. Then, its parahoric subgroups are the connected stabilizers of a self-dual chain of lattices as explained in [BT1, BT2] (see also [RZ, Appendix to Chapter 3]) and the local models are given as flat closures in certain corresponding linked Grassmannians. We refer the reader to the survey [PRS] for precise definitions and more references.

In [PZ1], we provided a general group theoretic definition of local models that is not tied to a particular representation of the group. This approach, inspired by constructions from the geometric Langlands program, allows us to make significant progress and resolve several open questions. Our local models are constructed starting from the "local Shimura data", i.e triples $(G, K, \{\mu\})$, where G is a (connected) reductive group over $\mathbb{Q}_p$, $K \subset G(\mathbb{Q}_p)$ a parahoric "level" subgroup, and $\{\mu\}$ a geometric conjugacy class of one parameter subgroups $\mu : \mathbb{G}_{m/\overline{\mathbb{Q}}_p} \to G_{\overline{\mathbb{Q}}_p}$ over an algebraic closure $\overline{\mathbb{Q}}_p$. Denote by E the field of definition of the conjugacy class $\{\mu\}$. This is the *local reflex field*; it is a finite extension of $\mathbb{Q}_p$ and is contained in a splitting field for G. Let $\mathcal{O}_E$ be the ring of integers of E and k_E its residue field. We will assume that G splits over a tamely ramified extension of $\mathbb{Q}_p$. Then from the data (G, K), we first construct a smooth affine group scheme $\mathcal{G}$ over $X = \mathrm{Spec}\mathcal{O}[u]$. Let $\mathrm{Gr}_{\mathcal{G},X}$ be the affine Grassmannian of $\mathcal{G}$ over X. Then $\{\mu\}$ determines a closed subscheme X_μ of $\mathrm{Gr}_{\mathcal{G},X} \otimes_{\mathcal{O}[u],u\mapsto p} E$, which is isomorphic to the homogeneous space for G_E that corresponds to μ. Then by definition, the (generalized) local model $M_{\mathcal{G},\mu}$ is the flat closure of X_μ in $\mathrm{Gr}_{\mathcal{G},X}$. We refer to §2 for the detailed construction.

The local model defined as above is by taking the flat closure of its generic fiber in some ambient scheme. Usually, an integral model defined in this way is not useful since the moduli interpretation is lost and therefore it is very difficult to study the special fiber. However, as observed by Pappas and Rapoport in [PR3], the brute force definition of the local model by taking the flat closure is not totally out of control as one might think. Namely, by its definition the special fibers of the local models are closed subschemes in the (partial) affine flag varieties and that their reduced subschemes are a union of Schubert varieties. Then it is a question to describe which Schubert varieties will appear in the special fibers and whether the special fibers are reduced. These questions are reduced to the coherence conjecture [PR3, PR4, Z1], which characterizes the dimension of the spaces of global sections of certain ample line bundles on certain union of Schubert varieties. The coherence conjecture was proved in [Z1]. As a result, one can apply known properties of affine Schubert varieties as developed in [Fa, PR3] to the understanding of the geometry of local models. Among other things, we mention that the local models are normal, with reduced special fibers. We refer to §3 for more details.

As mentioned at the beginning, local models were invented to understand the local structure of integral models of Shimura varieties. When the triple $(G, K, \{\mu\})$ is from a local Shimura datum, we conjecture that $\mathrm{M}(G, \{\mu\})_K := M_{\mathcal{G},\mu}$ describes the étale local structure of an integral model of the Shimura variety. In the case when $(G, \{\mu\})$ is of Hodge type (and G is tamely ramified over $\mathbb{Q}_p$ for $p > 2$), this is the case by [PZ1, KP]. In particular, the integral models of these Shimura varieties are normal with reduced special fibers. Another consequence is that the calculation of the sheaves of nearby cycles of the integral models of Shimura varieties reduces to calculations on the corresponding local models. The strategy of using local models to determine the (semi-simple) Hasse-Weil zeta function of Shimura varieties at places of (tame parahoric reduction) was first suggested by Rapoport [R]. It has since been advanced by Kottwitz, Haines and Ngô [HN], Haines and others. A centerpiece of this approach is a conjecture of Kottwitz that states that, in the case that G is split, the semi-simple trace of Frobenius on the sheaf of nearby cycles gives a central function in the parahoric Hecke algebra. This was proven by Haines and Ngô [HN] in split types A and C by following an argument of Gaitsgory [Ga]. Gaitsgory proved a stronger statement for general split groups in the function field case; he showed that the perverse sheaf of nearby cycles satisfies a commutativity constraint with respect to the convolution product. This implies Kottwitz's conjecture in the function field case. The main tools in Gaitsgory's approach are various versions of the global affine Grassmannian of Beilinson-Drinfeld. Our group theoretical definition of local models and their connections with affine Grassmannians like $\mathrm{Gr}_{\mathcal{G},X}$ allow us to generalize and simplify the approaches of both Gaitsgory and Haines-Ngô. In addition, we can determine the action of monodromy (i.e of inertia) on the sheaves of nearby cycles. We refer to §4 for more details.

Certainly, this survey does not cover all aspects of the theory of local models. For example, we do not discuss possible moduli interpretations of certain local models. Some progresses were made by [Sm1, Sm2, Sm3], following early works of [PR1, PR2, PR4]. We refer to another excellent survey paper [PRS] for this aspect. In some sense, our survey paper can be used as a complement of [PRS]. Another aspect we do not cover is the application of local models to the study of certain local deformation ring of Galois representations, started from [Ki] and recently further developed in [Le].

Acknowledgements. The author thanks the organizers for the invitation to speak on the 2013 ICCM. He thanks G. Pappas for the collaboration and X. He, T. Haines, G. Pappas and M. Rapoport for the discussion of the theory of local models.

2 Definition of local models

In this section, we review the group theoretical definition of local models, following [PZ1]. Let F be a p-adic field, $\mathcal{O}$ its ring of integers, ϖ a uniformizer, and $k = \mathcal{O}/\varpi$

the residue field. Let G be a (connected) reductive group over F, splits over a *tamely* ramified extension $\tilde{F}$ of F. Let H denote the split Chevalley form of G over $\mathcal{O}$.

2.a

Let $\mathcal{P}$ be a parahoric group scheme of G over $\mathcal{O}$. The first theorem we need is

Theorem 2.1. *There is a smooth connected affine group scheme $\mathcal{G}$ over $\mathcal{O}[u]$, such that*

(a) *$\mathcal{G}[u^{-1}]$ is (connected) reductive.*

(b) *The specialization of $\mathcal{G}$ along $u = \varpi$ is isomorphic to $\mathcal{P}$.*

(c) *The restriction of $\mathcal{G}$ to $\kappa[[u]]$, where $\kappa = F$ or k, is a parahoric group scheme.*

Remark 2.2. (i) In a forthcoming work [PZ2], we will show that the above group scheme $\mathcal{G}$ is unique up to isomorphism. This will follow from our study of the Bruhat-Tits group schemes over $\mathcal{O}[[u]]$, which allows us to classify reductive groups over $\mathcal{O}((u))$ and $\mathcal{O}[u, u^{-1}]$.

(ii) Let $G^\flat := \mathcal{G} \times_X k((u))$. Then $G \mapsto G^\flat$ gives a bijection between the isomorphism classes of tamely ramified reductive groups over F and those over $k((u))$. In addition, $\mathcal{P} \mapsto \mathcal{P}^\flat := \mathcal{G} \times_X k[[u]]$ gives a bijection between the isomorphism classes of parahorics group schemes over $\mathcal{O}$ and those over $k[[u]]$.

We continue with the same assumptions, but for a little while we allow $\mathcal{G}$ to be any smooth affine group scheme over $X = \mathbb{A}^1_{\mathcal{O}}$ with connected fibers.

Let $S \in (\mathrm{Sch}/X)$, with structure morphism $y : S \to X$. We will denote by $\Gamma_y \subset X \times S$ the closed subscheme given by the graph of y. We define the global affine Grassmannian of $\mathcal{G}$ over X to be the functor on (Sch/X) given by

$$\mathrm{Gr}_{\mathcal{G},X}(S) = \left\{ \text{iso-classes of pairs } (\mathcal{E}, \beta) \;\middle|\; \begin{array}{c} \mathcal{E} \text{ a } \mathcal{G}\text{-torsor on } X \times S, \\ \beta \text{ a trivialization of } \mathcal{E}|_{(X\times S)\setminus \Gamma_y} \end{array} \right\}. \tag{2.1}$$

Here and everywhere else the fiber products are over $\mathrm{Spec}(\mathcal{O})$. It is known that $\mathrm{Gr}_{\mathcal{G},X}$ is represented by an ind-scheme, ind of finite type over X.

Suppose now that $\mathcal{G}$ is as in Theorem 2.1. Then it was shown in [PZ1, Proposition 5.5] that $\mathrm{Gr}_{\mathcal{G},X}$ is further more ind-projective. Let κ be either the fraction field F or the residue field k of $\mathcal{O}$. Let $x : \mathrm{Spec}(\kappa) \to X$, where κ is as above and identify the completed local ring $\widehat{\mathcal{O}}_x$ of $X \times \mathrm{Spec}(\kappa)$ with $\kappa[[t]]$, using the local parameter $t = u - x$. Let

$$\mathcal{G}_{\kappa,x} := \mathcal{G} \times_{\mathrm{Spec}(\mathcal{O}[u])} \mathrm{Spec}(\kappa[[t]]). \tag{2.2}$$

(i) Suppose that x factors through $0 : \mathrm{Spec}(\mathcal{O}) \to X$. Recall that by Theorem 2.1 $P_\kappa := \mathcal{G}_{\kappa,0}$ is a parahoric group scheme over the dvr $\kappa[[t]] = \kappa[[u]]$ (note P_k is also denoted by $\mathcal{P}^\flat$).

Let L^+P_κ be the affine group scheme over $\mathrm{Spec}(\kappa)$ representing the functor on κ-algebras

$$R \mapsto L^+P_\kappa(R) = P_\kappa(R[[t]])\,,$$

and LP_κ the ind-group scheme over $\mathrm{Spec}(\kappa)$ representing the functor

$$R \mapsto LP_\kappa(R) = P_\kappa\big(R((t))\big) = G_\kappa\big(R((t))\big)\,.$$

Here $G_\kappa = P_\kappa[t^{-1}]$ is the connected reductive group over $\kappa((t))$ which is obtained by base changing $\mathcal{G} \to \mathrm{Spec}(\mathcal{O}[u])$ along $\mathcal{O}[u] \to \kappa((t))$, $u \mapsto t$. There is an ind-proper ind-scheme Gr_{P_κ} over κ which represents the quotient LP_κ/L^+P_κ of fpqc-sheaves on κ-schemes. More precisely, Gr_{P_κ} is the ind-scheme representing

$$R \mapsto \mathrm{Gr}_{P_\kappa}(R) = \left\{ \text{iso-classes of pairs } (\mathcal{E},\beta) \;\middle|\; \begin{array}{l} \mathcal{E} \text{ a } P_\kappa\text{-torsor on } \mathrm{Spec}R[[t]], \\ \beta \text{ a trivialization of } \mathcal{E}|_{\mathrm{Spec}R((t))} \end{array} \right\}.$$

The base change $\mathrm{Gr}_{P_\kappa} \times_{\mathrm{Spec}(\kappa)} \mathrm{Spec}(\bar\kappa)$ is an affine flag variety as in [PR3].

(ii) Suppose $x : \mathrm{Spec}(\kappa) \to X$ does not factor through $0 : \mathrm{Spec}(\mathcal{O}) \to X$. Then by Theorem 2.1 (a), the base change $\mathcal{G}_{\kappa,x}$ is a reductive group scheme which is a form of H. We can see that

$$\mathcal{G}_{\kappa,x} \times_{\mathrm{Spec}(\kappa)} \mathrm{Spec}(\kappa') \simeq H \times_{\mathrm{Spec}(\mathcal{O})} \mathrm{Spec}(\kappa'[[t]]) \tag{2.3}$$

for a finite κ'/κ. As above, we also have the affine Grassmannian $\mathrm{Gr}_{\mathcal{G}_{\kappa,x}}$ over κ; by the above, we can see that $\mathrm{Gr}_{\mathcal{G}_{\kappa,x}} \times_{\mathrm{Spec}(\kappa)} \mathrm{Spec}(\bar\kappa)$ can be identified with the usual affine Grassmannian Gr_H over $\bar\kappa$ for the split reductive group H. To summarize:

Proposition 2.3. *Let $x : \mathrm{Spec}(\kappa) \to X$, where κ is either the residue field k of $\mathcal{O}$, or the fraction field F of $\mathcal{O}$, and identify the completed local ring $\widehat{\mathcal{O}}_x$ of $X \times \mathrm{Spec}(\kappa)$ with $\kappa[[t]]$, using the local parameter $t = u - x$. Then restricting $\mathcal{G}$-bundles from $\mathcal{O}[u] \otimes_{\mathcal{O}} R$ to $R[[t]]$, $u \mapsto t + x$, induces an isomorphism over* $\mathrm{Spec}(\kappa)$,

$$i_x^*\colon \mathrm{Gr}_{\mathcal{G},X} \times_{X,x} \mathrm{Spec}(\kappa) \xrightarrow{\sim} \mathrm{Gr}_{\mathcal{G}_{\kappa,x}}\,.$$

Here $\mathrm{Gr}_{\mathcal{G}_{\kappa,x}} = L\mathcal{G}_{\kappa,x}/L^+\mathcal{G}_{\kappa,x}$ denotes the affine Grassmannian over κ as above; this is isomorphic to either Gr_{P_κ} if x maps to 0, or to Gr_H over $\bar\kappa$ otherwise.

2.b

We continue with the above assumptions and notations. Let $\overline{F}$ denote an algebraic closure of F that contains the field $\tilde F$ (that splits G), and let $\breve F$ be the completion of the maximal unramified extension of F in $\overline{F}$. Let A be a maximal F-split torus of G. Also, let S be a maximal $\breve F$-split torus in G which contains A and is defined over F. Such a torus exists by [BT2, 5.1.12]. Let $T = Z_G(S)$, which is a maximal torus of G which is defined over F and splits over $\tilde F$.

Suppose now that $\{\mu\}$ is a geometric conjugacy class of one parameter subgroups of G, defined over $\overline{F}$. Let E be the field of definition of $\{\mu\}$, a finite extension of F contained in $\overline{F}$ (the reflex field of the pair $(G, \{\mu\})$). We would like to define a projective scheme $M_{\mathcal{G},\mu}$ over $\mathcal{O}_E$ which we might view as a generalized local model.

By enlarge the field $\tilde{F}$ if necessary, we can assume that G is quasi-split over the maximal unramified extension $\tilde{F}_0$ of F in $\tilde{F}$, and therefore there exists a representative of $\{\mu\}$ defined over $E' = E\tilde{F}_0$ [Ko1, Lemma 1.1.3], which factors $\mu : \mathbb{G}_{\mathrm{m}E'} \to T_{E'} \to G_{E'}$. Notice that μ gives an $E'[z, z^{-1}]$-valued point of $G_{E'}$, therefore an $E'((z))$-valued point of $G_{E'}$, therefore an E'-valued point of the loop group LG. By Theorem 2.1 (c), $\mathcal{G}_{F,\varpi} \simeq G$ and therefore

$$G(F((z))) \xrightarrow{\sim} \mathcal{G}_{F,\varpi}(F((u-\varpi))) = \mathcal{G}_{F,\varpi}(F((t))).$$

We denote by s_μ the corresponding E'-valued point in $L\mathcal{G}_{F,\varpi}$. Consider the $L^+\mathcal{G}_{F,\varpi}$-orbit $(L^+\mathcal{G}_{F,\varpi})_{E'} \cdot [s_\mu]$ of the corresponding point $[s_\mu]$ in the affine Grassmannian $\mathrm{Gr}_{\mathcal{G}_{E',\varpi}}$. This orbit is contained in a projective subvariety of $\mathrm{Gr}_{\mathcal{G}_{E',\varpi}}$. Since the conjugacy class of $\mu : \mathbb{G}_{\mathrm{m}E'} \to G_{E'}$ is defined over E, the same is true for the orbit $(L^+\mathcal{G}_{F,\varpi})_{E'} \cdot [s_\mu]$: There is a closed E-subvariety X_μ of $\mathrm{Gr}_{\mathcal{G}_{E,\varpi}}$ such that $X_\mu \times_E E' = \overline{(L^+\mathcal{G}_{F,\varpi})_{E'} \cdot [s_\mu]}$.

We regard $\mathrm{Gr}_{\mathcal{G}_{E,\varpi}}$ as the generic fiber of $\mathrm{Gr}_{\mathcal{G},X} \otimes_{\mathcal{O}[u],u\mapsto\varpi} \mathcal{O}_{E'} \to \mathrm{Spec}(\mathcal{O}_{E'})$.

Definition 2.4. *The generalized local model (or mixed characteristic Schubert variety) $M_{\mathcal{G},\mu}$ is the reduced scheme over* $\mathrm{Spec}(\mathcal{O}_E)$ *which underlies the Zariski closure of the orbit X_μ in the ind-scheme* $\mathrm{Gr}_{\mathcal{G},\mathcal{O}_E} = \mathrm{Gr}_{\mathcal{G},X} \times_{\mathrm{Spec}(\mathcal{O}[u]),u\mapsto\varpi} \mathrm{Spec}(\mathcal{O}_E)$.

Since $\mathrm{Gr}_{\mathcal{G},X}$ is ind-projective over X, $M_{\mathcal{G},\mu}$ is also projective over $\mathrm{Spec}(\mathcal{O}_E)$.

Remark 2.5. In [PZ1], an important assumption is that the group G splits over a tamely ramified extension of F. This is due to Theorem 2.1. In fact, one can easily see that the tameness assumption is necessary for this theorem.

However, it is expected that the construction of local models (and its applications) can be extended to the following situation: if we write the adjoint form of G as $G_{\mathrm{ad}} = \prod_i \mathrm{Res}_{F_i/F} H_i$, where H_i is absolutely simple over F_i, then H_i splits over a tamely ramified extension of F_i. See the thesis of Levin [Le] for some progress in this direction. Note that this weaker assumption always holds if the residue characteristic p of F is > 3.

2.c

We give two examples in the above general construction.

The case of GL_N. Suppose $W = \mathcal{O}[u]^N$ and set $\overline{W} = W \otimes_{\mathcal{O}[u],u\mapsto 0} \mathcal{O}$. Write $\overline{W} = \oplus_{j=0}^{r-1} V_i$ and consider the parabolic subgroup $Q \subset \mathrm{GL}(\overline{W})$ which is the stabilizer of the flag $F_i = \oplus_{j\geq i} V_j$ given by the V_i. Denote by d_i the $\mathcal{O}$-rank of V_i. For $0 \leq i \leq r-1$, set W_i for the preimage of F_i under $W \to \overline{W}$ so that

$$uW \subset W_{r-1} \subset \cdots \subset W_1 \subset W_0 = W.$$

Extend the index set by requiring $W_{i+k\cdot r} = u^k W_i$ for $k \in \mathbb{Z}$. Denote by $\iota_i : W_{i+1} \to W_i$ the inclusion. We have a natural identification $\mathrm{GL}(W_i) = \mathrm{GL}(W_{i+r})$ given by conjugating by u.

The dilatation $\mathcal{G} = \mathrm{GL}(W)_Q$ of $\mathrm{GL}(W)$ along Q is isomorphic to the closed subgroup scheme in $H = \prod_{i=0}^{r-1} \mathrm{GL}(W_i) = \prod_{i\in\mathbb{Z}/r\mathbb{Z}} \mathrm{GL}(W_i)$ of tuples that commute with the maps $\iota_i : W_i \to W_{i+1}$. It is isomorphic to the group scheme over $\mathcal{O}[u]$ obtained by Theorem 2.1 applied to $G = \mathrm{GL}_N$ and the parahoric $\mathcal{P}$ corresponding to the lattice chain $\{\mathcal{L}_i := W_i \otimes_{\mathcal{O}[u], u\mapsto\varpi} \mathcal{O}\}_i$.

Consider the functor $\mathfrak{L}$ on (Sch/X) which to an X-scheme $y : S \to X$, associates the set of isomorphism classes of collections $(\mathcal{E}_i, \psi_i, \alpha_i)_{i\in\mathbb{Z}}$ where, for each $i \in \mathbb{Z}$, $\mathcal{E}_i$ are locally free coherent $\mathcal{O}_{X\times S}$-sheaves on $X \times S$ of rank N, $\psi_i : \mathcal{E}_{i+1} \to \mathcal{E}_i$ are $\mathcal{O}_{X\times S}$-module homomorphisms, and α_i are $\mathcal{O}_{X\times S-\Gamma_y}$-module isomorphisms $\alpha_i : W_i \otimes_{\mathcal{O}[u]} \mathcal{O}_{X\times S-\Gamma_y} \xrightarrow{\sim} \mathcal{E}_i \otimes_{\mathcal{O}_{X\times S}} \mathcal{O}_{X\times S-\Gamma_y}$ that satisfy the following conditions:

(a) the data are periodic of period r, $(\mathcal{E}_i, \psi_i, \alpha_i) = (\mathcal{E}_{i+r}, \psi_{i+r}, \alpha_{i+r})$, for all $i \in \mathbb{Z}$,

(b) we have $\alpha_{i+1} \cdot \psi_i = \alpha_i \cdot \iota_i$, for all $i \in \mathbb{Z}$,

(c) Each composition of r successive ψ_i is given by multiplication by u, i.e

$$\prod_{k=0}^{r-1} \psi_{i-k} = u : \mathcal{E}_{i+r} = \mathcal{E}_i \to \mathcal{E}_i,$$

for all $i \in \mathbb{Z}$, and,

(d) for each $i \in \mathbb{Z}$, the cokernel $\mathcal{E}_i/\psi_i(\mathcal{E}_{i+1})$ is a locally free $\mathcal{O}_S$-module of rank r_i.

We can see that $\mathfrak{L}$ is an fpqc sheaf on (Sch/X).

When $S = \mathrm{Spec}(R)$ is affine, and $y : \mathrm{Spec}(R) \to X = \mathrm{Spec}(\mathcal{O}[u])$ is given by $u \mapsto y$, we have $X \times S - \Gamma_y = \mathrm{Spec}(R[u][(u-y)^{-1}])$. Since $u - y$ is not a zero divisor in $R[u]$ we can use α_i to identify $\mathcal{E}_i$ with the sheaf corresponding to a $R[u]$-locally free rank N submodule E_i of $R[u][(u-y)^{-1}]^N$.

One can show that

Proposition 2.6. *There is a natural equivalence of functors* $\mathrm{Gr}_{\mathcal{G},X} \xrightarrow{\sim} \mathfrak{L}$ *where* $\mathcal{G}$ *is the group scheme as above.*

Now if $\mu : \mathbb{G}_{\mathrm{m}} \to \mathrm{GL}_N$ is the minuscule coweight given by

$$a \mapsto \mathrm{diag}(a^{(d)}, 1^{(N-d)})$$

and $\mathcal{O} = W(\mathbb{F}_p)$, we can see, using Proposition 2.6, that in this situation, the local models $M_{\mathcal{G},\mu}$ agree with the Rapoport-Zink local models for GL_N and μ considered in [RZ]. Indeed, in this case, by [Go1], the local models of [RZ] are flat over $\mathcal{O}$ and so they agree with the $M_{\mathcal{G},\mu}$ above.

The case of GSp_{2n}. Consider $W = \oplus_{i=1}^{2n}\mathcal{O}[u] \cdot e_i$ with the perfect $\mathcal{O}[u]$-bilinear alternating form $h : W \times W \to \mathcal{O}[u]$ determined by $h(e_i, e_{2n+1-j}) = \delta_{ij}$, $h(e_i, e_j) = h(e_{2n+1-i}, e_{2n+1-j}) = 0$ for $1 \le i, j \le n$. Let us fix a chain of $\mathcal{O}[u]$-submodules

$$uW \subset W_{r-1} \subset \cdots \subset W_1 \subset W_0 \subset W \tag{2.4}$$

such that

(i) $W_i^\vee = u^{-1}W_{r-i-1}$, for $0 \le i \le r-1$,
(ii) $W_i/W_{i+1} \simeq \mathcal{O}^{d_i}$.

Again, extend the index set by periodicity by setting $W_{i+k\cdot r} = u^k W_i$ so that the form h gives $W_i^\vee = W_{-i-a}$, with $a = 0$ or 1, and set $\iota_i : W_{i+1} \to W_i$ as before. Consider the group scheme $\mathcal{G}$ over $\mathcal{O}[u]$ of similitude automorphisms of the "polarized" system $(W_i, h_i)_{i\in\mathbb{Z}}$; more precisely this is the subgroup scheme of $\mathbb{G}_{\mathrm{m}} \times \prod_{i=0}^{r-1} \mathrm{GL}(W_i) = \mathbb{G}_{\mathrm{m}} \times \prod_{i\in\mathbb{Z}/r\mathbb{Z}} \mathrm{GL}(W_i)$ consisting of $(c, (g_i))$ such that

$$h(g_i(x), g_{-i-a}(y)) = c\cdot h(x,y), \quad \text{for all } i\in\mathbb{Z}.$$

As in [RZ, Appendix to Chapter 3], we can see that $\mathcal{G}$ is smooth over $\mathcal{O}[u]$; it is isomorphic to the group scheme obtained by Theorem 2.1 applied to $G = \mathrm{GSp}_{2n}$ and the parahoric $\mathcal{P}$ corresponding to the *self-dual* lattice chain

$$\{\mathcal{L}_i := W_i \otimes_{\mathcal{O}[u], u\mapsto\varpi} \mathcal{O}\}_i.$$

Consider the functor $\mathfrak{LSP}$ on (Sch/X) which to an X-scheme $y: S\to X$, associates the set of isomorphism classes of collections $(\mathcal{E}_i, \psi_i, \alpha_i, h_i)_{i\in\mathbb{Z}}$ where $(\mathcal{E}_i, \psi_i, \alpha_i)_{i\in\mathbb{Z}}$ give an object of $\mathfrak{L}$ and in addition

$$h_i : \mathcal{E}_i \times \mathcal{E}_{-i-a} \to \mathcal{O}_{X\times S}$$

are perfect $\mathcal{O}_{X\times S}$-bilinear forms that satisfy

(a) $h_i(\psi_i(x), y) = h_{i+1}(x, \psi_{-i-1-a}(y))$, for $x\in\mathcal{E}_{i+1}$, $y\in\mathcal{E}_{-i-a}$.

(b) There is $c\in\mathcal{O}^*_{X\times S-\Gamma_y}$, such that $h_i\cdot(\alpha_i, \alpha_{-i-a}) = c\cdot h$ for all $i\in\mathbb{Z}$ (as forms $W_i\otimes\mathcal{O}_{X\times S-\Gamma_y}\times W_{-i-a}\otimes\mathcal{O}_{X\times S-\Gamma_y}\to\mathcal{O}_{X\times S-\Gamma_y}$).

A similar argument as above, (cf. [RZ, Appendix to Chapter 3, Proposition A. 21]) now gives

Proposition 2.7. *There is a natural equivalence of functors* $\mathrm{Gr}_{\mathcal{G},X} \xrightarrow{\sim} \mathfrak{LSP}$ *where* $\mathcal{G}$ *is the (symplectic) group scheme as above.*

Again as a result of the above, combined with the flatness result of [Go2], we can see that if $\mu: \mathbb{G}_{\mathrm{m}} \to \mathrm{GSp}_{2n}$ is the standard minuscule coweight given by $a\mapsto \mathrm{diag}(a^{(n)}, 1^{(n)})$ and $\mathcal{O} = W(\mathbb{F}_p)$, then the local models $M_{\mathcal{G},\mu}$ in this situation agree with the local models for GSp_{2n} considered in [RZ].

Remark 2.8. One can find a similar interpretation of $\mathrm{Gr}_{\mathcal{G},X}$ as moduli spaces of chains of bundles with additional structure given by suitable forms in more cases, for example when G is an orthogonal group or a (ramified) unitary group. We will leave the details to the reader. A corresponding statement comparing the local models $M_{\mathcal{G},\mu}$ with the local models in the theory of PEL Shimura varieties [RZ, Go1, Go2, PR1, PR2, PR4, PRS] will be explained in later section.

3 Properties of local models

In this section, we will discuss some basic properties we know about the local models.

3.a

We keep to notations and assumptions as in the previous section. The following theorem summarizes the main geometric properties of local models.

Theorem 3.1. *Suppose that* $p \nmid |\pi_1(G_{\rm der})|$. *Then the scheme* $M_{\mathcal{G},\mu}$ *is normal. In addition, the special fiber* $\overline{M}_{\mathcal{G},\mu}$ *is reduced, and each geometric irreducible component of* $\overline{M}_{\mathcal{G},\mu}$ *is normal, Cohen-Macaulay and Frobenius split.*

Remark 3.2. It is conjectured that the whole scheme $\overline{M}_{\mathcal{G},\mu}$ is Cohen-Macaulay (which is equivalent to the Cohen-Macaulayness of $M_{\mathcal{G},\mu}$). Currently, only a few cases of this conjecture are known: (i) If G is general and $\mathcal{P}$ is a *very* special parahoric [Z1, PZ1]. We refer to [Z2, §6] (or §4) for the definition of very special parahorics. (ii) If G is unramified and $\mathcal{P}$ is an Iwahori group scheme (and more generally a parahoric that admits a map to a hyperspecial parahoric), and μ is minuscule [He].

The above theorem is an easy consequence of Theorem 3.3 below and the general theory of affine Schubert varieties developed in [Fa, PR3]. First, note that by definition

$$\overline{M}_{\mathcal{G},\mu} \otimes_{k_E} \bar{k} \subset {\rm Gr}_{P_{\bar{k}}} = LP_{\bar{k}}/L^+P_{\bar{k}}$$

is a closed subscheme of the affine flag variety ${\rm Gr}_{P_{\bar{k}}}$. It is not difficult to see that $\overline{M}_{\mathcal{G},\mu} \otimes_{k_E} \bar{k}$ is invariant under the left translation by $L^+P_{\bar{k}}$. Therefore, the reduced subscheme of $\overline{M}_{\mathcal{G},\mu} \otimes_{k_E} \bar{k}$ is a union of affine Schubert varieties. To describe which affine Schubert varieties are contained in $\overline{M}_{\mathcal{G},\mu} \otimes_{k_E} \bar{k}$, we need to introduce the μ-admissible set.

We write $F' = \bar{k}((u))$, $\mathcal{O}' = \bar{k}[[u]]$, and let $G' = \mathcal{G} \times_X \bar{k}((u)) = G^\flat_{F'}$ be the connected reductive group over F', which splits over a tamely ramified extension $\tilde{F}'$ of F'.

Let S' be a maximal F'-split torus of G' and $T' = Z_{G'}(S')$ a maximal torus. Let $I = {\rm Gal}(\tilde{F}'/F')$. We assume that $S' \subset G'$ extends to a closed embedding of the Néron model of S' into the parahoric group scheme $P_{\bar{k}}$. The orbits of the $L^+P_{\bar{k}}$-action on ${\rm Gr}_{P_{\bar{k}}}$ then are parameterized by $W^{P_{\bar{k}}}\backslash\widetilde{W}'/W^{P_{\bar{k}}}$, where $\widetilde{W}'$ is the Iwahori-Weyl group of G' and $W^{P_{\bar{k}}}$ is the Weyl group of $P_{\bar{k}} \otimes k$. For $w \in W^{P_{\bar{k}}} \backslash \widetilde{W}'/W^{P_{\bar{k}}}$, let $S_w^{P_{\bar{k}}} \subset {\rm Gr}_{P_{\bar{k}}}$ denote the corresponding Schubert variety, i.e. the closure of the $L^+P_{\bar{k}}$-orbit through w.

Let us also recall the structure of the Iwahori-Weyl group $\widetilde{W}'$ of G'. It is defined via the exact sequence

$$1 \to T'(\mathcal{O}') \to N'(F') \to \widetilde{W}' \to 1 \tag{3.1}$$

(where $T'(\mathcal{O}')$ is the group of $\mathcal{O}'$-valued points of the unique Iwahori group scheme for the torus T') and acts on $\mathcal{A}(G', S', F')$ via affine transformations. In addition, there is a short exact sequence

$$1 \to \mathbb{X}_\bullet(T')_I \to \widetilde{W}' \to W_0' \to 1, \tag{3.2}$$

where $W_0' = N'(F')/T'(F')$ is the relative Weyl group of G' over F'. In what follows, we use t_λ to denote the translation element in $\widetilde{W}'$ given by $\lambda \in \mathbb{X}_\bullet(T')_I$ from the above map (3.2)[1]. But occasionally, if no confusion is likely to arise, we will also use λ itself to denote this translation element. A choice of a special vertex v of $\mathcal{A}(G', S', F')$ gives a splitting of the above exact sequence and then we can write $w = t_\lambda w_f$ for $\lambda \in \mathbb{X}_\bullet(T')_I$ and $w_f \in W_0'$.

Let us choose a rational Borel subgroup B' of G' containing T'. This determines a set of positive roots $\Phi^+ = \Phi(G', S')^+$ for G'. There is a natural map $\mathbb{X}_\bullet(T')_I \to \mathbb{X}_\bullet(S')_\mathbb{R}$. We define

$$\mathbb{X}_\bullet(T')_I^+ = \{\lambda \mid (\lambda, a) \geq 0 \text{ for } a \in \Phi^+\}. \tag{3.3}$$

Observe that the chosen special vertex v and the rational Borel B' determine a unique alcove C in $\mathcal{A}(G', S', F')$. Namely, we identify $\mathcal{A}(G', S', F')$ with $\mathbb{X}_\bullet(S')_\mathbb{R}$ by v and then C is the unique alcove whose closure contains v, and is contained in the finite Weyl chamber determined by B'.

Let W'_{aff} be the affine Weyl group of G', i.e. the Iwahori-Weyl group of G'_{sc}, the simply-connected cover of G'_{der}. This is a Coxeter group. One has

$$1 \to \mathbb{X}_\bullet(T'_{\mathrm{sc}})_I \to W'_{\mathrm{aff}} \to W_0' \to 1, \tag{3.4}$$

where T'_{sc} is the inverse image of T' in G'_{sc}. One can write $\widetilde{W}' = W'_{\mathrm{aff}} \rtimes \Omega'$, where Ω' is the subgroup of $\widetilde{W}'$ that fixes the chosen alcove C. This gives $\widetilde{W}'$ a quasi Coxeter group structure; it makes sense to talk about the length of an element $w \in \widetilde{W}'$ and there is a Bruhat order on $\widetilde{W}'$. Namely, if we write $w_1 = w_1'\tau_1, w_2 = w_2'\tau_2$ with $w_i' \in W'_{\mathrm{aff}}, \tau_i \in \Omega'$, then $\ell(w_i) = \ell(w_i')$ and $w_1 \leq w_2$ if and only if $\tau_1 = \tau_2$ and $w_1' \leq w_2'$.

Now let us recall the definition of the *μ-admissible set* in the Iwahori-Weyl group. We continue with the above notations. Let $\bar{W}'$ be the absolute Weyl group of G', i.e. the Weyl group for $(G'_{\tilde{F}'}, T'_{\tilde{F}'})$. Let $\mu : (\mathbb{G}_m)_{\tilde{F}'} \to G' \otimes_{F'} \tilde{F}'$ be a geometrical conjugacy class of 1-parameter subgroups. It determines a $\bar{W}'$-orbit in $\mathbb{X}_\bullet(T')$. One can associate to μ a W_0'-orbit Λ in $\mathbb{X}_\bullet(T')_I$ as follows. Choose a Borel subgroup of G' containing T', and defined over F'. This gives a unique element in this $\bar{W}'$-orbit, still denoted by μ, which is dominant with respect to this Borel subgroup. Let $\bar{\mu}$ be its image in $\mathbb{X}_\bullet(T')_I$, and let $\Lambda = W_0'\bar{\mu}$. It turns out that Λ does not depend on the choice of the rational Borel subgroup of G', since any two such F'-rational Borel subgroups that contain T' are conjugate to each other by an element in W_0'. For $\mu \in \mathbb{X}_\bullet(T')$, define the admissible set

$$\mathrm{Adm}(\mu) = \{w \in \widetilde{W}' \mid w \leq t_\lambda, \text{ for some } \lambda \in \Lambda\}, \tag{3.5}$$

and more generally,

$$\mathrm{Adm}^{P_{\bar{k}}}(\mu) = W^{P_{\bar{k}}} \mathrm{Adm}(\mu) W^{P_{\bar{k}}}. \tag{3.6}$$

---

[1]Note that under the sign convention of the Kottwitz homomorphism in [Ko2], t_λ acts on $\mathcal{A}(G', S', F')$ by $v \mapsto v - \lambda$.

If $\mathrm{char}\, k \nmid |\pi_1(G'_{\mathrm{der}})|$, let us define a reduced closed subvariety of $\mathrm{Gr}_{P_{\bar{k}}}$ whose underlying set is given by

$$\mathcal{A}^{P_{\bar{k}}}(\mu) = \bigcup_{w \in \mathrm{Adm}(\mu)} L^+P_{\bar{k}} w L^+P_{\bar{k}}/L^+P_{\bar{k}} = \bigcup_{w \in W^{P_{\bar{k}}} \backslash \mathrm{Adm}^{P_{\bar{k}}}(\mu)/W^{P_{\bar{k}}}} S_w^{P_{\bar{k}}}.$$

There is an isomorphism between the Iwahori-Weyl groups $\widetilde{W}$ of $G \otimes_F \breve{F}$ and $\widetilde{W}'$ of G'. The constructions above can also be applied to $G \otimes_F \breve{F}$ and $\{\mu\}$ to produce a W_0-orbit Λ in $\mathbb{X}_\bullet(T)_I \subset \widetilde{W}$ and a subset $\mathrm{Adm}(\mu) \subset \widetilde{W}$. Using this identification, we will view Λ and $\mathrm{Adm}(\mu)$ also as, respectively, a W'_0-orbit in $\mathbb{X}_\bullet(T')_I \subset \widetilde{W}'$, and a subset $\mathrm{Adm}(\mu) \subset \widetilde{W}'$.

Theorem 3.3. *Suppose that $p \nmid |\pi_1(G_{\mathrm{der}})|$. Then we have*

$$\mathcal{A}^{P_{\bar{k}}}(\mu) = \overline{M}_{\mathcal{G},\mu} \otimes_{k_E} \bar{k}$$

as closed subschemes of $\mathrm{Gr}_{P_{\bar{k}}}$.

According to [Fa, PR3], if $\mathrm{char}\, k \nmid |\pi_1(G'_{\mathrm{der}})|$, where G'_{der} is the derived group of G', then $S_w^{P_{\bar{k}}}$ is normal, has rational singularities and is Frobenius-split if $\mathrm{char}\, k > 0$. These facts, together with Theorem 3.3 then imply Theorem 3.1.

Observe that this theorem will follow from: (i) $(\overline{M}_{\mathcal{G},\mu} \otimes_{k_E} \bar{k})_{red} = \mathcal{A}^{P_{\bar{k}}}(\mu)$; (ii) $\overline{M}_{\mathcal{G},\mu} \otimes_{k_E} \bar{k}$ is reduced. The first step (i) is a conjecture of Kottwitz-Rapoport, and its proof is based on the calculation of the sheaf of nearby cycles on $M_{\mathcal{G},\mu}$ (see Remark 4.4). Assuming (i), we now explain the second step (ii). The key ingredient is the coherence conjecture of Pappas and Rapoport [PR3] which we now explain.

3.b

We will just state a special form of the coherence conjecture (which turns out to be equivalent to the more general forms stated in [PR3, Z1]).

We continue to assume that $p \nmid |\pi_1(G_{\mathrm{der}})|$. In addition, we assume that G_{ad} is absolutely simple and $\mathcal{P}$ is an Iwahori group scheme of G. For every $w \in \widetilde{W}'$ with $\ell(w) = 1$, the corresponding affine Schubert variety $S_w^{P_{\bar{k}}}$ is isomorphic to $\mathbb{P}^1$. It is known (by combining [PR3, §6, §10]) that there is a unique line bundle $\mathcal{L}_s$ on the reduced subscheme $(\mathrm{Gr}_{P_{\bar{k}}})_{red}$ of $\mathrm{Gr}_{P_{\bar{k}}}$, such that its restriction to each of these $S_w^{P_{\bar{k}}}$ is isomorphic to $\mathcal{O}(1)$. On the other hand, all connected components of Gr_H are isomorphic to each other and let $\mathcal{L}_b$ be the line bundle that restricts to the ample generator of the Picard group of each connected component. Recall that $X_\mu \otimes \bar{F} \subset \mathrm{Gr}_{\mathcal{G}_{\bar{F},\varpi}} \simeq \mathrm{Gr}_H \otimes \bar{F}$ and therefore $\mathcal{L}_b$ restricts to a line bundle on X_μ.

Remark 3.4. If μ is a minuscule cocharacter, then $X_\mu \otimes \bar{F}$ is a homogenous space of $G_{\bar{F}}$, and $\mathcal{L}_b$ is just the ample generator of its Picard group.

The Coherence Conjecture. *For every $n \geq 0$,*

$$\dim \Gamma(\mathcal{A}^{P_{\bar{k}}}(\mu), \mathcal{L}_s^n) = \dim \Gamma(X_\mu \otimes \bar{F}, \mathcal{L}_b^{nh^\vee}),$$

where $h^\vee$ is the dual Coxeter number of H.

Let us briefly explain why the coherence conjecture implies Theorem 3.3. First, it is not difficult to see that $\mathcal{A}^{P_{\bar{k}}} \subset M_{\mathcal{G},\mu} \otimes k_E$. Second, there exists an ample line bundle $\mathcal{L}$ on $M_{\mathcal{G},\mu}$ such that $\mathcal{L} \otimes k_E \simeq \mathcal{L}_s$ and $\mathcal{L} \otimes \bar{F} \simeq \mathcal{L}_b^{h^\vee}$. Then Theorem 3.3 follows from

$$\dim \Gamma(\mathcal{A}^{P_{\bar{k}}}, \mathcal{L}_s^n) = \dim \Gamma(X_\mu \otimes \bar{F}, \mathcal{L}_b^{nh^\vee}) = \dim \Gamma(M_{\mathcal{G},\mu} \otimes k_E, \mathcal{L}_s^n).$$

Next, we give some ideas for the proof of the coherence conjecture, following [Z1]. The first observation is that the local models $M_{\mathcal{G},\mu}$ admits equal characteristic counterparts, which we denoted by $M^\flat_{\mathcal{G},\mu}$. (In [Z1], they were denoted by $\overline{\mathrm{Gr}}_{\mathcal{G},\mu}$ and were called global Schubert varieties.) Namely, $\{\mu\}$ can be also regarded as a geometric conjugacy class of one parameter subgroups of $G^\flat$. Let $E^\flat$ be its field of definition, which is a finite extension of $k((u))$. Then by the similar construction, it determines a subvariety $X^\flat_\mu \subset \mathrm{Gr}_{\mathcal{G},X} \times_X E^\flat$, and $M^\flat_{\mathcal{G},\mu}$ is its closure in $\mathrm{Gr}_{\mathcal{G},X}$. By the similar argument, the coherence conjecture will imply

Theorem 3.5. *Suppose that $p \nmid |\pi_1(G_{\mathrm{der}})|$. Then we have*

$$\mathcal{A}^{P_{\bar{k}}}(\mu) = \overline{M}^\flat_{\mathcal{G},\mu} \otimes_{k_{E^\flat}} \bar{k}$$

as closed subschemes of $\mathrm{Gr}_{P_{\bar{k}}}$.

But in fact, with some extra work, this above theorem is equivalent to the coherence conjecture [Z1, §4]. So it is enough to prove the above theorem.

At the first sight, it seems that the idea is circular. However, it is not the case. The reason is that $M^\flat_{\mathcal{G},\mu}$ now is a scheme over k and we have many more tools to attack the problem. Again the theorem is equivalent to show that: (i) $(\overline{M}^\flat_{\mathcal{G},\mu} \otimes_{k_{E^\flat}} \bar{k})_{red} = \mathcal{A}^{P_{\bar{k}}}(\mu)$; (ii) $\overline{M}^\flat_{\mathcal{G},\mu} \otimes_{k_{E^\flat}} \bar{k}$ is reduced.

Part (i) is essentially a conjecture of Kottwitz-Rapoport, and can be achieved by the calculation of the nearby cycle $\mathcal{Z}_\mu = R\Psi^{M^\flat_{\mathcal{G},\mu}}(\mathrm{IC})$ of the intersection cohomology sheaf of X_μ. We refer to §4.b (in particular Remark 4.4) for more detailed discussion. We turn to Part (ii), which is more difficult. Here is the first advantage for us to prove Theorem 3.5 rather than Theorem 3.3: As $M^\flat_{\mathcal{G},\mu}$ is defined over $\mathrm{char}\, k > 0$, we can use the powerful technique of Frobenius splitting [MR, BK]. To prove that $\overline{M}^\flat_{\mathcal{G},\mu} \otimes \bar{k}$ is reduced, it is enough to prove that it is Frobenius split. To achieve this goal, we embed $M^\flat_{\mathcal{G},\mu}$ into a larger scheme $M^\flat_{\mathcal{G},\mu,\lambda}$ over $\bar{k}[[u]]$, which is a closed subscheme of a version of the Beilinson-Drinfeld Grassmannian. The scheme $M^\flat_{\mathcal{G},\mu,\lambda}$ is normal and its special fiber is reduced. Then to prove that

$$\overline{M}^\flat_{\mathcal{G},\mu} \otimes \bar{k} = M^\flat_{\mathcal{G},\mu} \cap \overline{M}^\flat_{\mathcal{G},\mu,\lambda} \otimes \bar{k}$$

is Frobenius split, it is enough to construct a Frobenius splitting of $M^\flat_{\mathcal{G},\mu,\lambda}$, compatible with $M^\flat_{\mathcal{G},\mu}$ and with $\overline{M}^\flat_{\mathcal{G},\mu,\lambda} \otimes \bar{k}$. Since $M^\flat_{\mathcal{G},\mu,\lambda}$ is normal, it is enough to prove this for some nice open subscheme $U \subset M^\flat_{\mathcal{G},\mu,\lambda}$, such that $M^\flat_{\mathcal{G},\mu,\lambda} - U$

has codimension two. In particular, the open subscheme U will not intersect with $\overline{M}^{\flat}_{\mathcal{G},\mu}\otimes\bar{k}$, which is our primary interest. The second advantage of the equal characteristic version is as follows: The schemes $M^{\flat}_{\mathcal{G},\mu}$ and $M^{\flat}_{\mathcal{G},\mu,\lambda}$ can be naturally extended to $\mathbb{A}^1_{\bar{k}}=\mathrm{Spec}(\bar{k}[u])$ and can be even extended to $\mathbb{P}^1_{\bar{k}}$. This allows us to use the moduli stack of bundles on $\mathbb{P}^1_{\bar{k}}$ to construct a Frobenius splitting of U. We refer to [Z1] for details.

4 Applications to Shimura varieties

4.a Integral model of Shimura varieties

The direct application of local models is to study the local structure of the integral models of Shimura varieties.

Let (G,X) be a Shimura datum, i.e. G is a connected reductive group over $\mathbb{Q}$, and X is a $G_{\mathbb{R}}$ conjugacy class of certain homomorphisms $h:\mathbb{C}^\times\to G_{\mathbb{R}}$ satisfying Deligne's axioms:

(i) The adjoint representation $\mathrm{ad}\,h:\mathbb{C}^\times\to\mathrm{GL}(\mathfrak{g}_{\mathbb{R}})$ defines a rational Hodge structure of type $(-1,1),(0,0),(1,-1)$ on the Lie algebra $\mathfrak{g}$ of G;

(ii) $\mathrm{ad}\,h(i)$ induces a Cartan involution on G_{ad};

(iii) G_{ad} does not have a $\mathbb{Q}$-factor whose real points form a compact Lie group.

Attached to a Shimura datum, there is a number field $E\subset\mathbb{C}$, called the "reflex field", which is the field of definition of the geometric conjugacy class of one parameter subgroups

$$\mu_h:\mathbb{C}^\times\stackrel{(z,1)}{\to}\mathbb{C}^\times\times\mathbb{C}^\times\stackrel{h_{\mathbb{C}}}{\to}G_{\mathbb{C}}.$$

Let K a compact open subgroup of $G(\mathbb{A}_f)$, small enough, where $\mathbb{A}_f$ is the ring of finite adèle. Attaching to these data, there is a quasi projective variety $\mathrm{Sh}_K(G,X)$ defined over E, whose set of complex points are

$$\mathrm{Sh}_K(G,X)=G(\mathbb{Q})\backslash X\times G(\mathbb{A}_f)/K.$$

For example, let (V,ψ) be a non-degenerate symplectic space over $\mathbb{Q}$. Then one can define the symplectic similitudes group $\mathrm{GSp}(V,\psi)$ attached to (V,ψ), and the Siegel double space $S^{\pm}$ of homomorphisms $h:\mathbb{C}^\times\to\mathrm{GSp}(V,\psi)_{\mathbb{R}}$ such that $(V,h,\pm\psi)$ form a polarized rational Hodge structure of type $(-1,0)$ and $(0,-1)$. The pair $(\mathrm{GSp}(V,\psi),S^{\pm})$ is a Shimura datum, and $\mathrm{Sh}_K(G,X)$, sometimes called a Siegel moduli space (or a Mumford space), is a quasi-projective variety over $\mathbb{Q}$ classifying principally polarized abelian varieties with K-level structures. Recall that a Shimura datum (G,X) is said of Hodge type if there is an embedding $(G,X)\to(\mathrm{GSp}(V,\psi),S^{\pm})$.

For many arithmetic applications, it is desirable to have a reasonable extension of $\mathrm{Sh}_K(G,X)$ as a scheme over the ring of integers $\mathcal{O}_E$. So fix a prime v of E and let $p\mid v$. Write $K=K_pK^p\subset G(\mathbb{Q}_p)G(\mathbb{A}_f^p)$. Roughly speaking, the larger the group K_p, the better behavior of $\mathrm{Sh}_K(G,X)$ at v. For most p, one can choose K_p to be a hyperspecial subgroup of $G(\mathbb{Q}_p)$. For any p, one can choose K_p to be a parahoric subgroup, i.e. $K_p = \mathcal{P}(\mathbb{Z}_p)$, where $\mathcal{P}$ is a parahoric group scheme of $G_{\mathbb{Q}_p}$. We fix such a K_p and let K^p to be small enough.

The following theorem is the combination of the work of [PZ1] and the forthcoming work [KP].

Theorem 4.1. *Let (G,X) be a Shimura datum of Hodge type. Assume that G is connected and is tamely ramified at $p > 2$. Let $K = K_pK^p \subset G(\mathbb{A}_f)$ be as above. Then there is an integral model $\mathcal{S}_K$ of $\mathrm{Sh}_K(G,X)$, defined over $\mathcal{O}_{E,(v)}$ such that*

(a) *the inclusion $G \to \mathrm{GSp}(V)$ induces a finite morphism from $\mathcal{S}_K$ to the corresponding integral Siegel moduli scheme with parahoric level structures.*

(b) *The special fiber of $\mathcal{S}_K$ is reduced. Its geometrically irreducible components are normal, Cohen-Macaulay, and have rational singularities.*

If the Shimura datum is of PEL-type, the integral model $\mathcal{S}_K$ can be defined as the flat closure of the generic fiber of the moduli scheme defined in [RZ] (the naive model). In particular when $(G,X) = (\mathrm{GSp}(V,\psi), S^{\pm})$ and K_p is given by a self-dual lattice chain $\{\mathcal{L}_i\}_i$ as in §2.c, by the work of [Go2], $\mathcal{S}_K$ coincides with the naive model, which classifies principally polarized (in appropriate sense) chains of abelian varieties, with level structures K^p (see [RZ, Ha1] for the precise definition).

We call $\mathcal{S}_K$ an integral Siegel moduli scheme. If (G,X) is a general Shimura datum of Hodge type, one first embeds $\mathrm{Sh}_K(G,X)$ into an appropriate Siegel moduli space, and then defines $\mathcal{S}_K$ as the normalization of its flat closure in the corresponding integral Siegel moduli scheme.

Now let E_v be the completion of E at v. Then E_v is the local reflex field associated to the triple $(G_{\mathbb{Q}_p}, K_p, \{\mu_h\})$. Let $\mathrm{M}(G,\{\mu\})_K := M_{\mathcal{G},\mu}$ be the local model constructed from this triple as in §2. To prove the second point of the theorem, one constructs the local model diagram

$$\begin{array}{ccccc} & & \widetilde{\mathcal{S}}_K & & \\ & \stackrel{\pi}{\swarrow} & & \stackrel{\widetilde{\varphi}}{\searrow} & \\ \mathcal{S}_K \otimes \mathcal{O}_{E_v} & & & & \mathrm{M}(G,\{\mu\})_K \end{array} \tag{4.1}$$

of $\mathcal{O}_{E_v}$-schemes, in which:

- π is a torsor under the group $\mathcal{P}_{\mathcal{O}_{E_v}} = \mathcal{P} \otimes_{\mathbb{Z}_p} \mathcal{O}_{E_v}$.
- $\widetilde{\varphi}$ is $\mathcal{P}_{\mathcal{O}_{E_v}}$-equivariant and smooth of relative dimension $\dim G$.

Clearly, Theorem 4.1 will follow from Theorem 3.1 and the existence of such diagram. We refer to [RZ, PZ1, KP] for the construction of this diagram in various generalities.

4.b Sheaf of nearby cycles and Kottwitz' conjecture

The Langlands-Kottwitz method is a way to understand the cohomology of $\mathrm{Sh}_K(G, X)$ by comparing the Lefschetz trace formula and the Arthur-Selberg trace formula. The starting point is the expression

$$\mathrm{tr}^{\mathrm{ss}}(f\Phi^r \mid \mathrm{H}^*(\mathrm{Sh}_K(G,X))) = \sum_{\substack{(\gamma_0,\gamma,\delta)\\ \alpha(\gamma_0,\gamma,\delta)=1}} c(\gamma_0,\gamma,\delta)\mathrm{O}_\gamma(f)\mathrm{TO}_\delta(z_\mu),$$

where

- Φ is a lifting of the r-Frobenius element to the local Galois group $\mathrm{Gal}(\overline{\mathbb{Q}}_p/E_v)$.
- $f^p \in C_c^\infty(G(\mathbb{A}_f^p))$ is a test function away from p.
- z_μ is certain function on $G(\mathbb{Q}_{p^r})$.
- The sum is taken over Kottwitz triples $(\gamma_0, \gamma, \delta)$.
- $\mathrm{O}_\gamma(f)$ is the orbital integral of f, $\mathrm{TO}_\delta(z_\mu)$ is the twisted orbital integral of z_μ and $c(\gamma_0, \gamma, \delta)$ is certain volume factor.
- $\mathrm{tr}^{\mathrm{ss}}$ is the semisimple trace.

We refer to [R, HN, Ha1, Ha2] for the explanation of undefined terms and for more details. The most mysterious part of the above formula is the function z_μ. In this case K_p is an Iwahori subgroup (and G is unramified), Kottwitz conjectured an explicit form of z_μ. Let $Z(H_{K_p})$ be center of the Iwahori Hecke algebra, Sph be the spherical Hecke algebra, and $\mathrm{Rep}(^LG)$ the representation ring of the Langlands dual group ^LG. Recall that there are canonical isomorphisms

$$Z(H_{K_p}) \simeq \mathrm{Sph} \simeq \mathrm{Rep}(^LG),$$

where the first isomorphism is the Bernstein isomorphism, and the second isomorphism is the Satake isomorphism. Then Kottwitz conjectured that:

Conjecture.

(a) z_μ can be chosen in $Z(H_{K_p})$.

(b) For an explicit constant c_μ, $c_\mu z_\mu$ corresponds to the irreducible representation of the Langlands dual group ^LG of highest weight μ under the above isomorphism.

Remark 4.2. Recently, Haines [Ha2] gave a far reaching generalization of this conjecture, allowing arbitrary groups and arbitrary level structures.

One possible way to verify the above conjecture is to calculate the sheaves of nearby cycles of integral models $\mathcal{S}_K$ considered above (we refer to [Il, HN] for the background of nearby cycles). As observed by Rapoport, the sheaf of nearby cycles of $\mathcal{S}_K$ can be understood by the sheaf of nearby cycles of the corresponding local model. Recall that the local model diagram can be rewritten as a smooth morphism of algebraic stacks

$$\varphi : \mathcal{S}_K \otimes \mathcal{O}_{E_v} \to [\mathcal{P}_{\mathcal{O}_{E_v}} \backslash \mathrm{M}(G,\mu)_K].$$

Therefore,

$$R\Psi^{\mathcal{S}_K \otimes \mathcal{O}_{E_v}} = \varphi^* R\Psi^{[\mathcal{P}_{\mathcal{O}_{E_v}} \backslash \mathrm{M}(G,\mu)_K]}.$$

It is enough to understand $R\Psi^{\mathrm{M}(G,\mu)_K}$ as a $(\mathcal{P} \otimes k_{E_v})$-equivariant perverse sheaf. More generally, we would like to understand $R\Psi^{M_{\mathcal{G},\mu}} := R\Psi^{M_{\mathcal{G},\mu}}(\mathrm{IC}_\mu)$ as an $(L^+\mathcal{P} \otimes \bar{k})$-equivariant perverse sheaf on $\mathrm{Gr}_{P_{\bar{k}}}$, where IC_μ is the intersection cohomology sheaf of X_μ. Recall that given two $(L^+\mathcal{P} \otimes \bar{k})$-equivariant ℓ-adic complexes $\mathcal{F}_1$ and $\mathcal{F}_2$, Lusztig define the convolution $\mathcal{F}_1 \star \mathcal{F}_2$ (e.g. see [Ga, Z1, PZ1] for the definition), which is again an $(L^+\mathcal{P} \otimes \bar{k})$-equivariant complex on $\mathrm{Gr}_{P_{\bar{k}}}$. The main theorem is

Theorem 4.3. *For $w \in \widetilde{W}$, let $\mathrm{IC}_{w,\bar{k}}$ denote the intersection cohomology sheaf of $S_w^{P_{\bar{k}}}$. Then there is a canonical isomorphism $\mathrm{IC}_{w,\bar{k}} \star R\Psi^{M_{\mathcal{G},\mu}} \simeq R\Psi^{M_{\mathcal{G},\mu}} \star \mathrm{IC}_{w,\bar{k}}$.*

We refer to [Ha1, Ha2] for the discussion why the above theorem should lead to the Kottwitz conjecture. The proof of this theorem is based on some similar ideas as in [Ga]. In fact, Gaitsgory's work was motivated by Kottwitz' conjecture on the nearby cycles of Shimura varieties. However, for many years, it was not clear how to systematically apply Gaitsgory's techniques to prove the original Kottwitz' conjecture in general (except the special cases considered in [HN]). The main new ingredient is the definition of various types of affine Grassmannians over $\mathcal{O}$, starting with $\mathrm{Gr}_{\mathcal{G},X} \times_{\mathcal{O}[u],u\mapsto p} \mathcal{O}$.

Remark 4.4. Using some idea of [AB], the above theorem also implies Kottwitz-Rapoport's conjecture on the topological space of the special fibers of the local models, which is the first step towards the proof of Theorem 3.3. We refer to [Z1] for details.

One can also obtain results for quasi-split ramified groups when the level subgroup is *very special.* Here and elsewhere, we follow [Z2] to say that $K \subset G(\mathbb{Q}_p)$ is very special, if the corresponding parahoric subgroup of $G(\mathbb{Q}_p^{\mathrm{unr}})$ is special in the sense of Bruhat-Tits; then K itself is also special. In this case, we give a characterization of the function $\tau^{\mathrm{ss}}_{\mathrm{R}\Psi}$ (the semisimple trace of the sheaf of nearby cycles $R\Psi^{M_{\mathcal{G},\mu}}$) by identifying its trace on unramified representations; this also agrees with the prediction of a conjecture of Haines and Kottwitz. Corresponding to μ we have a minuscule coweight for G' which for simplicity we still denote by μ. The conjugacy class $\{\mu\}$ defines an algebraic $\overline{\mathbb{Q}}_\ell$-representation V_μ of the Langlands dual group ${}^LG = H^\vee \rtimes \mathrm{Gal}_{\mathbb{F}_q((u))}$ of $G^\flat$ over $\mathbb{F}_q((u))$. (Here we recall H is the Chevalley split form of G.) The inertia invariants V_μ^I give a representation of $(H^\vee)^I \rtimes \mathrm{Gal}(\overline{\mathbb{F}}_q/\mathbb{F}_q)$. If π is an irreducible smooth representation of $G^\flat(\mathbb{F}_q((u)))$ with a $K^\flat(\mathbb{F}_q[[u]])$-fixed vector, one can define its Langlands-Satake parameter $\mathrm{Sat}(\pi) : W \to {}^LG$: Among other properties, one can show that if $\Phi_q \in W$ is a (geometric) Frobenius element in the Weil group, then $\mathrm{Sat}(\pi)(\Phi_q)$ is a semi-simple element in $(H^\vee)^I \times \mathrm{Frob}_q^{-1}$, which is well-defined up to $(H^\vee)^I$-conjugacy and completely determines $\mathrm{Sat}(\pi)$. Our characterization of the Frobenius trace function is the identity

$$\mathrm{tr}(\pi(\tau^{\mathrm{ss}}_{\mathrm{R}\Psi})) = \mathrm{tr}(\mathrm{Sat}(\pi)(\Phi_q), V_\mu^I) \tag{4.2}$$

for all π as above. The proof relies on results from [Z2].

Finally, we have several results about the action of monodromy (i.e of inertia) on the sheaves of nearby cycles. These also imply corresponding results for Shimura varieties as explained above. Recall that we assume that G splits over a tamely ramified extension $\tilde{F}$. Let $I_{\tilde{F}} \subset \mathrm{Gal}(\bar{F}/\tilde{F})$ denote the inertia subgroup.

Theorem 4.5. *The action of $I_{\tilde{F}}$ on the sheaf of nearby cycles $\mathrm{R}\Psi^{M_{\mathcal{G},\mu}}$ is purely unipotent. If $\mathcal{P}$ is a very special it acts trivially on $\mathrm{R}\Psi^{M_{\mathcal{G},\mu}}$.*

Remark 4.6. The above theorem has some interesting consequence: We consider the Picard modular surface, which is the Shimura variety for some unitary group of signature $(2,1)$ associated to an imaginary quadratic extension $E = \mathbb{Q}(\sqrt{\Delta})$, where Δ is the discriminant of E. Assume that $p > 2$ and $p \mid \Delta$, and K_p is very special, then the Albanese of the corresponding Picard modular surface is trivial.

References

[AB] Arkhipov, S., Bezrukavnikov, R,: *Perverse sheaves on affine flags and Langlands dual group*, Israel Math. **170**, no. 1, 135–183.

[BK] Brion, M., Kumar, S.: *Frobenius Splitting Methods in Geometry and Representation Theory*, Progress in Mathematics, Vol. **231**. Birkhäuser, Boston, 2005.

[BT1] Bruhat, F., Tits, J.: *Groupes réductifs sur un corps local:I. Données radicielles valuées*, IHES Publ. Math. **41** (1972), 5–251.

[BT2] Bruhat, F., Tits, J.: *Groupes réductifs sur un corps local: II. Schémas en groupes. Existence d'une donné radicielle valuée*, IHES Publ. Math. **60** (1984), 197–376.

[CN] Chai, C. L., Norman, P.: *Singularities of the $\Gamma_0(p)$-level structure*, J. Alg. Geom. **1** (1992), 251–278.

[dJ] de Jong, A,: *The moduli spaces of principally polarized abelian varieties with $\Gamma_0(p)$-level structure*, J Algebraic Geom. **2** (1993), 667–688.

[DP] Deligne, P., Pappas, G: *Singularités des espaces de modules de Hilbert, en les caractéristiques divisants le discriminant*, Compositio Math. **90** (1994), 59–79.

[Fa] Faltings, G.: *Algebraic loop groups and moduli spaces of bundles*, J. Eur. Math. Soc. **5** (2003), 41–68.

[Ga] Gaitsgory, D.: *Construction of central elements in the affine Hecke algebra via nearby cycles*, Invent. Math. **144** (2001), no. 2, 253–280.

[Go1] Görtz, U.: *On the flatness of models of certain Shimura varieties of PEL-type*, Math. Ann. **321** (2001), 689–727.

[Go2] Görtz, U.: *On the flatness of local models for the symplectic case*, Adv. Math. **176** (2003), 89–115.

[Ha1] Haines, T.: *Introduction to Shimura varieties with bad reduction of parahoric type*, Harmonic analysis, the trace formula, and Shimura varieties, Clay Math. Proc., Vol. **4**, Amer. Math. Soc., Providence, RI, 2005, pp. 583–642.

[Ha2] Haines, T.: *The stable Bernstein center and test functions for Shimura varieties*, arXiv:1304.6293.

[HN] Haines, T., Ngô, B.C.: *Nearby cycles for local models of some Shimura varieties*, Compositio Math. **133** (2002), 117–150.

[He] He, X.: *Normality and Cohen-Macaulayness of local models of Shimura varieties*, Duke Math. J. **162** (2013), no. 13, 2509–2523.

[Il] Illusie, L.: *Autour du théorème de monodromie locale*, Astérisque (1994), no. 223, 9–57, Périodes p-adiques (Bures-sur-Yvette, 1988).

[Ki] Kisin, M.: *Moduli of finite flat group schemes, and modularity*, Ann. Math. 2nd series, Vol. **170**, no. 3, 2009.

[KP] Kisin, M., Pappas, G.: *Integral models for Shimura varieties with parahoric level structure*, in preparation.

[Ko1] Kottwitz,R.: *Shimura varieties and twisted orbital integrals*, Math. Ann. **269** (1984), no. 3, 287–300.

[Ko2] Kottwitz, R.: *Isocrystals with additional structure II*, Compositio Mathematica **109** (1997), 255–339.

[Le] Levin, B.: *G-valued flat deformations and local models*, available at http://www.math.ias.edu/ bwlevin/LevinThesis.pdf.

[MR] Metha, V.B., Ramanathan, A.: *Frobenius splitting and cohomology vanishing for Schubert varieties*, Ann. of math. (2) **122** (1985), no. 1, 27–40.

[Pa] Pappas, G.: *On the arithmetic moduli schemes of PEL Shimura varieties*, J. Algebraic Geom. **9** (2000), no. 3, 577–605.

[PR1] Pappas, G., Rapoport, M.: *Local models in the ramified case. I. The EL-case*, J. Algebraic Geom. **12** (2003), no. 1, 107–145.

[PR2] Pappas, G., Rapoport, M.: *Local models in the ramified case. II. Splitting models*, Duke Math. J. **127** (2005), no. 2, 193–250.

[PR3] Pappas, G., Rapoport, M.: *Twisted loop groups and their affine flag varieties*, Adv. in Math. **219** (2008), 118–198.

[PR4] Pappas, G., Rapoport, M.: *Local models in the ramified case. III. Unitary groups*, J. Inst. Math. Jussieu **8** (2009), no. 3, 507–564.

[PR5] Pappas, G., Rapoport, M.: *Some questions about $\mathcal{G}$-bundles on curves*, Algebraic and arithmetic structures of moduli spaces (Sapporo 2007), 159–171, Adv. Stud. Pure Math., **58**, Math. Soc. Japan, Tokyo, 2010.

[PRS] Pappas, G., Rapoport, M., Smithling, B.D.: *Local models of Shimura varieties, I. Geometry and combinatorics*, Handbook of Moduli, Vol. III, Higher Education Press and International Press, 2013, pp. 135–217.

[PZ1] Pappas, G., Zhu, X.: *Local models of Shimura varieties and a conjecture of Kottwitz*, Invent. Math. **194** (2013), 147–254.

[PZ2] Pappas, G., Zhu, X.: *Bruhat-Tits group schemes over some two-dimensional base*, in preparation.

[R] Rapoport, M.: *A guide to the reduction modulo p of Shimura varieties*, Astérisque **298** (2005), 271–318.

[RZ] Rapoport, M., Zink, T.: *Period spaces for p-divisible groups*, Ann. Math. Studies **141**, Princeton University Press, 1996.

[Sm1] Smithling, B: *Topological flatness of orthogonal local models in the split, even case. I*, Math. Ann. **350** (2011), no. 2, 381–416.

[Sm2] Smithling, B.: *Topological flatness of local models for ramified unitary groups. I. The odd dimensional case*, Adv. Math. **226** (2011), no. 4, 3160–3190.

[Sm3] Smithling, B: *Topological flatness of local models for ramified unitary groups. II. The even dimensional case*, J. Inst. Math. Jussieu **13** (2014), no. 2, 303–393.

[Z1] Zhu, X.: *On the coherence conjecture of Pappas and Rapoport*, Ann. Math **180** (2014), Issue 1, 1–85.

[Z2] Zhu, X.: *Geometric Satake correspondence for ramified groups*, to appear in Annales scientifiques de l'ENS **48** (2015), 409-451.

Department of Mathematics, California Institute of Technology
Pasadena, CA 91125, USA
E-mail address: xzhu@caltech.edu

Invited Lectures

Number Theory, Automorphic Forms and Arithmetic Geometry

Proceedings of the 6th International Congress of Chinese Mathematicians
ALM 36, pp. 271–280

New Estimates on Incomplete Character Sums

Mei-Chu Chang

Abstract

In this note we give a survey on various results on short incomplete character sums, their applications and some of the ingredients used.

Mathematics Subject Classification: 11L40, 11M06.

Keywords and Phrases: Character sums, zero free regions.

A function $\chi : \mathbb{Z}/q\mathbb{Z} \to \{z \in \mathbb{C} : |z| = 1\} \cup \{0\}$ is a multiplicative character mod q if it satisfies the properties that $\chi(mn) = \chi(m)\chi(n)$, and $\chi(n) = 0$ if $\gcd(n, q) \neq 1$. We are interested in bounding non-trivially the character sum of χ over an interval of length $N \leq q$. More precisely, we want to study the following problem.

Problem 1. *Assuming χ is non-principal and $q \gg 0$, how small $N = N(q)$ can be such that*

$$\Big|\sum_{n=a+1}^{a+N} \chi(n)\Big| < N^{1-\varepsilon}\ ? \tag{1}$$

In 1918, Polya and Vinogradov (Theorem 12.5 in [IK]) had the estimate that $N \gg \sqrt{q}\log q$. Forty four years later Burgess [Bu1] improved it to $N > q^{\frac{1}{4}+\delta}$, first for prime moduli, then for cube-free moduli.

At present time, Burgess' estimate is still the best. Davenport and Lewis generalized it to higher dimensions by replacing the interval by a box $B \subset \mathbb{F}_{p^n}$.

Let $\{\omega_1, \ldots, \omega_n\}$ be an arbitrary basis for $\mathbb{F}_{p^n}$ over $\mathbb{F}_p$. Then for any $x \in \mathbb{F}_{p^n}$, there is a unique representation of x in terms of the basis

$$x = x_1\omega_1 + \cdots + x_n\omega_n.$$

Supported in part by NSF grant DMS 1301608.

A *box* $B \subset \mathbb{F}_{p^n}$ is a set such that for each j, the coefficients x_j form an interval,

$$B = \Big\{ \sum_{j=1}^n x_j \omega_j : x_j \in [a_j + 1, a_j + H_j], \quad \forall i \Big\} = \prod_{j=1}^n [a_j + 1, a_j + H_j]. \tag{2}$$

Davenport and Lewis had the following nontrivial estimate for character sums over boxes. (See [Bu2, Ka] for work on special boxes.)

Theorem DL. ([DL], Theorem 2) *Let $H_j = H$ for $j = 1, \dots, n$, with*

$$H > p^{\frac{n}{2(n+1)}+\delta} \text{ for some } \delta > 0 \tag{3}$$

and let $p > p_1(\delta)$. Then, with B defined as above

$$\Big|\sum_{x\in B} \chi(x)\Big| < (p^{-\delta_1} H)^n,$$

where $\delta_1 = \delta_1(\delta) > 0$.

For $n = 1$ (i.e. $\mathbb{F}_q = \mathbb{F}_p$) this is Burgess' result. But as n increases, the exponent in (3) tends to $\frac{1}{2}$.

We improve Theorem DL for $n > 4$ ([C1, C4]). Our bound is also uniform in n.

Theorem 1. *Let χ be a nontrivial multiplicative character of $\mathbb{F}_{p^n}$, and let $\varepsilon > 0$ be given. If*

$$B = \prod_{j=1}^n [a_j + 1, a_j + H_j]$$

is a box satisfying

$$\prod_{j=1}^n H_j > p^{(\frac{2}{5}+\varepsilon)n},$$

then for $p > p(\varepsilon)$

$$\Big|\sum_{x\in B} \chi(x)\Big| \ll_n p^{-\frac{\varepsilon^2}{4}} |B|,$$

unless n is even and $\chi|_{F_2}$ is principal, where F_2 is the subfield of size $p^{n/2}$, in which case

$$\Big|\sum_{x\in B} \chi(x)\Big| \le \max_{\xi} |B \cap \xi F_2| + O_n(p^{-\frac{\varepsilon^2}{4}} |B|).$$

Theorem 1 was improved by Konyagin [K] for regular boxes.

Theorem K. *Let χ be a nontrivial multiplicative character of $\mathbb{F}_{p^n}$, and let B be given as in (2) with*

$$H > p^{\frac{1}{4}+\epsilon}, \quad \forall j.$$

Then

$$\Big|\sum_{x\in B}\chi(x)\Big| \ll_n p^{-\delta}|B|,$$

where $\delta=\delta(\epsilon)>0$.

As in [DL] (see p.131), Theorem 1 has the following application to the distribution of primitive roots of $\mathbb{F}_{p^n}$.

Corollary 2. *Let* $B\subset\mathbb{F}_{p^n}$ *be as in Theorem* 1 *and satisfying* $\max_\xi\big|B\cap\xi F_2\big| < p^{-\varepsilon}|B|$ *if* n *even. The number of primitive roots of* $\mathbb{F}_{p^n}$ *belonging to* B *is*

$$\frac{\varphi(p^n-1)}{p^n-1}|B|(1+o(p^{-\tau'})),$$

where $\tau'=\tau'(\varepsilon)>0$ *and assuming* $n\ll\log\log p$.

The proof of Corollary 2 is by combining character sums estimate with the following

$$\frac{\phi(p^n-1)}{p^n-1}\Big\{1+\sum_{\substack{d|p^n-1\\ d>1}}\frac{\mu(d)}{\phi(d)}\sum_{\operatorname{ord}(\chi)=d}\chi(x)\Big\}=\begin{cases}1 & \text{if } x \text{ is primitive,}\\ 0 & \text{otherwise.}\end{cases}$$

For special moduli, the condition on N in Burgess' theorem can be improved. One can proved (1) for much smaller intervals. There are two classical results, based on quite different arguments by Graham-Ringrose ([IK], Corollary 12.15) and Iwaniec ([IK], Theorem 12.16).

First, for the modulus q, we set up the following notations for the largest prime divisor $\mathcal{P}$ of q, and the *core* k of q,

$$\mathcal{P}=\max_{p|q}p \quad\text{and}\quad k=\prod_{p|q}p.$$

Theorem GR. *Let* χ *be a primitive character of modulus* $q\geq 3$ *with* q *square free. Then, for*

$$N=|I|\geq q^{\frac{4}{\sqrt{\log\log q}}}+\mathcal{P}^9$$

we have

$$\Big|\sum_{n\in I}\chi(n)\Big|\ll Ne^{-\sqrt{\log q}}.$$

The purpose of the work of Graham and Ringrose is to study the following least quadratic nonresidue problem.

Problem 2. *Let* p *be a prime, and let* $n(p)$ *be the least quadratic nonresidue mod* p. *Find a lower bound on* $n(p)$.

Using sieving, Burgess' result implies that $n(p)\leq p^{\frac{1}{4\sqrt{e}}+\varepsilon}$. Independently, Friedlander [F] and Salié [Sa] showed that $n(p)=\Omega(\log p)$, i.e. there are infinitely many p such that $n(p)>c\log p$ for some absolute constant c. By assuming the Generalized Riemann Hypothesis, in 1971 Montgomery [M] proved that $n(p) = \Omega(\log p\ \log\log p)$. Using Theorem GR, Graham and Ringrose showed that $n(p) = \Omega(\log p\ \log\log\log p)$ unconditionally.

In 1974, Iwaniec [I] proved the following theorem by generalizing Postnikov's theorem [P].

Theorem PI. *Let χ be a primitive character of modulus q, $2 \nmid q$. Then, for*

$$k^{100} < N < N' \leq 2N \tag{4}$$

we have

$$\Big|\sum_{N<n\leq N'} \chi(n)\Big| \leq C^{s(\log s)^2} N^{1-\frac{c}{s^2 \log s}},$$

where $s = \frac{\log q}{\log N}$.

Theorem PI is good for q powerful. A related problem is about the digital aspects of the primes. Let $x < N = 2^n$. Write

$$x = x_0 + 2x_1 + 2^2x_2 + \cdots + 2^{n-1}x_{n-1} \quad \text{with} \quad x_0, x_1, \ldots, x_{n-1} \in \{0, 1\}.$$

For $A \subset \{1, \ldots, n\}$, given $\{\alpha_j \in \{0,1\}\}_{j\in A}$, one expects

$$|\{p = x < N : x_j = \alpha_j,\ \forall j \in A\}| \sim \frac{N}{\log N}\, 2^{-|A|}.$$

Problem 3. *How large can A be?*

For related work, see Sierpinski [S], Harman and Katai [HK], Bourgain [B].

Theorem GR is for q with small prime factors (smooth moduli), for which one assumes

$$\log N \gg \log \mathcal{P} + \frac{\log q}{\sqrt{\log\log q}}. \tag{5}$$

On the other hand, Theorem PI is for q with small core. Condition (4) implies that

$$\log N \gg \log k. \tag{6}$$

If fix k, to get nontrivial result, in Theorem PI, one needs to assume

$$\log N \gg \log k + (\log q)^{\frac{3}{4}+\epsilon}. \tag{7}$$

Both are special cases of the following theorem in [C3].

Denote

$$K = \frac{\log q}{\log k}.$$

Theorem 3. *Assume N satisfies*

$$q > N > \mathcal{P}^{10^3}$$

and

$$\log N > (\log q)^{\frac{9}{10}} + 10^3 \frac{\log 2K}{\log\log q} \log k\ . \tag{8}$$

Let χ be a primitive multiplicative character modulo q and I an interval of size N. Then

$$\Big|\sum_{x\in I} \chi(x)\Big| \ll N e^{-(\log N)^{3/5}}. \tag{9}$$

Remark 1. In the same spirit as assumptions (5) and (7), assumption (8) can be replaced by the stronger and friendlier assumption

$$\log N \gg \log \mathcal{P} + \frac{\log q}{\log\log q}. \tag{10}$$

(The second term of condition (10) is clearly bigger than either term of (8).)

Remark 2. This result is completely general and gives bounds for very short character sums as soon as $\log \mathcal{P}$ is small compared with $\log q$.

Remark 3. To compare Theorem 3 with Theorem GR, we note also that condition (10) gives a better result than (5). Moreover, condition (5) is restricted to square free q. As for comparing Theorem 3 with Theorem PI, we let k be square free and $q = k^m$, where m is large but not too large. More precisely, we assume $\log m = o(\log\log k)$. Theorem PI certainly requires that $\log N > C \log k$ while condition (8) in Theorem 3 becomes $\log N > (m \log k)^{\frac{9}{10}} + C\frac{\log m}{\log\log q} \log k$, which is clearly better.

Remark 4. We did not try to optimize the power of $\log q$ in the first term in (8) nor the saving in (9).

The following is a mixed character sum version of Theorem 3.

Theorem 4. *Under the assumptions of Theorem 3,*

$$\Big|\sum_{x\in I} \chi(x) e^{if(x)}\Big| < N e^{-\sqrt{\log N}}$$

assuming $f(x) \in \mathbb{R}[x]$ of degree at most $(\log N)^c$ for some $c > 0$.

Corollary 5. *Let $T > 0$. Assume N satisfies*

$$q > N > \mathcal{P}^{10^3}$$

and q satisfies

$$\log N > (\log qT)^{\frac{9}{10}} + 10^3 \frac{\log 2K}{\log\log q} \log k.$$

Then for χ primitive, we have

$$\Big|\sum_{n\in I} \chi(n) n^{it}\Big| < N e^{-\sqrt{\log N}} \quad \textit{for } |t| < T.$$

Following the classical arguments going back to Hadamand, de-la-Vallee-Poissin, and Landau, the above estimates lead to zero-free regions for the corresponding Dirichlet L-functions. Denote

$$|L(s,\chi)| = \sum_n \chi(n)n^{-s}, \qquad s = \rho + it.$$

Iwaniec [I] obtained the following results.

Theorem I. *Assume* $|L(s,\chi)| < M$ *for* $\rho > 1-\eta$, $|t| < T^2$. *Then* $L(s,\chi)$ *has no zeros in the region* $\rho > 1 - \frac{\eta}{400 \log M}$, $|t| < T$, *except for possible Siegel zeros.*

Corollary I. *Assume* Y *and* $\gamma > 0$ *satisfy*

$$\Big|\sum_{n\sim N} \chi(n)n^{it}\Big| < N^{1-\gamma} \quad \textit{for } N > Y, |t| < T.$$

Then

$$\rho > 1 - \frac{c}{\log Y + \frac{1}{\gamma}\log\frac{1}{\gamma} + \frac{1}{\gamma}\log\log\frac{qT}{Y}}.$$

From Corollary 5 and Corollary I, one derives bounds on the Dirichlet L-function $L(s,\chi)$ and zero-free regions the usual way. For a detailed argument, see for instance Lemmas 8–11 in [I]. This leads to the following theorem.

Theorem 6. *Let* χ *be a primitive multiplicative character with modulus* q. *For* $T > 0$, *let*

$$\theta = c\min\left\{\frac{1}{\log \mathcal{P}}, \frac{\log\log k}{(\log k)\log 2K}, \frac{1}{(\log qT)^{9/10}}\right\}.$$

Then the Dirichlet L-function $L(s,\chi) = \sum_n \chi(n)n^{-s}, s = \rho + it$ *has no zeros in the region* $\rho > 1-\theta$, $|t| < T$, *except for possible Siegel zeros.*

It follows in particular that $\theta \log qT \to \infty$ if $\frac{\log \mathcal{P}}{\log q} \to 0$.

In certain range of k, this improves Iwaniec's bound (See [IK], Theorem 8.29)

$$\theta = \min\left\{c\frac{1}{(\log T)^{\frac{2}{3}}(\log\log T)^{\frac{1}{3}}}, \frac{1}{\log k}\right\}.$$

A well-known application of zero-free region is to primes in arithmetic progressions. In 1944, Linnik proved the following theorem about the least prime in an arithmetic progression.

Theorem L. *There exists* c *such that if* $(a,q) = 1$, *then there is a prime* $p < q^c$ *such that* $p \equiv a$ mod q.

For explicit value of c in Theorem L, Xylouris [X] has the best estimate $c = 5.2$ by using Heath-Brown [HB1] (who obtained $c = 5.5$) proposed improvements.

Theorem 6 gives a lower c for q smooth.

Theorem 7. *In Theorem L, one may take* $c = \frac{12}{5} + \epsilon$, *assuming* $\log \mathcal{P} < \frac{\log q}{\log\log q}$.

The proof of Theorem 7 uses the result of Heath-Brown. One distinguishes the following two cases.

(1) Either there is no Siegel zero or there is a Siegel zero ρ_1 satisfying

$$1 - \rho_1 > \frac{\epsilon}{\log q}.$$

In this case we use Theorem 6.

(2) There is a Siegel zero ρ_1 satisfying $1-\rho_1 \leq \frac{\epsilon}{\log q}$. In this case, Heath-Brown's result implies $c < 2 + \epsilon'$.

Estimates on short character sums are also related to Polya-Vinogradov's theorem. Based on the work of Granville and Soundararajan [GS] (which characterizes when character sums are large), Goldmakher [Go] used Theorem GR to improve Polya-Vinogradov's bound on $\sum_{n<x} \chi(n)$ and obtained the following

Theorem G. *Let* χ mod q *be a primitive character. Then*

$$\Big|\sum_{n<x} \chi(n)\Big| \ll \sqrt{q}\, \log q\, \sqrt{\log^3 q}\, \Big(\frac{1}{\log^2 q} + \frac{\log \mathcal{P}}{\log q}\Big)^{\frac{1}{4}}.$$

When q *is square free,*

$$\Big|\sum \chi(n)\Big| \ll \sqrt{q}\, \frac{\log q}{(\log\log q)^{\frac{1}{4}}}.$$

Instead of Theorem GR, we use Theorem 3 (and Granville-Soundararajan) and obtain the following.

Corollary 8. *Let* χ mod q *be a primitive character, and let*

$$M = (\log q)^{\frac{9}{10}} + (\log 2K)\frac{\log k}{\log\log k} + \log \mathcal{P}.$$

Then

$$\Big|\sum_{n<x} \chi(n)\Big| \ll \sqrt{q}\, \sqrt{\log q}\, \sqrt{M}\, \sqrt{\log\log\log q}.$$

When q *is square free,*

$$\Big|\sum_{n<x} \chi(n)\Big| \ll \sqrt{q}\, \frac{\log q}{(\log\log q)^{\frac{1}{2}}}.$$

We conclude the paper with some comments on the proof of Theorem 3. The main ideas include the following:

(1) Use Postnikov to replace multiplicative characters by additive characters with polynomial argument.

(2) Use mixed character sum estimates to remove additive characters.

(3) Use Graham-Ringrose-Weyl-type differencing scheme to reduce the modulus.

Here are some of the ingredients.

(a) The following is a simple example of the mixed character sum estimate. (See [C2].)

Theorem. *Let χ be a nontrivial multiplicative character mod p, and let $P(x) \in \mathbb{R}[X]$ be a polynomial of degree $d \geq 1$. Then for $|I| > p^{\frac{1}{4}+\kappa}$ with $\kappa > 0$,*

$$\Big|\sum_{n\in I} \chi(n)e^{iP(n)}\Big| < |I|\, p^{-c\kappa^2/d^2}.$$

(b) Postnikov's formula

Let χ be a primitive multiplicative character mod q^m. Then

$$\chi(1+qx) = e_{q^m}(F(qx)),$$

where $F(x) \in \mathbb{Z}[x]$,

$$F(x) = B\Big(x - \frac{x^2}{2} + \cdots \pm \frac{x^{m'}}{m'}\Big) \quad (m' > 2m)$$

with $B \in \mathbb{Z}$, $(B,q) = 1$.

References

[B] J. Bourgain, *Prescribing the binary digits of primes*, arXiv:1105.3895.

[Bu1] D.A. Burgess, *On character sums and primitive roots*, Proc LMS (3) **12** (1962), 179–192.

[Bu2] D.A. Burgess, *Character sums and primitive roots in finite fields*, Proc. London Math. Soc (3) **37** (1967), 11–35.

[C1] M.-C. Chang, *On a question of Davenport and Lewis and new character sum bounds in finite fields*, Duke Math. J. **145** (2008), no. 3, 409–442.

[C2] M.-C. Chang, *An estimate of incomplete mixed character sums*, In: An Irregular Mind. Editors: Brny, Imre, Solymosi, Jozsef. Bolyai Society Mathematical Studies. Budapest. **21** (2010), 243–250.

[C3] M.-C. Chang, *Short character sums for composite moduli*, preprint (2011).

[C4] M.-C. Chang, *Character sums in finite fields*, AMS Contemporary Mathematics. (2009), 19p.

[DL] H. Davenport, D. Lewis, *Character sums and primitive roots in finite fields*, Rend. Circ. Materm. Palermo-Serie II-Tomo XII-Anno (1963), 129–136.

[F] V.R. Friedlander, *On the least n-th power non-residue*, Dokl. Akad. Nauk. SSSR **66** (1949), 351–352.

[Ga] P.X. Gallagher, *Primes in progressions to prime-power modulus*, Invent. Math. **16** (1972), 191–201.

[Go] L. Goldmakher, *Character sums to smooth moduli are small*, Canad. J. Math. **62** (2010), 1099–1115.

[GR] S.W. Graham, C.J. Ringrose, *Lower bounds for least quadratic nonresidues*, In: Analytic Number Theory (Allerton Park, IL, 1989), Progr. Math., **85**, Birkhauser, Boston, MA, 1990, 269–309.

[GS] A. Granville, K. Soundararajan, *Large character sums: Pretentious characters and the Pólya-Vinogradov theorem*, J. Amer. Math. Soc. **20** (2007), 357–384.

[HK] G. Harman, I. Katai, *Primes with preassigned digits II*, Acta. Arith. **133** (2008), no. 2, 171–184.

[HB1] D. R. Heath-Brown, *Zero-free regions for Dirichlet L-functions, and the least prime in an arithmetic progression*, Proc. London Math. Soc. (3) **64** (1992), no. 2, 265–338.

[HB2] D.R. Heath-Brown, *Siegel zeros and the least prime in an arithmetic progression*, Quart. J. Math. Oxford Ser. (2) **41** (1990), 405–418.

[H] M.N. Huxley, *Large values of Dirichlet polynomials, III*, Acta Arith. **26** (1974), 435–444.

[IK] H. Iwaniec, E. Kowalski, *Analytic number theory*, Amer. Math. Soc., Providence RI, 2004.

[I] H. Iwaniec, *On zeros of Dirichlet's L series*, Invent. Math. **23** (1974), 97–104.

[Ka] A.A. Karacuba, *A certain arithmetic sum*, Soviet Math. Dokl. **12** (1971), no. 4, 1172–1174.

[K] S.V. Konyagin, *Estimates of character sums in finite fields*, Matematicheskie Zametki **88** (2010), no. 4, 529–542.

[L1] Y.V. Linnik, *On the least prime in an arithmetic progression I. The basic theorem*, Rec. Math. (Mat. Sbornik) N.S. **15** (1944), no. 57, 139–178.

[L1] Y.V. Linnik, *On the least prime in an arithmetic progression II. The Deuring-Heilbronn phenomenon*, Rec. Math. (Mat. Sbornik) N.S. **15** (1944), no. 57, 347–368.

[M] H.L. Montgomery, *Topics in Multiplicative Number Theory*, Lecture Notes in Math. **227**, Springer-Verlag, New York, 1971.

[P] A.G. Postnikov, *On Dirichlet L-series with the character modulus equal to the power of a prime number*, J. Indian Math. Soc. (N.S.) **20** (1956), 217–226.

[Sa] H.Salié, *Uber den kleinsten positiven quadratischen Nichtrest nach einer Primzahl*, Math. Nachr. **3** (1949), 7–8.

[S] W. Sierpinski, *Sur un problème concernment les nombres* $k \cdot 2n + 1$, Elem. Math. **15** (1960), 63–74.

[X] T. Xylouris, *On Linnik's constant*, 2009, arXiv:0906.2749.

Department of Mathematics, University of California,
Riverside, CA 92521, USA
E-mail address: mcc@math.ucr.edu

Proceedings of the 6th International
Congress of Chinese Mathematicians
ALM 36, pp. 281–295

Gelfand-Kapranov-Zelevinsky Hypergeometric Sheaves

Lei Fu

Abstract

We briefly recall the Gelfand-Kapranov-Zelevinsky (GKZ) hypergeometric system of differential equations and find an integral representation for a solution of this system in the space of generalized functions. An arithmetic analogue of this integral is the hypergeometric function over finite field introduced by Gelfand and Graev. The GKZ hypergeometric sheaf is a perverse sheaf in the derived category of ℓ-adic sheaves on the affine space. The traces of Frobenius elements at its stalks are given by the hypergeometric function over finite field. We study the weight filtration of GKZ hypergeometric sheaf.

Mathematics Subject Classification: Primary 14F20; Secondary 11T23, 14G15, 33C70.

Keywords and Phrases: ℓ-adic GKZ hypergeometric sheaf, Deligne-Fourier transformation, toric variety.

1 Classical hypergeometric functions

The *Gauss hypergeometric function* is defined by

$$_2F_1(a,b;c;z)=\sum_{n=0}^{\infty}\frac{(a)_n(b)_n}{(c)_n}\frac{z^n}{n!},$$

where c is not a negative integer, and

$$(q)_n=\begin{cases}1 & \text{if } n=0,\\ q(q+1)\cdots(q+n-1) & \text{if } n\geq 1.\end{cases}$$

Research is supported by the NSFC.

The series converges for $|z| < 1$, and can be analytically continued to $|z| \geq 1$ along any path avoiding the points 0 and 1. ${}_2F_1(a, b; c; z)$ is a solution of the *Gauss hypergeometric differential equation*

$$z(1-z)\frac{d^2y}{dz^2} + (c - (a+b+1)z)\frac{dy}{dz} - aby = 0.$$

This differential equation has regular singularities at $z = 0, 1, \infty$. One can construct from the Gauss hypergeometric function a basis for the space of the solutions of the Gauss hypergeometric differential equation. For example, near the point $z = 0$, two independent solutions of the Gauss hypergeometric equations are ${}_2F_1(a, b; c; z)$ and $z^{1-c}{}_2F_1(1+a-c, 1+b-c; 2-c; z)$.

Recall that an ordinary differential equation has *regular singularity* at $z = 0$ if it can be written in the form

$$\partial^n y + \sum_{i=1}^{n} a_i(z)\partial^{n-i}y = 0,$$

where $\partial = z\frac{d}{dz}$ and $a_i(z)$ are holomorphic at 0. Riemann proves that any second order differential equation on $\mathbb{C} \cup \{\infty\}$ with three regular singularities can be converted to the above equation by a change of variable.

Generalized hypergeometric series is defined to be

$${}_pF_q(a_1, \ldots, a_p; b_1, \ldots, b_q; z) = \sum_{n=0}^{\infty} \frac{(a_1)_n \cdots (a_p)_n}{(b_1)_n \cdots (b_q)_n} \frac{z^n}{n!}.$$

It satisfies the differential equation

$$z\prod_{i=1}^{p}\left(z\frac{d}{dz} + a_i\right)y = z\frac{d}{dz}\prod_{j=1}^{q}\left(z\frac{d}{dz} + b_j - 1\right)y.$$

A *confluent hypergeometric equation* is a degenerate form of the Gauss hypergeometric differential equation where two of the three regular singularities merge into an irregular singularity. One can check ${}_2F_1(a, b, c; b^{-1}z)$ satisfies the differential equation

$$z\left(1 - \frac{z}{b}\right)\frac{d^2y}{dz^2} + \left(c - (a+b+1)\frac{z}{b}\right)\frac{dy}{dz} - ay = 0,$$

which has regular singularities at 0, b, ∞. When $b \to \infty$, one gets the *Kummer confluent hypergeometric equation*

$$z\frac{d^2y}{dz^2} + (c-z)\frac{dy}{dz} - ay = 0,$$

which has a regular singularity at 0, and an irregular singularity at ∞. A solution of this equation is the *Kummer confluent hypergeometric function*

$$\lim_{b\to\infty} {}_2F_1(a, b, c; b^{-1}z) = {}_1F_1(a, c, z) = \sum_{n=0}^{\infty} \frac{(a)_n}{(c)_n} \frac{z^n}{n!}.$$

2 Gelfand-Kapranov-Zelevinsky hypergeometric systems

Let $A = \{\mathbf{w}_1, \ldots, \mathbf{w}_N\}$ be a subset of $\mathbb{Z}^n$ so that $\mathbf{w}_1, \ldots, \mathbf{w}_N$ generate $\mathbb{Q}^n$ over $\mathbb{Q}$. Write $\mathbf{w}_j = \begin{pmatrix} w_{1j} \\ \vdots \\ w_{nj} \end{pmatrix}$. Denote also by A the matrix $A = \begin{pmatrix} w_{11} & \ldots & w_{1N} \\ \vdots & & \vdots \\ w_{n1} & \ldots & w_{nN} \end{pmatrix}$. Define an action of the n-dimensional torus $\mathbb{T}^n_{\mathbb{Z}} = \operatorname{Spec} \mathbb{Z}[t_1^{\pm 1}, \ldots, t_n^{\pm 1}]$ on the N-dimensional affine space $\mathbb{A}^N_{\mathbb{Z}} = \operatorname{Spec} \mathbb{Z}[x_1, \ldots, x_N]$ by

$$\begin{gathered} \mathbb{T}^n_{\mathbb{Z}} \times \mathbb{A}^N_{\mathbb{Z}} \to \mathbb{A}^N_{\mathbb{Z}}, \\ \Big((t_1, \ldots, t_n), (x_1, \ldots, x_N)\Big) \mapsto (t_1^{w_{11}} \cdots t_n^{w_{n1}} x_1, \ldots, t_1^{w_{1N}} \cdots t_n^{w_{nN}} x_N). \end{gathered}$$

We say that A satisfies the *nonconfluence condition* if there exist rational numbers $c_1, \ldots, c_n \in \mathbb{Z}$ such that

$$\sum_{i=1}^{n} c_i w_{ij} = 1 \quad (j = 1, \ldots, N).$$

Let $(\gamma_1, \ldots, \gamma_n) \in \mathbb{C}^n$ be a fixed parameter. In [7], Gelfand, Kapranov and Zelevinsky introduce the system of differential equations

$$\begin{gathered} \sum_{j=1}^{N} w_{ij} x_j \frac{\partial f}{\partial x_j} + \gamma_i f = 0 \quad (i = 1, \ldots, n), \\ \prod_{a_j > 0} \left(\frac{\partial}{\partial x_j}\right)^{a_j} f = \prod_{a_j < 0} \left(\frac{\partial}{\partial x_j}\right)^{-a_j} f, \end{gathered}$$

where for the second system of equations, $(a_1, \ldots, a_N) \in \mathbb{Z}^N$ goes over the family of integral linear relations

$$\sum_{j=1}^{N} a_j \mathbf{w}_j = 0$$

among $\mathbf{w}_1, \ldots, \mathbf{w}_N$. We call this system the *GKZ hypergeometric system*. Holomorphic solutions of the GKZ hypergeometric systems are called *GKZ hypergeometric functions*.

Solutions of

$$\sum_{j=1}^{N} w_{ij} x_j \frac{\partial f}{\partial x_j} + \gamma_i f = 0 \quad (i = 1, \ldots, n)$$

have the property

$$f(t_1^{w_{11}} \cdots t_n^{w_{n1}} x_1, \ldots, t_1^{w_{1N}} \cdots t_n^{w_{nN}} x_N) = t_1^{-\gamma_1} \cdots t_n^{-\gamma_n} f(x_1, \ldots, x_N),$$

that is, $f(x_1, \ldots, x_N)$ is homogeneous of degree $(-\gamma_1, \ldots, -\gamma_n)$ with respect to the torus action. To proves this, let

$$g(t_1, \ldots, t_n) = f(t_1^{w_{11}} \cdots t_n^{w_{n1}} x_1, \ldots, t_1^{w_{1N}} \cdots t_n^{w_{nN}} x_N).$$

One can verify

$$t_i \frac{\partial g}{\partial t_i} + \gamma_i g = 0.$$

So we have $g(t_1, \ldots, t_n) = t_1^{-\gamma_1} \cdots t_n^{-\gamma_n} g(1, \ldots, 1)$.

Example 2.1. Let $(\gamma_1, \gamma_2, \gamma_3) = (c-1, a, b)$, let

$$A = \begin{pmatrix} 1 & -1 & 0 & 0 \\ 1 & 0 & 1 & 0 \\ 1 & 0 & 0 & 1 \end{pmatrix},$$

and let $\mathbf{w}_1, \ldots, \mathbf{w}_4$ be the column vectors of A. Note that A satisfies the nonconfluence condition

$$-w_{1j} + w_{2j} + w_{3j} = 1 \quad (j = 1, 2, 3, 4).$$

Integral linear relations among $\mathbf{w}_1, \ldots, \mathbf{w}_4$ are multiples of

$$\mathbf{w}_1 + \mathbf{w}_2 - \mathbf{w}_3 - \mathbf{w}_4 = 0.$$

The GKZ hypergeometric system is

$$\begin{aligned} x_1 \frac{\partial f}{\partial x_1} - x_2 \frac{\partial f}{\partial x_2} + (c-1) f &= 0, \\ x_1 \frac{\partial f}{\partial x_1} + x_3 \frac{\partial f}{\partial x_3} + a f &= 0, \\ x_1 \frac{\partial f}{\partial x_1} + x_4 \frac{\partial f}{\partial x_4} + b f &= 0, \\ \frac{\partial^2 f}{\partial x_1 \partial x_2} &= \frac{\partial^2 f}{\partial x_3 \partial x_4}. \end{aligned}$$

Applying $\frac{\partial}{\partial x_1}$ to the first equation, we get

$$\frac{\partial^2 f}{\partial x_1 \partial x_2} = x_2^{-1} \Big(x_1 \frac{\partial^2 f}{\partial x_1^2} + c \frac{\partial f}{\partial x_1} \Big).$$

From the second and the third equations, we get

$$\begin{aligned} \frac{\partial^2 f}{\partial x_3 \partial x_4} &= \frac{\partial}{\partial x_3} \Big(x_4^{-1} \Big(-x_1 \frac{\partial}{\partial x_1} - b \Big) f \Big) \\ &= \Big(x_4^{-1} \Big(-x_1 \frac{\partial}{\partial x_1} - b \Big) \Big) \frac{\partial f}{\partial x_3} \\ &= \Big(x_4^{-1} \Big(-x_1 \frac{\partial}{\partial x_1} - b \Big) \Big) \Big(x_3^{-1} \Big(-x_1 \frac{\partial}{\partial x_1} - a \Big) \Big) f. \end{aligned}$$

Combined with the fourth equation, we get

$$x_2^{-1} \Big(x_1 \frac{\partial^2 f}{\partial x_1^2} + c \frac{\partial f}{\partial x_1} \Big) = \Big(x_4^{-1} \Big(-x_1 \frac{\partial}{\partial x_1} - b \Big) \Big) \Big(x_3^{-1} \Big(-x_1 \frac{\partial}{\partial x_1} - a \Big) \Big) f.$$

This implies that $y = f(z,1,1,1)$ satisfies the Gauss hypergeometric equation

$$z(1-z)\frac{d^2y}{dz^2} + (c-(a+b+1)z)\frac{dy}{dz} - aby = 0.$$

Using the fact that f is homogeneous with respect to the torus action, one can show $f(x_1, x_2, x_3, x_4)$ is completely determined by $f(z,1,1,1)$.

Example 2.2. ([8]) Let $A_1, \ldots, A_m$ be finite subsets of $\mathbb{Z}^k$. To each $\mathbf{w} = (w_1, \ldots, w_k) \in \mathbb{Z}^k$, we associate the Laurent monomial $\mathbf{t}^{\mathbf{w}} = t_1^{w_1} \cdots t_k^{w_k}$. We regard $\mathbb{C}^{A_i}$ as the space of Laurent polynomials of the form $P_i(t_1, \ldots, t_k) = \sum_{\mathbf{w} \in A_i} x_{\mathbf{w}} \mathbf{t}^{\mathbf{w}}$. Let $(\alpha_1, \ldots, \alpha_m) \in \mathbb{C}^m$, $(\beta_1, \ldots, \beta_k) \in \mathbb{C}^k$. For $(P_1, \ldots, P_m) \in \prod_i \mathbb{C}^{A_i}$, consider

$$F(P_1, \ldots, P_m) = \int_\Sigma \prod_{i=1}^m P_i(t_1, \ldots, t_k)^{\alpha_i} t_1^{\beta_1} \cdots t_k^{\beta_k} \frac{dt_1}{t_1} \cdots \frac{dt_k}{t_k}.$$

This is an integration of a multivalued function on a cycle Σ. $F(P_1, \ldots, P_m)$ is a solution of the GKZ hypergeometric system for the parameter

$$(\gamma_1, \ldots, \gamma_{m+k}) = (-\alpha_1, \ldots, -\alpha_m, \beta_1, \ldots, \beta_k)$$

and the set

$$A = \bigcup_{i=1}^m (\{e_i\} \times A_i) \subset \mathbb{Z}^m \times \mathbb{Z}^k = \mathbb{Z}^{m+k}.$$

where $e_1, \ldots, e_m$ are the standard basis of $\mathbb{Z}^m$.

Example 2.3. In particular, let Σ be a cycle in $\mathbb{T}^n$. The period integral

$$\int_\Sigma \frac{1}{\sum_{j=1}^N x_j t_1^{w_{1j}} \cdots t_n^{w_{nj}}} \frac{dt_1}{t_1} \cdots \frac{dt_k}{t_k}$$

is a solution of the GKZ hypergeometric system.

Take $f(\psi, \mathbf{t}) = \psi - t_1 - t_2 - t_3 - t_4 - \frac{1}{t_1 t_2 t_3 t_4}$. Consider the period integral

$$\Pi_0(\psi) = \frac{1}{(2\pi i)^4} \int_\Sigma \frac{\psi}{f(\psi, \mathbf{t})} \frac{dt_1}{t_1} \cdots \frac{dt_4}{t_4},$$

where $\Sigma = \{(t_1, \ldots, t_4) | |t_i| = 1\}$. One can show $\Pi_0(\psi) = \sum_{k=0}^\infty \frac{(5k)!}{(k!)^5 (\psi)^{5k}}$. This example is used by Candelas, de la Ossa, Green, and Parkes in their work on the mirror symmetry for the quintic 3-fold.

3 An integral representation of a solution of the GKZ hypergeometric system

Let $C_c^\infty(\mathbb{R}^N)$ be the the space of complex valued smooth functions on $\mathbb{R}^N$ with compact support. For any $\phi(\xi)$ in $C_c^\infty(\mathbb{R}^N)$, consider its Fourier transform

$$\hat{\phi}(x) = \frac{1}{(2\pi)^{\frac{N}{2}}} \int_{\mathbb{R}^N} \phi(\xi) e^{i\xi \cdot x} d\xi.$$

It is an entire function in the complex variable $x \in \mathbb{C}^N$. If $\phi(\xi)$ is supported in $\{(\xi_1, \ldots, \xi_N) \in \mathbb{R}^N | |\xi_j| \leq a_j\}$, then for any nonnegative integers $q_1, \ldots, q_N$, there exists a constant C_q depending on $(q_1, \ldots, q_N)$ and $\phi(\xi)$ such that

$$|x_1^{q_1} \cdots x_N^{q_N} \hat{\phi}(x)| \leq C_q e^{a_1|\mathrm{Im}(x_1)| + \cdots + a_N|\mathrm{Im}(x_N)|}.$$

Conversely, any entire function with this property is the Fourier transform of a smooth function on $\mathbb{R}^N$ supported in $\{(\xi_1, \ldots, \xi_N) | |\xi_j| \leq a_j\}$. Let $Z(\mathbb{C}^N)$ be the space of all such entire functions, that is, the space of Fourier transforms of complex valued smooth functions on $\mathbb{R}^N$ with compact support. A *distribution* is a continuous linear functional on $C_c^\infty(\mathbb{R}^N)$. For any distribution $g(\xi)$, its Fourier transform $\hat{g}(x)$ is the linear functional on $Z(\mathbb{C}^N)$ defined by

$$(\hat{g}(x), \hat{\phi}(x)) = (g(\xi), \phi(-\xi))$$

for any $\phi(\xi) \in C_c^\infty(\mathbb{R}^N)$. Continuous linear functionals on $Z(\mathbb{C}^N)$ are called *generalized functions*. We can take the inverse Fourier transform of a generalized function to get a distribution. For more properties on generalized functions and the Fourier transform, see [9, Chapter III]. Intuitively, the above definition for the Fourier transform is motivated by the following calculation:

$$\begin{aligned}
(g(\xi), \phi(-\xi)) &= \int_{\mathbb{R}^N} g(\xi)\phi(-\xi)d\xi \\
&= \frac{1}{(2\pi)^{\frac{N}{2}}} \int_{\mathbb{R}^N} g(\xi)\Big(\int_{\mathbb{R}^N} \hat{\phi}(x)e^{i\xi\cdot x}dx\Big)d\xi \\
&= \frac{1}{(2\pi)^{\frac{N}{2}}} \int_{\mathbb{R}^N} \Big(\int_{\mathbb{R}^N} g(\xi)e^{i\xi\cdot x}d\xi\Big)\hat{\phi}(x)dx \\
&= (\hat{g}(x), \hat{\phi}(x)).
\end{aligned}$$

For any generalized function $f(x)$, its partial derivative $\frac{\partial f}{\partial x_j}$ is the generalized function defined by

$$\left(\frac{\partial f}{\partial x_j}, \psi(x)\right) = -\left(f(x), \frac{\partial \psi}{\partial x_j}\right)$$

for any $\psi(x) \in Z(\mathbb{C}^N)$. For any distribution $g(\xi)$, we have

$$\frac{\partial \hat{g}}{\partial x_j} = \widehat{(i\xi_j g)}, \quad x_j\hat{g} = \widehat{\left(i\frac{\partial g}{\partial \xi_j}\right)}.$$

Suppose $f(x)$ is a generalized function solution of the GKZ hypergeometric system. Let $g(\xi)$ be a distribution so that $\hat{g}(x) = f(x)$. Taking the inverse Fourier transform of the GKZ hypergeometric system, we see that $g(\xi)$ satisfies the system

$$\sum_{j=1}^N w_{ij}\frac{\partial}{\partial \xi_j}(\xi_j g) - \gamma_i g = 0,$$

$$\Big(\prod_{a_j>0}(i\xi_j)^{a_j} - \prod_{a_j<0}(i\xi_j)^{-a_j}\Big)g = 0 \quad \text{whenever } \sum_{j=1}^N a_j\mathbf{w}_j = 0.$$

The last equation means that $g(\xi)$ is supported on

$$\left\{(\xi_1,\ldots,\xi_N)\in\mathbb{R}^N\middle|\prod_{a_j>0}(i\xi_j)^{a_j}-\prod_{a_j<0}(i\xi_j)^{-a_j}=0 \text{ whenever } \sum_{j=1}^N a_j\mathbf{w}_j=0\right\}.$$

Over the complex field, this set is the Zariski closure of the image of the morphism

$$\iota:\mathbb{T}^n\to\mathbb{A}^N,\quad (t_1,\ldots,t_n)\mapsto(-it_1^{w_{11}}\cdots t_n^{w_{n1}},\ldots,-it_1^{w_{1N}}\cdots t_n^{w_{nN}}).$$

Suppose there exists a function $F(t_1,\ldots,t_n)$ on $\mathbb{T}^n$ and an n-dimensional cycle Σ contained in the set

$$\{(t_1,\ldots,t_n)\in(\mathbb{C}^*)^n|(-it_1^{w_{11}}\cdots t_n^{w_{n1}},\ldots,-it_1^{w_{1N}}\cdots t_n^{w_{nN}})\in\mathbb{R}^N\}$$

such that for any $\phi(\xi)\in C_c^\infty(\mathbb{R}^N)$, we have

$$(g(\xi),\phi(\xi))=\int_\Sigma F(t_1,\ldots,t_n)\phi(-it_1^{w_{11}}\cdots t_n^{w_{n1}},\ldots,-it_1^{w_{1N}}\cdots t_n^{w_{nN}})dt_1\cdots dt_n.$$

This is a distribution supported on $\overline{\mathrm{im}(\iota)}$. By the first equation for $g(\xi)$, we must have

$$\begin{aligned}
&0\\
&=\Big(\sum_{j=1}^N w_{ij}\frac{\partial}{\partial\xi_j}(\xi_j g)-\gamma_i g,\phi\Big)\\
&=\Big(g(\xi),-\sum_{j=1}^N w_{ij}\xi_j\frac{\partial\phi}{\partial\xi_j}-\gamma_i\phi(\xi)\Big)\\
&=\int_\Sigma F(t_1,\ldots,t_n)\\
&\quad\cdot\Bigg(-\sum_{j=1}^N w_{ij}(-it_1^{w_{1j}}\cdots t_n^{w_{nj}})\frac{\partial\phi}{\partial\xi_j}(-it_1^{w_{11}}\cdots t_n^{w_{n1}},\ldots,-it_1^{w_{1N}}\cdots t_n^{w_{nN}})\\
&\qquad-\gamma_i\phi(-it_1^{w_{11}}\cdots t_n^{w_{n1}},\ldots,-it_1^{w_{1N}}\cdots t_n^{w_{nN}})\Bigg)dt_1\cdots dt_n\\
&=\int_\Sigma F(t_1,\ldots,t_n)\Big(-t_i\frac{\partial}{\partial t_i}-\gamma_i\Big)\phi(-it_1^{w_{11}}\cdots t_n^{w_{n1}},\ldots,-it_1^{w_{1N}}\cdots t_n^{w_{nN}})dt_1\cdots dt_n\\
&=\int_\Sigma\Big(\frac{\partial}{\partial t_i}(t_iF)-\gamma_iF\Big)\phi(-it_1^{w_{11}}\cdots t_n^{w_{n1}},\ldots,-it_1^{w_{1N}}\cdots t_n^{w_{nN}})dt_1\cdots dt_n,
\end{aligned}$$

where for the last equality, we use the Stokes formula. So we should choose $F(t_1,\ldots,t_n)$ with the property

$$\frac{\partial}{\partial t_i}(t_iF)-\gamma_iF=0.$$

Let's take

$$F(t_1, \ldots, t_n) = t_1^{\gamma_1 - 1} \cdots t_n^{\gamma_n - 1}.$$

We then have

$$(g(\xi), \phi(\xi)) = \int_\Sigma t_1^{\gamma_1} \cdots t_n^{\gamma_n} \phi(-it_1^{w_{11}} \cdots t_n^{w_{n1}}, \ldots, -it_1^{w_{1N}} \cdots t_n^{w_{nN}}) \frac{dt_1}{t_1} \cdots \frac{dt_n}{t_n}.$$

We have

$$\begin{aligned}
&(f(x), \hat{\phi}(x)) = (\hat{g}(x), \hat{\phi}(x)) = (g(\xi), \phi(-\xi)) \\
&= \int_\Sigma t_1^{\gamma_1} \cdots t_n^{\gamma_n} \phi(it_1^{w_{11}} \cdots t_n^{w_{n1}}, \ldots, it_1^{w_{1N}} \cdots t_n^{w_{nN}}) \frac{dt_1}{t_1} \cdots \frac{dt_n}{t_n} \\
&= \frac{1}{(2\pi)^{\frac{N}{2}}} \int_\Sigma t_1^{\gamma_1} \cdots t_n^{\gamma_n} \int_{\mathbb{R}^N} \hat{\phi}(x_1, \ldots, x_N) e^{\sum_{j=1}^N x_j t_1^{w_{1j}} \cdots t_n^{w_{nj}}} dx_1 \cdots dx_N \frac{dt_1}{t_1} \cdots \frac{dt_n}{t_n}.
\end{aligned}$$

So we have

$$\begin{aligned}
(f(x), \hat{\phi}(x)) = &\frac{1}{(2\pi)^{\frac{N}{2}}} \int_{\mathbb{R}^N} \hat{\phi}(x_1, \ldots, x_N) \\
&\times \Big(\int_\Sigma t_1^{\gamma_1} \cdots t_n^{\gamma_n} e^{\sum_{j=1}^N x_j t_1^{w_{1j}} \cdots t_n^{w_{nj}}} \frac{dt_1}{t_1} \cdots \frac{dt_n}{t_n} \Big) dx_1 \cdots dx_N,
\end{aligned}$$

and hence

$$f(x_1, \ldots, x_N) = \frac{1}{(2\pi)^{\frac{N}{2}}} \int_\Sigma t_1^{\gamma_1} \cdots t_n^{\gamma_n} e^{\sum_{j=1}^N x_j t_1^{w_{1j}} \cdots t_n^{w_{nj}}} \frac{dt_1}{t_1} \cdots \frac{dt_n}{t_n}.$$

This is an integral representation of a generalized function solution of the GKZ system, where Σ can be taken to be any n-dimensional cycle in $\mathbb{T}^n$.

4 GKZ hypergeometric sheaves

Let $\mathbb{F}_q$ be a finite field of characteristic p with q elements, ℓ a prime number distinct from p, $\chi_1, \ldots, \chi_n : \mathbb{F}_q^* \to \overline{\mathbb{Q}}_\ell^*$ multiplicative characters, and $\psi : \mathbb{F}_q \to \overline{\mathbb{Q}}_\ell^*$ a nontrivial additive character. A typical choice for ψ is

$$\psi(x) = e^{\frac{2\pi i}{p} \mathrm{Tr}_{\mathbb{F}_q/\mathbb{F}_p}(x)}.$$

An analogue of the integral

$$f(x_1, \ldots, x_N) = \int_\Sigma t_1^{\gamma_1} \cdots t_n^{\gamma_n} e^{\sum_{j=1}^N x_j t_1^{w_{1j}} \cdots t_n^{w_{nj}}} \frac{dt_1}{t_1} \cdots \frac{dt_n}{t_n}$$

in the end of the previous section is

$$\mathrm{Hyp}_\psi(\chi_1, \ldots, \chi_n; x_1, \ldots, x_n) = \sum_{t_1, \ldots, t_n \in \mathbb{F}_q^*} \chi_1(t_1) \cdots \chi_n(t_n) \psi\Big(\sum_{j=1}^N x_j t_1^{w_{1j}} \cdots t_n^{w_{nj}} \Big).$$

This is the hypergeometric function over finite field introduced by Gelfand-Graev ([5], [6]).

We have a canonical isomorphism

$$\mathrm{Gal}\Big(\big(\mathbb{F}_q(t)[T]/(T^{q-1}-t)\big)/\mathbb{F}_q(t)\Big) \xrightarrow{\cong} \mathbb{F}_q^*, \quad \sigma \mapsto \frac{\sigma(T)}{T}.$$

For any multiplicative character $\chi : \mathbb{F}_q^* \to \overline{\mathbb{Q}}_\ell^*$, consider the Galois representation

$$\mathrm{Gal}\Big(\overline{\mathbb{F}_q(t)}/\mathbb{F}_q(t)\Big) \to \mathrm{Gal}\Big(\big(\mathbb{F}_q(t)[T]/(T^{q-1}-t)\big)/\mathbb{F}_q(t)\Big) \xrightarrow{\cong} \mathbb{F}_q^* \xrightarrow{\chi^{-1}} \overline{\mathbb{Q}}_\ell^*.$$

This Galois representation is unramified on $\mathbb{T}^1 = \mathbb{P}^1 - \{0, \infty\}$, and defines the *Kummer sheaf* $\mathscr{K}_\chi$ on $\mathbb{T}^1$. For any $\mathbb{F}_{q^k}$-points x on $\mathbb{T}^1$, we have

$$(\mathrm{Frob}_x, (\mathscr{K}_\chi)_{\bar{x}}) = \chi(\mathrm{Norm}_{\mathbb{F}_{q^k}/\mathbb{F}_q}(x)).$$

We have a canonical isomorphism

$$\mathrm{Gal}\Big(\big(\mathbb{F}_q(t)[T]/(T^q - T - t)\big)/\mathbb{F}_q(t)\Big) \xrightarrow{\cong} \mathbb{F}_q, \quad \sigma \mapsto \sigma(T) - T.$$

For the additive character $\psi : \mathbb{F}_q \to \overline{\mathbb{Q}}_\ell^*$, consider the Galois representation

$$\mathrm{Gal}\Big(\overline{\mathbb{F}_q(t)}/\mathbb{F}_q(t)\Big) \to \mathrm{Gal}\Big(\big(\mathbb{F}_q(t)[T]/(T^q - T - t)\big)/\mathbb{F}_q(t)\Big) \xrightarrow{\cong} \mathbb{F}_q \xrightarrow{\psi^{-1}} \overline{\mathbb{Q}}_\ell^*.$$

This Galois representation is unramified on $\mathbb{A}^1 = \mathbb{P}^1 - \{\infty\}$, and defines the *Artin-Schreier sheaf* $\mathscr{L}_\psi$ on $\mathbb{A}^1$. For any $\mathbb{F}_{q^k}$-points x on $\mathbb{A}^1$, we have

$$(\mathrm{Frob}_x, (\mathscr{L}_\psi)_{\bar{x}}) = \psi(\mathrm{Tr}_{\mathbb{F}_{q^k}/\mathbb{F}_q}(x)).$$

Let $\mathbb{A}^N$ be the N-dimensional affine space over $\mathbb{F}_q$, $\mathbb{T}^n$ the n-dimensional torus over $\mathbb{F}_q$, $\mathscr{K}_\chi$ the sheaf $p_1^*\mathscr{K}_{\chi_1} \otimes \cdots \otimes p_n^*\mathscr{K}_{\chi_n}$ on $\mathbb{T}^n \cong \mathbb{T}^1 \times \cdots \times \mathbb{T}^1$, F the morphism

$$F : \mathbb{T}^n \times \mathbb{A}^N \to \mathbb{A}^1, \quad (t_1, \ldots, t_n, x_1, \ldots, x_N) \mapsto \sum_{j=1}^N x_j t_1^{w_{1j}} \cdots t_n^{w_{nj}},$$

and $\pi_1 : \mathbb{T}^n \times \mathbb{A}^N \to \mathbb{T}^n$, $\pi_2 : \mathbb{T}^n \times \mathbb{A}^N \to \mathbb{A}^N$ the projections. We define the *ℓ-adic GKZ hypergeometric sheaf* associated to the Kummer sheaf $\mathscr{K}_\chi$ and the subset $A = \{\mathbf{w}_1, \ldots, \mathbf{w}_N\}$ of $\mathbb{Z}^n$ to be

$$\mathrm{Hyp}_\psi(\chi) = R\pi_{2!}\Big(\pi_1^*\mathscr{K}_\chi \otimes F^*\mathscr{L}_\psi\Big)[n+N].$$

It is an object in the derived category $D_c^b(\mathbb{A}^N, \overline{\mathbb{Q}}_\ell)$ of ℓ-adic sheaves on $\mathbb{A}^N$. Let $x = (x_1, \ldots, x_N)$ be a $\mathbb{F}_q$-rational point of $\mathbb{A}^N$. By the Grothendieck trace formula, we have

$$\begin{aligned}
&\mathrm{Tr}\Big(\mathrm{Frob}_x, (\mathrm{Hyp}_\psi(\chi))_{\bar{x}}\Big)\\
&= (-1)^{n+N}\mathrm{Tr}\Big(\mathrm{Frob}_x, \Big(R\pi_{2!}\Big(\pi_1^*\mathscr{K}_\chi\otimes F^*\mathscr{L}_\psi\Big)\Big)_{\bar{x}}\Big)\\
&= (-1)^{n+N}\sum_{t=(t_1,\ldots,t_n)}\Big(\mathrm{Frob}_{(t,x)}, (\pi_1^*\mathscr{K}_\chi\otimes F^*\mathscr{L}_\psi)_{\overline{(t,x)}}\Big)\\
&= (-1)^{n+N}\sum_{t=(t_1,\ldots,t_n)}\Big(\mathrm{Frob}_{(t,x)}, (\pi_1^*\mathscr{K}_\chi)_{\overline{(t,x)}}\Big)\Big(\mathrm{Frob}_{(t,x)}, (F^*\mathscr{L}_\psi)_{\overline{(t,x)}}\Big)\\
&= (-1)^{n+N}\sum_{t=(t_1,\ldots,t_n)}\Big(\mathrm{Frob}_{\pi_1(t,x)}, (\mathscr{K}_\chi)_{\overline{\pi_1(t,x)}}\Big)\Big(\mathrm{Frob}_{F(t,x)}, (\mathscr{L}_\psi)_{\overline{F(t,x)}}\Big)\\
&= (-1)^{n+N}\sum_{t=(t_1,\ldots,t_n)}\Big(\mathrm{Frob}_t, (\mathscr{K}_\chi)_{\bar{t}}\Big)\Big(\mathrm{Frob}_{F(t,x)}, (\mathscr{L}_\psi)_{\overline{F(t,x)}}\Big)\\
&= (-1)^{n+N}\sum_{t_1,\ldots,t_n\in\mathbb{F}_q^*}\chi_1(t_1)\cdots\chi_n(t_n)\psi\Big(\sum_{j=1}^N x_j t_1^{w_{1j}}\cdots t_n^{w_{nj}}\Big)\\
&= (-1)^{n+N}\mathrm{Hyp}_\psi(x_1,\ldots,x_N;\chi_1,\ldots,\chi_n).
\end{aligned}$$

So traces of Frobenius elements on stalks of $\mathrm{Hyp}_\psi(\chi)$ give rise to the hypergeometric function over finite field.

Recall the following theorem.

Theorem 4.1 (Adolphson [1], Gelfand-Zelevinsky-Kapranov [7]). *Suppose* $\mathbf{w}_1,\ldots,\mathbf{w}_N$ *generate* $\mathbb{Z}^n$.

(i) *The GKZ hypergeometric system is holonomic.*

(ii) *Suppose* $(\gamma_1,\ldots,\gamma_n)$ *satisfies the so-called semi-nonresonance condition. Then the dimension of the space of GKZ hypergeometric functions at a generic point is* $n!\mathrm{vol}(\Delta)$, *where* Δ *is the convex hull of* $\{0,\mathbf{w}_1,\ldots,\mathbf{w}_N\}$.

(iii) *Suppose* $(\gamma_1,\ldots,\gamma_n)$ *satisfies the so-called non-resonance condition. Then the sheaf of GKZ hypergeometric functions defines an irreducible local system on a Zariski dense open subset of* $\mathbb{C}^N$.

(iv) *If A satisfies nonconfluence condition, then the GKZ hypergeometric system is regular holonomic.*

In [3], we prove the following theorem.

Theorem 4.2.

(i) $\mathrm{Hyp}_\psi(\chi)$ *is a mixed perverse sheaf on* $\mathbb{A}^N$ *of weights* $\leq n+N$.

(ii) *The rank of* $\mathrm{Hyp}_\psi(\chi)$ *is* $(-1)^N n!\mathrm{vol}(\Delta)$, *where the rank of an object in* $D_c^b(\mathbb{A}^N,\overline{\mathbb{Q}}_\ell)$ *is the dimension of the stalk of the object at the generic point.*

(iii) *Suppose* $\mathbf{w}_1,\ldots,\mathbf{w}_N$ *generate* $\mathbb{Z}^n$, *and suppose* $\chi_1,\ldots,\chi_n$ *satisfy the non-resonance condition defined below. Then* $\mathrm{Hyp}_\psi(\chi)$ *is an irreducible pure perverse sheaf of weight* $n+N$.

(iv) *If A satisfies nonconfluence condition, then the GKZ hypergeometric sheaf is tame.*

(v) *Let*

$$P(\mathrm{Hyp}_\psi(\chi)) = \sum_{w\in\mathbb{Z}} e_w T^w,$$

where e_w *is the rank of the weight* w *sub-quotient of the weight filtration for the mixed perverse sheaf* $\mathrm{Hyp}_\psi(\chi)$. *Then we have*

$$\begin{aligned} P(\mathrm{Hyp}_\psi(\chi)) &= E(\Delta,\chi),\\ e_{n+N} &= e(\Delta,\chi), \end{aligned}$$

where $E(\Delta,\chi)$ *and* $e(\Delta,\chi)$ *are defined combinatorially below.*

(vi) *Let* U *be a Zariski open subset of* $\mathbb{A}^N$ *such that for any geometric point* $a = (a_1,\ldots,a_N)$ *of* U, *the Laurent polynomial* $F_a(t_1,\ldots,t_n) = \sum_{j=1}^N a_j t_1^{w_{1j}}\cdots t_n^{w_{nj}}$ *is non-degenerate with respect to* Δ *in the sense that for any face* τ *of* Δ *not containing* 0, *the locus*

$$\frac{\partial F_{a,\tau}}{\partial t_1} = \cdots = \frac{\partial F_{a,\tau}}{\partial t_n} = 0$$

in $\mathbb{T}^n$ *is empty, where* $F_{a,\tau} = \sum_{\mathbf{w}_j\in\tau} a_j t_1^{w_{1j}}\cdots t_n^{w_{nj}}$. *Then* $\mathrm{Hyp}_\psi(\chi)|_U = \mathscr{H}[n]$ *for some lisse sheaf* $\mathscr{H}$ *on* U.

Note that from Theorem 4.2 (i), (ii) and (vi), one can recover [2, Theorem 1.3] and [4, Proposition 0.1]. It shows that if $\sum_{j=1}^N a_j t_1^{w_{1j}}\cdots t_n^{w_{nj}}$ with $a_j\in\mathbb{F}_q$ is non-degenerate with respect to Δ, then we have the estimate

$$\Big|\sum_{t_1,\ldots,t_n\in\mathbb{F}_q^*} \chi_1(t_1)\cdots\chi_n(t_n)\psi\Big(\sum_{j=1}^N a_j t_1^{w_{1j}}\cdots t_n^{w_{nj}}\Big)\Big| \le n!\mathrm{vol}(\Delta)q^{\frac{n}{2}}.$$

For any convex polyhedral cone δ in $\mathbb{R}^n$ with 0 being a face, define the convex polytope $\mathrm{poly}(\delta)$ to be the intersection of δ with a hyperplane in $\mathbb{R}^n$ which does not contain 0 and intersects each one dimensional face of δ. Note that $\mathrm{poly}(\delta)$ is defined only up to combinatorial equivalence. For any convex polytope Δ in $\mathbb{R}^n$ and any face Γ of Δ, define $\mathrm{cone}_\Delta(\Gamma)$ to be the cone generated by $u'-u$ ($u'\in\Delta$, $u\in\Gamma$), and define $\mathrm{cone}^\circ_\Delta(\Gamma)$ to be the image of $\mathrm{cone}_\Delta(\Gamma)$ in $\mathbb{R}^n/\mathrm{span}(\Gamma-\Gamma)$. We define polynomials $\alpha(\delta)$ and $\beta(\Delta)$ in one variable T inductively by the following formulas:

$$\begin{aligned} \alpha(\{0\}) &= 1,\\ \beta(\Delta) &= (T^2-1)^{\dim(\Delta)} + \sum_{\Gamma \text{ face of } \Delta,\ \Gamma\neq\Delta} (T^2-1)^{\dim(\Gamma)}\alpha(\mathrm{cone}^\circ_\Delta(\Gamma)),\\ \alpha(\delta) &= \mathrm{trunc}_{\le \dim(\delta)-1}((1-T^2)\beta(\mathrm{poly}(\delta))), \end{aligned}$$

where $\mathrm{trunc}_{\le d}(\cdot)$ denotes taking the degree $\le d$ part of a polynomial. These polynomials are first introduced by Stanley. Note that $\alpha(\delta)$ and $\beta(\Delta)$ depend only on the combinatorial types of δ and Δ.

Let Δ be the convex hull of $\{0,\mathbf{w}_1,\ldots,\mathbf{w}_N\}$, let δ be the convex polyhedral cone generated by $\mathbf{w}_1,\ldots,\mathbf{w}_N$, and let T be the set of faces τ of δ so that $\tau\neq\delta$

and $\mathscr{K}_\chi \cong p_\tau^*\mathscr{K}_{\chi_\tau}$ for a Kummer sheaf $\mathscr{K}_{\chi_\tau}$ on the torus $\mathbb{T}_\tau = \operatorname{Spec} k[\mathbb{Z}^n \cap \operatorname{span}\tau]$, where

$$p_\tau : \mathbb{T}_k^n = \operatorname{Spec} k[\mathbb{Z}^n] \to \mathbb{T}_\tau = \operatorname{Spec} k[\mathbb{Z}^n \cap \operatorname{span}\tau]$$

is the morphism induced by the inclusion $k[\mathbb{Z}^n \cap \operatorname{span}\tau] \hookrightarrow k[\mathbb{Z}^n]$. $e(\Delta,\chi)$ and $E(\Delta,\chi)$ in Theorem 4.2 (v) are defined by

$$e(\Delta,\chi) = (-1)^N n!\mathrm{vol}(\Delta) + \sum_{\tau\in T}(-1)^{n-\dim(\tau)+N}(\dim(\tau))!\mathrm{vol}(\Delta\cap\tau)\alpha(\mathrm{cone}_\delta^\circ(\tau))(1),$$

$$E(\Delta,\chi) = e(\Delta,\chi)T^{n+N} - \sum_{\tau\in T}(-1)^{n-\dim(\tau)+N-N_\tau}T^{N-N_\tau}E(\Delta\cap\tau,\chi_\tau)\alpha(\mathrm{cone}_\delta^\circ(\tau)),$$

where N_τ is the number of those $\mathbf{w}_j$ ($j = 1,\ldots,N$) lying in τ, $\mathrm{cone}_\delta(\tau)$ is the cone generated by $u' - u$ ($u' \in \delta$, $u \in \tau$), and $\mathrm{cone}_\delta^\circ(\tau)$ is the image of $\mathrm{cone}_\delta(\tau)$ in $\mathbb{R}^n/\mathrm{span}(\tau)$. We say $\chi_1,\ldots,\chi_n$ satisfy the *non-resonance condition* if $\mathscr{K}_\chi \not\cong p_\tau^* K_{\chi_\tau}$ for any proper face τ of δ, that is, the set T is empty.

One of the main tool to prove Theorem 4.2 is the Deligne-Fourier transform. Fix notation by the following diagram:

$$\begin{array}{ccccc} \mathbb{A}^N & \stackrel{p_1}{\leftarrow} & \mathbb{A}^N\times\mathbb{A}^N & \stackrel{p_2}{\rightarrow} & \mathbb{A}^N \\ & & \langle\,,\rangle\downarrow & & \\ & & \mathbb{A}^1 & & \end{array}$$

where $\langle\,,\rangle : \mathbb{A}^N\times\mathbb{A}^N \to \mathbb{A}^1$ is the morphism defined by $\langle\xi,x\rangle = \sum_{j=1}^N \xi_j x_j$. The Deligne-Fourier transform is the functor

$$\mathscr{F}_\psi : D_c^b(\mathbb{A}^N,\overline{\mathbb{Q}}_\ell) \to D_c^b(\mathbb{A}^N,\overline{\mathbb{Q}}_\ell),\quad \mathscr{F}_\psi(K) = Rp_{2!}(p_1^*K\otimes\langle\,,\rangle^*\mathscr{L}_\psi)[N].$$

For any $\mathbb{F}_q$-rational points ξ and x of $\mathbb{A}^N$, let $f(\xi) = \mathrm{Tr}(\mathrm{Frob}_\xi, K_{\bar\xi})$ and let $\mathscr{F}(f)(x) = \mathrm{Tr}(\mathrm{Frob}_x,(\mathscr{F}_\psi(K))_{\bar x})$. Then we have

$$\mathscr{F}(f)(x) = \sum_\xi f(\xi)\psi(\langle\xi,x\rangle).$$

The Deligne-Fourier transform is an exact functor. It preserves the category of perverse sheaves, the weight filtration, the Verdier duality, and we have a Fourier inversion formula. See [10] for a complete treatment of the Deligne-Fourier transform. Let ι be the morphism

$$\iota : \mathbb{T}^n \to \mathbb{A}^N,\quad (t_1,\ldots,t_n)\mapsto (t_1^{w_{11}}\cdots t_n^{w_{n1}},\ldots,t_1^{w_{1N}}\cdots t_n^{w_{nN}}).$$

We have the following.

Proposition 4.3.

(i) *The morphism ι is quasi-finite and affine, and* $\mathrm{Hyp}_\psi(\chi) \cong \mathscr{F}_\psi(\iota_!\mathscr{K}_\chi[n])$.

(ii) *The weight $n+N$ subquotient of the weight filtration for the mixed perverse sheaf* $\mathrm{Hyp}_\psi(\chi)$ *is isomorphic to* $\mathscr{F}_\psi(\iota_{!*}(\mathscr{K}_\chi[n]))$.

(iii) *Suppose $\mathbf{w}_1,\ldots,\mathbf{w}_N$ generates $\mathbb{Z}^n$. Then ι is an immersion. Suppose furthermore that $\chi_1,\ldots,\chi_n$ satisfy the nonresonance condition. Then* $\iota_!(\mathscr{K}_\chi[n]) \cong \iota_{!*}(\mathscr{K}_\chi[n])$.

Let's explain how part of Theorem 4.2 follows from Proposition 4.3. From the fact that ι is quasi-finite and affine, one concludes that $\iota_!\mathscr{K}_\chi[n]$ is a mixed perverse sheaf of weight $\leq n$. It follows that $\mathscr{F}_\psi(\iota_!\mathscr{K}_\chi[n])$ is a mixed perverse sheaf of weights $\leq n+N$. Theorem 4.2 (i) then follows from Proposition 4.3 (i). Suppose $\mathbf{w}_1,\ldots,\mathbf{w}_N$ generates $\mathbb{Z}^n$ and $\chi_1,\ldots,\chi_n$ satisfy the nonresonance condition. Then ι is an immersion and $\iota_!(\mathscr{K}_\chi[n]) \cong \iota_{!*}(\mathscr{K}_\chi[n])$ is an irreducible perverse sheaf pure of weight n. It follows that $\mathrm{Hyp}_\psi(\chi) \cong \mathscr{F}_\psi(\iota_!\mathscr{K}_\chi[n])$ is irreducible and pure of weight $n+N$. This proves Theorem 4.2 (iii).

Again let δ be the convex polyhedral cone in $\mathbb{R}^n$ generated by $\mathbf{w}_1,\ldots,\mathbf{w}_N$, let T be the set of proper faces τ of δ so that $\mathscr{K}_\chi \cong p_\tau^*\mathscr{K}_{\chi_\tau}$ for a Kummer sheaf $\mathscr{K}_{\chi_\tau}$ on $\mathbb{T}_\tau$ under the morphism

$$p_\tau : \mathbb{T}^n = \operatorname{Spec}\mathbb{F}_q[\mathbb{Z}^n] \to \mathbb{T}_\tau = \operatorname{Spec}\mathbb{F}_q[\mathbb{Z}^n \cap \operatorname{span}\tau],$$

let N_τ be the number of those $\mathbf{w}_j$ ($j=1,\ldots,N$) lying in τ, let $\mathrm{Hyp}_\psi(\chi_\tau)$ be the GKZ-hypergeometric sheaf associated to the Kummer sheaf χ_τ and the vectors $(\mathbf{w}_j)_{\mathbf{w}_j\in\tau}$ in $\mathbb{Z}^n\cap\operatorname{span}\tau$, and let q_τ be the projection

$$q_\tau : \mathbb{A}^N \to \mathbb{A}^{N_\tau}, \quad (x_1,\ldots,x_N) \mapsto (x_j)_{\mathbf{w}_j\in\tau}.$$

Define $\operatorname{cone}_\delta(\tau)$ to be the cone generated by $x'-x$ ($x'\in\delta, x\in\tau$), and $\operatorname{cone}^\circ_\delta(\tau)$ the image of $\operatorname{cone}_\delta(\tau)$ under the canonical homomorphism $\mathbb{R}^n \to \mathbb{R}^n/\operatorname{span}\tau$. Let $x_\tau : \operatorname{Spec}\mathbb{F}_q \to \operatorname{Spec}\mathbb{F}_q[(\mathbb{Z}^n/\mathbb{Z}^n\cap\operatorname{span}\tau)\cap\operatorname{cone}^\circ_\delta(\tau)]$ be the point defined by the $\mathbb{F}_q$-homomorphism

$$F_q[(\mathbb{Z}^n/\mathbb{Z}^n\cap\operatorname{span}\tau)\cap\operatorname{cone}^\circ_\delta(\tau)] \to \mathbb{F}_q, \quad u \mapsto \begin{cases} 1 & \text{if } u=0, \\ 0 & \text{if } u\neq 0 \end{cases}$$

for any $u\in(\mathbb{Z}^n/\mathbb{Z}^n\cap\operatorname{span}\tau)\cap\operatorname{cone}^\circ_\delta(\tau)$, and let $j_\tau : \operatorname{Spec}k[\mathbb{Z}^n/\mathbb{Z}^n\cap\operatorname{span}\tau] \hookrightarrow \operatorname{Spec}\mathbb{F}_q[(\mathbb{Z}^n/\mathbb{Z}^n\cap\operatorname{span}\tau)\cap\operatorname{cone}^\circ_\delta(\tau)]$ be the immersion of the open dense torus in the affine toric scheme $\operatorname{Spec}\mathbb{F}_q[(\mathbb{Z}^n/\mathbb{Z}^n\cap\operatorname{span}\tau)\cap\operatorname{cone}^\circ_\delta(\tau)]$. The proof of Theorem 4.2 (v) relies on the following theorem and an induction argument.

Theorem 4.4. *In the K-group $K(D_c^b(\mathbb{A}^N,\overline{\mathbb{Q}}_\ell))$ of $D_c^b(\mathbb{A}^N,\overline{\mathbb{Q}}_\ell)$, we have*

$$\begin{aligned}&\Big[\mathrm{Hyp}_\psi(\chi)\Big]\\ &= \Big[\mathscr{F}_\psi(\iota_{!*}(\mathscr{K}_\chi[n]))\Big] - \sum_{\tau\in T}\Big[q_\tau^*\mathrm{Hyp}_\psi(\chi_\tau)[N-N_\tau]\otimes x_\tau^*(j_{\tau !*}\overline{\mathbb{Q}}_\ell[n-\dim\tau])\Big].\end{aligned}$$

We sketch the proof of Theorem 4.4. The morphism

$$\iota : \mathbb{T}^n \to \mathbb{A}^N, \quad (t_1,\ldots,t_N) \mapsto (t_1^{w_{11}}\cdots t_n^{w_{n1}},\ldots,t_1^{w_{1N}}\cdots t_n^{w_{nN}})$$

is induced by the $\mathbb{F}_q$-homomorphism

$$\iota^\natural : \mathbb{F}_q[\xi_1,\ldots,\xi_N] \to \mathbb{F}_q[t_1,\ldots,t_n], \quad \xi_j \mapsto t_1^{w_{1j}}\cdots t_n^{w_{nj}}.$$

The scheme theoretic image $\overline{\mathrm{im}\,\iota}$ of ι is the closed subscheme $\mathrm{Spec}\,(\mathbb{F}_q[\xi_1,\ldots,\xi_N]/\ker\iota^\natural)$ of $\mathbb{A}^N$. We have

$$\mathbb{F}_q[\xi_1,\ldots,\xi_N]/\ker\iota^\natural \cong \mathrm{im}\,\iota^\natural \cong \mathbb{F}_q[S],$$

where S is the sub-semigroup of $\mathbb{Z}^n$ generated by $\mathbf{w}_1,\ldots,\mathbf{w}_N$. So $\overline{\mathrm{im}\,\iota}$ is isomorphic to $\mathrm{Spec}\,\mathbb{F}_q[S]$, which is an affine toric variety. The normalization of $\mathrm{Spec}\,\mathbb{F}_q[S]$ is the toric variety $X(\Sigma(\delta)) = \mathrm{Spec}\,\mathbb{F}_q[\mathbb{Z}^n\cap\delta]$. Let g be the composite

$$X(\Sigma(\delta)) \to \overline{\mathrm{im}\,\iota} \to \mathbb{A}^N,$$

which is a finite morphism, and let $j:\mathbb{T}^n\to X(\Sigma(\delta))$ be the immersion of the dense torus. We have $\iota = g\circ j$. Recall the following result [4, Lemma 1.3]:

Lemma 4.5. *Let τ be a face of δ, and let O_τ be the orbit of torus action of $\mathbb{T}^n$ on $X(\Sigma(\delta))$ corresponding to τ.*

(i) *If $\mathscr{K}_\chi \not\cong p_\tau^* K_{\chi_\tau}$ for any Kummer sheaf $\mathscr{K}_{\chi_\tau}$, then*

$$(j_{!*}(\mathscr{K}_\chi[n]))|_{O_\tau} = 0.$$

(ii) *If $\mathscr{K}_\chi \cong p_\tau^*\mathscr{K}_{\chi_\tau}$, then*

$$(j_{!*}(\mathscr{K}_\chi[n]))|_{O_\tau} \cong (\mathscr{K}_{\chi_\tau}[\dim(\tau)])\otimes x_\tau^*(j_{\tau!*}(\overline{\mathbb{Q}}_\ell[n-\dim(\tau)])).$$

Let's continue the proof of Theorem 4.4. Let $\kappa_\tau : O_\tau \to X(\Sigma(\delta))$ be the immersion. By Lemma 4.5, in the K-group $K(D_c^b(X(\Sigma(\delta)),\overline{\mathbb{Q}}_\ell))$, we have

$$\Big[j_{!*}(\mathscr{K}_\chi[n])\Big] = \Big[j_!(\mathscr{K}_\chi[n])\Big] + \sum_{\tau\in T}\Big[\kappa_{\tau!}(\mathscr{K}_{\chi_\tau}[\dim\tau])\otimes x_\tau^*(j_{\tau!*}(\overline{\mathbb{Q}}_\ell[n-\dim\tau]))\Big].$$

Applying $g_! = Rg_*$ to the above equality, we get that in the K-group $K(D_c^b(\mathbb{A}'^N,\overline{\mathbb{Q}}_\ell))$, we have

$$\Big[\iota_{!*}(\mathscr{K}_\chi[n])\Big] = \Big[\iota_!(\mathscr{K}_\chi[n])\Big] + \sum_{\tau\in T}\Big[(g\kappa_\tau)_!(\mathscr{K}_{\chi_\tau}[\dim\tau])\otimes x_\tau^*(j_{\tau!*}(\overline{\mathbb{Q}}_\ell[n-\dim\tau]))\Big].$$

Applying the Deligne-Fourier transform to this equality, we get

$$\begin{aligned}&\Big[\mathscr{F}_\psi(\iota_{!*}(\mathscr{K}_\chi[n]))\Big]\\ &= \Big[\mathrm{Hyp}_\psi(\chi)\Big] + \sum_{\tau\in T}\Big[q_\tau^*\mathrm{Hyp}_\psi(\chi_\tau)[N-N_\tau]\otimes x_\tau^*(j_{\tau!*}\overline{\mathbb{Q}}_\ell[n-\dim\tau])\Big].\end{aligned}$$

This proves Theorem 4.4.

Also note that Proposition 4.3 (iii) follows from Lemma 4.5 (i).

We end this paper with a question.

Question 4.6. Define a mixed (possibly irregular) Hodge module structure on the GKZ hypergeometric D-module corresponding to the GKZ hypergeometric system. Study the pure piece corresponding to $\mathscr{F}_\psi(\iota_{!*}(\mathscr{K}_\chi[n]))$. Study the variation of Hodge structure corresponding to this pure piece.

References

[1] A. Adolphson, Hypergeometric functions and rings generated by monomials, *Duke Math. J.* **73** (1994), 269–290.

[2] J. Denef and F. Loeser, Weight of exponential sums, intersection cohomology and Newton polyhedra, *Invent. Math.* **106** (1991), 275–294.

[3] L. Fu, ℓ-adic GKZ hypergeometric sheaves and exponential sums, arXiv: alge.geom/1208.1373, to appear in *Adv. Math.*.

[4] L. Fu, Weight of twisted exponential sums, *Math. Zeit.* **262** (2009), 449–472.

[5] I. M. Gelfand, M. I Graev, Hypergeometric functions over finite fields, English translation, *Doklady Math.* **64** (2001), 402–406.

[6] I. M. Gelfand, M. I. Graev, V. S. Retakh, Hypergeometric functions over an arbitrary field, English translation, *Russian Math. Surveys* **59** (2004), 831–905.

[7] I. M. Gelfand, A. V. Zelevinsky, M. M. Kapranov, Hypergeometric functions and toric varieties, English translation, *Functional Anal. Appl.* **23** (1989), 94–106; Correction to the paper "Hypergeometric functions and toric varieties", English translation, *Functional. Anal. Appl.* **27** (1995), 295.

[8] I. M. Gelfand, M. M. Kapranov, A. V. Zelevinsky, Generalized Euler integrals and A-hypergeometric functions, *Adv. Math.* **84** (1990), 255–271.

[9] I. M. Gelfand, G. E. Shilov, Spaces of fundamental and generalized functions, *Generalized Functions*, Volume 2, English translation, Academic Press 1968.

[10] G. Laumon, Transformation de Fourier, constantes d'équations fontionnelles, et conjecture de Weil, *Publ. Math. IHES* **65** (1987), 131–210.

Chern Institute of Mathematics and LPMC, Nankai University, Tianjin 300071, China.

E-mail address: leifu@nankai.edu.cn

Proceedings of the 6th International Congress of Chinese Mathematicians
ALM 36, pp. 297–307

Note on Affine Deligne-Lusztig Varieties

Xuhua He

Abstract

This note is based on my talk at ICCM 2013, Taipei. We give an exposition of the group-theoretic method and recent results on the questions of non-emptiness and dimension of affine Deligne-Lusztig varieties in affine flag varieties.

Mathematics Subject Classification: 14L05, 20G25.

Keywords and Phrases: Affine Weyl groups, σ-conjugacy classes, affine Deligne-Lusztig varieties.

Introduction

Affine Deligne-Lusztig varieties were first introduced by Rapoport in [18], generalizing Deligne and Lusztig's classical construction. Understanding the emptiness/non-emptiness pattern and dimension of affine Deligne-Lusztig varieties is fundamental in the study of reduction of Shimura varieties with parahoric level structures.

In this note, we will discuss the new group-theoretic method to study affine Deligne-Lusztig varieties in affine flag varieties and anwser the above questions in terms of class polynomials of affine Hecke algebras. The two key ingredients are the Deligne-Lusztig reduction [2] and the combinatorial properties of affine Weyl groups [15].

1 Affine Deligne-Lusztig varieties

1.1

Let $\mathbb{F}_q$ be a finite field with q elements and $\mathbf{k}$ be an algebraic closure of $\mathbb{F}_q$. We consider the field $L = \mathbf{k}((\epsilon))$ and its subfield $F = \mathbb{F}_q((\epsilon))$. The Frobenius

The author is partially supported by HKRGC grant 602011.

automorphism σ of $\mathbf{k}/\mathbb{F}_q$ can be extended in the usual way to an automorphism of L/F such that $\sigma(\sum a_n \epsilon^n) = \sum \sigma(a_n)\epsilon^n$.

Let G be a split connected reductive group over $\mathbb{F}_q$ and let I be a σ-stable Iwahori subgroup of $G(L)$. The affine Deligne-Lusztig variety associated with w in the extended affine Weyl group $\tilde{W} \cong I \backslash G(L)/I$ and $b \in G(L)$ is defined to be

$$X_w(b) = \{gI \in G(L)/I; g^{-1}b\sigma(g) \in IwI\}.$$

Here we embed $\tilde{W}$ set-theoretically into $G(L)$.

This is a locally closed sub-ind scheme of the affine flag variety $G(L)/I$. It is a finite dimensional $\mathbf{k}$-scheme, locally of finite type over $\mathbf{k}$.[1]

1.2

To illustrate some difficulties in the study of affine Deligne-Lusztig varieties, let us begin with (classical) Deligne-Lusztig varieties. Let B be a Borel subgroup of G defined over $\mathbb{F}_q$, and W be the associate finite Weyl group. For any $w \in W$ and $b \in G(\mathbf{k})$, the classical Deligne-Lusztig variety $X_{w,b}$ is defined to be

$$X_{w,b} = \{gB \in G(\mathbf{k})/B(\mathbf{k}); g^{-1}b\sigma(g) \in B(\mathbf{k})wB(\mathbf{k})\}.$$

Note that in both the classical case and the affine cases, if b and b' are in the same σ-conjugacy class, then the associated classical/affine Deligne-Lusztig varieties are isomorphic. In other words, the classical/affine Deligne-Lusztig varieties depends on the finite/affine Weyl group element w and the σ-conjugacy class $[b]$.

However, in the classical case, $G(\mathbf{k})$ is a single σ-conjugacy class. Therefore, we may omit b in the definition and simply write

$$X_w = \{gB \in G(\mathbf{k})/B(\mathbf{k}); g^{-1}\sigma(g) \in B(\mathbf{k})wB(\mathbf{k})\}.$$

This is what appears in the literature. The classical Deligne-Lusztig variety X_w is always nonempty and is a smooth variety of dimension $\ell(w)$.

The loop group $G(L)$, on the other hand, contains infinitely many σ-conjugacy classes. The extra parameter $[b]$ makes the study of affine Deligne-Lusztig varieties much harder than that of the classical ones.

From a different point of view, the two parameters w and $[b]$ in the definition of affine Deligne-Lusztig varieties appear naturally. There are two important stratifications on the special fiber of a Shimura variety with Iwahori level structure: One is the Kottwitz-Rapoport stratification whose strata are indexed by specific elements $w \in \tilde{W}$; The other is the Newton stratification whose strata are indexed by specific σ-conjugacy classes $[b] \subseteq G(L)$. There is a close relation between the affine Deligne-Lusztig variety $X_w(b)$ and the intersection of the Newton stratum associated with $[b]$ with the Kottwitz-Rapoport stratum associated with w (see [9] and [23]).

---

[1] One may consider any connected reductive group over F that splits over a tamely ramified extension of L. As discussed for example in [12, Sections 6 and 7], the questions we'll discuss in this note can be reduced to quasi-split unramified groups. For simplicity, we only consider split groups in this note.

2 Conjugacy classes of $\tilde{W}$

2.1

To understand the affine Deligne-Lusztig varieties $X_w(b)$, one needs to understand the σ-conjugacy classes of $G(L)$ and its relation with the extended affine Weyl group $\tilde{W}$. We will relate the conjugacy classes of $\tilde{W}$ with the σ-conjugacy classes of $G(L)$. Based on the decomposition $G(L) = \sqcup_{w \in \tilde{W}} IwI$, this sounds plausible. However, the naive map sending IwI to the σ-conjugacy class of w does not work. The reasons are as follows:

(1) Given two elements w and w' in the same conjugacy class of $\tilde{W}$, the set IwI and $Iw'I$ may not be σ-conjugated to each other.

(2) One σ-conjugacy class of $G(L)$ may contain elements of different conjugacy classes in $\tilde{W}$. For example, in the classical case, $G(\mathbf{k})$ is a single σ-conjugacy class and W contains several conjugacy classes.

The ideas to overcome the difficulties is to use the minimal length elements in the conjugacy classes of $\tilde{W}$. We will recall some properties of the minimal length elements in this section and discuss the applications to σ-conjugacy classes of $G(L)$ in the next section.

2.2

We have the semidirect product

$$\tilde{W} = P \rtimes W = \{t^\lambda w; \lambda \in P, w \in W\},$$

where P is the coweight lattice of a split maximal torus T of G. Let Q be the coroot lattice of T. Then $W_a = Q \rtimes W \subseteq \tilde{W}$ is an affine Weyl group. The length function and Bruhat order on W_a extends in a natural way to $\tilde{W}$ and $\tilde{W} = W_a \rtimes \Omega$, where

$$\Omega = \{w \in \tilde{W}; \ell(w) = 0\} \cong P/Q.$$

Let $\kappa : \tilde{W} \to \tilde{W}/W_a$ be the natural projection. We call κ the Kottwitz map.

Let $P_{\mathbb{Q}} = P \otimes_{\mathbb{Z}} \mathbb{Q}$ and $P_{\mathbb{Q}}/W$ be the quotient of $P_{\mathbb{Q}}$ by the natural action of W. We may identify $P_{\mathbb{Q}}/W$ with $P_{\mathbb{Q},+}$, the set of dominant rational coweights. For any $w \in \tilde{W}$, $w^{n_0} = t^\lambda$ for some $\lambda \in P$, where $n_0 = \sharp(W)$. Let ν_w be the unique element in $P_{\mathbb{Q},+}$ that lies in the W-orbit of λ/n_0. We call the map

$$\tilde{W} \to P_{\mathbb{Q},+}, \qquad w \mapsto \nu_w$$

the Newton map.

Define $f : \tilde{W} \to \tilde{W}/W_a \times P_{\mathbb{Q},+}$ by $w \mapsto (\kappa(w), \nu_w)$. It is constant on each conjugacy class of $\tilde{W}$. We denote the image of f by $B(\tilde{W})$. Note that in general, a fiber of f contains more than one conjugacy class of $\tilde{W}$.

2.3

We call an element $w \in \tilde{W}$ straight if $\ell(w) = \langle \nu_w, 2\rho \rangle$, where ρ is the half sum of all the positive roots. It is easy to see that w is straight if and only if $\ell(w^n) = n\ell(w)$ for all $n \in \mathbb{N}$. We call a conjugacy class of $\tilde{W}$ straight if it contains some straight element. Note that a straight element is automatically of minimal length in its conjugacy class.

By [15, Proposition 3.2], a conjugacy class is straight if and only if it contains a length-zero element in the extended affine Weyl group $\tilde{W}_M$ for some standard Levi subgroup M of G.

By [15, Theorem 3.3], the map $f : \tilde{W} \to \tilde{W}/W_a \times P_{\mathbb{Q},+}$ induces a bijection from the set of straight conjugacy classes to $B(\tilde{W})$.[2]

2.4

Now we discuss the minimal length elements in a conjugacy class $\mathcal{O}$ of $\tilde{W}$. The following result is obtained in [15], motivated by previous results of Geck and Pfeiffer [8] for finite Coxeter groups.

Theorem 2.1. *Let $\mathcal{O}$ be a conjugacy class of $\tilde{W}$. Then*

(1) *Each element of $\mathcal{O}$ can be brought to a minimal length element by conjugation by simple reflections which reduce or keep constant the length.*

(2) *Any two minimal length elements in $\mathcal{O}$ are conjugate in the associated Braid group.*

(3) *If moreover, $\mathcal{O}$ is straight, then any two minimal length elements in $\mathcal{O}$ are conjugate by cyclic shifts.*

Let $\mathcal{O}$ be a straight conjugacy class and $\mathcal{O}'$ be another conjugacy class such that $f(\mathcal{O}) = f(\mathcal{O}')$. Then the minimal length elements in $\mathcal{O}'$ are related to the straight elements in $\mathcal{O}$ in the sense of [15, Theorem 3.4]. This property is used to overcome the difficulty §2.1 (2).

3 σ-conjugacy classes of $G(L)$

3.1

Recall that $G(L) = \sqcup_{w \in \tilde{W}} IwI$. Let $w \in \tilde{W}$ and $s \in \tilde{W}$ be a simple reflection. Then

$$IsIwI = \begin{cases} IswI, & \text{if } sw > w; \\ IswI \sqcup IwI, & \text{if } sw < w. \end{cases}$$

$$IwIsI = \begin{cases} IwsI, & \text{if } ws > w; \\ IwsI \sqcup IwI, & \text{if } ws < w. \end{cases}$$

[2] In [15, Section 3], we assume that G is adjoint. However, [15, Theorem 3.3] holds for any reductive group and the proof is the same as in loc. cit.

Therefore,

$$G(L) \cdot_\sigma IwI = \begin{cases} G(L) \cdot_\sigma IswsI, & \text{if } \ell(sws) = \ell(w); \\ G(L) \cdot_\sigma IswsI \cup G(L) \cdot_\sigma IswI, & \text{if } \ell(sws) < \ell(w). \end{cases}$$

Here $\cdot_\sigma$ denotes the σ-conjugation action.

This equality, together with Theorem 2.1, gives a reduction method in the study of $G(L) \cdot_\sigma IwI$ and allows us to reduce the general case to the case where w of minimal length in its conjugacy class. This is used to overcome the difficulty §2.1 (1).

We have the following parameterisation of σ-conjugacy classes of $G(L)$.

Theorem 3.1. *There is a canonical bijection between*
(a) *The set of σ-conjugacy classes of $G(L)$;*
(b) *The set of straight conjugacy classes of $\tilde{W}$;*
(c) *The image of $f : \tilde{W} \to \tilde{W}/W_a \times P_{\mathbb{Q},+}$.*

Here the bijection between (a) and (c) follows from Kottwitz's classification of σ-conjugacy classes [17] together with the fact that any σ-conjugacy class is represented by an element in $\tilde{W}$ [7, Corollary 7.2.2]. The bijection between (b) and (c) is discussed in §2.3. The bijection between (a) and (b) (and a new proof of the classification of σ-conjugacy classes) is obtained in [12, Section 3] using the strategy in §3.1.

3.2

Now we introduce partial orders on the three sets in Theorem 3.1.

Let C, C' be σ-conjugacy classes of $G(L)$. We write $C \leqslant C'$ if C is contained in the closure of C'.

Let $\mathcal{O}, \mathcal{O}'$ be straight conjugacy classes of $\tilde{W}$. We write $\mathcal{O} \leqslant \mathcal{O}'$ if for some w' of minimal length in $\mathcal{O}'$, there exists w of minimal length in $\mathcal{O}$ such that $w \leqslant w'$ with respect to the Bruhat order in $\tilde{W}$. It is proved in [10, §4.7] and [11, Corollary 7.5] that $\leqslant$ is a partial order on the set of straight conjugacy classes.

Let $(k, \nu), (k', \nu') \in \tilde{W}/W_a \times P_{\mathbb{Q},+}$. We write $(k, \nu) \leqslant (k', \nu')$ if $k = k'$ and $\nu' - \nu \in \sum_\alpha \mathbb{R}_{\geqslant 0}\alpha$, where α runs over all the simple roots. This partial order is studied in detail by Chai in [1].

Theorem 3.2. *Let C, C' be σ-conjugacy classes of $G(L)$ and $\mathcal{O}, \mathcal{O}'$ the corresponding straight conjugacy classes in $\tilde{W}$. Then the following conditions are equivalent:*
(1) $C \leqslant C'$;
(2) $\mathcal{O} \leqslant \mathcal{O}'$;
(3) $f(\mathcal{O}) \leqslant f(\mathcal{O}')$.

Here (1) $\Rightarrow$ (3) is proved by Rapoport and Richartz in [19, Theorem 3.6], (3) $\Rightarrow$ (1) is proved by Viehmann in [21, Theorem 2] and the equivalence between (1) and (2) is proved in [11, Section 11] and [13]. It is easy to show that (2) $\Rightarrow$ (3).

It is interesting to give a direct (combinatorial proof) that (3) $\Rightarrow$ (2).[3]

4 "Dimension=degree" theorem

4.1

We first recall the reduction method of Deligne and Lusztig [2, Theorem 1.6].

Theorem 4.1. *Let $w \in \tilde{W}$, and let s be a simple affine reflection.*

(1) If $\ell(sws) = \ell(w)$, then there exists a universal homeomorphism

$$X_w(b) \to X_{sws}(b).$$

(2) If $\ell(sws) = \ell(w) - 2$, then $X_w(b)$ can be written as a disjoint union $X_w(b) = X_1 \sqcup X_2$ where X_1 is closed and X_2 is open, and such that there exist morphisms $X_1 \to X_{sws}(b)$ and $X_2 \to X_{sw}(b)$ which are compositions of a Zariski-locally trivial fiber bundle with one-dimensional fibers and a universal homeomorphism.

The reduction method, together with Theorem 2.1, in principle, reduce the study of $X_w(b)$ to the case where w is of minimal length in its conjugacy class. The latter one, is studied in detail in [14] and [12]. In particular,

Theorem 4.2. *Let $w \in \tilde{W}$ be an element of minimal length in its conjugacy class and $b \in G(L)$.*

(1) If b and w are not in the same σ-conjugacy class of $G(L)$, then $X_w(b) = \emptyset$.

(2) If b and w are in the same σ-conjugacy class of $G(L)$, then $\dim X_w(b) = \ell(w) - \langle \nu_w, 2\rho \rangle$.

The emptiness/nonemptiness pattern and dimension formula for affine Deligne-Lusztig varieties is obtained if we can keep track of the reduction step from an arbitrary element to a minimal length element. This is accomplished via the class polynomials of affine Hecke algebras.

4.2

Let $\tilde{H}$ be the Hecke algebra associated with $\tilde{W}$, i.e., $\tilde{H}$ is the associated $A = \mathbb{Z}[v, v^{-1}]$-algebra with basis T_w for $w \in \tilde{W}$ and multiplication is given by

$$T_w T_{w'} = T_{ww'}, \quad \text{if } \ell(w) + \ell(w') = \ell(ww');$$
$$(T_s - v)(T_s + v^{-1}) = 0, \quad \text{for any simple reflection } s.$$

It is proved in [15] that for any $w \in \tilde{W}$ and conjugacy class $\mathcal{O}$ of $\tilde{W}$, there exists a unique polynomial $f_{w,\mathcal{O}} \in \mathbb{N}[v - v^{-1}]$ such that for any finite dimensional representation V of $\tilde{H}$,

$$Tr(T_w, V) = \sum_{\mathcal{O}} f_{w,\mathcal{O}} Tr(T_{w_{\mathcal{O}}}, V),$$

where $w_{\mathcal{O}}$ is a minimal length element in $\mathcal{O}$.

[3] In fact, (3) $\Leftrightarrow$ (1) $\Rightarrow$ (2) holds for any tamely ramified group. We expect that (2) $\Rightarrow$ (1) also holds for any tamely ramified group. As explained in the beginning of this note, it suffices to prove it for quasi-split unramified groups.

4.3

Now we state the main result in [12], which relates the dimension of affine Delgine-Lusztig varieties with the degree of the class polynomials.

Theorem 4.3. *Let $b \in G(L)$ and $w \in \tilde{W}$. Then*

$$\dim(X_w(b)) = \max_{\mathcal{O}} \frac{1}{2}(\ell(w) + \ell(\mathcal{O}) + \deg(f_{w,\mathcal{O}})) - \langle \bar{\nu}_b, 2\rho \rangle,$$

here $\mathcal{O}$ runs over conjugacy classes of $\tilde{W}$ with $f(\mathcal{O}) = f(b)$ and $\ell(\mathcal{O})$ is the length of any minimal length element in $\mathcal{O}$.

Here we use the convention that the dimension of an empty variety and the degree of a zero polynomial are both $-\infty$. So in particular, $X_w(b) \neq \emptyset$ if and only if $f_{w,\mathcal{O}} \neq 0$ for some conjugacy class $\mathcal{O}$ of $\tilde{W}$ with $f(\mathcal{O}) = f(b)$.

5 Affine Deligne-Lusztig varieties for basic b

5.1

We say that an element $b \in G(L)$ is basic if $\langle \nu_b, \alpha \rangle = 0$ for any root α of G. For basic b, we are able to give a more explicit description of the emptiness/nonemptiness pattern and dimension formula, as conjectured by Görtz, Haines, Kottwitz and Reuman in [7, Conjectures 1.1.1 and 1.1.3]. We first discuss the emptiness/nonemptiness pattern. It is given in terms of the P-alcoves introduced in [7, Definition 2.1.1].

Let $P = MN$ be a semistandard parabolic subgroup of G and $w \in \tilde{W}$. We say that w is a P-alcove element if $w \in \tilde{W}_M$ and $N(L) \cap wIw^{-1} \subseteq N(L) \cap I$.

Theorem 5.1. *Let $w \in \tilde{W}$ and $b \in G(L)$ be a basic element. Then $X_w(b) \neq \emptyset$ if and only if for any semistandard parabolic subgroup $P = MN$ such that w is a P-alcove element, $\kappa_M(b) \in \kappa_M([b] \cap M(L))$.*

The "only if" part is proved in [7, Theorem 1.1.2] as a consequence of the Hodge-Newton decomposition for affine Deligne-Lusztig varieties [7, Theorem 1.1.4]. The "if" part is proved in [5, Theorem A] by showing that the notion of P-alcoves is compatible with the Deligne-Lusztig reduction. An algebraic proof of the analogy of Hodge-Newton decomposition for affine Hecke algebras and a new proof of the "only if" part of the Theorem 5.1 is obtained in [16].

5.2

Define

- $\eta_1 : \tilde{W} = P \rtimes W \to W$, the projection map.
- $\eta_2 : \tilde{W} \to W$ such that $\eta_2(w)^{-1}w$ lies in the dominant Weyl chamber.
- $\eta(w) = \eta_2(w)^{-1}\eta_1(w)\eta_2(w)$.

As a consequence of Theorem 5.1, there is a simpler criterion for emptiness/nonemptiness if w lies in the shrunken Weyl chamber.

Corollary 5.2. *Assume that the Dynkin diagram of G is connected. Let $b \in G(L)$ be a basic element and $w \in \tilde{W}$ lies in the shrunken Weyl chamber. Then $X_w(b) \neq \emptyset$ if and only if $\kappa(b) = \kappa(w)$ and $\eta(w)$ is not in any proper parabolic subgroup of W.*

5.3

The following dimension formula is conjectured in [7, Conjecture 1.1.3] and proved in [12, Theorem 12.1].

Theorem 5.3. *Let $b \in G(L)$ be a basic element and $w \in \tilde{W}$ lie in the shrunken Weyl chamber. If $X_w(b) \neq \emptyset$, then*

$$\dim X_w(b) = \frac{1}{2}(\ell(w) + \ell(\eta(w)) - \mathit{def}(b)),$$

where def(b) is the defect of b.

Here the lower bound is obtained by constructing an explicit sequence from w to a minimal length element w' in some conjugacy class of $\tilde{W}$ (in most cases, w and w' are not in the same conjugacy class). The upper bound is obtained by combining the "partial conjugation" method in [10] and the dimension formula for affine Deligne-Lusztig varieties in affine Grassmannian in [6, Theorem 2.15.1] and [20, Theorem 1.1]. If w lies in the shrunken Weyl chamber, then the lower bound and the upper bound coincide and the theorem is proved.[4]

It is a challenging problem to give a close formula of $\dim X_w(b)$ for w in the critical strips.

6 Affine Deligne-Lusztig varieties for nonbasic b

6.1

An element $w \in \tilde{W}$ can be written in a unique way as $xt^\mu y$, where $x, y \in W$ and $\mu \in P$ such that $t^\mu y$ sends the fundamental alcove to an alcove in the dominant Weyl chamber. In this case, $\eta(xt^\mu y) = yx$.

Let w_0 be the longest element in W. For any $\mu \in P_+$, $w_0 t^\mu$ is the unique maximal element in its $W \times W$-coset. For such element, there is a complete answer for the emptiness/nonemptiness pattern and dimension formula.

Theorem 6.1. *Let $b \in G(L)$ and $\mu \in P_+$. Then $X_{w_0 t^\mu}(b) \neq \emptyset$ if and only if $f(b) \leqslant f(t^\mu)$. In this case,*

$$\dim X_{w_0 t^\mu}(b) = \langle \mu - \nu_b, \rho \rangle + \ell(w_0) - \frac{1}{2}\mathit{def}(b).$$

[4] I learned from Viehmann [22] that her student Paul Hamacher recently proved the conjectural dimension formula for affine Deligne-Lusztig varieties in affine Grassmannian for unramified groups. Combining this result, with the proof in [12] as we outlined above, and [5, Proposition 2.5.1], we are able to generalize the above theorem to any tamely ramified groups.

The affine Deligne-Lusztig variety $X_{w_0 t^\mu}(b)$ in the affine flag and the affine Deligne-Lusztig variety $X_\mu(b)$ in the affine Grassmannian are related as in [12, Theorem 10.1]. For affine Deligne-Lusztig varieties in the affine Grassmannian, the emptiness/nonemptiness is obtained by Gashi in [3, Theorem 1.1] and the dimension formula is obtained by Görtz, Haines, Kottwitz and Reuman [6, Theorem 2.15.1] and Viehmann [20, Theorem 1.1].

6.2

For other w, not much is known. Görtz, Haines, Kottwitz and Reuman made the following conjecture [7, Conjecture 9.5.1 (b)] which relates the affine Deligne-Lusztig variety $X_w(b)$ with $X_w(b')$ for basic b'.

Conjecture 6.2. *Let $b \in G(L)$ and b' be a basic element in $G(L)$ such that $\kappa(b) = \kappa(b')$. Then for $w \in \tilde{W}$ with sufficiently large length, $X_w(b) \neq \emptyset$ if and only if $X_w(b') \neq \emptyset$. In this case,*

$$\dim X_w(b) = \dim X_w(b') - \langle \nu_b, \rho \rangle + \frac{1}{2}(\mathit{def}(b') - \mathit{def}(b)).$$

Many numerical evidence for group of small rank is obtained by computer in support of this conjecture. Another evidence for split b is obtained in [13].

Theorem 6.3. *Assume that the Dynkin diagram of G is connected. Let $\mu \in P_+$ and $\lambda \in Q$ be a dominant and regular coweight. Then for any $x, y \in W$, $X_{xt^{\mu+\lambda}y}(t^\mu) \neq \emptyset$ if and only if yx is not in any proper parabolic subgroup of W. In this case,*

$$\dim X_{xt^{\mu+\lambda}y}(t^\mu) = \langle \lambda, \rho \rangle + \frac{1}{2}(\ell(x) + \ell(y) + \ell(yx)).$$

Acknowledgements. We are grateful to M. Rapoport and E. Viehmann for their comments on this note.

References

[1] C.-L. Chai, *Newton polygons as lattice points*, Amer. J. Math. **122** (2000), 967–990.

[2] P. Deligne and G. Lusztig, *Representations of reductive groups over finite fields*, Ann. of Math. (2) **103** (1976), no. 1, 103–161.

[3] Q. Gashi, *On a conjecture of Kottwitz and Rapoport*, Ann. Sci. École Norm. Sup. (4) **43** (2010), no. 6, 1017–1038.

[4] U. Görtz and X. He, *Dimension of affine Deligne-Lusztig varieties in affine flag varieties*, Documenta Math. **15** (2010), 1009–1028.

[5] U. Görtz, X. He, and S. Nie, *P-alcoves and nonemptiness of affine Deligne-Lusztig varieties*, arXiv:1211.3784.

[6] U. Görtz, T. Haines, R. Kottwitz, and D. Reuman, *Dimensions of some affine Deligne-Lusztig varieties*, Ann. Sci. École Norm. Sup. (4) **39** (2006), 467–511.

[7] U. Görtz, T. Haines, R. Kottwitz, and D. Reuman, *Affine Deligne-Lusztig varieties in affine flag varieties*, Compos. Math. **146** (2010), no. 5, 1339–1382.

[8] M. Geck and G. Pfeiffer, *On the irreducible characters of Hecke algebras*, Adv. Math. **102** (1993), no. 1, 79–94.

[9] T. Haines, *Introduction to Shimura varieties with bad reduction of parahoric type*, Clay Math. Proc. **4** (2005), 583–642.

[10] X. He, *Minimal length elements in some double cosets of Coxeter groups*, Adv. Math. **215** (2007), no. 2, 469–503.

[11] X. He, *Minimal length elements of extended affine Weyl group, I*, arXiv:1004.4040, preprint.

[12] X. He, *Geometric and homological properties of affine Deligne-Lusztig varieties*, Ann. of Math., to appear.

[13] X. He, unpublished note.

[14] X. He and G. Lusztig, *A generalization of Steinberg's cross-section*, J. Amer. Math. Soc. **25** (2012), 739–757.

[15] X. He and S. Nie, *Minimal length elements of extended affine Weyl group, II*, arXiv:1112.0824.

[16] X. He and S. Nie, *P-alcoves, parabolic subalgebras and cocenters of affine Hecke algebras*, preprint.

[17] R. Kottwitz, *Isocrystals with additional structure, II*, Compositio Math. **109** (1997), no. 3, 255–339.

[18] M. Rapoport, *A guide to the reduction modulo p of Shimura varieties*, Astérisque **298** (2005), 271–318.

[19] M. Rapoport and M. Richartz, *On the classification and specialization of F-isocrystals with additional structure*, Compositio Math. **103** (1996), 153–181.

[20] E. Viehmann. *The dimension of affine Deligne-Lusztig varieties*, Ann. Sci. École Norm. Sup. (4) **39** (2006), 513–526.

[21] E. Viehmann, *Newton strata in the loop group of a reductive group*, Amer. J. Math. **135** (2013), no. 2, 499–518.

[22] E. Viehmann, *Private communication*, 08/20/2013.

[23] E. Viehmann and T. Wedhorn, *Ekedahl-Oort and Newton strata for Shimura varieties of PEL type*, Math. Ann. **356** (2013), no. 4, 1493–1550.

Department of Mathematics and Institute for Advanced Study
The Hong Kong University of Science and Technology
Clear Water Bay, Kowloon, Hong Kong, China
E-mail address: maxhhe@ust.hk

Department of Mathematics, University of Maryland,
College Park 20742, USA
E-mail address: xuhuahe@math.umd.edu

Proceedings of the 6th International Congress of Chinese Mathematicians
ALM 36, pp. 309–325

On the Local Langlands Conjecture and Related Problems over p-adic Local Fields

Dihua Jiang and Chufeng Nien

Abstract

We discuss two issues related to the Local Langlands Conjecture over p-adic local fields. One is about the uniqueness of the local Langlands reciprocity map, which is a consequence of the local converse theorem; and the other is to establish the Local Langlands Conjecture for quasi-split classical groups based on the result for the tempered case, which is obtained by J. Arthur for symplectic and orthogonal groups and by C.-P. Mok for unitary groups.

Mathematics Subject Classification: Primary 11F70, 22E50; Secondary 11F85, 22E55.

Keywords and Phrases: Local gamma factors, Local Langlands Conjecture, local converse theorem, admissible representations.

1 Introduction

Let F be a p-adic local field of characteristic zero. The Local Langlands Conjecture for $\mathrm{GL}_n(F)$, which is proved by Harris and Taylor ([HT01]) and by Henniart ([H00]), is a correspondence between admissible n-dimensional complex representations of the Weil-Deligne group of F and admissible complex representations of $\mathrm{GL}_n(F)$, with certain constraints. It is regarded as the non-abelian extension of the classical Artin reciprocity law in the local abelian classfield theory. For a brief review of the progress towards the complete proof of the Local Langlands Conjecture for $\mathrm{GL}_n(F)$, we refer the readers to [HT01]. In this paper, we pursue two issues related to the Local Langlands Conjecture.

The work of the firsted named author is supported in part by NSF grants DMS–1001672 and DMS–1301567; and the work of the second named author is supported by NSC grants 102-2115-M-006-003.

One is about the uniqueness of the local Langlands reciprocity map, which follows from the local converse theorem for irreducible admissible generic representations of F-quasi-split reductive algebraic groups. In this aspect, we will discuss how to refine the local converse theorem, and in particular, discuss the recent progress towards the conjecture of H. Jacquet.

Another is about the Local Langlands Conjecture for classical groups. In order to extend the Local Langlands Conjecture from $\mathrm{GL}_n(F)$ to other classical groups, one may take an approach using the local Langlands functorial transfer from classical groups to the general linear group, via the standard embedding of the corresponding dual groups. We will discuss the problems involved in this approach.

In Section 2, we review briefly the Local Langlands Conjecture for $\mathrm{GL}_n(F)$, which is a theorem of Harris and Taylor ([HT01]) and of Henniart ([H00]). To discuss the issue on the uniqueness of the local Langlands reciprocity map, we continue with Section 3 on the local converse problem, where we explain the formulation of the local converse problem and our recent progress, joint with S. Stevens, on the Jacquet conjecture. The implication of the any positive answer to the local converse problem towards the uniqueness of the local Langlands reciprocity map is also discussed there. In Section 4, we discuss the second issue, i.e. the Local Langlands Conjecture for classical groups. First, we recall the tempered case from the work of Arthur ([Ar13]) and of Mok ([Mk15]), and then show that the general case follows from the tempered case by the Langlands classification theory. Section 5 discusses more detailed issues on how to use the local Langlands functorial transfer from classical groups to the general linear groups. In addition to the establishment of the local Langlands functorial transfer, one has to characterize the image of the transfer and show the subjectivity of the local reciprocity map. We explain all the issues for SO_{2n+1}, which should be applicable to all other classical groups.

Finally, we would like to thank the Mathematics Division of "the National Center for Theoretical Sciences", Taiwan, China, for funding support which makes parts of the collaborating work of the authors over the years possible.

2 The Local Langlands Conjecture

Let $\overline{F}$ be the algebraic closure of the local field F and $\Gamma_F = \mathrm{Gal}(\overline{F}/F)$ be the absolute Galois group. Let F^{ur} be the maximal unramified extension of F and denote by $\Gamma_F^{\mathrm{ur}} = \mathrm{Gal}(F^{\mathrm{ur}}/F)$ the corresponding Galois group. Consider the exact sequence

$$1 \to I_F \to \Gamma_F \to \Gamma_F^{\mathrm{ur}} \to 1,$$

with I_F the inertia group. The Galois group $\Gamma_F^{\mathrm{ur}} \cong \hat{\mathbb{Z}}$ is generated by the Frobenius element Frob. The Weil group $\mathcal{W}_F$ is defined through the following exact sequence

$$1 \to I_F \to \mathcal{W}_F \to \mathbb{Z} =< \mathrm{Frob} > \to 1$$

where $\mathbb{Z}$ is equipped with the discrete topology, which yields the topology on the Weil group $\mathcal{W}_F$. The Weil-Deligne group is defined to be the product $\mathcal{W}_F \times \mathrm{SL}_2(\mathbb{C})$

(a popular version). A Frobenius semisimple representation ρ of the Weil-Deligne group $\mathcal{W}_F \times \mathrm{SL}_2(\mathbb{C})$ is a representation satisfying the following two conditions:

1. $\rho|_{\mathcal{W}_F}$ is continuous and $\rho(\mathrm{Frob})$ is semisimple;

2. $\rho|_{\mathrm{SL}_2(\mathbb{C})}$ is algebraic.

Define $\Phi(\mathrm{GL}_n)$ to be the set of equivalence classes of n-dimensional Frobenius semisimple representations of $\mathcal{W}_F \times \mathrm{SL}_2(\mathbb{C})$.

Let (V, π) be a representation of $\mathrm{GL}_n(F)$, where V is a $\mathbb{C}$-vector space with $\dim_{\mathbb{C}} V \leq \infty$. A $\mathbb{C}$-representation V_π of $\mathrm{GL}_n(F)$ is called *smooth* if $\forall v \in V_\pi$, the stabilizer of v in $\mathrm{GL}_n(F)$

$$\mathrm{Stab}_{\mathrm{GL}_n(F)}(v) := \{g \in \mathrm{GL}_n(F) \mid \pi(g)(v) = v\}$$

is an open subgroup of $\mathrm{GL}_n(F)$. A smooth representation V_π of $\mathrm{GL}_n(F)$ is called *admissible* if for any open compact subgroup K of $\mathrm{GL}_n(F)$, the subspace K-invariants

$$V_\pi^K := \{v \in V_\pi \mid \pi(k)(v) = v, \forall k \in K\}$$

is of finite-dimension. Let $\Pi(\mathrm{GL}_n)$ be the set of equivalence classes of irreducible admissible representations of $\mathrm{GL}_n(F)$.

The Local Langlands Conjecture for $\mathrm{GL}_n(F)$ is proved by Harris and Taylor ([HT01]) and by Henniart ([H00]) and can be stated as follows.

Theorem 2.1. (Harris-Taylor and Henniart) *For each integer $n \geq 1$, there is a unique bijection*

$$\mathrm{rec}_{F,n} \; : \; \Pi(\mathrm{GL}_n) \longrightarrow \Phi(\mathrm{GL}_n)$$

such that for $\pi \in \Pi(\mathrm{GL}_n)$, $\pi' \in \Pi(\mathrm{GL}_{n'})$ and $\chi \in \Pi(\mathrm{GL}_1)$, the following hold:

1. *the Artin reciprocity law holds*

$$\mathrm{rec}_{F,1}(\chi) = \chi \circ \mathrm{Art}_F^{-1},$$

where Art_F is the Artin map from $F^\times$ to $\mathcal{W}_F^{\mathrm{ab}}$;

2. *the local factors are preserved*

$$L(s, \pi \times \pi') = L(s, \mathrm{rec}_{F,n}(\pi) \otimes \mathrm{rec}_{F,n'}(\pi'))$$

and

$$\epsilon(s, \pi \times \pi', \psi_F) = \epsilon(s, \mathrm{rec}_{F,n}(\pi) \otimes \mathrm{rec}_{F,n'}(\pi'), \psi_F)$$

for a given nontrivial additive character ψ_F of F;

3. *the twisting by character is preserved*

$$\mathrm{rec}_{F,n}(\pi \otimes (\chi \circ \det)) = \mathrm{rec}_{F,n}(\pi) \otimes \mathrm{rec}_{F,1}(\chi);$$

4. *the central character condition holds*

$$\det \circ \mathrm{rec}_{F,n}(\pi) = \mathrm{rec}_{F,n}(\omega_\pi),$$

where ω_π is the central character of π; and

5. *the duality holds*

$$\mathrm{rec}_{F,n}(\pi^\vee) = \mathrm{rec}_{F,n}(\pi)^\vee,$$

where $\vee$ denotes the contragredient.

The uniqueness of the local Langlands reciprocity $\mathrm{rec}_{F,n}$ follows from the local converse theorem for irreducible admissible representations of $\mathrm{GL}_n(F)$, which will be discussed with detail in the next section.

3 The local converse problem

Recall that the local L-factor $L(s, \pi \times \pi')$ and the local ϵ-factor $\epsilon(s, \pi \times \pi', \psi_F)$ in Theorem 2.1 were defined by Jacquet, Piatetski-Shapiro, and Shalika in [JPSS83]. The corresponding local gamma factor of the Rankin-Selberg type is defined by

$$\gamma(s, \pi \times \pi', \psi_F) = \epsilon(s, \pi \times \pi', \psi_F) \cdot \frac{L(1-s, \pi^\vee \times \pi'^\vee)}{L(s, \pi \times \pi')}.$$

On the Galois side, for $\phi, \phi' \in \Phi(\mathrm{GL}_n)$, Tate, following Deligne and Langlands, defined in [T79] the local gamma factors of Artin type by

$$\gamma(s, \phi \times \phi', \psi_F) = \epsilon(s, \phi \times \phi', \psi_F) \cdot \frac{L(1-s, \phi^\vee \times \phi'^\vee)}{L(s, \phi \times \phi')}.$$

Theorem 2.1 also implies that the following identity holds:

$$\gamma(s, \pi \times \pi', \psi_F) = \gamma(s, \mathrm{rec}_{F,n}(\pi) \times \mathrm{rec}_{F,n'}(\pi'), \psi_F)$$

for $\pi \in \Pi(\mathrm{GL}_n)$ and $\pi' \in \Pi(\mathrm{GL}_{n'})$.

It is clear that for $\pi \in \Pi(\mathrm{GL}_n)$, the local gamma factors for pair, $\gamma(s, \pi \times \tau, \psi_F)$, with $\tau \in \Pi(\mathrm{GL}_m)$ and with all $m = 1, 2, \ldots$, are invariants attached to π. The question is: *Is this family of invariants enough to determine π, up to equivalence?*

It is easy to show that if $m \geq n$, the family $\gamma(s, \pi \times \tau, \psi_F)$ with $\tau \in \Pi(\mathrm{GL}_m)$ determines π completely. Hence it remains to ask the following question:

Local converse problem. Find $n_0 < n$ such that the family $\gamma(s, \pi \times \tau, \psi_F)$ with $\tau \in \Pi(\mathrm{GL}_m)$ for $m = 1, 2, \ldots, n_0$ determine π completely.

We remark that any positive answer to the *local converse problem* implies the uniqueness of the collection of the reciprocity maps in the Local Langlands Conjecture for GL_n, i.e. in Theorem 2.1.

In fact, this can be verified as follows. If there are two reciprocity s σ_n and σ'_n from $\Phi(\mathrm{GL}_n)$ to $\Pi(\mathrm{GL}_n)$, then by the compatibility of local gamma factors, one has

$$\gamma(s, \sigma_n(\phi) \times \tau, \psi_F) = \gamma(s, \sigma'_n(\phi) \times \tau, \psi_F)$$

for all $\tau \in \Pi(\mathrm{GL}_m)$ with $m = 1, 2, \ldots, n-1$. Hence $\sigma_n(\phi) = \sigma'_n(\phi)$ holds for $\forall \phi \in \Phi(\mathrm{GL}_n)$ and $\forall n > 1$.

The progress towards the best answer to the *local converse problem* is briefly recalled here. In 1970, Jacquet and Langlands ([JL70]) show that for $n = 2$ and $n_0 = 1$, the *local converse problem* has a positive answer. Then in 1979, Jacquet, Piatetski-Shapiro and Shalika ([JPSS79]) show that for $n = 3$ and $n_0 = 1$, the *local converse problem* has a positive answer. After fourteen years, Henniart in [H93] proves that for general $n > 1$ with $n_0 = n - 1$, the *local converse problem* has a positive answer. This general result is refined by Chen in his 1996 Ph.D. thesis at Yale University and show that for general $n > 1$, it is enough to take $n_0 = n - 2$. This paper is published in 2006 ([Ch06]). On the other hand, Cogdell and Piatetski-Shapiro use the global method to prove in [CPS99] that the same result holds.

In the early 1980s, Jacquet made his conjecture on the possible answer to the *local converse problem.*

Conjecture 3.1. (Jacquet) *The* **local converse problem** *should have a positive answer for general $n > 1$ with $n_0 = [\frac{n}{2}]$, that is, for any $\pi_1, \pi_2 \in \Pi(\mathrm{GL}_m)$, if the identity for the local gamma factors for pairs*

$$\gamma(s, \pi_1 \times \tau, \psi_F) = \gamma(s, \pi_2 \times \tau, \psi_F)$$

holds for all $\tau \in \Pi(\mathrm{GL}_m)$ with $m = 1, 2, \ldots, [\frac{n}{2}]$, then π_1 must be isomorphic to π_2.

It is clear that the progress to the *local converse problem* accumulated over the years is still far away from the expectation of the Jacquet conjecture for general $n > 4$.

Recently, we started to investigate the finite field analogy of the Jacquet conjecture. The second named author is able to give a complete proof over finite fields of the Jacquet conjecture, based on the thesis work of E.-A. Roditty under the supervision of David Soudry at Tel Aviv University. The main idea and the new ingredient in her proof is to find special Bessel function with extra symmetry in the space of Whittaker models of irreducible cuspidal representations of GL_n over finite fields. This appears in [N14].

This idea has been carried over to general p-adic local fields. To be more precise, we recall from [JNS13] the basic notions and facts.

Let U_n be the unipotent radical of the standard Borel subgroup B_n of GL_n, which consists of all upper-triangular matrices. Denote by P_n the mirabolic subgroup of $\mathrm{GL}_n(F)$, consisting of matrices with last row equal to $(0, \ldots, 0, 1)$. We also fix a standard non-degenerate character ψ_n of U_n, so that all Whittaker functions are implicitly ψ_n-Whittaker functions.

Let π be an irreducible unitarizable supercuspidal representation of $\mathrm{GL}_n(F)$. Take K to be an open subgroup of $\mathrm{GL}_n(F)$, which is compact-mod-center. A Whittaker function W_π for π is called K-*special* if it satisfies:

$$W_{\pi_i}(g^{-1}) = \overline{W_{\pi_i}(g)} \text{ for all } g \in \mathrm{K},$$

and $\mathrm{Supp}W_\pi \subset \mathrm{U}_n\mathrm{K}$, where $^-$ denotes complex conjugation. Further, let π_1 and π_2 be irreducible unitarizable supercuspidal representations of $\mathrm{GL}_n(F)$ with the same central character and let W_{π_1} and W_{π_2} be Whittaker functions for π_1 and π_2, respectively. We call (W_{π_1}, W_{π_2}) a *special pair* (of Whittaker functions) for (π_1, π_2) if there exists a compact-mod-centre open subgroup K of $\mathrm{GL}_n(F)$ such that W_{π_1} and W_{π_2} are both K-special and

$$W_{\pi_1}(p) = W_{\pi_2}(p) \text{ for all } p \in \mathrm{P}_n.$$

In [JNS13], the Jacquet conjecture to the *local converse problem* was first proved under the assumption that for any pair of irreducible unitarizable supercuspidal representations (π_1, π_2) of $\mathrm{GL}_n(F)$, a special pair of Whittaker functions exists for (π_1, π_2), by using a refinement of the argument in [Ch96] and [Ch06]. In other words, we prove in [JNS13] the following theorem.

Theorem 3.2. *Let π_1 and π_2 be irreducible unitarizable supercuspidal representations of $\mathrm{GL}_n(F)$. Assume that a special pair (W_{π_1}, W_{π_2}) exists for (π_1, π_2). If the local gamma factors for pairs, $\gamma(s, \pi_1 \times \tau, \psi)$ and $\gamma(s, \pi_2 \times \tau, \psi)$, are equal as functions in the complex variable s, for all irreducible supercuspidal representations τ of $\mathrm{GL}_r(F)$ with $r = 1, \ldots, [\frac{n}{2}]$, then $W_{\pi_1} = W_{\pi_2}$ and π_1 and π_2 are equivalent as representations of $\mathrm{GL}_n(F)$.*

It remains critical to prove the existence of special pairs for irreducible unitarizable supercuspidal representations of $\mathrm{GL}_n(F)$. By using the construction of supercuspidal representations in terms of maximal simple types of Bushnell and Kutzko ([BK93]) and the explicit construction of Bessel functions of supercuspidal representations due to Paskunas and Stevens ([PS08]), we are able to handle this issue under certain constraints, which can be explained as follows. Given an irreducible supercuspidal representation π of $\mathrm{GL}_n(F)$, one of the invariants associated to it, by Bushnell and Henniart in [BH96], is its *endo-class* $\Theta(\pi)$. We prove:

Proposition 3.3. *Let π_1, π_2 be irreducible unitarizable supercuspidal representations of $\mathrm{GL}_n(F)$ with the same endo-class. Then there is a special pair (W_{π_1}, W_{π_2}) for (π_1, π_2).*

Theorem 3.2 with Proposition 3.3 implies, for example, that the Jacquet conjecture holds for two level zero irreducible unitarizable supercuspidal representations π_1, π_2 of $\mathrm{GL}_n(F)$. In fact, this is a special case of a more general result proved in [JNS13], as follows.

Attached to an irreducible supercuspidal representation π of $\mathrm{GL}_n(F)$, via its endo-class $\Theta(\pi)$, is an invariant which we call its *degree* $\deg(\pi)$. The degree is an integer dividing n: for example, $\deg(\pi) = 1$ if and only if π is a twist of a level zero representation. By using results on the conductor of pairs of supercuspidal representations from [BHK98, BH03], we obtain the following corollary.

Corollary 3.4. *Let π_1 and π_2 be irreducible unitarizable supercuspidal representations of $\mathrm{GL}_n(F)$ and suppose that $\deg(\pi_1) < n$. If the twisted local gamma factors $\gamma(s, \pi_1 \times \tau, \psi)$ and $\gamma(s, \pi_2 \times \tau, \psi)$ are equal as functions in the complex variable s, for all irreducible supercuspidal representations τ of $\mathrm{GL}_r(F)$ with $r = 1, \ldots, [\frac{n}{2}]$, then π_1 and π_2 are equivalent as representations of $\mathrm{GL}_n(F)$.*

Hence it remains to show the existence of special pairs for irreducible unitarizable supercuspidal representations π_1 and π_2 of $\mathrm{GL}_n(F)$ with the condition that $\deg(\pi_1) = \deg(\pi_2) = n$. In this aspect, Moshe Adrian and Baiying Liu has recently checked that special pairs for irreducible unitarizable supercuspidal representations π_1 and π_2 of $\mathrm{GL}_n(F)$ exists when both π_1 and π_2 are of epipelagic type, which is a case of $\deg(\pi_1) = \deg(\pi_2) = n$.

4 The Local Langlands Conjecture for classical groups

For simplicity, we may assume that the classical groups discussed here are F-split, and hence they are either SO_{2n+1}, Sp_{2n}, or SO_{2n}, which are denoted by G_n. The Langlands dual group is ${}^L G_n := G_n^\vee(\mathbb{C}) \times \Gamma_F$, where $G_n^\vee(\mathbb{C})$ is the complex dual group given by the following diagram.

An F-split of G_n is given by a pair (B, T), where B is an F-Borel subgroup and T is the maximal F-torus contained in B. This pair (B, T) determines the root data $(X, \Delta; X^\vee, \Delta^\vee)$ for G_n. The dual of $(X, \Delta; X^\vee, \Delta^\vee)$ is $(X^\vee, \Delta^\vee; X, \Delta)$, which determines the *complex dual group* $G_n^\vee(\mathbb{C})$ of G_n as displayed in the following diagram:

$$\begin{array}{ccc} G_n & \Longleftrightarrow & (X, \Delta; X^\vee, \Delta^\vee) \\ \updownarrow & & \updownarrow \\ G_n^\vee(\mathbb{C}) & \Longleftrightarrow & (X^\vee, \Delta^\vee; X, \Delta) \end{array} \tag{4.1}$$

For examples, one has that $\mathrm{GL}_n^\vee(\mathbb{C}) = \mathrm{GL}_n(\mathbb{C})$, $\mathrm{SO}_{2n+1}(\mathbb{C}) = \mathrm{Sp}_{2n}(\mathbb{C})$, $\mathrm{Sp}_{2n}^\vee(\mathbb{C}) = \mathrm{SO}_{2n+1}(\mathbb{C})$, and $\mathrm{SO}_{2n}^\vee(\mathbb{C}) = \mathrm{SO}_{2n}(\mathbb{C})$.

We refer to [Ar13, Mk15, JZ14, GGP12] for more explicit description of the Langlands dual groups of F-quasisplit groups.

The local Langlands parameters for $G_n(F)$ can be described as follows. Let $\iota : G_n^\vee(\mathbb{C}) \to \mathrm{GL}_{n^*}(\mathbb{C})$ be the natural embedding, where $\mathrm{GL}_{n^*} = \mathrm{GL}_{2n}$ if G_n is an orthogonal group; and $\mathrm{GL}_{n^*} = \mathrm{GL}_{2n+1}$ if G_n is Sp_{2n}. The set $\Phi(G_n)$ of local Langlands parameters for $G_n(F)$ consists of all local Langlands parameters $\phi \in \Phi(\mathrm{GL}_{n^*})$ which factor through $G_n^\vee(\mathbb{C})$:

$$\begin{array}{ccc} & & \mathrm{GL}_{n^*}(\mathbb{C}) \\ & \nearrow & \\ \mathcal{W}_F \times \mathrm{SL}_2(\mathbb{C}) & & \uparrow \\ & \searrow & \\ & & G_n^\vee(\mathbb{C}) \end{array} \tag{4.2}$$

The **Local Langlands Conjecture** asserts that there is a reciprocity map

$$\mathrm{rec}_{F,G_n} : \Pi(G_n) \longrightarrow \Phi(G_n) \tag{4.3}$$

with finite fibres and with certain constraints, where $\Pi(G_n)$ is the set of equivalence classes of irreducible admissible representations of $G_n(F)$. The main problems are

1. How to define the reciprocity rec_{F,G_n}? and
2. How to construct the fibres, i.e. local L-packets?

4.1 The Local Langlands Conjecture for classical groups: tempered case

Let $\Pi_{\mathrm{temp}}(G_n)$ be the subset of $\Pi(G_n)$ consisting of all irreducible admissible representations π whose matrix coefficients are $L^{2+\epsilon}$ on $G_n(F)$ modulo the center Z_{G_n}. Let $\Phi_{\mathrm{bdd}}(G_n)$ be the subset of $\Phi(G_n)$ consisting of all local Langlands parameters ϕ such that the image $\phi(\mathcal{W}_F \times \mathrm{SU}_2)$ in $G^\vee(\mathbb{C})$ is bounded modulo the center, here $\mathrm{SL}_2(\mathbb{C})$ is replaced by SU_2.

Conjecture 4.1. (Local Langlands Conjecture (tempered case)) *The local Langlands reciprocity map is given by*

$$\mathrm{rec}^t_{F,G_n} \;:\; \Pi_{\mathrm{temp}}(G_n) \longrightarrow \Phi_{\mathrm{bdd}}(G_n)$$

and the fibre: $\mathrm{rec}^{-1}_{F,G_n}(\phi)$ *is called the* **local tempered L-packet** *attached to* ϕ, *and denoted by* $\Pi_{G_n}(\phi)$, *which is parameterized by the dual of the component group of the centralizer of* ϕ *in* $G^\vee(\mathbb{C})$, *and hence it is nonempty. Moreover,* $\Pi_{\mathrm{temp}}(G_n)$ *is a disjoint union of* $\Pi_{G_n}(\phi)$, *for* $\forall\phi \in \Phi_{\mathrm{bdd}}(G_n)$.

This conjecture has been proved by Arthur in [Ar13] for quasi-split orthogonal groups and symplectic groups; and by Mok in [Mk15] for quasi-split unitary groups. We sketch here how to establish the reciprocity map rec^t_{F,G_n}.

One first needs to establish the local Langlands functorial transfer

$$\ell \;:\; \Pi_{\mathrm{temp}}(G_n) \longrightarrow \Pi_{\mathrm{temp}}(\mathrm{GL}_{n^*})$$

through the global method, i.e. by comparing the stable trace formula on G_n to the stable, twisted trace formula on GL_{n^*} and using the fundamental lemma proved by B.-C. Ngô and the local transfer established by J.-L. Waldspurger. Then by the Local Langlands Conjecture for GL_{n^*} with tempered representations

$$\mathrm{rec}_{F,\mathrm{GL}_*} \;:\; \Pi_{\mathrm{temp}}(\mathrm{GL}_{n^*}) \longrightarrow \Phi_{\mathrm{bdd}}(\mathrm{GL}_{n^*})$$

where $\phi \in \Phi_{\mathrm{bdd}}(\mathrm{GL}_{n^*})$ is of form:

$$\phi = (\rho_1, b_1) \boxplus (\rho_2, b_2) \boxplus \cdots \boxplus (\rho_r, b_r)$$

with ρ_i irreducible with bounded image in $\mathrm{GL}_{a_i}(\mathbb{C})$ modulo the center and b_i denoting the b_i-dimensional algebraic representation of $\mathrm{SL}_2(\mathbb{C})$, and $n^* = \sum_{i=1}^r a_i b_i$. It follows that for any irreducible admissible tempered representation $\pi \in \Pi_{\mathrm{temp}}(G_n)$, there exists a local Langlands parameter $\phi \in \Phi_{\mathrm{bdd}}(\mathrm{GL}_{n^*})$ which corresponds to π. One has to show that this local Langlands parameter ϕ actually belongs to the set $\Phi_{\mathrm{bdd}}(\mathrm{GL}_{n^*})$. This is done in [Ar13] and in [Mk15].

Next, one has to show that the local Langlands reciprocity map, established above,

$$\mathrm{rec}^t_{F,G_n} \;:\; \Pi_{\mathrm{temp}}(G_n) \longrightarrow \Phi_{\mathrm{bdd}}(G_n)$$

is actually surjective.

Let $\Pi^{\mathrm{gen}}(G_n)$ be the subset of $\Pi(G_n)$ consisting of all irreducible admissible representations which are generic, i.e. have nonzero Whittaker model; and let

$\Pi^{\rm gen}_{\rm temp}(G_n)$ be the intersection of $\Pi^{\rm gen}(G_n)$ and $\Pi_{\rm temp}(G_n)$. By the local descent method and the classification of irreducible generic representations of $G_n(F)$, it is proved in [JS03] and [JS04] (see also [JNQ10]), in [Liu11], and in [JtL13] that the map

$$\mathrm{rec}^{\rm gen}_{F,G_n} \ : \ \Pi^{\rm gen}_{\rm temp}(G_n) \longrightarrow \Phi_{\rm bdd}(G_n)$$

is surjective for all split classical groups G_n, and the local factors are preserved:

$$\begin{aligned} L(s, \pi \times \pi') &:= L(s, \mathrm{rec}_{F,G_n}(\pi) \otimes \mathrm{rec}_{F,G_n}(\pi')); \\ \gamma(s, \pi \times \pi', \psi_F) &:= \gamma(s, \mathrm{rec}_{F,G_n}(\pi) \otimes \mathrm{rec}_{F,G_n}(\pi'), \psi_F). \end{aligned}$$

We remark that this should also be true for quasi-split classical groups.

By the Shahidi conjecture, which is a consequence of the local descent method ([JS12]), for $\forall \phi \in \Phi_{\rm bdd}(G_n)$, the subset $\Pi^{\rm gen}_{G_n}(\phi)$ of the local tempered L-packet $\Pi_{G_n}(\phi)$is not empty. By Arthur's definition of the local factors of the local tempered L-packets ([Ar13] and [Mk15]), these two reciprocity maps are the same and hence the local reciprocity map

$$\mathrm{rec}^{t}_{F,G_n} \ : \ \Pi_{\rm temp}(G_n) \longrightarrow \Phi_{\rm bdd}(G_n)$$

is surjective. This establishes the reciprocity map in the Local Langlands Conjecture for split classical groups G_n. Note that the surjectivity of the reciprocity map for the tempered case is part of the theorem in [Ar13] and [Mk15]. Of course, it is impossible at least for the moment to define the tempered local L-packets in general without using the trace formula method.

4.2 The Local Langlands Conjecture for split classical groups: general case

We are going to deduce the general case of the Local Langlands Conjecture for split classical groups from the tempered case via the Langlands classification theory.

The Langlands classification of irreducible admissible representations of p-adic reductive algebraic groups in terms of essentially tempered representations, for any $\pi \in \Pi(G_n)$, there exists a pair (P, σ), where $P = MN$ is a parabolic subgroup of G_n and σ is essentially tempered on $M(F)$, s.t.

$$\mathrm{Ind}^{G_n(F)}_{P(F)}(\sigma) \to \pi$$

gives the unique irreducible quotient, up to Weyl group action. More precisely, we write

$$M = \mathrm{GL}_{n_1} \times \cdots \times \mathrm{GL}_{n_t} \times G_{n_0},$$

a Levi subgroup of G_n, and write

$$\sigma = \sigma_1 |\det|^{s_1} \otimes \cdots \otimes \sigma_t |\det|^{s_t} \otimes \sigma_0$$

with σ_i for $i = 1, 2, \ldots, t$ being all tempered and s_i being all real, so that $s_1 \geq \cdots \geq s_t$. The local Langlands transfer from G_n to GL_{n^*} is given by the following diagram:

$$\begin{array}{ccc} \Pi_{\mathrm{temp}}(M) & \longrightarrow & \Pi_{\mathrm{temp}}(M') \\ \downarrow & & \downarrow \\ \Pi(G) & \longrightarrow & \Pi(\mathrm{GL}_*) \end{array}$$

where M' is the Levi subgroup GL_{n^*} corresponding to the Levi subgroup M of G_n, which is given by

$$\mathrm{GL}_{n_1} \times \cdots \times \mathrm{GL}_{n_t} \times \mathrm{GL}_{n_0^*} \times \mathrm{GL}_{n_t} \times \cdots \times \mathrm{GL}_{n_1}.$$

Take $\tau_0 = \ell(\sigma_0)$ to be the local Langlands transfer of σ_0 from G_{n_0} to $\mathrm{GL}_{n_0^*}$ and define

$$\sigma' := \sigma_1 |\det|^{s_1} \otimes \cdots \otimes \sigma_r |\det|^{s_t} \otimes \tau_0 \otimes \sigma_t |\det|^{-s_t} \otimes \cdots \otimes \sigma_1 |\det|^{-s_1}.$$

The local Langlands functorial transfer $\ell(\pi)$ of π is given by

$$\mathrm{Ind}_{P'(F)}^{\mathrm{GL}_{n^*}(F)}(\sigma') \longrightarrow \ell(\pi)$$

which is the unique irreducible quotient. The conditions on local factors are satisfied: for $\forall \pi \in \Pi(G_n)$ and $\forall \tau \in \Pi(\mathrm{GL}_m)$, define

$$\begin{aligned} L(s, \pi \times \tau) &:= L(s, \ell(\pi) \times \tau); \\ \gamma(s, \pi \times \tau, \psi_F) &:= \gamma(s, \ell(\pi) \times \tau, \psi_F). \end{aligned}$$

The right hand side is defined by Jacquet, Piatetski-Shapiro and Shalika (1983). It is clear that

$$\mathrm{rec}_{F,G_n} = \mathrm{rec}_{F,n} \circ \ell \ : \ \Pi(G_n) \to \Phi(G_n)$$

is a local Langlands reciprocity map in general. The surjectivity of the local Langlands reciprocity map has been checked for split classical groups through the work ([JS03, JS04, Liu11, JtL13]). This is also expected to be true for quasi-split classical groups. For general reductive algebraic groups over F, we refer to a recent work [ABPS14].

5 Related problems

We explain here some interesting issues related to the establishment of the reciprocity map in the Local Langlands Conjecture for split classical groups G_n as sketched in the previous section. We treat the case when $G_n = \mathrm{SO}_{2n+1}$ and make comments on the other classical groups, including the quasi-split unitary groups.

As explained before, one general strategy to establish the Local Langlands Conjecture for $G_n(F)$ is to go with the local Langlands functorial transfer from $\Pi(G_n)$ to $\Pi(\mathrm{GL}_{n^*})$. For $G_n = \mathrm{SO}_{2n+1}$, the complex dual group is $G_n^\vee(\mathbb{C}) = \mathrm{Sp}_{2n}(\mathbb{C})$. In this case, we have that $n^* = 2n$. For a local Langlands parameter $\forall\phi \in \Phi(\mathrm{SO}_{2n+1})$, one has the following structure:

$$\begin{array}{ccc} \mathrm{Sp}_{2n}(\mathbb{C}) & \longrightarrow & \mathrm{GL}_{2n}(\mathbb{C}) \\ \nwarrow & & \nearrow \\ & \mathcal{W}_F \times \mathrm{SL}_2(\mathbb{C}) & \end{array} \tag{5.1}$$

The *local Langlands functorial transfer* asserts that the following diagram is commutative:

$$\begin{array}{ccc} \Pi(\mathrm{SO}_{2n+1}) & \longrightarrow & \Pi(\mathrm{GL}_{2n}) \\ \downarrow & & \downarrow \\ \Phi(\mathrm{SO}_{2n+1}) & \longrightarrow & \Phi(\mathrm{GL}_{2n}) \end{array} \tag{5.2}$$

In order to carry out this strategy with details, one has to deal with the following problems:

1. **Local Langlands Functorial Transfer:**
$$\ell \ : \ \Pi(\mathrm{SO}_{2n+1}) \longrightarrow \Pi(\mathrm{GL}_{2n});$$
2. **Characterization of Image:**
$$\ell(\Pi(\mathrm{SO}_{2n+1})) \subset \Pi(\mathrm{GL}_{2n});$$
3. **Surjectivity of the Reciprocity Maps:**
$$\ell(\Pi(\mathrm{SO}_{2n+1})) = \mathrm{rec}_{F,\mathrm{GL}_{2n}}^{-1}(\iota(\Phi(\mathrm{SO}_{2n+1})));$$
4. **Uniqueness of the Reciprocity Maps:**
$$\mathrm{rec}_{F,\mathrm{SO}_{2n+1}} \ : \ \Pi(\mathrm{SO}_{2n+1}) \longrightarrow \Phi(\mathrm{SO}_{2n+1}).$$

In the process, one has to check that the local reciprocity map is compatible with the local factors and goes with other constraints. We will discuss them with some details in the following subsections. Note that the local Langlands functorial transfer is already discussed before.

5.1 Characterization of the image

We consider the local Langlands functorial transfer:

$$\ell \ : \ \Pi(\mathrm{SO}_{2n+1}) \longrightarrow \Pi(\mathrm{GL}_{2n})$$

and ask for the characterization of the image of this transfer. When an irreducible admissible representation $\tau \in \Pi(\mathrm{GL}_{2n})$ is supercuspidal, we have the following characterization which is proved in our paper joint with Y.-J. Qin based on previous work of many others. We refer to [JNQ08] for detailed discussion on the background of this theorem and what was proved in [JNQ08]. However, when τ is not supercuspidal, much more work needs to be done.

Theorem 5.1. ([JNQ08]) *Let τ be an irreducible supercuspidal representation of $\mathrm{GL}_{2n}(F)$. Then the following are equivalent:*

1. *τ has a nonzero Shalika model.*
2. *The local exterior square L-factor $L(s,\tau,\wedge^2)$ has a pole at $s=0$.*
3. *The square γ-factor $\gamma(s,\tau,\wedge^2,\psi_F)$ has a pole at $s=1$.*
4. *τ is a local Langlands functorial transfer from $\mathrm{SO}_{2n+1}(F)$.*

If one of the above holds for τ, then τ is self-dual.

Remark 5.2. *By the classification theory for $\mathrm{GL}_{2n}(F)$, it is enough to describe the image of ℓ explicitly. In principle, this can be established for other classical groups.*

It remains in progress to extend the above characterization to more general representations in $\Pi(\mathrm{GL}_{2n})$, in order to study other relevant problems.

5.2 Surjectivity of the reciprocity map

In the case of $G_n = \mathrm{SO}_{2n+1}$, one has to show that for any local Langlands parameter $\phi \in \Phi(\mathrm{GL}_{2n})$, which is of symplectic type, i.e. the image $\phi(\mathcal{W}_F \times \mathrm{SL}_2(\mathbb{C}) \subset \mathrm{Sp}_{2n}(\mathbb{C})$, in other words, $\phi \in \Phi(\mathrm{SO}_{2n+1})$, one has to find an irreducible admissible representation $\pi \in \Pi(\mathrm{SO}_{2n+1})$, which has the local Langlands parameter ϕ. It is a theorem of the first named author and David Soudry in [JS04] which is stated as below.

Theorem 5.3. ([JS04]) *For any local parameter $\phi \in \Phi(\mathrm{SO}_{2n+1})$, there exists (can be explicitly constructed) a member $\pi \in \Pi(\mathrm{SO}_{2n+1})$ such that*

$$\begin{aligned} L(s,\pi\times\tau) &:= L(s,\mathrm{rec}_{F,\mathrm{GL}}^{-1}(\phi)\times\tau); \\ \gamma(s,\pi\times\tau,\psi_F) &:= \gamma(s,\mathrm{rec}_{F,\mathrm{GL}}^{-1}(\phi)\times\tau,\psi_F). \end{aligned}$$

Hence the reciprocity map

$$\mathrm{rec}_{F,\mathrm{SO}_{2n+1}} : \Pi(\mathrm{SO}_{2n+1}) \longrightarrow \Phi(\mathrm{SO}_{2n+1})$$

is surjective.

We remark that Liu in [Liu11] extended the above theorem to Sp_{2n}, based on the local descent theory of the first named author and Soudry, which is published in [JS12]. Liu with Jantzen in [JtL13] extended the above theorem to split SO_{2n}. The result should also be true for quasi-split unitary groups, but it still remains to be checked.

5.3 Uniqueness of the reciprocity map

In general, one has to show that the local Langlands reciprocity map established above

$$\mathrm{rec}_{F,\mathrm{SO}_{2n+1}} : \Pi(\mathrm{SO}_{2n+1}) \longrightarrow \Phi(\mathrm{SO}_{2n+1})$$

is in fact unique. From the deduction of the general case of the Local Langlands Conjecture to the tempered case of the Local Langlands Conjecture, it is enough to prove the uniqueness of the reciprocity map for

$$\mathrm{rec}^t_{F,\mathrm{SO}_{2n+1}} : \Pi_{\mathrm{temp}}(\mathrm{SO}_{2n+1}) \longrightarrow \Phi_{\mathrm{bdd}}(\mathrm{SO}_{2n+1}).$$

Now by the Arthur classification of the tempered spectrum of $\mathrm{SO}_{2n+1}(F)$ in [Ar13] and by the local descent theory of the first named author and Soudry in [JS03] and [JS12], the Shahidi conjecture holds. This means that for any local Langlands parameter $\phi \in \Phi_{\mathrm{bdd}}(\mathrm{SO}_{2n+1})$, the local tempered L-packet $\Pi_{\mathrm{SO}_{2n+1}}(\phi)$ has the property that the subset $\Pi_G^{\mathrm{gen}}(\phi)$ of generic members in the packet is nonempty. Hence it is enough to show the uniqueness of the reciprocity map

$$\mathrm{rec}^{\mathrm{gen}}_{F,\mathrm{SO}_{2n+1}} : \Pi^{\mathrm{gen}}_{\mathrm{temp}}(\mathrm{SO}_{2n+1}) \longrightarrow \Phi_{\mathrm{bdd}}(\mathrm{SO}_{2n+1})$$

with the property:

$$\gamma(s, \pi \times \tau, \psi_F) = \gamma(s, \mathrm{rec}^{\mathrm{gen}}_{F,\mathrm{SO}_{2n+1}}(\pi) \otimes \mathrm{rec}_{F,\mathrm{GL}}(\tau), \psi_F)$$

for all $\tau \in \Pi(\mathrm{GL}_m/F)$ and for all $m \geq 1$.

As in the case of GL_n, the uniqueness follows from the a positive solution to the **local converse problem** for GL_n. Hence the uniqueness of the local Langlands reciprocity map follows from a positive solution to the **local converse problem** for G_n, which will be discussed in the next section.

6 Local converse problems in general

In [J06], the first named author formulated the local converse problem for general reductive groups over p-adic local fields. We will specialize the general formulation in [J06] to the split classical groups and make the problem more precise. As remarked in the previous sections, we consider the local converse problem for irreducible generic representations of $G_n(F)$.

To recall the notion of Whittaker models, we fix an F-Borel subgroup $B = TU$ with a maximal split torus T, which defines the root system $\Phi(G_n, T)$ with the positive roots Φ^+ determined by U and gives the set of the simple roots Δ. Take a basis vector X_α in the one-dimensional F-root space of $\alpha \in \Delta$, which gives the following isomorphism

$$U/[U,U] \cong \oplus_{\alpha \in \Delta} F \cdot X_\alpha \tag{6.1}$$

as abelian groups. Any character ψ of $U(F)$ factors through the quotient $U(F)/[U(F), U(F)]$. By (6.1), a character ψ of $U(F)$ is called *generic* if ψ is nontrivial at each of the simple root α. By the Pontryagin duality, such characters of $U(F)$ is parameterized by n-tuples

$$\underline{a} = (a_1, \ldots, a_n) \in (F^\times)^n.$$

An irreducible admissible representation (π, V_π) of $G_n(F)$ is called generic or ψ-generic if the following space

$$\mathrm{Hom}_{U(F)}(V_\pi, \psi) \cong \mathrm{Hom}_{G_n(F)}(V_\pi, \mathrm{Ind}_{U(F)}^{G_n(F)}(\psi))$$

is nonzero. Any nonzero functional $\ell_\psi \in \mathrm{Hom}_{U(F)}(V_\pi, \psi)$ yields a $G_n(F)$-equivariant homomorphism

$$v \in V_\pi \mapsto W_v^\psi(g) = \ell_\psi(\pi(g)(v)).$$

The subspace $\{W_v^\psi(g) \mid v \in V_\pi\}$ is called the ψ-Whittaker model associated to π. It is of multiplicity one in the space $\mathrm{Ind}_{U(F)}^{G_n(F)}(\psi)$ because of the uniqueness of local Whittaker models ([Shl74]).

For $t \in T(F)$, we have that $t \circ \psi(u) = \psi(t^{-1}ut)$, via the adjoint action. It follows that a $\pi \in \Pi(G_n)$ is ψ-generic if and only if π is $t \circ \psi$-generic for any $t \in T(F)$. In other words, the genericity of π depends on the $T(F)$-orbit of the generic characters. An easy exercise leads to

Proposition 6.1. ([K02]) *The set of $T(F)$-orbits the generic characters of $U(F)$ is in one-to-one correspondence with $H^1(\Gamma_F, Z_{G_n})$, where Γ_F is the absolute Galois group of F and Z_{G_n} is the center of G_n.*

For a $\pi \in \Pi^{\mathrm{gen}}(G_n)$, we define

$$\mathcal{F}(\pi) = \{\psi \mid \pi \textit{ is } \psi\textit{-generic}\}, \tag{6.2}$$

which is clearly $T(F)$-stable. Hence the genericity of π is completely described by the $T(F)$-orbits $\mathcal{F}(\pi)/T(F)$. Moreover, we recall from [J06] the following refinement of the Shahidi conjecture.

Conjecture 6.2. ([J06]) *In a generic local L-packet $\Pi(\phi)$, for any generic members $\pi_1, \pi_2 \in \Pi(\phi)$, the sets $\mathcal{F}(\pi_1)$ and $\mathcal{F}(\pi_2)$ are disjoint and the union of $T(F)$-orbits of generic characters over the subset $\Pi^{\mathrm{gen}}(\phi)$ of $\Pi(\phi)$ consisting of all generic members in $\Pi(\phi)$, i.e.*

$$\bigcup_{\pi \in \Pi^g(\phi)} \mathcal{F}(\pi)/T(F)$$

is in one-to-one correspondence with $H^1(\Gamma_F, Z_G)$.

Based on Conjecture 6.2, we recall from [J06] a general conjecture on the local converse theorem.

Conjecture 6.3. (local converse theorem) *For $\pi_1, \pi_2 \in \Pi^{\mathrm{gen}}(G_n)$, assume that*

1. *the intersection of $\mathcal{F}(\pi_1)$ and $\mathcal{F}(\pi_2)$ is not empty, and*

2. *the twisted local* γ*-factors are equal, i.e.*

$$\gamma(s, \pi_1 \times \tau, \psi) = \gamma(s, \pi_2 \times \tau, \psi)$$

holds for all supercuspidal $\tau \in \Pi(\mathrm{GL}_l)$ *with* $l = 1, 2, \ldots, n$.

Then $\pi_1 \cong \pi_2$.

In [JS03], the first named author with Soudry proved a weak version of Conjecture 6.3 for $G_n = \mathrm{SO}_{2n+1}$.

Theorem 6.4. ([JS03]) *For* $\pi, \pi' \in \Pi^{\mathrm{gen}}(\mathrm{SO}_{2n+1})$, *assume that the twisted local gamma factors* $\gamma(\pi \times \tau, s, \psi)$ *and* $\gamma(\pi' \times \tau, s, \psi)$ *are the same, i.e.*

$$\gamma(\pi \times \tau, s, \psi) = \gamma(\pi' \times \tau, s, \psi)$$

for all supercuspidal $\tau \in \Pi(\mathrm{GL}_l)$ *with* $l = 1, 2, \ldots, 2n-1$. *Then the representations* π *and* π' *are equivalent.*

Note that for $\pi, \pi' \in \Pi^{\mathrm{gen}}(\mathrm{SO}_{2n+1})$, it is easy to check that $\mathcal{F}(\pi) = \mathcal{F}(\pi')$. We refer to [J06] for more detailed discussion on relevant topics.

References

[Ar13] Arthur, J. *The endoscopic classification of representations: orthogonal and symplectic groups.* Colloquium Publication, Vol. **61**, American Mathematical Society, 2013.

[ABPS14] Aubert, A.-M.; Baum, P.; Plyen, R.; Solleveld, M. *On the local Langlands correspondence for non-tempered representations.* Münster J. Math. **7** (2014), no. 1, 27–50.

[BH96] Bushnell, C. J.; Henniart, G. M. *Local tame lifting for* GL(N). *I. Simple characters.* Inst. Hautes Études Sci. Publ. Math. **83** (1996), 105–233.

[BH03] Bushnell, C. J.; Henniart, G. M. *Local tame lifting for* GL(n). *IV. Simple characters and base change*, Proc. London Math. Soc. (3) **87** (2003), no. 2, 337–362.

[BHK98] Bushnell, C. J.; Henniart, G. M.; Kutzko, P. C. *Local Rankin-Selberg convolutions for* GL_n*: explicit conductor formula.* J. Amer. Math. Soc. **11** (1998), no. 3, 703–730.

[BK93] Bushnell, C. J.; Kutzko, P. C. *The admissible dual of* GL(N) *via compact open subgroups.* Annals of Mathematics Studies, **129**, Princeton University Press, Princeton, NJ, 1993.

[Ch96] Chen, J. -P. J. *Local Factors, Central Characters, and Representations of the General Linear Group over Non-Archimedean Local Fields.* Thesis, Yale University, 1996.

[Ch06] Chen, J. -P. J. *The $n \times (n-2)$ local converse theorem for* $\mathrm{GL}(n)$ *over a p-adic field.* Journal of Number Theory **120** (2006), Issue 2, 193–205.

[CPS99] Cogdell, J.; Piatetski-Shapiro, I. *Converse theorems for* GL_n. *II.* J. Reine Angew. Math. **507** (1999), 165–188.

[GGP12] Gan, W. T.; Gross, B. H.; Prasad, D. *Symplectic local root numbers, central critical L-values, and restriction problems in the representation theory of classical groups.* Astérisque **346** (2012), 1–109.

[HT01] Harris, M.; Taylor, R. *The geometry and cohomology of some simple Shimura varieties.* Annals of Math. Studies, **151**, Princeton Unversity Press, 2001.

[H93] Henniart, G. *Caractérisation de la correspondance de Langlands locale par les facteurs ϵ de paires.* (French) Invent. Math. **113** (1993), no. 2, 339–350.

[H00] Henniart, G. *Une preuve simple des conjectures de Langlands pour* $\mathrm{GL}(n)$ *sur un corps p-adique.* (French) Invent. Math. **139** (2000), no. 2, 439–455.

[JL70] Jacquet, H.; Langlands, R. P. *Automorphic forms on* $\mathrm{GL}(2)$. Lecture Notes in Mathematics, Vol. **114**, Springer-Verlag, Berlin-New York, 1970.

[JPSS79] Jacquet, H.; Piatetski-Shapiro, I.; Shalika, J. *Automorphic forms on* $\mathrm{GL}(3)$. *I and II.* Ann. of Math. **109** (1979), no. 1, 169–212 and no. 2, 213–258.

[JPSS83] Jacquet, H.; Piatetski-Shapiro, I.; Shalika, J. *Rankin-Selberg convolutions.* Amer. J. Math. **105** (1983), no. 2, 367–464.

[JtL13] Jantzen, C; Liu, B. *The generic dual of p-Adic split* SO_{2n} *and local Langlands parameters.* Israel Journal of Mathematics **204** (2014), Issue 1, 199–260.

[J06] Jiang, D. *On local gamma factors.* Arithmetic Geometry and Number Theory. Series on Number Theory and Applications, Vol **1**, 1–28, World Scientific, 2006.

[JNQ08] Jiang, D.; Nien, C.; Qin, Y. *Local Shalika models and functoriality.* Manuscripta Math. **127** (2008), no. 2, 187–217.

[JNQ10] Jiang, D.; Nien, C.; Qin, Y. *Symplectic supercuspidal representations of* $\mathrm{GL}(2n)$ *over p-adic fields.* Pacific J. Math. **245** (2010), no. 2, 273–313.

[JNS13] Jiang, D.; Nien, C.; Stevens, S. *Towards the Jacquet conjecture on the local converse problem for p-adic* $\mathrm{GL}(n)$. Journal of the European Mathematical Society **17** (2015), 991–1007.

[JS03] Jiang, D.; Soudry, D. *The local converse theorem for* $\mathrm{SO}(2n+1)$ *and applications.* Ann. of Math. **157** (2003), 743–806.

[JS04] Jiang, D.; Soudry, D. *Generic Representations and the Local Langlands Reciprocity Law for p-adic* $\mathrm{SO}(2n+1)$. Contributions to automorphic forms, geometry, and number theory, 457–519, Johns Hopkins Univ. Press, Baltimore, MD, 2004.

[JS12] Jiang, D.; Soudry, D. *On the local descent from* $\mathrm{GL}(n)$ *to classical groups.* Amer. J. Math. **134** (2012), no. 3, 767–772.

[JZ14] Jiang, D.; Zhang, L. *A product of tensor product L-functions of quasi-split classical groups of Hermitian type.* Geom. Funct. Anal. **24** (2014), no. 2, 552–609.

[K02] Kuo, W. *Principal nilpotent orbits and reducible principal series.* Represent. Theory **6** (2002), 127–159.

[L70] Langlands, R. P. *Problems in the theory of automorphic forms.* Lectures in modern analysis and applications, III, pp. 18–61. Lecture Notes in Math., Vol. **170**, Springer, Berlin, 1970.

[Liu11] Liu, B. *Genericity of representations of p-adic* Sp_{2n} *and local Langlands parameters.* Canad. J. Math. **63** (2011), 1107–1136.

[Mk15] Mok, C. P. *Endoscopic classification of representations of quasi-split unitary groups.* Mem. Amer. Math. Soc. **235** (2015), no. 1108, vi+248 pp.

[N14] Nien, C. *A proof of the finite field analogue of Jacquet's conjecture.* Amer. J. Math. **136** (2014), no. 3, 653–674.

[PS08] Paskunas, V.; Stevens, S. *On the realization of maximal simple types and epsilon factors of pairs.* Amer. J. Math. **130** (2008), no. 5, 1211–1261.

[Shl74] Shalika, J. *The multiplicity one theorem for* GL_n. Ann. of Math. (2) **100** (1974), 171–193.

[T79] Tate, J. *Number theoretic background.* Proc. Sympos. Pure Math., **33**, Automorphic forms, representations and L-functions, Part 2, pp. 3–26, Amer. Math. Soc., 1979.

[Z80] Zelevinsky, A. *Induced representations of reductive p-adic groups. II. On irreducible representations of* $\mathrm{GL}(n)$. Ann. Sci. École Norm. Sup. (4) **13** (1980), no. 2, 165–210.

School of Mathematics, University of Minnesota
Minneapolis, MN 55455, USA
E-mail address: dhjiang@math.umn.edu

Department of Mathematics, Cheng Kung University
& "National Center for Theoretical Sciences(South)"
Tainan 701, Taiwan, China
E-mail address: nienpig@mail.ncku.edu.tw

Proceedings of the 6th International Congress of Chinese Mathematicians
ALM 36, pp. 327–335

The Möbius Disjointness Conjecture for Distal Flows

Jianya Liu and Peter Sarnak

Abstract

We summarize main results in our joint paper "The Möbius function and distal flows", where we establish the Möbius Disjointness Conjecture of the second author for various distal flows. In addition we give a direct proof, with a rate, of the result that the Möbius function is disjoint from Furstenberg's irregular systems.

Mathematics Subject Classification: 11L03, 37A45, 11N37.

Keywords and Phrases: The Möbius function, distal flow, affine linear map, skew product, nilmanifold.

1 The Möbius disjointness conjecture

Let $\mathscr{X} = (T, X)$ be a flow, namely X is a compact topological space and $T : X \to X$ a continuous map. The sequence $\xi(n)$ is observed in $\mathscr{X}$ if there is an $f \in C(X)$ and an $x \in X$, such that $\xi(n) = f(T^n x)$. Let $\mu(n)$ be the Möbius function, that is $\mu(n)$ is 0 if n is not square-free, and is $(-1)^t$ if n is a product of t distinct primes. We say that μ is linearly disjoint from $\mathscr{X}$ if

$$\frac{1}{N}\sum_{n\le N}\mu(n)\xi(n) \to 0, \quad \text{as } N\to\infty, \tag{1.1}$$

for every observable ξ of $\mathscr{X}$. The Möbius Disjointness Conjecture of the second author ([16, 17]) states the following

Conjecture 1.1. (The Möbius Disjointness Conjecture) *The Möbius function μ is linearly disjoint from every $\mathscr{X}$ whose entropy is* 0.

This Conjecture has been established for many flows $\mathscr{X}$ (see [5, 14, 9, 3, 2]) however all of these flows are quasi-regular (or rigid) in the sense that the Birkhoff averages

$$\frac{1}{N}\sum_{n\leq N}\xi(n) \tag{1.2}$$

exist for every ξ observed in $\mathscr{X}$. In [13] we establish some new cases of the Disjointness Conjecture and in particular for irregular flows $\mathscr{X}$, that is ones for which (1.2) fails. These flows are complicated in terms of the behavior of their individual orbits but they are distal and of zero entropy, so that the disjointness is still expected to hold.

2 Results

In this section we summarize the results we have established in [13]. The first result in [13] is concerned with certain regular flows, namely affine linear maps of a compact abelian group X. Such a flow (T, X) is given by

$$T(x) = Ax + b \tag{2.1}$$

where A is an automorphism of X and $b \in X$ (see [10, 11]).

Theorem 2.1. *Let $\mathscr{X} = (T, X)$ be an affine linear flow on a compact abelian group which is of zero entropy. Then μ is linearly disjoint from $\mathscr{X}$.*

Theorem 2.1 actually holds with a rate of convergence. We first reduce to the torus case and then handle the torus case by Fourier analysis and classical results of Davenport [5] and Hua [12] on exponential sums concerning the Möbius function.

The flows in Theorem 2.1 are distal, and our main result in [13] is concerned with nonlinear distal flows on such spaces. We restrict to $X = \mathbb{T}^2$ the two dimensional torus $\mathbb{R}^2/\mathbb{Z}^2$ and consider nonlinear smooth (or even analytic) skew products as discussed in Furstenberg [6]. $T : \mathbb{T}^2 \to \mathbb{T}^2$ is given by

$$T(x, y) = (ax + \alpha, cx + dy + h(x)) \tag{2.2}$$

where $a, c, d \in \mathbb{Z}, ad = \pm 1, \alpha \in \mathbb{R}$ and h is a smooth periodic function of period 1. The affine linear part is in the form

$$\begin{bmatrix} a & 0 \\ c & d \end{bmatrix} \in GL_2(\mathbb{Z}),$$

ensuring that T has zero entropy (and it can always be brought into this form). The flow $(T, \mathbb{T}^2)$ is distal and this skew product is a basic building block (with $e(h(x))$ continuous) in Furstenberg's classification theory of minimal distal flows [7]. If α is diophantine, that is

$$\left|\alpha - \frac{a}{q}\right| \geq \frac{c}{q^B}$$

for some $c > 0, B < \infty$ and all a/q rational, then T can be conjugated by a smooth map of $\mathbb{T}^2$ to its affine linear part

$$(x, y) \mapsto (ax + \alpha, cx + dy + \beta) \tag{2.3}$$

where

$$\beta = \int_0^1 h(x)dx$$

(see [15]). Hence the disjointness of μ from $\mathscr{X} = (T, \mathbb{T}^2)$ for a T with a diophantine α, follows from Theorem 2.1. However if α is not diophantine the dynamics of the flow $(T, \mathbb{T}^2)$ can be very different from an affine linear flow. For example, as Furstenberg shows it may be irregular (i.e. the limits in (1.2) fail to exist for certain observables). Our main result is that these nonlinear skew products are linearly disjoint from μ, at least if h satisfies some further technical hypothesis. Firstly we assume that h is analytic, namely that if

$$h(x) = \sum_{m \in \mathbb{Z}} \hat{h}(m)e(mx) \tag{2.4}$$

then

$$\hat{h}(m) \ll e^{-\tau|m|} \tag{2.5}$$

for some $\tau > 0$. Secondly we assume that there is $\tau_2 < \infty$ such that

$$|\hat{h}(m)| \gg e^{-\tau_2|m|}. \tag{2.6}$$

This is not a very natural condition being an artifact of our proof. However it is not too restrictive and the following applies rather generally (and most importantly there is no condition on α).

Theorem 2.2. *Let $\mathscr{X} = (T, \mathbb{T}^2)$ be of the form* (2.2), *with h satisfying* (2.5) *and* (2.6). *Then μ is linearly disjoint from $\mathscr{X}$.*

The assertion of Theorem 2.2 holds for all α, and so we have to consider all diophantine possibilities of α. The tool is the Bourgain-Sarnak-Ziegler [3] finite version of the Vinogradov method, incorporated with various analytic methods such as Poisson's summation and stationary phase. Furstenberg [6] gives examples of skew product transformations of the form (2.2) which are not regular in the sense of (1.2). Many of the flows $\mathscr{X}$ in Theorem 2.2 have this property and we show in [13] that Furstenberg's examples are smoothly conjugate to such $\mathscr{X}$'s. In particular his examples are linearly disjoint from μ.

Theorem 2.1 deals with the affine linear distal flows on the n-torus. A different source of homogeneous distal flows are the affine linear flows on nilmanifold $X = G/\Gamma$ where G is a nilpotent Lie group and Γ a lattice in G. For $\mathscr{X} = (T, G/\Gamma)$ where $T(x) = \alpha x\Gamma$ with $\alpha \in G$, i.e. translation on G/Γ, the linear disjointness of μ and $\mathscr{X}$ is proven in [8] and [9]. Using the classification of zero entropy (equivalently distal) affine linear flows on nilmanifolds [4], and Green and Tao's results we establish in [13] the following

Theorem 2.3. *Let $\mathscr{X} = (T, G/\Gamma)$ where T is an affine linear map of the nilmanifold G/Γ of zero entropy. Then μ is linearly disjoint from $\mathscr{X}$.*

The above results for $\mu(n)$ can be proved in the same way for similar multiplicative functions such as $\lambda(n) = (-1)^{\tau(n)}$ where $\tau(n)$ is the number of prime factors of n.

3 Disjointness of μ from Furstenberg's systems

As a consequence of Theorem 2.2, it is proved in [13] that μ is linearly disjoint from Furstenberg's systems. But no rate of convergence is obtained there since Theorem 2.2 in general offers no rate. In this section we show that, for Furstenberg's systems, rate of convergence is actually available if we work on these systems directly rather than appeal to Theorem 2.2.

3.1 The continued fraction expansion of α

We assume that α is irrational, and our argument will depend on the continued fraction expansion of α. Every real number α has its continued fraction representation

$$\alpha = a_0 + \cfrac{1}{a_1 + \frac{1}{a_2 + \cdots}} \tag{3.1}$$

where $a_0 = [\alpha]$ is the integral part of α, and $a_1, a_2, \ldots$ are positive integers. The expression (3.1) is infinite since $\alpha \notin \mathbb{Q}$. We write $[a_0; a_1, a_2, \ldots]$ for the expression on the right-hand side of (3.1), which is the limit of the finite continued expressions

$$[a_0; a_1, a_2, \ldots, a_k] = a_0 + \cfrac{1}{a_1 + \frac{1}{a_2 + \cdots + \frac{1}{a_k}}} \tag{3.2}$$

as $k \to \infty$. Writing

$$\frac{l_k}{q_k} = [a_0; a_1, a_2, \ldots, a_k],$$

we have $l_0 = a_0, l_1 = a_0 a_1 + 1, q_0 = 1, q_1 = a_1$, and for $k \geq 2$,

$$l_k = a_k l_{k-1} + l_{k-2}, \quad q_k = a_k q_{k-1} + q_{k-2}.$$

Since α is irrational we have $q_{k+1} \geq q_k + 1$ for all $k \geq 1$. An induction argument gives the stronger assertion that $q_k \geq 2^{(k-1)/2}$ for all $k \geq 2$, and thus q_k increases at least like an exponential function of k. The irrationality of α also implies that, for all $k \geq 2$,

$$\frac{1}{2 q_k q_{k+1}} < \left| \alpha - \frac{l_k}{q_k} \right| < \frac{1}{q_k q_{k+1}}. \tag{3.3}$$

3.2 Furstenberg's examples

Furstenberg [6] gave examples of smooth transformation $T : \mathbb{T}^2 \to \mathbb{T}^2$ such that the ergodic averages do not all exist. Let α be as above such that

$$q_{k+1} \geq e^{q_k} \tag{3.4}$$

for all positive k. Define $q_{-k} = -q_k$ and set

$$h(x) = \sum_{k \neq 0} \frac{e(q_k \alpha) - 1}{|k|} e(q_k x). \tag{3.5}$$

It follows from (3.3) and (3.4) that $h(x)$ is a smooth function. We also have $h(x) = g(x + \alpha) - g(x)$ where

$$g(x) = \sum_{k \neq 0} \frac{1}{|k|} e(q_k x) \tag{3.6}$$

so that $g(x) \in L^2(0, 1)$ and in particular defines and measurable function. But $g(x)$ cannot correspond to a continuous function, as shown in Furstenberg [6].

3.3 Disjointness of μ from Furstenberg's systems

In the following we consider slightly more general h's with

$$h(x) = \sum_{k \neq 0} c_k (1 - e(q_k \alpha)) e(q_k x) \tag{3.7}$$

where α satisfies (3.4) and, for all positive k, the coefficients c_k satisfy

$$c_k = c_{-k}, \quad |c_k| \leq C \tag{3.8}$$

for some positive constant C which we may assume to be greater than 1. We note that if c_k is not bounded then $g(x)$ will not be L^2. Now what we want to estimate is essentially

$$S(N) := \sum_{n \leq N} \mu(n) e\left(\sum_{j=0}^{n-1} h(j\alpha) \right), \tag{3.9}$$

and our result is the following

Proposition 3.1. *Let $S(N)$ be as in (3.9). Then*

$$S(N) \ll N \log^{-A} N \tag{3.10}$$

where the implied constant depends on A, but is independent of α.

Proof. By (3.7) we have

$$\sum_{j=0}^{n-1} h(j\alpha) = \sum_{k\neq 0} c_k(1-e(q_k\alpha)) \sum_{j=0}^{n-1} e(q_k j\alpha) = \sum_{k\neq 0} c_k(1-e(nq_k\alpha)).$$

We should cut the last sum at some point. Let K be such that $q_{K-1} < 2\log N \leq q_K$. Then for $k \geq K$ we have

$$|q_k\alpha - l_k| \leq \frac{1}{q_{k+1}} \leq e^{-q_k}, \tag{3.11}$$

so that

$$\sum_{|k|\geq K} |c_k(1-e(nq_k\alpha))| \ll C \sum_{|k|\geq K} ne^{-q_k} \ll CNe^{-q_K} \ll CN^{-1}$$

where the implied constant does not depend on C. It follows that

$$S(N) = \sum_{n\leq N} \mu(n) e\Big\{ \sum_{1\leq|k|\leq K-1} c_k(1-e(nq_k\alpha)) + O\Big(\frac{C}{N}\Big)\Big\}.$$

The O-term as well as $\sum\limits_{1\leq|k|\leq K-1} c_k$ above are harmless, and so from now on we concentrate on

$$\begin{aligned} \widetilde{S}(N) &:= \sum_{n\leq N} \mu(n) e\Big\{ \sum_{1\leq|k|\leq K-1} c_k e(nq_k\alpha)\Big\} \\ &= \sum_{n\leq N} \mu(n) e\Big\{ \sum_{1\leq|k|\leq K-1} c_k e(n\theta_k)\Big\} \end{aligned} \tag{3.12}$$

on writing $\theta_k = \|q_k\alpha\|$. Note that by (3.11) we have $\theta_k \leq e^{-q_k}$.

We have a sequence

$$q_1 < \exp(q_1) \leq q_2 < \exp(q_2) \leq q_3 < \cdots,$$

and therefore $q_k > \exp\cdots\exp(q_2)$ with $k-2$ repeated exp's. Taking $k = K-1$ and noting that $q_2 \geq 2$ give

$$\exp\cdots\exp(2) \leq \exp\cdots\exp(q_2) < q_{K-1} \leq 2\log N \tag{3.13}$$

where on the left-hand side there are $K-3$ exp's.

Now let

$$\phi(x) = 2c_1\cos(2\pi x), \quad x \in [\theta_1, \theta_1 N]. \tag{3.14}$$

Then $e(\phi(x))$ is a smooth periodic function and hence can expanded into Fourier series

$$e(\phi(x)) = \sum_{l\in\mathbb{Z}} a_l(c_1)e(lx), \tag{3.15}$$

where

$$a_l(c_1) = \int_0^1 e(\phi(x))e(-lx)dx. \tag{3.16}$$

We must compute the dependence of $a_l(c_1)$ on c_1. By partial integration for $l \neq 0$ we have

$$\begin{aligned} a_l(c_1) &= -\frac{1}{2\pi il}\int_0^1 e(\phi(x))de(-lx) \\ &= -\frac{1}{l}\int_0^1 e(\phi(x))\phi'(x)e(-lx)dx \\ &= \frac{1}{2\pi il^2}\int_0^1 \frac{d[e(\phi(x))\phi'(x)]}{dx}e(-lx)dx. \end{aligned}$$

Since

$$\phi'(x) = -4\pi c_1 \sin(2\pi x), \quad \phi''(x) = -8\pi^2 c_1 \cos(2\pi x),$$

we have

$$\left|\frac{d[e(\phi(x))\phi'(x)]}{dx}\right| = |e(\phi(x))[\phi''(x) + 2\pi i\phi'(x)\phi'(x)]| \ll |c_1| + |c_1|^2 \ll C^2$$

and consequently

$$a_l(c_1) \ll \frac{C^2}{l^2} \tag{3.17}$$

for all $l \neq 0$. The above $\ll$constant is absolute. Of course the above l^2 can be improved to l^A for arbitrary $A > 0$, but we are not going to use this.

The sum $\widetilde{S}(N)$ in (3.12) is of the form

$$\begin{aligned} \widetilde{S}(N) &= \sum_{n\leq N} \mu(n)e(\phi(n\theta_1) + F(n)) \\ &= \sum_{n\leq N} \mu(n)e(F(n))\sum_{l\in\mathbb{Z}} a_l(c_1)e(ln\theta_1) \\ &= \sum_{l\in\mathbb{Z}} a_l(c_1)\sum_{n\leq N} \mu(n)e(ln\theta_1 + F(n)), \end{aligned}$$

where $F(n)$ stands for the remaining terms in the $e(\cdot)$ in (3.12). It follows from this and (3.17) that

$$\begin{aligned} |\widetilde{S}(N)| &\leq \sum_{l\in\mathbb{Z}} |a_l(c_1)|\left|\sum_{n\leq N} \mu(n)e(ln\theta_1 + F(n))\right| \\ &\ll C^2\sum_{l\in\mathbb{Z}} \frac{1}{l^2+1}\left|\sum_{n\leq N} \mu(n)e(ln\theta_1 + F(n))\right| \\ &\ll C^2 \sup_{l_1}\left|\sum_{n\leq N} \mu(n)e(nl_1\theta_1 + F(n))\right|. \end{aligned}$$

Repeating this procedure in (3.12), we get

$$|\widetilde{S}(N)| \ll C^{2(K-1)} \sup_{l_1,\ldots,l_{K-1}} \left| \sum_{n\le N} \mu(n)e(nl_1\theta_1 + \cdots + nl_{K-1}\theta_{K-1}) \right|.$$

The inner sum involving μ can be estimated by classical results of Davenport [5] and Hua [12] as

$$\ll N\log^{-A} N$$

where $A > 0$ is arbitrary, and the implied constant depends at most on A, i.e. it does not depend on the coefficients $l_1, l_2, \ldots, l_{K-1}$ or $\theta_1, \theta_2, \ldots, \theta_{K-1}$. By (3.13) we have $C^{2(K-1)} \le \log N$. The proposition is proved. □

Acknowledgements. The first author is supported by the 973 Program, NSFC grant 11031004, and IRT1264 from the Ministry of Education. The second author is supported by an NSF grant.

References

[1] N. Aoki, Topological entropy of distal affine transformations on compact abelian groups, J. Math. Soc. Japan **23**(1971), 11–17.

[2] J. Bourgain, On the correlation of the Möbius function with random rank one systems (2011), arXiv:1112.1031.

[3] J. Bourgain, P. Sarnak, and T. Ziegler, Disjointness of Möbius from horocycle flows, From Fourier analysis and number theory to radon transforms and geometry, 67–83, Dev. Math. **28**, Springer, New York, 2013.

[4] S. G. Dani, Dynamical systems on homogeneous spaces, in Chapter 10, Dynamical Systems, Ergodic Theory and Applications, Encyclopedia of Mathematical Sciences, Vol. **100**, Springer.

[5] H. Davenport, On some infinite series involving arithmetical functions II, Quart. J. Math. **8**(1937), 313–350.

[6] H. Furstenberg, Strict ergodicity and transformation of the torus, Amer. J. Math. **83**(1961), 573–601.

[7] H. Furstenberg, The structure of distal flows, Amer. J. Math. **85**(1963), 477–515.

[8] B. Green and T. Tao, The quatitative behaviour of polynomial obits on nilmanifolds, Ann. Math. (2) **175**(2012), 465–540.

[9] B. Green and T. Tao, The Möbius function is strongly orthorgonal to nilsequences, Ann. Math. (2) **175**(2012), 541-566.

[10] F. J. Hahn, On affine transformations of compact abelian groups, Amer. J. Math. **85**(1963), 428–446; Errata: Amer. J. Math. **86**(1964), 463–464.

[11] H. Hoare and W. Parry, Affine transformations with quasi-discrete spectrum I, J. London Math. Soc. **41**(1966), 88–96.

[12] L. K. Hua, Additive theory of prime numbers, AMS Translations of Mathematical Monographs, Vol. **13**, Providence, R.I. 1965.

[13] J. Liu and P. Sarnak, The Möbius function and distal flows, Duke Math. J. **164** (2015), 1353–1399.

[14] C. Mauduit and J. Rivat, Sur un problème de Gelfond: la somme des chiffres des nombres premiers, Ann. Math. (2) **171**(2010), 1591–1646.

[15] N. M. dos Santos and R. Urzúa-Luz, Minimal homeomorphisms on low-dimensional tori, Ergodic Theory Dynam. Systems **29**(2009), 1515–1528.

[16] P. Sarnak, *Three lectures on the Möbius function, randomness and dynamics*, IAS Lecture Notes, 2009; http://publications.ias.edu/sites/default/files/MobiusFunctionsLectures(2).pdf.

[17] P. Sarnak, Möbius randomness and dynamics, Not. S. Afr. Math. Soc. **43**(2012), 89-97.

[18] P. Sarnak and A. Ubis, The horocycle at prime times, J. Math. Pures Appl. **103**(2015), 575–618.

School of Mathematics, Shandong University
Jinan, Shandong 250100, China
E-mail address: jyliu@sdu.edu.cn

Department of Mathematics, Princeton University
& Institute for Advanced Study
Princeton, NJ 08544-1000, USA
E-mail address: sarnak@math.princeton.edu

Proceedings of the 6th International Congress of Chinese Mathematicians
ALM 36, pp. 337–350

Survey on Automorphic Period for General Bessel Model

Yifeng Liu

Abstract

In this survey, we introduce the conjecture of explicit formulas for Bessel periods, which refines the Gan–Gross–Prasad conjecture. We also summarize some recent progresses and approaches toward this problem.

Mathematics Subject Classification: Primary: 11F67; Secondary: 11F30, 11F55.

Keywords and Phrases: Gan–Gross–Prasad conjecture, Bessel model, period integral, automorphic L-function.

1 Automorphic period

Let F be a number field with ring of adèles $\mathbb{A}$ and the set of places Σ. Fix a nontrivial character $\psi\colon F\backslash\mathbb{A} \to \mathbb{C}^\times$. Throughout the note, for a linear algebraic group G over F, we always use the Tamagawa measure $\mathrm{d}g$ on $G(\mathbb{A})$. In particular, the volume of $G(F)\backslash G(\mathbb{A})$ is 1 if G is unipotent.

Consider a reductive group G over F and a closed F-subgroup H. Denote $\mathcal{A}(G)$ the space of automorphic forms on $G(\mathbb{A})$ and $\mathcal{A}_{\mathrm{cusp}}(G)$ the subspace of cusp forms, both being representations of $G(\mathbb{A})$ via the right translation. Let

$$\nu\colon H(F)\backslash H(\mathbb{A}) \to \mathbb{C}^\times$$

be an automorphic character.

Definition 1. For $\varphi \in \mathcal{A}(G)$, we define the following integral

$$\mathcal{P}_{H,\nu}(\varphi) = \int_{H(F)\backslash H(\mathbb{A})} \varphi(h)\nu(h)^{-1}\mathrm{d}h$$

The author is partially supported by NSF Grant DMS #1302000.

to be the (H, ν)-*period* of φ, as long as it is absolutely convergent.

In what follows, we will mainly focus on the subspace $\mathcal{A}_{\mathrm{cusp}}(G)$ of cusp forms and the integral defining $\mathcal{P}_{H,\nu}(\varphi)$ is always absolutely convergent.

Definition 2. Let $\pi \subset \mathcal{A}_{\mathrm{cusp}}(G)$ be an irreducible (cuspidal automorphic) representation of $G(\mathbb{A})$. We say π is (globally) (H, ν)-*distinguished* if the linear functional $\mathcal{P}_{H,\nu}$ is nontrivial on π; in other words, $\mathcal{P}_{H,\nu}(\varphi) \neq 0$ for some $\varphi \in \pi$.

It is clear that if π is (H, ν)-distinguished, then

$$\mathrm{Hom}_{H(\mathbb{A})}(\pi, \nu) \neq \{0\},$$

where π is regarded as a representation of $H(\mathbb{A})$ by restriction.

Example 3. (Global genericity) Let G be a quasi-split reductive group over F. Choose a Borel subgroup B of G (defined over F) with the unipotent radical U. Let $\nu\colon U(F)\backslash U(\mathbb{A}) \to \mathbb{C}^\times$ be a generic character. Then $\mathcal{P}_{U,\nu}(\varphi)$ is absolutely convergent for $\varphi \in \mathcal{A}(G)$, which we usually call the (U, ν)-*Whittaker–Fourier coefficient* of φ. For an irreducible cuspidal representation $\pi \subset \mathcal{A}_{\mathrm{cusp}}(G)$, we say π is *globally generic* with respect to the pair (U, ν) if it is (U, ν)-distinguished.

Example 4. (Rankin–Selberg convolution) Take $n, m, r \geq 0$ with $r \leq n - m - 1$. Put $G = \mathrm{GL}(n)_F \times \mathrm{GL}(m)_F$. Consider the subgroup $H_r \subset \mathrm{GL}(n)_F$ consisting of matrices of the form

$$h = \begin{pmatrix} g_m & & & & & \star & \cdots & \star \\ \star & 1 & u_{1,2} & \cdots & \star & \star & \cdots & \star \\ \vdots & & \ddots & \ddots & \vdots & \vdots & \ddots & \vdots \\ \star & & & 1 & u_{r,r+1} & \star & \cdots & \star \\ & & & & 1 & u_{r+1,r+2} & \cdots & \star \\ & & & & & 1 & \ddots & \vdots \\ & & & & & & \ddots & u_{n-m-1,n-m} \\ & & & & & & & 1 \end{pmatrix}$$

with $g_m \in \mathrm{GL}(m)_F$. We may view H_r as a subgroup of G by sending h to (h, g_m). Define

$$\nu(h) = \psi(u_{1,2} + \cdots + u_{n-m-1,n-m})$$

to be an automorphic character of $H_r(\mathbb{A})$. Then $\mathcal{P}_{H_r,\nu}(\varphi)$ is absolutely convergent for $\varphi \in \mathcal{A}_{\mathrm{cusp}}(G)$. When $r = 0$, it is simply the Rankin–Selberg integral studied in [JPSS83]. In general, we prove in [Liua] that for an irreducible cuspidal representation $\pi \subset \mathcal{A}_{\mathrm{cusp}}(G)$, π is (H_r, ν)-distinguished if and only if $L(1/2, \pi, \mathrm{RS}) \neq 0$, where $\mathrm{RS}\colon {}^L G \to {}^L\mathrm{GL}(nm)_F$ is the tensor product homomorphism.

Example 5. (Unitary base change) Let E/F be an étale algebra of degree 2. Put $G = \mathrm{Res}_{E/F}\,\mathrm{GL}(n)_E/\mathbf{G}_{m,F}$ and $H = \mathrm{GL}(n)_F/\mathbf{G}_{m,F} \subset G$, where $\mathbf{G}_{m,F}$ is identified with the subgroup of F-rational scalar matrices. Denote $\eta_{E/F}\colon F^\times\backslash\mathbb{A}^\times \to \{\pm 1\}$ the quadratic character associated to E/F via global class field theory. Put

$$\nu = \eta_{E/F}^{n+1} \circ \det\colon H(F)\backslash H(\mathbb{A}) \to \mathbb{C}^\times$$

which is well-defined. Then $\mathcal{P}_{H,\nu}(\varphi)$ is absolutely convergent for $\varphi \in \mathcal{A}_{\mathrm{cusp}}(G)$. Moreover, Flicker [Fli88] proved that for an irreducible cuspidal representation $\pi \subset \mathcal{A}_{\mathrm{cusp}}(G)$, π is (H,ν)-distinguished if and only if the partial L-function $L^S(s,\pi,\mathrm{As}^{(-1)^{n+1}})$ has a (simple) pole at $s=1$, where $\mathrm{As}^{\pm}\colon {}^LG \to {}^L\mathrm{GL}(n^2)_F$ is the Asai (resp. twisted Asai) representation. Moreover, one expects that π is (H,ν)-distinguished if and only if it is the standard base change of a (cuspidal) automorphic representation of the quasi-split unitary group of n-variables defined by E/F.

Example 6. (Shalika period) Put $G = \mathrm{GL}(2n)_F$. Let $H \subset G$ be the subgroup consisting of matrices

$$h = \begin{pmatrix} g & \\ X & g \end{pmatrix}, \quad g \in \mathrm{GL}(n)_F, X \in \mathrm{Mat}(n)_F.$$

Define the character $\nu\colon H(F)\backslash H(\mathbb{A}) \to \mathbb{C}^\times$ by

$$\nu(h) = \psi(\mathrm{tr}\, X).$$

Then $\mathcal{P}_{H,\nu}(\varphi)$ is absolutely convergent for $\varphi \in \mathcal{A}_{\mathrm{cusp}}(G)$. By a well-known theorem of Jacquet and Shalika [JS90], for an irreducible cuspidal representation $\pi \subset \mathcal{A}_{\mathrm{cusp}}(G)$, π is (H,ν)-distinguished if and only if the partial L-function $L^S(s,\pi,\wedge^2)$ has a (simple) pole at $s=1$, where $\wedge^2\colon {}^L\mathrm{GL}(2n)_F \to {}^L\mathrm{GL}(n(2n-1))_F$ is the wedge square representation. Moreover, one expects that π is (H,ν)-distinguished if and only if it is the functorial lift of a (cuspidal) automorphic representation of $\mathrm{SO}(n,n+1)_F$.

2 Bessel period and multiplicity one

Let E/F be a field extension of degree at most 2. Denote c the unique automorphism of E such that $E^{c=1} = F$. An E-hermitian space is an E-vector space V equipped with a nondegenerate pairing $(\cdot,\cdot)\colon V \times V \to E$ (the hermitian form) that is E-linear in the first variable and satisfies $(u,v) = (v,u)^c$ for all $u, v \in V$. Denote $\mathrm{U}(V)$ the neutral connected component of the group of isometries of $V, (\cdot,\cdot)$, which is a connected reductive group over F. More explicitly, $\mathrm{U}(V)$ is the special orthogonal (resp. unitary) group attached to $V, (\cdot,\cdot)$ when $E = F$ (resp. $[E:F] = 2$). We will fix an E-hermitian space $L = \mathrm{Span}_E\{x_0\}$ of dimension 1.

Consider an E-hermitian space $V, (\cdot,\cdot)$ of dimension $m \geq 0$. Let $r \geq 0$. Put

$$X = \mathrm{Span}_E\{x_1, \ldots, x_r\}, \quad X^* = \mathrm{Span}_E\{x_1^*, \ldots, x_r^*\}$$

and equip $X \oplus X^*$ with a hermitian form such that $(x_i, x_j^*) = \delta_{ij}$ for $1 \leq i,j \leq r$. Put

$$V^\sharp = L \oplus V \oplus X \oplus X^*$$

as an E-hermitian space whose hermitian form is denoted by $(\cdot,\cdot)^\sharp$. In particular, the dimension of $V^\sharp$ is $n = m + 2r + 1$. The group $G_0 = \mathrm{U}(V)$ is naturally a closed subgroup of $G_1 = \mathrm{U}(V^\sharp)$. Put $G = G_1 \times G_0$. Now we are going to define the subgroup $H = H_r$ of G. Consider the following complete flag of E-subspaces of X:

$$\{0\} = X_0 \subset X_1 \subset \cdots \subset X_r = X, \tag{2.1}$$

where $X_i = \mathrm{Span}_E\{x_r, \ldots, x_{r+1-i}\} \subset X$ for $i = 0, \ldots, r$. Let $P_r \subset G_1$ be the parabolic subgroup stabilizing the flag (2.1) and U_r its unipotent radical. The group G_0 acts on U_r by conjugation in G_1. Define

$$H_r = U_r \rtimes G_0$$

to be the r-*th Bessel subgroup* of G_1. In what follows, we will view H_r as a subgroup of $G = G_1 \times G_0$ via the quotient homomorphism to the second component. We have the following homomorphism

$$\begin{aligned} \rho\colon H_r &\to \mathrm{Res}_{E/F}\, \mathbf{G}^r_{a,E}, \\ h &\mapsto \left((hx_{r-1}, x_r^*)^\sharp, \ldots, (hx_0, x_1^*)^\sharp\right). \end{aligned} \tag{2.2}$$

Define a character $\nu_r\colon H_r(F)\backslash H_r(\mathbb{A}) \to \mathbb{C}^\times$ by

$$\nu_r(h) = \psi(\mathrm{Tr}_{E/F}\, \rho(h)).$$

Example 7. Consider the case where $G_0 \simeq \mathrm{SO}(2,2)_F$ and $G_1 \simeq \mathrm{SO}(4,3)_F$ (in particular, $E = F$ and $r = 1$). We identify G_i with the group

$$\{g \in \mathrm{GL}(3i+4)_F \mid {}^t g \mathrm{J}_{3i+4} g = \mathrm{J}_{3i+4}\}$$

for $i = 0, 1$, where J_n is the n-by-n matrix with 1 on the non-principal diagonal and 0 elsewhere. We regard G_0 as a subgroup of G_1 via the embedding

$$\begin{pmatrix} g_{11} & g_{12} & g_{13} & g_{14} \\ g_{21} & g_{22} & g_{23} & g_{24} \\ g_{31} & g_{32} & g_{33} & g_{34} \\ g_{41} & g_{42} & g_{43} & g_{44} \end{pmatrix} \mapsto \begin{pmatrix} 1 &&&&&& \\ & g_{11} & g_{12} && g_{13} & g_{14} & \\ & g_{21} & g_{22} && g_{23} & g_{24} & \\ &&& 1 &&& \\ & g_{31} & g_{32} && g_{33} & g_{34} & \\ & g_{41} & g_{42} && g_{43} & g_{44} & \\ &&&&&& 1 \end{pmatrix}.$$

Then H_1 consists of matrices in G_1 of the form

$$h = \begin{pmatrix} 1 & u_1 & u_2 & u_0 & u_3 & u_4 & 0 \\ & g_{11} & g_{12} && g_{13} & g_{14} & u_4^* \\ & g_{21} & g_{22} && g_{23} & g_{24} & u_3^* \\ &&& 1 &&& -u_0 \\ & g_{31} & g_{32} && g_{33} & g_{34} & u_2^* \\ & g_{41} & g_{42} && g_{43} & g_{44} & u_1^* \\ &&&&&& 1 \end{pmatrix},$$

where $u_i \in \mathbf{G}_{a,F}$ for $i = 0, 1, 2, 3, 4$,

$$\begin{pmatrix} g_{11} & g_{12} & g_{13} & g_{14} \\ g_{21} & g_{22} & g_{23} & g_{24} \\ g_{31} & g_{32} & g_{33} & g_{34} \\ g_{41} & g_{42} & g_{43} & g_{44} \end{pmatrix} \in G_0$$

and $u_{5-i}^* = -(g_{i1}u_4 + g_{i2}u_3 + g_{i3}u_2 + g_{i4}u_1)$ for $i = 1, 2, 3, 4$. The homomorphism $\rho\colon H_1 \to \mathbf{G}_{a,F}$ is given by $h \mapsto u_0$.

Note that the only split torus that G_0 could be is $\mathbf{G}_{m,F}$, in which case we must have $E = F$ and $m = 2$. In what follows, we will assume that G_0 is *not* $\mathbf{G}_{m,F}$, and $n + [E : F] \geq 4$ (equivalently, G_1 is not abelian). Then $\mathcal{P}_{H_r,\nu_r}(\varphi)$ is absolutely convergent for $\varphi \in \mathcal{A}_{\mathrm{cusp}}(G)$.

Example 8. When $r = 0$, that is, $n = m + 1$. The unipotent group U_0 is trivial and we have $H_0 = G_0$, diagonally embedded into G. Moreover, the character ν_0 is trivial, and

$$\mathcal{P}_{H_0,\nu_0}(\varphi) = \int_{G_0(F)\backslash G_0(\mathbb{A})} \varphi(g)\mathrm{d}g.$$

When $m + [E : F] \leq 2$, the group G_0 is trivial and G_1 is quasi-split. The group U_r is a maximal unipotent subgroup of G_1 and ν_r is generic. Therefore,

$$\mathcal{P}_{H_r,\nu_r}(\varphi) = \int_{U_r(F)\backslash U_r(\mathbb{A})} \varphi(u)\nu(u)^{-1}\mathrm{d}u$$

is simply the Whittaker–Fourier coefficient of φ. This is a special case of Example 3.

Let $\pi \simeq \otimes'_{v\in\Sigma}\pi_v \subset \mathcal{A}_{\mathrm{cusp}}(G)$ be an irreducible representation of $G(\mathbb{A})$. The assignment $\varphi \in \pi \mapsto \mathcal{P}_{H_r,\nu_r}(\varphi)$ defines an element in

$$\mathrm{Hom}_{H_r(\mathbb{A})}(\pi, \nu) = \bigotimes_{v\in\Sigma} \mathrm{Hom}_{H_r(F_v)}(\pi_v, \nu_v).$$

The following is the *Multiplicity One Theorem* for the Bessel subgroup H_r.

Theorem 9. ([AGRS10, GGP12, JSZ10, SZ12, Wal12b]) *Let v be a place of F. For every irreducible smooth representation π_v of $G(F_v)$,*

$$\dim_{\mathbb{C}} \mathrm{Hom}_{H_r(F_v)}(\pi_v, \nu_v) \leq 1.$$

In particular, the dimension of the space $\mathrm{Hom}_{H_r(\mathbb{A})}(\pi, \nu)$ is at most 1. For a place v of F, it is known that

$$\dim_{\mathbb{C}} \mathrm{Hom}_{H_r(F_v)}(\pi_v, \nu_v) = \dim_{\mathbb{C}} \mathrm{Hom}_{H_r(F_v)}(\tilde{\pi}_v, \nu_v^{-1}),$$

where $\tilde{\pi}_v$ is the contragredient of π_v. In the next section, we will construct a candidate for the generator of the space

$$\mathrm{Hom}_{H_r(F_v)}(\pi_v, \nu_v) \otimes_{\mathbb{C}} \mathrm{Hom}_{H_r(F_v)}(\tilde{\pi}_v, \nu_v^{-1}),$$

whose dimension is at most 1, under the assumption that π_v is tempered.

3 Local invariant functional

In this section, we will suppress the place v from notation. In particular, F is a local field; E/F is an étale algebra of degree at most 2; $G = G(F_v)$ and similarly for other groups. Let π be an irreducible smooth (resp. Casselman–Wallach) representation of G if F is non-archimedean (resp. archimedean), and $\tilde{\pi}$ its contragredient. Recall that we may associate to a pair of elements $\varphi \in \pi$ and $\tilde{\varphi} \in \tilde{\pi}$ a smooth function $\Phi_{\varphi,\tilde{\varphi}}(g) = \langle \pi(g)\varphi, \tilde{\varphi}\rangle$ on G, called a *matrix coefficient* of π, where $\langle \cdot, \cdot \rangle \colon \pi \times \tilde{\pi} \to \mathbb{C}$ is the canonical pairing. We say π is *tempered* if for every K-finite φ and $\tilde{\varphi}$, $\Phi_{\varphi,\tilde{\varphi}}$ belongs to $L^{2+\varepsilon}(G)$ for any $\varepsilon > 0$. The function $\Phi_{\varphi,\tilde{\varphi}}$ may be viewed as a local analogue of automorphic form. Thus if one would like to mimic the definition of automorphic periods in the global case, the natural thought would be the following integral

$$\int_{H_r} \Phi_{\varphi,\tilde{\varphi}}(h)\nu(h)^{-1}\mathrm{d}h. \tag{3.1}$$

Formally, it does define an element in $\mathrm{Hom}_{H_r}(\pi, \nu) \otimes_{\mathbb{C}} \mathrm{Hom}_{H_r}(\tilde{\pi}, \nu^{-1})$. Unfortunately, unlike the global case, the above integral is not absolutely convergent in general, except when $r = 0$ (which is studied originally by Ichino and Ikeda [II10] in the orthogonal case and later by Harris [Har] in the unitary case). Therefore, we need to regularize the integral (3.1).

We will provide two methods for such regularization. The first one works only for F non-archimedean which we now describe. Thus assume F is non-archimedean. The following lemma is proved in [Liub].

Lemma 10. *Let π be an irreducible smooth representation of G. For every $\varphi \in \pi$ and $\tilde{\varphi} \in \tilde{\pi}$, there exists a compact open subgroup $\mathcal{U}$ of U_r such that for all other compact open subgroups $\mathcal{U}'$ of U_r containing $\mathcal{U}$,*

$$\int_{\mathcal{U}'} \Phi_{\varphi,\tilde{\varphi}}(u)\nu(u)^{-1}\mathrm{d}u = \int_{\mathcal{U}} \Phi_{\varphi,\tilde{\varphi}}(u)\nu(u)^{-1}\mathrm{d}u.$$

If we denote the integral on the right-hand side in the above proposition by

$$\int_{U_r}^{\mathrm{st}} \Phi_{\varphi,\tilde{\varphi}}(u)\nu(u)^{-1}\mathrm{d}u,$$

then one easily sees that it is independent of $\mathcal{U}$ that satisfies the property in the proposition. For $g \in G$, put

$$\mathcal{F}_\nu(\Phi_{\varphi,\tilde{\varphi}})(g) = \int_{U_r}^{\mathrm{st}} \Phi_{\pi(g)\varphi,\tilde{\varphi}}(u)\nu(u)^{-1}\mathrm{d}u$$

which is a smooth function on G. We have the following proposition.

Proposition 11. ([Liub, Theorem 2.1]) *Assume π is tempered. Then the integral*

$$\alpha(\varphi, \tilde{\varphi}) := \int_{G_0} \mathcal{F}_\nu(\Phi_{\varphi,\tilde{\varphi}})(g)\mathrm{d}g$$

is absolutely convergent, where G_0 *is diagonally embedded into* G *and* $\mathrm{d}h = \mathrm{d}u \cdot \mathrm{d}g$. *In particular, the functional* $(\varphi, \tilde{\varphi}) \mapsto \alpha(\varphi, \tilde{\varphi})$ *defines an element in*

$$\mathrm{Hom}_{H_r}(\pi, \nu) \otimes_{\mathbb{C}} \mathrm{Hom}_{H_r}(\tilde{\pi}, \nu^{-1}).$$

The second way works for an arbitrary local field F of characteristic 0. Define U_r° to be the kernel of $\rho|_{U_r}$ where ρ is defined in (2.2). Then ρ induces an isomorphism $U_r/U_r^\circ \xrightarrow{\sim} E^r$. One may view ν as an element in the dual space $\widehat{U_r/U_r^\circ} \simeq \widehat{E^r}$. In fact, it is in the regular locus of $\widehat{E^r}$, that is, those characters that are nontrivial on each component of E^r. We fix a decomposition $\mathrm{d}h = \mathrm{d}u \cdot \mathrm{d}u^\circ \cdot \mathrm{d}g$ of the Haar measure $\mathrm{d}h$ on H_r into Haar measures on U_r/U_r°, U_r° and G_0, respectively.

Proposition 12. *Assume* π *is tempered. For any* K*-finite vectors* $\varphi \in \pi$ *and* $\tilde{\varphi} \in \tilde{\pi}$,

1. *the integral*

$$\mathfrak{a}_{\varphi,\tilde{\varphi}}(u) := \int_{U_r^\circ G_0} \Phi_{\varphi,\tilde{\varphi}}(uu^\circ g)\mathrm{d}u^\circ \mathrm{d}g$$

   is absolutely convergent for every $u \in U_r/U_r^\circ$;

2. *the measure* $\mathfrak{a}_{\varphi,\tilde{\varphi}}(u)\mathrm{d}u$ *is a tempered distribution on* U_r/U_r°;

3. *the Fourier transform* $\widehat{\mathfrak{a}_{\varphi,\tilde{\varphi}}}$, *as a tempered distribution on* $\widehat{U_r/U_r^\circ} \simeq \widehat{E^r}$, *is smooth on the regular locus. In particular,*

$$\alpha(\varphi, \tilde{\varphi}) := \widehat{\mathfrak{a}_{\varphi,\tilde{\varphi}}}(\nu)$$

   is well-defined.

The functional $(\varphi, \tilde{\varphi}) \mapsto \alpha(\varphi, \tilde{\varphi})$ *defines an element in the space*

$$\mathrm{Hom}_{H_r}(\pi, \nu) \otimes_{\mathbb{C}} \mathrm{Hom}_{H_r}(\tilde{\pi}, \nu^{-1}).$$

In [Wal12a], Waldspurger also provided a regularization of the integral (3.1) when F is non-archimedean, using a way different from the previous two. It is not hard to see that in fact, three ways give rise to the same $\alpha(\varphi, \tilde{\varphi})$ when F is non-archimedean. We have the following properties about $\alpha(\varphi, \tilde{\varphi})$, in which (1) is proved by Waldspurger [Wal12a] and (2), (3) are proved in [Liub].

Theorem 13. *Let* π *be tempered and* $\alpha(\varphi, \tilde{\varphi})$ *as defined above.*

1. *Assume* $E = F$ *that is non-archimedean. If* $\dim_{\mathbb{C}} \mathrm{Hom}_{H_r}(\pi, \nu) = 1$, *then* $\alpha(\varphi, \tilde{\varphi}) \neq 0$ *for some* v *and* $\tilde{\varphi}$.

2. *Assume* π *is unitary, that is,* $\tilde{\pi} \simeq \bar{\pi}$. *Then for* $\varphi \in \pi$,

$$\alpha(\varphi, \bar{\varphi}) \geq 0.$$

3. *Assume F is non-archimedean and G is unramified. Choose a hyperspecial maximal subgroup K_G of G and assume that φ and $\tilde{\varphi}$ are both fixed by K_G with $\langle \varphi, \tilde{\varphi} \rangle = 1$; ψ has the conductor $U_r \cap K_G$; and the volume of $K_G \cap H_r$ under $\mathrm{d}h$ is 1. Then*

$$\alpha(\varphi, \tilde{\varphi}) = \frac{\Delta_{G_1} L(1/2, \pi, \mathrm{Std})}{L(1, \pi, \mathrm{Ad})},$$

where Δ_{G_1} is certain L-value of the Gross motive of G_1 [Gro97], *whose explicit formula can be found in* [Liub, *Section 2*].

We conjecture that part (1) in the above theorem holds in all other cases as well, especially when F is archimedean.

4 Distinguished packet

Let us go back to the global situation and thus F is a number field. Let $\mathfrak{V}$ be the set of all isomorphism classes of E-hermitian spaces of dimension m (and a fixed discriminant if $E = F$), which we call a *pure inner class.* For $\beta \in \mathfrak{V}$, denote V_β the corresponding E-hermitian space. We have groups $G_0^\beta = \mathrm{U}(V_\beta)$, $G_1^\beta = \mathrm{U}(V_\beta^\sharp)$, $G^\beta = G_1^\beta \times G_0^\beta$, $H_r^\beta \subset G^\beta$, and the character ν_r^β. Note that $G^\beta \simeq G^{\beta'}$ if and only if $\beta = \beta'$. For G^β, denote $\Pi(G^\beta)_{\mathrm{cusp}}^{\mathrm{temp}}$ the set of isomorphism classes of irreducible tempered cuspidal automorphic representations[1] of $G^\beta(\mathbb{A})$. By Arthur's conjecture, the multiplicity of any member of $\Pi(G^\beta)_{\mathrm{cusp}}^{\mathrm{temp}}$ in $\mathcal{A}_{\mathrm{cusp}}(G^\beta)$ is 1.

The Langlands dual group ${}^L G^\beta$ depends only on the pure inner class $\mathfrak{V}$, which we will denote by ${}^L G_{\mathfrak{V}}$ by abuse of notation. There is a conjectural Langlands group $\mathcal{L}_F$ associated to the field F. A (global) *tempered elliptic Arthur parameter*[2] for the pure inner class $\mathfrak{V}$ is an equivalence class of homomorphisms

$$\phi \colon \mathcal{L}_F \to {}^L G_{\mathfrak{V}}$$

such that the image is bounded, and

$$\mathcal{S}_\phi := \mathrm{Cent}_{{}^L G_{\mathfrak{V}}^0}(\mathrm{Im}\,\phi), \tag{4.1}$$

the centralizer of $\mathrm{Im}\,\phi$ in the neutral connected component ${}^L G_{\mathfrak{V}}^0$ of ${}^L G_{\mathfrak{V}}$, is finite. In fact, $\mathcal{S}_\phi$ will then be a finite 2-group in our case. Denote $\Phi(\mathfrak{V})_{\mathrm{ell}}^{\mathrm{temp}}$ the set of equivalent classes of tempered elliptic Arthur parameters for $\mathfrak{V}$. A part of Arthur's conjecture predicts a natural finite-to-one surjective map

$$\mathrm{rec} \colon \coprod_{\mathfrak{V}} \Pi(G^\beta)_{\mathrm{cusp}}^{\mathrm{temp}} \to \Phi(\mathfrak{V})_{\mathrm{ell}}^{\mathrm{temp}}$$

[1] An irreducible cuspidal automorphic representation $\pi \simeq \otimes'_{v \in \Sigma} \pi_v$ is tempered if π_v is tempered for all v.

[2] See [GGP12] for a formulation without using the Langlands group $\mathcal{L}_F$.

such that $\pi \in \mathrm{rec}^{-1}(\phi)$ if and only if they have the same partial (unramified) L-function

$$L^S(s,\pi,r) = L^S(s,\phi,r)$$

for any L-homomorphism $r\colon {}^LG_{\mathfrak{V}} \to {}^L\mathrm{GL}(N)_F$. Moreover, for any $v \in \Sigma$, $L(s,\phi_v,\mathrm{Std})L(2s,\phi_v,\mathrm{Ad})^{-1}$ is expected to be holomorphic and nonzero at $s=1/2$. In what follows, we are going to assume all these conjectures. The inverse image $\Pi_\phi := \mathrm{rec}^{-1}(\phi)$ for an element $\phi \in \Phi(\mathfrak{V})^{\mathrm{temp}}_{\mathrm{ell}}$ is called a *tempered elliptic Vogan packet* for (the pure inner class) $\mathfrak{V}$.

Definition 14. Let Π_ϕ be a tempered elliptic Vogan packet for $\mathfrak{V}$. We say Π_ϕ is (globally) *Bessel distinguished* if there is $\beta \in \mathfrak{V}$ and $\varphi_\beta \in \pi_\beta \in \Pi_\phi \cap \Pi(G^\beta)^{\mathrm{temp}}_{\mathrm{cusp}}$ such that $\mathcal{P}_{H^\beta_r,\nu^\beta_r}(\varphi_\beta) \neq 0$.

The following conjecture is known as the global Gan–Gross–Prasad conjecture for Bessel periods.

Conjecture 15. ([GGP12]) *Let Π_ϕ be a tempered elliptic Vogan packet for $\mathfrak{V}$. Then Π_ϕ is Bessel distinguished if and only if $L(1/2,\phi,\mathrm{Std}) \neq 0$.*

The following result is a consequence of the recent work of Waldspurger [Wal10, Wal12a] (resp. Beuzart-Plessis [BP]) on the local Gan–Gross–Prasad conjecture for special orthogonal (resp. unitary) groups in the non-archimedean case.

Proposition 16. *Let Π_ϕ be a tempered elliptic Vogan packet for $\mathfrak{V}$. If there are two members π_β and $\pi'_{\beta'}$ that are (H^β_r,ν^β_r)- and $(H^{\beta'}_r,\nu^{\beta'}_r)$-distinguished, respectively, then their finite components are isomorphic; in other words, $(\pi_\beta)^\infty \simeq (\pi'_{\beta'})^\infty$ (in particular, $G^\beta(\mathbb{A}^\infty) \simeq G^{\beta'}(\mathbb{A}^\infty)$).*

We expect that $\pi_\beta \simeq \pi'_{\beta'}$ (in particular, $\beta=\beta'$), which will follow by a similar result of Waldspurger and Beuzart-Plessis in the archimedean local case.

5 Refined Gan–Gross–Prasad conjecture and examples

Let $\pi \simeq \otimes'_{v\in\Sigma}\pi_v \subset \mathcal{A}_{\mathrm{cusp}}(G)$ be an irreducible tempered cuspidal automorphic representation of $G(\mathbb{A})$. Then π is unitary and $\bar{\pi} = \{\bar{\varphi} \mid \varphi \in \pi\}$. For each $v \in \Sigma$, choose a nontrivial $G(F_v)$-invariant hermitian paring $\langle\cdot,\cdot\rangle_v\colon \pi_v \times \bar{\pi}_v \to \mathbb{C}$. Motivated by Theorem 13 (3), we put

$$\alpha^\natural_v = \frac{L(1,\pi_v,\mathrm{Ad})}{\Delta_{G_1,v}L(1/2,\pi_v,\mathrm{Std})}\alpha_v,$$

where $\alpha_v \in \mathrm{Hom}_{H_r(F_v)}(\pi_v,\nu_v) \otimes_{\mathbb{C}} \mathrm{Hom}_{H_r(F_v)}(\bar{\pi}_v,\nu_v^{-1})$ is the invariant functional defined in Proposition 12 using the paring $\langle\cdot,\cdot\rangle_v$ and the local Haar measure $\mathrm{d}h_v$ on $H_r(F_v)$. We assume that the product measure $\prod_{v\in\Sigma}\mathrm{d}h_v$ on $H_r(\mathbb{A})$ coincides with the Tamagawa measure, and $\otimes_{v\in\Sigma}\langle\cdot,\cdot\rangle_v$ gives the Petersson inner product on $\pi\times\bar{\pi}$ (defined by the Tamagawa measure on $G(\mathbb{A})$). The product

$$\alpha^{\natural}_{\mathbb{A}}=\prod_{v\in\Sigma}\alpha^{\natural}_v$$

is a well-defined element in $\mathrm{Hom}_{H_r(\mathbb{A})}(\pi,\nu)\otimes_{\mathbb{C}}\mathrm{Hom}_{H_r(\mathbb{A})}(\bar{\pi},\nu^{-1})$.

Assume that Theorem 13 (1) holds in all cases. Then

$$\mathrm{Hom}_{H_r(\mathbb{A})}(\pi,\nu)\otimes_{\mathbb{C}}\mathrm{Hom}_{H_r(\mathbb{A})}(\bar{\pi},\nu^{-1})=\mathrm{Span}_{\mathbb{C}}\{\alpha^{\natural}_{\mathbb{A}}\}.$$

On the other hand, for $\varphi_1\in\pi,\bar{\varphi}_2\in\bar{\pi}$,

$$\beta(\varphi_1,\bar{\varphi}_2)=\mathcal{P}_{H_r,\nu_r}(\varphi_1)\mathcal{P}_{H_r,\nu_r^{-1}}(\bar{\varphi}_2)\tag{5.1}$$

also defines an element in $\mathrm{Hom}_{H_r(\mathbb{A})}(\pi,\nu)\otimes_{\mathbb{C}}\mathrm{Hom}_{H_r(\mathbb{A})}(\bar{\pi},\nu^{-1})$. Thus by Multiplicity One Theorem 9, there exists a constant c_π depending only π, such that

$$\beta=c_\pi\cdot\alpha^{\natural}_{\mathbb{A}}.\tag{5.2}$$

The following conjecture, proposed originally by Ichino and Ikeda [II10] in the case $E=F$ and $r=0$ and then extended by N. Harris [Har] in the case $[E:F]=2$ and $r=0$, is formulated in [Liub] for all Bessel periods. This is often called the *refined global Gan–Gross–Prasad conjecture.*

Conjecture 17. ([Liub]) *Let $\pi\subset\mathcal{A}_{\mathrm{cusp}}(G)$ be an irreducible tempered cuspidal automorphic representation of $G(\mathbb{A})$ occurred with multiplicity one. We have*

$$c_\pi=\frac{1}{|\mathcal{S}_{\mathrm{rec}(\pi)}|}\frac{\Delta_{G_1}L(1/2,\pi,\mathrm{Std})}{L(1,\pi,\mathrm{Ad})},$$

where $\mathcal{S}_{\mathrm{rec}(\pi)}$ is the finite 2-group defined in (4.1).

In the Whittaker–Fourier case, that is, $m+[E:F]\le 2$, the above conjecture is also formulated by Lapid and Mao in [LM], as special cases of their conjecture on Whittaker–Fourier coefficients for more general reductive groups.

The above conjecture is fully known when

- $E=F$, $m=2$ and $r=0$, which is essentially the Waldspurger formula [Wal85];
- $E=F$, $m=3$ and $r=0$, which is known as the triple product formula, finally settled by Ichino [Ich08];
- $[E:F]=2$, $m=1$ and $r=0$, which can be deduced from the Waldspurger formula.

Moreover, if $E=F$ and $r=0$, we have some other known special cases when m is small. See [II10] for a summary of all other examples in this case. Toward G of higher ranks, we have the following celebrated result recently obtained by W. Zhang.

Theorem 18. ([Zhab]) *We assume certain standard hypotheses from Arthur's conjecture. Suppose* $[E:F]=2$, $m\geq 2$ *and* $r=0$. *Denote* Σ_s *(resp.* Σ_i*) the subset of* Σ *consisting of places of* F *that are split (resp. non-split) in* E. *Let* $\pi\subset\mathcal{A}_{\mathrm{cusp}}(G)$ *be an irreducible (tempered) cuspidal automorphic representation of* $G(\mathbb{A})$. *If*

1. Σ_s *contains all archimedean places;*

2. *there exists* $v_0\in\Sigma_s$ *such that* π_{v_0} *is supercuspidal;*

3. *at any place* $v\in\Sigma_i$ *(which is non-archimedean by (1)), either* $G_0(F_v)$ *is compact, or* π_v *is supercuspidal, or* π_v *is unramified and the Jacquet–Rallis fundamental lemma for* G *holds at* F_v,

then Conjecture 17 *holds for such* π.

Remark 19. In the above theorem, condition (2) will force π to be tempered and moreover $|\mathcal{S}_{\mathrm{rec}(\pi)}|=4$, under certain hypothesis from Arthur's conjecture. The Jacquet–Rallis fundamental lemma is known by the work of Yun [Yun11] providing the residue characteristic of F_v greater that $c(m)$, a constant depending only on m.

We have two more examples in the case $r=1$, which are new.

Theorem 20. ([Liub]) *Assume* $E=F$, $m=2$ *and* $r=1$. *Let* $\pi=\pi_1\boxtimes\pi_0$ *be an irreducible cuspidal automorphic representation of* $G(\mathbb{A})=G_1(\mathbb{A})\times G_0(\mathbb{A})$. *Suppose that* $G_1\simeq\mathrm{SO}(2,3)_F$ *and* π_1 *is an endoscopic Yoshida lift. Then Conjecture* 17 *holds for such* π.

In the above theorem, G_0 is a non-split torus and hence π_0 is simply an automorphic character of $G_0(\mathbb{A})$. The representation π_1 is expected to be tempered, but we do *not* assume this. In particular, we have a way to define $\alpha^\natural_{\mathbb{A}}$ without using the assumption that π_1 is tempered. Finally, $|\mathcal{S}_{\mathrm{rec}(\pi)}|=8$.

Theorem 21. ([Liub]) *Assume* $E=F$, $m=3$ *and* $r=1$. *Let* $\pi=\pi_1\boxtimes\pi_0$ *be an irreducible tempered cuspidal automorphic representation of* $G(\mathbb{A})=G_1(\mathbb{A})\times G_0(\mathbb{A})$. *Suppose that* G_1 *is quasi-split but not split and* π_1 *is a certain endoscopic tempered cuspidal automorphic representation. Assume that Conjecture* 17 *holds for* $m=2$, $r=0$ *and* E *being the discriminant field of the (6-dimensional) quadratic space* $V^\sharp$. *Then Conjecture* 17 *holds for such* π.

In fact, in the above theorem, G_1 is naturally isomorphic to a subgroup of $\mathrm{PGU}(2,2)_E$ whose quotient is abelian, where the quasi-split unitary group is associated to the quadratic extension E/F with E the discriminant field of the quadratic space $V^\sharp$. The representation π_1 we consider comes from the global theta lift of $\mathrm{U}(W)$ where W is a 3-dimensional E-hermitian space.

6 Relative trace formula

In the last section, we would like to introduce an approach toward Conjecture 17 in the unitary case, namely, the *relative trace formula*. Several years ago, Jacquet and Rallis [JR11] proposed such approach in the case $[E:F]=2$ and $r=0$. After that, significant progress has been achieved, including the work of Yun on the proof of the corresponding (relative) fundamental lemma [Yun11], of Zhang on the proof of the corresponding smooth matching [Zhaa] and finally Theorem 18 in [Zhab]. On the other hand, the author extended such relative trace formula to all r (and even the case of Fourier–Jacobi periods, which are not discussed in this note) in [Liua]. In the rest of the section, we would like to explain its basic idea. Before doing so, let us remark that in certain orthogonal case – when $E=F$, $m=2$ and $r=1$, there are also some relative trace formulas and related results, mainly due to M. Furusawa, K. Martin and J. Shalika. See [FM] for a summary.

Let us now assume $[E:F]=2$ and fix a pure inner class $\mathfrak{V}$. Put $\mathbf{G}=\mathrm{Res}_{E/F}(\mathrm{GL}(n)_E\times\mathrm{GL}(m)_E)$ and one may similarly define a subgroup $\mathbf{H}_r\subset\mathbf{G}$ like H_r. In addition, put $\mathbf{G}'=\mathrm{GL}(n)_F\times\mathrm{GL}(m)_F$ which is a subgroup of $\mathbf{G}$. We define two distributions: $\mathcal{I}$ on the space $\mathcal{C}_c^\infty(G(\mathbb{A}))$ and $\mathcal{J}$ on the space $\mathcal{C}_c^\infty(\mathbf{G}(\mathbb{A}))$. Very roughly speaking, $\mathcal{I}$ encodes the information of the linear functional β (5.1), that is, the left-hand side of (5.2), and $\mathcal{J}$ encodes the information on the special L-value, that is, the right-hand side of (5.2).

Formally, both distributions have two sorts of decompositions: the *spectral decomposition*

$$\mathcal{I}=\sum_\pi\mathcal{I}_\pi,\quad \mathcal{J}=\sum_\Pi\mathcal{J}_\Pi,$$

and the *orbital decomposition*

$$\mathcal{I}=\sum_{H_r(F)\backslash G(F)/H_r(F)}\mathcal{I}_\gamma,\quad \mathcal{J}=\sum_{\mathbf{H}_r(F)\backslash\mathbf{G}(F)/\mathbf{G}'(F)}\mathcal{J}_\delta.$$

Inspired by Proposition 16, we should consider all $\beta\in\mathfrak{V}$ at the same time. Our goal is to compare the spectral side, that is, $\mathcal{I}_\pi$ and $\mathcal{J}_\Pi$, which can be transferred to the comparison on the orbital side. The following lemma is the first step toward such comparison.

Lemma 22. ([Liua]) *Denote* $(\mathbf{H}_r(F)\backslash\mathbf{G}(F)/\mathbf{G}'(F))_{\mathrm{reg}}$ *the subset of*

$$\mathbf{H}_r(F)\backslash\mathbf{G}(F)/\mathbf{G}'(F)$$

consisting of orbits whose stabilizer is trivial, and similarly for

$$(H_r^\beta(F)\backslash G^\beta(F)/H_r^\beta(F))_{\mathrm{reg}}.$$

Then they are Zariski dense (resp. v-adically dense) if F is a number field (resp. F is viewed as its localization at v). Moreover, we have a natural bijection

$$(\mathbf{H}_r(F)\backslash\mathbf{G}(F)/\mathbf{G}'(F))_{\mathrm{reg}}\simeq\coprod_{\beta\in\mathfrak{V}}(H_r^\beta(F)\backslash G^\beta(F)/H_r^\beta(F))_{\mathrm{reg}}.$$

We also propose the conjectures on the corresponding relative fundamental lemma and smooth matching. The fundamental lemma is open unless $r = 0$ or $m = 0$ (which follows from a slightly modified argument of Ngô [NBC99] in the proof of the Jacquet–Ye fundamental lemma). In general, it is hopeful that one may attack it by an algebro-geometric method that is a combination of [NBC99] and [Yun11]. The smooth matching is widely open unless $r = 0$, even in the case $m = 0$.

References

[AGRS10] A. Aizenbud, D. Gourevitch, S. Rallis, and G. Schiffmann, *Multiplicity one theorems*, Ann. of Math. (2) **172** (2010), no. 2, 1407–1434.

[BP] R. Beuzart-Plessis, *La conjecture locale de Gross-Prasad pour les représentations tempérées des groupes unitaires.* arXiv:1205.2987.

[Fli88] Y. Z. Flicker, *Twisted tensors and Euler products*, Bull. Soc. Math. France **116** (1988), no. 3, 295–313 (English, with French summary).

[FM] M. Furusawa and K. Martin, *On central critical values of the degree four L-functions for* GSp(4): *a simple trace formula*, Math. Z. **277** (2014), no. 1-2, 149–180.

[GGP12] W. T. Gan, B. H. Gross, and D. Prasad, *Symplectic local root numbers, central critical L-values, and restriction problems in the representation theory of classical groups*, Astérisque **346** (2012), 1–109.

[Gro97] B. H. Gross, *On the motive of a reductive group*, Invent. Math. **130** (1997), no. 2, 287–313.

[GP92] B. H. Gross and D. Prasad, *On the decomposition of a representation of* SO_n *when restricted to* SO_{n-1}, Canad. J. Math. **44** (1992), no. 5, 974–1002.

[GP94] B. H. Gross and D. Prasad, *On irreducible representations of* $SO_{2n+1} \times SO_{2m}$, Canad. J. Math. **46** (1994), no. 5, 930–950.

[Har] R. N. Harris, *The refined Gross-Prasad conjecture for unitary groups*, Int. Math. Res. Not. (2014), no. 2, 303–389.

[Ich08] A. Ichino, *Trilinear forms and the central values of triple product L-functions*, Duke Math. J. **145** (2008), no. 2, 281–307.

[II10] A. Ichino and T. Ikeda, *On the periods of automorphic forms on special orthogonal groups and the Gross-Prasad conjecture*, Geom. Funct. Anal. **19** (2010), no. 5, 1378–1425.

[JPSS83] H. Jacquet, I. Piatetski-Shapiro, and J. A. Shalika, *Rankin-Selberg convolutions*, Amer. J. Math. **105** (1983), no. 2, 367–464.

[JR11] H. Jacquet and S. Rallis, *On the Gross–Prasad conjecture for unitary groups*, On certain L-functions, Clay Math. Proc., Vol. **13**, Amer. Math. Soc., Providence, RI, 2011, pp. 205–264.

[JS90] H. Jacquet and J. Shalika, *Exterior square L-functions*, Automorphic forms, Shimura varieties, and L-functions, Vol. **II** (Ann Arbor, MI, 1988), Perspect. Math., Vol. **11**, Academic Press, Boston, MA, 1990, pp. 143–226.

[JSZ10] D. Jiang, B. Sun, and C.- B. Zhu, *Uniqueness of Bessel models: the Archimedean case*, Geom. Funct. Anal. **20** (2010), no. 3, 690–709.

[LM] E. Lapid and Z. Mao, *A conjecture on Whittaker–Fourier coefficients of cusp forms*, J. Number Theory **146** (2015), 448–505.

[Liua] Y. Liu, *Relative trace formulae toward Bessel and Fourier-Jacobi periods of unitary groups*, Manuscripta Math. **145** (2014), no. 1-2, 1–69.

[Liub] Y. Liu, *Refined Gan-Gross-Prasad conjecture for Bessel periods*, J. reine angew. Math., to appear.

[NBC99] B. C. Ngô, *Le lemme fondamental de Jacquet et Ye en caractéristique positive*, Duke Math. J. **96** (1999), no. 3, 473–520.

[SZ12] B. Sun and C. -B. Zhu, *Multiplicity one theorems: the Archimedean case*, Ann. of Math. (2) **175** (2012), no. 1, 23–44.

[Wal85] J. -L. Waldspurger, *Sur les valeurs de certaines fonctions L automorphes en leur centre de symétrie*, Compositio Math. **54** (1985), no. 2, 173–242 (French).

[Wal10] J. -L. Waldspurger, *Une formule intégrale reliée à la conjecture locale de Gross-Prasad*, Compos. Math. **146** (2010), no. 5, 1180–1290 (French, with English summary).

[Wal12a] J. -L. Waldspurger, *Une formule intégrale reliée à la conjecture locale de Gross-Prasad, 2e partie: extension aux représentations tempérées*, Astérisque **346** (2012), 171–312.

[Wal12b] J. -L. Waldspurger, *Une variante d'un résultat de Aizenbud, Gourevitch, Rallis et Schiffmann*, Asterisque **346** (2012), 313–318.

[Yun11] Z. Yun, *The fundamental lemma of Jacquet and Rallis*, Duke Math. J. **156** (2011), no. 2, 167–227 (With an appendix by Julia Gordon).

[Zhaa] W. Zhang, *Fourier transform and the global Gan-Gross-Prasad conjecture for unitary groups*, available at http://www.math.columbia.edu/~wzhang/, preprint.

[Zhab] W. Zhang, *Automorphic period and the central value of Rankin-Selberg L-function*, J. Amer. Math. Soc. **27** (2014), no. 2, 541–612.

Department of Mathematics, Massachusetts Institute of Technology
Cambridge, MA 02139, USA
E-mail address: liuyf@math.mit.edu

Proceedings of the 6th International Congress of Chinese Mathematicians
ALM 36, pp. 351–359

Counting Integral Points in Certain Homogeneous Spaces

Fei Xu

Abstract

The asymptotic formula of the number of integral points in non-compact symmetric homogeneous spaces of semi-simple simply connected algebraic groups is given by the average of the product of the number of local solutions twisted by the Brauer-Manin obstruction. The similar result is also true for homogeneous spaces of reductive groups with some restriction. As application, we will give the explicit asymptotic formulae of the number of integral points of certain norm equations and prove that the asymptotic formula of the number of integral matrices with a fixed irreducible characteristic polynomial over $\mathbb{Z}$ studied by Eskin-Mozes-Shah is equal to the product of the number of local integral solutions over all primes although the density function defined by Borovoi and Rudnick is not trivial in general. We also answer a question raised by Borovoi for comparing the number of integral symmetric matrices with the given determinant with the product of the number of all local solutions.

Mathematics Subject Classification: 11E72, 11G35, 11R37, 14F22, 14G25, 20G30, 20G35.

Keywords and Phrases: Integral point, homogeneous space, Tamagawa measure.

1 Introduction

Let $\mathbf{X}$ be a separated scheme of finite type over $\mathbb{Z}$ with a fixed coordinates (closed immersion)

$$X_{\mathbb{Q}} = \mathbf{X} \times_{\mathbb{Z}} \mathbb{Q} \hookrightarrow \mathrm{Spec}\mathbb{Q}[x_1, \dots, x_n].$$

Our basic question is to find the asymptotic formula for

$$N(\mathbf{X}, T) = \sharp\{(x_i) \in \mathbf{X}(\mathbb{Z}) : |x_i| \le T \quad \text{for all } 1 \le i \le n\}$$

as $T \to \infty$.

• The Hardy-Littlewood circle method (if it can be applied) gives

$$N(\mathbf{X},T) \sim (\prod_{p \text{ primes}} N_p(\mathbf{X})) \cdot N_\infty(X_\mathbb{Q},T)$$

as $T \to \infty$, where

$$N_p(\mathbf{X}) = \lim_{k\to\infty} \frac{\sharp\mathbf{X}(\mathbb{Z}/(p^k))}{p^{k\cdot \dim X_\mathbb{Q}}}$$

and

$$N_\infty(X_\mathbb{Q},T) = \mathrm{vol}(\{(x_1,\dots,x_n) \in X_\mathbb{Q}(\mathbb{R}) : |x_i| \le T,\ 1 \le i \le n\}).$$

Once the Hardy-Littlewood circle method can be applied, one has that the local-global principle holds for $\mathbf{X}$, i.e.

$$X_\mathbb{Q}(\mathbb{R}) \times \prod_{p \text{ primes}} \mathbf{X}(\mathbb{Z}_p) \neq \emptyset \quad \Leftrightarrow \quad \mathbf{X}(\mathbb{Z}) \neq \emptyset.$$

• Recent progress on strong approximation with Brauer-Manin obstruction on various homogeneous spaces has been studied extensively in [4, 9, 5, 35, 36, 2], which provides many examples that the local-global principle does not hold. It is natural to ask how the asymptotic formula of the number of integral solutions will be related to Brauer-Manin obstruction for such homogeneous spaces.

• Borovoi and Rudnick have studied such problem for homogeneous spaces of semi-simple groups in [3], where they measured the difference between the number of integral solutions and that of Hardy-Littlewood expectation by introducing so called the density functions. These density functions were described by using Kottwitz invariant and were not determined locally in [3].

2 Main results

Let $\mathbf{X}$ be a separated scheme of finite type over $\mathbb{Z}$ such that

$$X_\mathbb{Q} = \mathbf{X} \times_\mathbb{Z} \mathbb{Q} \cong G/H$$

where G and H are reductive groups over $\mathbb{Q}$.

• For any

$$\xi \in Br(X_\mathbb{Q}) = H^2_{et}(X_\mathbb{Q}, \mathbb{G}_m),$$

one can regarded ξ as a locally constant function over $X_\mathbb{Q}(\mathbb{A}_\mathbb{Q})$ and $X_\mathbb{Q}(\mathbb{Q}_p)$ by setting

$$\xi : X_\mathbb{Q}(\mathbb{A}_\mathbb{Q}) \to \mu_\infty \subset \mathbb{C}^\times; \quad (x_p) \mapsto \prod_{p\le\infty} \mathrm{inv}_p(\xi)$$

and

$$\xi : X_\mathbb{Q}(\mathbb{Q}_p) \to \mu_\infty \subset \mathbb{C}^\times; \quad x_p \mapsto \mathrm{inv}_p(\xi)$$

with a fixed identification $\mathbb{Q}/\mathbb{Z} \cong \mu_\infty$, where $\mathbb{A}_\mathbb{Q}$ is the adeles ring of $\mathbb{Q}$ and $\mathbb{Q}_p$ is the p-adic completion of $\mathbb{Q}$ and $\mathbb{Q}_\infty = \mathbb{R}$ and μ_∞ is the set of all roots of unity.

• Define

$$N_p(\mathbf{X}, \xi) = \int_{\mathbf{X}(\mathbb{Z}_p)} \xi d_p$$

for any prime p and

$$N_\infty(X, T, \xi) = \int_{X_\mathbb{Q}(\mathbb{R}, T)} \xi d_\infty$$

for $T > 0$, where $\prod_{p \le \infty} d_p$ is the Tamagawa measure and

$$X_\mathbb{Q}(\mathbb{R}, T) = \{(x_i) \in X(\mathbb{R}) : \ |x_i| \le T \ \text{ with } \ 1 \le i \le n\}.$$

• **Remark.** If $\xi = 1$, then

$$N_p(\mathbf{X}, 1) = \int_{\mathbf{X}(\mathbb{Z}_p)} d_p = \lim_{k \to \infty} \frac{\sharp \mathbf{X}(\mathbb{Z}/(p^k))}{p^{k \cdot \dim X_\mathbb{Q}}} = N_p(\mathbf{X})$$

and

$$N_\infty(X, T, 1) = \int_{X_\mathbb{Q}(\mathbb{R}, T)} d_\infty = \mathrm{vol}(X_\mathbb{Q}(\mathbb{R}, T))$$

for $T \gg 0$, which are the same as the local solutions in the circle method.

• **Theorem.** If G is simply connected and almost simple such that $G(\mathbb{R})$ is not compact and $X_\mathbb{Q}$ is symmetric, then

$$N(\mathbf{X}, T) \sim \sum_{\xi \in (Br(X)/Br(\mathbb{Q}))} \Big(\prod_{p \text{ primes}} N_p(\mathbf{X}, \xi)\Big) \cdot N_\infty(X, T, \xi)$$

as $T \to \infty$ for any separated scheme $\mathbf{X}$ of finite type over $\mathbb{Z}$ with $X = \mathbf{X} \times_\mathbb{Z} \mathbb{Q}$.

• For general reductive groups, one can have the similar result with certain restriction.

3 Sketch of proof for semi-simple case

Step 1: Choose a sufficiently large finite set S of primes containing ∞ such that there are the group schemes $\mathbf{G}$ and $\mathbf{H}$ of finite type over $\mathbb{Z}_S$ such that

$$\mathbf{G} \times_{\mathbb{Z}_S} \mathbb{Q} = G, \quad \mathbf{H} \times_{\mathbb{Z}_S} \mathbb{Q} = H \quad \text{and} \quad \mathbf{X}_S = \mathbf{X} \times_\mathbb{Z} \mathbb{Z}_S \cong \mathbf{G}/\mathbf{H}$$

where $\mathbb{Z}_S$ is the sub-ring of $\mathbb{Q}$ such that the denominator of any element in $\mathbb{Z}_S$ is divided only by the primes in S.

Define

$$St(\mathbf{X}(\mathbb{Z}_p)) = \begin{cases} \mathbf{G}(\mathbb{Z}_p) & p \notin S, \\ \{g \in \mathbf{G}_p(\mathbb{Z}_p) : \ g \cdot \mathbf{X}(\mathbb{Z}_p) = \mathbf{X}(\mathbb{Z}_p)\} & p \in S \setminus \infty, \\ G(\mathbb{R}) & p = \infty \end{cases}$$

where $\mathbf{G}_p$ is a fixed group scheme of finite type over $\mathbb{Z}_p$ such that $\mathbf{G}_p \times_{\mathbb{Z}_p} \mathbb{Q}_p = G \times_{\mathbb{Q}} \mathbb{Q}_p$ for $p \in S \setminus \infty$.

Let

$$St(\mathbf{X}) = \prod_{p \leq \infty} St(\mathbf{X}(\mathbb{Z}_p)) \quad \text{and} \quad \Gamma = G(\mathbb{Q}) \cap St(\mathbf{X}).$$

Then Γ acts on $\mathbf{X}(\mathbb{Z})$ with finite orbits.

Step 2: Partition of the orbits.

• Define the equivalent relation " $\sim$ " on $\mathbf{X}(\mathbb{Z})$.

For any $x, y \in \mathbf{X}(\mathbb{Z})$, we define the equivalent relation on $\mathbf{X}(\mathbb{Z})$

$$x \sim y \quad \Leftrightarrow \quad x = s_{\mathbb{A}} \cdot y$$

for some $s_{\mathbb{A}} \in St(\mathbf{X})$. The set of the equivalent classes is denoted by $\mathbf{X}(\mathbb{Z})/\sim$.

Proposition 3.1. *If G is semi-simple and simply connected such that $G'(\mathbb{R})$ is not compact for any simple factor of G, then the diagonal map*

$$\mathbf{X}(\mathbb{Z})/\sim \ \xrightarrow{\cong} \ St(\mathbf{X})\backslash(\prod_{p \leq \infty} \mathbf{X}(\mathbb{Z}_p))^{Br(X_{\mathbb{Q}})}$$

is bijective, where

$$(\prod_{p \leq \infty} \mathbf{X}(\mathbb{Z}_p))^{Br(X_{\mathbb{Q}})} = \Big\{(x_p) \in \prod_{p \leq \infty} \mathbf{X}(\mathbb{Z}_p) : \prod_{p \leq \infty} \xi(x_p) = 1, \quad \forall\, \xi \in Br(X_{\mathbb{Q}})\Big\}.$$

Therefore

$$\mathbf{X}(\mathbb{Z}) = \bigcup_i (\mathbf{X}(\mathbb{Z}) \cap St(\mathbf{X})x_i)$$

where $x_i \in \mathbf{X}(\mathbb{Z})$ are the representatives of orbits of $St(\mathbf{X})$.

• Define the further equivalent relation $\sim_G$ on $\mathbf{X}(\mathbb{Z}) \cap St(\mathbf{X})x_i$.

For $y, z \in \mathbf{X}(\mathbb{Z}) \cap St(\mathbf{X})x_i$, we can define

$$y \sim_G z \quad \Leftrightarrow \quad y = g \cdot z$$

for some $g \in G(\mathbb{Q})$.

Proposition 3.2. *With the same condition as that in Proposition* 3.1, *one has*

$$\sharp((\mathbf{X}(\mathbb{Z}) \cap St(\mathbf{X})x_i)/\sim_G) = \sharp(\ker(H^1(\mathbb{Q}, H) \to \prod_{p \leq \infty} H^1(\mathbb{Q}_p, H))).$$

Therefore

$$\mathbf{X}(\mathbb{Z}) \cap St(\mathbf{X})x_i = \bigcup_j (G(\mathbb{Q})y_j^{(i)} \cap St(\mathbf{X})x_i)$$

where $y_j^{(i)} \in \mathbf{X}(\mathbb{Z})$ are representatives of such equivalent classes.

• Γ acts on $G(\mathbb{Q})y_j^{(i)} \cap St(\mathbf{X})x_i$.

Proposition 3.3. *With the same condition as that in Proposition* 3.1, *there is a bijection*

$$\Gamma\backslash(G(\mathbb{Q})y_j^{(i)} \cap St(\mathbf{X})x_i) \xrightarrow{\cong} H_{ij}(\mathbb{Q})\backslash(H_{ij}(\mathbb{A}_\mathbb{Q})/(H_{ij}(\mathbb{A}_\mathbb{Q}) \cap St(\mathbf{X}))$$

where H_{ij} is the stabilizer of $y_j^{(i)}$.

Therefore

$$G(\mathbb{Q})y_j^{(i)} \cap St(\mathbf{X})x_i = \bigcup_k \Gamma z_k^{(i,j)}$$

with $z_k^{(i,j)} \in \mathbf{X}(\mathbb{Z})$.

Step 3. Counting via equi-distribution.
Decompose

$$\mathbf{X}(\mathbb{Z}) = \bigcup_i \bigcup_j \bigcup_k \Gamma z_k^{(i,j)}.$$

We need the following equi-distribution property.

Theorem 3.4. (*Duke-Rudnick-Sanark, Eskin-Mozes-Shah*) *If G is almost simple and simply connected such that $G(\mathbb{R})$ is not compact and $X_\mathbb{Q}$ is symmetric, then*

$$\sharp\{y \in \Gamma x : |y| \le T\} \sim \frac{\nu_\infty(\Gamma_{H_x}\backslash H_x(\mathbb{R}))}{\lambda_\infty(\Gamma\backslash G(\mathbb{R}))} d_\infty(G(\mathbb{R})x \cap X_\mathbb{Q}(\mathbb{R}, T))$$

as $T \to \infty$, where H_x is the stabilizer of x in G,

$$\Gamma_{H_x} = H_x(F) \cap \Gamma,$$

ν_∞, λ_∞ and d_∞ are the Tamagawa measures over $H_x(\mathbb{R})$, $G(\mathbb{R})$ and $X_\mathbb{Q}(\mathbb{R})$ respectively which match together algebraically.

By this theorem, one has

$$N(\mathbf{X}, T) \sim \sum_i \sum_j \sum_k \frac{\nu_\infty(\Gamma_{i,j,k}\backslash H_{i,j,k}(\mathbb{R}))}{\lambda_\infty(\Gamma\backslash G(\mathbb{R}))} d_\infty(G(\mathbb{R})z_k^{(i,j)} \cap X_\mathbb{Q}(\mathbb{R}, T))$$

as $T \to \infty$, where $H_{i,j,k}$ is the stabilizer of $z_k^{(i,j)}$ in G, $\Gamma_{i,j,k} = \Gamma \cap H_{i,j,k}(\mathbb{Q})$.

By Proposition 3.3, one has

$$N(\mathbf{X}, T) \sim \sum_i \frac{d_\infty(G(\mathbb{R})x_i \cap X_\mathbb{Q}(\mathbb{R}, T))}{\lambda_\infty(\Gamma\backslash G(\mathbb{R}))} \cdot \Big(\sum_j \tau(H_{i,j}) \prod_{p \text{ primes}} \nu_p^{(i,j)}(H_{i,j}(\mathbb{Q}_p) \cap St(\mathbf{X}(\mathbb{Z}_p)))^{-1}\Big)$$

as $T \to \infty$, where $\tau(H_{i,j})$ is the Tamagawa number of $H_{i,j}$.

By Proposition 3.2, one has

$$N(\mathbf{X}, T) \sim \sharp\mathrm{Pic}(H) \sum_i \frac{d_\infty(G(\mathbb{R})x_i \cap X_\mathbb{Q}(\mathbb{R}, T)) \prod_{p \text{ primes}} d_p(St(\mathbf{X}(\mathbb{Z}_p)x_i)}{\lambda_\infty(\Gamma\backslash G(\mathbb{R})) \prod_{p \text{ primes}} \lambda_p(St(\mathbf{X}(\mathbb{Z}_p)))}$$

as $T \to \infty$

By Proposition 3.1, one has

$$N(\mathbf{X},T) \sim \sharp(Br(X_{\mathbb{Q}})/Br(\mathbb{Q})) \int_{(\prod_{p \text{ primes}} \mathbf{X}(\mathbb{Z}_p)\times X_{\mathbb{Q}}(\mathbb{R},T))^{Br(X_{\mathbb{Q}})}} d_{X_{\mathbb{Q}}}$$

$$= \sum_{\xi\in(Br(X_{\mathbb{Q}})/Br(\mathbb{Q}))} \int_{\prod_{p \text{ primes}} \mathbf{X}(\mathbb{Z}_p)\times X_{\mathbb{Q}}(\mathbb{R},T)} \xi \, d_{X_{\mathbb{Q}}}$$

$$= \sum_{\xi\in(Br(X_{\mathbb{Q}})/Br(\mathbb{Q}))} (\prod_{p \text{ primes}} N_p(\mathbf{X},\xi))N_\infty(X_{\mathbb{Q}},T,\xi)$$

as $T \to \infty$.

4 An example

As application, we will explain that the asymptotic formula of the number of integral solutions in Theorem 1.1 of [8] is the same as the Hardy-Littlewood expectation although the density function defined in [3] is not trivial in general by §6.6 in [3]. It should be pointed out that such a result only holds over $\mathbb{Q}$ and is no longer true over general number fields (see Example 6.3 in [37]).

Example 4.1. *Let $p(\lambda)$ be an irreducible monic polynomial of degree $n \geq 2$ over $\mathbb{Z}$ and $\mathbf{X}$ be the scheme over $\mathbb{Z}$ defined by the following equations with variables $x_{i,j}$ for $1 \leq i,j \leq n$*

$$\det(\lambda I_n - (x_{i,j})) = p(\lambda)$$

and $X_{\mathbb{Q}} = \mathbf{X} \times_{\mathbb{Z}} \mathbb{Q}$. Then

$$N(\mathbf{X},T) \sim (\prod_{p<\infty} \int_{\mathbf{X}(\mathbb{Z}_p)} d_p) \cdot \int_{\mathbf{X}(\mathbb{R},T)} d_\infty$$

as $T \to \infty$.

In order to match the result in [8], we modify

$$X_{\mathbb{Q}}(\mathbb{R},T) = \left\{(x_{i,j}) \in X_{\mathbb{Q}}(\mathbb{R}) : \sqrt{\sum_{i,j} x_{i,j}^2} < T\right\}$$

and

$$N(\mathbf{X},T) = \sharp[\mathbf{X}(\mathbb{Z}) \cap X_{\mathbb{Q}}(\mathbb{R},T)]$$

for $T > 0$. One can recover the following result by explicit computation of local factors.

Corollary 4.2. (*Eskin-Mozes-Shah*) *If $p(\lambda)$ splits completely over $\mathbb{R}$ and $\mathbb{Z}[\theta]$ is the ring of integers of $\mathbb{Q}(\theta)$ for a root θ of $p(\lambda)$, then*

$$N(\mathbf{X},T) \sim \frac{2^{n-1}hR\omega_n}{\sqrt{D}\prod_{i=2}^n \Lambda(\frac{i}{2})} T^{\frac{n(n-1)}{2}}$$

as $T \to \infty$, *where* h *is the class number of* $\mathbb{Z}[\theta]$, R *is the regulator of* $\mathbb{Q}(\theta)$, D *is the discriminant of* $p(\lambda)$, ω_n *is the volume of the unit ball in* $\mathbb{R}^{\frac{n(n-1)}{2}}$ *and* $\Lambda(s) = \pi^{-s}\Gamma(s)\zeta(2s)$.

The detail or more general results can be found in [37]. This is a joint work with Dasheng Wei.

References

[1] Y.Benoist and H.Oh, *Effective equidistribution of S-integral points on symmetric varieties*, preprint.

[2] M.Borovoi and C.Demarche, *Manin obstruction to strong approximation for homogeneous spaces*, to appear in Commentarii Mathematici Helvetici.

[3] M.Borovoi and Z.Rudnick, *Hardy-Littlewood varieties and semisimple groups*, Invent. Math. 119 (1995), 37–66.

[4] J.-L.Colliot-Thélène and F.Xu, *Brauer-Manin obstruction for integral points of homogeneous spaces and representations by integral quadratic forms*, Compositio Math. **145** (2009), 309–363.

[5] C. Demarche, *Le défaut d'approximation forte dans les groupes linéaires connexes*, Proc.London Math.Soc. **102** (2011), no. 3, 563–597.

[6] W.Duke, Z.Rudnick, and P.Sanark, *Density of integer points on affine homogeneous varieties*, Duke Math. J. **71** (1993), 143–179.

[7] A.Eskin and C.McMullen, *Mixing, counting, and equidistribution in Lie groups*, Duke Math. J. **71** (1993), 181–209.

[8] A.Eskin, S.Mozes, and N.Shah, *Unipotent flows and counting lattice points on homogeneous varieties*, Annals of Math. **143** (1996), 253–199.

[9] D. Harari, *Le défaut d'approximation forte pour les groupes algébriques commutatifs*, Algebra and Number Theory **2** (2008), no. 5, 595–611.

[10] R. Hartshorne, *Algebraic Geometry*, GTM, Vol. **52**, Springer-Verlag, 1977.

[11] Y. Kitaoka, *Arithmetic of Quadratic Forms*, Cambridge Tracts in Mathematics, Vol. **106**, Cambridge University Press, 1993.

[12] M.Kneser, *Darstellungsmasse indefiniter quadratischer Formen*, Math. Z. **77** (1961), 188–194.

[13] M.-A. Knus, A.Merkurjev, M.Rost, and J.-P. Tignol, *The Book of Involutions*, Colloquium Publication, Vol. **44**, American Mathematical Society, 1998.

[14] R.E.Kottwitz, *Stable trace formula: Cuspidal tempered terms*, Duke Math. J. **51** (1984), no. 3, 611–650.

[15] R. E. Kottwitz, *Tamagawa numbers*, Annals of Math. **127** (1988), 629–646.

[16] C.G. Latimer and C. C. MacDuffee, *A correspondence between classes of ideals and classes of matrices*, Annals of Math. **34** (1933), 313–316.

[17] J. S. Milne, *Etale Cohomology*, Princeton Press, 1980.

[18] J. S. Milne, *Arithmetic Duality Theorems*, Perspectives in Mathematics, Vol. **1**, Academic Press, 1986.

[19] J.Neukirch, *Algebraic Number Fields*, Grundlehren, Vol. **322**, Springer, 1999.

[20] J.Neukirch, A.Schmidt, and K.Wingberg, *Cohomology of Number Fields*, Grundlehren, Vol. **323**, Springer, 2000.

[21] O.T.O'Meara, *Introduction to Quadratic Forms*, Springer, 1973.

[22] T.Ono, *On the Tamagawa number of algebraic tori*, Ann. of Math. **78** (1963), 47–73.

[23] V. P. Platonov and A. S. Rapinchuk, *Algebraic Groups and Number Theory*, Academic Press, 1994.

[24] V.L.Popov and E.B.Vinberg, *II. Invariant Theory*, Algebraic Geometry IV: Linear Algebraic Groups, Invariant Theory; Edited by A.N.Parshin and I.R.Shafarevich, Springer-Verlag, Berlin Heidelberg, 1994.

[25] J.-J. Sansuc, *Groupe de Brauer et arithmétique des groupes algébriques linéaires sur un corps de nombres*, J.reine angew. Math. **327** (1981), 12–80.

[26] W.M. Schmidt, *The density of integer points on homogeneous varieties*, Acta Math. **154** (1985), no. 3–4, 243–296.

[27] R. Schulze-Pillot, *Darstellungsmasse von Spinorgeschlechtern ternärer quadratischer Formen*, J. reine angew. Math. **352** (1984), 114–132.

[28] J. P. Serre, *Cohomologie Galoisienne*, Lecture Notes in Mathematics, Vol. **5**, Springer, Berlin, 1965.

[29] J.M. Shyr, *A generalization of Dirichlet's unit theorem*, J. Number Theory **9** (1977), no. 2, 213–217.

[30] A.N. Skorobogatov, *Torsors and Rational Points*, Cambridge Tracts in Mathematics, Vol. **144**, Cambridge University Press, 2001.

[31] V.E.Voskresenskii, *Algebraic Groups and Their Birational Invariants*, Translations of Mathematical Monographs, Vol. **179**, AMS, 1998.

[32] A. Weil, *Sur la theórie des formes quadratiques*, Colloque sur la théorie des groupes algébriques, C.B.R.M. Bruxelles (1962), 9–22.

[33] A. Weil, *Basic Number Theory*, Springer, Berlin-Heidelberg-New York, 1974.

[34] A. Weil, *Adeles and Algebraic Groups*, Birkhäuser, Boston, 1982.

[35] D. Wei and F. Xu, *Integral points for multi-norm tori*, Proc. London Math. Soc. **104** (2012), no. 5, 1019–1044.

[36] D. Wei and F. Xu, *Integral points for groups of multiplicative type*, Adv. Math. **232** (2013), no. 1, 36–56.

[37] D. Wei and F. Xu, *Counting integral points in certain homogeneous spaces*, arXiv:1211.2286[math.NT] (2012).

[38] F. Xu, *Representation masses of spinor genera*, Duke Math. J. **110** (2001), no. 2, 279–307.

School of Mathematical Sciences, Capital Normal University
Beijing 100048, China
E-mail address: xufei@math.ac.cn

Proceedings of the 6th International Congress of Chinese Mathematicians
ALM 36, pp. 361–379

On Existence and Density of the Ordinary Locus of Certain Shimura Varieties

Chia-Fu Yu

Abstract

The main purpose of this note is to discuss the problems of the existence and density of the ordinary locus of the reduction modulo p of certain Shimura varieties.

Mathematics Subject Classification: 14G35,11G18.

Keywords and Phrases: Ordinary locus, PEL-type Shimura varieties, Hilbert-Siegel moduli spaces, KR stratification.

The present are the notes for the author's talk given in the 6th International Congress for Chinese Mathematicians held in July 14–19, 2013. The main purpose is to discuss the problems of the existence and density of the ordinary locus of the reduction modulo p of certain Shimura varieties. This is a survey article where methods and ideas of proofs are indicated. Details of new results present here would be published elsewhere.

1 Introduction (general questions)

Starting with a PEL-datum $\mathcal{D} = (B, *, V, \psi)$ with possible additional data (see Sections 2.1 and 2.2 for details), one associates a moduli space $\mathbf{M}_{\mathcal{D}}$ over $O_{E,\mathfrak{p}}$ (or over $O_{E,\mathfrak{p}}[\zeta_N]$ for a choice of a primitive Nth root of unity ζ_N in $\overline{\mathbb{Q}} \subset \mathbb{C}$) of PEL-type. Here p is a prime number, E a number field in $\mathbb{C}$ (the reflex field), O_E the ring of integers in E, and $O_{E,\mathfrak{p}}$ the localization at a prime $\mathfrak{p}|p$ (corresponding to a fixed embedding $\overline{\mathbb{Q}} \hookrightarrow \overline{\mathbb{Q}}_p$). The moduli scheme $\mathbf{M}_{\mathcal{D}}$ parametrizes equivalence classes of polarized abelian varieties $(A, \lambda, \iota, \dots)$ together with an endomorphism structure $\iota : O_B \to \mathrm{End}(A)$ and with a prime-to-p level structure that satisfy certain conditions, where O_B is an order of B, which is part of additional data added to $\mathcal{D}$. Then one can consider the reduction modulo p of the moduli scheme: $\mathcal{M}_{\mathfrak{D}} := \mathbf{M}_{\mathfrak{D}} \otimes \overline{\mathbb{F}}_p$, which again parameterizes equivalence classes of the same geometric objects but in characteristic $p > 0$. Let $\mathcal{M}_{\mathfrak{D}}^{\mathrm{ord}}$ denote the ordinary locus of the moduli space $\mathcal{M}_{\mathfrak{D}}$, which parametrizes objects $(A, \lambda, \iota, \dots)$ in $\mathcal{M}_{\mathfrak{D}}$ whose underlying abelian variety A is ordinary. Recall that an abelian variety A over a field k of characteristic $p > 0$ is said to be *ordinary* if $A[p](\bar{k}) \simeq (\mathbb{Z}/p\mathbb{Z})^{\dim A}$, where $\bar{k}$ is an algebraic closure of k.

Basic questions:

(1) Is the ordinary locus $\mathcal{M}_{\mathfrak{D}}^{\mathrm{ord}}$ of $\mathcal{M}_{\mathfrak{D}}$ non-empty?

(2) Suppose that the ordinary locus $\mathcal{M}_{\mathfrak{D}}^{\mathrm{ord}}$ is non-empty. Is $\mathcal{M}_{\mathfrak{D}}^{\mathrm{ord}}$ Zariski dense in the moduli space $\mathcal{M}_{\mathfrak{D}}$?

(3) In any case, what are the maximal Newton strata in $\mathcal{M}_{\mathfrak{D}}$? Another way to rephrase this question is to find the notion of generalized ordinarity, which plays the similar role as the naive ordinarity of Siegel moduli spaces does. For example, the generalized ordinary locus should be dense in the moduli space $\mathcal{M}_{\mathfrak{D}}$.

In the case of good reduction (thus the moduli space $\mathcal{M}_{\mathfrak{D}}$ is smooth), the problems (1), (2) and (3) are solved by Wedhorn [34]. Wedhorn defined the notion of μ-ordinarity by group theory and showed that the μ-ordinary locus $\mathcal{M}_{\mathfrak{D}}^{\mu\text{-ord}}$ is Zariski dense in $\mathcal{M}_{\mathfrak{D}}$. He also showed that in the cases of type A or C (see Section 2.4) Question (1) has affirmative answer if and only if so does Question (2). Moreover, this is also equivalent to the following condition: (see [34, Theorem 1.6.3])

(*) The p-adic completion $E_{\mathfrak{p}}$ of E at $\mathfrak{p}$ is equal to $\mathbb{Q}_p$.

In the case of type C, the reflex field E is just $\mathbb{Q}$, and hence the ordinary locus is always non-empty and Zariski dense. Then one uses the criterion (*) only in the type A case.

The equivalence of affirmative answers to Questions (1) and (2) (in the smooth case) may be viewed as a consequence of a more involved result on the existence of smooth compactifications; see Rapoport [31], Faltings-Chai [9], Larsen [22] and Lan [21]. The latter furnishes a natural bijection between the set of irreducible components of the geometric special fiber $\mathcal{M}_{\mathfrak{D}}$ and that of the geometric generic fiber $\mathbf{M} \otimes \overline{\mathbb{Q}}_p$. This bijection also respects prime-to-p Hecke correspondences. Therefore, prime-to-p Hecke correspondences operate transitively on the set of components of $\mathcal{M}_{\mathfrak{D}}$ if and only if it is so for $\mathbf{M} \otimes \overline{\mathbb{Q}}_p$.

The only new part of our work is to investigate the cases where p is *ramified* in the PEL-datum $\mathfrak{D}$. We shall assume $p \neq 2$ whenever the datum $\mathfrak{D}$ contains a factor of type D (see Section 2.4). Moreover, we will restrict ourselves only to the naive notion of ordinarity.

In the case of bad reduction one may also consider the moduli space $\mathcal{M}_{\mathfrak{D},\mathcal{P}}$ over $\overline{\mathbb{F}}_p$ with level structure at p furnished by a parahoric subgroup $\mathcal{P}$. On this moduli space one has the so called *Kottwitz-Rapoport* (KR) stratification (see Kottwitz-Rapoport [20], Ngô-Genestier [24] and Haines [15], also see Görtz-Yu [13, 14] and cf. [47] and Section 4 for a special case).

A more general question is:

(4) What are the Newton vectors of the generic points of each maximal KR stratum of $\mathcal{M}_{\mathcal{D},\mathcal{P}}$?

Under the assumption of existence of a good compactification $\overline{\mathcal{M}}_{\mathcal{D},\mathcal{P}} \subset \overline{\mathbf{M}}_{\mathcal{D},\mathcal{P}}$, one may expect that there is a natural bijection between the set of *connected* components of the geometric special fiber $\mathcal{M}_{\mathcal{D},\mathcal{P}}$ and that of the geometric generic fiber $\mathbf{M}_{\mathcal{D},\mathcal{P}} \otimes \overline{\mathbb{Q}}_p$. Moreover, this bijection should be also compatible with prime-to-p Hecke correspondences. However, the existence of a naive ordinary point in one connected component of $\mathcal{M}_{\mathcal{D},\mathcal{P}}$ does not conclude the density of the ordinary locus in the connected component. Therefore, in the bad reduction case one does not expect for the equivalence of affirmative answers to Questions (1) and (2) in general.

For example, consider the Hilbert modular surface $\mathcal{M}_{F,I}$ with Iwahori level structure at p associated to a real quadratic field F where p is inert in F Stamm [33] showed that the ordinary locus $\mathcal{M}_{F,I}^{\mathrm{ord}}$ is non-empty and not Zariski dense. It is easier to understand the geometry of $\mathcal{M}_{F,I}$ through the KR stratification. The moduli space $\mathcal{M}_{F,I}$ is stratified by 9 KR strata: There is one unique minimal KR stratum of dimension zero which consists of chains of superspecial points, and four maximal KR strata of dimension two. Two of maximal KR strata contain only chains of ordinary points, while another two strata contain only chains of supersingular points.

Question (4) is closely related to a general question concerning the non-emptiness of the affine Deligne-Lusztig varieties $X_w(b)$ (which is a set in the mixed characteristic case) which may be interpreted as "the intersection of the Newton stratum associated to b and the KR stratum associated to w". At least in the function field case, this is known in a rather great generality due to Görtz-Haines-Kottwitz-Reumann [12] (in the split case) and Görtz-He (in the tamely ramified case).

For Shimura varieties, Question (4) is known at least in the following cases:

- Siegel moduli spaces with Iwahori level structure at p, due to Ngô-Genestier [24]. They showed that every maximal KR stratum contains only chains of ordinary abelian varieties.

- Unitary Shimura varieties of signature (n, m) associated to an imaginary quadratic field K, together with Iwahori level structure at p due to Haines [15], where p splits in K. Haines showed a similar result that every maximal KR stratum contains only chains of ordinary abelian varieties.

- Hilbert-Blumenthal moduli spaces with Iwahori level structure at p due to the author [44], where p is assumed to be unramified in the totally real field F. In this case, the ordinary locus is not necessarily Zariski dense. However, this is the case exactly when p splits completely in the totally real number field F.

As an application, the answer to Question (4) allows one to predict the number of irreducible components of the moduli space $\mathcal{M}_{\mathfrak{D},\mathcal{P}}$. We illustrate the ideas and methods around. It suffices to count the number of components of maximal KR strata as the union of them is a open and Zariski dense subvariety. First one divides maximal KR strata into two classes: basic KR strata and non-basic KR strata. The basic locus is the minimal Newton polygon stratum, which is a generalization of the supersingular locus in Siegel moduli spaces. Basic KR strata are those that are entirely contained in the basic locus. For the case of non-basic KR strata, a method due to Chai [3] is to show that the associated ℓ-adic monodromy is as big as possible. Then one may investigate whether or not in each connected component of $\mathcal{M}_{\mathfrak{D},\mathcal{P}}$ the set of connected (=irreducible) components of each KR stratum is transitive under prime-to-p Hecke correspondences. If this is the case, then a result of Chai [3] also allows one to determine the number of components of this maximal KR stratum if it is not contained in the basic locus. For basic maximal KR strata, the set of the components is usually parametrized by a double coset space of an algebraic group G' (or possibly a union of them) arising from basic points. However, it is very difficult to determine explicitly what these double coset spaces are. Once this difficult problem is solved, the standard method for computing these class numbers is to establish related geometric mass formulas. The reader is referred to [8, 2, 39, 41, 43, 44, 45] for some special cases.

2 On non-emptiness of ordinary loci

We shall first recall PEL-data, p-integral PEL-data and then define the associated moduli scheme $\mathbf{M}_{\mathfrak{D}}$ and its reduction $\mathcal{M}_{\mathfrak{D}} := \mathbf{M}_{\mathfrak{D}} \otimes \overline{\mathbb{F}}_p$ modulo p. Then we investigate Question (1) for the moduli space $\mathcal{M}_{\mathfrak{D}}$. We shall treat the case with minimal level structure at p. The defining algebraic group G (see Section 2.1) here needs not to be unramified at p.

2.1 PEL-data

Definition 2.1. A *PEL datum* $\mathfrak{D}$ is a 4-tuple $(B, *, V, \psi)$, where

- B is a finite-dimensional semi-simple algebra over $\mathbb{Q}$;
- $*$ is a positive involution on B, that is, one has $\mathrm{tr}_{B/\mathbb{Q}}(bb^*) > 0$ for all $b \neq 0 \in B$;
- V is a finite faithful B-module;
- $\psi : V \times V \to \mathbb{Q}$ is a non-degenerate alternating $\mathbb{Q}$-valued pairing such that $\psi(bx, y) = \psi(x, b^*y)$ for all $x, y \in V$ and $b \in B$.

To each PEL datum $\mathcal{D}$ one associates reductive algebraic $\mathbb{Q}$-groups G and G_1 as follows. The groups of $\mathbb{Q}$-points of these groups are defined respectively by

$$G(\mathbb{Q}) = \{x \in \mathrm{End}_B(V)^\times \mid x'x \in \mathbb{Q}^\times \},$$

$$G_1(\mathbb{Q}) = \{x \in \mathrm{End}_B(V)^\times \mid x'x = 1 \},$$

where x' is the adjoint of x with respect to the pairing ψ.

Let $h : \mathbb{C} \to \mathrm{End}_B(V_\mathbb{R})$ be an $\mathbb{R}$-linear algebra homomorphism such that $h(z)' = h(\bar{z})$ and the symmetric bilinear form $(x, y) := \psi(x, h(i)y)$ is positive definite (or negative definite). Such a homomorphism exists and it is unique up to conjugate by elements of $G_1(\mathbb{R})$; see Kottwitz [19, Section 4]. One usually adds h into $\mathcal{D}$ as part of the datum $\mathcal{D}$ though it is determined by $(B, *, V, \psi)$.

Let $X \subset \mathrm{Hom}_{\mathbb{R}-\mathrm{gps}}(\mathbb{C}^\times, G_\mathbb{R})$ be the $G(\mathbb{R})$-conjugacy class of h. It is shown that the $G(\mathbb{R})$-homogeneous space X admits a natural structure of a finite union of Hermitian symmetric domains.

The pair (G, X) satisfies the axioms given by Deligne [7] for defining Shimura varieties. In particular the map $\mathrm{Int}\, h(i)$ induces a Cartan involution on $G^{\mathrm{ad}}_\mathbb{R}$. For each open compact subgroup $U \subset G(\mathbb{A}_f)$, one puts

$$\mathrm{Sh}_U(G, X)_\mathbb{C} := G(\mathbb{Q})\backslash X \times G(\mathbb{A}_f)/U.$$

By the work of Baily and Borel, the analytic variety $\mathrm{Sh}_U(G, X)_\mathbb{C}$ admits a canonical structure of algebraic variety over $\mathbb{C}$, denoted also by $\mathrm{Sh}_U(G, X)_\mathbb{C}$, which is normal and quasi-projective. Let

$$\mathrm{Sh}(G, X)_\mathbb{C} = \{\mathrm{Sh}_U(G, X)_\mathbb{C}\}_{U \subset G(\mathbb{A}_f)}$$

be the tower of the family indexed by open compact subgroups U; this is called the (complex) Shimura variety associated to the Shimura datum (G, X).

Each $\mathrm{Sh}_U(G, X)_\mathbb{C}$ has a modular interpretation as a moduli space of abelian varieties with a polarization, a B-action and a level structure. Moreover using the modular interpretation, each $\mathrm{Sh}_U(G, X)_\mathbb{C}$ is the base change to $\mathbb{C}$ of a quasi-projective scheme $\mathrm{Sh}_U(G, X)$ defined over a number field $E(G, X) \subset \mathbb{C}$, called the *reflex field*. The model $\mathrm{Sh}_U(G, X)$ is characterized by certain properties and is unique up to a canonical isomorphism. Moreover, the action of $G(\mathbb{A}_f)$ on the tower $\mathrm{Sh}(G, X)$, known as the Hecke symmetries, is defined over $E(G, X)$. Note that the reflex field $E(G, X)$ only depends on (G, X) but not on the level subgroup U. However, each connected component of $\mathrm{Sh}_U(G, X)_\mathbb{C}$ may be defined over an abelian extension of $E(G, X)$.

We recall the definition of the reflex field $E(G, X)$ for the reader's convenience. The element $h(i)$ gives rise to a complex structure on $V_\mathbb{R}$ which commutes with the B-action. Thus one has the decomposition

$$V_\mathbb{C} = V^{-1,0} \oplus V^{0,-1} \tag{2.1}$$

of eigenspaces with eigenvalues i and $-i$, respectively. The reflex field $E(G, X)$ then is defined to be the field of definition of the complex algebra representation $V^{-1,0}$

of B. Alternatively, if let $\mu_h(z) := h_{\mathbb{C}}(z,1) : \mathbb{G}_{m\mathbb{C}} \to G_{\mathbb{C}}$ be the Hodge cocharacter and let $Y \subset \mathrm{Hom}_{\mathbb{C}}(\mathbb{G}_{m\mathbb{C}}, G_{\mathbb{C}})$ be the $G^0(\mathbb{C})$-conjugacy class of μ_h, then the reflex field $E(G,X)$ also equals the field of definition of the smooth complex subvariety Y.

2.2 p-integral PEL-datum

Let p be a prime number. A *p-integral PEL-datum* is a tuple

$$\mathcal{D} := (B, *, V, \psi, O_B, L_0, h),$$

where

- $(B, *, V, \psi, h)$ is a (rational) PEL-datum;
- $O_B \subset B$ is an order which is stable under $*$ and is maximal at p, that is, $O_B \otimes \mathbb{Z}_p$ is a maximal order in $B \otimes \mathbb{Q}_p$;
- $L_0 \subset V$ is an O_B-lattice such that $\psi(L_0, L_0) \subset \mathbb{Z}$ and the $O_B \otimes \mathbb{Z}_p$-lattice $L_0 \otimes_{\mathbb{Z}} \mathbb{Z}_p$ is self-dual with respect to the pairing ψ.

We remark that it is not always possible, for given any semi-simple $\mathbb{Q}$-algebra $(B, *)$ with positive involution, to admit an order O_B which is both stable under $*$ and maximal at p. We need to assume that such an order exists. Similarly the existence of the O_B-lattice L_0 as above also requires some conditions. However, one can modify ψ without changing other data so that such a lattice L_0 exists. We will assume that $p \neq 2$ whenever the group G is not connected.

With a p-integral PEL-datum $\mathcal{D}$ and assumptions as above, we shall define a moduli scheme $\mathbf{M}_{\mathcal{D}}$. Fix an embedding $\overline{\mathbb{Q}} \hookrightarrow \overline{\mathbb{Q}}_p$, where $\overline{\mathbb{Q}}_p$ is an algebraic closure of $\mathbb{Q}_p$. This fixes a prime $\mathfrak{p}$ over p of the ring of integers O_E in the reflex field E. Let $O_{E,\mathfrak{p}}$ denote the localization at $\mathfrak{p}$. The lattice L_0 defines an open compact subgroup group $G(\hat{\mathbb{Z}})$. Let $U = U_p U^p \subset G(\hat{\mathbb{Z}})$ be an open compact subgroup. We assume that $U_p = G(\mathbb{Z}_p)$ and $U^p \subset G(\hat{\mathbb{Z}}^{(p)})$ is sufficiently small. Finally put $g := 1/2 \dim_{\mathbb{Q}} V$. Let $\mathbf{M}_{\mathcal{D}}$ be the moduli scheme over $O_{E,\mathfrak{p}}$ of equivalence classes of the objects $(A, \lambda, \iota, \bar{\eta})_S$, where

- S is a locally Noetherian connected $O_{E,\mathfrak{p}}$-scheme;
- A is an abelian scheme of relative dimension g;
- $\lambda : A \to A^t$ is a prime-to-p degree polarization;
- $\iota : O_B \to \mathrm{End}_S(A)$ is a ring monomorphism such that $\lambda \circ \iota(b^*) = \iota(b)^t \circ \lambda$ for all $b \in O_B$;
- $\bar{\eta}$ is a $\pi_1(S, \bar{s})$-invariant U^p-orbit of $O_B \otimes \hat{\mathbb{Z}}^{(p)}$-linear isomorphisms

$$\eta : L_0 \otimes \hat{\mathbb{Z}}^{(p)} \simeq T^{(p)}(A_{\bar{s}}) = \prod_{\ell \neq p} T_\ell(A_{\bar{s}})$$

  which preserve the pairings up to a scalar.

that satisfy the Kottwitz determinant condition (see Kottwitz [19, Section 5], Rapoport-Zink [32], or cf. (3.2) for a special case).

Let $\mathcal{M}_{\mathcal{D}} := \mathbf{M}_{\mathcal{D}} \otimes \overline{\mathbb{F}}_p$ denote the reduction modulo p.

2.3 Morita reduction step

Using the Wedderburn theorem, one can decompose a semi-simple algebra B into a product of simple algebras. If B is equipped with a positive involution $*$, then each simple factor in the decomposition is stable under $*$. As a result, one has a decomposition of semi-simple algebras with positive involutions:

$$(B, *) \simeq \prod_{i=1}^{r} \mathrm{Mat}_{n_i}(B_i, *_i),$$

where $(B_i, *_i)$ is a finite-dimensional simple $\mathbb{Q}$-algebra with a positive involution. We also have the decomposition

$$(V, \psi) \simeq \bigoplus_{i=1}^{r} (V_i, \psi_i)^{\oplus n_i},$$

where (V_i, ψ_i) is a skew-Hermitian B_i-module. For each $i = 1, \ldots, r$, we choose an order O_{B_i} of B_i and an O_{B_i}-lattice L_i to make a p-integral PEL-datum $\mathcal{D}_i = (B_i, *_i, V_i, \psi_i, O_{B_i}, L_i, h_i)$. Then one can show that up to a finite etale morphism the moduli space $\mathcal{M}_{\mathcal{D}}$ is equivalent to the product of the moduli spaces $\mathcal{M}_{\mathcal{D}_i}$:

$$\mathcal{M}_{\mathcal{D}} \sim \prod_{i=1}^{r} \mathcal{M}_{\mathcal{D}_i}.$$

It follows that Question (1) or (2) for $\mathcal{M}_{\mathcal{D}}$ has an affirmative answer if and only if it has an affirmative answer for $\mathcal{M}_{\mathcal{D}_i}$ for all $i = 1, \ldots, r$. Moreover, one can reduce the description for the answer to Question (3) to that for each $\mathcal{M}_{\mathcal{D}_i}$. Without loss of generality we may and will assume that B is a *division* $\mathbb{Q}$-algebra with a positive involution.

2.4 Albert's classification

We recall the classification of finite-dimensional division $\mathbb{Q}$-algebras with positive involutions due to Albert. A convenience reference for this is Mumford [23, Section 21]. Let $(B, *)$ be a finite-dimensional division $\mathbb{Q}$-algebra with a positive involution. Then we have the following four classes.

(I) $B = F$ is a totally real number field and the involution $*$ is trivial.

(II) B is a totally indefinite quaternion algebra over a totally real number field F and $x^* = \alpha \bar{x} \alpha^{-1}$ for some traceless element $\alpha \in B^\times$ such that $a = \alpha^2 \in F$ is a totally negative element, where $x \mapsto \bar{x}$ is the canonical involution.

(III) B is a totally definite quaternion algebra over a totally real number field F and $*$ is the canonical involution, which is the unique positive involution on B.

(IV) B is a central division algebra over a CM field K and $*$ is of the second kind, i.e. the restriction to K is nontrivial. There is a condition for invariants of B:

$$\begin{array}{ll} \mathrm{inv}_w(B) + \mathrm{inv}_{\bar{w}}(B) = 0, & \text{if } v = w\bar{w} \text{ splits in } K, \\ \mathrm{inv}_w(B) = 0, & \text{if } w = \bar{w}. \end{array} \tag{2.2}$$

Moreover, the necessary condition (2.2) is also sufficient in order for a central division K-algebra B to equip with a positive involution.

For the first three cases the involution $*$ is of the first kind; while it is of the second kind for the case (IV). In terms of classification by group theory, for the case of (I) or (II), each simple factor of the derived group G_{der} is of type C in the Dynkin diagram. For the case of type (III), it is of type D and while the last case is of type A.

2.5 Non-emptiness of ordinary loci

Let $\mathcal{D} = (B, *, V, \psi, O_B, L_0, h)$ be a simple p-integral PEL-datum, i.e. B is a division $\mathbb{Q}$-algebra. Let $\mathcal{M}_{\mathcal{D}}$ be the associated moduli space over $\overline{\mathbb{F}}_p$ of PEL-type. Put

$$F := Z(B), \quad \text{if } (B,*) \text{ is of type (I), (II) or (III)},$$
$$K := Z(B) \text{ and } F := K^{*=id}, \quad \text{if } (B,*) \text{ is of type (IV)},$$
$$d := [F:\mathbb{Q}], \quad q := \sqrt{[B:Z(B)]},$$
$$m := \begin{cases} 1/2 \dim_F V & \text{if } B = F \text{ (of type (I))}, \\ \dim_B V & \text{otherwise}, \end{cases}$$
$$g := \frac{1}{2}\dim_{\mathbb{Q}} V = \begin{cases} md & \text{in type (I) case}, \\ 2md & \text{in type (II) or (III) cases}, \\ mdq^2 & \text{in type (IV) case}. \end{cases}$$

Proposition 2.2. *Notations as above. Assume that $F = Z(B)$. Then $\mathcal{M}_{\mathcal{D}}^{\mathrm{ord}}$ is non-empty if and only if*

(1) $(B,*)$ *is of type* (I), *or*

(2) $(B,*)$ *is of type* (II) *or* (III), *and one of the following conditions holds:*

(i) *m is even;*

(ii) $\mathrm{inv}_v(B) = 0$ *for all places v of F over p.*

2.6 Sketch of the proof of Proposition 2.2

The most essential part of the proof is to construct a pair (A, ι) over $\overline{\mathbb{F}}_p$, where A is a g-dimensional *ordinary* abelian variety over $\overline{\mathbb{F}}_p$ and $\iota : B \to \mathrm{End}^0(A) = \mathrm{End}(A) \otimes \mathbb{Q}$ is an embedding of $\mathbb{Q}$-algebras.

Case of type (I). Take an ordinary elliptic curve E over $\overline{\mathbb{F}}_p$, and put $A := (O_F \otimes_{\mathbb{Z}} E)^{\oplus m}$ (noting that $g = md$). Here $O_F \otimes_{\mathbb{Z}} E$ stands for Serre's tensor construction for abelian varieties. Then one gets an embedding $F \hookrightarrow \mathrm{End}^0(A)$ of $\mathbb{Q}$-algebras.

Case of type (II) or (III). One has $g = 2md$.

(i) Suppose that m is even. Take $A := (O_F \otimes_{\mathbb{Z}} E)^{\oplus 2m}$ where E is as above. Then there is an embedding $\mathrm{Mat}_{2m}(F) \hookrightarrow \mathrm{End}^0_F(A)$ of F-algebras. As m is even, one has the following embeddings of F-algebras

$$B \hookrightarrow \mathrm{Mat}_4(F) \hookrightarrow \mathrm{Mat}_{2m}(F) \hookrightarrow \mathrm{End}^0_F(A).$$

Therefore, we construct an embedding $\iota : B \to \mathrm{End}^0(A)$ of $\mathbb{Q}$-algebras.

(ii) Suppose that $\mathrm{inv}_v(B) = 0$ for all places v of F over p. We shall first construct a pair (A_1, ι_1) over $\overline{\mathbb{F}}_p$ where A_1 is an ordinary abelian variety of dimension $2d$ and $\iota_1 : B \hookrightarrow \mathrm{End}^0(A_1)$ is an embedding of $\mathbb{Q}$-algebras.

Take an ordinary elliptic curve E over $\overline{\mathbb{F}}_p$ so that the imaginary quadratic field $K_1 := \mathrm{End}^0(A)$ has the following property:

(∗) for any place u of F with $\mathrm{inv}_u(B) \neq 0$, the completion $K_1 \otimes_{\mathbb{Q}} \mathbb{Q}_q$ is a field extension over $\mathbb{Q}_q$, where q is the prime below the place u.

Put $L := FK_1$, which is a quadratic CM extension of F. The property (∗) ensures that L_u can be embedded into B_u for all places u of F. By the local-global principle, the CM field L can be embedded as a maximal subfield of B. As a result, $B \otimes_F L \simeq \mathrm{Mat}_2(L)$. Put $A_1 := (O_F \otimes_{\mathbb{Z}} E)^{\oplus 2}$, then $\mathrm{End}^0_F(A_1) \simeq \mathrm{Mat}_2(L)$. Thus, one has an embedding of F-algebras $\iota_1 : B \hookrightarrow B \otimes_F L \simeq \mathrm{Mat}_2(L) \simeq \mathrm{End}^0_F(A_1)$. As a result, we have constructed an ordinary abelian variety A_1 and an embedding $\iota_1 : B \hookrightarrow \mathrm{End}^0(A_1)$ of $\mathbb{Q}$-algebras.

Take $A := A_1^{\oplus m}$ and $\iota : B \hookrightarrow \mathrm{End}^0(A_1) \subset \mathrm{End}^0(A) = \mathrm{Mat}_m(\mathrm{End}^0(A_1))$. This finishes the construction of the pair (A, ι) over $\overline{\mathbb{F}}_p$.

We describe the idea of constructing an ordinary point in the moduli space $\mathcal{M}_{\mathcal{D}}$. Let (A, ι) be an ordinary abelian B-variety as above. By a lemma of Kottwitz [19, Lemma 9.2] there is a B-linear polarization λ on (A, ι). Replacing A by one in the isogeny class of A, we may assume that λ has prime-to-p degree and that ι maps O_B into $\mathrm{End}(A)$ (i.e. $\iota : O_B \hookrightarrow \mathrm{End}(A)$).

In the case of type (I) or (II), one can show that for any prime $\ell \neq p$ there is a $B \otimes \mathbb{Q}_\ell$-linear isomorphism

$$\eta_\ell : L_0 \otimes \mathbb{Q}_\ell \simeq T_\ell(A) \otimes \mathbb{Q}_\ell \tag{2.3}$$

which preserves the pairings up to a scalar. So we can furnish a prime-to-p level structure $\bar{\eta}$ and get an ordinary point $(A, \lambda, \iota, \bar{\eta})$ in $\mathcal{M}_{\mathfrak{D}}$.

In the case of type (III) the isomorphism η_ℓ in (2.3) may not exist. However, we can adjust the prime-to-p part of the polarization λ so that the ordinary points (A, λ, ι) lies in $\mathcal{M}_{\mathfrak{D}}$ "up to finite ambiguity" in the following sense (this is a weaker result than Proposition 2.2).

There are p-integral PEL-data $\mathfrak{D}_1, \dots, \mathfrak{D}_r$ with $\mathfrak{D}_i = (B, *, V, \psi_i, O_B, L_i, h_i)$ for suitable choices of ψ_i and L_i (noting that $\mathfrak{D}_i$ share the common data B, $*$, V and O_B) where the integer r only depends on $B, *$ and g such that one can adjust the prime-to-p part of λ so that the point $(A, \lambda, \iota, \eta) \in \mathcal{M}_{\mathfrak{D}_i}$ for some $i = 1, \dots, r$. In other words, one shows a weaker result

$$\bigcup_{i=1}^{r} \mathcal{M}_{\mathfrak{D}_i}^{\mathrm{ord}} \neq \emptyset. \tag{2.4}$$

To improve this result, we need to work more on Galois cohomology as in Kottwitz [19].

2.7 Non-emptiness of ordinary loci: case of type (IV) (type A)

Recall $B \supset K = Z(B) \supset F$ and $[B : K] = q^2$. Write $F \otimes \mathbb{Q}_p = \prod_{v|p} F_v$ and $K \otimes \mathbb{Q}_p = \prod_{w|p} K_w$, where F_v (resp. K_w) denotes the completion of F (resp. K) at the place v (resp. w).

Put

$$\Sigma_F := \mathrm{Hom}_{\mathbb{Q}}(F, \mathbb{C}) = \mathrm{Hom}_{\mathbb{Q}}(F, \overline{\mathbb{Q}}) = \mathrm{Hom}_{\mathbb{Q}_p}(F \otimes \mathbb{Q}_p, \overline{\mathbb{Q}}_p) = \coprod_{v|p} \Sigma_{F_v},$$

$$\Sigma_K := \mathrm{Hom}_{\mathbb{Q}}(K, \mathbb{C}) = \mathrm{Hom}_{\mathbb{Q}}(K, \overline{\mathbb{Q}}) = \mathrm{Hom}_{\mathbb{Q}_p}(K \otimes \mathbb{Q}_p, \overline{\mathbb{Q}}_p) = \coprod_{w|p} \Sigma_{K_w},$$

where

$$\Sigma_{F_v} := \mathrm{Hom}_{\mathbb{Q}_p}(F_v, \overline{\mathbb{Q}}_p) \quad \text{and} \quad \Sigma_{K_w} := \mathrm{Hom}_{\mathbb{Q}_p}(K_w, \overline{\mathbb{Q}}_p).$$

Remind that we have fixed an embedding $\overline{\mathbb{Q}} \hookrightarrow \overline{\mathbb{Q}}_p$. Let $\mathrm{res} : \Sigma_K \to \Sigma_F$ be the restriction map to F.

Definition 2.3. For elements $\tau \in \Sigma_K$ and $\sigma \in \Sigma_F$, and places v and w of F and of K over p respectively, we say τ (resp. σ) is *associated to* w (resp. to v) if $\tau \in \Sigma_{K_w}$ (resp. $\sigma \in \Sigma_{F_v}$).

The PEL-datum $\mathfrak{D}$ gives rise to a tuple $(n_\tau)_{\tau \in \Sigma_K}$ of non-negative integers n_τ indexed by Σ_K as follows. One can also write this as $((n_\tau, n_{\bar{\tau}}))_{\sigma \in \Sigma_F}$, where $\tau, \bar{\tau}$ are the elements in Σ_K over σ, i.e. $\mathrm{res}^{-1}(\sigma) = \{\tau, \bar{\tau}\}$. The map $h : \mathbb{C} \to \mathrm{End}_B(V_{\mathbb{R}})$ induces a complex structure on $V_{\mathbb{R}}$, which yields the decomposition $V_{\mathbb{C}} = V^{-1,0} \oplus V^{0,-1}$ where $h(z)$ acts by z on $V^{-1,0}$. The subspace $V^{-1,0}$ is a

$B \otimes \mathbb{C}$-submodule. As $\dim_{\mathbb{Q}} V = 2dq^2m$, the subspace $V^{-1,0}$ has dimension dq^2m over $\mathbb{C}$. Restricting to the K-linear action, one has an isomorphism of $K \otimes \mathbb{C}$-modules

$$V^{-1,0} \simeq \bigoplus_{\sigma\in\Sigma_F} (qn_\tau \cdot \mathbb{C}_\tau \oplus qn_{\bar{\tau}} \cdot \mathbb{C}_{\bar{\tau}}) \tag{2.5}$$

for non-negative integers n_τ and $n_{\bar{\tau}}$ with $n_\tau + n_{\bar{\tau}} = qm$ for all $\tau \in \Sigma_K$. Here $\mathbb{C}_\tau$ is the one-dimensional complex representation of K on which K acts through the embedding map $\tau : K \hookrightarrow \mathbb{C}$. The invariant $((n_\tau, n_{\bar{\tau}}))_{\sigma\in\Sigma_F}$ is called the *signature* of the PEL-datum $\mathcal{D}$.

Proposition 2.4. *Let $\mathcal{D}$ be a simple p-integral PEL-datum of type* (IV) *with signature $((n_\tau, n_{\bar{\tau}}))_{\sigma\in\Sigma_F}$, and let $\mathcal{M}_{\mathcal{D}}$ be the associated moduli space of PEL-type over $\overline{\mathbb{F}}_p$. Then the ordinary locus $\mathcal{M}_{\mathcal{D}}^{\mathrm{ord}}$ is non-empty if and only if the following conditions hold.*

(1) *For any place v of F over p such that $K_v := K \otimes_F F_v$ is a field (i.e. v is inert or ramified in K), one has $n_\tau = n_{\bar{\tau}}$ for all elements $\tau \in \Sigma_K$ with* $\mathrm{res}(\tau)$ *associated to v;*

(2) *For any place v of F over p such that $v = w\bar{w}$ splits in K, one has $\delta_w | n_\tau$ for all $\tau \in \Sigma_K$ with* $\mathrm{res}(\tau)$ *associated to v, where δ_w is the index of the central simple K_w-algebra $B_w = B \otimes_K K_w$, which is also the order of the class $[B_w]$ in the Brauer group* $\mathrm{Br}(K_w)$.

Note that $\delta_w = \delta_{\bar{w}}$. So we can also rephrase the condition in (2) by $\delta_w | n_\tau$ for all τ associated to w.

The idea of the proof of Proposition 2.4 is similar to the previous case. However, the construction of ordinary abelian varieties with the additional structures is more involved.

3 Density of the ordinary locus of Hilbert-Siegel moduli spaces

The second part of this article discusses Question (2) for Hilbert-Siegel moduli spaces.

3.1 Statement of the main result

Let F be a totally real number field of degree d and let O_F denote the ring of integers. Recall that a *polarized abelian O_F-variety* is a triple (A, λ, ι), where A is an abelian variety, $\iota : O_F \to \mathrm{End}(A)$ is a ring monomorphism and $\lambda : A \to A^t$ is a polarization such that $\lambda \circ \iota(a) = \iota(a)^t \circ \lambda$ for all $a \in O_F$.

Let m be a positive integer and put $g := dm$. Let (V, ψ) be vector space over F of dimension $2m$, together with a non-degenerate alternating pairing $\psi : V \times V \to \mathbb{Q}$ such that $\psi(ax, y) = \psi(x, ay)$ for all $a \in F$ and $x, y \in V$. This is unique up to isomorphism. Choose an O_F-lattice L_0 in V so that $\psi(L_0, L_0) \subset \mathbb{Z}$ and the

$O_F \otimes \mathbb{Z}_p$-lattice $L_0 \otimes_{\mathbb{Z}} \mathbb{Z}_p$ is self-dual with respect to the pairing ψ. Let $N \geq 3$ be a prime-to-p integer. Put

$$U := \ker(G(\hat{\mathbb{Z}}) \to G(\mathbb{Z}/N\mathbb{Z})),$$

where G is the reductive algebraic $\mathbb{Q}$-group defined in Section 2.1. Write $U = U_p U^p$ where $U_p = G(\mathbb{Z}_p)$ and $U^p \subset G(\mathbb{A}_f^{(p)})$ is an open compact subgroup.

The prime-to-p level structure on (A, λ, ι) is an U^p-orbit $\bar{\eta}$ of $O_F \otimes \hat{\mathbb{Z}}^{(p)}$-linear isomorphisms

$$\eta : L_0 \otimes_{\mathbb{Z}} \hat{\mathbb{Z}}^{(p)} \simeq T^{(p)}(A) := \prod_{\ell \neq p} T_\ell(A) \tag{3.1}$$

that preserve the pairings up to a scalar. The Kottwitz determinant condition for an object $(A, \lambda, \iota)_S$ over a $\mathbb{Z}_{(p)}$-scheme S is an equality of the following two characteristic functions:

$$\mathrm{char}(a|\,\mathrm{Lie}(A/S)) = \mathrm{char}(a|V^{-1,0}) \quad \text{in } O_S[\underline{X}] \tag{3.2}$$

for all $a \in O_F$, where $V^{-1,0} \subset V_{\mathbb{C}}$ is a subspace defined in Section 2.1.

Consider the integral model of Hilbert-Siegel moduli space. This is the moduli scheme $\mathbf{M}$ over $\operatorname{Spec} \mathbb{Z}_{(p)}$ of g-dimensional prime-to-p degree polarized abelian O_F-varieties $(A, \lambda, \iota, \bar{\eta})$ together with a prime-to-p level structure and satisfying the Kottwitz determinant condition as (3.1) and (3.2).

Let $\mathcal{M} := \mathbf{M} \otimes \overline{\mathbb{F}}_p$ be the reduction modulo p of the Hilbert-Siegel moduli scheme $\mathbf{M}$. Let $\mathcal{M}^{\mathrm{ord}} \subset \mathcal{M}$ be the ordinary locus of $\mathcal{M}$, which parameterizes the objects $(A, \lambda, \iota, \bar{\eta})$ in $\mathcal{M}$ whose underlying abelian variety A is *ordinary*.

Theorem 3.1. *The ordinary locus $\mathcal{M}^{\mathrm{ord}}$ is open and dense in $\mathcal{M}$.*

Remark 3.2. In the case where the prime p does not divide the discriminant $\Delta_{F/\mathbb{Q}}$ of F over $\mathbb{Q}$, that is, p is unramified in F, the moduli space $\mathbf{M}$ is smooth over $\operatorname{Spec} \mathbb{Z}_{(p)}$ and Theorem 3.1 is known due to T. Wedhorn. The only new part of this theorem deals with the case where p divides the discriminant $\Delta_{F/\mathbb{Q}}$. In this case the moduli space has singularities. In the special case where $m = 1$, that is, $\mathbf{M}$ is a Hilbert moduli scheme, Theorem 3.1 is proved in [37].

3.2 Ingredients of the proof of Theorem 3.1

The proof of Theorem 3.1 is divided into two parts.

1. We study the Kottwitz-Rapoport (KR) stratification on the moduli space $\mathcal{M}$. From this we show that the smooth locus $\mathcal{M}^{\mathrm{sm}} \subset \mathcal{M}$ is open and dense in $\mathcal{M}$.

2. Deform any point x in the smooth locus $\mathcal{M}^{\mathrm{sm}}$ to a point y in the ordinary locus $\mathcal{M}^{\mathrm{ord}}$. That is, the ordinary locus $\mathcal{M}^{\mathrm{ord}}$ is open and dense in the smooth locus $\mathcal{M}^{\mathrm{sm}}$.

We point out that the step (1) may be a nontrivial fact. This of course follows immediately if we knew that the special fiber $\mathcal{M}$ is reduced. However, the latter is not known yet under the author's knowledge. Nevertheless, the result (1) shows that an open dense subscheme of the special fiber is reduced. U. Görtz showed the following result, which is in the content of the Rapoport-Zink conjecture [32, p. 95] about the flatness of local models .

Theorem 3.3. (Görtz [11]) *The structure morphism* $\mathbf{M} \to \operatorname{Spec} \mathbb{Z}_{(p)}$ *is topologically flat. That is, every generic points in the special fiber can be lifted to characteristic zero.*

Theorem 3.3 provides an evidence that the integral model $\mathbf{M}$ could be *flat* over $\operatorname{Spec} \mathbb{Z}_{(p)}$. It follows from Theorem 3.3 that the integral model $\mathbf{M} \to \operatorname{Spec} \mathbb{Z}_{(p)}$ would be flat provided that the special fiber $\mathcal{M}$ is reduced. However, the flatness of an integral model usually does not say anything about whether the special fiber is reduced. There are already some examples given by curves: The special fiber of the minimal regular integral model for a projective smooth curve over a complete discrete valuation field K may have non-reduced irreducible components.

Let $\mathbf{M}^{\mathrm{can}}$ denote the scheme-theoretic closure of the general fiber $\mathbf{M}_{\mathbb{Q}}$ in $\mathbf{M}$, and let $\mathcal{M}^{\mathrm{can}} := \mathbf{M} \otimes_{\mathbb{Z}_{(p)}} \overline{\mathbb{F}}_p$ denote the reduction modulo p of $\mathbf{M}^{\mathrm{can}}$. The following is a special case of results of Pappas and Zhu [30].

Theorem 3.4. (Pappas-Zhu) *The reduced subscheme* $(\mathcal{M}^{\mathrm{can}})_{\mathrm{red}}$ *of* $\mathcal{M}^{\mathrm{can}}$ *is a union of normal subvarieties.*

In fact, they proved much more results concerning the closure of each KR stratum. The methods of He in [17] may lead to conclude that $\mathbf{M}^{\mathrm{can}}$ is Cohen-Macaulay and normal, which is still under investigation. Results proved in [17] are for the cases where $G_{\mathbb{Q}_p}$ is unramified but the level group $U_p \subset G(\mathbb{Q}_p)$ is a parahoric subgroup.

4 The Kottwitz-Rapoport stratification

4.1 Local models

Now we come to explain the first part of the proof of Theorem 3.1. We show this through studying the Kottwitz-Rapoport (KR) stratification. Let F, V, ψ, L_0 be as in Section 3. Put

$$\mathcal{O}_p := O_F \otimes \mathbb{Z}_p = \prod_{v|p} \mathcal{O}_v. \tag{4.1}$$

Let e_v and f_v be the ramification index and inert degree of v, respectively. Put $\Lambda := L_0 \otimes_{\mathbb{Z}} \mathbb{Z}_p$; this is a free $\mathcal{O}_p$-module of rank $2m$ together with a $\mathbb{Z}_p$-valued skew-Hermitian form $\psi : \Lambda \times \Lambda \to \mathbb{Z}_p$. Let $\mathcal{G} := \mathrm{GU}(\Lambda, \psi)$ be the automorphism group scheme of (Λ, ψ) over $\mathbb{Z}_p$; this is a group scheme over $\mathbb{Z}_p$ which represents the following functor: For any $\mathbb{Z}_p$-algebra R, $\mathcal{G}(R) := \mathrm{GU}(\Lambda \otimes_{\mathbb{Z}_p} R, \psi)$, which consists of $\mathcal{O}_p \otimes R$-automorphisms on $V \otimes R$ that preserve the pairing ψ up to a scalar in $R^\times$. Denote by $\mathbf{M}_\Lambda$ the associated local model. This is a projective scheme over $\operatorname{Spec}\mathbb{Z}_p$ of finite type which represents the following functor. For any $\mathbb{Z}_p$-scheme S, $\mathbf{M}_\Lambda(S)$ is the set of all locally free $\mathcal{O}_S$-submodules $\mathcal{F} \subset \Lambda \otimes \mathcal{O}_S$ of rank md such that

- locally in the Zariski topology on S, $\mathcal{F}$ is a direct summand of $\Lambda \otimes_{\mathbb{Z}_p} \mathcal{O}_S$;
- $\mathcal{F}$ is $\mathcal{O}_p$-invariant and is isotropic with respect to the pairing ψ;
- $\mathcal{F}$ satisfies the determinant condition; cf. (3.2).

Let $\mathcal{M}_\Lambda := \mathbf{M}_\Lambda \otimes \overline{\mathbb{F}}_p$ the reduction modulo p of the local model $\mathbf{M}_\Lambda$.

4.2 Local model diagrams

Let $\widetilde{\mathcal{M}}$ denote the moduli space over $\overline{\mathbb{F}}_p$ parameterizing equivalence classes of objects $(\underline{A}, \xi)$, where $\underline{A} = (A, \lambda, \iota, \bar{\eta})$ is an object in $\mathcal{M}$ and $\xi : H_1^{\mathrm{DR}}(A/S) \simeq \Lambda \otimes \mathcal{O}_S$ is an $\mathcal{O}_p \otimes_{\mathbb{Z}_p} \mathcal{O}_S$-linear isomorphism that preserves the pairings up to a scalar in $\mathcal{O}_S^\times$. We have the local model diagram (see Rapoport-Zink [32]):

$$\begin{array}{ccccc} & & \widetilde{\mathcal{M}} & & \\ & \swarrow^{\varphi^{\mathrm{mod}}} & & \searrow^{\varphi^{\mathrm{loc}}} & \\ \mathcal{M} & & & & \mathcal{M}_\Lambda \end{array} \tag{4.2}$$

In the above diagram φ^{loc} is the morphism that sends each object $(\underline{A}, \xi)$ to the image $\xi(\omega')$ of the Hodge submodule $\omega' \subset H_1^{\mathrm{DR}}(A)$, and φ^{mod} is the morphism that forgets the trivialization ξ.

The special fiber $\mathcal{G} \otimes \overline{\mathbb{F}}_p$ of $\mathcal{G}$ acts on $\widetilde{\mathcal{M}}$ and on $\mathcal{M}_\Lambda$ from the left. Furthermore,

- the morphism φ^{loc} is $\mathcal{G} \otimes \overline{\mathbb{F}}_p$-equivalent, surjective and smooth, and
- the morphism $\varphi^{\mathrm{mod}} : \widetilde{\mathcal{M}} \to \mathcal{M}$ is a $\mathcal{G} \otimes \overline{\mathbb{F}}_p$-torsor.

4.3 Kottwitz-Rapoport stratification

Let

$$\mathcal{M}_\Lambda = \coprod_{\underline{e} \in \mathrm{Adm}(\mu)} \mathcal{M}_{\Lambda, \underline{e}} \tag{4.3}$$

be the decomposition of $\mathcal{M}_\Lambda$ into the $\mathcal{G} \otimes \overline{\mathbb{F}}_p$-orbits, where $\mathrm{Adm}(\mu)$ is the finite index set for the orbit spaces. The set $\mathrm{Adm}(\mu)$ is also called the μ-admissible set, which is originally defined by group theory and shown to agree with the set of geometric orbits; see Section 4.4. For this moment, we just regard it as the orbit set, which is defined by geometry. Pull back the $\mathcal{G} \otimes \overline{\mathbb{F}}_p$-orbits in (4.3) to the moduli space $\widetilde{\mathcal{M}}$ and get a stratification of $\widetilde{\mathcal{M}}$ into locally closed subsets:

$$\widetilde{\mathcal{M}} = \coprod_{\underline{e} \in \mathrm{Adm}(\mu)} \widetilde{\mathcal{M}}_{\underline{e}}. \tag{4.4}$$

Since φ^{mod} is a $\mathcal{G} \otimes \overline{\mathbb{F}}_p$-torsor, the stratification of $\widetilde{\mathcal{M}}$ descends to a stratification of $\mathcal{M}$:

$$\mathcal{M} = \coprod_{\underline{e} \in \mathrm{Adm}(\mu)} \mathcal{M}_{\underline{e}}, \tag{4.5}$$

which is called the *KR stratification.* Notice that the morphism φ^{loc} is surjective if and only if each KR stratum $\mathcal{M}_{\underline{e}}$ is non-empty for $\underline{e} \in \mathrm{Adm}(\mu)$.

4.4 Description of $\mathrm{Adm}(\mu)$

Let k be an algebraically closed field of characteristic p, $W = W(k)$ the ring of Witt vectors over k and $L := \mathrm{Frac}(W)$ the fraction field. Let $\mathcal{O}_v^{\mathrm{nr}}$ be the etale extension of $\mathbb{Z}_p$ in $\mathcal{O}_v$ and $\mathcal{O}_p^{\mathrm{nr}} := \prod_{v|p} \mathcal{O}_v^{\mathrm{nr}}$. Set

$$\Sigma := \mathrm{Hom}_{\mathbb{Z}_p}(\mathcal{O}_p^{\mathrm{nr}}, W) = \coprod_{v|p} \Sigma_v, \quad \text{and} \quad \Sigma_v := \mathrm{Hom}_{\mathbb{Z}_p}(\mathcal{O}_p^{\mathrm{nr}}, W).$$

Regard $G(L) \subset \prod_{v|p} \prod_{\alpha \in \Sigma_v} \mathrm{GSp}_{2m}(F_v L)$ as a subgroup. Let $T \subset B \subset G$ be the diagonal maximal torus over $\mathbb{Q}_p$ and the upper triangular Borel subgroup. Denote by $X_*(T)_L$ the group of co-characters defined over L. Let $X_*(T)_{L,+} \subset X_*(T)_L$ be the subset of dominant co-characters with respect to B. The Cartan decomposition gives the following natural bijection

$$X_*(T)_{L,+} \xrightarrow{\mathrm{ev}} G(W)\backslash G(L)/G(W), \quad t \mapsto [(t(\varpi_v))_{v|p, \alpha \in \Sigma_v})], \tag{4.6}$$

where ϖ_v is an uniformizer of $\mathcal{O}_v \subset F_v$.

Let $\mu := \mu_h : \mathbb{G}_{m\mathbb{C}} \to G_{\mathbb{C}}$ be the Hodge co-character defined by h. The $G(\mathbb{C})$-conjugacy class of μ defines a dominant co-character still denoted by $\mu \in X_*(T)_+$. Then there is an unique element $t_\mu \in X_*(T)_{L,+}$ such that $\mathrm{ev}(t_\mu) = [\mu(p)]$. Explicitly,

$$t_\mu = (\underline{e}_\alpha)_\alpha, \quad \underline{e}_\alpha = (e_v, \dots, e_v, 0, \dots, 0) \quad \text{(each multiplicity } m),$$

where α ranges elements in Σ and v is the place for which Σ_v contains α.

The original group-theoretic definition of the μ-admissible set is given by

$$\mathrm{Adm}(\mu) := \{\underline{e} \in X_*(T)_{L,+} \,|\, \underline{e} \preceq t_\mu\}, \tag{4.7}$$

where $\preceq$ denotes the Bruhat order. It is shown in Haines and Ngô [16, Theorems 1, 4 and Proposition 5] (also see [11, Theorem 7.2]) that it coincides with the set $\mathrm{Adm}(\mu)$ defined in (4.3) by geometry.

We have the following result.

Theorem 4.1.

1. *For each element $\underline{e} \in \mathrm{Adm}(\mu)$, the KR stratum $\mathcal{M}_{\underline{e}}$ is non-empty. If $\overline{\mathcal{M}_{\underline{e}}}$ denotes the Zariski closure of $\mathcal{M}_{\underline{e}}$ in $\mathcal{M}$, then one has*

$$\overline{\mathcal{M}_{\underline{e}}} = \bigcup_{\underline{e}' \preceq \underline{e}} \mathcal{M}_{\underline{e}'}. \tag{4.8}$$

2. *Each KR stratum* $\mathcal{M}_{\underline{e}}$ *is smooth, equi-dimensional, and*

$$\dim \mathcal{M}_{\underline{e}} = \langle 2\rho, \underline{e} \rangle, \tag{4.9}$$

where ρ *is the half sum of all positive roots.*

3. *There is a unique minimal KR stratum and a unique maximal KR stratum* $\mathcal{M}_{t_\mu}$. *Moreover, the maximal KR stratum* $\mathcal{M}_{t_\mu}$ *is exactly the smooth locus* $\mathcal{M}^{\text{sm}}$.

Proof. See the proof in [48, Section 6].

An immediate consequence of Theorem 4.1 is the following

Corollary 4.2. *The smooth locus* $\mathcal{M}^{\text{sm}} \subset \mathcal{M}$ *is open and dense.*

This finishes the first step of the proof. For the second step, we construct the universal deformation for every point x in the smooth locus $\mathcal{M}^{\text{sm}}$. Then we show that the generic fiber of the universal deformation is indeed ordinary. For details, we refer to the forthcoming paper.

This finishes the sketch of the proof of our main result Theorem 3.1.

References

[1] J. D. Achter, Hilbert-Siegel moduli spaces in positive characteristic. *Rocky Mountain J. Math.* **33** (2003), 1–25.

[2] T. Ibukiyama, T. Katsura and F. Oort, Supersingular curves of genus two and class numbers. *Compositio Math.* **57** (1986), 127–152.

[3] C.-L. Chai, Monodromy of Hecke-invariant subvarieties. Special issue in memory of A. Borel. *Pure Appl. Math. Q.* **1** (2005), 291–303.

[4] C.-L. Chai and P. Norman, Singularities of the $\Gamma_0(p)$-level structure. *J. Algebraic Geom.* **1** (1992), 251–278.

[5] A.J. de Jong, The moduli spaces of polarized abelian varieties. *Math. Ann.* **295** (1993), 485–503.

[6] A.J. de Jong, The moduli spaces of principally polarized abelian varieties with $\Gamma_0(p)$-level structure. *J. Algebraic Geom.* **2** (1993), 667–688.

[7] P. Deligne, Variétés de Shimura: interprétation modulaire, et techniques de construction de modèles canoniques. *Automorphic forms, representations and L-functions* (*Proc. Sympos.Pure Math., Oregon State Univ. Corvallis,* 1977), *Part* 2, 247–289. Proc. Sympos. Pure Math., **XXXIII**, *Amer. Math. Soc.*, 1979.

[8] T. Ekedahl, On supersingular curves and supersingular abelian varieties. *Math. Scand.* **60** (1987), 151–178.

[9] G. Faltings and C.-L. Chai, *Degeneration of Abelian Varieties.* Ergebnisse der Mathematik und ihrer Grenzgebiete (3), **22**. Springer-Verlag, Berlin, 1990.

[10] A. Grothendieck, *Groupes de Barsotti-Tate et Cristaux de Dieudonné.* Les Presses de l'Université de Montréal, 1974.

[11] U. Görtz, Topological flatness of local models in the ramified case. *Math. Z.* **250** (2005), 775–790.

[12] U. Görtz, T. Haines, R. E. Kottwitz and D. C. Reumann, Dimensions of some affine Deligne-Lusztig varieties. *Ann. Sci,. École Norm. Sup.* (4) **39** (2006), 467–511.

[13] U. Görtz and C.-F. Yu, Supersingular Kottwitz-Rapoport strata and Deligne-Lusztig varieties. *J. Inst. Math. Jussieu* **9** (2010), no. 2, 357–390.

[14] U. Görtz and C.-F. Yu, The supersingular locus in Siegel modular varieties with Iwahori level structure. *Math. Ann.* **353** (2012), 465–498.

[15] T. Haines, Introduction to Shimura varieties with bad reduction of parahoric type. *Harmonic Analysis, the Trace Formula, and Shimura Varieties*, 583–642, Clay Math. Proc., **4**, Amer. Math. Soc., 2005.

[16] T. Haines and B. C. Ngô, Alcoves associated to special fibers of local models. *Amer. J. Math.* **124** (2002), 1125–1152.

[17] X. He, Normality and Cohen-Macaulayness of local models of Shimura varieties. *Duke Math. J.* **162** (2013), no. 13, 2509–2523.

[18] N. Koblitz, p-adic variant of the zeta-function of families of varieties defined over finite fields. *Compositio Math.* **31** (1975), 119–218.

[19] R. E. Kottwitz, Points on some Shimura varieties over finite fields. *J. Amer. Math. Soc.* **5** (1992), 373–444.

[20] R. E. Kottwitz and M. Rapoport, Minuscule alcoves for GL_n and GSp_{2n}. *Manuscripta Math.* **102** (2000), 403–428.

[21] K. -W. Lan, Arithmetic compactifications of PEL-type Shimura varieties. Thesis (Ph.D.), Harvard University, 2008.

[22] M. J. Larsen, Arithmetic compactification of some Shimura surfaces. *The Zeta Functions of Picard Modular Surfaces*, 31–45, Univ. Montréal, Montreal, QC, 1992.

[23] D. Mumford, *Abelian Varieties.* Oxford University Press, 1974.

[24] B.C. Ngô and A. Genestier, Alcôves et p-rang des variétés abéliennes. *Ann. Inst. Fourier* (*Grenoble*) **52** (2002), 1665–1680.

[25] B. C. Ngô and P. Polo, Résolutions de Demazure affines et formule de Casselman-Shalika géométrique. *J. Algebraic Geom.* **10** (2001), no. 3, 515–547.

[26] G. Pappas, On the arithmetic moduli schemes of PEL Shimura varieties. *J. Algebraic Geom.* **9** (2000), 577–605.

[27] G. Pappas and M. Rapoport, Local models in the ramified case I. The EL-case. *J. Algebraic Geom.* **12** (2003), 107–145.

[28] G. Pappas and M. Rapoport, Local models in the ramified case II. Splitting models. *Duke Math. J.* **127** (2005), 193–250.

[29] G. Pappas and M. Rapoport, Local models in the ramified case. III. Unitary groups. *J. Inst. Math. Jussieu* **8** (2009), no. 3, 507–564.

[30] G. Pappas and X. Zhu, Local models of Shimura varieties and a conjecture of Kottwitz, *Invent. Math.* **194** (2013), no. 1, 147–254.

[31] M. Rapoport, Compactifications de l'espaces de modules de Hilbert-Blumenthal. *Compositio Math.* **36** (1978), 255–335.

[32] M. Rapoport and Th. Zink, *Period Spaces for p-divisible Groups*. Ann. Math. Studies **141**, Princeton University Press, 1996.

[33] H. Stamm, On the reduction of the Hilbert-Blumenthal-moduli scheme with $\Gamma_0(p)$-level structure. *Forum Math.* **4** (1997), 405–455.

[34] T. Wedhorn, Ordinariness in good reductions of Shimura varieties of PEL-type. *Ann. Sci,. École Norm. Sup.* (4) **32** (1999), 575–618.

[35] C.-F. Yu, On the supersingular locus in Hilbert-Blumenthal 4-folds. *J. Algebraic Geom.* **12** (2003), 653–698.

[36] C.-F. Yu, Lifting abelian varieties with additional structures. *Math. Z.* **242** (2002), 427–441.

[37] C.-F. Yu, On reduction of Hilbert-Blumenthal varieties. *Ann. Inst. Fourier (Grenoble)* **53** (2003), 2105–2154.

[38] C.-F. Yu, Irreducibility of the Siegel moduli spaces with parahoric level structure. *Int. Math. Res. Not.* **2004**, no. 48, 2593–2597.

[39] C.-F. Yu, On the mass formula of supersingular abelian varieties with real multiplications. *J. Australian Math. Soc.* **78** (2005), 373–392.

[40] C.-F. Yu, On the slope stratification of certain Shimura varieties. *Math. Z.* **251** (2005), 859–873.

[41] C.-F. Yu, The supersingular loci and mass formulas on Siegel modular varieties. *Doc. Math.* **11** (2006), 449–468.

[42] C.-F. Yu, Irreducibility and p-adic monodromies on the Siegel moduli spaces. *Adv. Math.* **218** (2008), 1253–1285.

[43] C.-F. Yu, An exact geometric mass formula. *Int. Math. Res. Not.* **2008**, Article ID rnn113, 11 pages.

[44] C.-F. Yu, Irreducibility of Hilbert-Blumenthal moduli spaces with parahoric level structure. *J. Reine Angew. Math.* **635** (2009), 187–211.

[45] C.-F. Yu, Simple mass formulas on Shimura varieties of PEL-type. *Forum Math.* **22** (2010), no. 3, 565–582.

[46] C.-F. Yu, Geometry of the Siegel modular threefold with paramodular level structure. *Proc. Amer. Math. Sci.* **139** (2011), no. 9, 3181–3190.

[47] C.-F. Yu, Kottwitz-Rapoport strata in the Siegel moduli spaces. *Taiwanese J. Math.* **14** (2010), no. 6, 2343–2364.

[48] C.-F. Yu, Notes on density of the ordinary locus. *Bull. Inst. Math. Acad. Sin.* (*N.S.*), **9** (2014), no. 4, 711–733.

Institute of Mathematics, "Academia Sinica" and NCTS (Taipei Office)
6th Floor, Astronomy Mathematics Building
No. 1, Roosevelt Rd. Sec. 4
Taipei 10617, Taiwan, China
E-mail address: chiafu@math.sinica.edu.tw

Discrete Mathematics, Combinatorics

Proceedings of the 6th International Congress of Chinese Mathematicians
ALM 36, pp. 383–392

Bounding Some Ramsey Functions

Yusheng Li

Abstract

For positive integers k and n, define Ramsey number $r(k,n)$ to be the smallest integer N such that any red-blue edge-coloring of K_N contains a red K_k or a blue K_n, and van der Waerden number $w(k,n)$ to be the smallest N such that any red-blue coloring of $\{1,\dots,N\}$ contains a red k-AP or a blue n-AP, where a k-AP is a nontrivial arithmetic progression of k terms. We introduce the best known bounds for these important functions, and describe the proofs in outline, including some of ours. The most of methods are non-constructive.

Mathematics Subject Classification: 05C55.

Keywords and Phrases: Ramsey number, Van der Waerden number, non-constructive method.

1 Definitions of two Ramsey functions

A typical result in Ramsey theory states that if a large mathematical object is partitioned into finite many parts, then one of the parts must contain a sub-object of particular property. The smallest size of the large object to guarantee the property is called Ramsey number. We only discuss two of important Ramsey numbers. In combinatorics, a partition of a set is called a coloring of the members in the set, or a coloring of the set simply.

Definition 1. *For positive integers k and n, let $r(k,n)$ be the smallest N such that any red-blue edge-coloring of complete graph K_N contains either a red K_k or a blue K_n.*

Let us write a nontrivial arithmetic progression of t terms as t-AP, and write the set $\{1,2,\dots,N\}$ for $N \geq 1$ as $[N]$.

Supported in part by NSFC.

Definition 2. *For positive integers k and n, let $w(k,n)$ be the smallest N such that any red-blue coloring of $[N]$ contains either a red k-AP or a blue n-AP.*

We will call $r(n,n)$ and $w(n,n)$ as diagonal Ramsey number and diagonal van der Waerden number, respectively.

2 Upper bounds for Ramsey functions

From the definitions, we see that if N is an integer that satisfies the property in the definition, then any large integer must do. Thus the argument for existence of a Ramsey number is often associated with an upper bound. Generally, it is hard to obtain good bounds for Ramsey functions.

Ramsey theory is named after the paper of Ramsey [19], but in fact, two famous problems, namely Schur number [20] and van der Waerden number [29], were studied before Ramsey's paper, which are now called the Ramsey numbers on integers.

The Ramsey number was popular after Erdős and Szekeres [11] rediscovered it and gave the upper bound

$$r(k,n) \le \binom{k+n-2}{k-1}. \tag{1}$$

This bound implies that for large n,

$$r(n,n) \le \binom{2n-2}{n-1} \sim \frac{4^n}{4\sqrt{\pi n}},$$

and for fixed k and large n,

$$r(k,n) \le \binom{k+n-2}{k-1} \sim \frac{n^{k-1}}{(k-1)!}.$$

An improvement of Thomason [28] for upper bounds of diagonal case in (1) is as follows

$$r(n,n) \le \frac{\exp(c\sqrt{\log n})}{\sqrt{n}}\binom{2n-2}{n-1},$$

where $\frac{\exp(c\sqrt{\log n})}{\sqrt{n}} = \frac{1}{n^{1/2-o(1)}}$. Conlon [7] did it furthermore as

$$r(n,n) \le \frac{1}{n^{c\log n/\log\log n}}\binom{2n-2}{n-1}.$$

Note that the two improvements did not change the upper bound in the form $(4+o(1))^n$. It is believed that $(2-o(1))^n \le r(n,n) \le (2+o(1))^n$.

For the off-diagonal case, a significant improvement of Ajtai, Komlós and Szemerédi [2] is

$$r(k,n) \le (5000)^k \frac{n^{k-1}}{\log^{k-2} n}, \tag{2}$$

where $k \geq 2$ is fixed and n is large. For $k = 3$, Shearer [21] improved the above bound as

$$r(3,n) \leq (1+o(1))\frac{n^2}{\log n}.$$

For general $k \geq 2$, Li, Rousseau and Zang [16, 17] gave

$$r(k,n) \leq (1+o(1))\frac{n^{k-1}}{\log^{k-2} n}. \tag{3}$$

Kim [15] showed that the order of $r(3,n)$ is $n^2/\log n$ by giving the correct order of the lower bound, and thus we cannot expect to improve the upper bound of $r(3,n)$ in (1) more than a multiplicative constant. However, we may expect that the asymptotic formula of $r(3,n)$ is $n^2/\log n$. Note that (3) is an asymptotical equality for $k = 2$ as $r(2,n) = n$. Let us further compare the known values of $r(3,n)$ and $n^2/\log n$ in the following table.

n	3	4	5	6	7	8	9	10	11
$r(3,n)$	6	9	14	18	23	28	36	40/43	46/51
$n^2/\log n$	8.2	11.5	15.5	20.1	25.2	30.8	36.9	43.4	50.5
Error(%)	36.5	28.2	11.0	11.6	9.5	9.9	2.4		

How did the improvement (2) come?

Let us explain a bit more on a basic problem. For a graph G, its independence number $\alpha(G)$ is an important parameter. It is well-known for any graph with average degree d,

$$\alpha(G) \geq \frac{N}{1+d},$$

which is called Turán's bound as it can be implied from Turán theorem on maximum number of edges of graphs with a forbidden cliques. Let $\omega(G)$ be the clique number of G. Then the statement G contains no K_k is equivalent to $\omega(G) < k$.

Turán's bound is associated with Ramsey $r(k,n)$ tightly. It is easy to see

$$r(k,n) - 1 = \max\{|V(G)| : \alpha(G) < k, \omega(G) < k\}.$$

Since the Turán bound is for general graphs so that we could not employ it to obtain any good bound. We are more interested with this kind bound for graphs that have a small fixed forbidden graph.

Ajtai, Erdős, Komlós and Szemerédi [1] showed

$$\alpha(G) \geq cN\frac{\log d}{d}$$

for triangle-free graph G of order N with average degree $d > 0$, where $c > 0$ is a constant. In the proof, they invented a method called "semi-random method" or "nibble method" . This bound improves the Turán's bound significantly when d is large. Shearer [21] improved the above result as

$$\alpha(G) \geq Nf(d)$$

where

$$f(x) = \frac{x\log x - x + 1}{(x-1)^2}.$$

We should have some similar result for more general graphs instead of only triangle-free graphs. It is showed in [16, 17] that

$$\alpha(G) \geq N f_m(d)$$

for any graph G of average degree d, in which each subgraph induced by a neighborhood has a maximum degree less than the integer $m \geq 1$, where

$$f_m(x) = \int_0^1 \frac{(1-t)^{1/m}}{m + (x-m)t} dt.$$

A simple calculation shows that $f_1(x)$ is the function used by Shearer for triangle-free graphs. The function $f_m(x)$ satisfies a property as follows.

$$\frac{\log(x/m)}{x-m} > f_m(x) > \frac{\log(x/m) - 1}{x}, \tag{4}$$

for $x > m \geq 1$. Let us give a proof for (3) which simplifies that in [17].

Proof of (3). We shall prove (3) by induction on $k \geq 2$ starting at $r(2, n) = 2$. The proof for $k = 3$ is a special case of the inductive step. Let $k \geq 4$ and let G be a graph of order $N = r(k, n) - 1$ that contains no K_k and $\alpha(G) < n$. Then the average degree $d \leq \Delta \leq r(k-1, n) - 1$ and each subgraph induced by a neighborhood has maximum degree at most $r(k-2, n) - 1$, and thus it contains no $K_{1,m}$ with $m = r(k-2, n)$. Thus, from the induction hypothesis on $r(k-1, n)$ and $r(k-2, n)$ and the fact that the right-hand side of (4) is decreasing on each of d and m, we have (3) for general k. □

Note that an application of (4) on information theory was done by Gao, Yang and Ge [12] to obtain an improvement for Gilbert-Varshamov bound.

3 Lower bounds for Ramsey functions

A remarkable result of Erdős [8] showed

$$r(n, n) \geq \frac{n}{e\sqrt{2}} 2^{n/2}. \tag{5}$$

In the proof, Erdős counted all graphs of order n. This proof, together with some other results by so called probabilistic method, bringed to the birth of random graphs [10]. In addition to probabilistic method, random graphs are one of most important topic for graph theory and other areas.

In earlier years, Erdős employed the random graph to disprove some conjectures by showing that "almost all" graphs were counterexamples.

After a long time, Spencer [23] improved (5) by proving

$$r(n,n) \geq (1-o(1))\frac{\sqrt{2}n}{e}2^{n/2},$$

which only improved (5) by an asymptotic multiplicative constant 2. In the proof, Spencer used Lovász local lemma [9]. In another paper [24], Spencer generalize the local lemma and showed

$$r(k,n) \geq c\left(\frac{n}{\log n}\right)^{(k+1)/2} \tag{6}$$

for fixed $k \geq 3$ and large n. Since then, the local lemma is known as a powerful tool in probabilistic method. Now original form of local lemma is called symmetric form and the generalized is called general form.

Let $A_1, \ldots, A_n$ be "bad" random events. We wish that no "bad" event happens, that is to say, we wish

$$\Pr\left(\bigcap \overline{A}_i\right) > 0.$$

This can be guaranteed if

$$\sum_{i=1}^{n} \Pr(A_i) < 1,$$

or if $A_1, \ldots, A_n$ are mutually independent and $\Pr(A_i) < 1$.

The Lovász local lemma may be understood in term of taking advantage of *partial independence* of the events $A_1, \ldots, A_n$ and far weaker bounds on

$$\sum \Pr(A_i) < 1.$$

A graph D on vertices $[n] = \{1, 2, \ldots, n\}$ (the set of indices for the events A_i) is called *dependency graph* for events if for every i, the event A_i is mutually independent of all A_j with $ij \notin E(D)$ and $j \neq i$.

The original of the local lemma is as follows.

Theorem 1. *If* $\Pr(A_i) \leq p$, *each vertex in the dependence graph has degree at most* $d \geq 1$, *and if*

$$4dp \leq 1,$$

then $\Pr(\cap_{i=1}^{n} \overline{A}_i) > 0$.

The following is the general form which appeared in [24].

Theorem 2. *Suppose that* $0 < x_i < 1$ *and*

$$\Pr(A_i) \leq x_i \prod_{j:\ ij \in E(D)} (1 - x_j).$$

Then $\Pr(\cap_{i=1}^{n} \overline{A}_i) \geq \Pi_{i=1}^{n}(1 - x_i) > 0$.

Then, the original form of local lemma can be strengthened as follows, which is also called as symmetric form.

Corollary 1. *Suppose that each vertex in dependency graph has a degree at most* d, $\Pr(A_i) \le p$, *and*

$$e(d+1)p \le 1,$$

then $\Pr(\cap_{i=1}^n \overline{A}_i) > 0$.

An event B is said to be *positive* (*negative*) to A if $\Pr(A|B) \ge \Pr(A)$ ($\Pr(A|B) \le \Pr(A)$). When the inequality holds strictly, then B is said to be *strictly positive* (*strictly negative*) to A. For a set S of indices of events, we write

$$\mathcal{C}_S = \bigcap_{j \in S} \overline{A}_j.$$

A graph F on $[n]$ (the set of indices for the events) is called *negative graph* of events $A_1, \ldots, A_n$ if for every i and any set $S \subseteq [n] \setminus N[i]$, the event $\mathcal{C}_S$ is negative to A_i.

Let us remark that the dependency graph in Theorem 2, Theorem 2 and Corollary 1 can be replaced by a negative graph. As a negative graph can have less edges than dependency graph, the replacement often gives some improvement or simplicity of the proof.

After around twenty years of Spencer's lower bound (6), Kim [15] proved

$$r(3,n) \ge c\frac{n^2}{\log n}. \tag{7}$$

So the order of $r(3,n)$ is $n^2/\log n$. Using different probabilistic approach, Bohman [4] proved Kim's bound again. Furthermore, Bohman and Keevash [5] have the following result, which improves the lower bound (6) by a factor $\log^{1/(k-2)} n$ as

$$r(k,n) \ge c\left(\frac{n}{\log n}\right)^{(k+1)/2} \log^{1/(k-2)} n.$$

In [4, 5, 15], they employed triangle-free process, which we shall explain a bit. It is often difficult to show the existence of small events. The local lemma is a tool for such proof that improved most lower bounds from basic probabilistic method. The key for the proof of the local lemma itself is conditional probability. A revolutionary idea for finding the small events is also "conditional". If we know a certain condition in which the event is likely to appear, then the probability for event is large under the condition. In other words, we try to switch a small event to be a large one conditionally. However, we may encounter the difficulty to finger the conditional probability out.

Obtaining the right order of magnitude of $r(m,n)$ even $r(3,n)$ was certainly a challenge in decades. The celebrated result (7) of Kim showed that of $r(3,n)$ is $n^2/\log n$, which was obtained again by Bohman. They used different analysis on the same random graph process, called the triangle-free process.

The triangle-free process can be described as follows. We begin with the empty graph, denoted by G_0, on N vertices. At step i we form the graph G_i by adding a new edge to G_{i-1} chosen uniformly at random from the collection of pairs of vertices that neither appear as edges in G_{i-1} nor form triangles when added as edges to G_{i-1}. The process terminates at a maximal triangle-free graph G_M, for which the random variable M is the number of edges of G_M. Note that a maximal triangle-free graph is connected and the number of edges in a triangle-free graph of order N is at most $N^2/4$ (see Chapter 8), we have

$$N-1 \le M \le \frac{N^2}{4}.$$

However, Bohman proved that almost surely

$$c_1 N^{3/2}\sqrt{\log N} \le M \le c_2 N^{3/2}\sqrt{\log N}.$$

From a result in Chapter 3, we have that the independence numbers of such graphs are at least $\Omega(\sqrt{N\log N})$. Remarkably, Kim and Bohman show that a.s. independence numbers of such graphs are at most $O(\sqrt{N\log N})$, which implies that $r(3,n) \ge \Omega(n^2/\log n)$.

Let us talk a bit more on the process. For a set V, let $V^{(2)}$ be the set of all pairs u, v of V, which is the edge set of complete graph on V. The vertex set of our complete graph of order N is on $[N] = \{1, 2, \ldots, N\}$. In the evolution of the triangle-free process, we shall track the some random sets. Recall that G_i is the graph given by the first i edges selected by the process. The graph G_i partitions $[N]^{(2)}$ into three parts: E_i, O_i and C_i. The set E_i is simply the edge set of G_i. A pair of $[N]^{(2)}$ is open, and in the set O_i, if it can still be added as an edge without violating the triangle-free condition. A pair of $[N]^{(2)}$ is closed, and in the set C_i, if it is neither an edge in the graph nor open; that is, a pair $e = \{u, v\}$ is in C_i if $e \notin E_i \cup O_i$ and there exists a vertex w such that $\{u, w\}, \{v, w\} \in E_i$. Note that e_{i+1} is chosen uniformly at random from O_i. That is to say, each edge of O_i has the same probability $1/|O_i|$ to be chosen as e_{i+1}. We do not express this as $\Pr(e_{i+1} \in O_i) = 1/|O_i|$ as only edges in random set O_i are available.

4 Van der Waerden functions

Let us discuss the bounds for van der Waerden function $w(k, n)$. The current best upper bound for $w(n, n)$ is due to Gowers [13], who proved

$$w(n,n) \le \left. 2^{2^{\cdot^{\cdot^{\cdot^{n+1}}}}} \right\} 6,$$

which is a tower of height 6. Even for $n = 1$, this bound is such a huge number that is beyond the capacity of any modern computer. However, this proves the bound of Shelah [22] significantly, which was

$$w(n,n) \le s(n),$$

where $s(n)$ is called "wowzer", which is a tower of height $s(n-1)$. How huge was the original bound [29] for $w(n,n)$ made by van der Waerden? It is hard to say what is like, and the bound, denoted by $A(n)$, is called Ackerman function, witch increases fastest among all functions used in history.

The lower bounds for $w(n,n)$ are more "reasonable". Berlekamp [3] showed if 2^p-1 is a (Mersenne) prime, then

$$w(p+1,p+1) \geq p(2^p-1).$$

Note that a big open problem in number theory is asking that there are how many Mersenne primes. Are they infinitely many as conjectured?

Using the local lemma, Szabó [25] proved $w(n,n) \geq \frac{2^n}{2en}$. If we use the lemma with negative graph for events, then we can have

$$w(n,n) \geq \frac{2^n}{en}.$$

Let us turn to the off-diagonal van der Waerden function $w(k,n)$. It is showed that

$$c_1\left(\frac{n}{\log n}\right)^2 \leq w(3,n) \leq n^{c_2n^2}.$$

The upper bound is in Bourgain [6], and the lower bound is in Li and Shu [18], which is a special case of the general one as for fixed $k \geq 3$ and large n,

$$w(k,n) \geq c\left(\frac{n}{\log n}\right)^{k-1}.$$

Let us talk more on AP in a dense set of integers. For a set A of positive integers, Erdős-Turán conjectured that if $\overline{\lim}_n \frac{|A\cap[n]|}{n} > 0$, then A contains arbitrarily long AP. It was Szemerédi [26] who proved the conjecture completely. His proof is a deep and complicated combinatorial argument, and he used a result called the Regularity Lemma in bipartite version. In 1978, he proved the full lemma in [27]. The lemma has now become a totally new tool in extremal graph theory. We here just describe the regularity lemma briefly.

Theorem 3. (Regularity Lemma) *For fixed $\epsilon > 0$ and $m \geq 1$, there exists $N = N(\epsilon,m) > m$ such that any graph of order $n \geq N$ can be partition into k parts $C_1, C_2, \ldots, C_k$ with $m \leq k \leq N$ such that $\Big||C_i|-|C_j|\Big| \leq 1$ and all pair (C_i,C_j) except $\epsilon\binom{k}{2}$ of them are ϵ-regular.*

It is a long history that people conjectured that there are arbitrarily long AP of primes, which has been proved by Green and Tao [14]. Another conjecture on AP of integers of Erdős-Turán is that a set $A=\{a_i\}$ with $\sum 1/a_i = \infty$ contains arbitrarily long AP. So the result of Green and Tao is an important case as $\sum_{p\leq n} 1/p \sim \log\log n$.

We shall propose a notation to conclude the paper. Define $T(n)$ is the minimum integer N such that $[N]$ contains an n-AP of primers. It must be interesting and difficult to estimate $T(n)$.

References

[1] M. Ajtai, P. Erdős, J. Komlós and E. Szemerédi, On Turán theorem for sparse graphs, *Combinatorica*, **1** (1981), 313–317.

[2] M. Ajtai, J. Komlós and E. Szemerédi, A note on Ramsey numbers, *J. Combin. Theory Ser. A*, **29** (1980), 354–360.

[3] E. Berlekamp, A construction for partitions which avoid long arithmetic progressions, *Canad. Math. Bull.*, **11** (1968), 409–414.

[4] T. Bohman, The triangle-free process, *Adv. Math.*, **221** (2009), 1653–1677.

[5] T. Bohman and P. Keevash, The early evolution of H-free process, *Invent. Math.*, **181** (2010), 291–336.

[6] J. Bourgain, On triples in arithmetic progression, *Geom. Funct. Anal.*, **9** (1999), 968–984.

[7] D. Conlon, A new upper bound for diagonal Ramsey numbers, *Ann. of Math.*, **170** (2009), 941–960.

[8] P. Erdős, Some remarks on the theory of graphs, *Bull. Amer. Math. Soc.*, **53** (1947), 292–294.

[9] P. Erdős and L. Lovász, Problems and results on 3-chromatic hypergraph and some related questions, in: *Infinite and Finite Sets* (*to Paul Erdős on His* 60*th Birthday* 1973) *II, A. Hajnal, R. Rado and V. Sos Eds.*, Colloq. Math. Soc. Janos Bolyai, North-Holland, Amsterdam/London, 1975.

[10] P. Erdős and A. Rényi, On the evolution of random graphs, *Publ. Math. Inst. Hungar. Acad. Sci.*, **5** (1960), 17–61.

[11] P. Erdős and G. Szekeres, A combinatorial problem in geometry, *Compos. Math.* **2** (1935), 464–470.

[12] F. Gao, Y. Yang and G. Ge, An improvement to Gilbert-Varshamov bound for permutation, *IEEE Trans Inform Theory*, to appear.

[13] W. T. Gowers, A new proof of Szemerédi's theorem, *Geom. Funct. Anal.*, **11** (2001), 465–588.

[14] B. Green and T. Tao, The primes contain arbitrarily long arithmetic progressions, *Ann. of Math.*, **167** (2008), 481–547.

[15] J. Kim, The Ramsey number $R(3,t)$ has order of magnitude $t^2/\log t$, *Random Structures Algorithms*, **7** (1995), 173–207.

[16] Y. Li and C. Rousseau, On book-complete graph Ramsey numbers, *J. Combin. Theory Ser. B*, **68** (1996), 36–44.

[17] Y. Li, C. Rousseau and W. Zang, Asymptotic upper bounds for Ramsey functions, *Graphs Combin.*, **17** (2001), 123–128.

[18] Y. Li and J. Shu, A lower bound for off-diagonal van der Waerden numbers, *Adv. in Appl. Math.*, **44** (2010), 243–247.

[19] F. Ramsey, On a problem of formal logic, *Proc. Lond. Math. Soc. Ser.* 2, **30** (1930), 264–286.

[20] I. Schur, Über die Kongruenz $x^m + y^m = z^m \pmod{p}$, *Jahresber. Deutsch. Math.-Verein.*, **25** (1916), 114–117.

[21] J. Shearer, A note on the independence number of triangle-free graphs, *Discrete Math.*, **46** (1983), 83–87.

[22] S. Shelah, Primitive recursive bounds for van der Waerden numbers, *J. Amer. Math. Soc.*, **1** (1988), 683–697.

[23] J. Spencer, Ramsey's theorem—a new lower bound, *J. Combin. Theory Ser. A*, **18** (1975), 108–115.

[24] J. Spencer, Asymptotic lower bound for Ramsey functions, *Discrete Math.*, **20** (1977), 69–76.

[25] T. Szabó, An application of Lovász' local lemma—a new lower bound for the van der Waerden numbers, *Random Structures Algorithms*, **1** (1990), 343–360.

[26] E. Szemerédi, On sets of integers containing no k elements in arithmetic progression, *Acta Arith.*, **27** (1975), 199–245.

[27] E. Szemerédi, Regular partitions of graphs, Problémes combinatories et théorie des graphs, Orsay 1976, J.-C. Bermond, J.-C. Fournier, M. Las Vergnas and D. Scotteau (Editors), *Colloq. Internat. CNRS* 260, Paris, 1978, 399–401.

[28] A. Thomason, An upper bound for some Ramsey numbers, *Europ. J. Combin.*, **12** (1988), 509–517.

[29] B. L. van der Waerden, Beweis einer Baudetschen Vermutung, *Nieuw Arch. Wisk.*, **15** (1927), 257–271.

Department of Mathematics, Tongji University
Shanghai 200092, China
E-mail address: li_yusheng@tongji.edu.cn

Proceedings of the 6th International
Congress of Chinese Mathematicians
ALM 36, pp. 393–405

Some Results and Problems About Circumferences of Graphs

Xingxing Yu

Abstract

The circumference of a graph is the length of a longest cycle in that graph. In general it is NP-hard to determine the circumference of a graph. In this article, we discuss results and problems about circumferences of graphs with certain structures, including graphs on a fixed surface, graphs with forbidden minors, and graphs with bounded maximum degree.

Mathematics Subject Classification: 05C38, 05C10, 05C40, 05C45, 05C63, 05C83.

Keywords and Phrases: Circumference of graphs, long cycles, planar graphs, graphs on surfaces, infinite graphs, Hamilton circles, graph connectivity, graph minors, graphs with bounded degree, cubic graphs.

1 Introduction

Let V be a set and let E be a set of subsets of V; then the pair (V, E) is a *hypergraph*. An element of V is called a *vertex*, and an element of E is called an *edge*; so V, E are the *vertex set*, *edge set* of the hypergraph, respectively. An edge e is *incident* with a vertex v if $v \in e$. When each member of E is a 2-element subset of V, (V, E) is a *graph*. A graph (V, E) is *finite* (respectively, *infinite*) if $|V|$ is finite (respectively, infinite).

In this article, graphs are finite except when mentioned explicitly as infinite graphs. We will often use uppercase letters to denote graphs. When we say G is a graph, we understand that it has vertex set $V(G)$ and edge set $E(G)$. A graph H is a *subgraph* of a graph G if $V(H) \subseteq V(G)$ and $E(H) \subseteq E(G)$.

Let $n \geq 3$ be an integer. A *cycle of length* n is a graph isomorphic to the graph with vertex set $\{v_1, \ldots, v_n\}$ and edge set $\{\{v_i, v_{i+1}\} : 1 \leq i \leq n-1\} \cup \{\{v_1, v_n\}\}$.

Partially supported by NSA grant ID-111008 and NSF grant DMS-1265564.

The *circumference* of a graph G, denoted by $c(G)$, is the length of a longest cycle in G.

Let $n \geq 0$ be an integer. A *path of length* n is the graph isomorphic to the graph with vertex set $\{v_1, \ldots, v_{n+1}\}$ and edge set $\{\{v_i, v_{i+1}\} : 1 \leq i \leq n\}$. (Such a path is also said to be *between* v_1 *and* v_{n+1}.) Path and cycle are fundamental concepts in graph theory. A graph G is *connected* if $|V(G)| \geq 2$ and for any distinct $x, y \in V(G)$, there is a path in G between x and y. The maximal connected subgraphs of a graph G are the *components* of G. For any positive integer k, a graph G is k*-connected* if $|V(G)| \geq k+1$ and removing at most $k-1$ vertices (as well as their incident edges) results in a connected graph.

For standard graph theory terminology not defined here, we refer the reader to Bollobás [7] and Diestel [24].

A cycle (respectively, path) in a graph G containing all vertices of G is said to be a *Hamilton* cycle (respectively, path). The *Hamilton Cycle Problem*, which is NP-complete (see [29]), asks whether a given graph contains a Hamilton cycle. Karger, Motwani and Ramkumar [40] showed that unless $P = NP$ it is impossible to find, in polynomial time, a cycle of length $n - n^{\epsilon}$ (for any $\epsilon < 1$) in an n-vertex Hamiltonian graph. Thus, in view of approximation algorithms, the problem of determining the circumference of a graph is one of the "harder" NP-hard problems. Therefore, it is meaningful to find good lower bounds on the circumferences of graphs with natural properties.

There is also a connection between the Hamilton Cycle Problem and the Four Color Problem (now a theorem). A graph is *planar* if it has a drawing in the plane with no edges crossing. Attempting to solve the Four Color Problem, Tait [62] conjectured in 1880 that every cubic 3-connected planar graph contains a Hamilton cycle. (A graph is *cubic* if each vertex is incident with exactly three edges.) If this were true, then we would have an easy proof of the Four Color Theorem. In 1946, Tutte [70] found a counterexample to Tait's conjecture. There has since been extensive research concerning circumferences of cubic 3-connected planar graphs, as well as circumferences of planar graphs, graphs embedded on a fixed surface, cubic 3-connected graphs, 3-connected graphs with bounded degree, and 3-connected graphs with certain forbidden minors.

The author will mention a number of results and open problems about circumferences of these graphs. In Section 2, we deal with planar graphs. In Section 3, we consider graphs on surfaces. In Section 4, we handle infinite planar graphs.

2 Planar graphs

Tutte [70] gave examples of cubic 3-connected planar graphs without Hamilton cycles. For 4-connected planar graphs the story is different. Whitney [75] showed that every 4-connected planar triangulation contains a Hamilton cycle. Tutte [71] in 1956 extended Whitney's theorem to all 4-connected planar graphs. (See [22] for an interesting application of Tutte's theorem to length of thick knots.)

Theorem 2.1. (Tutte [71]) *Every 4-connected planar graph contains a Hamilton cycle.*

In fact, Thomassen [67] proved that for any 4-connected planar graph G and any distinct $x, y \in V(G)$, there is a Hamilton path in G between x and y. Thus, 4-connected planar graphs are *Hamilton connected.* Since it is difficult to deal with 4-connected graphs, the approach taken by Thomassen [67] is to prove that every 2-connected planar graph G contains a "Tutte path" which must be a Hamilton path if G is in fact 4-connected.

Let G be a graph and P a subgraph of G. A *P-bridge* of G is a subgraph of G which is induced by a component D of $G - V(P)$ and all edges of G with one incident vertex in P and another in D. For a P-bridge B of G, the vertices of $B \cap P$ are the *attachments* of B on P. We say that P is a *Tutte subgraph* of G if every P-bridge of G has at most three attachments on P. A *Tutte path* (respectively, *Tutte cycle*) in a graph G is a path (respectively, cycle) which is a Tutte subgraph of G. We can now state a slightly stronger version of Thomassen's result in [67], see [59].

Theorem 2.2. (Sanders [59]) *Let G be a 2-connected plane graph, and let $x, y \in V(G)$ be distinct. Then for any edge $e \in E(G)$, G contains a Tutte-path between x and y and through e.*

As an immediate consequence of Theorem 2.2, every 4-connected planar graph is Hamilton connected. Using Thomassen's technique, Chiba and Nishizeki [18] gave a linear-time algorithm for finding a Hamilton cycle in a 4-connected planar graph.

Let G be a 4-connected n-vertex planar graph. Then Theorem 2.1 says that G contains a cycle of length n. It was observed by Nelson (see [67]) that the Tutte path technique can be used to show that G also contains a cycle of length $n-1$. Plummer [57] conjectured that G should contain a cycle of length $n-2$, which was established by Thomas and Yu [64], again using the Tutte path technique. Theorem 2.2 implies that G also contains a cycle of length $n-3$. The following is a conjecture made by Malkevitch [48].

Conjecture 2.3. (Malkevitch [48]) *Let G be a 4-connected n-vertex planar graph. If G contains a cycle of length 4, then G contains a cycle of length k for all $3 \leq k \leq n$.*

The condition that G contains a cycle of length 4 is necessary in the sense that there are 4-connected planar graphs without cycles of length 4. (For example, line graphs of cyclically 4-edge-connected cubic planar graphs do not contain 4-cycles.) Chen, Fan and Yu [12] observed that the Tutte path technique employed in [71, 67, 64] alone is not adequate to establish Conjecture 2.3. By combining a "contractible edge" argument, it is shown in [12] that G also contains cycles of lengths $n-4, n-5$, and $n-6$. An edge e in a k-connected graph G is said to be *k-contractible* if G/e, the graph obtained from G by contracting e, is also k-connected. It is easy to see that for $k = 1, 2$, every k-connected graph other than K_{k+1} has a contractible edge. A theorem of Tutte [72] establishes the same result for $k = 3$. However, there are two infinite families of 4-connected graphs which do not contain any contractible edge, see [26, 49].

3 Graphs on surfaces

Grünbaum [32] conjectured that every 4-connected graph embeddable in the projective plane contains a Hamilton cycle. This conjecture was proved by Thomas and Yu [64], extending Theorem 2.1 to the projective plane. Dean [20] conjectured that such graphs are Hamilton connected. Recently, Kawarabayashi and Ozeki [44] announced a proof of Dean's conjecture.

To extend Theorem 2.1 to graphs embedded in the torus, Grünbaum [32] and, independently, Nash-Williams [55] made the following conjecture. A *toroidal* graph is a graph embeddable in the torus.

Conjecture 3.1. (Grünbaum [32], Nash-Williams [55]) *Every 4-connected toroidal graph contains a Hamilton cycle.*

Conjecture 3.1 is established in [1] for 6-connected toroidal graphs. Brunet and Richter [11] proved Conjecture 3.1 for 5-connected triangulations of the torus. Thomas and Yu [65] proved Conjecture 3.1 for all 5-connected toroidal graphs. In 2005, Thomas, Zang and Yu [66] proved that every 4-connected toroidal graph contains a Hamilton path.

The Tutte subgraph technique plays important roles in [65, 66]. In addition, one needs a simple closed curve in the surface that intersects the embedded graph minimum number of times (this minimum number is called *face-width*, or *representativity* of that graph). When the face-width of a graph is large, one simply apply graph reductions so that the new graph has smaller face-width. When the face-width is small (3 or less), one cuts the surface and graph simultaneously, and reduces the problem to a problem about disjoint Tutte paths in planar graphs.

Recently, using different techniques, Nakamoto and Ozeki [53] proved Conjecture 3.1 for bipartite toroidal quadrangulations of the torus, and Fujisawa, Nakamoto and Ozeki [27] proved Conjecture 3.1 for graphs with toughness exactly one.

Very recently, Kawarabayashi and Ozeki [43] announced a proof of Conjecture 3.1 for 4-connected toroidal triangulations, and it appears that the Tutte path technique is used again. Nevertheless, Conjecture 3.1 remains open. Perhaps, one reason that Conjecture 3.1 is still unresolved is because there are 4-connected toroidal graphs that are not Hamilton connected. (Take a toroidal grid $C_{2m} \times C_{2n}$, and take two nonadjacent vertices from a 4-cycle that bounds a face. We see that there is no Hamilton path joining these two vertices.)

We now turn to graphs embeddable in other surfaces. Brunet, Nakamoto and Negami [10] showed that every 5-connected triangulation of the Klein bottle contains a Hamilton cycle. Archdeacon, Hartsfield and Little [2] showed that for each integer $n \geq 3$ there is a triangulation of some surface which is n-connected and has face-width at least n, and in which every spanning tree has maximum degree at least n (so no Hamilton path or cycle exists). Yu [76] proved that if a 5-connected triangulation of a surface has large face-width, then it contains a Hamilton cycle, establishing a conjecture of Thomassen [69]. Kawarabayashi [41] observed that the same technique actually proves that such graphs are Hamilton connected. Mohar [51], Thomassen[69], and Zhao [82] independently conjectured that the triangulation requirement is not necessary.

Conjecture 3.2. (Mohar [51], Thomassen[69], Zhao [82]) *Every* 5*-connected graph in a surface with large face-width contains a Hamilton cycle.*

4 Infinite planar graphs

A *ray* or 1*-way infinite path* is a graph which is isomorphic to the graph with vertex set $\{v_i : i = 1, 2, \ldots\}$ and edge set $\{v_i v_{i+1} : i = 1, 2, \ldots\}$. A *double ray* or 2*-way infinite path* is a graph which is isomorphic to the graph with vertex set $\{v_i : i = \ldots, -2, -1, 0, 1, 2, \ldots\}$ and edge set $\{v_i v_{i+1} : i = \ldots, -2, -1, 0, 1, 2, \ldots\}$.

In an attempt to generalize Theorem 2.1 to infinite graphs, Nash-Williams [54, 55] made two conjectures: A 4-connected infinite planar graph contains a spanning ray iff it is 2*-indivisible* (i.e., the deletion of any finite set of vertices results in at most one infinite component); a 4-connected infinite planar graph contains a spanning double ray iff it is 3*-indivisible* (i.e., the deletion of any finite set of vertices results in at most two infinite components). Dean, Thomas and Yu [21] proved the first conjecture, and Yu [77, 78, 79, 80, 81] proved the second conjecture.

The indivisibility conditions in Nash-Williams' conjectures show in a sense that spanning rays and spanning double rays are not the right infinite version of a Hamilton cycle. Diestle [23] introduced the notion of "circle", which seems to be the right definition for "infinite cycle". He uses the Freudenthal compactification of an infinite locally finite graph.

An *end* of an infinite graph is an equivalence class of rays, where two rays are *equivalent* if there are infinitely many vertex disjoint paths between them. An end may be thought of as a point at infinity to which its rays converge. Viewing an infinite graph as a 1-complex and compactifying it by adding its ends as points at infinity, Freudenthal defined a topological space, which is compact and Hausdorff when the graph is locally finite. This space is called the *Freudenthal compactification* of that infinite graph. Diestel [23] defined a *circle* in a graph G (finite or infinite) as a homeomorphic image of the unit circle in the Freudenthal compactification of G. A *Hamilton circle* is then defined to be a circle containing all vertices and ends. With this definition, Bruhn (see [23]) made the following

Conjecture 4.1. (Bruhn, see [23]) *Every* 4*-connected infinite locally finite planar graph admits a Hamilton circle.*

The results in [77, 78] show that Conjecture 4.1 holds for 2-indivisible graphs. Bruhn and Yu [9] proved Conjecture 4.1 for 6-connected graphs with finitely many ends. Cui, Wang and Yu [19] proved Conjecture 4.1 for graphs containing no dividing cycles. (A cycle in an infinite plane graph is *dividing* if each closed region of the plane bounded by this cycle contains infinitely many vertices of the graph.)

5 3-connected planar graphs

Theorem 2.1 no longer holds for 3-connected graphs, as shown by Tutte [70]. Indeed, there are many 3-connected planar graphs containing no Hamilton cycles, see [34]. Hence, we are interested in bounding the circumference of such graphs. Jackson and Wormald [38] showed that if G is a 3-connected planar n-vertex graph then $c(G) = \Omega(n^{\beta})$, where $\beta \approx 0.2$, improving previous bounds in [3, 8, 37]. Gao and Yu [28] improved β to 0.4.

Chen and Yu [16] further improved β to $\log_3 2$, which establishes an old conjecture (implicit in Moon and Moser [52] and explicit in Grünbaum and Walther [33]). The exponent $\log_3 2$ is best possible as is evident from the class of graphs obtained from K_4 by repeatedly applying the following operation: For each face F, insert a vertex v_F inside F, and add an edge between v_F and each vertex incident with F.

In fact, Chen and Yu [16] proved that if G is a 3-connected graph embeddable in the torus or the Klein bottle then $c(G) = \Omega(n^{\log_3 2})$. Sheppardson and Yu [61] generalized this result to all 3-connected graphs with large face-width in an orientable surface. Kawarabayashi [42] did the same for non-orientable surfaces. Böhme, Mohar and Thomassen [6] showed that if G is a 3-connected n-vertex graph embedded in a surface of Euler genus g then $c(G) \geq \alpha(g)n^{\log_3 2}$, where $\alpha(g)$ is a constant dependent on g.

We conclude this section with two interesting conjectures about Hamilton cycles in cubic 3-connected planar graphs. Recall that Tutte [70] gave a cubic 3-connected planar graph that does not contain a Hamilton cycle. That example is not bipartite. Tutte conjectured that every cubic 3-connected bipartite graph is Hamiltonian, but this was disproved by Horton [36]. Barnette [4] proposed the following conjecture which is a weakened combination of Tait's and Tutte's conjectures; some related work may be found in Goodey [30, 31].

Conjecture 5.1. (Barnette [4]) *Every cubic 3-connected bipartite planar graph has a Hamilton cycle.*

A graph is *cyclically* 4-edge-connected if for every set S of at most three edges, at most one component of $G - S$ contains a cycle. The following conjecture is due to Bondy (see [37]).

Conjecture 5.2. (Bondy, see [37]) *There is a constant $\alpha > 0$ such that every cubic 3-connected cyclically 4-edge-connected planar n-vertex graph has a cycle of length αn.*

6 Graphs without $K_{3,t}$-minors

Planar graphs are characterized by the well-known theorem of Kuratowski [45]. Wagner [74] gave an alternative characterization of planar graphs using forbidden minors. A graph H is said to be a *minor* of another graph G if H is isomorphic to a graph that can be obtained from a subgraph of G by contracting edges. Thus, Wagner's theorem states that a graph is planar iff it does not contain $K_{3,3}$ or K_5

as a minor. For a 3-connected graph $G \neq K_5$, G is planar, or G does not contain a $K_{3,3}$-minor. Thus we know from [16] that 3-connected n-vertex graphs without $K_{3,3}$-minor have circumference $\Omega(n^{\log_3 2})$.

It is natural to study the circumference for 3-connected graphs without $K_{3,t}$-minor (for $t \geq 4$). A structural characterization of all graphs with no $K_{3,t}$-minors is not known and is probably difficult to obtain. Oporowski, Oxley, and Thomas [56] proved that if a 3-connected graph contains no $K_{3,t}$-minor, then it must contain a large wheel. Motivated by this result, Thomas [63] conjectured that there exist functions $\alpha(t) > 0$ and $\beta(t) > 0$ such that, for any integer $t \geq 3$ and any 3-connected n-vertex graph G with no $K_{3,t}$-minor, $c(G) \geq \alpha(t)n^{\beta(t)}$. This conjecture was proved by Chen, Sheppardson, Yu and Zang [14].

Seymour and Thomas [60] further conjectured that there exist function $\alpha(t) > 0$ and constant $\beta > 0$ such that, for any integer $t \geq 3$ and any 3-connected n-vertex graph G with no $K_{3,t}$-minor, $c(G) \geq \alpha(t)n^{\beta}$. In [17], Chen, Yu and Zang prove this conjecture with $\alpha(t) = (1/2)^{t(t-1)}$ and $\beta = \log_{1729} 2$. We believe that the base 1729 is far from being best possible and that $\beta = \log_3 2$ might be the truth.

Conjecture 6.1. (Chen, Yu and Zang [17]) *There exists function $\alpha(t) > 0$ such that, for any integer $t \geq 3$ and any 3-connected n-vertex graph G with no $K_{3,t}$-minor, $c(G) \geq \alpha(t)n^{\log_3 2}$.*

We also speculate that the following may be true.

Conjecture 6.2. (Chen, Yu and Zang [17]) *There exists a function $\alpha(t) > 0$ such that, for any integer $t \geq 4$ and any 4-connected n-vertex graph G with no $K_{3,t}$-minor, $c(G) \geq \alpha(t)n$.*

We mention that the arguments in [14, 17] use a result of Tutte [73] and Maclaine [47] about the decomposition of 2-connected graphs into 3-connected "components". Hopcroft and Tarjan [35] gave a linear time algorithm for obtaining such a decomposition. These results are also used for bounding the circumference of graphs with bounded maximum degree.

7 Graphs with bounded maximum degree

Karger, Motwani and Ramkumar [40] showed that unless $P = NP$ it is impossible to find, in polynomial time, a cycle of length $n - n^{\epsilon}$ (for any $\epsilon < 1$) in an n-vertex Hamiltonian graph. They conjectured that this is also the case when restricted to graphs with bounded maximum degree.

Conjecture 7.1. (Karger, Motwani and Ramkumar [40]) *Unless $P = NP$ it is impossible to find, in polynomial time, a cycle of length $n - n^{\epsilon}$ (for any $\epsilon < 1$) in an n-vertex Hamiltonian graph with bounded maximum degree.*

On the positive side, Feder, Motwani and Subi [25] gave a polynomial time algorithm for finding long cycles in 3-connected cubic graphs. Jackson and Wormald [39] showed that if G is a 3-connected n-vertex graph with maximum degree at most d, then $c(G) \geq \frac{1}{2}n^r + 1$, where $r = \log_{6d^2} 2$. Chen, Xu and Yu [15] gave a cubic algorithm which, given any 3-connected n-vertex graph with maximum degree at most d, finds a cycle in G of length $\Omega(n^{\log_\beta 2})$, where $\beta = d^2 - d + 1$. Later, Chen, Gao, Yu and Zang [13] improved β to $\max\{64, 4d + 1\}$. (The arguments used in these papers depend on a decomposition theorem of Tutte [73] and Maclaine [47], as well as an algorithm of Hopcroft and Tarjan [35] for obtaining such a decomposition.) Jackson and Wormald [39] showed that when $d \geq 4$ the best possible value for β could be $d - 1$, and made the following conjecture.

Conjecture 7.2. (Jackson and Wormald [39]) *If $d \geq 4$ is an integer and G is a 3-connected n-vertex graph with maximum degree at most d, then $c(G) = \Omega(n^{\log_{d-1} 2})$.*

8 Cubic graphs

Barnette [3] showed that every cubic 3-connected graph of order n has circumference $\Omega(\log n)$. Bondy and Simonovits [8] improved this bound to $\exp(\Omega(\sqrt{\log n}))$ and conjectured that it can be improved further to $\Omega(n^\alpha)$ for some constant $0 < \alpha < 1$. This conjecture was confirmed by Jackson [37], with $\alpha = \log_2(1+\sqrt{5})-1 \approx 0.694$. Bondy and Simonovits [8] used the Petersen graph to construct an infinite family of 3-connected cubic graphs with circumference $\Theta(n^{\log_9 8}) \approx \Theta(n^{0.946})$. Thus, there is still a considerable gap between the two exponents.

Bilinski, Jackson, Ma and Yu [5] showed that every 3-connected cubic graph of order n has circumference $\Omega(n^\alpha)$, where $\alpha \approx 0.753$ is the real root of $4^{1/x} - 3^{1/x} = 2$. This is proved by reducing the problem to one about Eulerian subgraphs in 3-edge-connected graphs. Recently, Liu, Yu and Zhang [46] proved that if G is a 3-connected n-vertex cubic graph then $c(G) = \Omega(n^r)$, where $r \approx 0.8$ is the real root of $8.956^r + 1.036^r = 10.992^r$. Here, we used a more direct approach by studying the structure of a possible minimum counterexample to this result. Such structural information makes us believe that the following should be true.

Conjecture 8.1. (Liu, Yu and Zhang [46]) *If G is a cubic 3-connected n-vertex graph then $c(G) = \Omega(n^{\log_5 4})$.*

However, the approach in [5] has its own merit. Let G be a cubic graph. We obtain a claw-free cubic graph G' from G by replacing each vertex with a triangle in a natural way. Then, $c(G) = \Theta(n^\alpha)$ iff $c(G') = \Theta(n^\alpha)$. We then obtain a line graph $L(H)$ (of a graph H) from G' by performing the closure operation. Using a result of Ryjáček [58], we reduce the problem of finding a long cycle in G' to the problem of finding a large Eulerian subgraph in H. The closure operation was used by Ryjáček [58] to show that the following two conjectures are equivalent.

Conjecture 8.2. (Mathews and Sumner [50]) *Every 4-connected claw-free graph contains a Hamilton cycle.*

Conjecture 8.3. (Thomassen [68]) *Every 4-connected line graph contains a Hamilton cycle.*

References

[1] A. Altshuler, Hamiltonian circuits in some maps on the torus, *Discrete Math.* **1** (1972), 299–314.

[2] D. Archdeacon, N. Hartsfield, and C. H. C. Little, Non-hamiltonian triangulations with large connectivity and representativity, *J. Combin. Theory Ser. B* **68** (1996), 45–55.

[3] D. Barnette, Trees in polyhedral graphs, *Canad. J. Math.* **18** (1966), 731–736.

[4] D. W. Barnette, Conjecture 5, in *Recent Progress in Combinatorics: Proceedings of the Third Waterloo Conference on Combinatorics* (Ed. W. T. Tutte), May 1968, Academic Press, New York.

[5] M. Bilinski, B. Jackson, J. Ma, and X. Yu, Circumference of 3-connected claw-free graphs and large Eulerian subgraphs of 3-edge connected graphs, *J. Combin. Theory Ser. B* **101** (2011), 214–236.

[6] T. Böhme, B. Mohar, and C. Thomassen, Long cycles in graphs on a fixed surface, *J. Combin. Theory Ser. B* **85** (2002), 338–347.

[7] B. Bollobás, *Modern Graph Theory*, Graduate Texts in Mathematics, Vol. **184**, Springer, 1998.

[8] J. A. Bondy and M. Simonovits, Longest cycles in 3-connected cubic graphs, *Canad. J. Math.* **32** (1980), 987–992.

[9] H. Bruhn and X. Yu, Hamilton circles in planar locally finite graphs, *SIAM J. Discrete Math.* **22** (2008), 1381–1392.

[10] R. Brunet, A. Nakamoto, and S. Negami, Every 5-connected triangulations of the Klein bottle is Hamiltonian, *Yokohama Mathematical Journal* **47** (1999), 239–244.

[11] R. Brunet and R. B. Richter, Hamiltonicity of 5-connected toroidal triangulations, *J. Graph Theory* **20** (1995), 267–286.

[12] G. Chen, G. Fan, and X. Yu, Cycles in 4-connected planar graphs, *Europ. J. Combin.* **25** (2004), 763–780.

[13] G. Chen, Z. Gao, X. Yu, and W. Zang, Approximating longest cycles in graphs with bounded degree, *SIAM J. Computing* **36** (2006), 635–656.

[14] G. Chen, L. Sheppardson, X. Yu, and W. Zang, Circumferences of 3-connected graphs with no $K_{3,t}$-minors, *J. Combin. Theory Ser. B* **96** (2006), 822–845.

[15] G. Chen, J. Xu, and X. Yu, Circumferences of graphs with bounded degree, *SIAM J. Computing* **33** (2004), 1136–1170.

[16] G. Chen and X. Yu, Long cycles in 3-connected graphs, *J. Combin. Theory Ser. B* **86** (2002), 80–99.

[17] G. Chen, X. Yu, and W. Zang, The circumference of a graph with no $K_{3,t}$-minor, II, *J. Combin. Theory Ser. B* **102** (2012), 1211–1240.

[18] N. Chiba and T. Nishizeki, The Hamiltonian cycle problem is linear-time solvable for 4-connected planar graphs, *J. Algorithms* **10** (1989), 187–211.

[19] Q. Cui, J. Wang, and X. Yu, Hamilton circles in infinite planar graphs, *J. Combin. Theory Ser. B* **99** (2009), 110–138.

[20] N. Dean, Lecture at Twenty-First Southeastern Conference on Combinatorics, Graph Theory and Computing, Boca Raton, Florida, 1990.

[21] N. Dean, R. Thomas, and X. Yu, Spanning paths in infinite planar graphs, *J. Graph Theory* **23** (1996), 163–177.

[22] Y. Diao, C. Ernst, and X. Yu, Hamilton knot projections and lengths of thick knots, *Topology and its applications* **28** (2004), 7–36.

[23] R. Diestel, The cycle space of an infinite graph, *Combinatorics, Probability and Computing* **14** (2005), 59–79.

[24] R. Diestel, *Graph Theory*, Graduate Texts in Mathematics, Vol. **173**, Springer-Verlag, Heidelberg, 2010.

[25] T. Feder, R. Motwani, and C. Subi, Approximating the longest cycle problem in sparse graphs, *SIAM J. Comput.* **31** (2002), 1596–1607.

[26] M. Fontet, Graphes 4-essential, *C. R. Acad. Sci. Paris* **287** (1978), 289–290.

[27] J. Fujisawa, A. Nakamoto, and K. Ozeki, Hamiltonian cycles in bipartite toroidal graphs with a partite set of degree four vertices, *J. Combin. Theory Ser. B* **103** (2013), 46–60.

[28] Z. Gao and X. Yu, Convex programming and circumference of 3-connected graphs of low genus, *J. Combin. Theory Ser. B* **69** (1997), 39–51.

[29] M. R. Garey and D. S. Johnson, *Computers and Intractability: A Guide to the Theory of NP-Completeness*, A Series of Books in the Mathematical Sciences, W. H. Freeman and Co., 1979.

[30] P. R. Goodey, Hamiltonian circuits on simple polytopes, *J. London Math. Soc.* (2) **5** (1972), 504–510.

[31] P. R. Goodey, Hamiltonian circuits in polytopes with even sided faces, *Israel J. Math.* **22** (1975), 52–56.

[32] B. Grünbaum, Polytopes, graphs, and complexes, *Bull. Amer. Math. Soc.* **76** (1970), 1131–1201.

[33] B. Grünbaum and H. Walther, Shortness exponents of families of graphs, *J. Combin. Theory Ser. A* **14** (1973), 364–385.

[34] D. A. Holton and B. D. McKay, The smallest non-hamiltonian 3-connected cubic planar graphs have 38 vertices, *J. Combin. Theory Ser. B* **45** (1988), 305–319.

[35] J. E. Hopcroft and R. E. Tarjan, Dividing a graph into triconnected components, *SIAM J. Computing* **2** (1973), 135–158.

[36] J. D. Horton, On two-factors of bipartite regular graphs, *Discrete Mathematics* **41** (1982), 35–41.

[37] B. Jackson, Longest cycles in 3-connected cubic graphs, *J. Combin. Theory Ser. B* **41** (1986), 17–26.

[38] B. Jackson and N. Wormald, Longest cycles in 3-connected planar graphs, *J. Combin. Theory Ser. B* **54** (1992), 291–321.

[39] B. Jackson and N. C. Wormald, Longest cycles in 3-connected graphs of bounded maximum degree, in *Graphs, Matrices, and Designs* (R. S. Rees, editor), Marcel and Dekker, Inc (1993), 237–254.

[40] D. Karger, R. Motwani, and G. D. S. Ramkumar, On approximating the longest path in a graph, *Algorithmica* **18** (1997), 82–98.

[41] K. Kawarabayashi, Private communication, 2004.

[42] K. Kawarabayashi, On longest cycles in 3-connected graphs in a fixed surface, preprint, 2005.

[43] K. Kawarabayashi and K. Ozeki, Hamilton cycles in 4-connected toroidal triangulations, preprint.

[44] K. Kawarabayashi and K. Ozeki, 4-connected projective-planar graphs are hamiltonian-connected, preprint.

[45] K. Kuratowski, Sur le probléme des courbes gauches en topologie, *Fund. Math.* (in French) **15** (1930), 271–283.

[46] Q. Liu, X. Yu, and Z. Zhang, Circumference of 3-connected cubic graphs, submitted.

[47] S. Maclaine, A structural characterization of planar combinatorial graphs, *Duke Math. J.* **3** (1937), 460–472.

[48] J. Malkevitch, Polytopal graphs, in *Selected Topics in Graph Theory*, Vol. **3** (Beineke and Wilson, eds.), Academic Press, London, 1988, 169–188.

[49] N. Martinov, Uncontractible 4-connected graphs, *J. Graph Theory* **6** (1982), 343–344.

[50] M. M. Mathews and D. P. Sumner, Hamiltonian results in $K_{1,3}$-free graphs, *J. Graph Theory* **8** (1984), 139–146.

[51] B. Mohar and C. Thomassen, *Graphs and Surfaces*, Johns Hopkins University Press, Baltimore, 2001.

[52] J. W. Moon and L. Moser, Simple paths on polyhedra, *Pacific J. Math.* **13** (1963), 629–631.

[53] A. Nakamoto and K. Ozeki, Hamilton cycles in bipartite quadrangulations on the torus, *J. Graph Theory* **69** (2012), 143–151.

[54] C. St. J. A. Nash-Williams, Hamiltonian lines in infinite graphs with few vertices of small valency, *Aequations Math.* **7** (1972), 59–81.

[55] C. St. J. A. Nash–Williams, Unexplored and semi-explored territories in graph theory, in *New Directions in Graph Theory*, Academic Press, New York, 1973, 169–176.

[56] B. Oporowski, J. Oxley, and R. Thomas, Typical subgraphs of 3- and 4-connected graphs, *J. Combin. Theory Ser. B* **57** (1993), 239–257.

[57] M. D. Plummer, Problems, in *Infinite and Finite Sets*, Colloq. Math. Soc. J. Bolyai 10, Vol. **III** (A. Hajnal, R. Rado and V. T. Sós, Eds.), North Holland, Amsterdam, 1975, 1549–1550.

[58] Z. Ryjáček, On a closure concept in claw-free graphs, *J. Combin. Theory Ser. B* **70** (1997), 217–224.

[59] D. P. Sanders, On paths in planar graphs, *J. Graph Theory* **24** (1997), 341–345.

[60] P. D. Seymour and R. Thomas, Private communication, 1992.

[61] L. Sheppardson and X. Yu, Long cycles in 3-connected graphs in orientable surfaces, *J. Graph Theory* **41** (2002), 69–84.

[62] P. G. Tait, Remarks on colouring maps, *Proc. Roy. Soc. Edinburgh Ser. A* **10** (1880), 729.

[63] R. Thomas, Private communication, 1992.

[64] R. Thomas and X. Yu, 4-connected projective-planar graphs are hamiltonian, *J. Combin. Theory Ser. B* **62** (1994), 114–132.

[65] R. Thomas and X. Yu, 5-connected toroidal graphs are Hamiltonian, *J. Combin. Theory Ser. B* **69** (1997), 79–96.

[66] R. Thomas, W. Zang, and X. Yu, Hamilton paths in toroidal graphs, *J. Combin. Theory Ser. B* **94** (2005), 214–236.

[67] C. Thomassen, A theorem on paths in planar graphs, *J. Graph Theory* **7** (1983), 169–176.

[68] C. Thomassen, Reflections on graph theory, *J. Graph Theory* **10** (1986), 309–324.

[69] C. Thomassen, Trees in triangulations, *J. Combin. Theory Ser. B* **60** (1994), 56–62.

[70] W. T. Tutte, On Hamilton circuits, *J. London Math. Soc.* **21** (1946), 98–101.

[71] W. T. Tutte, A theorem on planar graphs, *Trans. Amer. Math. Soc.* **82** (1956), 99–116.

[72] W. T. Tutte, A theory of 3-connected graphs, *Indag. Math.* **23** (1961), 441–455.

[73] W. T. Tutte, *Connectivity in Graphs*, University of Toronto Press, 1966.

[74] K. Wagner, Über eine Eigenschaft der ebenen Komplexe, *Math. Ann.* **114** (1937), 570–590.

[75] H. Whitney, A theorem on graphs, *Ann. of Math.* **32** (1931), 378–390.

[76] X. Yu, Disjoint paths, planarizing cycles, and spanning walks, *Trans. Amer. Math. Soc.* **349** (1997), 1333–1358.

[77] X. Yu, Infinite paths in planar graphs I, graphs with radial nets, *J. Graph Theory* **47** (2004), 147–162.

[78] X. Yu, Infinite paths in planar graphs II, structures and ladder nets, *J. Graph Theory* **48** (2005), 247–266.

[79] X. Yu, Infinite paths in planar graphs III, 1-way infinite paths, *J. Graph Theory* **51** (2006), 175–198.

[80] X. Yu, Infinite paths in planar graphs IV, dividing cycles, *J. Graph Theory* **53** (2006), 173–195.

[81] X. Yu, Infinite paths in planar graphs V, 3-indivisible graphs, *J. Graph Theory* **57** (2008), 275–312.

[82] Y. Zhao, Private communication, 1994.

School of Mathematics, Georgia Institute of Technology
Atlanta, GA 30332-0160, USA
E-mail address: yu@math.gatech.edu

Proceedings of the 6th International Congress of Chinese Mathematicians
ALM 36, pp. 407–424

Extremal Graph Theory for Degree Sequences

Xiao-Dong Zhang

Abstract

This paper surveys some recent results and progress on the extremal problems in a given set consisting of all simple connected graphs with the same graphic degree sequence. In particular, we study and characterize the extremal graphs having the maximum (or minimum) values of graph invariants such as (Laplacian, p-Laplacian, signless Laplacian) spectral radius, the first Dirichlet eigenvalue, the Wiener index, the Harary index, the number of subtrees and the chromatic number etc, in given sets with the same tree, unicyclic, graphic degree sequences. Moreover, some conjectures are included.

Mathematics Subject Classification: 05C50, 05C05, 05C07, 05C12, 05C35.

Keywords and Phrases: Graphic degree sequence, majorization, tree, unicyclic graphs, spectral radius, Wiener index.

1 Introduction

A nonincreasing sequence of nonnegative integers $\pi = (d_0, d_1, \ldots, d_{n-1})$ with $d_0 \geq d_1 \geq \cdots \geq d_{n-1}$ is called *graphic* if there exists a simple graph having π as its vertex degree sequence and such a graph is called a *realization* of π. Erdös and Gallai [17] characterized all graphic degree sequences.

Theorem 1.1. ([17]) *A nonincreasing sequence of nonnegative integers* $\pi = (d_0, d_1, \ldots, d_{n-1})$ *is graphic degree sequence if and only if* $\sum_{i=0}^{n-1} d_i$ *is even and*

$$\sum_{i=0}^{r} d_i \leq r(r+1) + \sum_{i=r+1}^{n-1} \min\{r+1, d_i\},$$

Supported by National Natural Science Foundation of China (No.11271256) and Innovation Program of Shanghai Municipal Education Commission (No:14ZZ016).

for $r = 0, 1, \ldots, n-2$.

Moreover, Senior [41] and Hakimi[24] characterized all graphic degree sequences with at least a realization being connected.

Theorem 1.2. ([41, 24]) *A nonincreasing sequence of nonnegative integers* $\pi = (d_0, d_1, \ldots, d_{n-1})$ *is graphic sequence with at least a realization being connected if and only if* $\sum_{i=0}^{n-1} d_i$ *is even and*

$$\sum_{i=0}^{r} d_i \leq r(r+1) + \sum_{i=r+1}^{n-1} \min\{r, d_i\},$$

for $r = 0, 1, \ldots, n-2$, $\sum_{i=0}^{n-1} d_i = \frac{n(n-1)}{2}$ *and* $d_{n-1} \geq 1$.

Let π be graphic degree sequence with at least a realization being connected. Denote by $\mathcal{G}(\pi)$ the set of all connected graphs which are realizations of π, i.e.,

$$\mathcal{G}(\pi) = \{G \mid G \text{ is connected and has } \pi \text{ as its degree sequence}\}. \tag{1}$$

Sierksma and Hoogeveen [42] surveyed seven criteria for a nonnegative integer sequence being graphic. For graphic sequences, Ferrara [19] surveyed recent research progress and new results on graphic sequences and presented a number of approachable open problems.

Extremal graph theory is an extremal important branch of graph theory. One is interested in relations among the various graph invariants, such as order, size, connectivity, chromatic number, diameter and eigenvalues, and also in the values of these invariants which ensure that the graph has certain properties. Bollobás published an excellent book [7] and recent survey in [8]. Erdös, Jacobson and Lehel in [18] proposed the problem on the possible clique number attained by graphs with the same degree sequence, i.e., determine the smallest integer number M so that each graphic sequence $\pi = (d_0, \ldots, d_{n-1})$ with $d_0 \geq d_1 \geq \cdots \geq d_{n-1} \geq 1$ and $\sum_{i=0}^{n-1} d_i \geq M$ has a *realization* G having given clique number. Li and Yin [32] surveyed extremal graph theory and degree sequences, in particular for the above problem and its variants.

On the other hand, many graphs coming from the real world have power law degree distribution: the number of vertices with degree k is proportional for $k^{-\beta}$ for some exponent $\beta \geq 1$. Chung and Lu (for example, see [11, 12]) proposed the famous Chung-Lu's model: random graphs with given expected degree sequences and studied their properties such as diameters, eigenvalues, the average distance, etc. In this survey, we just focus on some extremal properties of graphs with given degree sequences. Graph invariants such as the spectral radius, the Dirichlet eigenvalues, the Wiener index, the Harary index, the number of subtrees, etc., will be considered. Graphic degree sequences such as tree degree sequences, unicyclic degree sequences, etc., will be involved. Moreover, the majorization theorem on two different degree sequences are obtained. Therefore, we may propose the following general problem.

Problem 1.3. *For a given graphic degree sequence* π, *let*

$$\mathcal{G}_\pi = \{G \mid G \text{ is connected with } \pi \text{ as its degree sequence}\}.$$

For some graph invariants, such as spectral radius, the Wiener index, etc., find the maximum (minimum) value of these graph invariants in $\mathcal{G}_\pi$ *and characterize all extremal graphs which attain these values.*

2 Preliminary

In this section, we introduce some notions and properties. For a graph $G = (V, E)$ with a root v_0, denote by $\text{dist}(v, v_0)$ the distance between vertices v and v_0. Moreover, the distance $\text{dist}(v, v_0)$ is called the *height* $h(v) = \text{dist}(v, v_0)$ of a vertex v.

Definition 2.1. ([4, 61]) *Let* $G = (V, E)$ *be a graph with root* v_0. *A well-ordering* $\prec$ *of the vertices is called a breadth-first search ordering (BFS-ordering for short) if the following holds for all vertices* $u, v \in V$:
(1) $u \prec v$ *implies* $h(u) \le h(v)$;
(2) $u \prec v$ *implies* $d(u) \ge d(v)$;
(3) *let* $uv \in E(G)$, $xy \in E(G)$, $uy \notin E(G), xv \notin E(G)$ *with* $h(u) = h(x) = h(v) - 1 = h(y) - 1$. *If* $u \prec x$, *then* $v \prec y$.
We call a graph that has a BFS-ordering of its vertices a BFS-graph.

For example, for a given degree sequence

$$\pi = (4, 4, 3, 3, 3, 3, 2, 1, 1, 1, 1, 1, 1, 1, 1, 1, 1),$$

T_π^* is the tree of order 17 (see Fig.1). There is a vertex v_{01} in layer 0; four vertices $v_{11}, v_{12}, v_{13}, v_{14}$ in layer 1; nine vertices $v_{21}, v_{22}, \ldots, v_{29}$ in layer 2; three vertices v_{31}, v_{32}, v_{33} in layer 3.

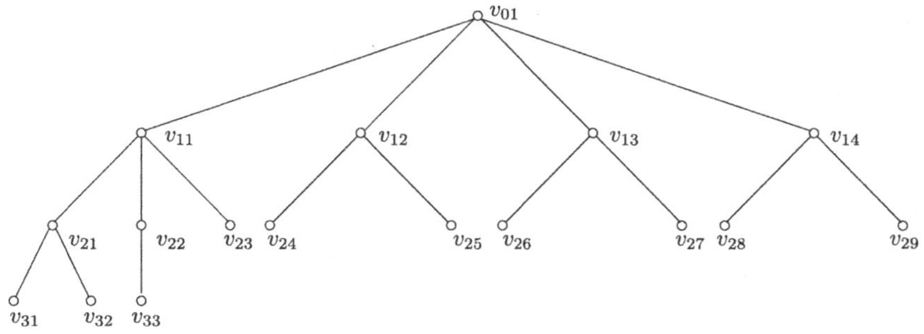

Fig. 1

It is easy to see that every tree has an ordering such that it satisfies the conditions (1) and (3) by using breadth-first search, but not every tree has a BFS-ordering. For example, the following tree T of order 17 does not have a

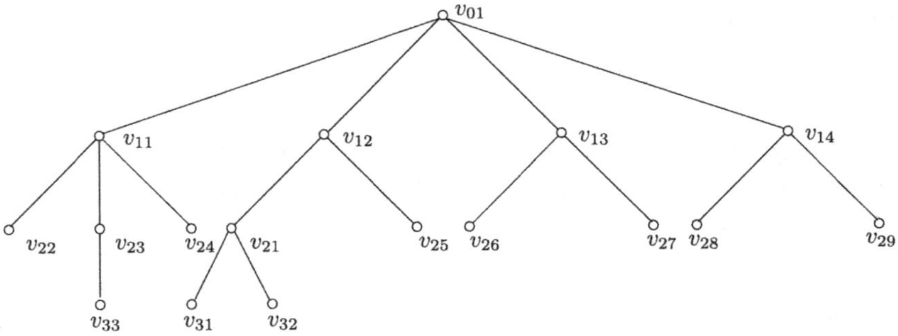

Fig. 2

BFS-ordering with degree sequence $\pi = (4, 4, 3, 3, 3, 3, 2, 1, 1, 1, 1, 1, 1, 1, 1, 1, 1)$ (see Fig.2).

Moreover, it is easy to see [4] that there may be many BFS graphs for a given graphic degree sequence. Another importation notion is majorization. Let $x = (x_1, x_2, \ldots, x_p)$ and $y = (y_1, y_2, \ldots, y_p)$ be two nonnegative integers. We arrange the entries of x and y in nonincreasing order $x_\downarrow = (x_{[1]}, \ldots, x_{[p]})$ and $y_\downarrow = (y_{[1]}, \ldots, y_{[p]})$. If

$$\sum_{i=1}^{k} x_{[i]} \geq \sum_{i=1}^{k} y_{[i]}, \quad \text{for } k = 1, \ldots, p, \tag{2}$$

x is said to *weakly majorize* y and denoted $x \rhd^w y$ or $y \lhd^w x$. Further, if $y \lhd^w x$ and

$$\sum_{i=1}^{p} x_{[i]} = \sum_{i=1}^{p} y_{[i]}, \tag{3}$$

x is said to *majorize* y and denoted by $x \rhd y$ or $y \lhd x$. For details, the readers are referred to the book of Marshall and Olkin [35]. It is well known that the following result holds (see [17] or [53]).

Proposition 2.2. ([17, 53]) *Let $\pi = (d_0, \ldots, d_{n-1})$ and $\pi' = (d'_0, \ldots, d'_{n-1})$ be two different nonincreasing graphic degree sequences. If $\pi \lhd \pi'$, then there exists a series graphic degree sequences $\pi_1, \ldots, \pi_k$ such that $\pi \lhd \pi_1 \lhd \cdots \lhd \pi_k \lhd \pi'$, and only two components of π_i and π_{i+1} are different from 1.*

Let $G = (V, E)$ be a simple graph with vertex set $V(G) = \{v_1, \ldots, v_n\}$ and edge set $E(G)$. There are many matrices associated with a graph. Let $A(G) = (a_{ij})$ be the $(0, 1)$ *adjacency matrix* of G, where $a_{ij} = 1$ if v_i and v_j are adjacent, and 0 elsewhere. The spectral radius of $A(G)$ is denoted by $\rho(G)$ and called the *spectral radius* of G. Moreover, there exists a nonnegative unit eigenvector of $A(G)$ corresponding to $\rho(G)$. By the Perron-Frobenius theorem, f is unique and positive for connected graph G. So f is called *Perron vector* of G. In addition, denote by $d(v_i)$ the *degree* of vertex v_i and $D(G) = \text{diag}(d(u), u \in V)$ the diagonal matrix of vertex degrees of G, thus the matrix $L(G) = D(G) - A(G)$ is called the *Laplacian matrix* of a graph G. The spectral radius of $L(G)$ is equal to the largest eigenvalue of $L(G)$ and denoted by $\lambda(G)$. Moreover, $\lambda(G)$ is called the *Laplacian spectral radius* of G. $\mathcal{L}(G) = D^{-1/2}L(G)D^{-1/2}$ is call the *normal Laplacian matrix* of a graph. The spectral radius of $\mathcal{L}(G)$ is denoted by $\mu(G)$ and called the *normal Laplacian spectral radius.* In addition, the matrix $Q(G) = D(G) + A(G)$ is called the *signless Laplacian matrix* of G. The spectral radius of $Q(G)$ is denoted by $q(G)$ and called the *signless Laplacian spectra radius.* Let g be the unit eigenvector of $Q(G)$ corresponding to eigenvalue $q(G)$ and be called *signless Laplacian Perron vector* of G. The adjacency, Laplacian, normal Laplacian and signless Laplacian matrices play a key role in spectral graph theory. The book [13] by Cvetković, Doob abd Sachs focuses on the adjacency matrix. The book [37] by Molitierno studies many properties of Laplacian matrices. The book [10] by Chung focuses on normal Laplacian matrix. Moreover, there are several surveys on the topic of signless Laplacian matrices (see [14] and the references therein).

3 Eigenvalues with given tree degree sequences

If a nonnegative sequence $\pi = (d_0, \ldots, d_{n-1})$ is graphic and $\sum_{i=0}^{n-1} d_i = 2(n-1)$, π is called *tree degree sequence* and denoted by $\mathcal{T}_\pi$ the set of all trees having degree sequence π.

For a tree degree sequence $\pi = (d_0, d_1, \ldots, d_{n-1})$ with $d_0 \geq d_1 \geq \cdots \geq d_{n-1}$, with the aid of breadth-first search method, we can define a special tree T_π^* with degree sequence π as follows. Assume that $d_m > 1$ and $d_{m+1} = \cdots = d_{n-1} = 1$ for $0 \leq m < n-1$. Put $s_0 = 0$. Select a vertex v_{01} as a root and begin with v_{01} in layer 0. Put $s_1 = d_0$ and select s_1 vertices $\{v_{11}, \ldots, v_{1s_1}\}$ in layer 1 such that they are adjacent to v_{01}. Thus $d(v_{01}) = s_1 = d_0$. We continue to construct all other layers by recursion. In general, put $s_t = d_{s_0+s_1+\cdots+s_{t-2}+1} + \cdots + d_{s_0+s_1+\cdots+s_{t-2}+s_{t-1}} - s_{t-1}$ for $t \geq 2$ and assume that all vertices in layer t have been constructed and are denoted by $\{v_{t1}, \ldots, v_{ts_t}\}$ with $d(v_{t-1,1}) = d_{s_0+\cdots+s_{t-2}+1}, \ldots, d(v_{t-1,s_{t-1}}) = d_{s_0+\cdots+s_{t-1}}$. Now using the induction hypothesis, we construct all vertices in layer $t+1$. Put $s_{t+1} = d_{s_0+\cdots+s_{t-1}+1} + \cdots + d_{s_0+\cdots+s_t} - s_t$. Select s_{t+1} vertices $\{v_{t+1,1}, \ldots, v_{t+1,s_{t+1}}\}$ in layer $t+1$ such that $v_{t+1,i}$ is adjacent to v_{tr} for $r = 1$ and $1 \leq i \leq d_{s_0+\cdots+s_{t-1}+1} - 1$ and for $2 \leq r \leq s_t$ and $d_{s_0+\cdots+s_{t-1}+1} + d_{s_0+\cdots+s_{t-1}+2} + \cdots + d_{s_0+\cdots s_{t-1}+r-1} - r + 2 \leq i \leq d_{s_0+\cdots+s_{t-1}+1} + d_{s_0+\cdots+s_{t-1}+2} + \cdots + d_{s_0+\cdots+s_{t-1}+r} - r$. Thus $d(v_{tr}) = d_{s_0+\cdots+s_{t-1}+r}$ for $1 \leq r \leq s_t$. Assume that $m = s_0 + \cdots + s_{p-1} + q$. Put $s_{p+1} = d_{s_0+\cdots+s_{p-1}+1} + \cdots + d_{s_0+\cdots+s_{p-1}+q} - q$ and select s_{p+1} vertices $\{v_{p+1,1}, \ldots, v_{p+1,s_{p+1}}\}$ in layer $p+1$ such that $v_{p+1,i}$ is adjacent to v_{pr} for $1 \leq r \leq q$ and $d_{s_0+\cdots+s_{t-1}+1} + \cdots + d_{s_0+\cdots+s_{p-1}+2} + \cdots + d_{s_0+\cdots s_{p-1}+r-1} - r + 2 \leq i \leq d_{s_0+\cdots+s_{p-1}+1} + \cdots + d_{s_0+\cdots+s_{p-1}+2} + \cdots + d_{s_0+\cdots+s_{p-1}+r} - r$. Thus $d(v_{p,i}) = d_{s_0+\cdots+s_{p-1}+i}$ for $1 \leq i \leq q$. In this way, we obtain a tree T_π^*. It is easy to see that T_π^* has degree sequence π. Moreover, this special tree T_π^* is called *greedy tree* with a given tree degree sequence π. In addition, it is easy to show the following assertion holds.

Proposition 3.1. ([4, 60]) *Given a tree degree sequence π, there exists a unique tree T_π^* with degree sequence π having a BFS-ordering. In other words, the greedy tree with degree sequence π is only tree having a BFS-ordering in the set $\mathcal{T}_\pi$. More-*

over, any two trees with the same degree sequences and having BFS-ordering are isomorphic.

3.1 The (Laplacian, signless Laplacian) spectral radius

Biyikoğlu and Leydold [4] characterized all extremal trees having the largest spectral radius in the set $\mathcal{T}_\pi$ consisting of all trees with a given degree sequence π. While, Zhang [60] characterized all extremal trees having the largest (signless) Laplacian spectral radius in the set $\mathcal{T}_\pi$ consisting of all trees with a given degree sequence π, since the signless Laplacian radius of a tree T is always equal to the Laplacian spectral radius. It is not a surprise that the two extremal graphs are coincident.

Theorem 3.2. ([4, 60]) *For a given tree degree sequence π, the greedy tree T_π^* is the only tree having the largest spectral radius (resp. signless Laplacian radius) in $\mathcal{T}_\pi$. Moreover, The BFS-ordering of the greedy tree is consistent with the Perron vector f of G in such a way that $f(u) > f(v)$ (resp. $g(u) > g(v)$) implies $u \prec v$.*

Theorem 3.3. ([4, 60]) *Let π and τ be two different tree degree sequences with the same order. Let T_π^* and T_τ^* have the largest spectral radii (resp. Laplacian, signless Laplacian spectral radii) in $\mathcal{T}_\pi$ and $\mathcal{T}_\tau$, respectively. If $\pi \lhd \tau$, then $\rho(T_\pi^*) < \rho(T_\tau^*)$, $\lambda(T_\pi^*) < \lambda(T_\tau^*)$ and $q(T_\pi^*) < q(T_\tau^*)$.*

With the aid of Theorems 3.2 and 3.3, we may deduce extremal graphs with the largest (resp. signless Laplacian) spectral radius in some class of graphs.

Corollary 3.4. ([4, 60]) *Let $\mathcal{T}_{n,s}^{(1)}$ be the set of all trees of order n with s leaves and let T_π^* be the greedy tree with $\pi = (t, 2, \ldots, 2, 1, \ldots, 1)$ (i.e., if $n-1 = sq+t, 0 \le t < s$, then T_π^* is obtained from t paths of order $q+2$ and $s-t$ paths of order $q+1$ by identifying one end of the s paths). Then $\rho(T) \le \rho(T_\pi^*)$ (resp. $\lambda(T) \le \lambda(T_\pi^*)$, $q(T) \le q(T_\pi^*)$) for any tree $T \in \mathcal{T}_{n,s}^{(1)}$, with equality if and only if $T = T_\pi^*$.*

Corollary 3.5. ([60]) *Let $\mathcal{T}_{n,\Delta}^{(2)}$ be the set of all trees of order n with the maximum degree $\Delta \ge 3$ and let T_π^* be the greedy tree with π which is defined as follows: Denote $p = \lceil \log_{(\Delta-1)} \frac{n(\Delta-2)+2}{\Delta} \rceil - 1$ and $n - \frac{\Delta(\Delta-1)^p-2}{\Delta-2} = (\Delta-1)r + q$ for $0 \le q < \Delta - 1$. If $q = 0$, put $\pi = (\Delta, \ldots, \Delta, 1, \ldots, 1)$ with the number $\frac{\Delta(\Delta-1)^{p-1}-2}{\Delta-2} + r$ of degree Δ. If $1 \le q$, put $\pi = (\Delta, \ldots, \Delta, q, 1, \ldots, 1)$ with the number $\frac{\Delta(\Delta-1)^{p-1}-2}{\Delta-2} + r$ of degree Δ. Then $\rho(T) \le \rho(T_\pi^*)$ (resp. $\lambda(T) \le \lambda(T_\pi^*)$, $q(T) \le q(T_\pi^*)$) for any tree $T \in \mathcal{T}_{n,\Delta}^{(2)}$, with equality if and only if $T = T_\pi^*$.*

Corollary 3.6. ([59, 60]) *Let $\mathcal{T}_{n,\alpha}^{(3)}$ be the set of all trees of order n with the independence number α and let T_π^* be the greedy tree with $\pi = (\alpha, 2, \ldots, 2, 1, \ldots, 1)$ the numbers $n - \alpha - 1$ of 2 and α of 1. Then $\rho(T) \le \rho(T_\pi^*)$ (resp. $\lambda(T) \le \lambda(T_\pi^*)$, $q(T) \le q(T_\pi^*)$) for any tree $T \in \mathcal{T}_{n,\alpha}^{(3)}$, with equality if and only if $T = T_\pi^*$.*

Corollary 3.7. ([22, 60]) *Let $\mathcal{T}_{n,\beta}^{(4)}$ be the set of all trees of order n with the matching number β and T_π^* be the greedy tree with $\pi = (n-\beta, 2, \ldots, 2, 1, \ldots, 1)$ and* *the number* $n - \beta$ *of* 1. *Then* $\rho(T) \leq \rho(T_\pi^*)$ (*resp.* $\lambda(T) \leq \lambda(T_\pi^*)$, $q(T) \leq q(T_\pi^*)$) *for any tree* $T \in \mathcal{T}_{n,\beta}^{(4)}$, *with equality if and only if* $T = T_\pi^*$.

On the minimum (Laplacian, signless Laplacian) spectral radius of with a given tree degree sequence, it seems to be more difficult. The related results are referred to [6].

3.2 The spectral radius of p-Laplacian of weighted trees

Now we turn to consider p-Laplacian eigenvalues of weighted graphs. Let $G^W = (V(G), E(G), W(G))$ be a *weighted graph* with vertex set $V(G) = \{v_0, v_1, \ldots, v_{n-1}\}$, edge set $E(G)$ and weight set $W(G) = \{w_k > 0,\ k = 1,\ 2,\ \ldots,\ |E(G)|\}$. If $uv \in E(G)$, denote by $w_G(uv)$ the weight of an edge uv; if $uv \notin E(G)$, define $w_G(uv) = 0$. The weight of a vertex u, denoted by $w_G(u)$, is the sum of weights of all edges incident to u in G.

For $p > 1$, the *discrete p-Laplacian* $\Delta_p(G)$ of a function f on $V(G)$ is defined to be

$$\Delta_p(G)f(u) = \sum_{v,uv\in E(G)} (f(u) - f(v))^{[p-1]} w_G(uv), \tag{4}$$

where $x^{[q]} = \text{sign}(x)|x|^q$. When $p = 2$, $\Delta_2(G)$ is the well-known graph Laplacian (see [5]), i.e., $\Delta_2(G) = L(G) = D(G) - A(G)$, where $A(G) = (w_G(v_iv_j))_{n\times n}$ denotes the weighted adjacency matrix of G and $D(G) = \text{diag}(w_G(v_0),\ w_G(v_1),\ \ldots,\ w_G(v_{n-1}))$ denotes the weighted diagonal matrix of G.

A real number λ is called an *eigenvalue* of $\Delta_p(G)$ if there exists a function $f \neq 0$ on $V(G)$ such that for $u \in V(G)$,

$$\Delta_p(G)f(u) = \lambda f(u)^{[p-1]}. \tag{5}$$

The function f is called the *eigenfunction* corresponding to λ. The largest eigenvalue of $\Delta_p(G)$, denoted by $\lambda^p(G)$, is called the *p-Laplacian spectral radius*.

Given a tree degree sequence π and a positive set W, denote by $\mathcal{T}_{\pi,W}$ the set of trees with a given tree degree sequence π and a positive weight set W. Thus the un-weighted greedy tree T_π^* has a BFS-ordering $v_0 \prec v_2 \prec \cdots \prec v_{n-1}$. Hence the unique weighted greedy tree $T_{\pi,W}^*$ is obtained from the un-weighted greedy tree T_π^* whose edge weighted has the following property" for any two edges v_iv_j and v_kv_l with $i < j, k < l$, weight $w(v_iv_j) \geq w(v_kv_l)$ if $i < k$ or $i = k$ and $j < l$". Zhang and Zhang [55] proved the following results.

Theorem 3.8. ([55]) *For a given tree degree sequence* π *and a positive weight set* W, *the weighted greedy tree* $T_{\pi,W}^*$ *is the unique weighted tree with the largest p-Laplacian spectral radius in* $\mathcal{T}_{\pi,W}$, *which is independent of* p.

Theorem 3.9. ([55]) *Let* π *and* τ *be two different tree degree sequences. Let* $T_{\pi,W}^*$ *and* $T_{\tau,W}^*$ *be two weighted greedy trees in* $\mathcal{T}_{\pi,W}$ *and* $\mathcal{T}_{\tau,W}$ *respectively. If* $\pi \triangleleft \tau$, *then* $\lambda^p(T_{\pi,W}^*) < \lambda^p(T_{\tau,W}^*)$.

Remark. If the positive set W consists of all 1, Theorems 3.8 and 3.9 become Biyikoğlu, Hellmuth and Leydold's result [5]. If $p = 2$, Theorems 3.8 and 3.9

become Tan's result [48]. If the positive set W consists of all 1 and $p = 2$, Theorems 3.8 and 3.9 become Zhang' result[60]. In addition, Corollaries 3.4, 3.5, 3.6 and 3.7 can be easily generalized to the weighted trees, respectively. Moreover, the p-adjacency eigenvalue may be similarly defined and Theorems 3.8 and 3.9 can be translated to the p-adjacency eigenvalue.

3.3 The Dirichlet eigenvalue of trees with boundary

In this subsection, we consider the Dirichlet eigenvalues of trees with boundary. A *graph with boundary* $G = (V_0 \cup \partial V,\ E_0 \cup \partial E)$ consists of interior vertex set V_0, boundary vertex set ∂V, interior edge set E_0 that connect interior vertices, and boundary edge set ∂E that join interior vertices with boundary vertices (for example, see [10] or [21]). Throughout this subsection, we always assume that the degree of any boundary vertex is 1 and the degree of any interior vertex is at least 2.

A real number λ^D is called a *Dirichlet eigenvalue* of G if there exists a function $f \neq 0$ such that they satisfy the Dirichlet eigenvalue problem:

$$\begin{cases} L(G)f(u) = \lambda f(u) & u \in V_0; \\ f(u) = 0 & u \in \partial V. \end{cases}$$

The function f is called an *eigenfunction* corresponding to λ^D. Recently, Biyikoğlu and Leydold [3] proposed the following problem:

Problem 3.10. ([3]) *Give a characterization of all graphs in a given class $\mathcal{C}$ with the Faber-Krahn property, i.e., characterize those graphs in $\mathcal{C}$ which have minimal first Dirichlet eigenvalue for a given "volume".*

In order to study the above problem, Biyikoğlu and Leydold [3] extended the concept of an SLO*-ordering for describing the trees with the Faber-Krahn property, which was introduced by Pruss (see [38]). The notation of an SLO*-ordering may be extended for any connected graphs.

Definition 3.11. ([3]) *Let $G = (V_0 \cup \partial V, E_0 \cup \partial E)$ be a connected graph with root v_0. Then a well-ordering $\prec$ of the vertices is called spiral-like (SLO*-ordering for short) if the following holds for all vertices $u, v, x, y \in V_0 \cup \partial V$:*

(1) *$v \prec u$ implies $h(v) \leq h(u)$, where $h(v)$ denotes the distance between v and v_0.*

(2) *Let $uv \in E(G)$, $xy \in E(G), uy \notin E(G)$, $xv \notin E(G)$ with $h(u) = h(v) - 1$ and $h(x) = h(y) - 1$. If $u \prec x$, then $v \prec y$.*

(3) *If $v \prec u$ and $v \in \partial V$, then $u \in \partial V$.*

(4) *If $v \prec u$ and $v, u \in V_0$, then $d(v) \leq d(u)$.*

Clearly, if G is a tree, an SLO*-ordering of G is consistent with the definition of an SLO*-ordering in [3]. Moreover, if there exists a positive integer r such that the number of vertices v with $h(v) = i+1$ is not less than the number of vertices v with $h(v) = i$ for $i = 1, \ldots, r-1$, and $h(u) \in \{r, r+1\}$ for any boundary vertex $u \in \partial V$, G is called a *ball approximation*. Given a tree degree sequence π, denote by $\mathcal{T}_{\pi,B}$

the set consisting of all trees with boundary and degree sequence π. Biyikoğlu and Leydold [3] proved that $\mathcal{T}_{\pi,B}$ contains an SLO*-tree that is uniquely determined up to isomorphism. This special tree in $\mathcal{T}_{\pi,B}$ is denoted by $T^*_{\pi,B}$.

Theorem 3.12. ([3]) *For a given tree degree sequence* π,

$$\lambda^D(T) \geq \lambda^D(T^*_{\pi,B}) \quad \text{for } T \in \mathcal{T}_{\pi,B}$$

with equality if and only if $T = T^*_{\pi,B}$.

Theorem 3.13. ([3]) *Let* π *and* τ *be two different tree degree sequences. If* $\pi \triangleleft \tau$, *then* $\lambda^D(T^*_{\pi,B}) > \lambda^D(T^*_{\tau,B})$.

4 Chemical indices for tree degree sequences

There are many indices which are used to describe molecules and molecular compounds in chemical graph theory. One of the most widely known topological descriptor is the *Wiener index* which is named after chemist Wiener [54] who firstly considered it. The *Wiener index* of a graph G is defined as

$$W(G) = \sum_{\{v_i,v_j\}\subseteq V(G)} d(v_i, v_j). \tag{6}$$

Let $T = (V, E)$ be a rooted tree with root r. For each vertex u, let $T(u)$ be the subtree of the rooted tree T induced by u and all its successors in T. In other words, if u is not the root r of tree T and v is the parent of u, then $T(u)$ is the connected component of T obtained from T by deleting the edge uv such that the component does not contain the root r; if u is the root r, then $T(u)$ is the tree T. Let $\phi_T(u) = |T(u)|$ be the number of vertices in $T(u)$ and denote $\phi(T) = (\phi_T(u), u \in V(T))$. We prove the following results.

Theorem 4.1. ([64]) *Let* T *be a rooted tree in* $\mathcal{T}_\pi$. *Then the following conditions are equivalent:*
(1) T *has a BFS-ordering;*
(2) $\phi(T)_\downarrow = \phi(T^*_\pi)_\downarrow$;
(3) T *is isomorphic to* T^*_π.

Further, this result can be used to characterize the trees having the minimum Wiener index among all trees with a given degree sequence. Wang [52] independently proved the same result by a different approach.

Theorem 4.2. ([52, 64]) *Given a tree degree sequence* π, *the greedy tree* T^*_π *is a unique tree with the minimum Wiener index in* $\mathcal{T}_\pi$.

Theorem 4.3. ([64]) *Let* π *and* τ *be two different tree degree sequences. Let* T^*_π *and* T^*_τ *be two greedy trees in* $\mathcal{T}_\pi$ *and* $\mathcal{T}_\tau$, *respectively. If* $\pi \triangleleft \tau$, *then* $W(T^*_\pi) > W(T^*_\tau)$.

On the other hand, it is natural to investigate the extremal trees which attain the maximum Wiener index among all trees with given degree sequences. But this problem seems to be difficult, since it is discovered that the extremal trees are not unique and depend on the values of components of degree sequences. However, there are some partial results. A *caterpillar* is a tree with the property that a path remains if all leaves are deleted and this path is called *spine*.

Theorem 4.4. ([43]) *If a tree $T_\pi^\sharp$ has the maximum Wiener index among all trees in $\mathcal{T}_\pi$ for a given tree degree sequence π, then $T_\pi^\sharp$ is a caterpillar.*

Theorem 4.5. ([63]) *Let $\pi = (d_0, \ldots, d_{n-1})$ with $d_0 \geq \cdots \geq d_{k-1} \geq 2 \geq d_{k+1} = \cdots = d_{n-1} = 1$. If a caterpillar $T_\pi^\sharp$ has the maximum Wiener index among all trees in $\mathcal{T}_\pi$ and the spine is $v_0 v_1 \cdots v_{k-1}$ with $d(v_0) \geq d(v_{k-1})$, then there exists a $1 \leq t \leq k-2$ such that $d(v_0) \geq d(v_1) \geq \cdots \geq d(v_t)$ and $d(v_t) \leq d(v_{t+1}) \leq \cdots \leq d(v_{k-1})$.*

Further, Schmuck, Wagner and Wang [40] proposed and studied a new graph invariant which are based on distance.

Theorem 4.6. ([40]) *Let $\pi = (d_0, d_1, \ldots, d_{n-1})$ be a tree degree sequence, let r be an arbitrary positive integer. If $p_r(T)$ is the number of pairs (u, v) of vertices such that $d(u, v) \leq r$ for a tree $T \in \mathcal{T}_\pi$, then $p_r(T) \leq p_r(T_\pi^*)$ with equality for all $r = 1, \ldots, n-1$ if and only if T is the greedy tree T_π^*.*

Recently, Schmuck, Wagner and Wang [40] proposed a general graph invariant

$$W_\psi(T) = \sum_{\{u,v\} \subseteq V(G)} \psi(d(u, v)) \tag{7}$$

for any nonnegative function ψ. Further, they proved the following result.

Theorem 4.7. ([40, 51]) *Let π be a tree degree sequence and $\psi(x)$ be any nonnegative and nondecreasing (resp. nonincreasing) function on x. Then for any tree $T \in \mathcal{T}_\pi$,*

$$W_\psi(T) \geq \text{(resp. } \leq \text{)}\ W_\psi(T_\pi^*). \tag{8}$$

Further, if $\psi(x)$ is strictly increasing function, then equality holds if and only if T is the greedy tree T_π^*.

Remark. If $\psi(x) = x$, then $W_\psi(T)$ is the classical Wiener index. If $\psi(x) = \frac{x(x+1)}{2}$, then $W_\psi(T)$ is the hyper-Wiener index [29]. If $\psi(x) = \frac{1}{x}$, then $W_\psi(T)$ is the Harary index. Moreover, it is easy to deduce the following result from [40] and [51].

Theorem 4.8. *Let π and τ be two different tree degree sequences with $\pi \triangleleft \tau$. If $\psi(x)$ is any nonnegative and strictly increasing (resp. decreasing) function on x, then*

$$W_\psi(T_\pi^*) > \text{(resp. } < \text{)}\ W_\psi(T_\tau^*). \tag{9}$$

Proof. Let $p_r(T)$ be the number of pairs $\{u, v\}$ of vertices with $d(u, v) \le r$. By Theorem in [51], $p_r(T_\pi^*) \le p_r(T_\tau^*)$ for $r = 1, \ldots, n$ with strict inequality for at least one r. Observe that

$$W_\psi(T_\pi^*) = \sum_{k \ge 0} (\psi(k+1) - \psi(k)) \left(\frac{n(n-1)}{2} - p_k(T_\pi^*) \right)$$

and

$$W_\psi(T_\tau^*) = \sum_{k \ge 0} (\psi(k+1) - \psi(k)) \left(\frac{n(n-1)}{2} - p_k(T_\tau^*) \right).$$

If $\psi(x)$ is nonnegative and strictly increasing, then $W_\psi(T_\pi^*) > W_\psi(T_\tau^*)$. □

Theorem 4.9. ([40]) *Let π be a tree degree sequence and $\psi(x)$ be strictly increasing and convex function on x. If $T_\pi^\sharp$ is a tree that maximizes $W_\psi(T)$ among all trees in the set $\mathcal{T}_\pi$, then $T_\pi^\sharp$ is a caterpillar.*

In order to study total π-electron energy on molecular structure, Gutman and Trinajstić [23] proposed the *second Zagreb index*, which is defined to be

$$ZA(T) = \sum_{uv \in E(T)} d(u)d(v), \tag{10}$$

where $d(u)$ and $d(v)$ are the degrees of vertices u and v, respectively.

Theorem 4.10. ([40]) *Let π be a tree degree sequence. Then the greedy tree T_π^* is the only tree having the maximum second zagreb index among all trees in $T \in \mathcal{T}_\pi$, i.e., for any tree $T \in \mathcal{T}_\pi$,*

$$ZA(T) \le ZA(T_\pi^*) \tag{11}$$

with equality if and only if T is the greedy tree T_π^*.

Let G be a simple graph. Denoted by $m(G, k)$ the number of k-matchings in G and $m(G, 0) = 1$. The *Hosoya index*, named after Haruo Hosoya, is defined to be

$$Z(G) = \sum_{k \ge 0} m(G, k). \tag{12}$$

Denote by $i(G, k)$ the number of k-independent sets in G and $i(G, 0) = 1$. The *Merrifield-Simmons index* is defined to be

$$MS(G) = \sum_{k \ge 0} i(G, k). \tag{13}$$

The *energy of graph* is defined by the sum of the absolute values of all the eigenvalues of $A(G)$, and denoted by $En(G)$. Andriantiana [1] proved that the greedy tree has minimum energy, Hosoya index and maximum Merrifield-Simmons index.

Theorem 4.11. ([1]) *Let π be a tree degree sequence. Then the greedy tree T_π^* is the only tree having minimum energy, minimum Hosoya index and maximum Merrifield-Simmons indices, respectively among all trees in $T \in \mathcal{T}_\pi$, i.e., for any tree $T \in \mathcal{T}_\pi$,*

$$Z(T) \geq Z(T_\pi^*), \quad En(T) \geq En(T_\pi^*), \quad MS(T) \leq MS(T_\pi^*), \tag{14}$$

with any one equality if and only if T is the greedy tree T_π^*.

Theorem 4.12. ([1]) *Let π and τ be two different tree degree sequences. Then the greedy trees T_π^* and T_τ^* are the minimum energy, minimum Hosoya index and maximum Merrifield-Simmons indices in $\mathcal{T}_\pi$ and $\mathcal{T}_\tau$ respectively. If $\pi \lhd \tau$, then*

$$En(T_\pi^*) > En(T_\tau^*), \quad Z(T_\pi^*) > Z(T_\tau^*), \quad MS(T_\pi^*) < MS(T_\tau^*). \tag{15}$$

In addition, there are the following results on the number of subtrees of a tree.

Theorem 4.13. ([65]) *With a given degree sequence π, T_π^* is the unique tree with the maximum number of subtrees in $\mathcal{T}_\pi$.*

Theorem 4.14. ([65]) *Given two different degree sequences π and τ. If $\pi \lhd \tau$, then the number of subtrees of T_π^* is less than the number of subtrees of T_τ^*.*

5 Unicyclic degree sequences

For a given nonincreasing degree sequence $\pi = (d_0, d_1, \ldots, d_{n-1})$ of a unicyclic graph with $n \geq 3$, we construct a special unicyclic graph U_π^* as follows: If $d_0 = 2$, denote by U_π^* the cycle of order n. If $d_0 \geq 3$, denote by U_π^* is the graph obtained from a cycle C_3 of order 3 and by breadth-first search ordering, i.e., U_π^* is BFS graph with the first three vertices, each pair of which is adjacent. Clearly, there exists only one such unicyclic graph U_π^* for a given unicyclic degree sequence π, which is called *the greedy unicyclic graph* with a given unicyclic degree sequence π.

Proposition 5.1. ([24, 41]) *Let $\pi = (d_0, \ldots, d_{n-1})$ be a positive nonincreasing integer sequence with $n \geq 3$. Then π is unicyclic graphic if and only if $\sum_{i=0}^{n-1} d_i = 2n$ and $d_2 \geq 2$.*

Theorem 5.2. ([33, 61]) *Let $\pi = (d_0, d_1, \ldots, d_{n-1})$ be a unicyclic degree sequence. Then the greedy unicyclic graph U_π^* is the only unicyclic graph having the largest (signless Laplacian) spectral radius in $\mathcal{U}_\pi$, i.e.,*

$$\rho(G) \leq \rho(U_\pi^*) \quad (q(G) \leq q(U_\pi^*)) \tag{16}$$

with equality if and only if G is the greedy unicyclic graph U_π^*.

Theorem 5.3. ([33, 61]) *Let π and τ be two different degree sequences of unicyclic graphs with the same order. If $\pi \lhd \tau$ then $\rho(U_\pi^*) < \rho(U_\tau^*)$ and $q(U_\pi^*) < q(U_\tau^*)$.*

In order to present the results on the Dirichlet eigenvalues, we need the following notion. Let $\pi = (d_0, \ldots, d_{k-1}, 1, \ldots, 1)$ be a unicyclic degree sequence with $d_0 \geq d_1 \geq \cdots \geq d_{k-1}$. If $d_2 \geq 3$, then let $U^*_{\pi,B}$ have SLO*-ordering with each pair of the first three vertices being adjacent; if $d_0 = \cdots = d_{m-1} = 2$ and $d_m = 3$ for $3 \leq m \leq k-1$, then let $U^*_{\pi,B}$ be obtained from by identifying one end vertex of a path P_m with one vertex of triangle and the other end vertex of P_m with the root of $T^*_{\tau,B}$ having SLO*-ordering with $\tau = (d_{m+1} - 1, d_{m+2}, \ldots, d_{n-1})$; if $d_0 = \cdots = d_{m-1} = 2$ and $d_m \geq 4$ for $3 \leq m \leq k-1$, then let $U^*_{\pi,B}$ be obtained from by identifying one vertex of a cycle C_m and the root of $T^*_{\tau,B}$ having SLO*-ordering with $\tau = (d_m - 2, d_{m+2}, \ldots, d_{n-1})$. Clearly, $U^*_{\pi,B}$ is uniquely determined by π. Then we have the following results.

Theorem 5.4. ([56]) *For a given graphic unicyclic degree sequence $\pi = (d_0, d_1, \ldots, d_{n-1})$, with $3 \leq d_0 \leq \cdots \leq d_k$ and $d_{k+1} = \cdots = d_{n-1} = 1$, let $G = (V_0 \cup \partial V, E_0 \cup \partial E)$ be a graph with the Faber-Krahn property in $\mathcal{U}_\pi$. Then G has an SLO*-ordering consistent with the first eigenfunction f of G in such a way that $v \prec u$ implies $f(v) \geq f(u)$.*

Theorem 5.5. ([56]) *For a given graphic unicyclic degree sequence $\pi = (d_0, d_1, \ldots, d_{n-1})$ with $3 \leq d_0 \leq \cdots \leq d_k$ and $d_{k+1} = \cdots = d_{n-1} = 1$, then $\lambda^D(G) \geq \lambda^D(U^*_{\pi,B})$ with equality if and only if $G = U^*_{\pi,B}$. In other words, $U^*_{\pi,B}$ is the unique uncyclic graph with the Faber-Krahn property in $\mathcal{U}_\pi$, which can be regarded as ball approximation.*

For $d_0 = 2$, Zhang and Zhang proposed the following conjecture.

Conjecture 5.6. ([56]) *Let $\pi = (d_0, d_1, \ldots, d_{k-1}, 1, \ldots, 1)$ be a graphic unicyclic degree sequence with $2 \leq d_0 \leq d_1 \leq \cdots \leq d_{k-1}$ and $d_k = \cdots = d_{n-1} = 1$. Then $\lambda^D(G) \geq \lambda^D(U^*_{\pi,B})$ with equality if and only if $G = U^*_{\pi,B}$. In other words, $U^*_{\pi,B}$ is the unique graph with the Faber-Krahn property in $\mathcal{U}_\pi$.*

Theorem 5.7. ([58]) *For a given positive integer $k \leq n-3$, let $U^*_{n,k}$ be unicyclic graph obtained from by identifying one end vertex of a path P_{n-k-1} and one vertex of a triangle and the other end vertex of P_{n-k-1} and the center of star $K_{1,k}$. Then for any unicyclic graph G of order n with k pendant vertices, $\lambda^D(G) \geq \lambda^D(U^*_{n,k})$ with equality if and only if $G = U^*_{n,k}$. In other words, U^* is the only unicyclic graph with the Faber-Krahn property in the set $\mathcal{U}_k$ of all unicyclic graphs of order n with k pendant vertices, which can be regarded a ball.*

6 Graphic degree sequences

In this section, we discuss some extremal properties of $\mathcal{G}_\pi$.

Theorem 6.1. ([4, 61]) *For a given graphic degree sequence π, if G is a simple connected graph that has the largest (resp. signless Laplacian) spectral radius in $\mathcal{G}_\pi$, then G has a BFS-ordering, i.e., G is a BFS graph.*

Moreover, all extremal graphs having the maximum (signless Laplacian) spectral radius for bicyclic (tricyclic) degree sequence have been characterized (see [26, 28, 33]).

On the Dirichlet eigenvalues for bicyclic degree sequences, we have the following results. For given a graphic bicyclic degree sequence $\pi = (d_0, \ldots, d_{n-1})$ with $3 \leq d_0 \leq \cdots \leq d_{k-1}$ and $d_k = \cdots = d_{n-1} = 1$, let $G^*_{\pi,B}$ be the unique graph having SLO* ordering $v_0 v_1 \cdots v_{n-1}$ with the induced subgraph by $\{v_0, v_1, v_2, v_3\}$ being $K_4 - v_2 v_3$. Then

Theorem 6.2. ([57]) *Let $\pi = (d_0, d_1, \ldots, d_{k-1}, 1, 1, \ldots, 1)$ be a graphic bicyclic degree sequence with $3 \leq d_0 \leq d_1 \leq \cdots \leq d_{k-1}$ for $k \geq 4$. Then for any bicyclic graph G with degree sequence π, $\lambda^D(G) \geq \lambda^D(G_{\pi,B})$ with equality if and only if $G = G^*_{\pi,B}$. In other words, $G^*_{\pi,B}$ is the only extremal graph with the smallest first Dirichlet eigenvalue in $\mathcal{G}_\pi$.*

Theorem 6.3. ([57]) *Let $\pi = (d_0, d_1, \ldots, d_{k-1}, 1, 1, \ldots, 1)$ and $\tau = (d'_0, d'_1, \ldots, d'_{k-1}, 1, 1, \ldots, 1)$ be two different graphic bicyclic degree sequences. If $\pi \lhd \tau$ with $d_0 \geq 3$, $d'_0 \geq 3$, then $\lambda^D(G_{\pi,B}) > \lambda^D(G_{\tau,B})$.*

A graph H is a *minor* of a graph G if H can be obtained from a subgraph of G by contracting edges. An H minor is a minor isomorphic to H. If H is a complete graph, we say that G contains a *clique minor* of size $|H|$. For a graph G, the Hadwiger number $h(G)$ of G is the maximum integer k such that G contains a clique minor of size k. Moreover, denote by $\chi(G)$ the chromatic number of G. Let

$$h(\pi) = \max\{h(G) : G \in \mathcal{G}_\pi\}, \quad \chi(\pi) = \max\{\chi(G) : G \in \mathcal{G}_\pi\}. \tag{17}$$

Conjecture 6.4. (Hadwigers Conjecture for Degree Sequences, [39]) *For every graphic degree sequence π, $\chi(\pi) \leq h(\pi)$.*

Recently, Dvořák and Mohar [16] proved a stronger result than the above conjecture.

Theorem 6.5. ([16]) *For every graphic degree sequence π,*

$$\chi(\pi) \leq h'(\pi) \leq h(\pi), \tag{18}$$

where $h'(G)$ is the maximum k such that G has a K_k-topological minor and $h'(\pi) = \max\{h'(G) : G \in \mathcal{G}_\pi\}$.

References

[1] E. O. D. Andriantiana, Energy, Hosoya index and Merrifield-Simmons index of trees with prescribed degree sequence, *Discrete Appl. Math.*, **161** (2013), 724–741.

[2] F. Belardo, E. M. Marzi, S. K. Simć, and J. F. Wang, On the spectral radius of unicyclic graphs with presribed degree sequence, *Linear Algebra Appl.*, **432**(2010), 2323–2334.

[3] T. Biyikoğlu and J. Leydold, Faber-Krahn type inequalities for trees, *J. Combin. Theory Ser. B,* **97**(2007), 159–174.

[4] T. Biyikoğlu and J. Leydold, Graphs with given degree sequence and maximal spectral radius. *Electron. J. Combin.*, **15**(2008), Paper 119.

[5] T. Biyikoğlu, M. Hellmuth, and J. Leydold. Largest eigenvalues of the discrete p-Laplacian of trees with degree sequence. *Electron. J. Linear Algebra,* **18**(2009), 202–210.

[6] T. Biyikoğlu and J. Leydold, Semiregular trees with minimal Laplacian spectral radius, *Linear Algebra Appl.*, **432**(2010), 2335–2341.

[7] B. Bollobás, *Extremal Graph Theory,* Academic Press, New York, 1978.

[8] B. Bollobás, Extremal graph theory. *Handbook of Combinatorics,* Vol. **1**, **2**, 1231–1292, Elsevier, Amsterdam, 1995.

[9] J. A. Bondy and U. S. R. Murty, *Graph Theory with Applications,* Macmillan Press, New York, 1976.

[10] F. R. K. Chung, *Spectral Graph Theory,* AMS Publications, 1997.

[11] F. Chung and L.Lu, The average distance in random graphs with given expected degree, *Proceeding of National Academy of Science*, **99**(2002), 15879–15882.

[12] F. Chung and L. Lu, *Complex Graphs and Networks,* AMS Publications, Providence, 2006.

[13] D. Cvetković, M. Doob, and H. Sachs, *Spectra of Graphs-Theory and Applications,* Academic Press, New York, 1980. 3rd edition, 1995.

[14] D. Cvetković, P. Rowlinson, and S. Simić, Signless Laplacian of finite graphs, *Linear Algebra Appl.,* **423**(2007), 155–171.

[15] A. A. Dobrynin, R. Entringer, and I. Gutman, Wiener index of trees: theory and applications, *Acta Appl. Math.*, **66**(2001), 211–249.

[16] Z. Dvořák and B. Mohar, Chromatic number and complete graph substructures for degree sequences, Combinatorica, **33**(2013), 513–529.

[17] P. Erdös and T. Gallai, Graphs with prescribed degrees of vertices (Hungarian) *Mat. Lapok,* **11**(1960), 264–274.

[18] P. Erdös, M. S. Jacobson, and J. Lehel, Graphs realizing the same degree sequences and their respective clique numbers, *Graph Theory, Combinatorics, and Applications,* Vol. **1** (Kalamazoo, MI, 1988), 439–449, Wiley-Intersci. Publ., Wiley, New York, 1991.

[19] M. Ferrara, Some problems on graphic sequences, *Graph Theory Notes N. Y.*, **64**(2013), 19–25.

[20] M. Fischermann, A. Hoffmann, D. Rautenbach, L. Szekely, and L. Volkmann, Wiener index versus maximum degree in trees, *Discrete Appl. Math.*, **122**(2002), 127–137.

[21] J. Friedman, Some geometric aspects of graphs and their eigenfunctions, *Duke Math. J.*, **69**(1993), 487–525.

[22] J. M. Guo, On the Laplacian spectral radius of a tree, *Linear Algebra Appl.*, **368**(2003), 379–385.

[23] I. Gutman and N. Trinajstić, Graph theory and molecular orbitals. Total π-electron energy of alternant hydrocarbons, *Chem. Phys. Lett.*, **17**(1972), 535–538.

[24] S. L. Hakimi, On realizability of a set of integers as degrees of the vertices of a linear graph, *SIAM Appl. Math.*, **10**(1962), 496–506.

[25] H. Hosoya, Topological index, a newly proposed quantity characterizing the topological nature of structural isomers of saturated hydrocarbons, *Bull. Chem. Soc. Jpn.*, **4**(1971), 2332–2339.

[26] Y. F. Huang, B. L. Liu, and Y. L. Liu, The signless Laplacian spectral radius of bicyclic graphs with prescribed degree sequences, *Discrete Math.*, **311**(2011), 504–511.

[27] F. Jelen and E. Triesch, Superdominance order and distance of trees with bounded maximum degree, *Discrete Appl. Math.*, **125**(2003), 225–233.

[28] X. Jiang, Y. Liu, and B. L. Liu, A further result on majorization theorem, *Linear Multilinear Algebra*, **59**(2011), 957–967.

[29] D. J. Klein, I. Lukovits, and I. Gutman, On the definition of the hyper-Wiener index for cycle-containing structures, *J. Chem. Inf. Comput. Sci.* , **35**(1995), 50–52.

[30] J. Leydold, A Faber-Krahn-type inequality for regular trees, *Geom. Funct. Anal.*, **7**(1997), 364–378.

[31] J. Leydold, The geometry of regular trees with the Faber-Krahn property, *Discrete Math.*, **245**(2002), 155–172.

[32] J. S. Li and J. H. Yin, Extremal graph theory and degree sequences (Chinese), *Adv. Math.* (*China*), **33**(2004), 273–283.

[33] M. H. Liu, The (signless Laplacian) spectral radii of connected graphs with prescribed degree sequences, *Electron. J. Combin.*, **19**(2012), no. 4, Paper 35, 19 pp.

[34] L. Lovász, *Combinatorial Problems and Exercises*, 2nd edition, North-Holland, Amsterdam, 1993.

[35] A. W. Marshall and I. Olkin, *Inequalities: Theory of Majorization and Its Applications*, Mathematics in Science and Engineering, Vol. **143**, Academic Press, New York, 1979.

[36] R. Merris, Laplacian matrices of graphs: a survey, *Linear Algebra Appl.*, **197–198**(1994), 143–176.

[37] J. J.Molitierno, *Applications of Combinatorial Matrix Theory to Laplacian Matrices of Graphs,* Discrete Mathematics and Its Applications (Boca Raton). CRC Press, Boca Raton, FL, 2012.

[38] A. R. Pruss, Discrete convolution-rearrangement inequalities and the Faber-Krahn inequality on regular trees, *Duke Math. J.*, **91**(1998), no. 3, 463–514.

[39] N. Robertson and Z. X. Song, Hadwiger number and chromatic number for near regular degree sequences,*J. Graph Theory*, **64**(2010), 175-183.

[40] N. Schmuck, S. Wagner, and H. Wang, Greedy trees, caterpillars, and Wiener-type graph invariants, *MATCH Commun. Math. Comput. Chem.*, **68**(2012), 273–292.

[41] J. K. Senior, Partitions and their representative graphs, *Amer. J. Math.*, **73**(1951), 663–689.

[42] G. Sierksma and H. Hoogeveen, Seven criteria for integer sequences being graphic, *J. Graph Theory,* **15**(1991), 223–231.

[43] R. Shi, The average distance of trees, *Systems Science and Mathematical Sciences*, **6**(1993), no. 1, 18–24.

[44] D. Stevanović, Bounding the largest eigenvalue of trees interms of the largest vertex degree, *Linear Algebra Appl.*, **360**(2003), 35–42.

[45] L. A. Székely and H. Wang, On subtrees of trees, *Adv. in Appl. Math.*, **34**(2005), 138–155.

[46] L. A. Székely, H. Wang, and T. Y. Wu, The sum of the distances between the leaves of a tree and the "semi-regular" property, *Discrete Math.*, **311**(2011), 1197–1203.

[47] H. Takeuchi. The spectrum of the p-Laplacian and p-Harmonic morphisms on graphs, *Illinois Journal of Mathematics,* **47**(2003), 939–955.

[48] S. W. Tan. On the weighted trees with given degree sequence and positive weight set, *Linear Algebra Appl.*, **433**(2010), 380–389.

[49] A. Vince and H. Wang, The average order of a subtree of a tree, *J. Combin. Theory Ser. B.*, **100**(2010), 161–170.

[50] S. G. Wagner, Correlation of graph-theoretical indeces, *SIAM J. Discrete Math.*, **21**(2007), 33–46.

[51] S. G. Wagner, H. Wang, and X.-D. Zhang, Distance-based graph invariants of trees and the Harary index, *Filomat*, **27**(2013), 41–50.

[52] H. Wang, The extremal values of the Wiener index of a tree with given degree sequence, *Discrete Appl. Math.*, **156**(2009), 2647–2654.

[53] W. D. Wei, The class $\Re(R, S)$ of (0,1)-matrices, *Discrete Math.*, **39**(1982), 201–205.

[54] H. Wiener, Structural determination of paraffin boiling points, *J. Amer. Chem. Soc.*, **69**(1947), 17–20.

[55] G. J. Zhang and X. D. Zhang, The p-Laplacian spectral radius of weighted trees with a degree sequence and a weight set, *Electron. J. Linear Algebra,* **22**(2011), 267–276.

[56] G. J. Zhang, J. Zhang, and X. D. Zhang, Faber-Krahn type inequality for unicyclic graphs. *Linear Multilinear Algebra,* **60**(2012), 1355–1364.

[57] G. J. Zhang and X. D. Zhang, The first Dirichlet eigenvalue of bicyclic graphs, *Czech. Math. J.*, **62**(2012), 441–451.

[58] G. J. Zhang and X. D. Zhang, Faber-Krahn type inequalities for unicyclic graphs with the same boundaries, *Chinese Journal of Contemporary Mathematics,* to appear.

[59] X. D. Zhang, On the two conjectures of Graffiti, *Linear Algebra Appl.*, **385**(2004), 369–379.

[60] X.-D. Zhang, The Laplacian spectral radii of trees with degree sequences, *Discrete Math.*, **308**(2008), 3143–3150.

[61] X.-D. Zhang, The signless Laplacian spectral radius of graphs with given degree sequences, *Discrete Appl. Math.*, **157**(2009), no. 13, 2928–2937.

[62] X. D. Zhang and J. S. Li, The Laplacian spectrum of mixed graphs, *Linear Algebra Appl.*, **351**(2002), 11–20.

[63] X.-D. Zhang, Y. Liu, and M.-X. Han, Maximum Wiener index of trees with given degree sequence, *MATCH Commun. Math. Comput. Chem.*, **64**(2010), 661–682.

[64] X. D. Zhang, Q. Y. Xiang, L. Q. Xu, and R. Y. Pan, The Wiener index of trees with given degree sequences, *MATCH Commun. Math. Comput. Chem.*, **60**(2008), 623–644.

[65] X.-M. Zhang, X.-D. Zhang, D. Gray, and H. Wang, The number of subtrees with given degree sequence, *Journal of Graph Theory,* **73**(2013), 280–295.

Department of Mathematics and MOE-LSC
Shanghai Jiao Tong University
800 Dongchuan road, Shanghai 200240, China
E-mail address: xiaodong@sjtu.edu.cn

Algebraic Geometry, Complex Geometry

Proceedings of the 6th International Congress of Chinese Mathematicians
ALM 36, pp. 427–435

On the Toda Systems of VHS Type

Chen-Yu Chi

Abstract

We consider the Toda systems of VHS type with singular sources and provide a criterion for the existence of solutions with prescribed asymptotic behaviour near singularities. We also prove the uniqueness of solution. Our approach uses Simpson's theory of constructing Higgs-Hermitian-Yang-Mills metrics from stability.

Mathematics Subject Classification: 32L25, 35J60, 53C07, 53C28.

Keywords and Phrases: Toda systems, Higgs bundles, variation of Hodge structures.

1 Introduction

Let M be a compact Riemann surface and $g = ds^2$ a (smooth) hermitian metric on M. A Toda system on M is a system of the following form with unknown functions $(u_1, \ldots, u_n)$:

$$\frac{1}{4}\Delta_g \begin{pmatrix} u_1 \\ u_2 \\ \vdots \\ u_n \end{pmatrix} = \epsilon C \begin{pmatrix} e^{u_1} \\ e^{u_2} \\ \vdots \\ e^{u_n} \end{pmatrix} + \text{ perturbation terms}, \tag{1.1}$$

where Δ_g is the Laplacian associated to g, C is the Cartan matrix of certain Lie algebra, and $\epsilon = \pm 1$. We call it a Toda system of usual or reversed type depending on $\epsilon = -1$ or 1.

Toda systems have been arising from many aspects in mathematical sciences, ranging from the study of water waves to high energy physics. From the mathematical point of view, Toda systems of reversed type can be viewed as a very special and important type of Higgs bundles, as introduced by Hitchin [2] and further developed by Simpson [6], which generalize the notion of bundles with hermitian-Yang-Mills connections/metrics and are related to notions of stability. On the other hand, each Higgs-hermitian-Yang-Mills equation has a "partner" obtained by reversing the sign of the curvature, as studied by Hitchin [3], and a Toda systems of usual type is simply the partner of the Higgs bundle corresponding to some Toda system of reversed type. The studies of these "reversed" partners are quite different in nature from that of the original Higgs bundles in that, instead of stability of bundles, they are more related to harmonic maps. A typical example first appeared in [4], where a special kind of Toda systems on (domains in) the two dimensional plane are studied by establishing a one-to-one correspondence between the solutions and smooth harmonic maps to specific projective spaces.

The n-th *Toda system of VHS type* is the following system of PDEs

$$\frac{1}{4}\Delta_g \begin{pmatrix} u_1 \\ u_2 \\ \vdots \\ u_n \end{pmatrix} - \frac{K_g}{2} \begin{pmatrix} 1 \\ 1 \\ \vdots \\ 1 \end{pmatrix} = \begin{pmatrix} 2 & -1 & & \\ -1 & 2 & & \\ & & \ddots & -1 \\ & & -1 & 2 \end{pmatrix} \begin{pmatrix} e^{u_1} \\ e^{u_2} \\ \vdots \\ e^{u_n} \end{pmatrix} \tag{1.2}$$

where K_g is the Gauss curvature associated to g. The Toda systems we consider here are usually called type A Toda systems, manifesting its relation to the Lie algebra A_n. Our original motivation was to generalize the work of [4] to the case with singularities and to general Riemann surfaces, which amounts to counting harmonic maps with prescribed singularities. If the Riemann surface is equipped with a Riemann metric g, the corresponding equation becomes

$$\frac{1}{4}\Delta_g \begin{pmatrix} u_1 \\ u_2 \\ \vdots \\ u_n \end{pmatrix} - \frac{K_g}{2} \begin{pmatrix} 1 \\ 1 \\ \vdots \\ 1 \end{pmatrix} = -\begin{pmatrix} 2 & -1 & & \\ -1 & 2 & & \\ & & \ddots & -1 \\ & & -1 & 2 \end{pmatrix} \begin{pmatrix} e^{u_1} \\ e^{u_2} \\ \vdots \\ e^{u_n} \end{pmatrix}. \tag{1.3}$$

In their seminal work [5], Lin and Wang delt with some cases of one single equation with one singularities on the flat torus. So far we have not been able to solve the problem completely due to some technical difficulty, and we turn to study its reversed partner, which turns out to be the system (1.2).

Definition 1.1. (1) *For any n-tuple $\mu = (\mu_1, \ldots, \mu_n)$ of real valued functions on M, we define*

$$S_\mu := \{p \in M : \mu(p) \neq (0, \ldots, 0)\}.$$

μ is called an assignment of singular strengths if S_μ is finite.

(2) *For any assignment of singular strengths μ, an n-tuple $u = (u_1, \ldots, u_n) \in \mathcal{C}^\infty(M \setminus S_\mu)$ is called μ-admissible if for all $p \in S_\mu$ there exists a holomorphic chart (U, z) centered at p and a bounded function $v_j \in \mathcal{C}^\infty(U \setminus \{p\})$ such that*

$$u_j = 2\mu_j(p) \log |z| + v_j$$

on $U \setminus \{p\}$, $j = 1, \ldots, n$.

(3) *Let μ be an assignment of singular strengths and $p \in M$. We define*

$$d_{k,p}(\mu) := \frac{-1}{n+1}\sum_{j=1}^{n}(n+1-j)\mu_j(p) + \sum_{j=1}^{k}\mu_j(p)$$

and

$$d_k(\mu) := \sum_{p\in M} d_{k,p}(\mu),$$

$k = 1, \ldots, n$.

Our main result is the following theorem (Theorem 4.2).

Theorem. *There exists a μ-admissible solution to* (1.2) *if and only if*

$$d_{n-m+1}(\mu) + \cdots + d_n(\mu) < m(n-m+1)(\text{genus}(M)-1),$$

$m = 1, \ldots, n$. *In addition, the μ-admissible solution is unique if it exists.*

Our main tool in getting the criterion for existence of solutions is the theory of getting Higgs-Hermitian-Yang-Mills metrics (which are flat in our case) from stability, developed by Hitchin [2] and Simpson [6]. The presence of singularities is the main difficulty to be overwhelmed. In order to deal with singularities, we have to use the results of [6] (see also [7]) for quasiprojective curves.

2 The Toda systems of VHS type and polarized complex variation of Hodge structure

Let X be a complex manifold.

Definition 2.1. (1) *A complex pre-variation of Hodge structure* ($\mathbf{C}$-*pVHS for short*) $(\{V^{r,s}\}, \nabla)$ *of weight an integer n on X consists of*

(i) $\mathcal{C}^\infty$ $\mathbf{C}$*-vector bundles $V^{r,s}$ over X, $r, s \in \mathbf{Z}$, $r+s=n$, with $V^{r,s}=0$ for all but finitely many (r,s) and*

(ii) *a connection $\nabla : \mathcal{A}^0_X(V) \to \mathcal{A}^1_X(V)$ on $V := \oplus_{r+s=n}V^{r,s}$ which only has components of total degree $(1,0)$ and $(0,1)$. In other words, $\nabla = \oplus_{r,s}\nabla^{r,s}$ with*

$$\nabla^{r,s} : \mathcal{A}^0_X(V^{r,s}) \to \mathcal{A}^{1,0}_X(V^{r-1,s+1}) \oplus \mathcal{A}^1_X(V^{r,s}) \oplus \mathcal{A}^{0,1}_X(V^{r+1,s-1})$$

for each pair (r,s). (If we write $\nabla = \theta + \nabla^{\text{Hodge}} + \theta'$ where

$$\nabla^{\text{Hodge}} = \bigoplus_{r+s=n}\{\nabla^{\text{Hodge},r,s} : \mathcal{A}^0_X(V^{r,s}) \longrightarrow \mathcal{A}^1_X(V^{r,s})\},$$

$$\theta := \bigoplus_{r+s=n}\{\theta^{r-1,s+1} : \mathcal{A}^0_X(V^{r,s}) \longrightarrow \mathcal{A}^{1,0}_X(V^{r-1,s+1})\},$$

and

$$\theta' := \bigoplus_{r+s=n}\{\theta'^{r+1,s-1}\mathcal{A}^0_X(V^{r,s}) \longrightarrow \mathcal{A}^{0,1}_X(V^{r+1,s-1})\},$$

then it is clear that ∇^{Hodge} *is also a connection and* θ *and* θ' *are* $\mathcal{A}_X$*-linear.* θ *is called the Hodge field of the* **C**-*pVHS.)*

(2) *A* **C**-*pVHS* $(\{V^{r,s}\},\nabla)$ *over* X *is called diagonal* (Δ *for short) if the curvature* F^∇ *of* ∇ *is diagonally valued, i.e.*

$$F^\nabla(Y_1,Y_2)(V_x^{r,s}) \subset V_x^{r,s}, \text{ for all } (r,s), Y_1, Y_2 \in T_xX, x \in X.$$

(3) *A* **C**-*pVHS* $(\{V^{r,s}\},\nabla)$ *over* X *is called a special linear* (*SL for short) if its induced connection on* $\det V$ *is flat.*

(4) *A* **C**-*pVHS* $(\{V^{r,s}\},\nabla)$ *is a* **C**-*VHS if* ∇ *is flat.*

Definition 2.2. *A polarization of a* **C**-*pVHS* $(\{V^{r,s}\},\nabla)$ *of weight* n *consists of a family* $\{h^{r,s}\}$ *of smooth hermitian metrics on* $V^{r,s}$ *such that if* $h := \oplus_{r+s=n} h^{r,s}$ *then* $h(C^{-1}\cdot,\cdot)$ *is preserved by* ∇, *where* C *is the Weil operator defined by* $C|_{V^{r,s}} := i^{r-s}\mathrm{id}_{V^{r,s}} = i^n(-1)^s\mathrm{id}_{V^{r,s}}$.

More precisely, this means that

$$X\langle\sigma,\tau\rangle = \langle\nabla_X\sigma,\tau\rangle + \langle\sigma,\nabla_{\overline{X}}\tau\rangle,$$

for all $X \in T_xX$, $\sigma,\tau \in \mathcal{A}^0_X(V)_x$, and $\langle\cdot,\cdot\rangle = h(C\cdot,\cdot)$. If this is the case, we say that $(\{V^{r,s}\},\nabla)$ is polarized by $\{h^{r,s}\}$ (or by h for short). It is not hard to see that $\{h^{r,s}\}$ is a polarization if and only if ∇^{Hodge} preserves h and $\theta' = \theta^*$, where

$$\theta^* : \mathcal{A}^0_X(V) \longrightarrow \mathcal{A}^{0,1}_X(V)$$

is the adjoint of θ with respect to h.

Now let (X,g) be a Riemann surface with a smooth metric, L a smooth line bundle on X such that $L^{n+1} \simeq K_X^{\frac{n(n+1)}{2}}$. We set $V^{n-k,k} := L \otimes K_X^{-k}$ if $k = 0,\ldots,n$ and 0 otherwise, and $h_g = \oplus_k h^{n-k,k}$ be the metric on $V := \oplus_k V^{n-k,k}$ induced by g. In this situation, we have a bijection between the set of "potential Hodge fields" and $\mathcal{C}^\infty(X)^n$ since

$$\mathcal{A}^{1,0}_X(\mathrm{Hom}(V^{n-k+1,k-1},V^{n-k,k})) \simeq \mathcal{A}^0_X.$$

More precisely, let $\varphi = \lambda dz$ ($\lambda > 0$) be a unitary frames of K_X and e_L a unitary frames of L such that $e_L^{n+1} = \varphi^{\frac{n(n+1)}{2}}$ under the isomorphism $L^{n+1} \simeq K_X^{\frac{n(n+1)}{2}}$. Then $e_k := e_L \otimes \varphi^{-k}$ is a unitary frame of $V^{k,n-k}$ and the correspondence is given by

$$\{\theta^{n-k,k} = b_k\varphi \otimes e_k \otimes e^*_{k-1}\}_{k=0,\ldots,n} \longleftrightarrow (b_1,\ldots,b_n) \in \mathcal{C}^\infty(X)^n.$$

Question. For a given $(b_1,\ldots,b_n) \in \mathcal{C}^\infty(X)^n$, do there exist connections

$$(\nabla^{\text{Hodge}})^{n-k,k}$$

on $V^{n-k,k}$, $k = 0,\ldots,n$, such that $\nabla := \theta + \nabla^{\text{Hodge}} + \theta^*$ defines a **C**-VHS polarized by h_g?

We have the following result.

Proposition 2.1. ([1]) *We have a bijection between the set*

$$\{(b_1, \dots, b_n) \in \mathcal{C}^\infty(X)^n : \textit{All } b_k \textit{ are nonvanishing}\}$$

and the set

$$\left\{ \begin{array}{c} \Delta\text{-}SL\text{-}\mathbf{C}\text{-}p\textit{VHSs which are underlied by } \{V^{r,s}\} \\ \textit{and polarized by } h_g \textit{ with all } \theta^{r,s} \textit{ nonvanishing} \end{array} \right\}$$

Under this bijection, those n-tuples $(b_1, \dots, b_n)$ with $(\log|b_1|^2, \dots, \log|b_n|^2)$ solving (1.2) correspond exactly to $\mathbf{C}$-VHSs.

3 The associated Higgs bundles

Let X be a complex manifold.

Definition 3.1. *A Higgs bundle (E, Φ) on X consists of a holomorphic vector bundle E on X and a holomorphic $\mathrm{End}(E)$-valued $(1,0)$-form Φ, called the Higgs field.*

Now let M be a compact Riemann surface equipped with a smooth hermitian metric g, μ an assignment of singular strengths, and $X = M \setminus S_\mu$. We fix from now on a holomorphic spinor bundle $K_M^{\frac{1}{2}}$ on M in our discussion and let $L := (K_M^{\frac{1}{2}})^n$. Set

$$\widetilde{E} := \bigoplus_k \widetilde{E}^{n-k,k} \text{ and } \widetilde{\Phi} := \bigoplus_k \Phi_k$$

where

$$\widetilde{E}^{n-k,k} := L \otimes K_M^{-k}$$

and

$$\Phi_k \longleftrightarrow 1 \in \Gamma\big(M, K_M \otimes \mathrm{Hom}(L \otimes K_M^{-k+1}, L \otimes K_M^{-k})\big),$$

$k = 1, \dots, n$. Then $(\widetilde{E}, \widetilde{\Phi})$ is a Higgs bundle on M. $(E, \Phi) := (\widetilde{E}, \widetilde{\Phi})|_X$ is called the *associated Higgs bundle* on X.

Definition 3.2. *A smooth hermitian metric H on E is called of type μ if $H = \oplus_k H_k$ for some smooth hermitian metrics H_k on $\widetilde{E}^{n-k,k}|_X$ such that for any $p \in M$ and any sufficiently small coordinate chart (U, z) centered at p, the function $|z|^{2d_{k,p}(\mu)}$ and the expression of H_k with respect to any local trivialization of $\widetilde{E}^{n-k,k}$ are mutually bounded multiplicatively.*

The key correspondence is given by the following result.

Proposition 3.1. *There is a bijection between*

$$\{(b_1, \dots, b_n) \in \mathcal{C}^\infty(X)^n | b_k > 0 \textit{ with } (\log b_1^2, \dots, \log b_n^2)\ \mu\textit{-admissible}\}$$

and

$$\{\textit{smooth hermitian metrics } H \textit{ on } E \textit{ of type } \mu \textit{ with } \det H = 1\}.$$

Under this bijection, n-tuples $(b_1, \dots, b_n)$ with $(\log b_1^2, \dots, \log b_n^2)$ solving (1.2) correspond exactly to metrics H such that the connection $\Phi + \nabla^H + \Phi^*$ is flat, where ∇^H denotes the metrical connection and Φ^* denotes the H-adjoint of Φ.

4 Stability and Higgs-Hermitian-Yang-Mills metrics

In this section we provide the criterion for the existence of solutions to Toda systems of VHS type for compact Riemann surfaces. Our main ingredient is Simpson's theory of constructing Higgs-Hermitian-Yang-Mills metrics on Higgs bundles from stability [6]. Let (X, g) be a Kähler manifold satisfying suitable assumptions (cf. [6], Section 2, Assumptions 1,2 and 3), which will be fulfilled in our situation (cf. [6], Proposition 2.2 or 2.4).

Definition 4.1. ([6], p.877,878) (1) *A sub-Higgs sheaf of a Higgs bundle (E, Φ) is an analytic subsheaf $\mathcal{V} \subset \mathcal{O}(E)$ such that $\Phi : \mathcal{V} \longrightarrow \mathcal{O}(K_X) \otimes \mathcal{V}$. (If $\mathcal{V}$ is saturated, outside a set of codimension 2 it is the coherent sheaf associated to a subbundle of E.)*

(2) *For a saturated subsheaf $\mathcal{V}$ and a smooth metric K on E such that*

$$\sup_X |\Lambda_g F_K^{\text{Higgs}}|_K < \infty,$$

$$\deg(\mathcal{V}, K) := i \int_X \text{Tr } \Lambda_g F_K^{\text{Higgs}}.$$

(This is either a real number or $-\infty$.)

Definition 4.2. *Let (E, Φ) be a Higgs bundle on X and A be a group acting on X by biholomorphic maps preserving the metric g and acting on E compatibly by automorphisms $a : E \longrightarrow E$ preserving the metric K and changing Φ by homotheties $a\Phi a^{-1} = \lambda(a)\Phi$. (E, Φ, K) is stable with respect to the A-action if for every proper saturated sub-Higgs sheaf $\mathcal{V}$ preserved by A,*

$$\frac{\deg(\mathcal{V}, K)}{\text{rk}(\mathcal{V})} < \frac{\deg(E, K)}{\text{rk}(E)}.$$

For a Higgs bundle (E, Φ) on X polarized by a metric H, we denote the curvature of the connection $\Phi + \nabla^H + \Phi^*$ by F_H^{Higgs}.

Definition 4.3. *H is a Higgs-Yang-Mills metric if the trace-free part of $\Lambda_g F_H^{\text{Higgs}}$ vanishes. (Here Λ_g denotes the adjoint operator of the Lefschetz operator $L = \omega \wedge$ on spaces of bundle-valued differential forms, where ω is the Kähler form.)*

Recall the following result due to Simpson.

Theorem 4.1. (Simpson) *If (E, Φ, K) is stable with respect to the A-action, then there exists a smooth A-invariant Higgs-Hermitian-Yang-Mills metric H with H and K mutually bounded, $\det H = \det K$, and $\overline{\partial} h + [\Phi, h] \in L^2_{g,K}$, where h is the unique endomorphism of E such that*

$$(\cdot, \cdot)_H = (h(\cdot), \cdot)_K.$$

If furthermore $\Phi \wedge \Phi = 0$, the first Chern form $c_1(E, K) = 0$, and $\int_X c_2(E, K) \wedge \omega_g^{n-2} = 0$, then the connection $\nabla = \Phi + \nabla^H + \Phi^*$ is flat. Conversely, if there exists an A-invariant Higgs-Hermitian-Yang-Mills metric, then

$$\frac{\deg(\mathcal{V}, K)}{\mathrm{rk}(\mathcal{V})} \leq \frac{\deg(E, K)}{\mathrm{rk}(E)}$$

for every proper saturated sub-Higgs sheaf $\mathcal{V}$ preserved by A and equality holds only if $E = \mathcal{V} \oplus \mathcal{V}^{\perp}$ is an orthogonal direct sum of Higgs subbundles.

Now we are ready to prove our main result mentioned in the introduction.

Theorem 4.2. *There exists a μ-admissible solution to* (1.2) *if and only if*

$$d_{n-m+1}(\mu) + \cdots + d_n(\mu) < m(n - m + 1)(\mathrm{genus}(M) - 1), \tag{4.1}$$

$m = 1, \ldots, n$. *In addition, the μ-admissible solution is unique if it exists.*

Proof. We will omit the proof of the part of uniqueness, which is a special case of more general results about Higgs-Yang-Mills metrics. Let (E, Φ) be the associated Higgs bundle and A be the group $U(1)^{\times(n+1)} \cap SU(n + 1)$ acting on X trivially and on E in the obvious diagonal manner. We equip E with a metric $K := \oplus_k K_k$ of type μ such that $\det K = 1$. Such a metric can be constructed as follows. Choose a covering of M by coordinate disks (U_α, z_α) and a partition of unity $\{\rho_\alpha\}$ subordinate to $\{U_\alpha\}$. We may assume that

(1) each $s \in S_\mu$ is contained in U_α for exactly one α, denoted as $\alpha(s)$, and $z_{\alpha(s)}(s) = 0$;

(2) for every $s \in S_\mu$, there exists an open neighborhood $W_{\alpha(s)} \subset U_{\alpha(s)}$ of s such that $\rho_{\alpha(s)}|_{W_{\alpha(s)}} \equiv 1$;

(3) every $\widetilde{E}^{n-k,k}$ is trivialized on U_α by a holomorphic local frame $e_{k,\alpha}$.

Let σ be an element in the fibre of E_k above $p \in M \setminus S$. We write

$$\sigma = \sigma_\alpha e_{k,\alpha}(p)$$

if $p \in U_\alpha$. For $p \in U_\alpha$, we define

$$f_\alpha(p) := \begin{cases} |z_\alpha(p)|^{2d_{k,s}(\mu)}, & \text{if } U_\alpha \cap S_\mu = \{s\}; \\ 1, & \text{otherwise.} \end{cases}$$

The metric K_k on E_k for $k = 1, \ldots, n$ is defined as follows:

$$|\sigma|^2_{K_k} := \sum_{\alpha : p \in U_\alpha} \rho_\alpha(p) f_\alpha(p) |\sigma_\alpha|^2.$$

Finally, we let K_0 on E_0 be defined so that $\det K = 1$. Then $c_1(E, K) = 0$ by the construction of K. $\Phi \wedge \Phi = 0$ and $c_2(E, K) = 0$ automatically on a Riemann surface.

Since dim $M = 1$, proper saturated sub-Higgs sheaves are exactly proper holomorphic subbundles of E preserved by Φ. By the form of Φ (a nilpotent string), it is clear that such kind of subbundles are exactly

$$F^m := E_{n-m+1} \oplus \cdots \oplus E_n,$$

$m = 1, \ldots, n$.

$$\deg(F^m, K) = \sum_{k=n-m+1}^{n} \deg(E_k, K_k).$$

Note that for each k the smooth metric K_k on E_k can be viewed as a singular metric on $\widetilde{E}^{n-k,k}$, whose curvature current represents the first Chern class of $\widetilde{E}^{n-k,k} = L \otimes K_M^{\otimes -k}$. By the Poincaré–Lelong formula

$$\deg \widetilde{V}^{n-k,k} = \deg(E_k, K_k) - d_k(\mu).$$

Therefore

$$\deg(E_k, K_k) = (\text{genus}(M) - 1)(n - 2k) + d_k(\mu),$$

and hence

$$\deg(F^m, K) = (\text{genus}(M) - 1)m(m - n - 1) + d_{n-m+1}(\mu) + \cdots + d_n(\mu).$$

It is clear that $\deg(E, K) = 0$. If (4.1) holds for $m = 1, \ldots, n$, then (E, Φ, K) is stable. By the first part of Simpson's theorem we obtain a μ-admissible solution to (1.2). Conversely, if there exists such a solution, by the second part of Simpson's theorem, we have

$$d_{n-m+1}(\mu) + \cdots + d_n(\mu) \leq m(n - m + 1)(\text{genus}(M) - 1),$$

$m = 1, \ldots, n$. Actually, all inequalities above are strict since the orthogonal complement of F^m is not a sub-Higgs sheaf, $m = 1, \ldots, n$. Therefore (4.1) holds for $m = 1, \ldots, n$. □

References

[1] C.-Y. Chi; *On the Toda systems of VHS type*, arXiv:1302.1287.

[2] N. Hitchin; *The self-duality equations on a Riemann surface*, Proc. London Math. Soc. (3) **55** (1987), no. 1, 59–126.

[3] N. Hitchin; *Harmonic maps from a 2-torus to the 3-sphere*, J. Differential Geom. **31** (1990), no. 3, 627–710.

[4] J. Jost and G. Wang; *Classification of solutions of a Toda system in R^2*, Int. Math. Res. Not. 2002, no. 6, 277–290.

[5] C.-S. Lin and C.-L. Wang; *Elliptic functions, Green functions and the mean field equations on tori*, Ann. of Math. (2) **172** (2010), no. 2, 911–954.

[6] C. Simpson; *Constructing variations of Hodge structure using Yang-Mills theory and applications to uniformization*, J. Amer. Math. Soc. **1** (1988), no. 4, 867–918.

[7] C. Simpson; *Harmonic bundles on noncompact curves*, J. Amer. Math. Soc. **3** (1990), no. 3, 713–770.

Department of Mathematics and Taida Institute for Mathematical Sciences
Taiwan University, Taipei, Taiwan, China
E-mail address: chi@tims.ntu.edu.tw, chi1@ntu.edu.tw

Proceedings of the 6th International
Congress of Chinese Mathematicians
ALM 36, pp. 437–451

Classification of Isoparametric Hypersurfaces

Quo-Shin Chi

Abstract

We survey chronologically the classification of isoparametric hypersurfaces.

Mathematics Subject Classification: 53C40, 13F20.

Keywords and Phrases: Ideal, isoparametric.

1 Early history of isoparametric hypersurfaces

Wikipedia. In physics, a *wavefront* is the locus of points having the same phase: a line or curve in 2-d or a surface for a wave propagating in 3-d.

A typical example is the crests of ocean waves forming wave fronts. A skillful surfer, on the other hand, knows how to ride a wavefront below the crest.

Question 1. (Laura, 1918 [16]) *What are the wavefronts whose front speed remains constant on each front surface?*

The wave equation is

$$\Delta\phi = \frac{\partial^2\phi}{\partial t^2}.$$

Wave fronts are *level surfaces* of ϕ, at each moment, which propagate along the normal directions of the level surfaces. That the front speed remains constant on each level surface means

$$|\nabla\phi| = \text{ change per unit length of } \phi \text{ along the normal} = a(\phi),$$
$$ds/dt = b(\phi),$$

for some smooth functions a and b, where s is the distance a wavefront travels. Therefore,

$$\frac{\partial\phi}{\partial t} = \frac{\partial\phi}{\partial s}\frac{ds}{dt} = a(\phi)b(\phi) := c(\phi),$$
$$\Delta\phi = \frac{\partial^2\phi}{\partial t^2} = c'(\phi)c(\phi).$$

Definition 2. *A smooth function f over $\mathbb{R}^3$ is transnormal if*

$$|\nabla f| = A(f)$$

for some smooth functions A. A transnormal function is isoparametric if

$$\Delta f = B(f).$$

Let c be a regular value of an isoparametric function f. The level surface $f^{-1}(c)$ is called an isoparametric surface.

Theorem 3. (Somigliana, 1918–1919 [28]) *A transnormal function f is isoparametric if and only if each regular level surface of f has constant mean curvature.*

Theorem 4. (Somigliana) *The regular level surfaces of an isoparametric function must be either all spheres, all cylinders or all planes.*

This theorem was rediscovered later by Segre [26] in 1924 and Levi-Civita [17] in 1937. Levi-Civita's approach lent its way to the subsequent generalization to higher dimensions.

Lemma 5. (Levi-Civita, 1937 [17]) *A transnormal f is isoparametric if and only if the two principal curvatures of each regular level surface are constant.*

Theorem 6. (Segre, 1938 [27]) *The same conclusion of Somigliana holds on $\mathbb{R}^n$. That is, an isoparametric hypersurface, which is a regular level hypersurface of an isoparametric function f over $\mathbb{R}^n$ satisfying*

$$|\nabla f| = A(f), \quad \Delta f = B(f),$$

is either a hypersphere, a hyperplane, both are totally umbilic (one principal curvature), or a cylinder $S^k \times \mathbb{R}^{n-1-k}$.

Theorem 7. (Cartan, 1938 [3]) *The same conclusion of Segre holds on the hyperbolic space H^n of constant curvature -1. That is, an isoparametric hypersurface in H^n must be either a sphere, a hyperbolic H^{n-1}, a Euclidean $\mathbb{R}^{n-1}$ (called a horosphere), all three are totally umbilic, or a cylinder $S^k \times H^{n-k-1}$.*

IDEA: Cartan first showed that a transnormal function f (in the standard space forms) is isoparametric if and only if each regular level hypersurface of f has *constant mean curvature*, if and only if the shape operator of each regular level hypersurface has *constant principal curvatures* (*values*), counting multiplicities.

In the hyperbolic case, he then showed again that there are at most two principal values of the shape operator. Indeed, let $\lambda_1, \ldots, \lambda_{n-1}$ be the principal values of an isoparametric hypersurface in a standard space form of dimension n with constant curvature C. Then we have

$$\sum_{j \neq k} m_j \frac{C + \lambda_k \lambda_j}{\lambda_k - \lambda_j} = 0, \quad \text{summed on } j, \tag{1}$$

referred to by Cartan as the "Fundamental Formula", which was proved by Segre in the Euclidean case and by Cartan in general. Here, m_j is the multiplicity of λ_j and $\lambda_i \neq \lambda_j$ if $i \neq j$.

Remark 8. *Say, $C = 1$. That is, the ambient space is the unit sphere in which the isoparametric hypersurface sits. Write*

$$\lambda_j = \cot(\theta_j).$$

Then the fundamental formula is nothing but

$$\sum_{j \neq k} \cot(\theta_k - \theta_j) = 0,$$

which carries a significant geometric meaning. Namely, in the spherical case, wavefronts, that is, the 1-parameter family of isoparametric hypersurfaces, eventually degenerate to two subspaces of much smaller dimensions whose mean curvatures are zero.

The case $C = 1$ is remarkably deep! At this point, 95 years after Laura first investigated isoparametric surfaces, there remains the last case (out of infinitely many) to be classified:

Classify the isoparametric hypersurfaces in S^{31} with four principal values of multiplicities $7, 7, 8, 8$.

Different fields of mathematics, such as differential geometry, algebraic geometry, algebraic topology, homotopy theory, K-theory, representation theory, etc., interplay in this arena.

Definition 9. *g is the number of principal values of an isoparametric hypersurface in S^n.*

Theorem 10. (Cartan, 1939–1940 [4, 6])

$g = 1$. *This is the 1-parameter family of parallel hyperspheres degenerating to the North and South Poles, called the focal submanifolds of the family.*

$g = 2$. *This is the 1-parameter family of generalized tori $S^k \times S^{n-k-1}$, whose points are*

$$(x_0, \ldots, x_k, x_{k+1}, \ldots, x_n),\ x_0^2 + \cdots + x_k^2 = r^2,\ x_{k+1}^2 + \cdots + x_n^2 = s^2,\ r^2 + s^2 = 1,$$

which degenerates to two focal submanifolds S^k and S^{n-k-1} of radius 1 as r approaches 0 or 1.

$g = 3$. I. *The three principal values have equal multiplicity* $m = 1, 2, 4,$ *or* 8.

II. *In the ambient Euclidean space* $\mathbb{R}^{n+1} \supset S^n$, *there is a homogeneous polynomial* F *of degree* 3, *satisfying*

$$|\nabla F|^2 = 9r^2, \quad r \text{ is the Euclidean radial distance, } \quad \text{and,}$$
$$\Delta F = 0,$$

whose restriction to S^n *is exactly the isoparametric function* f. *The range of* f *is* $[-1, 1]$ *and* ± 1 *are the only critical values. Thus* $f^{-1}(c)$, $-1 < c < 1$, *form a* 1*-parameter family of isoparametric hypersurfaces that degenerates to the two focal submanifolds* $f^{-1}(1)$ *and* $f^{-1}(-1)$.

III. *The two focal submanifolds are the real, complex, quaternionic, or octonion projective plane corresponding to the principal multiplicity* $m = 1, 2, 4,$ *or* 8. *Each isoparametric hypersurface in the family is a tube around the projective plane.*

IV. *Let* $\mathbb{F}$ *be one of the normed algebras* $\mathbb{R}, \mathbb{C}, \mathbb{H}$ *and* $\mathbb{O}$. *Let* $X, Y, Z \in \mathbb{F}$ *and* $a, b \in \mathbb{R}$. *Then*

$$F = a^3 - 3ab^2 + \frac{3a}{2}(X\overline{X} + Y\overline{Y} - 2Z\overline{Z}) + \frac{3\sqrt{3}b}{2}(X\overline{X} - Y\overline{Y}) + \frac{3\sqrt{3}}{2}((XY)Z + \overline{(XY)Z}). \tag{2}$$

$g = 4$. *He assumed equal multiplicity* m *and classified the cases when* $m = 1$ *or* 2.

Question 11. (Cartan, 1940 [6]) (i) *What are the possible* g?
(ii) *Is equal multiplicity of principal values always true?*
(iii) *Are all isoparametric hypersurfaces homogeneous?*

Technically, it is more convenient to work with the following definition of an isoparametric hypersurface, though it is equivalent to the above one, as we will see later.

Definition 12. *A hypersurface in* $\mathbb{R}^n, S^n$ *or* H^n, *is called isoparametric if its principal values are everywhere constant, counting multiplicities.*

Question 13. *Classify all isoparametric hypersurfaces in spheres.*

This is Problem 34 on Yau's list of important open problems in geometry proposed in 1992 [32].

2 Development in the early 1970s, the homogeneous case

Nomizu wrote two papers [23, 24] in the early 1970s that revived the interest in isoparametric hypersurfaces. At about the same time, Takagi and Takahashi [30] classified homogeneous isoparametric hypersurfaces in spheres. Takagi and Takahashi's work was based on the comprehensive work of Cartan on the classification of symmetric spaces, and Hsiang and Lawson's 1971 work [15] on the classification of homogeneous hypersurfaces in spheres.

Definition 14. *A connected hypersurface M in a smooth manifold X is called homogeneous if $I(X, M)$, the group of isometries of X leaving M invariant, acts transitively on M.*

It is clear that for such a hypersurface, the principal values of its shape operator are everywhere constant, counting multiplicities.

Theorems 6 and 7 classify all isoparametric hypersurfaces in $\mathbb{R}^n$ and H^n to be exactly the homogeneous hypersurfaces in these space forms.

What is interesting is then the spherical case.

We start with understanding the homogeneous ones:

Theorem 15. (Hsiang and Lawson, 1971 [15]) *A homogeneous hypersurface in a sphere is exactly a principal orbit of the isotropic representation of a rank* 2 *symmetric space.*

The following table is the collection of all symmetric spaces G/K of Types I and II whose isotropy representations give homogeneous (isoparametric) hypersurfaces M. There are at most two multiplicities $(m_1, m_2), m_1 \leq m_2$, for the g principal curvatures.

G	K	$\dim M$	g	(m_1, m_2)
$S^1 \times SO(n+1)$	$SO(n)$	n	1	$(1, 1)$
$SO(p+1) \times SO(n+1-p)$	$SO(p) \times SO(n-p)$	n	2	$(p, n-p)$
$SU(3)$	$SO(3)$	3	3	$(1, 1)$
$SU(3) \times SU(3)$	$SU(3)$	6	3	$(2, 2)$
$SU(6)$	$Sp(3)$	12	3	$(4, 4)$
E_6	F_4	24	3	$(8, 8)$
$SO(5) \times SO(5)$	$SO(5)$	8	4	$(2, 2)$
$SO(10)$	$U(5)$	18	4	$(4, 5)$
$SO(m+2), m \geq 3$	$SO(m) \times SO(2)$	$2m-2$	4	$(1, m-2)$
$SU(m+2), m \geq 2$	$S(U(m) \times U(2))$	$4m-2$	4	$(2, 2m-2)$
$Sp(m+2), m \geq 2$	$Sp(m) \times Sp(2)$	$8m-2$	4	$(4, 4m-5)$
E_6	$(Spin(10) \times SO(2))/\mathbb{Z}_4$	30	4	$(6, 9)$
G_2	$SO(4)$	6	6	$(1, 1)$
$G_2 \times G_2$	G_2	12	6	$(2, 2)$

3 Development in the early 1970s, the general case

Münzner [22] in 1980 proved a remarkable result that extended Cartan's investigation, recorded in Theorem 10, in a far-reaching manner:

Theorem 16. *Let M be any isoparametric hypersurfaces with g principal curvatures in S^n. Then we have the following.*

(1) *There is a homogeneous polynomial F, called Cartan-Münzner polynomial, of degree g over $\mathbb{R}^n$ satisfying*

$$|\nabla F|^2 = g^2 r^{2g-2}, \quad \Delta F = \frac{m_- - m_+}{2} g^2 r^{g-2},$$

where r is the radial function over $\mathbb{R}^{n+1}$.

(2) *Let $f := F|_{S^n}$. Then the range of f is $[-1,1]$. The only critical values of f are ± 1. Moreover, $M_\pm := f^{-1}(\pm 1)$ are connected submanifolds of codimension $m_\pm + 1$ in S^n, called focal submanifolds, whose principal curvatures are $\cot(k\pi/g), 1 \leq k \leq g-1$.*

(3) *For any $c \in (-1,1)$, $M_c := f^{-1}(c)$ is an isoparametric hypersurface with at most two multiplicities $m_\pm$ associated with the principal curvatures. In fact, if we order the principal curvatures $\lambda_1 > \cdots > \lambda_g$ with multiplicities $m_1, \ldots, m_g$, then $m_i = m_{i+2}$ with index modulo g; in particular, all multiplicities are equal when g is odd, and when g is even, there are at most two multiplicities precisely equal to $m_\pm$.*

(4) *Each of the 1-parameter isoparametric hypersurfaces is a tube around the two focal submanifolds, so that S^n is obtained by gluing two disk bundles over $M_\pm$ along the isoparametric hypersurface $M_0 := f^{-1}(0)$. As a consequence, algebraic topology implies that the only possible values of g are 1, 2, 3, 4, or 6.*

Corollary 17. *$M_\pm$ are minimal submanifolds of S^n. The minimality condition is exactly equation (1), the fundamental formula of Segre and Cartan.*

Corollary 18. *There is a unique minimal isoparametric hypersurface in the 1-parameter family M_c.*

Now that S^n is obtained by gluing two disk bundles over the focal submanifolds $M_\pm$ along an isoparametric hypersurface M, Münzner used algebraic topology to express the cohomology ring of M, with $\mathbb{Z}_2$ coefficients, as modules of those of $M_\pm$, whose module structures then give $g = 1, 2, 3, 4$, or 6.

Based on Münzner's work, Ozeki and Takeuchi [25] in 1975–1976 constructed two classes, each with infinitely many members, of *inhomogeneous* isoparametric hypersurfaces with $g = 4$. They also classified all isoparametric hypersurfaces with $g = 4$ when one of the multiplicities $m_\pm$ is 2, which are all homogeneous.

An important ingredient in their work is their expansion formula of the Cartan-

Münzner polynomial:

$$\begin{aligned}F(tx+y+w) =& t^4 + (2|y|^2 - 6|w|^2)t^2 + 8(\sum_{a=0}^{m_+} p_a w_a)t \\ &+ |y|^4 - 6|y|^2|w|^2 + |w|^4 - 2\sum_{a=0}^{m_+}(p_a)^2 + 8\sum_{a=0}^{m_+} q_a w_a \\ &+ 2\sum_{a,b=0}^{m_+} \langle \nabla p_a, \nabla p_b \rangle w_a w_b.\end{aligned}$$

Here, x is a point on M_+, y is tangent to M_+ at x, and w is normal to M_+ with coordinates w_i with respect to the chosen orthonormal normal basis $\mathbf{n}_0, \mathbf{n}_1, \ldots, \mathbf{n}_{m_+}$ at x. Moreover, $p_a(y)$ (resp., $q_a(y)$) is the a-th component of the 2nd (resp., 3rd) fundamental form of M_+ at x. Furthermore, p_a and q_a are subject to ten convoluted equations, to be seen later, of which the first three assert that, since $S_{\mathbf{n}}$, the 2nd fundamental matrix of M_+ in any unit normal direction $\mathbf{n}$, has eigenvalues $1, -1, 0$ with fixed multiplicities, it must be that $(S_{\mathbf{n}})^3 = S_{\mathbf{n}}$.

The expansion formula coupled with the ten identities are fundamentally important for the classification of isoparametric hypersurfaces with $g = 4$.

4 Development in the 1980s

The multiplicities of the principal values for $g = 4$ and $g = 6$ had remained undetermined until Abresch [2] in 1983 extended Münzner's work to show, by algebraic topology, that for $g = 6$ we have $m_+ = m_- = 1$ or 2. This is in agreement with the multiplicities of the homogeneous examples. Although he derived some constraints in the case $g = 4$, among which we have, for instance, $m_+ = m_-$ implies $m_+ = m_- = 1$ or 2, etc., the case remained open.

Meanwhile, Ferus, Karcher and Münzner [14] in 1981 generalized the inhomogeneous examples of Ozeki and Takeuchi to construct infinitely many classes, each with infinitely many members, of *inhomogeneous* isoparametric hypersurfaces with $g = 4$. Their construction can be best motivated by the example in Nomizu's 1973 paper mentioned earlier:

Consider $\mathbb{C}^k = \mathbb{R}^k \oplus \mathbb{R}^k$ and write $z \in \mathbb{C}^k$ as $z = x + \sqrt{-1}y$ accordingly. Define an homogeneous polynomial of degree 4 on $\mathbb{C}^k$ by

$$\tilde{F} = (|x|^2 - |y|^2)^2 + 4(\langle x, y \rangle)^2.$$

Then F is an isoparametric function with multiplicities $\{1, k-2\}$. In fact, the isoparametric hypersurfaces are the principal isotropy orbits of the symmetric spaces $SO(k+2)/S(2) \times SO(k)$.

Note that $\tilde{f} = \tilde{F}|_{S^{2k-1}}$ has range $[0, 1]$. So we normalize it by defining $f := 1 - 2\tilde{f}$, or rather, by setting

$$F := (|x|^2 + |y|^2)^2 - 2\tilde{F}.$$

F is an isoparametric function such that f has range $[-1,1]$. Let us set

$$P_0 := \begin{pmatrix} I & 0 \\ 0 & -I \end{pmatrix}, \quad P_1 := \begin{pmatrix} 0 & I \\ I & 0 \end{pmatrix}, \quad u := (x,y)^{tr},$$

where I is the k by k identity matrix. Then F can be rewritten as

$$F = |u|^4 - 2\sum_{i=0}^{1}(\langle P_i u, u\rangle)^2, \quad P_iP_j + P_jP_i = 2\delta_{ij}I.$$

Ferus, Karcher and Münzner's construction is a generalization of this.

Definition 19. *The skew-symmetric (resp., symmetric) Clifford algebra C_n (resp., C'_n) over $\mathbb{R}$ is the algebra generated by the standard basis $e_1, \ldots, e_n$ of $\mathbb{R}^n$ subject to the only constraint*

$$e_ie_j + e_je_i = -2\delta_{ij}I \quad (\textit{resp.}, \quad e_ie_j + e_je_i = 2\delta_{ij}I).$$

The classification of the Clifford algebras are known:

n	1	2	3	4	5	6	7	8
C_n	$\mathbb{C}$	$\mathbb{H}$	$\mathbb{H}\oplus\mathbb{H}$	$\mathbb{H}(2)$	$\mathbb{C}(4)$	$\mathbb{R}(8)$	$\mathbb{R}(8)\oplus\mathbb{R}(8)$	$\mathbb{R}(16)$
δ_n	1	2	4	4	8	8	8	8
C'_n	$\mathbb{R}\oplus\mathbb{R}$	$\mathbb{R}(2)$	$\mathbb{C}(2)$	$\mathbb{H}(2)$	$\mathbb{H}(2)\oplus\mathbb{H}(2)$	$\mathbb{H}(4)$	$\mathbb{C}(8)$	$\mathbb{R}(16)$
θ_n	2	4	8	8	16	16	16	16

Here, δ_n is the dimension of an irreducible module of C_{n-1}, and θ_n is the dimension of an irreducible module of C'_{n+1}. Moreover, C_n (resp., C'_n) is subject to the periodicity condition $C_{n+8} = C_n \otimes \mathbb{R}(16)$ (resp., $C'_{n+8} = C'_n \otimes \mathbb{R}(16)$). The generators $e_1, \ldots, e_n$ acting on each irreducible module of either C_n or C'_n in the table give rise to n skew-symmetric or symmetric orthogonal matrices $T_1, \ldots, T_n$ satisfying

$$T_iT_j + T_jT_i = \pm 2\delta_{ij}I,$$

a representation of C_n or C'_n on the irreducible module. Note that we have

$$\theta_n = 2\delta_n.$$

This is not fortuitous. It says that we can construct symmetric representations of C'_{m+1} from skew-symmetric representations of C_{m-1}, and vice versa. Indeed, let us be given k irreducible representations $V_1, \ldots, V_k$ of C_{m-1}. Set

$$V := V_1 \oplus \cdots \oplus V_k \simeq \mathbb{R}^l, \quad l = k\delta_m.$$

The representations of $e_1, \ldots, e_{m-1}$ on $V_1, \ldots, V_k$ give rise to $m-1$ skew-symmetric orthogonal matrices $E_1, \ldots, E_{m-1}$ on V. Set

$$P_0 := \begin{pmatrix} I & 0 \\ 0 & -I \end{pmatrix}, \quad P_1 := \begin{pmatrix} 0 & I \\ I & 0 \end{pmatrix}, \quad P_{1+i} = \begin{pmatrix} 0 & E_i \\ -E_i & 0 \end{pmatrix}, \ 1 \le i \le m-1.$$

Then

$$P_iP_j + P_jP_i = 2\delta_{ij}I.$$

$P_0, \dots, P_m$ give a representation of C'_{m+1} on $\mathbb{R}^{2l}$.

Ferus, Karcher and Münzner's examples, referred to as of OT-FKM type henceforth (OT short for Ozeki and Takeuchi), are

$$F := |u|^4 - 2\sum_{i=0}^{m}(\langle P_i u, u\rangle)^2, \quad u \in \mathbb{R}^{2l}, \quad l = k\delta_m.$$

Note that we recover Nomizu's example when $m = 1$.

By a straightforward calculation, we conclude

Proposition 20. *The two multiplicities $m_\pm$ of an isoparametric hypersurface of OT-FKM type are*

$$(m, k\delta_m - m - 1),$$

where $m, k \in \mathbb{N}$ to make the second entry positive.

We have the following table for the multiplicity pair $(m, k\delta_m - m - 1)$.

$\delta_m =$	1	2	4	4	8	8	8	8	16	$\cdots$
$k = 1$	–	–	–	–	$(5,2)$	$(6,1)$	–	–	$(9,6)$	$\cdots$
$k = 2$	–	$(2,1)$	$(3,4)$	$(4,3)$	$(5,10)$	$(6,9)$	$(7,8)$	$(8,7)$	$(9,22)$	$\cdots$
$k = 3$	$(1,1)$	$(2,3)$	$(3,8)$	$(4,7)$	$(5,18)$	$(6,17)$	$(7,16)$	$(8,15)$	$(9,38)$	$\cdots$
$k = 4$	$(1,2)$	$(2,5)$	$(3,12)$	$(4,11)$	$(5,26)$	$(6,25)$	$(7,24)$	$(8,23)$	$(9,54)$	$\cdots$
$k = 5$	$(1,3)$	$(2,7)$	$(3,16)$	$(4,15)$	$(5,34)$	$(6,33)$	$(7,32)$	$(8,31)$	$(9,70)$	$\cdots$
$\vdots$	$\vdots$	$\vdots$	$\vdots$	$\vdots$	$\vdots$	$\vdots$	$\vdots$	$\vdots$	$\vdots$	$\vdots$

Among other things, Ferus, Karcher and Münzner established

Theorem 21. (1) *OT-FKM type with multiplicities on the first, second, fourth columns, $(4,3)$ and $(9,6)$ are exactly the homogeneous examples, except for the two with multiplicities $\{2,2\}$ and $\{4,5\}$ not on the list.*

(2) *OT-FKM type with multiplicities on the third and seventh columns are exactly the inhomogeneous examples constructed by Ozeki and Takeuchi.*

So, except for the first, second and fourth columns, we have infinitely many families, each with infinitely many members, of inhomogeneous isoparametric hypersurfaces with four principal curvatures. Note that we also have the fact that OT-FKM type with multiplicities $(2, l)$ is congruent to the one with multiplicities $(l, 2)$. Note also that Cartan classified the cases when the multiplicities are $\{1,1\}$ and $\{2,2\}$, both being homogeneous.

Wang [31] investigated the topology of OT-FKM type by K-theory and showed that there are many pairs of minimal isoparametric hypersurfaces in spheres, of identical constant scalar curvature, which are diffeomorphic but noncongruent to each other.

Dorfmeister and Neher [12] in 1985 settled one of the two cases when $g = 6$:

Theorem 22. *An isoparametric hypersurface with $g = 6$ and multiplicities $m_\pm = 1$ is homogeneous.*

5 Development in the 1990s

Another remarkable result, via homotopy theory, was achieved in 1999 by Stolz [29], who classified all the possible multiplicity pairs (m_1, m_2), where $m_1 \leq m_2$, of isoparametric hypersurfaces with $g = 4$:

Theorem 23. *The multiplicity pairs $(m_1, m_2), m_1 \leq m_2$, of isoparametric hypersurfaces with four principal curvatures are exactly those in the above table for the OT-FKM type, except for the pairs $(2,2)$ and $(4,5)$ not in the table.*

He established that if $(m_1, m_2), m_1 \leq m_2$, is neither $(2,2)$ nor $(4,5)$, then $m_1 + m_2 + 1$ is a multiple of $2^{\phi(m_1-1)}$, where $\phi(n)$ denotes the number of natural numbers $s, 1 \leq s \leq n$, such that $s \equiv 0, 1, 2, 4 \pmod 8$. One can see easily that such pairs (m_1, m_2) are exactly those for the OT-FKM type in the above table.

His approach is reminiscent of the theorem of Adams [1]:

Theorem 24. *If there are k independent vector fields on S^n, then $n + 1$ is a multiple of $2^{\phi(k)}$.*

The core technique Adams developed for proving the above theorem on vector fields was to what Stolz reduced his proof.

6 Development in the 2000s

The author and his collaborators [7] and later the author [8, 10] established the following:

Theorem 25. *When $g = 4$, except possibly for the case with multiplicity pair $\{7, 8\}$, an isoparametric hypersurface is, up to congruence, either of OT-FKM type, or homogeneous of multiplicity pair $\{2, 2\}$ or $\{4, 5\}$.*

The proof utilizes commutative algebra, algebraic geometry and Stolz's multiplicity result.

Miyaoka [19, 21] recently settled the other case when $g = 6$:

Theorem 26. *An isoparametric hypersurface with $g = 6$ and multiplicities $m_\pm = 2$ is homogeneous.*

She also gave a simpler and more geometric proof of the theorem of Dorfmeister and Neher by the same technique [18, 20].

So as it stands now, only the case with $g = 4$ and $\{m_+, m_-\} = \{7, 8\}$ remains open, for which we know three inhomogeneous examples of OT-FKM type.

7 Idea of attack on the classification for $g = 4$

We outline the simpler and more powerful method employed in [8] and [10]. In the expansion formula

$$\begin{aligned}F(tx+y+w) &= t^4 + (2|y|^2 - 6|w|^2)t^2 + 8(\sum_{a=0}^{m_+} p_a w_a)t \\ &+ |y|^4 - 6|y|^2|w|^2 + |w|^4 - 2\sum_{a=0}^{m_+}(p_a)^2 + 8\sum_{a=0}^{m_+} q_a w_a \\ &+ 2\sum_{a,b=0}^{m_+} \langle \nabla p_a, \nabla p_b \rangle w_a w_b.\end{aligned}$$

of Ozeki and Takeuchi, the second components of the 2nd and 3rd fundamental forms are highly convoluted in ten equations. The first three are that the shape operator S_n satisfies $(S_n)^3 = S_n$. Set

$$< f, g >:= \langle \nabla(f), \nabla(g) \rangle, \quad G = \sum_a p_a^2.$$

Then we have the remaining seven equations:

$$\begin{aligned}&< p_a, q_a >= 0; \\ &< p_a, q_b > + < p_b, q_a >= 0, \quad a \neq b; \\ &<< p_a, p_b >, q_c > + << p_c, p_a >, q_b > + << p_b, p_c >, q_a >= 0, \quad a, b, c \text{ distinct}; \\ &\sum_a p_a q_a = 0; \\ &16(\sum_a q_a^2) = 16G(\sum_i y_i^2) - < G, G >; \\ &8 < q_a, q_a >= 8(< p_a, p_a > (\sum_i y_i^2) - p_a^2) + << p_a, p_a >, G > \\ &\qquad - 24G - 2\sum_b < p_a, p_b >^2; \\ &8 < q_a, q_b >= 8(< p_a, p_b > (\sum_i y_i^2) - p_a p_b) + << p_a, p_b >, G > \\ &\qquad - 2\sum_c < p_a, p_c >< p_b, p_c > .\end{aligned}$$

Here, y_i are the coordinates of y. They seem to be too overwhelming to solve! But,

every cloud has a silver lining.

The 4th of these gory equations is a syzygy equation:

$$\sum_{a=1}^{m_1} p_a q_a = 0.$$

What happens when the homogeneous $p_0, \ldots, p_{m_1}$ of degree 2 form a regular sequence, for which the syzygy is trivial, so that

$$q_a = \sum_b r_{ab} p_b,$$

where $r_{ab} = -r_{ba}$ are homogeneous of degree 1? (That is, we are looking into complete intersections.)

Moreover, what conditions guarantee that $p_0, \ldots, p_{m_1}$ form a regular sequence?

The answer to the second question is Serre's criterion of normal varieties [13]:

Theorem 27. (Special Case) *Over the complex numbers let $p_1, \ldots, p_s$ be a regular sequence of homogeneous polynomials in a polynomial ring, let V be the variety cut out by $p_1, \ldots, p_s$, and let J be the subvariety of V where the rank of the Jacobian of $p_1, \ldots, p_s$ is $< s$. If $\dim(J) \leq \dim(V) - 2$, then the ideal $(p_1, \ldots, p_s)$ is prime.*

Corollary 28. ([7]) *Over the complex numbers, let $p_1, \ldots, p_k, k \geq 2$, be linearly independent homogeneous polynomials of equal degree ≥ 1 in a polynomial ring. For $i \leq k$, let V_i be the variety cut out by $p_1, \ldots, p_i$, and let J_i be the subvariety of V_i where the rank of the Jacobian matrix of $p_1, \ldots, p_i$ is $< i$. If $\dim(J_i) \leq \dim(V_i) - 2$ for $1 \leq i \leq k-1$, then $p_1, \ldots, p_k$ form a regular sequence in the polynomial ring.*

The answer to the first question is then facilitated by the following:

Theorem 29. ([8]) *Let $\{m_1, m_2\}$ be the multiplicity pair of an isoparametric hypersurface with four principal values in the sphere. If $m_1 < m_2$ and $p_0, \ldots, p_{m_1}$ form a regular sequence, then the hypersurface is of OT-FKM type.*

Theorem 25 follows from Corollary 28 and Theorem 29.

In fact, when $m_2 \geq 2m_1 - 1$, one can show that indeed $p_0, \ldots, p_{m_1}$ do form a regular sequence by Corollary 28, so that the isoparametric hypersurface is of OT-FKM type, which had been proven in [7] by a considerably more complicated method that does not seem to extend to the remaining cases. By Stolz's multiplicity result, this takes care of exactly the multiplicity pairs except for the anomalous ones $\{3,4\}, \{4,5\}, \{6,9\}, \{7,8\}$, for which $p_0, \ldots, p_{m_1}, m_1 < m_2$, *do not* form a regular sequence in general. The zero locus of a nonregular sequence, even over complex numbers, can be wildly untamed. However, one can now employ the notion of *Condition A* of Ozeki and Takeuchi [25]:

Definition 30. *A focal submanifold of an isoparametric hypersurface with four principal values in the sphere is said to be of Condition A, if its shape operator has a fixed kernel in all normal directions at some point.*

The focal submanifold satisfying Condition A must be of codimension 2, 4, or 8 in the sphere [25], so that, in particular, the associated isoparametric hypersurface cannot be of multiplicity pair $\{4,5\}$ or $\{6,9\}$. On the other hand, the isoparametric hypersurfaces of OT-FKM type with multiplicity pair $\{3,4\}$ or $\{7,8\}$ do admit points of Condition A on the focal submanifold with codimension 4 or 8 [25, 14].

Conversely, it is proved in [11] that the existence of points of Condition A implies that the isoparametric hypersurface is of OT-FKM type (see also [9]).

With Condition A and Theorem 29, one can now conclude [8, 10] that the isoparametric hypersurface is either homogeneous in the $\{4,5\}$ case, or is of OT-FKM type in the $\{3,4\}$ and $\{6,9\}$ cases, for which the codimension 2 estimates in Corollary 28 is manageable. The essential point is that, for the three multiplicity pairs $\{3,4\}, \{4,5\}$, and $\{6,9\}$, nonexistence of points of Condition A (even locally) implies that either $p_0, \ldots, p_{m_1}, m_1 < m_2$, form a regular sequence, so that the isoparametric hypersurface is of OT-FKM type, or the 2nd and, hence, the 3rd fundamental forms of the hypersurface coincide with those of the homogeneous example.

References

[1] J. F. Adams, *Vector fields on spheres*, Ann. Math. **75** (1962), 603–632.

[2] U. Abresch, *Isoparametric hypersurfaces with four or six principal curvatures*, Math. Ann. **264** (1983), 283–302.

[3] E. Cartan, *Familles de surfaces isoparamétriques dans les espaces à courbure constante*, Ann. Mat. Pura Appl. (4) **17** (1938), 177–191.

[4] E. Cartan, *Sur des familles remarquables d'hypersurfaces isoparamétriques dans les espaces sphériques*, Math. Z. **45** (1939), 335–367.

[5] E. Cartan, *Sur quelque familles remarquables d'hypersurfaces*, C. R. Congr. Math. Liège (1939), 30–41.

[6] E. Cartan, *Sur des familles d'hypersurfaces isoparamétriques des espaces sphériques à 5 et à 9 dimensions*, Rev. Univ. Tucuman, Ser. A **1** (1940), 5–22.

[7] T. Cecil, Q.-S. Chi and G. Jensen, *Isoparametric hypersurfaces with four principal curvatures,* Ann. Math. **166** (2007), 1–76.

[8] Q.-S. Chi, *Isoparametric hypersurfaces with four principal curvatures*, II, Nagoya Math. J. **204** (2011), 1–18.

[9] Q. -S. Chi, *A new look at Condition A*, Osaka J. Math. **49** (2012), 133–166.

[10] Q. -S. Chi, *Isoparametric hypersurfaces with four principal curvatures*, III, J. Diff. Geom. **94** (2013), 469–504.

[11] J. Dorfmeister and E. Neher, *Isoparametric triple systems of algebra type*, Osaka J. Math. **20** (1983), 145–175.

[12] J. Dorfmeister and E. Neher, *Isoparametric hypersurfaces, case $g = 6, m = 1$*, Communications in Algebra **13** (1985), 2299–2368.

[13] D. Eisenbud, *Commutative Algebra with a View Toward Algebraic Geometry*, Springer-Verlag, New York, 1995.

[14] D. Ferus, H. Karcher and H.-F. Münzner, *Cliffordalgebren und neue isoparametrische Hyperfläschen,* Math. Z. **177** (1981), 479–502.

[15] W. Y. Hsiang and H. B. Lawson, *Minimal submanifolds with low cohomogeneity*, J. Diff. Geom. **5** (1971), 1–38.

[16] E. Laura, *Sopra la propagazione di onde in un mezzo indefinito*, Scritti matematici offerti ad Enrico D'Ovidio (1918), 253–278.

[17] T. Levi-Civita, *Famiglie di superficie isoparametrische nell'ordinario spacio euclideo*, Atti Accad. Naz. Lincei Cl. Sci. Fis. Mat. Natur. Rend. Lincei (9) Math. Appl. **26** (1937), 355–362.

[18] R. Miyaoka, *The Dorfmeister-Neher theorem on isoparametric hypersurfaces*, Osaka J. Math. **46** (2009), 695–715.

[19] R. Miyaoka, *Isoparametric hypersurfaces with* $(g, m) = (6, 2)$, Ann. Math. **177** (2013), 53–110.

[20] R. Miyaoka, *Remarks on the Dorfmeister-Neher theorem on isoparametric hypersurfaces*, preprint.

[21] R. Miyaoka, *Errata on isoparametric hypersurfaces with* $(g, m) = (6, 2)$, preprint.

[22] H. F. Münzner, *Isoparametrische Hyperflächen in Sphären*, I, Math. Ann. **251** (1980); II, **256** (1981).

[23] K. Nomizu, *Some results in E. Cartan's theory of isoparametric families of hypersurfaces*, Bull. Amer. Math. Soc. **79** (1973), 1184–1188.

[24] K. Nomizu, *Élie Cartan's work on isoparametric families of hypersurfaces*, Proc. Sympos. Pure Math. **27** (1975), 191–200.

[25] H. Ozeki and M. Takeuchi, *On some types of isoparametric hypersurfaces in spheres*, I, Tôhoku Math. J. **27** (1975), 515-559; II, **28** (1976), 7–55.

[26] B. Segre, *Una Proprietá caratteristixca di tre sistemi* ∞^1 *di superficie*, Atti Accad. Sci. Torino Cl. Sci. Fis. Mat. Natur. **29** (1924), 666–671.

[27] B. Segre, *Famiglie di ipersuperficie isoparametrische negli spazi euclidei ad un qualunque numero di demesioni*, Atti Accad. Naz. Lincie Cl. Sci. Fis. Mat. Natur. Rend. Lincei (9) Mat. Appl. **27** (1938), 203–207.

[28] C. Somigliana, *Sulle relazione fra il principio di Huygens e l'ottica geometrica*, Atti. Accad. Sci. Torino Cl. Sci. Fis. Mat. Natur. **24** (1918–1919), 974–979.

[29] S. Stolz, *Multiplicities of Dupin hypersurfaces*, Inven. Math. **138** (1999), 253–279.

[30] R. Takagi and T. Takahashi, *On the Principal Curvatures of Homogeneous Hypersurfaces in a Sphere*, Differential Geometry in honor of K. Yano, 469–481, Kinokuniya, Tokyo, 1972.

[31] Q. -M. Wang, *On the topology of Clifford isoparametric hypersurfaces*, J. Diff. Geom. **27** (1988), 55–66.

[32] S. T. Yau, Open problems in geometry, *Chern—A Great Geometer of the Twentieth Century*, International Press, 1992, 275-319.

DEPARTMENT OF MATHEMATICS, WASHINGTON UNIVERSITY
ST. LOUIS, MO 63017, USA
E-mail address: chi@math.wustl.edu

Proceedings of the 6th International Congress of Chinese Mathematicians
ALM 36, pp. 453–465

New Invariants for Higher Dimensional Singularities and the Complex Plateau Problem

Rong Du[†], Yun Gao[††], and Stephen Yau[*]

Abstract

Let X be a compact connected strongly pseudoconvex CR manifold of real dimension $2n-1$ in $\mathbb{C}^N$. It has been an interesting question to find an intrinsic smoothness criteria for the complex Plateau problem. For $n \geq 3$ and $N = n+1$, Yau found a necessary and sufficient condition for the interior regularity of the Harvey–Lawson solution to the complex Plateau problem by means of Kohn–Rossi cohomology groups on X in 1981. For $n = 2$ and $N \geq n+1$, the first and third authors introduced a new CR invariant $g^{(1,1)}(X)$ of X. The vanishing of this invariant will give the interior regularity of the Harvey-Lawson solution up to normalization. For $n \geq 3$ and $N > n+1$, the problem still remains open. In this paper, we survey some recent results on such problem.

Mathematics Subject Classification: 32V15, 32S05, 32S20.

Keywords and Phrases: Complex Plateau problem, CR manifold, Harvey–Lawson solution.

1 Introduction

One of the natural fundamental questions of complex geometry is to study the boundaries of complex varieties. For example, the famous classical complex Plateau

[†] The Research Sponsored by the National Natural Science Foundation of China (Grant No. 11471116), Science and Technology Commission of Shanghai Municipality (Grant No. 13dz2260400) and Shanghai Pujiang Program. The first author would also like thank Professor N. Mok for supporting his research when he was in the University of Hong Kong.

[††] The Research Sponsored by the National Natural Science Foundation of China and the Innovation Program of Shanghai Municipal Education Commission.

[*] Research partially supported by Department of Mathematical Sciences, Tsinghua University, Beijing, China.

problem asks which odd dimensional real sub-manifolds of $\mathbb{C}^N$ are boundaries of complex sub-manifolds in $\mathbb{C}^N$. In their beautiful seminal paper, Harvey and Lawson [Ha-La1] proved that for any compact connected CR manifold X of real dimension $2n-1$, $n \geq 2$, in $\mathbb{C}^N$, there is a unique complex variety V in $\mathbb{C}^N$ such that the boundary of V is X. In fact, Harvey and Lawson proved the following theorem.

Theorem. (Harvey–Lawson [Ha-La1, Ha-La2]) Let X be an embeddable strongly pseudoconvex CR manifold. Then X can be CR embedded in some $\mathbb{C}^{\widetilde{N}}$ and X bounds a unique Stein variety $V \subseteq \mathbb{C}^{\widetilde{N}}$ with at most isolated singularities.

The above theorem relates theory of strongly pseudoconvex CR manifolds on the one hand and the theory of isolated normal singularities on the other hand. We call V in the above theorem the Harvey–Lawson solution to X.

The next question is to determine when X is a boundary of a complex sub-manifold in $\mathbb{C}^N$, i.e., when V is smooth. In 1981, Yau [Ya] solved this problem for the case $n \geq 3$ by calculation of Kohn–Rossi cohomology groups $H^{p,q}_{KR}(X)$. More precisely, suppose X is a compact connected strongly pseudoconvex CR manifold of real dimension $2n-1$, $n \geq 3$, in the boundary of a bounded strongly pseudoconvex domain D in $\mathbb{C}^{n+1}$. Then X is a boundary of the complex sub-manifold $V \subset D - X$ if and only if Kohn–Rossi cohomology groups $H^{p,q}_{KR}(X)$ are zeros for $1 \leq q \leq n-2$ (see Theorem 5.1).

For $n = 2$, i.e. X is a 3-dimensional CR manifold, the intrinsic smoothness criteria for the complex Plateau problem remains unsolved for over a quarter of a century even for the hypersurface case. The main difficulty is that the Kohn–Rossi cohomology groups are infinite dimensional in this case. Let V be a complex variety with X as its boundary. In [Du-Ya], the first and the third authors introduced an new CR invariant $g^{(1,1)}(X)$ to solve the regularity problem of the Harvey–Lawson solution to the complex Plateau problem. More precisely, they showed that if X is a strongly pseudoconvex compact Calabi–Yau CR manifold of dimension 3 and X is contained in the boundary of a strongly pseudoconvex bounded domain D in $\mathbb{C}^N$ with holomorphic De Rham cohomology $H^2_h(X) = 0$ (this condition can be omitted (see [Du]), then X is a boundary of the complex sub-manifold up to normalization $V \subset D - X$ with boundary regularity if and only if $g^{(1,1)}(X) = 0$. In particular, if $N = 3$, then X is a boundary of the complex sub-manifold $V \subset D - X$ if and only if $g^{(1,1)}(X) = 0$.

For $n \geq 3$ and $N > n+1$, i.e., non-hypersurface type, the complex Plateau problem still remains open. Recently, in the paper [Du-Ga-Ya], we generalize the invariant $g^{(1,1)}(X)$ to higher dimension as $g^{(\Lambda^n 1)}(X)$ and show that if $g^{(\Lambda^n 1)}(X) = 0$, then the interior which X bounds has at most finite rational singularities. In particular, if X is Calabi–Yau of real dimension 5, i.e., $n = 3$, then the vanishing of this invariant is equivalent to give the interior regularity. The purpose of this paper is to give a self-contained survey of the results obtained in [Du-Ga-Ya].

2 Strongly pseudoconvex CR manifolds

Kohn–Rossi cohomology was first introduced by Kohn–Rossi. Following Tanaka [Ta], we reformulate the definition in a way independent of the interior manifold.

Definition 2.1. *Let X be a connected orientable manifold of real dimension $2n-1$. A CR structure on X is an $(n-1)$-dimensional subbundle S of $\mathbb{C}T(X)$ (complexified tangent bundle) such that*

1. $S\bigcap \bar{S} = \{0\}$,

2. *If L, L' are local sections of S, then so is $[L, L']$.*

Such a manifold with a CR structure is called a CR manifold. There is a unique subbundle $\mathcal{H}$ of $T(X)$ such that $\mathbb{C}\mathcal{H} = S\oplus\bar{S}$. Furthermore, there is a unique homomorphism $J : \mathcal{H} \longrightarrow \mathcal{H}$ such that $J^2 = -1$ and $S = \{v - iJv : v \in \mathcal{H}\}$. The pair $(\mathcal{H}, J)$ is called the real expression of the CR structure.

Let X be a CR manifold with structure S. For a complex valued C^∞ function u defined on X, the section $\bar{\partial}_b u \in \Gamma(\bar{S}^*)$ is defined by

$$\bar{\partial}_b u(\bar{L}) = \bar{L}(u), \quad L \in S.$$

The differential operator $\bar{\partial}_b$ is called the (tangential) Cauchy–Riemann operator, and a solution u of the equation $\bar{\partial}_b u = 0$ is called a holomorphic function.

Definition 2.2. *A complex vector bundle E over X is said to be holomorphic if there is a differential operator*

$$\bar{\partial}_E : \Gamma(E) \longrightarrow \Gamma(E \otimes \bar{S}^*)$$

satisfying the following conditions:

1. $\bar{\partial}_E(fu)(\bar{L}_1) = (\bar{\partial}_b f)(\bar{L}_1)u + f(\bar{\partial}_E u)(\bar{L}_1) = (\bar{L}_1 f)u + f(\bar{\partial}_E u)(\bar{L}_1)$,

2. $(\bar{\partial}_E u)[\bar{L}_1, \bar{L}_2] = \bar{\partial}_E(\bar{\partial}_E u(\bar{L}_2))(\bar{L}_1) - \bar{\partial}_E(\bar{\partial}_E u(\bar{L}_1))(\bar{L}_2)$,

where $u \in \Gamma(E)$, $f \in C^\infty(X)$ and L_1, $L_2 \in \Gamma(S)$.

The operator $\bar{\partial}_E$ is called the Cauchy–Riemann operator and a solution u of the equation $\bar{\partial}_E u = 0$ is called a holomorphic cross section.

A basic holomorphic vector bundle over a CR manifold X is the vector bundle $\widehat{T}(X) = \mathbb{C}T(X)/\bar{S}$. The corresponding operator $\bar{\partial} = \bar{\partial}_{\widehat{T}(X)}$ is defined as follows. Let p be the projection from $\mathbb{C}T(X)$ to $\widehat{T}(X)$. Take any $u \in \Gamma(\widehat{T}(X))$ and express it as $u = p(Z)$, $Z \in \Gamma(\mathbb{C}T(X))$. For any $L \in \Gamma(S)$, define a cross section $(\bar{\partial}u)(\bar{L})$ of $\widehat{T}(X)$ by $(\bar{\partial}u)(\bar{L}) = p([\bar{L}, Z])$. One can show that $(\bar{\partial}u)(\bar{L})$ does not depend on the choice of Z and that $\bar{\partial}u$ gives a cross section of $\widehat{T}(X) \otimes \bar{S}^*$. Furthermore one can show that the operator $u \longmapsto \bar{\partial}u$ satisfies (1) and (2) of Definition 2.2, using the Jacobi identity in the Lie algebra $\Gamma(\mathbb{C}T(X))$. The resulting holomorphic vector bundle $\widehat{T}(X)$ is called the holomorphic tangent bundle of X.

If X is a real hypersurface in a complex manifold M, we may identify $\widehat{T}(M)$ with the holomorphic vector bundle of all $(1,0)$ tangent vectors to M and $\widehat{T}(X)$ with the restriction of $\widehat{T}(M)$ to X. In fact, since the structure S of X is the bundle of all $(1,0)$ tangent vectors to X, the inclusion map $\mathbb{C}T(X) \longrightarrow \mathbb{C}T(M)$ induces a natural map $\widehat{T}(X) \xrightarrow{\phi} \widehat{T}(M)|_X$ which is a bundle isomorphism satisfying $\bar{\partial}(\phi(u))(\bar{L}) = \phi(\bar{\partial}u(\bar{L}))$, $u \in \Gamma(\widehat{T}(X))$, $L \in S$.

For a holomorphic vector bundle E over X, set

$$C^q(X,E) = E \otimes \wedge^q \bar{S}^*, \quad \mathscr{C}^q(X,E) = \Gamma(C^q(X,E))$$

and define a differential operator

$$\bar{\partial}_E^q : \mathscr{C}^q(X,E) \longrightarrow \mathscr{C}^{q+1}(X,E)$$

by

$$\begin{aligned}(\bar{\partial}_E^q\phi)(\bar{L}_1,\ldots,\bar{L}_{q+1}) &= \sum_i (-1)^{i+1}\bar{\partial}_E(\phi(\bar{L}_1,\ldots,\widehat{\bar{L}_i},\ldots,\bar{L}_{q+1}))(\bar{L}_i)\\ &\quad + \sum_{i<j}(-1)^{i+j}\phi([\bar{L}_i,\bar{L}_j],\bar{L}_1,\ldots,\widehat{\bar{L}_i},\ldots,\bar{L}_{q+1})\end{aligned}$$

for all $\phi \in \mathscr{C}^q(X,E)$ and $L_1,\ldots,L_{q+1} \in \Gamma(S)$. One shows by standard arguments that $\bar{\partial}_E^q\phi$ gives an element of $\mathscr{C}^{q+1}(X,E)$ and that $\bar{\partial}_E^{q+1}\bar{\partial}_E^q = 0$. The cohomology groups of the resulting complex $\{\mathscr{C}^q(X,E), \bar{\partial}_E^q\}$ is denoted by $H^q(X,E)$.

Let $\{\mathscr{A}^k(X), d\}$ be the De Rham complex of X with complex coefficients, and let $H^k(X)$ be the De Rham cohomology groups. There is a natural filtration of the De Rham complex, as follows. For any integer p and k, put $A^k(X) = \wedge^k(\mathbb{C}T(X)^*)$ and denote by $F^p(A^k(X))$ the subbundle of $A^k(X)$ consisting of all $\phi \in A^k(X)$ which satisfy the equality

$$\phi(Y_1,\ldots,Y_{p-1},\bar{Z}_1,\ldots,\bar{Z}_{k-p+1}) = 0$$

for all $Y_1,\ldots,Y_{p-1} \in \mathbb{C}T(X)_0$ and $Z_1,\ldots,Z_{k-p+1} \in S_0$, 0 being the origin of ϕ. Then

$$\begin{aligned}A^k(X) &= F^0(A^k(X)) \supset F^1(A^k(X)) \supset \cdots\\ &\supset F^k(A^k(X)) \supset F^{k+1}(A^k(X)) = 0.\end{aligned}$$

Setting $F^p(\mathscr{A}^k(X)) = \Gamma(F^p(A^k(X)))$, we have

$$\begin{aligned}\mathscr{A}^k(X) &= F^0(\mathscr{A}^k(X)) \supset F^1(\mathscr{A}^k(X)) \supset \cdots\\ &\supset F^k(\mathscr{A}^k(X)) \supset F^{k+1}(\mathscr{A}^k(X)) = 0.\end{aligned}$$

Since clearly $dF^p(\mathscr{A}^k(X)) \subseteq F^p(\mathscr{A}^{k+1}(X))$, the collection $\{F^p(\mathscr{A}^k(X))\}$ gives a filtration of the De Rham complex.

We denote by $H_{KR}^{p,q}(X)$ the groups $E_1^{p,q}(X)$ of the spectral sequence $\{E_r^{p,q}(X)\}$ associated with the filtration $\{F^p(\mathscr{A}^k(X))\}$. We call $H_{KR}^{p,q}(X)$ the Kohn–Rossi cohomology group of type (p,q). More explicitly, let

$$A^{p,q}(X) = F^p(A^{p+q}(X)), \quad \mathscr{A}^{p,q}(X) = \Gamma(A^{p,q}(X)),$$

$$C^{p,q}(X) = A^{p,q}(X)/A^{p+1,q-1}(X), \quad \mathscr{C}^{p,q}(X) = \Gamma(C^{p,q}(X)).$$

Since $d : \mathscr{A}^{p,q}(X) \longrightarrow \mathscr{A}^{p,q+1}(X)$ maps $\mathscr{A}^{p+1,q-1}(X)$ into $\mathscr{A}^{p+1,q}(X)$, it induces an operator $d'' : \mathscr{C}^{p,q}(X) \longrightarrow \mathscr{C}^{p,q+1}(X)$. $H^{p,q}_{KR}(X)$ are then the cohomology groups of the complex $\{\mathscr{C}^{p,q}(X), d''\}$.

Alternatively $H^{p,q}_{KR}(X)$ may be described in terms of the vector bundle $E^p = \wedge^p(\widehat{T}(X)^*)$. If for $\phi \in \Gamma(E^p)$, $u_1, \dots, u_p \in \Gamma(\widehat{T}(X))$, $Y \in S$, we define $(\bar\partial_{E^p}\phi)(\bar Y) = \bar Y\phi$ by

$$\bar Y\phi(u_1, \dots, u_p) = \bar Y(\phi(u_1, \dots, u_p)) + \sum_i (-1)^i \phi(\bar Y u_i, u_1, \dots, \widehat{u_i}, \dots, u_p)$$

where $\bar Y u_i = (\bar\partial_{\widehat{T}(X)} u_i)(\bar Y)$, then we easily verify that E^p with $\bar\partial_{E^p}$ is a holomorphic vector bundle. Tanaka [Ta] proves that $C^{p,q}(X)$ may be identified with $C^q(X, E^p)$ in a natural manner such that

$$d''\phi = (-1)^p \bar\partial_{E^p}\phi, \quad \phi \in \mathscr{C}^{p,q}(X).$$

Thus, $H^{p,q}_{KR}(X)$ may be identified with $H^q(X, E^p)$.

We denote by $H^k_h(X)$ the groups $E_2^{k,0}(X)$ of the spectral sequence $\{E_r^{p,q}(X)\}$ associated with the filtration $\{F^p(\mathscr{A}^k(X))\}$. We call $H^k_h(X)$ the holomorphic De Rham cohomology groups. The groups $H^k_h(X)$ are the cohomology groups of the complex $\{\mathscr{S}^k(X), d\}$, where we put $\mathscr{S}^k(X) = E_1^{k,0}(X)$ and $d = d_1 : E_1^{k,0} \longrightarrow E_1^{k+1,0}$. Recall that $\mathscr{S}^k(X)$ is the kernel of the following mapping:

$$d_0 : E_0^{k,0} = F^k\mathscr{A}^k = \mathscr{A}^{k,0}(X)$$
$$\rightarrow E_0^{k,1} = F^k\mathscr{A}^{k+1}/F^{k+1}\mathscr{A}^{k+1} = \mathscr{A}^{k,1}(X)/\mathscr{A}^{k+1,0}.$$

Note that $\mathscr{S}$ may be characterized as the space of holomorphic k-forms, namely holomorphic cross sections of E^k. Thus the complex $\{\mathscr{S}^k(X), d\}$ (respectively, the groups $H^k_h(X)$ will be called the holomorphic De Rham complex (respectively, the holomorphic De Rham cohomology groups).

Definition 2.3. *Let $L_1, \dots, L_{n-1}$ be a local frame of the CR structure S on X so that $\bar L_1, \dots, \bar L_{n-1}$ is a local frame of $\bar S$. Since $S \oplus \bar S$ has complex codimension one in $\mathbb{C}T(X)$, we may choose a local section N of $\mathbb{C}T(X)$ such that $L_1, \dots, L_{n-1}, \bar L_1, \dots, \bar L_{n-1}, N$ span $\mathbb{C}T(X)$. We may assume that N is purely imaginary. Then the matrix (c_{ij}) defined by*

$$[L_i, \bar L_j] = \sum_k a^k_{i,j} L_k + \sum_k b^k_{i,j} \bar L_k + c_{i,j} N$$

is Hermitian, and is called the Levi form of X.

Proposition 2.4. *The number of non-zero eigenvalues and the absolute value of the signature of (c_{ij}) at each point are independent of the choice of $L_1, \dots, L_{n-1}, N$.*

Definition 2.5. *X is said to be strongly pseudoconvex if the Levi form is positive definite at each point of X.*

Definition 2.6. *Let X be a CR manifold of real dimension $2n-1$. X is said to be Calabi–Yau if there exists a nowhere vanishing holomorphic section in $\Gamma(\wedge^n \widehat{T}(X)^*)$, where $\widehat{T}(X)$ is the holomorphic tangent bundle of X.*

Remark 2.7. 1. Let X be a CR manifold of real dimension $2n-1$ in $\mathbb{C}^n$. Then X is a Calabi–Yau CR manifold.

2. Let X be a strongly pseudoconvex CR manifold of real dimension $2n-1$ contained in the boundary of bounded strongly pseudoconvex domain in $\mathbb{C}^{n+1}$. Then X is a Calabi–Yau CR manifold.

3 Minimal discrepancy and 3-dimensional canonical Gorenstein singularities

Harvay–Lawson's theorem relates theory of strongly pseudoconvex CR manifolds and the theory of isolated normal singularities, so the complex Plateau problem becomes the problem of singularities in some sense. For $n \geq 3$, we need to consider higher dimensional singularities. There are two types of singularities, canonical and terminal singularities, being considered by many mathematicians for classifying algebraic varieties of higher dimension. Canonical singularities appear as singularities of the canonical model of a projective variety, and terminal singularities are special cases that appear as singularities of minimal models. They were introduced by Reid in 1980 ([Re]). Terminal singularities are important in the minimal model program because smooth minimal models do not always exist, and thus one must allow certain singularities, namely the terminal singularities.

Suppose that X is a normal variety such that its canonical class K_X is $\mathbb{Q}$-Cartier, and let $f : Y \to X$ be a resolution of the singularities of X. Then

$$K_Y = f^* K_X + \sum_i a_i E_i,$$

where the sum is over the irreducible exceptional divisors, and the a_i are rational numbers, called the discrepancies.

Then the singularities of X are called:

$$\begin{cases} \text{terminal} & a_i > 0 \quad \text{for all i,} \\ \text{canonical} & a_i \geq 0 \quad \text{for all i,} \\ \text{log-terminal} & a_i > -1 \quad \text{for all i,} \\ \text{log-canonical} & a_i \geq -1 \quad \text{for all i.} \end{cases} \tag{3.1}$$

Definition 3.1. *The minimal discrepancy of a variety X at 0, denoted by $Md_0(X)$ (or $Md(X)$ for short), is the minimum of all discrepancies of discrete valuations of $\mathbb{C}(X)$, whose center on X is 0.*

Remark 3.2. The minimal discrepancy only exists when X has log-canonical singularities (see, e.g. [C-K-M]). Whenever $Md(X)$ exists it is at least -1.

Shokurov conjectured that the minimal discrepancy is bounded above in term of the dimension of a variety.

Conjecture 3.3. (Shokurov [Sh]) *The minimal discrepancy $Md_0(X)$ of a variety X at 0 of dimension n is at most $n-1$. Moreover, if $Md_0(X) = n-1$, then $(X, 0)$ is nonsingular.*

The conjecture was confirmed for surfaces ([Al]) and 3-dimensional singularities after the explicit classification ([Re2]) of Gorenstein terminal 3-fold singularities with [E-M-Y] or [Ma]. If X is a local complete intersection, then the conjecture also holds (see [E-M] and [E-M-Y]).

In this paper, we are going to consider the 3-dimensional singularities. Mori ([Mo]), Cutkosky ([Cu]) and Brenton ([Br]) had some classification theorems about special 3-folds singularities. The following two theorems will be used to prove our main theorems.

Theorem 3.4. ([Re] Corollary 2.14, Corollary 2.12) *Let $0 \in V$ be a 3-dimensional rational Gorenstein singularity, then there exists a partial resolution $\pi : \widetilde{V} \to V$ such that $\pi^{-1}\{0\}$ is a union of nonsingular rational or elliptic ruled surfaces, $\widetilde{V}$ only has terminal singularities and $K_{\widetilde{V}} = \pi^* K_V$.*

Remark 3.5. The information of the irreducible and reduced components of the exceptional set can also be found in [Br].

Definition 3.6. *A singular point 0 is called compound Du Val (cDV) if for a general section H through 0, $0 \in H$ is a Du Val singularities.*

Remark 3.7. A cDV singularity is formally equivalent to the germ of a hypersurface singularity $(\{f = 0\}, 0)$ in $\mathbb{C}^4$, where

$$f(x_0, x_1, x_2, x_3) = f_{X_n}(x_0, x_1, x_2) + x_3 g(x_0, x_1, x_2, x_3), \tag{3.2}$$

where X_n stands for A_n, D_n or E_n, $g(0,0,0,0) = 0$ and f_{X_n} is one of the following polynomials:

$$f_{A_n} = x_0^2 + x_1^2 + x_2^{n+1} \quad (n \geq 1), \tag{3.3}$$

$$f_{D_n} = x_0^2 + x_1^2 x_2 + x_2^{n-1} \quad (n \geq 4), \tag{3.4}$$

$$f_{E_6} = x_0^2 + x_1^3 + x_2^4, \tag{3.5}$$

$$f_{E_7} = x_0^2 + x_1^3 + x_1 x_2^3, \tag{3.6}$$

$$f_{E_8} = x_0^2 + x_1^3 + x_2^5. \tag{3.7}$$

Theorem 3.8. ([Re2] Theorem 1.1) *Let $0 \in V$ be a 3-dimensional singularity, then 0 is an isolated cDV singularity if and only if 0 is Gorenstein terminal.*

4 New invariants of singularities and new CR invariants

Let V be a n-dimensional complex analytic subvariety in $\mathbb{C}^N$ with only isolated singularities. In [Ya2], Yau considered four kinds of sheaves of germs of holomorphic p-forms

1. $\bar{\Omega}_V^p := \pi_* \Omega_M^p$, where $\pi : M \longrightarrow V$ is a resolution of singularities of V.

2. $\bar{\bar{\Omega}}_V^p := \theta_* \Omega_{V \backslash V_{sing}}^p$ where $\theta : V \backslash V_{sing} \longrightarrow V$ is the inclusion map and V_{sing} is the singular set of V.

3. $\Omega_V^p := \Omega_{\mathbb{C}^N}^p / \mathscr{K}^p$, where $\mathscr{K}^p = \{f\alpha + dg \wedge \beta : \alpha \in \Omega_{\mathbb{C}^N}^p; \beta \in \Omega_{\mathbb{C}^N}^{p-1}; f, g \in \mathscr{I}\}$ and $\mathscr{I}$ is the ideal sheaf of V in $\mathbb{C}^N$.

4. $\widetilde{\Omega}_V^p := \Omega_{\mathbb{C}^N}^p / \mathscr{H}^p$, where $\mathscr{H}^p = \{\omega \in \Omega_{\mathbb{C}^N}^p : \omega|_{V \backslash V_{sing}} = 0\}$.

Clearly Ω_V^p, $\widetilde{\Omega}_V^p$ are coherent. $\bar{\Omega}_V^p$ is a coherent sheaf because π is a proper map. $\bar{\bar{\Omega}}_V^p$ is also a coherent sheaf by a theorem of Siu (cf. Theorem A of [Si]). In case V is a normal variety, the dualizing sheaf ω_V of Grothendieck is actually the sheaf $\bar{\bar{\Omega}}_V^n$.

Definition 4.1. *Let V be a n-dimensional Stein space with 0 as its only singular point. Let $\pi : (M, A) \to (V, 0)$ be a resolution of the singularity with A as exceptional set. The geometric genus p_g and the irregularity q of the singularity are defined as follows (cf. [Ya2], [St-St]):*

$$p_g := \dim \Gamma(M \backslash A, \Omega^n) / \Gamma(M, \Omega^n), \tag{4.1}$$

$$q := \dim \Gamma(M \backslash A, \Omega^{n-1}) / \Gamma(M, \Omega^{n-1}), \tag{4.2}$$

$$g^{(p)} := \dim \Gamma(M, \Omega_M^p) / \pi^* \Gamma(V, \Omega_V^p). \tag{4.3}$$

Let X be a compact connected strongly pseudoconvex CR manifold of real dimension $2n - 1$, in the boundary of a bounded strongly pseudoconvex domain D in $\mathbb{C}^N$. By a result of Harvey and Lawson mentioned in the introduction, there is a unique complex variety V in $\mathbb{C}^N$ such that the boundary of V is X. Let $\pi : (M, A_1, \ldots, A_k) \to (V, 0_1, \ldots, 0_k)$ be a resolution of the singularities with $A_i = \pi^{-1}(0_i)$, $1 \le i \le k$, as exceptional sets.

In order to solve the classical complex Plateau problem, we need to find some CR invariant which can be calculated directly from the boundary X and the vanishing of this invariant will give regularity of Harvey–Lawson solution to complex Plateau problem.

For this purpose, we define a new sheaf $\bar{\bar{\Omega}}_V^{1,1}$, new invariant of surface singularities $g^{(1,1)}$ and new CR invariant $g^{(1,1)}(X)$ in [Du-Ya]. Now, we are going to generalize them to higher dimension for dealing with general complex Plateau problem.

Definition 4.2. *Let $(V, 0)$ be a Stein germ of a n-dimensional analytic space with an isolated singularity at 0. Define a sheaf of germs $\bar{\bar{\Omega}}_V^{\Lambda^p 1}$ by the sheaf associated to the presheaf*

$$U \mapsto < \Lambda^p \Gamma(U, \bar{\bar{\Omega}}_V^1) >,$$

where U is an open set of V and $2 \le p \le n$.

Lemma 4.3. *Let V be a n-dimensional Stein space with 0 as its only singular point in $\mathbb{C}^N$. Let $\pi : (M, A) \to (V, 0)$ be a resolution of the singularity with A as exceptional set. Then $\bar{\bar{\Omega}}_V^{\Lambda^p 1}$ is coherent and there is a short exact sequence*

$$0 \longrightarrow \bar{\bar{\Omega}}_V^{\Lambda^p 1} \longrightarrow \bar{\bar{\Omega}}_V^p \longrightarrow \mathscr{G}^{(\Lambda^p 1)} \longrightarrow 0 \tag{4.4}$$

where $\mathscr{G}^{(\Lambda^p 1)}$ is a sheaf supported on the singular point of V. Let

$$G^{(\Lambda^p 1)}(M \backslash A) := \Gamma(M \backslash A, \Omega_M^p) / < \Lambda^p \Gamma(M \backslash A, \Omega_M^1) >, \tag{4.5}$$

then $\dim \mathscr{G}_0^{(\Lambda^p 1)} = \dim G^{(\Lambda^p 1)}(M \backslash A)$.

Thus, from Lemma 4.3, we can define a local invariant of a singularity which is independent of resolution.

Definition 4.4. *Let V be a n-dimensional Stein space with 0 as its only singular point. Let $\pi : (M, A) \to (V, 0)$ be a resolution of the singularity with A as exceptional set. Let*

$$g^{(\Lambda^p 1)}(0) := \dim \mathscr{G}_0^{(\Lambda^p 1)} = \dim G^{(\Lambda^p 1)}(M \backslash A). \tag{4.6}$$

We will omit 0 in $g^{(\Lambda^p 1)}(0)$ if there is no confusion from the context.

Let $\pi : (M, A_1, \ldots, A_k) \to (V, 0_1, \ldots, 0_k)$ be a resolution of the singularities with $A_i = \pi^{-1}(0_i)$, $1 \le i \le k$, as exceptional sets. In this case we still let

$$G^{(\Lambda^p 1)}(M \backslash A) := \Gamma(M \backslash A, \Omega_M^p) / < \Lambda^p \Gamma(M \backslash A, \Omega_M^1) >,$$

where $A = \cup_i A_i$.

Definition 4.5. *If X is a compact connected strongly pseudoconvex CR manifold of real dimension 3, in the boundary of a bounded strongly pseudoconvex domain D in $\mathbb{C}^N$. Suppose V in $\mathbb{C}^N$ such that the boundary of V is X. Let $\pi : (M, A = \cup_i A_i) \to (V, 0_1, \ldots, 0_k)$ be a resolution of the singularities with $A_i = \pi^{-1}(0_i)$, $1 \le i \le k$, as exceptional sets. Let*

$$G^{(\Lambda^p 1)}(M \backslash A) := \Gamma(M \backslash A, \Omega_M^p) / < \Lambda^p \Gamma(M \backslash A, \Omega_M^1) > \tag{4.7}$$

and

$$G^{(\Lambda^p 1)}(X) := \mathscr{S}^p(X) / < \Lambda^p \mathscr{S}^1(X) > \tag{4.8}$$

where $\mathscr{S}^q$ are holomorphic cross sections of $\wedge^q(\widehat{T}(X)^*)$. Then we set

$$g^{(\Lambda^p 1)}(M \backslash A) := \dim G^{(\Lambda^p 1)}(M \backslash A), \tag{4.9}$$

$$g^{(\Lambda^p 1)}(X) := \dim G^{(\Lambda^p 1)}(X). \tag{4.10}$$

Lemma 4.6. *Let X be a compact connected strongly pseudoconvex CR manifold of real dimension $2n - 1$ which bounds a bounded strongly pseudoconvex variety V with only isolated singularities $\{0_1, \ldots, 0_k\}$ in $\mathbb{C}^N$. Let $\pi : (M, A_1, \ldots, A_k) \to (V, 0_1, \ldots, 0_k)$ be a resolution of the singularities with $A_i = \pi^{-1}(0_i)$, $1 \le i \le k$, as exceptional sets. Then $g^{(\Lambda^p 1)}(X) = g^{(\Lambda^p 1)}(M \backslash A)$, where $A = \cup_i A_i$, $1 \le i \le k$.*

By Lemma 4.6 and Lemma 4.3, we can get the following lemma easily.

Lemma 4.7. *Let X be a compact connected strongly pseudoconvex CR manifold of real dimension 3, which bounds a bounded strongly pseudoconvex variety V with only isolated singularities $\{0_1, \dots, 0_k\}$ in $\mathbb{C}^N$. Then $g^{(\Lambda^p 1)}(X) = \sum_i g^{(\Lambda^p 1)}(0_i) = \sum_i \dim \mathscr{G}_{0_i}^{(\Lambda^p 1)}$.*

The following proposition is to show that $g^{(\Lambda^p 1)}$ is bounded above.

Proposition 4.8. *Let V be a n-dimensional Stein space with 0 as its only singular point. Then*

$$g^{(\Lambda^p 1)} \leq \begin{cases} g^{(p)}, & p \leq n-2; \\ g^{(n-1)} + q, & p = n-1; \\ g^{(n)} + p_g, & p = n. \end{cases}$$

The following theorem is the crucial part for solving the classical complex Plateau problem of real dimension 3.

Theorem 4.9. ([Du-Ya]) *Let V be a 2-dimensional Stein space with 0 as its only normal singular point with $\mathbb{C}^*$-action. Let $\pi : (M, A) \to (V, 0)$ be a minimal good resolution of the singularity with A as exceptional set, then $g^{(\Lambda^2 1)} \geq 1$.*

Remark 4.10. We also show that $g^{(\Lambda^2 1)}$ is strictly positive for rational singularities ([Du-Ga]) and minimal elliptic singularities ([Du-Ga2])and exact 1 for rational double points, triple points and quotient singularities ([Du-Lu-Ya]).

Similarly, the following theorem is the crucial part for solving the classical complex Plateau problem of real dimension 5.

Theorem 4.11. *Let V be a n-dimensional Stein space with 0 as its only non-rational singular point, where $n > 2$, then $g^{(\Lambda^n 1)} \geq 1$.*

Theorem 4.12. *Let V be a 3-dimensional Stein space with 0 as its only normal Gorenstein singular point, then $g^{(\Lambda^3 1)} \geq 1$.*

5 The classical complex Plateau problem

In 1981, Yau [Ya] solved the classical complex Plateau problem for the case $n \geq 3$.

Theorem 5.1. ([Ya]) *Let X be a compact connected strongly pseudoconvex CR manifold of real dimension $2n-1$, $n \geq 3$, in the boundary of a bounded strongly pseudoconvex domain D in $\mathbb{C}^{n+1}$. Then X is a boundary of the complex submanifold $V \subset D - X$ if and only if Kohn–Rossi cohomology groups $H_{KR}^{p,q}(X)$ are zeros for $1 \leq q \leq n-2$*

When $n = 2$, the Plateau problem remains unsolved for many years even there are no any criterion to judge whether X is a boundary of the complex manifold. In [Du-Ya], the first and the third authors used CR invariant $g^{(1,1)}(X)$ to give the sufficient and necessary condition for the variety bounded by a Calabi–Yau CR manifold X being smooth if $H_h^2(X) = 0$.

Theorem 5.2. ([Du-Ya]) *Let X be a strongly pseudoconvex compact Calabi–Yau CR manifold of dimension 3. Suppose that X is contained in the boundary of a strongly pseudoconvex bounded domain D in $\mathbb{C}^N$ with $H_h^2(X) = 0$. Then X is a boundary of the complex sub-manifold (up to normalization) $V \subset D - X$ with boundary regularity if and only if $g^{(1,1)}(X) = 0$.*

Theorem 5.3. ([Du-Ya]) *Let X be a strongly pseudoconvex compact CR manifold of dimension 3. Suppose that X is contained in the boundary of a strongly pseudoconvex bounded domain D in $\mathbb{C}^3$ with $H_h^2(X) = 0$. Then X is a boundary of the complex sub-manifold $V \subset D - X$ if and only if $g^{(1,1)}(X) = 0$.*

If $n > 2$ and X is not of hypersurface type, the situation is complicated. It is impossible to get the smooth Harvey–Lawson solution to X in general. To see the least, we ask if there exists a solution V which is smooth after normalization or has a good types of singularities. We will use the new CR invariant $g^{(\Lambda^n 1)}(X)$ to deal with this problem of when X is of general type.

Theorem 5.4. *Let X be a strongly pseudoconvex compact CR manifold of dimension $2n - 1$, where $n > 2$. Suppose that X is contained in the boundary of a strongly pseudoconvex bounded domain D in $\mathbb{C}^N$. Then X is a boundary of the variety $V \subset D - X$ with boundary regularity and the number of non-rational singularities is not great than $g^{(\Lambda^n 1)}(X)$. In particular, if $g^{(\Lambda^n 1)}(X) = 0$, then V has at most finite rational singularities.*

When X is a Calabi–Yau CR manifold of dimension 5, we give the following necessary and sufficient condition for the variety bounded by X being smooth up to normalization.

Theorem 5.5. *Let X be a strongly pseudoconvex compact Calabi–Yau CR manifold of dimension 5. Suppose that X is contained in the boundary of a strongly pseudoconvex bounded domain D in $\mathbb{C}^N$. Then X is a boundary of the complex sub-manifold (up to normalization) $V \subset D - X$ with boundary regularity if and only if $g^{(\Lambda^3 1)}(X) = 0$.*

Corollary 5.6. *Let X be a strongly pseudoconvex compact Calabi–Yau CR manifold of dimension 5. Suppose that X is contained in the boundary of a strongly pseudoconvex bounded domain D in $\mathbb{C}^4$. Then X is a boundary of the complex sub-manifold $V \subset D - X$ with boundary regularity if and only if $g^{(\Lambda^3 1)}(X) = 0$.*

References

[Al] Alexeev, V., *Two two-dimensional terminations*, Duke Math. J. **69** (1993), no. 3, 527–545.

[Br] Brenton, L., *On singular complex surfaces with negative canonical bundle, with applications to singular compactifications of $\mathbb{C}^2$ and to 3-dimensional rational singularities*, Math. Ann. **248** (1980), no. 2, 117–124.

[C-K-M] Clemens, H., Kollár, J. and Mori, S., *Higher dimensional complex geometry*, Astérisque **166** (1988).

[Cu] Cutkosky, S. *Elementary contractions of Gorenstein threefolds*, Math. Ann. **280** (1988), no. 3, 521–525.

[Du] Du, R., *Smooth Harvey-Lawson solution to 3 dimensional strictly pseudoconvex CR manifolds*, in preperation.

[Du-Ga] Du, R. and Gao, Y., *New invariants for complex manifolds and rational singularities*, Pacific J. of Math., to appear.

[Du-Ga2] Du, R. and Gao, Y., *Some remarks on Yau's conjecture and complex Plateau problem*, Methods and Applications of Analysis, to appear.

[Du-Ga-Ya] Du, R., Gao, Y. and Yau, S.S.-T., *On higher dimensional complex Plateau problem*, submitted.

[Du-Lu-Ya] Du, R., Luk, H.S. and Yau, S.S.-T., *New invariants for complex manifolds and isolated singularities*, Communications in Analysis and Geometry **19** (2011), no. 5, 991–1021.

[Du-Ya] Du, R. and Yau, S.S.-T., *Kohn–Rossi cohomology and its application to the complex Plateau problem, III*, Journal of Differential Geometry **90** (2012), no. 2, 251–266.

[E-M] Ein, L. and Mustaţă, M., *Inversion of adjunction for local complete intersection varieties*, Amer. J. Math. **126** (2004), no. 6, 1355–1365.

[E-M-Y] Ein, L., Mustaţă, M. and Yasuda, T., *Jet schemes, log discrepancies and inversion of adjunction*, Invent. Math. **153** (2003), no. 3, 519–535.

[E-V] Esnault, H. and Viehweg, E., *Lectures on Vanishing Theorems*, DMV Seminar, **20**, Birkhäuser Verlag, Basel, 1992.

[Gr] Greuel, G.-M., *Dualität in der lokalen Kohomologie isolierter Singularitäten*, Math. Ann. **250** (1980), 157–173.

[Ha-La1] Harvey, R. and Lawson, B., *On boundaries of complex analytic varieties*, I, Ann. of Math. **102** (1975), 233–290.

[Ha-La2] Harvey, R. and Lawson, B., *Addendum to Theorem* 10.4 *of [HL]*, arXiv: math/0002195.

[Ma] Markushevich, D., *Minimal discrepancy for a terminal cDV singularity is* 1, J. Math. Sci. Univ. Tokyo **3** (1996), no. 2, 445–456.

[Mo] Mori, S., *Threefolds whose canonical bundles are not numerically effective*, Ann. of Math. (2) **116** (1982), no. 1, 133–176.

[Re] Reid, M., *Canonical 3-folds*, Journées de Géometrie Algébrique d'Angers, Juillet 1979/Algebraic Geometry, Angers, 1979, pp. 273-310, Sijthoff & Noordhoff, Alphen aan den Rijn-Germantown, Md., 1980.

[Re2] Reid, M., *Minimal models of canonical 3-folds*, Algebraic varieties and analytic varieties (Tokyo, 1981), 131–180, Adv. Stud. Pure Math., **1**, North-Holland, Amsterdam, 1983.

[Sh] Shokurov, V.V., *Problems about Fano varieties*, Birational Geometry of Algebraic Varieties. Open Problems-Katata, 1988, 30–32.

[Si] Siu, Y.-T., *Analytic sheaves of local cohomology*, Trans. AMS **148** (1970), 347–366.

[St-St] Straten, D.V. and Steenbrink, J., *Extendability of holomorphic differential forms near isolated hypersurface singularities*, Abh. Math. Sem. Univ. Ham-burg **55** (1985), 97–110.

[Ta] Tanaka, N., *A Differential Geometry Study on Strongly Pseudoconvex Manifolds*, Lecture in Mathematics, **9**, Kyoto University, Kinokuniya Bookstore Co. Ltd, 1975.

[Ya] Yau, S.S.-T., *Kohn–Rossi cohomology and its application to the complex Plateau problem, I*, Ann. of Math. **113** (1981), 67–110.

[Ya2] Yau, S.S.-T., *Various numerical invariants for isolated singularities*, Amer. J. Math. **104** (1982), no. 5, 1063–1110.

Department of Mathematics, Shanghai Key Laboratory of PMMP
East China Normal University
Rm. 312, Math. Bldg, No. 500, Dongchuan Road
Shanghai 200241, China
E-mail address: rdu@math.ecnu.edu.cn

Department of Mathematics, Shanghai Jiao Tong University
Shanghai 200240, China
E-mail address: gaoyunmath@sjtu.edu.cn

Department of Mathematical Sciences, Tsinghua University
Beijing 100084, China
E-mail address: yau@uic.edu, syau@math.tsinghua.edu.cn

Proceedings of the 6th International Congress of Chinese Mathematicians
ALM 36, pp. 467–488

Relative Resolution and Its Applications

Yi Hu

Abstract

We present an introduction to the *derived and relative resolutions* of the moduli of stable maps. We discuss one application and mention a few problems.

Mathematics Subject Classification: 14D23, 14E15.

Keywords and Phrases: Relative resolution, moduli stack.

1 Introduction

We aim to provide a *relative resolution* of a singular moduli, preferably by geometric method, not by any algorithm. Our primary example is the Kontsevich moduli space $\overline{M}_g(\mathbb{P}^n, d)$ of degree d stable maps from genus g curves to the projective space $\mathbb{P}^n$. In this note, we will give a hopefully leisurely introduction to some recent works on the two kinds of "resolutions" of $\overline{M}_g(\mathbb{P}^n, d)$ ([17, 7, 8, 9]).

The first is the *derived resolution*. Its primary purpose is to define the *reduced* GW invariants which properly count the contributions of the main components of the moduli of stable maps. The derived resolution always exists and has a minimal one, unique up to isomorphism.

The second is the *relative resolution*. For a *smooth* stack $\mathfrak{M}$, we say that a blowup $\widetilde{\mathfrak{M}} \to \mathfrak{M}$ is smooth if $\widetilde{\mathfrak{M}}$ is smooth. Thus, smooth blowups include blowups along smooth closed centers. If a moduli space has several components, it usually comes equipped with a distinguished one, called the main component.

Definition 1.1. *Let X be a Deligne-Mumford stack with a main component X'. Let $\mathfrak{M}$ be a smooth Artin stack and $X \to \mathfrak{M}$ be a morphism. Suppose that we have a sequence of smooth blowups $\widetilde{\mathfrak{M}} \to \mathfrak{M}$ such that if let $\widetilde{X} = X \times_{\mathfrak{M}} \widetilde{\mathfrak{M}}$, then the main component $\widetilde{X}' = X' \times_{\mathfrak{M}} \widetilde{\mathfrak{M}}$ is smooth and the entire stack $\widetilde{X}$ has at worst normal crossing singularities. Then we say $\widetilde{X} \to X$ is a resolution of X relative* *to $\widetilde{\mathfrak{M}} \to \mathfrak{M}$. If in addition, every irreducible component of $\widetilde{X}$ is smooth, then we say it is a strong relative resolution.*

Here, a DM stack (not necessarily irreducible) is said to have at worst normal crossing singularities if at any singular point, it is locally equivalent to a union of coordinate subspaces of an affine space. A relative resolution is a natural extension of the ordinary resolution in that we allow $\widetilde{X}$ to be reducible but require normal crossing singularities for all irreducible components as well as for their intersections, which seems to be the best one can hope for.

For the moduli space $\overline{M}_g(\mathbb{P}^n, d)$, one should interpret a relative resolution as follows. The stack $\overline{M}_g(\mathbb{P}^n, d)$ fibers over a smooth stack $\mathfrak{M}$ such that all its obstructions lie in the fiber direction. A relative resolution is a process to remove all the obstructions in the fiber direction and at the mean time preserve smoothness of the base (whence the requirement that the blowup $\widetilde{\mathfrak{M}} \to \mathfrak{M}$ is smooth).

For many topological applications, lesser resolution may be sufficient. Let $E^\bullet$ be a two-term perfect derived object over the DM stack X. Suppose that $\widetilde{\mathfrak{M}} \to \mathfrak{M}$ is a blowup (not necessarily smooth) such that if let $\widetilde{X} = X \times_{\mathfrak{M}} \widetilde{\mathfrak{M}}$, then the pullback of $E^\bullet$ to $\widetilde{X}$ becomes *locally diagonalizable.* Then we say $\widetilde{X} \to X$ is a *derived resolution* for the object $E^\bullet$. We let $\pi : \mathcal{C} \to \overline{M}_g(\mathbb{P}^n, d)$ be the universal family and $\mathfrak{f} : \mathcal{C} \to \mathbb{P}^n$ be the universal map. The canonical derived objects $\mathbf{R}\pi_*\mathfrak{f}^*\mathscr{O}_{\mathbb{P}^n}(k)$ are of central importance in the Gromov-Witten theory.

Our main results may be summarized as follows. *We obtained a strong relative resolution for $\overline{M}_1(\mathbb{P}^n, d)$ ([17, 7]). We achieved relative resolutions for $\overline{M}_2(\mathbb{P}^n, d)$ ([9]). We constructed the derived resolutions for the canonical objects $\mathbf{R}\pi_*\mathfrak{f}^*\mathscr{O}_{\mathbb{P}^n}(k)$ over $\overline{M}_g(\mathbb{P}^n, d)$ for all g and $k > 0$ ([8]).* More details are as follows.

When $g = 1$, the relative resolution is achieved over a canonical smooth blowup of the smooth Artin stack of weighted curves ([7]); for $g = 2$, it is relative to a canonical smooth blowup of the smooth Artin stack of pairs of curve and line bundle ([9]). The resolution of the main component of $\overline{M}_1(\mathbb{P}^n, d)$ is originally due to Vakil and Zinger ([17]) and was applied to calculate the genus one reduce GW numbers by Zinger ([18]). Chang and Li used the relative resolution to give an algebro-geometric proof of the Li-Zinger formula for genus one GW invariants of quintics ([3]). It is expected that the relative resolution of $\overline{M}_2(\mathbb{P}^n, d)$ can be applied to prove the genus two version of the LZ formula. Higher genus cases are considerably harder.

We hope that this note will benefit researchers in the Gromov-Witten theory as well as other mathematicians wandering around. All the works on the derived resolution, reduced invariants, and relative resolutions of stable map moduli are jointly obtained with Li ([7, 8, 9]). Section 4 on the genus one LZ formula is due to a joint work of Chang and Li ([3]). I thank Jun Li for collaboration and Huai-Liang Chang for spending hours in a series of seminars explaining their works. This note is based upon my lectures at Taiwan University and at Hong Kong University of Science and Technology during the summer of 2013. I thank TIMS of Taiwan University for partial financial support and Chin-Lung Wang for his hospitality when I was there. I thank ICCM committee for its kind invitation to the sixth ICCM-Taiwan and for its subsequent invitation to writing up my lecture notes. I thank Huai-Liang Chang and Wei-Ping Li for the invitation to visit HKUST in the summer of 2013 and for their warm hospitality; the partial financial support from HKUST is also gratefully acknowledged. Some of the research work described in this article was partially supported by NSF grant DMS 0901136.

2 Derived and relative resolutions: an overview

2.1 A little background

The moduli space $\overline{M}_g(\mathbb{P}^n, d)$ in algebraic geometry was introduced by Kontsevich. Its original primary application is to laid a foundation for the Gromov-Witten theory.

The moduli space $\overline{M}_g(\mathbb{P}^n, d)$ is a *moduli* compactification of $M_g(\mathbb{P}^n, d)$, the open subset of degree d maps from smooth projective curves of genus g to the projective space $\mathbb{P}^n$. It is not a compactification in the usual sense since $\overline{M}_g(\mathbb{P}^n, d)$ may contain many components of various dimensions. The closure of $M_g(\mathbb{P}^n, d)$, is the main component, denoted $\overline{M}_g(\mathbb{P}^n, d)'$. It is of a central interest in the GW theory to understand the contribution to the GW invariants of each individual component. For this, it is helpful to understand the structures of the components and how they interact, to certain extent. But, when $n \geq 2$, the spaces $M_g(\mathbb{P}^n, d)$ (hence also $\overline{M}_g(\mathbb{P}^n, d)$) can contain arbitrary singularity types when g varies and $0 < d < 2g - 2$ according to a theorem of Vakil (based upon Mnëv's universality theorem). In addition, the main component $\overline{M}_g(\mathbb{P}^n, d)'$ needs not to be irreducible if $d \leq 2g - 2$. So, it looks rather hopeless to fully grasp the geometric structures of the moduli space.

However, thinking positively, the moduli spaces $\overline{M}_g(\mathbb{P}^n, d)$ provide us a single geometric setting to resolve all possible singularities in algebraic geometry, possibly by a uniform geometric method. Even if one is not interested in resolution of general singularities but is only interested in the GW invariant of hypersurfaces or more generally the GW invariants of complete intersections in $\mathbb{P}^n$, it is still desirable to have a resolution of the singularities of $\overline{M}_g(\mathbb{P}^n, d)$, to certain extent[1]. We will explain and hopefully convince the reader that a *partial* relative resolution of $\overline{M}_g(\mathbb{P}^n, d)$, in a suitable sense, can be applied to study the GW numbers of a quintic threefold and more generally to study the GW invariants of a hypersurface of $\mathbb{P}^n$.

2.2 Where are the singularites?

So, to get a "resolution" of singularities of the moduli space $\overline{M}_g(\mathbb{P}^n, d)$, first we need to understand where its singularities are.

A point of the moduli space $\overline{M}_g(\mathbb{P}^n, d)$ is a degree d stable morphism $u : C \to \mathbb{P}^n$, where C is a projective curve of genus g with at worst nodal singularities. Its stability means that its automorphism group is finite. We will often abbreviate a stable map $u : C \to \mathbb{P}^n$ as $[u, C]$. Take any point $[u_0 : C_0]$ in $\overline{M}_g(\mathbb{P}^n, d)$.

[1] E.g., in the sense of Definition 1.1, or some weaker ones to be explained in what follows.

Fix homogeneous coordinates $[x_0, \ldots, x_n]$ of $\mathbb{P}^n$ with $x_i \in H^0(\mathcal{O}_{\mathbb{P}^n}(1))$. Then we can write $u_0 = [s_0, \ldots, s_n]$ where $s_i = u^* x_i \in H^0(u_0^* \mathcal{O}_{\mathbb{P}^n}(1))$. Thus, we see that a deformation of $[u_0, C_0]$ is determined by a combined deformation of the curve C_0 and the sections $\{s_0, \ldots, s_n\}$. As the deformation of the (nodal) curve C_0 is unobstructed, all possible obstructions to deform the map $[u_0, C_0]$ come from the obstructions of deforming the sections $\{s_0, \ldots, s_n\}$. If in a neighborhood of $[u_0, C_0]$, the rank of $H^0(u^* \mathcal{O}_{\mathbb{P}^n}(1))$ remains constant, then one has a vector bundle over the neighborhood with fibers $H^0(u^* \mathcal{O}_{\mathbb{P}^n}(1))$. After shrinking the neighborhood, we may assume that this vector bundle is trivial; after a trivialization, one sees that there is no obstruction to choose a set of sections $\{S_0, \ldots, S_n\}$ of this trivialized bundle, extending the vectors $\{s_0, \ldots, s_n\}$ in $H^0(u_0^* \mathcal{O}_{\mathbb{P}^n}(1))$ to its nearby fibers $H^0(u^* \mathcal{O}_{\mathbb{P}^n}(1))$. Or else, if the rank of $H^0(u_0^* \mathcal{O}_{\mathbb{P}^n}(1))$ is higher than its nearby general fibers $H^0(u^* \mathcal{O}_{\mathbb{P}^n}(1))$, then some sections of $\{s_0, \ldots, s_n\}$ may experience obstructions when we try to extend them to nearby fibers. If this happens, we encounter a singularity of $\overline{M}_g(\mathbb{P}^n, d)$ at $[u_0, C_0]$.

Informally, we say that a stable map has two directions: the curve direction and the section direction; its curve direction is unobstructed but its section direction may experience obstructions.

Now, to formulate the above informal discussions in formal terms, we need the universal curve $\pi : \mathcal{C} \longrightarrow \overline{M}_g(\mathbb{P}^n, d)$ over $\overline{M}_g(\mathbb{P}^n, d)$ and the universal map $\mathfrak{f} : \mathcal{C} \longrightarrow \mathbb{P}^n$. For a point $[u, C] \in \overline{M}_g(\mathbb{P}^n, d)$, the fiber $\pi^{-1}([u, C]) \subset \mathcal{C}$ is just the curve C; the map f restricted to the fiber $\pi^{-1}([u, C]) = C$ is just the map $u : C \longrightarrow \mathbb{P}^n$. To capture the variation of vector space $H^0(u^* \mathcal{O}_{\mathbb{P}^n}(1))$, we have the direct image sheaf $\pi_* \mathfrak{f}^* \mathcal{O}_{\mathbb{P}^n}(1)$. In diagrams, this is

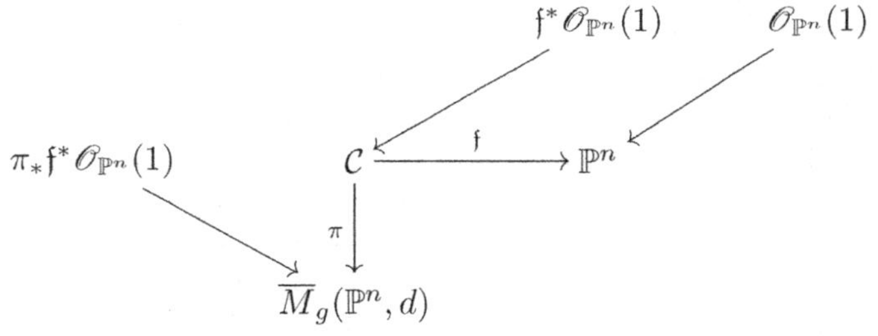

By the above discussions, we have the following important observation. The moduli space $\overline{M}_g(\mathbb{P}^n, d)$ is smooth at a point $[u, C]$ if and only if the sheaf $\pi_* \mathfrak{f}^* \mathcal{O}_{\mathbb{P}^n}$ (1) is locally free at $[u, C]$. Or, equivalently, the moduli space $\overline{M}_g(\mathbb{P}^n, d)$ is singular at a point $[u, C]$ if and only if the sheaf $\pi_* \mathfrak{f}^* \mathcal{O}_{\mathbb{P}^n}(1)$ is not locally free at $[u, C]$. (By this characterization, all smooth stable maps must lie in the main component of $\overline{M}_g(\mathbb{P}^n, d)$. The remainder components contain no smooth points.) Thus, intuitively, one would expect that *making the sheaf $\pi_* \mathfrak{f}^* \mathcal{O}_{\mathbb{P}^n}(1)$ locally free will have consequence on the resolution of the moduli space $\overline{M}_g(\mathbb{P}^n, d)$*. We will elaborate this the later sections.

Now, what about the other direct image sheaves $\pi_* \mathfrak{f}^* \mathcal{O}_{\mathbb{P}^n}(k)$, $k > 1$?

2.3 The canonical derived objects $\mathbf{R}\pi_*\mathfrak{f}^*\mathscr{O}_{\mathbb{P}^n}(k)$

Well, if we consider a hypersurface X of degree k in $\mathbb{P}^n$ (e.g., a quintic in $\mathbb{P}^4$), we may assume that it is given by the zeros of a section $s \in H^0(\mathscr{O}_{\mathbb{P}^n}(k))$. Then one checks that at least set-theoretically, we have that the moduli space $\overline{M}_g(X, d) \subset \overline{M}_g(\mathbb{P}^n, d)$ is given by the zeros of the section $\sigma = \pi_*\mathfrak{f}^*s$ of the sheaf $\pi_*\mathfrak{f}^*\mathscr{O}_{\mathbb{P}^n}(k)$. Thus, in an ideal situation such as if $\pi_*\mathfrak{f}^*\mathscr{O}_{\mathbb{P}^n}(k)$ were locally free and $\overline{M}_g(\mathbb{P}^n, d)$ were of pure and expected dimension, then $\pi_*\mathfrak{f}^*\mathscr{O}_{\mathbb{P}^n}(k)$ would possess the Euler class and the fundamental class of $\overline{M}_g(X, d)$ would be the intersection of the Euler class with the fundamental class of $\overline{M}_g(\mathbb{P}^n, d)$. This motivates that we should also make $\pi_*\mathfrak{f}^*\mathscr{O}_{\mathbb{P}^n}(k)$ locally free so that we can define Euler class and obtain invariants when restricted to the main component of $\overline{M}_g(\mathbb{P}^n, d)$.

Thus, for either purpose of desingularizing the moduli space $\overline{M}_g(\mathbb{P}^n, d)$ and making applications to the GW theory of hypersurfaces in $\mathbb{P}^n$, one would like to resolve the sheaves $\pi_*\mathfrak{f}^*\mathscr{O}_{\mathbb{P}^n}(k)$, $k \geq 1$. However, we should point out here that as it turns out, it is not very helpful to just consider the direct image sheaf $\pi_*\mathfrak{f}^*\mathscr{O}_{\mathbb{P}^n}(k)$, it is more natural and mathematically correct to also include the higher direct image sheaf $R^1\pi_*\mathfrak{f}^*\mathscr{O}_{\mathbb{P}^n}(k)$ and consider the derived object $\mathbf{R}\pi_*\mathfrak{f}^*\mathscr{O}_{\mathbb{P}^n}(k)$.

To stress the points, one should bear in mind that when $k = 1$ the structures of the object $\mathbf{R}\pi_*\mathfrak{f}^*\mathscr{O}_{\mathbb{P}^n}(1)$ implies structural properties of the underlying moduli space $\overline{M}_g(\mathbb{P}^n, d)$; when $k > 1$, the structures of the object $\mathbf{R}\pi_*\mathfrak{f}^*\mathscr{O}_{\mathbb{P}^n}(k)$ will have applications to the GW invariants of the hypersurfaces of degree k in $\mathbb{P}^n$. Our treatment for $\mathbf{R}\pi_*\mathfrak{f}^*\mathscr{O}_{\mathbb{P}^n}(k)$, $k \geq 1$, is uniform.

2.4 The quintic Calabi-Yau threefolds

We now spend a subsection to discuss the important case of quintic CY threefolds which is related to the object $\mathbf{R}\pi_*\mathfrak{f}^*\mathscr{O}_{\mathbb{P}^4}(5)$. Let $Q \subset \mathbb{P}^n$ be a smooth quintic threefold. The (virtual) numbers of curves in Q are central topics in the GW theory for the last two decades. To this date, only the cases of rational and elliptic curves are mathematically known. The case of higher genus is still out of reach.

Consider the moduli space $\overline{M}_g(Q, d)$ of degree d stable maps from genus g curves into the quintic Q. Its expected dimension is zero and possesses a virtual fundamental cycle $[\overline{M}_g(Q, d)]^{\mathrm{vir}}$ which is a zero cycle with rational coefficients. Its degree is a rational number, denoted $N_{g,d}$. This provides a virtual count of genus g, degree d curves in Q. Let us know examine the meanings of $N_{g,d}$ via the structures of the moduli space $\overline{M}_g(Q, d)$. The space $\overline{M}_g(Q, d)$ consists of stable maps classified as in Fig. 1. The domain of every stable map is a genus g curve. But in its image, one may only see a genus h curve if a genus $(g - h)$ subcurve is contracted (the dotted curve in the picture is contracted); such a stable map is a general point in its corresponding component $\overline{M}_g(Q, d)_h$ of $\overline{M}_g(Q, d)$. Intuitively, this indicates that the virtual number $N_{g,d}$ should contain the contribution of the component of $\overline{M}_g(Q, d)_h$; this contribution should come with two parts: one for $\overline{M}_h(Q, d)$, i.e., the genus h curves that we see in its images, which we denote by $N'_{h,d}$; the other comes from the contracted curves of genus $g - h$, and should be an invariant on the moduli space $\overline{M}_{g-h}$ of stable curves of genus $(g - h)$, which we denote by c_{g-h}.

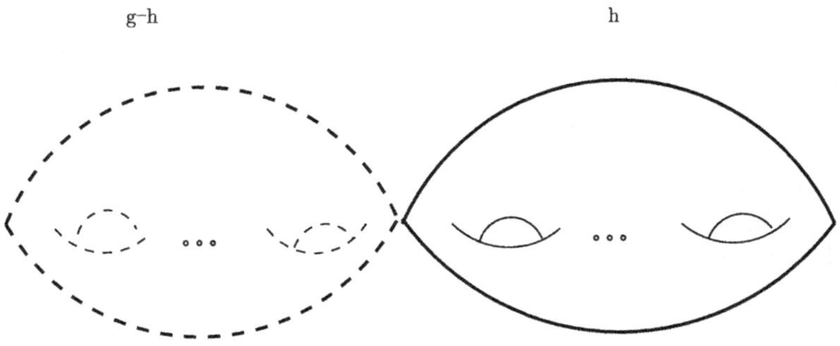

Fig. 1

There are other components, but conjecturally they do not contribute. This explains the idea behind Li-Zinger's conjecture:

$$N_{g,d} = N'_{g,d} + c_1 N'_{g-1,d} + \cdots + c_{g-1} N'_{1,d} + c_g N'_{0,d}. \tag{2.1}$$

What is $N'_{h,d}$? This should be the contribution of the main component $\overline{M}_h(Q,d)'$ of $\overline{M}_h(Q,d)$ whose general points are stable maps with smooth domain curves. How do we rigorously define $N'_{g,d}$? We will explain this in the next couple of sections.

2.5 The derived resolution

To rigorously define $N'_{g,d}$, we need to make the sheaf $\pi_* \mathfrak{f}^* \mathscr{O}_{\mathbb{P}^4}(5)$ locally free so that it can provide an Euler class that we can use. As pointed out earlier, it is more correct to consider the object $\mathbf{R}\pi_* \mathfrak{f}^* \mathscr{O}_{\mathbb{P}^4}(5)$ rather than just the sheaf $\pi_* \mathfrak{f}^* \mathscr{O}_{\mathbb{P}^4}(5)$ alone. We can do this in more general setting. So now we examine the properties of the object $\mathbf{R}\pi_* \mathfrak{f}^* \mathscr{O}_{\mathbb{P}^n}(k)$. First, it is a particularly nice kind of object: it is a two term perfect object, meaning locally, it can be presented as a two term complex $[E \xrightarrow{\varphi} F]$ of locally free sheaves E and F. (Even though, for $\mathbf{R}\pi_* \mathfrak{f}^* \mathscr{O}_{\mathbb{P}^n}(k)$, a global presentation can be found, we still prefer its local presentations, as in practice, local presentations come more naturally, and more importantly, possess geometric meanings.) Recall that one of our naive aim is to make the sheaf $\pi_* \mathfrak{f}^* \mathscr{O}_{\mathbb{P}^n}(k)$ locally free. As it turns out, the more functorial way to achieve this is to *locally diagonalize* the object $\mathbf{R}\pi_* \mathfrak{f}^* \mathscr{O}_{\mathbb{P}^n}(k)$ from which the local freeness of $\pi_* \mathfrak{f}^* \mathscr{O}_{\mathbb{P}^n}(k)$ comes as an immediate consequence. So, we explain now the very useful notion of *locally diagonalizable* derived object.

For any scheme X, a homomorphism $\varphi : \mathscr{O}_X^{\oplus p} \longrightarrow \mathscr{O}_X^{\oplus q}$ is said to be diagonalizable if we have direct sum decompositions by trivial sheaves

$$\mathscr{O}_X^{\oplus p} = G_0 \oplus \bigoplus_{i=1}^{l} G_i \text{ and } \mathscr{O}_X^{\oplus q} = H_0 \oplus \bigoplus_{i=1}^{l} H_i \tag{2.2}$$

with $\varphi(G_i) \subset H_i$ for all i such that

1. $\varphi|_{G_0} = 0$;

2. for every $1 \leq i \leq l$, $\varphi|_{G_i}$ equals to $p_i I_i$ for some $0 \neq p_i \in \Gamma(\mathscr{O}_X)$ where $I_i : G_i \to H_i$ is an isomorphism;

3. $\langle p_i \rangle \supsetneqq \langle p_{i+1} \rangle$.

A homomorphism $\varphi : E \to F$ between locally free sheaves of a scheme X is said to be locally diagonalizable if there are trivializations of E and F over some open covering of X such that $\varphi : E \to F$ is diagonalizable over every open subset.

Now if $E^\bullet$ is a two-term perfect derived object over a DM stack X, using any local presentation of $[E \xrightarrow{\varphi} F]$ of $E^\bullet$, we can introduce the notion of locally diagonalizable derived object. One can prove that the notion does not depend on the local presentation ([8]).

Not every homomorphism (or two-term perfect derived object) can be locally diagonalized. But if it can be, then there are some good implications. For example, suppose that a homomorphism $\varphi : E \to F$ (or an object $E^\bullet$) is locally diagonalizable, then for every irreducible component X' of X with the reduced scheme structure, $\ker(\varphi|_{X'})$ ($\mathscr{H}^0(E^\bullet)$) is locally free. Also, if $f : Y \to X$ is a morphism, then $f^*\varphi$ ($f^*E^\bullet$) is also locally diagonalizable, that is, "locally diagonalizable" has base change property. Note here that the rank of $\ker \varphi$ depends on the properties of the functions p_i, hence may not be constant over X. Further, when X' is not reduced, $\ker(\varphi|_{X'})$ needs not to be locally free. These technical issues lead us to use the "integral" assumption whenever we want to produce a locally free sheaf.

For any $0 \leq i \leq r := \max(\operatorname{rank} E, \operatorname{rank} F)$, we let $\mathscr{I}_{\varphi,i}$ be the i-th determinantal ideal sheaf of $\varphi : E \to F$. That is, $\mathscr{I}_{\varphi,i}$ is the ideal sheaf of the zero scheme of $\wedge^{i+1}E \to \wedge^{i+1}F$. Again, using local presentations of the object $E^\bullet$, one can introduce the i-th determinantal ideal sheaf $\mathscr{I}_{E^\bullet,i}$ of $E^\bullet$ and show that it is independent of the choice of the presentations. We proved the following.

Theorem 2.1. ([8]) *Suppose $E^\bullet$ is a two-term perfect derived object over a DM stack X. If we let $b : \tilde{X} \longrightarrow X$ be the blowup of X along the ideal sheaves $\mathscr{I}_{E^\bullet,0}, \ldots, \mathscr{I}_{E^\bullet,r}$ and so on, then the pullback $b^*E^\bullet$ becomes locally diagonalizable over $\tilde{X}$. In particular, $\mathscr{H}^0(b^*E^\bullet)$ is locally free provided X is integral. Further, $\tilde{X}$ is minimal in the sense that if there is another dominating morphism $g : Z \to X$ such that $g^*E^\bullet$ is locally diagonalizable, then g factors uniquely as $Z \to \tilde{X} \to X$.*

In fact, we can consider any perfect derived object $\mathscr{E}^\bullet$ in the bounded derived category $\mathrm{D}^b(M)$ of an integral DM stack M with cohomologies concentrated in the non-negative places. Using the above theorem, we can show

Theorem 2.2. ([8]) *Let $\mathscr{E}^\bullet$ be any perfect derived object over an integral DM stack X. Assume that $\mathscr{E}^\bullet$ can locally be represented by a complex of locally free sheaves of finite length supported only in non-negative degrees. Then there is another integral DM stack $\widetilde{X}$ and a surjective birational morphism $b : \widetilde{X} \to X$ such that $\mathscr{H}^0(b^*\mathscr{E}^\bullet)$ is locally free.*

The above allows us to define the Euler class of such a derived object. For this, we suppose further that the cohomologies in positive places $\mathscr{H}^{i>0}(\mathscr{E}^\bullet)$ are all torsion sheaves over X. Then, we define the Euler class $\mathbf{e}(\mathscr{E}^\bullet)$ in the Chow group A_*X of cycles on X by

$$\mathbf{e}(\mathscr{E}^\bullet) := b_*(c_r(\mathscr{H}^0(b^*\mathscr{E}^\bullet)) \cdot [\widetilde{X}]), \tag{2.3}$$

where $r = \operatorname{rank} \mathscr{H}^0(b^*\mathscr{E}^\bullet)$. One can show the Euler class $\mathbf{e}(\mathscr{E}^\bullet)$ is independent of the choice of the resolutions $b : \widetilde{X} \to X$.

2.6 The reduced GW numbers of quintics and LZ conjecture

The above, when applied to the derived object $\mathbf{R}\pi_*\mathfrak{f}^*\mathscr{O}_{\mathbb{P}^4}(5)$ restricted to the primary component $\overline{\mathfrak{M}}_g(\mathbb{P}^4, d)'$ of the moduli stack $\overline{\mathfrak{M}}_g(\mathbb{P}^4, d)$, enables us to construct the modular Euler class when $d > 2g - 2$. Here, the general points of $\overline{\mathfrak{M}}_g(\mathbb{P}^4, d)'$ are maps with smooth domains, and $\overline{\mathfrak{M}}_g(\mathbb{P}^4, d)'$ is irreducible and of the expected dimension when $d > 2g-2$. In this case, letting π' and f' be the restrictions of the projection of the universal family and universal map to $\overline{\mathfrak{M}}_g(\mathbb{P}^4, d)'$, respectively, then $R^1\pi'_*f'^*\mathscr{O}_{\mathbb{P}^4}(5)$ is a torsion sheaf over $\overline{\mathfrak{M}}_g(\mathbb{P}^4, d)'$. Hence, for $d > 2g - 2$, we can define the modular Euler class of $\mathbf{R}\pi'_*\mathfrak{f}^*\mathscr{O}_{\mathbb{P}^4}(5)$ over $\overline{\mathfrak{M}}_g(\mathbb{P}^4, d)'$ to be

$$\mathbf{e}(\mathbf{R}\pi'_*f'^*\mathscr{O}_{\mathbb{P}^4}(5)) \in A_*(\overline{\mathfrak{M}}_g(\mathbb{P}^4, d)'); \tag{2.4}$$

for any smooth Calabi-Yau manifold Q in $\mathbb{P}^4$, we define

$$N'_{g,d}(Q) = \deg \mathbf{e}(\mathbf{R}\pi'_*f'^*\mathscr{O}_{\mathbb{P}^4}(5)). \tag{2.5}$$

Using these numbers $N'_{g,d}(Q)$, in [8], we rigorously formulate the conjecture of Li and Zinger:

Conjecture 2.3. *Let $N_{g,d}(Q)$ be the genus g, degree d GW-invariants of the smooth quintic $Q \subset \mathbb{P}^4$. Then there are universal constants c_h such that for all $d > 2g - 2$,*

$$N_{g,d}(Q) = \sum_{0 \le h \le g} c_h N'_{h,d}(Q).$$

The $g = 0$ case is trivial: $N_{0,d} = N'_{0,d}$. The $g = 1$ version, $N_{1,d} = N'_{1,d} + \frac{1}{12}N'_{0,d}$, is proved by Li and Zinger (and later by Chang-Li, using algebro-geometric method). The $g \ge 2$ cases are open.

2.7 Relative resolutions: a general program

The derived resolutions, i.e., the blowup $b : \tilde{X} \longrightarrow X$ in Theorem 2.1 (or Theorem 2.2), are unfortunately very singular in general. This is so even when we start with a smooth scheme X. For various purpose, a resolution in the usual sense, which, at the mean time, can also serve as a derived resolution, is desirable. This is, of course, much hard to achieve in general.

For the case of $\overline{M}_g(\mathbb{P}^n, d)$, we hope to achieve a relative resolution. Recall that a stable map has two directions to deform, one is curve, the other is collection of sections. Intuitively, one could image that the derived resolution removes the obstructions of the section directions but at the cost of destroying the (original) smoothness of the curve direction. The difficulty is to resolve the obstructions of section direction and at the mean time preserve the smoothness of the curve direction. Below, we will make it clear this vague point.

To find out the base of relative resolution of $\overline{M}_g(\mathbb{P}^n, d)$, we should begin with analyzing a neighborhood U of a singular point $[u, C] \in \overline{M}_g(\mathbb{P}^n, d)$. We use $\pi_U : \mathcal{C}_U \to U$ to denote the restriction $\mathcal{C}|_U$ of the universal family and f_U the restriction $f|_{\mathcal{C}_U}$ of the universal map. So, in diagram, we have

$$\begin{array}{ccc} \mathcal{C}_U & \xrightarrow{f_U} & \mathbb{P}^n \\ {\scriptstyle \pi_U}\downarrow & & \\ U & & \end{array}$$

Suppose we plan to resolve $\mathbf{R}\pi_* \mathfrak{f}^* \mathscr{O}_{\mathbb{P}^n}(k)$ (just remember that the case $k = 1$ ties closely to the local structures of the underlying moduli). First, we represent $\mathscr{O}_{\mathbb{P}^n}(k)$ as $\mathscr{O}_{\mathbb{P}^n}(H)$ by picking a general hypersurface of degree k meeting the image curve $f(C)$ (which is of degree d) at $m := dk$ many *isolated* points. If we perturb the map $[u, C]$ slightly, the image curve still meets H at isolated m points. Putting all these neighboring maps together, this implies that by replacing U by a smaller neighborhood if necessary, we may assume that we have $f_U^* \mathscr{O}_{\mathbb{P}^n}(H) = \mathscr{O}_{\mathcal{C}_U}(D)$ with $D = D_1 + \cdots + D_m$ where each D_i is a section of the family $\mathcal{C}_U \to U$ and they are disjoint. If C' is a component of C such that the degree of $u|_{C'}$ is d', them there should be $d'k$ many sections passing through the curve $C' \subset \mathcal{C}_U$; if C' gets contracted by the map f, then C' meets none of the sections.

Observe that the choice of the general hypersurface H determines a morphism by assigning the map $[u, C]$ to the pair $(C, \delta_1 + \cdots + \delta_m)$ where $\delta_i = C \cap D_i$. For this purpose, we introduce the Artin stack $\mathfrak{M}_g^{\text{div}}$ of pairs $(C, \delta_1 + \cdots + \delta_m)$ where C is a prestable curve of genus g and $\delta_1, \ldots, \delta_m$ are disjoint smooth points on C. This stack is smooth. By the above, we have a morphism $U \to \mathcal{V} \subset \mathfrak{M}_g^{\text{div}}$ where $\mathcal{V}$ is a chart of $\mathfrak{M}_g^{\text{div}}$. (The morphism is not canonical because it depends on the choice of H.) The geometry of the pairs is easier to get hold of than that of the stable map. This leads to the idea that we should work over the chart $\mathcal{V}$ and then pull back the results to the neighborhood U. For this, the Artin stack admits a universal curve $\mathcal{C}_\mathcal{V}$ and a universal divisor $\mathcal{D} = \mathcal{D}_1 + \cdots + \mathcal{D}_m$ so that we have a fiber square

$$\begin{array}{ccc} \mathcal{C}_U & \xrightarrow{\alpha} & \mathcal{C}_\mathcal{V} \supset \mathcal{D} \\ {\scriptstyle \pi_U}\downarrow & & {\scriptstyle \rho_\mathcal{V}}\downarrow \\ U & \longrightarrow & \mathcal{V} \end{array}$$

and $D_i = \alpha^* \mathcal{D}_i$ and $D = \alpha^* \mathcal{D}$. This allows us to study the object $\mathbf{R}(\pi_U)_* \mathscr{O}_{\mathcal{C}_U}(D)$ via the object $\mathbf{R}(\rho_\mathcal{V})_* \mathscr{O}_{\mathcal{C}_\mathcal{V}}(\mathcal{D})$. That is, a derived resolution of $\mathbf{R}(\rho_\mathcal{V})_* \mathscr{O}_{\mathcal{C}_\mathcal{V}}(\mathcal{D})$ will induce a derived resolution of $\mathbf{R}\pi_* \mathscr{O}_{\mathcal{C}_U}(D)$ because of the base change property. The task now is: smoothly blow up $\mathcal{V}$ so that the object $\mathbf{R}(\rho_\mathcal{V})_* \mathscr{O}_{\mathcal{C}_\mathcal{V}}(\mathcal{D})$ becomes locally diagonalizable. This is what we mean earlier when we mention that to get a true resolution we should *resolve the obstructions of the section directions* *(i.e., to locally diagonalize* $\mathbf{R}(\rho_{\mathcal{V}})_*\mathscr{O}_{\mathcal{C}_{\mathcal{V}}}(\mathcal{D})$*) and preserve the smoothness of curve directions (i.e., smoothly blowup the chart* $\mathcal{V}$*).*

Unfortunately, we have no obvious global morphisms from $\overline{M}_g(\mathbb{P}^n, d)$ to $\mathfrak{M}_g^{\mathrm{div}}$. To overcome to this difficulty, one could in principle work over local charts $\mathcal{V}$ of $\mathfrak{M}_g^{\mathrm{div}}$ and then (cumbersomely) argue how to patch the data over the charts. Instead, we choose to introduce another Artin stack so that it admits a global morphism from $\overline{M}_g(\mathbb{P}^n, d)$ and not too much technical information is lost.

This substituting stack is the Artin stack $\mathfrak{P}_g$ of pairs (C, L) of a prestable curve of genus g and a line bundle over C. This again is a smooth stack. It comes with a universal curve

$$\rho_{\mathfrak{P}} : \mathcal{C}_{\mathfrak{P}} \longrightarrow \mathfrak{P}_g$$

and a universal line bundle $\mathscr{L}$. The relevant derived object here is $\mathbf{R}(\rho_{\mathfrak{P}})_*\mathscr{L}$. We have a (global) morphism (depending on k)

$$\varsigma_k : \overline{M}_g(\mathbb{P}^n, d) \longrightarrow \mathfrak{P}_g, \quad [u, C] \to (C, u^*\mathscr{O}_{\mathbb{P}^n}(k)).$$

Our earlier local morphism clearly factors through

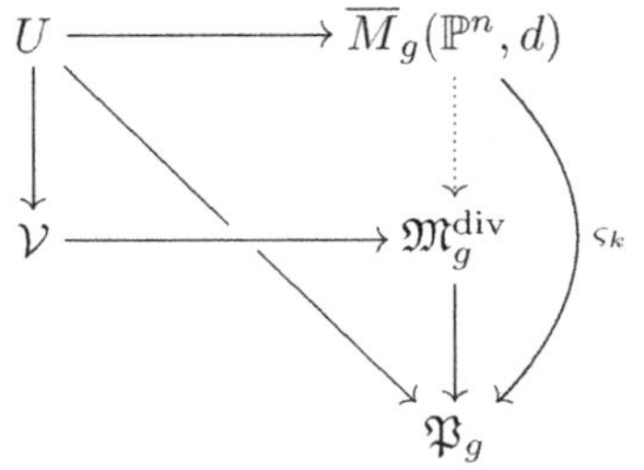

The idea now is that we can work over the chart $\mathcal{V}$ (which we found easy to cope with since it conveniently parameterizes a pair of curve and a divisor). Then, according the local smooth blowup of $\mathcal{V}$ we find, we may translate it to the stack $\mathfrak{P}_g$ to obtain the corresponding global smooth blowup, as desired. The goal is that we need to smoothly blow up $\mathcal{V}$ and hence also $\mathfrak{P}_g$ so that $\mathbf{R}(\rho_{\mathfrak{P}})_*\mathscr{L}$, hence also $\mathbf{R}\pi_*\mathfrak{f}^*\mathscr{O}_{\mathbb{P}^n}(k)$, becomes locally diagonalizable. the author excepts the following

Conjecture 2.4. *Fix $g > 0$ and $k > 0$. There is a canonical sequence of smooth blowups $\tilde{\mathfrak{P}}_g^k \to \mathfrak{P}_g$ of $\mathfrak{P}_g$ such that if we form the cartesian diagram*

$$\begin{array}{ccc} \widetilde{M}_g^k(\mathbb{P}^n, d) = \tilde{\mathfrak{P}}_g^k \times_{\varsigma_k, \mathfrak{P}_g} \overline{M}_g(\mathbb{P}^n, d) & \longrightarrow & \tilde{\mathfrak{P}}_g^k \\ \downarrow & & \downarrow \\ \overline{M}_g(\mathbb{P}^n, d) & \longrightarrow & \mathfrak{P}_d \end{array}$$

then the derived object $\mathbf{R}\pi_*\mathfrak{f}^*\mathscr{O}_{\mathbb{P}^n}(k)$, when pulling back to $\widetilde{M}_g^k(\mathbb{P}^n, d)$, becomes locally diagonalizable.

Such a *relative derived resolution* should be sufficient to apply to the GW theory[2] of hypersurfaces in $\mathbb{P}^n$. Because $\tilde{\mathfrak{P}}_g^k$ is smooth, we have a perfect relative

[2]Indeed, an easier partial relative derived resolution is sufficient. This will be explained in a forthcoming paper.

obstruction theory $\widetilde{M}_g^k(\mathbb{P}^n, d)/\tilde{\mathfrak{P}}_g^k$ by pulling back a perfect relative obstruction theory $\overline{M}_g(\mathbb{P}^n, d)/\mathfrak{P}_g$ (which is known to exist). We may apply this to $R\pi_* \mathfrak{f}^* \mathscr{O}_{\mathbb{P}^4}(5)$ to study the GW invariants of quintic threefolds.

To get a relative resolution of $\overline{M}_g(\mathbb{P}^n, d)$, we specialize the above to the case of $k = 1$.

Conjecture 2.5. *Fix $g > 0$. There is a canonical sequence of smooth blowups $\tilde{\mathfrak{P}}_g \to \mathfrak{P}_g$ of $\mathfrak{P}_g$ such that if we form the cartesian diagram*

$$\begin{array}{ccc} \widetilde{M}_g(\mathbb{P}^n, d) = \tilde{\mathfrak{P}}_g \times_{\mathfrak{P}_g} \overline{M}_g(\mathbb{P}^n, d) & \longrightarrow & \tilde{\mathfrak{P}}_g \\ \downarrow & & \downarrow \\ \overline{M}_g(\mathbb{P}^n, d) & \longrightarrow & \mathfrak{P}_d \end{array}$$

(we omit ς_1 from the fiber product), then the derived object $\mathbf{R}\pi_* \mathfrak{f}^* \mathscr{O}_{\mathbb{P}^n}(1)$, when pulling back to $\widetilde{M}_g^k(\mathbb{P}^n, d)$, becomes locally diagonalizable. Further, assume that $d > 2g - 2$, then we have

1. *the main component of $\widetilde{M}_g(\mathbb{P}^n, d)$ is smooth;*

2. *the entire stack $\widetilde{M}_g(\mathbb{P}^n, d)$ has at worst normal crossing singularities.*

The above is a resolution of $\overline{M}_g(\mathbb{P}^n, d)$ *relative to* $\mathfrak{P}_g$. This sort of *relative resolution* problem may be characterized as "*desingularization by removing relative obstructions, while preserving the smoothness of the base*". The author expects this to be true possibly after substituting $\mathfrak{P}_g$ by another moduli stack, or re-interpret the smoothness.

We can carry out this program for the cases of genus one and two. That is, Conjectures 2.4 and 2.5 hold in this two cases. In fact, our statements are slightly stronger than stated in the conjecture.

2.8 Relative resolution: the genus one case

Ideally, for technical convenience as we discussed earlier, we would prefer to describe a relative resolution of $\overline{M}_g(\mathbb{P}^n, d)$ with $\mathfrak{M}_g^{\mathrm{div}}$ as the base. But, we do not have a global morphism to achieve so. Thus we are forced to compromise to work over the stack $\mathfrak{P}_g$ downstairs. However, the stack $\mathfrak{P}_g$ is not that convenient to work with. Therefore, whenever possible, we should replace $\mathfrak{P}_g$ by another stack that is easier to work with. For genus one, this substituting stack is the smooth Artin stack $\mathfrak{M}_1^{\mathrm{wt}}$ of weighted curves. In general, the stack $\mathfrak{M}_g^{\mathrm{wt}}$ consists of pairs of pre-stable curve of genus g and a nonnegative integer (called weight). Forgetting the weight, we obtain a morphism from $\mathfrak{M}_g^{\mathrm{wt}}$ to the smooth Artin stack $\mathfrak{M}_g$ of pre-stable curves. This morphism is étale, showing the smoothness of $\mathfrak{M}_g^{\mathrm{wt}}$. There is a canonical morphism $\mathfrak{P}_g \to \mathfrak{M}_g^{\mathrm{wt}}$ by assigning a pair $(C, L) \in \mathfrak{P}_g$ to $(C, c_1(L)) \in \mathfrak{M}_g^{\mathrm{wt}}$. Thus, for every $k > 0$, we have

$$\overline{M}_g(\mathbb{P}^n, d) \to \mathfrak{P}_g \to \mathfrak{M}_g^{\mathrm{wt}}, \quad [u, C] \to (C, u^* \mathscr{O}_{\mathbb{P}^n}(k)) \to (C, c_1(u^* \mathscr{O}_{\mathbb{P}^n}(k))).$$

For genus one, the resolution can be achieved relative to the base $\mathfrak{M}_1^{\mathrm{wt}}$,

In [17], Vakil and Zinger provided a natural resolution of the main component of $\overline{M}_1(\mathbb{P}^n,d)$. We follow exactly their blowing up procedure, but instead blow up $\mathfrak{M}_1^{\mathrm{wt}}$, and then treat the entire moduli space by taking fiber product. In this case, we will only use the morphism

$$\varsigma_1 : \overline{M}_1(\mathbb{P}^n,d) \longrightarrow \mathfrak{M}_1^{\mathrm{wt}}, \quad [u,C] \to (C, c_1(u^*\mathscr{O}_{\mathbb{P}^n}(1))).$$

For any $i > 0$, let $\Theta_i \subset \mathfrak{M}_1^{\mathrm{wt}}$ be the close subset such its general points are of the form (C,w) such that C can be obtained from a smooth elliptic curve E by attaching (connected) trees of rational curves at i many disjoint points of E and the weight w restricted to E is zero. Every Θ_i is a smooth closed substack of codimension i.

Theorem 2.6. ([17, 7]) *We let $\widetilde{\mathfrak{M}}_1^{\mathrm{wt}} \longrightarrow \mathfrak{M}_1^{\mathrm{wt}}$ be the successive blowup of $\mathfrak{M}_1^{\mathrm{wt}}$ along the smooth closed substacks $\Theta_2, \Theta_3, \ldots$, and so on. If we let*

$$\widetilde{M}_1(\mathbb{P}^n,d) = \widetilde{\mathfrak{M}}_1^{\mathrm{wt}} \times_{\mathfrak{M}_1^{\mathrm{wt}}} \overline{M}_1(\mathbb{P}^n,d),$$

then we have

1. *every (including the main) irreducible component of $\widetilde{M}_1(\mathbb{P}^n,d)$ is smooth;*

2. *the entire DM stack $\widetilde{M}_1(\mathbb{P}^n,d)$ has at worse normal crossing singularities; further*

3. *the derived object $\mathbf{R}\pi_*\mathfrak{f}^*\mathscr{O}_{\mathbb{P}^n}(k)$, upon pulling back to $\widetilde{M}_1(\mathbb{P}^n,d)$, becomes locally diagonalizable for all $k > 0$; in particular*

4. *for any irreducible component N of $\widetilde{M}_1(\mathbb{P}^n,d)$, let $(\pi_N, \mathfrak{f}_N)$ be the pullback of the universal family $(\pi,\mathfrak{f})$ to N, then the direct image sheaf $(\pi_N)_*\mathfrak{f}_N^*\mathscr{O}_{\mathbb{P}^n}(k)$ is locally free for all $k > 0$.*

2.9 Relative resolutions: the genus two case

The genus two case is already substantially more complicated than the case of genus one. For one thing, unlike $g=1$, for $g=2$, there are more than one rounds of canonical sequence of blowups, and the case $k=1$ requires its own sequence of blowups; for another, the blowup centers are complicated to describe. Because it requires substantial preparation to describe them, we will have to be vague on the blowing up centers here. But, we will give a good hint near the end of the next section.

Theorem 2.7. ([9]) *There are three rounds of sequences of successive smooth blowups*[3] *$\widetilde{\mathfrak{M}}_2^{\mathrm{wt}} \longrightarrow \mathfrak{M}_2^{\mathrm{wt}}$ of $\mathfrak{M}_2^{\mathrm{wt}}$ such that if we let*

$$\widetilde{M}_2^k(\mathbb{P}^n,d) = \widetilde{\mathfrak{M}}_2^{\mathrm{wt}} \times_{\varsigma_k, \mathfrak{M}_2^{\mathrm{wt}}} \overline{M}_2(\mathbb{P}^n,d),$$

[3]We mention here that the first two rounds of sequences of smooth blowups are *topological* and the third round is *geometric*. We will be more specific about them in the next section. We believe that the sequences of topological blowups extend to all genera and are sufficient to study the GW numbers of quintics. This will appear in a forthcoming paper.

then we have

1. *the derived object* $\mathbf{R}\pi_*\mathfrak{f}^*\mathscr{O}_{\mathbb{P}^n}(k)$, *upon pulling back to* $\widetilde{M}_2^k(\mathbb{P}^n, d)$, *becomes locally diagonalizable for all* $k \geq 2$; *in particular*

2. *for any irreducible component* N *of* $\widetilde{M}_2^k(\mathbb{P}^n, d)$, *let* $(\pi_N, \mathfrak{f}_N)$ *be the pullback of the universal family* $(\pi, \mathfrak{f})$ *to* N, *then the direct image sheaf* $(\pi_N)_*\mathfrak{f}_N^*\mathscr{O}_{\mathbb{P}^n}(k)$ *is locally free for all* $k \geq 2$.

Unlike $g = 1$, some of the smooth blowups for $g = 2$ are not blowups along smooth closed centers.

The object $\mathbf{R}\pi_*\mathfrak{f}^*\mathscr{O}_{\mathbb{P}^n}(1)$ underpins the local structure of the underlying moduli space and requires one more sequence of blowups to resolve.

Theorem 2.8. ([9]) *There is a canonical sequence of smooth blowups of* $\widetilde{\mathfrak{P}}_2 \to \mathfrak{P}_2 \times_{\mathfrak{M}_2^{\mathrm{wt}}} \widetilde{\mathfrak{M}}_2^{\mathrm{wt}}$ *such that if we let*

$$\widetilde{M}_2(\mathbb{P}^n, d) = \widetilde{\mathfrak{P}}_2 \times_{\mathfrak{P}_2} \overline{M}_1(\mathbb{P}^n, d),$$

and suppose $d > 2$, *then we have*

1. *the main component of* $\widetilde{M}_2(\mathbb{P}^n, d)$ *is smooth;*

2. *the entire DM stack* $\widetilde{M}_2(\mathbb{P}^n, d)$ *has at worse normal crossing singularities; further*

3. *the derived object* $\mathbf{R}\pi_*\mathfrak{f}^*\mathscr{O}_{\mathbb{P}^n}(k)$, *upon pulling back to* $\widetilde{M}_2(\mathbb{P}^n, d)$, *becomes locally diagonalizable for all* $k \geq 1$; *in particular*

4. *for any irreducible component* N *of* $\widetilde{M}_2(\mathbb{P}^n, d)$, *let* $(\pi_N, \mathfrak{f}_N)$ *be the pullback of the universal family* $(\pi, \mathfrak{f})$ *to* N, *then the direct image sheaf* $(\pi_N)_*\mathfrak{f}_N^*\mathscr{O}_{\mathbb{P}^n}(k)$ *is locally free for all* $k \geq 1$.

3 Why does relative resolution work?

In this section, we provide some details to explain how we could potentially derive the modular resolutions for all g and why the cases $g = 1$ and 2 work.

We remind the reader that our goal is to smoothly blow up $\mathfrak{P}_g$ so that the derived object $\mathbf{R}(\rho_{\mathfrak{P}})_*\mathscr{L}$ becomes locally diagonalizable. As in Section 2.7, locally, we can factorize the canonical morphism $\overline{M}_g(\mathbb{P}^n, d) \to \mathfrak{P}_g$ as

$$U \longrightarrow \mathcal{V} \longrightarrow \mathfrak{P}_g$$

where U is an open neighborhood of a point $[u, C] \in \overline{M}_g(\mathbb{P}^n, d)$ and $\mathcal{V}$ is a chart over $\mathfrak{M}_g^{\mathrm{div}}$. Any blowup of $\mathfrak{P}_g$ induces a blowup of $\mathcal{V}$ by taking the fiber product. By now, we get the idea that to carry out our relative resolution program, it is important to know the local structures of the derived object $\mathbf{R}(\rho_{\mathcal{V}})_*\mathscr{O}_{\mathcal{C}_{\mathcal{V}}}(\mathcal{D})$ (which is related to $\mathbf{R}\pi_*\mathfrak{f}^*\mathscr{O}_{\mathbb{P}^n}(k)$ by Cohomology and Base Change).

So, we begin to analyze the derived object $\mathbf{R}(\rho_\mathcal{V})_*\mathscr{O}_{\mathcal{C}_\mathcal{V}}(\mathcal{D})$ and pin down a local presentation. This is known to admit a local presentation of the form $[E \xrightarrow{\psi} F]$ where E and F are locally free sheaves. But now, we are not content with such an inexplicit form (for the derived resolution, it suffices since we blow up along determinantal ideals whose explicit forms are not necessary), we need an explicit form so that we can extract the geometry behind the form. For this purpose, shrinking $\mathcal{V}$ if necessary, we can choose g many disjoint sections $\mathcal{A}_1, \ldots, \mathcal{A}_g$ of $\mathcal{C}_\mathcal{V} \to \mathcal{V}$ such that they are disjoint from the divisor $\mathcal{D}$ and for any $v \in \mathcal{V}$ and any irreducible subcurve C' of $C_v = \rho_\mathcal{V}^{-1}(v)$ of genus $h \leq g$, the set $\{\mathcal{A}_1 \cap C', \ldots, \mathcal{A}_g \cap C'\}$ consists of h many distinct points in general position (that is, they do not form a special divisor on the curve C'). Set $\mathcal{A} = \sum_i \mathcal{A}_i$. Then from the short exact sequence

$$0 \longrightarrow \mathscr{O}_{\mathcal{C}_\mathcal{V}}(\mathcal{D}) \longrightarrow \mathscr{O}_{\mathcal{C}_\mathcal{V}}(\mathcal{D}+\mathcal{A}) \longrightarrow \mathscr{O}_\mathcal{A}(\mathcal{A}) \longrightarrow 0$$

we have a long one

$$(\rho_\mathcal{V})_*\mathscr{O}_{\mathcal{C}_\mathcal{V}}(\mathcal{D}) \hookrightarrow (\rho_\mathcal{V})_*\mathscr{O}_{\mathcal{C}_\mathcal{V}}(\mathcal{D}+\mathcal{A}) \xrightarrow{\psi} (\rho_\mathcal{V})_*\mathscr{O}_\mathcal{A}(\mathcal{A}) \twoheadrightarrow R^1(\rho_\mathcal{V})_*\mathscr{O}_{\mathcal{C}_\mathcal{V}}(\mathcal{D})$$

because one shows that $R^1(\rho_\mathcal{V})_*\mathscr{O}_{\mathcal{C}_\mathcal{V}}(\mathcal{D}+\mathcal{A}) = 0$ by the assumptions on $\mathcal{A}$. Hence, $(\rho_\mathcal{V})_*\mathscr{O}_{\mathcal{C}_\mathcal{V}}(\mathcal{D}+\mathcal{A})$ is locally free of rank $d+1$ by Riemann-Roch. Also, it is clear that $(\rho_\mathcal{V})_*\mathscr{O}_\mathcal{A}(\mathcal{A})$ is locally free of rank g since $\mathcal{A}$ are disjoint sections. Thus, we have

$$\mathbf{R}(\rho_\mathcal{V})_*\mathscr{O}_{\mathcal{C}_\mathcal{V}}(\mathcal{D}) = [(\rho_\mathcal{V})_*\mathscr{O}_{\mathcal{C}_\mathcal{V}}(\mathcal{D}+\mathcal{A}) \xrightarrow{\psi} (\rho_\mathcal{V})_*\mathscr{O}_\mathcal{A}(\mathcal{A})].$$

We will call ψ a structural homomorphism; understanding this homomorphism is a key to our program. We can always choose frames of the free sheaves

$$E = (\rho_\mathcal{V})_*\mathscr{O}_{\mathcal{C}_\mathcal{V}}(\mathcal{D}+\mathcal{A}) \text{ and } F = (\rho_\mathcal{V})_*\mathscr{O}_\mathcal{A}(\mathcal{A})$$

so that the homomorphism ψ can be represented by a matrix. Then our task becomes to derive an explicit form for every entry in the matrix. Before we do it, we need some useful lemmas to decompose the homomorphism ψ.

First we choose another general section $\mathcal{B}$; this will help us to isolate a canonical trivial sheaf in the kernel of ψ.

Lemma 3.1. *We have a natural splitting of locally free sheaves*

$$(\rho_\mathcal{V})_*\mathscr{O}_{\mathcal{C}_\mathcal{V}}(\mathcal{D}+\mathcal{A}) = \mathscr{O}_\mathcal{V} \oplus (\rho_\mathcal{V})_*\mathscr{O}_{\mathcal{C}_\mathcal{V}}(\mathcal{D}+\mathcal{A}-\mathcal{B})$$

such that $\mathscr{O}_\mathcal{V} \subset \ker\psi$ *and* ψ *can be naturally decomposed as*

$$\mathscr{O}_\mathcal{V} \oplus (\rho_\mathcal{V})_*\mathscr{O}_{\mathcal{C}_\mathcal{V}}(\mathcal{D}+\mathcal{A}-\mathcal{B}) \xrightarrow{0\oplus\varphi} (\rho_\mathcal{V})_*\mathscr{O}_\mathcal{A}(\mathcal{A})$$

where $\varphi : (\rho_\mathcal{V})_*\mathscr{O}_{\mathcal{C}_\mathcal{V}}(\mathcal{D}+\mathcal{A}-\mathcal{B}) \to (\rho_\mathcal{V})_*\mathscr{O}_\mathcal{A}(\mathcal{A})$ *is the restriction homomorphism.*

Since $\psi = 0 \oplus \varphi$, we also call φ a structural homomorphism. Observe now that the rank of the locally free sheaf $(\rho_\mathcal{V})_*\mathscr{O}_{\mathcal{C}_\mathcal{V}}(\mathcal{D}+\mathcal{A}-\mathcal{B})$ coincides with the degree of $\mathcal{D} = \sum_{i=1}^m \mathcal{D}_i$. Indeed, if we let $\varphi_j : (\rho_\mathcal{V})_*\mathscr{O}_{\mathcal{C}_\mathcal{V}}(\mathcal{D}_j+\mathcal{A}-\mathcal{B}) \to (\rho_\mathcal{V})_*\mathscr{O}_\mathcal{A}(\mathcal{A})$ be the restriction homomorphism, then we have

Lemma 3.2. *We have the following natural decomposition of locally free sheaves*

$$(\rho_{\mathcal{V}})_* \mathscr{O}_{\mathcal{C}_{\mathcal{V}}}(\mathcal{D} + \mathcal{A} - \mathcal{B}) = \bigoplus_{j=1}^{m} (\rho_{\mathcal{V}})_* \mathscr{O}_{\mathcal{C}_{\mathcal{V}}}(\mathcal{D}_j + \mathcal{A} - \mathcal{B}).$$

Further, the homomorphism φ also decomposes as

$$\varphi = \bigoplus_{j=1}^{m} \varphi_j.$$

This reduces the study of the homomorphism ψ to each φ_i which can further be reduced to

$$\varphi_{ij} : (\rho_{\mathcal{V}})_* \mathscr{O}_{\mathcal{C}_{\mathcal{V}}}(\mathcal{D}_j + \mathcal{A} - \mathcal{B}) \to (\rho_{\mathcal{V}})_* \mathscr{O}_{\mathcal{A}_i}(\mathcal{A}_i) \cong \mathcal{V}$$

for all $1 \le i \le g$. Shrinking $\mathcal{V}$ if necessary, we can assume all sheaves $(\rho_{\mathcal{V}})_* \mathscr{O}_{\mathcal{C}_{\mathcal{V}}}(\mathcal{D}_j + \mathcal{A} - \mathcal{B})$ are isomorphic to the trivial sheaf $\mathscr{O}_{\mathcal{V}}$. Thus, upon fixing frames, we may regard φ_{ij} as a function in $\Gamma(\mathscr{O}_{\mathcal{V}})$. We found that this function vanishes at $v \in \mathcal{V}$ if and only if there are separable nodes between $\mathcal{A}_i \cap \mathcal{C}_v$ and $\mathcal{D}_j \cap \mathcal{C}_v$. Here a node q of a connected curve C is separable if $C \setminus q$ is disconnected; the separable node q is said to be between two points $a, \delta \in C$ if a and δ lie in different components of $C \setminus q$. For any two points $a, \delta \in C$, we let $N_{[a,\delta]}$ denote the set of all separable nodes between a and δ. Every node q of C is associated with a function ζ_q in $\Gamma(\mathscr{O}_{\mathcal{V}})$, called a node-smoothing parameter. Geometrically, at any point $v \in (\zeta_q = 0)$, the fiber curve $\mathcal{C}_v$ contains a node corresponding to q; at any point v away from the locus $(\zeta_q = 0)$, the curve $\mathcal{C}_v$ does not have a node corresponding to q (the node is smoothed out). This leads us to the following key proposition.

Proposition 3.3. *Upon fixing frames, we have $\varphi_{ij} = c_{ij} \prod_{q \in N_{[a_i,\delta_j]}} \zeta_q$ where $c_{ij} \in \Gamma(\mathscr{O}_{\mathcal{V}}^*)$.*

Putting all together, we obtain a matrix representation of the structural homomorphism φ by

$$\Phi = (c_{ij} \prod_{q \in N_{[a_i,\delta_j]}} \zeta_q)_{1 \le i \le g, 1 \le j \le m}.$$

This matrix is important because it provides a local equation of $\overline{M}_g(\mathbb{P}^n, d)$ at the point $[u, C]$ when $m = d$ (i..e, for $\pi_* \mathfrak{f}^* \mathscr{O}_{\mathbb{P}^n}(1)$). The idea is roughly that the kernel of the homomorphism φ gives rise to sections of $\pi_* \mathfrak{f}^* \mathscr{O}_{\mathbb{P}^n}(1)$ but sections of $\pi_* \mathfrak{f}^* \mathscr{O}_{\mathbb{P}^n}(1)$ determine stable maps. Note that Φ is a matrix of size $g \times d$ (for $\pi_* \mathfrak{f}^* \mathscr{O}_{\mathbb{P}^n}(1)$). We let $\mathbf{w}^i$ be the *column* vector $(w_1^i, \ldots, w_d^i), i = 1, \ldots, n$ with $w_j^i \in \mathbb{A}^1$.

Theorem 3.4. ([7]) *The local equation of a (small) neighborhood U of the point $[u, C] \in \overline{M}_g(\mathbb{P}^n, d)$ is given by the system of equations*

$$\Phi \cdot \mathbf{w}^i = 0, \quad i = 1, \ldots, n, \tag{3.1}$$

realizing it as a closed subset of the smooth space $\mathcal{V} \times \mathbb{A}^{dn}$.

The simplest kind of matrices are diagonal ones. In such a case, equations (3.1) become elementary. However, one can not hope in general to find a diagonal form of Φ simply by choosing frames. To make Φ diagonal in general, birational base change is necessary. A good thing is that the local equation (3.1) pulls back to provide a local equation under any base change.

Theorem 3.5. *Let $f : \mathcal{V}' \to \mathcal{V}$ be any morphism, $U' = U \times_{\mathcal{V}} \mathcal{V}'$ and Φ' a matrix representation of the pullback homomorphism $f^*\varphi$. Then U' is defined by the system of equations*

$$\Phi' \cdot \mathbf{w}^i = 0, \quad i = 1, \ldots, n, \tag{3.2}$$

realizing it as a closed subset of $\mathcal{V}' \times \mathbb{A}^{dn}$.

Now, *suppose* $f : \tilde{\mathcal{V}} \to \mathcal{V}$ is a smooth blowup of $\mathcal{V}$ such that the derived object $\mathbf{R}(\rho_{\mathcal{V}})_* \mathscr{O}_{\mathcal{C}_{\mathcal{V}}}(\mathcal{D})$ (related to $\mathbf{R}\pi_* \mathfrak{f}^* \mathscr{O}_{\mathbb{P}^n}(1)$) becomes diagonalizable upon pulling back to $\tilde{\mathcal{V}}$. Then we can find a diagonal form for its corresponding matrix $\tilde{\Phi}$. Write $\tilde{\Phi}$ as $\mathrm{diag}(z_1, \ldots, z_g, 0, \ldots, 0)$ (here, we assume $d > g$). Then

Corollary 3.6. *The open subset $\tilde{U} = U \times_{\mathcal{V}} \tilde{\mathcal{V}}$ is given by equations*

$$z_j \cdot \mathbf{w}_j^i = 0, \quad 1 \le j \le g, \; i = 1, \ldots, n, \tag{3.3}$$

realizing it as a closed subset of the smooth space $\tilde{\mathcal{V}} \times \mathbb{A}^{dn}$.

The main component of $\tilde{U}$ is found by requiring the non-vanishing of z_j over its generic points, hence it is given by

$$w_j^i = 0, \quad 1 \le j \le g, \; i = 1, \ldots, n,$$

which define a smooth closed subset of the smooth space $\tilde{\mathcal{V}} \times \mathbb{A}^{dn}$. This shows the smoothness of the main component. Since $f : \tilde{\mathcal{V}} \to \mathcal{V}$ is a smooth blowup of $\mathcal{V}$, we conclude that $(z_j = 0)$ defines a normal crossing divisor over $\mathcal{V}$. Any other irreducible component is defined by a mixed vanishing of prime factors of z_j and w_j^i. We conclude that every irreducible component admits at worst normal crossing singularities and they meet in normal crossing way. A priori, we cannot exclude the case when some prime divisor of irreducible component has self-intersection. If in addition, we know that every prime divisor of irreducible component is smooth, then we can conclude that every irreducible component is smooth. This is the case when $g = 1$, but not the case when $g = 2$, hence it should not be the case for all $g \ge 2$.

When $g = 1$, the matrix $\Phi = (c_{ij} \prod_{q \in N_{[a_i, \delta_j]}} \zeta_q)$ in the local equation in (3.1) has only one row. We can absorb all the invertible coefficients c_{ij} into the frames. Then, a simple analysis implies that all the singularities are caused by the presence of nodes. We call these singularities of topological type. These are straightforward to resolve as in Theorem 2.6.

When $g = 2$, the matrix $\Phi = (c_{ij} \prod_{q \in N_{[a_i, \delta_j]}} \zeta_q)$ in the local equation in (3.1) has two rows. We classify the singularities into two types. Topological type: the ones caused solely by the presence of nodes. Geometric type: the ones caused by vanishing of minor determinants of the matrix (c_{ij}). This classification works for all genera. In genus two, we prove that the singularities of geometric type means: Weierstrass and conjugate rational tails, double cover of rational curves (hyperelliptic). Thus, one may also call the geometric type the Brill-Noether type. These are not too hard to handle when $g = 2$. In Theorems 2.7 and 2.8, the first two rounds of sequences of smooth blowups are along centers of topological type. Roughly, the first round deals with (1×1) minors; the second round deals with (2×2) minors. Both rounds ignore singularities of geometric type. Then the third rounds resolve singularities of Weierstrass and conjugate rational tails. The fourth round resolves singularities caused by hyperelliptic maps.

For general g, it is not hard for the author to "see" how to resolve singularities of topological type. The remainder singularities of geometric type are fortunately of high codimension. I believe that a canonical partial relative resolution that resolves singularities of topological type will be sufficient to prove Conjecture 2.3 (the details will be in a forthcoming publication).

4 Applications to the GW theory

In this section, we simply summarize the approach of [2].

The GSW invariants originate from Guffin and Sharpe's work [6]. Let $\overline{M}_1(\mathbb{P}^4, d)^p$ be the DM stack of pairs $([u,C],p)$ with $[u,C] \in \overline{M}_1(\mathbb{P}^4,d)$ and $p \in H^0(C, u^*\mathscr{O}_{\mathbb{P}^4}(-5)\otimes\omega_C)$. This is the moduli of stale maps with p-fields. Chang and Li [2] showed that it admits with a perfect obstruction theory. We shall use the following shorthands

$$\mathcal{X} = \overline{M}_1(\mathbb{P}^4,d), \quad \mathcal{Y} = \overline{M}_1(\mathbb{P}^4,d)^p.$$

The polynomial $x_1^5 + \cdots + x_5^5$ induces a cosection of its obstruction sheaf $\sigma : \mathcal{O}b_{\mathcal{Y}} \longrightarrow \mathscr{O}_{\mathcal{Y}}$ whose degeneracy locus (where σ fails to be surjective) is

$$\overline{M}_1(Q,d) \subset \mathcal{Y}, \quad Q = (x_1^5 + \cdots + x_5^5 = 0) \subset \mathbb{P}^4,$$

which is proper. The cosection localized virtual class of Kiem-Li ([13]) defines a localized virtual cycle $[\mathcal{Y}]^{\mathrm{vir}}_\sigma \in A_0\overline{M}_1(Q,d)$. The GSW-invariant of the quintic Q is

$$N^p_{1,d} = \deg[\mathcal{Y}]^{\mathrm{vir}}_\sigma. \tag{4.1}$$

Theorem 4.1. ([2]) *The GSW-invariant coincides with the GW-invariant of the quintic Q up to a sign:*

$$N^p_{1,d} = (-1)^{5d}\cdot N_{1,d}$$

where $N_{1,d}$ is the (usual) genus 1, degree d GW invariant of the smooth quintic $Q \subset \mathbb{P}^4$.

This theorem translates the problem from over $\mathcal{X}$ to over $\mathcal{Y}$. To prove Conjecture 2.3, it requires to study the separation of the virtual cycle $[\mathcal{Y}]^{\mathrm{vir}}_\sigma$. For this, we need some structural results on its intrinsic normal cone. So, we let $(f_{\mathcal{X}},\pi_{\mathcal{X}}) : \mathcal{C}_{\mathcal{X}} \longrightarrow \mathbb{P}^4\times\mathcal{X}$ be the universal family of $\mathcal{X}$; let

$$(f_{\mathcal{Y}},\pi_{\mathcal{Y}}) : \mathcal{C}_{\mathcal{Y}} \longrightarrow \mathbb{P}^4\times\mathcal{Y}, \quad \psi_{\mathcal{Y}} \in \Gamma(\mathcal{Y},\mathcal{P}_{\mathcal{Y}}), \quad \mathcal{P}_{\mathcal{Y}} = f^*_{\mathcal{Y}}\mathscr{O}(-5)\otimes\omega_{\mathcal{C}_{\mathcal{Y}}/\mathcal{Y}}$$

be the universal family of $\mathcal{Y}$. A perfect obstruction theory of $\mathcal{X}$ relative to the Artin stack $\mathfrak{M}$ of genus g nodal curves is ([1])

$$\phi_{\mathcal{X}/\mathfrak{M}} : (E_{\mathcal{X}/\mathfrak{M}})^{\vee} \longrightarrow L^{\bullet}_{\mathcal{X}/\mathfrak{M}}, \quad E_{\mathcal{X}/\mathfrak{M}} := R^{\bullet}\pi_{\mathcal{X}*}f^*_{\mathcal{X}}T_{\mathbb{P}^4}.$$

Using the Euler sequence, a perfect obstruction theory of $\mathcal{X}$ relative to $\mathfrak{P}_1$ is (cf. [2])

$$\phi_{\mathcal{X}/\mathfrak{P}_1} : (E_{\mathcal{X}/\mathfrak{P}_1})^{\vee} \longrightarrow L^{\bullet}_{\mathcal{X}/\mathfrak{P}_1}, \; E_{\mathcal{X}/\mathfrak{P}_1} := R^{\bullet}\pi_{\mathcal{X}*}\mathcal{L}_{\mathcal{X}}^{\oplus 5}, \; \mathcal{L}_{\mathcal{X}} = f^*_{\mathcal{X}}\mathscr{O}(1).$$

Chang and Li also worked out a perfect obstruction theory of $\mathcal{Y}$ relative to $\mathfrak{P}_1$:

$$\phi_{\mathcal{Y}/\mathfrak{P}_1} : (E_{\mathcal{Y}/\mathfrak{P}_1})^{\vee} \longrightarrow L^{\bullet}_{\mathcal{Y}/\mathfrak{P}_1}, \; E_{\mathcal{Y}/\mathfrak{P}_1} := R^{\bullet}\pi_{\mathcal{Y}*}(\mathcal{L}_{\mathcal{Y}}^{\oplus 5} \oplus \mathcal{P}_{\mathcal{Y}}), \; \mathcal{L}_{\mathcal{Y}} = f^*_{\mathcal{Y}}\mathscr{O}(1).$$

The cohomology sheaf

$$\mathcal{O}b_{\mathcal{Y}/\mathfrak{P}_1} := H^1(E_{\mathcal{Y}/\mathfrak{P}_1}) = R^1\pi_{\mathcal{Y}*}(\mathscr{L}_{\mathcal{Y}}^{\oplus 5} \oplus \mathcal{P}_{\mathcal{Y}})$$

is the relative obstruction sheaf of $\phi_{\mathcal{Y}/\mathfrak{P}_1}$.

The following theorem will be useful.

Theorem 4.2. *We let* $\widetilde{\mathfrak{M}}_1^{\mathrm{wt}} \longrightarrow \mathfrak{M}_1^{\mathrm{wt}}$ *be the successive blowup of* $\mathfrak{M}_1^{\mathrm{wt}}$ *along the smooth closed substacks* $\Theta_2, \Theta_3, \ldots,$ *and so on. If we let*

$$\widetilde{M}_1(\mathbb{P}^4, d)^p = \widetilde{\mathfrak{M}}_1^{\mathrm{wt}} \times_{\mathfrak{M}_1^{\mathrm{wt}}} \overline{M}_1(\mathbb{P}^4, d)^p,$$

then we have

1. *every (including the main) irreducible component of* $\widetilde{M}_1(\mathbb{P}^4, d)^p$ *is smooth;*
2. *the entire DM stack* $\widetilde{M}_1(\mathbb{P}^4, d)^p$ *has at worse normal crossing singularities; further*
3. *the object* $R^{\bullet}\pi_{\mathcal{Y}*}(\mathcal{L}_{\mathcal{Y}}^{\oplus 5} \oplus \mathcal{P}_{\mathcal{Y}})$*, upon pulling back to* $\widetilde{M}_1(\mathbb{P}^4, d)^p$*, becomes locally diagonalizable.*

This is analogous to Theorem 2.6 and can be proved parallelly.

We let $\tilde{\mathcal{Y}}$ be $\widetilde{M}_1(\mathbb{P}^n, d)^p$ and $\tilde{\mathfrak{P}}_1 = \widetilde{\mathfrak{M}}_1^{\mathrm{wt}} \times_{\mathfrak{M}_1^{\mathrm{wt}}} \mathfrak{P}_1$. Then we have a canonical morphism $\tilde{\mathcal{Y}} \to \tilde{\mathfrak{P}}_1$. By working out the relative perfect obstruction theory of $\tilde{\mathcal{Y}} \to \tilde{\mathfrak{P}}_1$, we obtain its obstruction complex $E_{\tilde{\mathcal{Y}}/\tilde{\mathfrak{P}}_1}$. The intrinsic normal cone $\mathbf{C}_{\tilde{\mathcal{Y}}/\tilde{\mathfrak{P}}_1}$ of $\tilde{\mathcal{Y}} \to \tilde{\mathfrak{P}}_1$ is embedded in $h^1/h^0(E_{\tilde{\mathcal{Y}}/\tilde{\mathfrak{P}}_1})$.

The blowup $\tilde{\mathcal{X}}$ is a union of smooth DM stacks: one, denoted $\tilde{\mathcal{X}}'$, is the proper transform of the main component $\mathcal{X}'$ of $\mathcal{X}$, the rest are indexed by partitions μ of d. In formula,

$$\tilde{\mathcal{X}} = \tilde{\mathcal{X}}' \cup (\cup_{\mu \vdash d}\tilde{\mathcal{X}}_{\mu}).$$

Geometrically, general points of $\tilde{\mathcal{X}}'$ are stable morphisms with smooth domain curves; general points of $\tilde{\mathcal{X}}_{\mu}$ lie over stable morphisms whose domain curves are the union of a smooth genus one curve and connected trees of rational curves such that the genus one curve is contracted and the degrees of the stable morphism on the connected trees form the partition μ. The blowup $\tilde{\mathcal{Y}}$ has the similar decomposition

$$\tilde{\mathcal{Y}} = \tilde{\mathcal{Y}}' \cup (\cup_{\mu \vdash d} \tilde{\mathcal{Y}}_\mu).$$

Theorem 4.2 implies that

Lemma 4.3. *Let $\mathbf{C}_{\tilde{\mathcal{Y}}/\tilde{\mathfrak{P}}_1} \subset h^1/h^0(E_{\tilde{\mathcal{Y}}/\tilde{\mathfrak{P}}_1})$ be the intrinsic normal cone embedded in $h^1/h^0(E_{\tilde{\mathcal{Y}}/\tilde{\mathfrak{P}}_1})$ via the obstruction theory $\phi_{\tilde{\mathcal{Y}}/\tilde{\mathfrak{P}}_1}$. Then*

1. *away from $\cup_{\mu \vdash d} \tilde{\mathcal{Y}}_\mu$, $\mathbf{C}_{\tilde{\mathcal{Y}}/\tilde{\mathfrak{P}}_1}$ is the zero section of $h^1/h^0(E_{\tilde{\mathcal{Y}}/\tilde{\mathfrak{P}}_1})$;*

2. *away from $\tilde{\mathcal{Y}}'$, it is a rank two subbundle stack of $h^1/h^0(E_{\tilde{\mathcal{Y}}/\tilde{\mathfrak{P}}_1})$.*

Lemma 4.3 allows us to conclude that the cone $\mathbf{C}_{\tilde{\mathcal{Y}}/\tilde{\mathcal{D}}}$ can be separated as

$$[\mathbf{C}_{\tilde{\mathcal{Y}}/\tilde{\mathcal{D}}}] = [\mathbf{C}'] + \sum_{\mu \vdash d} [\mathbf{C}_\mu],$$

where $\mathbf{C}'$ is an irreducible cycle lying over $\tilde{\mathcal{Y}}_{\text{pri}}$; each $\mathbf{C}_\mu$ lies over $\tilde{\mathcal{Y}}_\mu$. (The cycles $\mathbf{C}_\mu$ need not to be irreducible.) Thus, applying Kiem-Li's cosection localized virtual class, we obtain

$$\deg 0^!_{\tilde{\sigma},\text{loc}}[\mathbf{C}_{\tilde{\mathcal{Y}}/\tilde{\mathcal{D}}}] = 0^!_{\tilde{\sigma},\text{loc}}[\mathbf{C}'] + \sum_{\mu \vdash d} 0^!_{\tilde{\sigma},\text{loc}}[\mathbf{C}_\mu], \tag{4.2}$$

where the cosection $\tilde{\sigma} : \mathcal{O}b_{\tilde{\mathcal{Y}}} \longrightarrow \mathscr{O}_{\tilde{\mathcal{Y}}}$ is induced from $\sigma : \mathcal{O}b_{\mathcal{Y}} \longrightarrow \mathscr{O}_{\mathcal{Y}}$.

Now, a general standard argument implies that

$$\deg 0^!_{\tilde{\sigma},\text{loc}}[\mathbf{C}'] = (-1)^{5d} N'_{1,d}.$$

By some accurate dimensional arguments, one finds the vanishing

$$\deg 0^!_{\tilde{\sigma},\text{loc}}[\mathbf{C}_\mu] = 0$$

for all $\mu \neq (d)$ where (d) be the partition of d into a single part (i.e., the non-partition of d). Finally, Chang and Li calculate to prove

$$\deg 0^!_{\tilde{\sigma},\text{loc}}[\mathbf{C}_{(d)}] = \frac{(-1)^{5d}}{12} N_{0,d}.$$

Putting these together, one gets $N_{1,d} = N'_{1,d} + \frac{1}{12} N_{0,d}$.

5 Modular resolution program

Suppose we have a moduli space $\mathfrak{M}$ and we let $\widetilde{\mathfrak{M}} \to \mathfrak{M}$ be a blowup. In general, it is not clear how to provide $\widetilde{\mathfrak{M}}$ a modular meaning. There are a few interesting cases where we know that the blowups have geometric meanings. Here we mention one class of such examples.

We let $\mathrm{Gr}(k,V)$ be the Grassmannian of k-dimensional subspaces in a vector space $V \cong \mathbf{k}^n$ where $\mathbf{k}$ is the base field $(0<k<n)$. Consider the variety $\mathcal{Q}_d$ of degree d maps from $\mathbf{P}^1$ to $\mathrm{Gr}(k,V)$. It comes with a naive compactification, the Grothendieck Quot schome $\overline{\mathcal{Q}}_d$ $(= \mathrm{Quot}^{d,n-k}_{V_{\mathbf{P}^1}/\mathbf{P}^1/\mathbf{k}})$. The compactification is smooth but the boundary $\overline{\mathcal{Q}}_d \setminus \mathcal{Q}_d$ has rather intricate singularities. It comes equipped with a natural filtration by closed subsets

$$Z_{d,0} \subset Z_{d,1} \subset \cdots \subset Z_{d,d-1} = \overline{\mathcal{Q}}_d \setminus \mathcal{Q}_d$$

where $Z_{d,r}$ consists of non-locally free sheaves whose torsion parts have degree at least $d-r$.

Theorem 5.1. ([11, 16]) *The singularities of $Z_{d,r}$ can be resolved by repeatedly blowing up $Z_{d,0}, Z_{d,1}, \ldots, Z_{d,r-1}$. Consequently, by successively blowing up the Quot scheme $\overline{\mathcal{Q}}_d$ along $Z_{d,0}, Z_{d,1}, \ldots, Z_{d,d-1}$, we obtain a smooth compactification $\widetilde{\mathcal{Q}}_d$ such that the boundary $\widetilde{\mathcal{Q}}_d \setminus \mathcal{Q}_d$ is a simple normal crossing divisor.*

Next, it is natural to ask whether $\widetilde{\mathcal{Q}}_d$ admits any modular interpretation. We provide an affirmative answer to this question. For any coherent sheaf $\mathcal{E}$ over $\mathbb{P}^1$, we let $\mathcal{E}_t$ denote the torsion subsheaf of $\mathcal{E}$ and $\mathcal{E}_f = \mathcal{E}/\mathcal{E}_t$ denote the locally free part of $\mathcal{E}$.

Definition 5.2. *A complete quotient of $V_{\mathbb{P}^1}$ consists of $(\mathcal{E}^0, \ldots, \mathcal{E}^\ell)$ where $\mathcal{E}^0$ is a coherent quotient sheaf of $V_{\mathbb{P}^1}$, and for every $1 \le i \le \ell$, $\mathcal{E}^i$ is a nonsplit extension of $\mathcal{E}^{i-1}_t$ by $\mathcal{E}^{i-1}_f$ such that the last sheaf $\mathcal{E}^\ell$ is the unique one that is locally free. (We allow $\ell = 0$.)*

Theorem 5.3. ([12]) *The variety $\widetilde{\mathcal{Q}}_d$ in Theorem 5.1 parameterizes complete quotients of $V_{\mathbb{P}^1}$ of degree d and rank $n-k$.*

There are further interesting questions. For $g=1$, the analogous of $\overline{\mathcal{Q}}_d$ is MOP's moduli $\overline{\mathrm{SQ}}_1(\mathrm{Gr}(r,V),d)$ of stable quotients. A smooth blowup

$$\widetilde{\mathrm{SQ}}_1(\mathrm{Gr}(r,V),d) \to \overline{\mathrm{SQ}}_1(\mathrm{Gr}(r,V),d),$$

resolving all the boundary singularities, is currently being worked out by Maienschein ([14]). The above should generalize to all high genera.

The author believes the following

Conjecture 5.4. *For every $g>0$, there is a smooth moduli stack $\mathfrak{M}$ admitting a dominating morphism $\overline{M}_g(\mathbb{P}^n,d) \to \mathfrak{M}$; further, there is also another smooth* **moduli** *stack $\widetilde{\mathfrak{M}}$ admitting a birational dominating morphism to $\mathfrak{M}$ such that if we let*

$$\widetilde{M}_g(\mathbb{P}^n,d) = \widetilde{\mathfrak{M}} \times_{\mathfrak{M}} \overline{M}_g(\mathbb{P}^n,d),$$

then $\widetilde{M}_g(\mathbb{P}^n,d)$ inherits moduli meaning. Further, suppose $d > 2g-2$, then we have

1. *the main component of $\widetilde{M}_g(\mathbb{P}^n,d)$ is smooth;*

2. *the entire stack* $\widetilde{M}_g(\mathbb{P}^n, d)$ *has at worst normal crossing singularities.*

This is a kind of reformulation of Conjecture 2.5. Here, the smoothness can subject to re-definition. For $g = 1$, it is believed that $\widetilde{\mathfrak{M}}$ can be taken to be the moduli of weighted genus one curves with certain admissible log structures (this is due to Qile Chen).

Beyond reach, the author ends this note with an unlimitedly wild speculation.

Conjecture 5.5. *Let* $\mathfrak{X}$ *be a singular moduli stack with the main component* $\mathfrak{X}'$. *Then under some reasonable hypotheses on* $\mathfrak{X}$, *there is another moduli stack* $\widetilde{\mathfrak{X}}$ *with a smooth main component* $\widetilde{\mathfrak{X}}'$, *at worst normal crossing singularities, and admitting a dominant morphism to* $\mathfrak{X}$ *such that it restricts to a birational morphism* $\widetilde{\mathfrak{X}}' \to \mathfrak{X}'$.

Derived from *Anna Karenina*,

Regularities are all alike; every singularity is singular in its own way.

The conjecture would say that every singularity can be resolved because there are geometric reasons to let us achieve it (not because we can implant an algorithm). de Jong's alteration has this flavor. We hope to furnish more interesting examples of this in the future publications.

It was pointed out to the author that the reference [9] misses a special case in the interior of the moduli of genus-2 stable maps. Thus, Theorem 2.8 (1) should be replaced by: the main component is smooth away from this special locus. This special case will soon be studied.

References

[1] K. Behrend and B. Fantechi. The intrinsic normal cone. *Invent. Math.* **128**, no. 1: 45–88, 1997.

[2] H.-L. Chang and J. Li. Gromov-Witten invariants of stable maps with fields. *International Mathematics Research Notices* **16**, no. 18: 4163–4217, 2011.

[3] H.-L. Chang and J. Li. An algebraic proof of the hyperplane property of the genus one GW-invariants of quintics. *Journal of Differential Geometry* **100**, no. 2, 2012.

[4] I. Ciocan-Fontanine, B. Kim and D. Maulik. Stable quasimaps to GIT quotients. *Journal of Geometry and Physics* **75**, no. 1: 17–47, 2014.

[5] A.J. de Jong. Smoothness, semi-stability and alterations. *Publications Mathematiques I.H.E.S.* **83**, 51–93, 1996.

[6] J. Guffin and E. Sharpe. A-twisted Landau-Ginzburg models. arXiv:0801.3836.

[7] Y. Hu and J. Li. Genus-one stable maps, local equations and vakil-zinger's desingularization. *Math. Ann.* **348**, no. 4: 929–963, 2010.

[8] Y. Hu and J. Li. Derived resolution property for stacks, Euler classes and applications. *Math. Research Letters* **18**, no. 4: 1–14, 2011.

[9] Y. Hu and J. Li. Genus-two stable maps, local equations and modular Resolutions, 2012. arXiv:1201. 2427.

[10] D. Huybrechts and M. Lehn. *The Geometry of Moduli Spaces of Sheaves.* Cambridge University Press, 2010.

[11] Y. Hu, J. Lin and Y. Shao. A compactification of the space of algebraic maps from $\mathbb{P}^1$ to $\mathbb{P}^n$. *Communications in Analysis and Geometry* **19**, no. 1: 1–30, 2011.

[12] Y. Hu and Y. Shao. The space of complete quotients. *Pure and Applied Mathematics Quarterly* **10**, no. 1, 2013.

[13] Y. -H. Kiem and J. Li, Localizing virtual cycles by cosections. arXiv:1007.3085.

[14] T. D. Maienschein. Desingularization of the boundary of the moduli space of genus one stable quotients. Ph.D Dissertation at the University of Arizona, 2014.

[15] A. Marian, D. Oprea and R. Pandharipande. The moduli space of stable quotients. *Geometry & Topology* **15**, 1651–1706, 2011.

[16] Y. Shao. A compactification of the space of parametrized rational curves in Grassmannians. Ph.D Dissertation at the University of Arizona, 2010. arXiv:1108.2299.

[17] R. Vakil and A. Zinger. A desingularization of the main component of the moduli space of genus-one stable maps into $\mathbb{P}^n$. *Geom. Topol.* **12**, no. 1: 1–95, 2008.

[18] A. Zinger. The reduced genus-one Gromov-Witten invariants of Calabi-Yau hypersurfaces. *Journal of AMS* **22**, no. 3: 691–737, 2007.

Department of Mathematics, University of Arizona
Tucson, AZ 85721-0089, USA
E-mail address: yhu@math.arizona.edu

Proceedings of the 6th International Congress of Chinese Mathematicians
ALM 36, pp. 489–503

Mirror Theorem, Seidel Representation, and Counting Holomorphic Disks in Toric Varieties

Hsian-Hua Tseng

To the memory of Professor Chao-Liang Shen

Abstract

There are three methods for computing the quantum cohomology ring of a projective toric manifold: mirror theorem [16, 28], Seidel representation [30], and disk enumerations [13]. We give a survey on these methods, and explain the relations between them recently established by González-Iritani [18, 19] and Chan-Lau-Leung-Tseng [6, 7].

Mathematics Subject Classification: 53D45, 14N35.

Keywords and Phrases: Open Gromov-Witten invariants, mirror symmetry, Seidel representation, toric manifolds.

1 Introduction

The quantum cohomology ring[1] $QH^*(X)$ of a projective toric manifold X is a well-studied topic. Batyrev [1] proposed a presentation of $QH^*(X)$ which can be viewed as a quantization of the Stanley-Reisner presentation of the ordinary cohomology ring of X. As a consequence of the toric mirror theorem [16, 28, 22], Batyrev's presentation is proven to be correct when X is Fano, and when X is not Fano, some corrections are needed. We briefly review this in Section 3.1.

Another way to obtain a presentation of $QH^*(X)$, due to [30], is to use a group homomorphism from $\pi_1(Ham(X))$ to $QH^*(X)^\times$ called Seidel representation to translate relations in $\pi_1(Ham(X))$ to relations in $QH^*(X)$. The toric manifold

[1] In this note we only consider small quantum cohomology.

X has many circle actions, giving rise to elements of $\pi_1(Ham(X))$ with relations described by the toric combinatorics of X. These relations suffice to give a presentation of $QH^*(X)$, which also recovers Batyrev's presentation when X is Fano. Some more details are discussed in Section 3.2.

More recently, Fukaya-Oh-Ohta-Ono [12] define open Gromov-Witten invariants which are virtual numbers of holomorphic disks in X with boundaries on a Lagrangian torus fiber of the moment map of X. These invariants assemble to a generating function W^{LF} called Lagrangian Floer superpotential of X, which should be understood as the mirror of X. As a justification, it is shown [13] that $QH^*(X)$ is isomorphic to the Jacobian ring of W^{LF}, see Section 3.3 for more details.

Qualitative and quantitative properties of open Gromov-Witten invariants of X are of great importance. However, the nature of these invariants render standard tools in Gromov-Witten theory (e.g. virtual localization) inapplicable to them. It is therefore natural and important to ask the following

Question 1.1. *How to compute these open Gromov-Witten invariants of X?*

On the other hand, having three methods to compute $QH^*(X)$ naturally motivates the following

Question 1.2. *How are these three methods related?*

In this note we address these Questions in the case when X is semi-Fano. Our discussion is based on the recent works of González-Iritani [18, 19] and Chan-Lau-Leung-Tseng [6, 7].

The rest of this note is organized as follows. In Section 2 we discuss some preparatory materials. In Section 3 we briefly review the three methods of computing $QH^*(X)$. Finally in Section 4 we explain the relations between these methods.

Acknowledgements. The author is grateful to K. Chan, C.-H. Cho, S.-C. Lau, and N. C. Leung for discussions and collaborations on this subject. The author is also grateful to the organizers of the International Congress of Chinese Mathematicians 2013 for an opportunity to present these works. The author is supported in part by Simons Foundation Collaboration Grant.

2 Preliminary materials

In this Section we review some basic materials.

2.1 Quantum cohomology

Let (M, ω) be a compact symplectic manifolds. The (small) quantum cohomology of (M, ω), denoted by $QH^*(M)$, is additively the ordinary cohomology of M,

$$QH^*(M) = H^*(M, \mathbb{C}) \otimes_{\mathbb{C}} \Lambda_{nov},$$

with coefficients in the *Novikov ring* Λ_{nov} of M, which is a suitable completion of the group ring $\mathbb{C}[H_2(M, \mathbb{Z})^{\mathrm{eff}}]$ associated to the semi-group $H_2(M, \mathbb{Z})^{\mathrm{eff}} \subset H_2(M, \mathbb{Z})$ of effective curve classes of M. The space $QH^*(M)$ is equipped with

1. the usual Poincaré pairing:

$$(a,b) := \int_M a \cup b, \quad a,b \in H^*(M,\mathbb{C});$$

2. an associative multiplication called *quantum cup product*, defined as follows: for $a, b \in H^*(M,\mathbb{C})$,

$$a * b := \sum_\alpha \sum_{\beta \in H_2(M,\mathbb{Z})^{\text{eff}}} \langle a, b, \phi_\alpha \rangle_{0,3,\beta}^M q^\beta \phi^\alpha,$$

where

(a) $q^\beta \in \Lambda_{nov}$ is the element corresponding to β;

(b) $\{\phi_\alpha\} \subset H^*(M)$ is an additive basis, and $\{\phi^\alpha\} \subset H^*(M)$ is the Poincaré dual basis;

(c) $\langle a, b, c\rangle_{0,3,\beta}^M$ is the Gromov-Witten invariant defined by integration against the virtual fundamental class of the moduli space $\overline{\mathcal{M}}_{0,3}(M,\beta)$ of 3-pointed genus 0 stable maps to M of degree β:

$$\langle a, b, c\rangle_{0,3,\beta}^M := \int_{[\overline{\mathcal{M}}_{0,3}(M,\beta)]^{vir}} ev_1^* a \cup ev_2^* b \cup ev_3^* c,$$

where $ev_i : \overline{\mathcal{M}}_{0,3}(M,\beta) \to M, i = 1, 2, 3$ are the evaluation maps.

Properties of Gromov-Witten invariants imply that $QH^*(M)$ admits some abstract structures, e.g. Frobenius manifold and D-module. For the purpose of this note we do not need the details of these structures and omit the discussions about them.

2.2 Toric manifolds

In this Section we collect some basic materials about toric manifolds. Excellent introductions to toric manifolds can be found in e.g. [14, 11].

Throughout this note we consider a projective toric manifold X of real dimension $2n$ equipped with a toric symplectic structure. Such a toric manifold can be constructed as symplectic quotients of torus actions on complex vector spaces. Alternatively, X can be defined by using the combinatorial data of a fan $\Sigma \subset N \otimes_{\mathbb{Z}} \mathbb{Q}$, where N is a lattice. Let

$$v_1, v_2, ..., v_n, v_{n+1}, ..., v_m \in N \tag{2.1}$$

be generators of rays of Σ, listed in such a way that $v_1, ..., v_n$ generate a top-dimensional cone of Σ and form a $\mathbb{Z}$-basis of N. Let

$$\nu_1, ..., \nu_n \in N^* \tag{2.2}$$

be the elements dual to $v_1, ..., v_n$, namely $(\nu_i, v_j) = \delta_{ij}$.

Also, let

$$D_1, \ldots, D_m \subset X$$

be the toric prime divisors of X corresponding to $v_1, \ldots, v_m$. We write $[D_1], \ldots, [D_m] \in H^2(X, \mathbb{Z})$ for the corresponding cohomology classes. It is known that $[D_{n+1}], \ldots, [D_m]$ form an additive basis of $H^2(X, \mathbb{Z})$. For each $j = 1, \ldots, m$ we may write $[D_j] = \sum_{i=1}^{m-n} m_{ij}[D_{n+i}]$ with $m_{ij} \in \mathbb{Z}$.

By definition, X is *semi-Fano* if the anti-canonical divisor $-K_X = \sum_{k=1}^{m} D_k$ is nef, i.e. $-K_X \cdot C \geq 0$ for every holomorphic curve $C \subset X$.

3 The three methods

In this Section we explain three known methods of computing quantum cohomology rings of toric manifolds.

3.1 Toric mirror theorem

We briefly review the toric mirror theorem of Givental [16] and Lian-Liu-Yau [28]. First we define the following combinatorial object:

Definition 3.1. *The* I-function *of X is defined to be*

$$I^X(\check{q}, z) := \exp\left(\frac{1}{z}\sum_{k=1}^{m-n}(\log \check{q}_k)[D_{n+k}]\right) \sum_{d \in H_2(X,\mathbb{Z})^{\mathrm{eff}}} \check{q}^d \prod_{l=1}^{m} \frac{\prod_{s=-\infty}^{0}(D_l + sz)}{\prod_{s=-\infty}^{D_l \cdot d}(D_l + sz)}. \tag{3.1}$$

Here $\check{q}^d := \prod_{k=1}^{m-n} \check{q}_k^{(d, D_{n+k})}$, and the expression is to be interpreted as a $1/z$-series with values in $H^*(X) \otimes \Lambda_{nov,\check{q}}$ by using the geometric series and the following convention: $D_l + sz = sz(1 + \frac{D_l}{sz})$.

The following generating function of closed Gromov-Witten invariants of X is of great importance.

Definition 3.2. *The* J-function *of X is defined to be*

$$J^X(q, z) := \exp\left(\frac{1}{z}\sum_{k=1}^{m-n}(\log q_k)[D_{n+k}]\right) \cdot \left(1 + \sum_a \sum_{d \in H_2(X,\mathbb{Z})^{\mathrm{eff}} \setminus \{0\}} q^d \left\langle 1, \frac{\phi_a}{z - \psi}\right\rangle_{0,2,d}^X \phi^a\right). \tag{3.2}$$

Here $q^d := \prod_{k=1}^{m-n} q_k^{(d, D_{n+k})}$, and $\{\phi_a\}, \{\phi^a\} \subset H^*(X, \mathbb{C})$ are two additive bases which are Poincaré dual to each other. The symbol $\langle \cdots \rangle^X$ are closed descendant Gromov-Witten invariants of X, and the expression is to be interpreted as a $1/z$-series with values in $\Lambda_{nov,q}$ via $\frac{1}{z-\psi} = \frac{1}{z}\frac{1}{1-\psi/z} = \frac{1}{z}\sum_{k \geq 0}(\frac{\psi}{z})^k$.

It follows immediately from the definition that the J-function admits the following expansion:

$$J^X(q,z) = 1 + \frac{1}{z}\sum_{k=1}^{m-n}(\log q_k)[D_{n+k}] + o\left(\frac{1}{z}\right). \tag{3.3}$$

If X is semi-Fano, then a direct calculation shows that the I-function admits the following expansion:

$$I^X(\check{q},z) = 1 + \frac{1}{z}\sum_{k=1}^{m-n}(\log \check{q}_k + f_k(\check{q}))[D_{n+k}] + o\left(\frac{1}{z}\right), \tag{3.4}$$

where $f_k(\check{q}), k = 1, ..., m-n$ are certain explicit series in $\check{q}$.

The toric mirror theorem in the semi-Fano case can be stated as follows.

Theorem 3.3. ([16, 28]) *For a projective semi-Fano toric manifold X, we have*

$$I^X(\check{q},z) = J^X(q(\check{q}),z),$$

after the mirror map $q = q(\check{q})$ *given by*

$$q_k = \check{q}_k \exp\left(f_k(\check{q})\right), \quad k = 1, ..., m-n. \tag{3.5}$$

Remark 3.4. When X is not semi-Fano, the expansion of I^X into a series of z contains positive powers of z. In this case a less explicit version of toric mirror theorem, formulated in terms of Givental's Lagrangian cone [17], is proven in [22].

Since the I-function of X and the mirror map $q = q(\check{q})$ are both explicitly given, it is clear that Theorem 3.3 gives a way to compute some closed Gromov-Witten invariants of X. Moreover, it is observed by Givental [15] that differential operators of certain forms that annihilate J^X give rise to relations in $QH^*(X)$. It can be directly checked (see [16]) from definition that the I-function satisfies the so-called GKZ system of differential equations. Together with Theorem 3.3, this yields a complete set of relations that determine $QH^*(X)$ for semi-Fano toric manifold X. In line with mirror symmetry, the presentation of $QH^*(X)$ obtained this way can be formulated as follows. Define the *Hori-Vafa superpotential* of X to be

$$W_{\check{q}}^{\mathrm{HV}} := \sum_{l=1}^{m} Z_l, \tag{3.6}$$

where

$$Z_l := \begin{cases} z_l, & 1 \le l \le n; \\ \check{q}_{l-n}\prod_{k=1}^{n} z_k^{(\nu_k, v_l)}, & n+1 \le l \le m. \end{cases} \tag{3.7}$$

Here $z_1, ..., z_n$ are variables, and $\nu_1, ..., \nu_n \in N^*$ are given in (2.2). Note that $W_{\check{q}}^{\mathrm{HV}}$ depends on $\check{q}$.

By definition, the Jacobian ring of $W_{\check{q}}^{\mathrm{HV}}$ is

$$\mathrm{Jac}(W_{\check{q}}^{\mathrm{HV}}) := \frac{\Lambda_{nov,\check{q}}[z_1^{\pm}, ..., z_n^{\pm}]}{\left\langle z_i \frac{\partial W_{\check{q}}^{\mathrm{HV}}}{\partial z_i} \middle| i = 1, ..., n\right\rangle}. \tag{3.8}$$

Theorem 3.5. ([16, 28]) *For a projective semi-Fano toric manifold X, we have*

$$QH^*(X) \simeq \mathrm{Jac}(W^{HV}_{\check{q}(q)}),$$

after the inverse mirror map $\check{q} = \check{q}(q)$ *given by inverting* (3.5).

3.2 Seidel representation in the toric case

Let (M, ω) be a compact symplectic manifold. The collection $Ham(M, \omega)$ of Hamiltonian diffeomorphisms of (M, ω) has the structure of a topological group. Elements in the fundamental group $\pi_1(Ham(M, \omega))$ are represented by loops of Hamiltonian diffeomorphisms of (M, ω). Let $\varphi : S^1 \times M \to M$ be a loop of Hamiltonian diffeomorphisms of (M, ω). Let D be a disk. One can use φ to glue two copies of $M \times D$ along the boundaries by identifying $(x, t) \in \partial(M \times D) = M \times S^1$ with $(\varphi_t(x), t) \in \partial(M \times D) = M \times S^1$. The resulting space E_φ has a natural map $E_\varphi \to S^2$ and can be equipped with a symplectic structure.

Definition 3.6. *The* Seidel element *associated to φ is defined to be*

$$S_\varphi := \sum_\alpha \sum_{\beta \in H_2(M,\mathbb{Z})^{eff}} \langle \iota_* \phi_\alpha \rangle^{E_\varphi}_{0,1,\sigma_\varphi + \beta} q^\beta \phi^\alpha.$$

Here $\sigma_\varphi \in H_2(E_\varphi, \mathbb{Z})^{eff}$ is the class of a section of $E_\varphi \to S^2$, $\iota : M \to E_\varphi$ is the inclusion of a fiber, $\{\phi_\alpha\} \subset H^*(M)$ is an additive basis, and $\{\phi^\alpha\} \subset H^*(M)$ is the Poincaré dual basis.

Evidently $S_\varphi \in QH^*(M)$. It can be shown that S_φ only depends on the homotopy class $[\varphi]$ of φ.

If φ arises from a $\mathbb{C}^*$-action on M, then $E_\varphi = (M \times (\mathbb{C}^2 \setminus \{0\}))/\mathbb{C}^*$ where $\mathbb{C}^*$ acts on $\mathbb{C}^2 \setminus \{0\}$ by $\lambda \cdot (u, v) = (\lambda u, \lambda v)$. The map $E_\varphi \to S^2$ is obtained from the projection $M \times (\mathbb{C}^2 \setminus \{0\}) \to \mathbb{C}^2 \setminus \{0\}$. In this case, the class σ_φ is taken to be the class of a section emanating from a $\mathbb{C}^*$-fixed point of M the normal bundle of which has negative $\mathbb{C}^*$ weights.

Let

$$QH^*(M)^\times \subset QH^*(M)$$

be the collection of elements invertible under quantum cup product. Clearly $QH^*(M)^\times$ is a group under quantum cup product. The main result concerning the construction of Seidel representation can be stated as follows.

Theorem 3.7. ([31, 29]) 1. $S_\varphi \in QH^*(M)^\times$.

2. *The map*

$$S : \pi_1(Ham(M, \omega)) \to QH^*(M)^\times, \quad [\varphi] \mapsto S_\varphi \tag{3.9}$$

is a group homomorphism.

The map (3.9) is called *Seidel representation* of (M, ω). This Theorem is proven for monotone symplectic manifolds in [31] and for general symplectic manifolds in [29]. See [32] for a treatment using Kuranishi structures.

The application of Seidel representation to quantum cohomology is based on the following very simple idea. Since (3.9) is a group homomorphism, a relation in $\pi_1(Ham(M,\omega))$ yields a (multiplicative) relation in $QH^*(M)$. Therefore, knowing $\pi_1(Ham(M,\omega))$ sufficiently well may produce enough relations to determine $QH^*(M)$.

This idea has been successfully carried out in the toric case. Let X be a projective toric manifold equipped with a toric symplectic structure, as in Section 2.2. It is clear that Hamiltonian S^1-actions on X give rise to loops of Hamiltonian diffeomorphisms of X. The toric manifold X has plenty of Hamiltonian S^1-actions, in fact every element in the lattice N gives such an S^1-action. Moreover, if $u_1, ..., u_k \in N$ satisfy a linear relation

$$a_1 u_1 + \cdots + a_k u_k = 0 \tag{3.10}$$

in N with $a_1, ..., a_k \in \mathbb{Z}$, then one obtains a relation for the corresponding loops of Hamiltonian diffeomorphisms and, via (3.9), a relation

$$S_{u_1}^{a_1} * \cdots * S_{u_k}^{a_k} = 1 \tag{3.11}$$

in $QH^*(X)$. Relations like (3.10) can be explicitly described via combinatorics, and there are lots of such relations. By computing the Seidel elements S_v corresponding to $v \in N$, a presentation of the quantum cohomology ring $QH^*(X)$ is obtained in [30]. This presentation is related to Batyrev's original proposal [1] and takes a particularly simple form for *semi-Fano* symplectic toric manifolds. In this presentation of $QH^*(X)$, the Seidel elements

$$S_{v_1}, ..., S_{v_m} \tag{3.12}$$

corresponding to $v_1, ..., v_m \in N$ in (2.1) give a set of generators. We omit the explicit description of this presentation and refer to [30] for more details.

Remark 3.8. The same idea works for *toric orbifolds* as well, see [33] for results in this direction.

3.3 Counting holomorphic disks in toric varieties

Let X be a projective toric manifold of real dimension $2n$, equipped with a toric symplectic structure and a Hamiltonian action by the torus $T := (S^1)^n$. Let $v : X \to \mathfrak{t}^*$ be the moment map of this T-action. The image $v(X) \subset \mathfrak{t}^*$ is a convex polytope. Let

$$L := v^{-1}(p) \subset X$$

be the fiber of v over a point p in the interior of $v(X)$. Then $L \simeq (S^1)^n$ is a Lagrangian submanifold of X.

Open Gromov-Witten invariants of the pair (X, L), as defined by [12], are (virtual) counts of stable disks in X with boundary conditions given by L. In what follows we sketch the construction. We assume that X is semi-Fano.

Let $\beta \in \pi_2(X, L)$ be a class representable by holomorphic disks. Consider the moduli space

$$\mathcal{M}_{l,k}^{op}(\beta)$$

which parametrizes holomorphic maps $f : (C, \partial C) \to (X, L)$ from boarded Riemann surfaces C to X together with marked points $p_1, ..., p_l$ on the interior of C and $q_1, ..., q_k \in \partial C$ such that

1. $\partial C \simeq S^1$ for generic f;
2. $f(\partial C) \subset L$;
3. f is stable;
4. $f_*[C] = \beta$.

It is shown in [12] that $\mathcal{M}^{op}_{l,k}(\beta)$ admits a T-equivariant Kuranishi structure and T-invariant multisections can be chosen to give a *virtual fundamental chain*

$$[\mathcal{M}^{op}_{l,k}(\beta)]_{vir},$$

which is a singular $\mathbb{Q}$-chain on $X^l \times L^k$ in dimension $\mathrm{vdim}\mathcal{M}^{op}_{l,k}(\beta) := n + \mu(\beta) + k + 2l - 3$. Here $\mu(\beta)$ is the Maslov index of β.

It is known that in toric case, those $\beta \in \pi_2(X, L)$ which are representable by holomorphic disks must have $\mu(\beta) \geq 2$. We consider the following two cases:

Case 1: When $\mu(\beta) = 2$, $k = 1, l = 0$, we have $\mathrm{vdim}\mathcal{M}^{op}_{0,1}(\beta) = n$. Since Maslov index is minimized, there is no real codimension 1 boundaries, as those are formed by disk bubbles. So $[\mathcal{M}^{op}_{0,1}(\beta)]_{vir}$ is a dimension n cycle in L. Define

$$n_1(\beta) := \int_{[\mathcal{M}^{op}_{0,1}(\beta)]_{vir}} [pt]_L \in \mathbb{Q}, \tag{3.13}$$

where $[pt]_L \in H^n(L, \mathbb{Q})$ is the class dual to a point.

Case 2: When $\mu(\beta) = 2$, $k = 1, l = 1$, we have $\mathrm{vdim}\mathcal{M}^{op}_{0,1}(\beta) = n + 2$. For the same reason, $[\mathcal{M}^{op}_{1,1}(\beta)]_{vir}$ is a dimension $n+2$ cycle in $X \times L$. If $D \subset X$ is a real codimension 2 toric cycle, then we define

$$n_{1,1}(\beta; D) := \int_{[\mathcal{M}^{op}_{1,1}(\beta)]_{vir}} [D] \times [pt]_L \in \mathbb{Q}. \tag{3.14}$$

The two invariants are related by the following open analogue of the divisor equation (see [8]):

$$n_{1,1}(\beta; D) = (D \cdot \beta) n_1(\beta). \tag{3.15}$$

According to [9], for X semi-Fano, a homotopy class $\beta \in \pi_2(X, L)$ of Maslov index 2 represented by holomorphic disks must be of the form $\beta = \beta_l + \alpha$, where

1. $\beta_l \in \pi_2(X, L)$ is the *basic disk class* associated to D_l, which is characterized by

$$D_k \cdot \beta_l = \begin{cases} 1, & k = l; \\ 0, & k \neq l. \end{cases} \tag{3.16}$$

Moreover, a explicit holomorphic disk in X representing β_l is described in [9].

2. $\alpha \in H_2(X,\mathbb{Z})^{\text{eff}}$ is such that $-K_X \cdot \alpha = 0$.

Given this, we can define the following generating functions:

$$\delta_l(q) := \sum_{\alpha \in H_2(X,\mathbb{Z})^{\text{eff}} \setminus \{0\}} n_1(\beta_l + \alpha) q^\alpha, \tag{3.17}$$

where $q^\alpha := \prod_{k=1}^{m-n} q_k^{(\alpha, D_{n+k})}$. Define the *Lagrangian Floer superpotential* of X by

$$W_q^{\text{LF}} := \sum_{l=1}^{m} (1 + \delta_l(q)) Z_l, \tag{3.18}$$

where

$$Z_l := \begin{cases} z_l, & 1 \le l \le n; \\ q_{l-n} \prod_{k=1}^{n} z_k^{(\nu_k, v_l)}, & n+1 \le l \le m. \end{cases} \tag{3.19}$$

Note that W_q^{LF} depends on q.

The Lagrangian Floer superpotential W_q^{LF} gives a construction of the mirror of the toric manifold X that is intrinsic to the symplectic geometry of X. The following result should be considered as a mirror symmetry statement.

Theorem 3.9. ([13]) *There is an isomorphism of rings*

$$QH^*(X) \xrightarrow{\simeq} \mathrm{Jac}(W_q^{LF}). \tag{3.20}$$

Remark 3.10. In fact the result proven in [13] is much more general: it is proven that for *any* compact toric manifold X there is a ring isomorphism between the *big* quantum cohomology ring of X and the Jacobian ring of the Lagrangian Floer superpotential defined using disk invariants *with bulks*.

4 Relations between the methods

In this Section we explain the relations between the three methods, and a geometric approach to establishing these relations.

4.1 Results

4.1.1 Mirror theorem and Seidel representation

The fact that both mirror theorem and Seidel representation compute quantum cohomology of a toric manifold motivates González-Iritani to investigate their relations. Here we summarize their results in [18].

Assume that X is a projective semi-Fano toric manifold as in Section 2.2. For each $j = 1, ..., m$ define the *Batyrev element* to be

$$[\widetilde{D}_j] = \sum_{i=1}^{m-n} m_{ij} \sum_{k=1}^{m-n} \frac{\partial \log q_k}{\partial \log \check{q}_i} [D_{n+k}].$$

Evidently the Batyrev elements are determined by the mirror map (3.5). The following result shows that Batyrev elements determine the Seidel elements.

Theorem 4.1. ([18], Theorem 1.1) *For each $j = 1, ..., m$ we have*

$$S_{v_j}(q) = \exp\left(-g_0^{(j)}(\check{q})\right)[\widetilde{D}_j] \tag{4.1}$$

after the mirror map $q = q(\check{q})$. Here

$$g_0^{(j)}(\check{q}) := \sum_{\substack{-K_X \cdot d=0, D_j \cdot d<0 \\ D_i \cdot d \geq 0 \text{ for } i \neq j}} \frac{(-1)^{D_j \cdot d}(-(D_j \cdot d) - 1)!}{\prod_{i \neq j}(D_i \cdot d)!} \check{q}^d. \tag{4.2}$$

It is also proven that Seidel elements $S_{v_1}, ..., S_{v_m}$ determine the mirror map (3.5), see [18, Theorem 1.2].

4.1.2 Seidel representation and disks

In [13], the map (3.20) is defined and proven to be injective and surjective. The proof of surjectivity does not give an explicit description of the inverse map. The following result achieves that in the semi-Fano case:

Theorem 4.2. ([7]) *For a projective semi-Fano toric manifold X, the isomorphism* (3.20) *maps the Seidel element S_{v_l} to Z_l, for each $l = 1, ..., m$.*

4.1.3 Mirror theorem and disks

The following result may be considered as an *open mirror theorem.*

Theorem 4.3. ([6, 7]) *For a projective semi-Fano toric manifold X, we have*

$$W_q^{LF} = W_{\check{q}(q)}^{HV}$$

after the inverse mirror map $\check{q} = \check{q}(q)$.

The mirror map (3.5) is combinatorially explicitly defined. Geometric meanings of the mirror map and its inverse is very desirable. The following result gives an enumerative meaning of the inverse mirror map $\check{q}_k = \check{q}_k(q)$.

Theorem 4.4. ([7]) *For a projective semi-Fano toric manifold X, its inverse mirror map $\check{q}_k = \check{q}_k(q)$ satisfies*

$$\check{q}_k(q) = q_k(1 + \delta_{n+k}(q)) \prod_{p=1}^{n} (1 + \delta_p(q))^{-(v_{n+k}, \nu_p)}, \quad k = 1, ..., m-n. \tag{4.3}$$

Remark 4.5. According to [10] and [34], the mirror map (3.5) may be interpreted as generating series of certain *quasimap invariants* or *gauged Gromov-Witten invariants* defined using a presentation of X as a GIT quotient.

4.1.4 Counting disks

The following result gives an explicit effective formula for open Gromov-Witten invariants of semi-Fano toric manifold X.

Theorem 4.6. ([6, 7]) *For a projective semi-Fano toric manifold X, we have*

$$1 + \delta_l(q) = \exp\left(g_0^{(j)}(\check{q})\right) \tag{4.4}$$

after the inverse mirror map $\check{q} = \check{q}(q)$. Here $g_0^{(j)}(\check{q})$ is given by (4.2).

4.2 A geometric approach

We now discuss a geometric approach to Theorems 4.2, 4.3, 4.4, and 4.6.

First of all, it is shown in [6, 7] are Theorems 4.2, 4.3, 4.4, and 4.6 are equivalent. For this reason, proving one of them will suffice. We choose to address Theorem 4.6.

Proving Theorem 4.6 essentially amounts to computing open Gromov-Witten invariants $n_1(\beta)$ of X. Because X is toric, many computational tools (e.g. virtual localization) are available to calculate *closed* Gromov-Witten invariants of X. However the situation is very different for open Gromov-Witten invariants $n_1(\beta)$: these invariants are defined using the geometry and analysis of moduli spaces of pseudo-holomorphic disk whose structures are very sophisticated, and virtual localization is not applicable in this situation, making these invariants very hard to compute.

By the classification result of holomorphic disks in toric manifolds [9], a stable disk of Maslov index two in a toric manifold X must have its domain being a union of a disk D and a collection of rational curves. If one can "cap off" the disk D by finding another disk D' and gluing D and D' together along boundaries to form a sphere, then one can hope to relate open Gromov-Witten invariants to closed Gromov-Witten invariants. This idea was first worked out successfully in [2] for toric manifolds of the form $X = \mathbb{P}(K_Y \oplus \mathcal{O}_Y)$ with Y a toric Fano manifold. In this case the disk D' used to cap off D arises from the $\mathbb{P}^1$-bundle structure of X. This is a very special case. In more general situation one must alter the toric manifold X in order to find the disk D'.

When X is a toric *Calabi-Yau* manifold (which is necessarily non-compact), capping-off disks D' can be found by suitably compactifying X to projective semi-Fano toric manifolds. In this case the idea of relating open and closed Gromov-Witten invariants by using capping-off disks works very well, see [23, 24, 5, 4] for progresses and [3] for a complete solution.

When X is a projective semi-Fano toric manifold, a novel idea in [7] that allows us to relate open Gromov-Witten invariants of X to closed Gromov-Witten invariants is the use of *Seidel spaces*. Let $v_l \in N$ be a primitive generator of a ray of the fan Σ. Let D_k be a toric prime divisor which is linearly equivalent to $ND_l + \tilde{D}$ for some $N \in \mathbb{Z}_{\geq 0}$ and effective $\tilde{D}$ such that $\tilde{D} - D_l$ is no longer effective. Let $\alpha \in H_2(X, \mathbb{Z}^{\text{eff}})$ be such that $-K_X \cdot \alpha = 0$. Then we have the following *open/closed equality:*

$$n_{1,1}^X(\beta_l+\alpha; ND_l+\tilde{D}) = \langle \iota_* D_k, [\mathrm{pt}]\rangle_{0,2,\sigma_l^-+\alpha}^{E_{-v_l}}, \tag{4.5}$$

where

1. $\iota : X \to E_{-v_l}$ is the inclusion of a fiber;
2. $\sigma_l^- \in H_2(E_{-v_l}, \mathbb{Z})^{\mathrm{eff}}$ is the class of a specific section of $E_{-v_l} \to S^2$ explained above (see also e.g. [7, Section 4]).

Note that since divisors D_k satisfying the above requirement span $H^2(X,\mathbb{Z})$, the divisor equation (3.15) implies that $n_{1,1}^X(\beta_l+\alpha)$ are determined by $n_{1,1}^X(\beta_l+\alpha; ND_l+\tilde{D})$.

We now explain the idea behind (4.5). The explicit description of basic disks in [9] implies that one can identify disks of class $\beta_l+\alpha$ in X with disks of suitable class in the Seidel space E_{-v_l}. Furthermore, this identification also equates the deformations/obstructions of disks and yields an identification of the relevant moduli spaces of disks as Kuranishi spaces. Again by the description of basic disks in [9], one can see that the basic disk of E_{-v_l} considered here naturally extends[2] to a holomorphic sphere $S^2 \to E_{-v_l}$ in class σ_l^-. This in turn identifies the relevant moduli space of disks in X with moduli space of stable maps to E_{-v_l} in class $\sigma_l^-+\alpha$ as Kuranishi spaces. (4.5) then follows.

Because of (4.5), the problem of computing open Gromov-Witten invariants of X is transformed into the problem of computing 2-point genus 0 closed Gromov-Witten invariants

$$\langle \iota_* D_k, [\mathrm{pt}]\rangle_{0,2,\sigma_l^-+\alpha}^{E_{-v_l}} \tag{4.6}$$

for all l, k, and α. It is easy to see that E_{-v_l} is a toric manifold. Hence the cohomology of E_{-v_l} is multiplicatively generated in degree 2. Because of the divisor equation of closed Gromov-Witten invariants, derivatives of the J-function of E_{-v_l} with respect to the Novikov variables give generating series of 2-point closed Gromov-Witten invariants of E_{-v_l}. If there is an explicit formula for the J-function of E_{-v_l}, then one can compute 2-point closed Gromov-Witten invariants of E_{-v_l} this way. Unfortunately this *does not* work: because E_{-v_l} is not semi-Fano, the toric mirror theorem (Theorem 3.3) does not apply to give an explicit formula for the J-function of E_{-v_l}. An alternative approach is required.

We now sketch the method of computing invariants (4.6) used in [7]. The starting point is the observation that invariants (4.6) appear in the quantum product $S_{-v_l} * D_k$ of the divisor D_k and the Seidel element S_{-v_l} associated to $-v_l$. Because $v_l + (-v_l) = 0$, the $\mathbb{C}^*$-actions associated to v_l and $-v_l$ are inverse to each other. A basic property (see [29]) implies that

$$S_{v_l} * S_{-v_l} = 1.$$

This property should be understood as an example of the degeneration formula of Gromov-Witten invariants [25, 20, 21, 26, 27]. Hence we have

$$S_{v_l} * (S_{-v_l} * D_k) = D_k. \tag{4.7}$$

[2]Roughly speaking, the direct of a section of E_{-v_l} provides the capping-off disks.

Recall that the Seidel elements S_{v_l} are explicitly computed in [18], as mentioned in Theorem 4.1 above. (4.7) may be viewed as a system of linear equations with (4.6) as part of the unknowns. This system can be solved in part to obtain (4.6) and consequently Theorem 4.6. Details are given in [7, Section 6].

Remark 4.7. 1. If one is willing to assume that the generating functions $\delta_l(q)$ of open Gromov-Witten invariants are convergent power series of q, then Theorem 4.3 can be derived by combining Theorems 3.5 and 3.9 and applying the theory of unfoldings of holomorphic functions. This also yields Theorems 4.2, 4.4, and 4.6 under the convergence assumption. We refer to [6] for more details.

2. We point out that González-Iritani [19] developed a beautiful approach to Theorem 4.2 based on a conjectural degeneration formula of disk invariants.

References

[1] V. Batyrev, *Quantum cohomology rings of toric manifolds*, Journées de Géométrie Algébrique d'Orsay (Orsay, 1992), Astérisque No. **218** (1993), 9–34.

[2] K. Chan, *A formula equating open and closed Gromov-Witten invariants and its applications to mirror symmetry*, Pacific J. Math. **254** (2011), no. 2, 275–293.

[3] K. Chan, C.-H. Cho, S.-C. Lau, H.-H. Tseng, *Gross fibrations, SYZ mirror symmetry, and open Gromov-Witten invariants for toric Calabi-Yau orbifolds*, arXiv:1306.0437.

[4] K. Chan, S.-C. Lau, *Open Gromov-Witten invariants and superpotentials for semi-Fano toric surfaces*, to appear in Internat. Math. Res. Notices, arXiv:1010.5287.

[5] K. Chan, S.-C. Lau, N. C. Leung, *SYZ mirror symmetry for toric Calabi-Yau manifolds*, J. Differential Geom. **90** (2012), no. 2, 177–250.

[6] K. Chan, S.-C. Lau, N. C. Leung, H.-H. Tseng, *Open Gromov-Witten invariants and mirror maps for semi-Fano toric manifolds*, arXiv:1112.0388.

[7] K. Chan, S.-C. Lau, N. C. Leung, H.-H. Tseng, *Open Gromov-Witten invariants and Seidel representations for toric manifolds*, arXiv:1209.6119.

[8] C.-H. Cho, *Products of Floer cohomology of torus fibers in toric Fano manifolds*, Comm. Math. Phys. **260** (2005), no. 3, 613–640.

[9] C.-H. Cho, Y.-G. Oh, *Floer cohomology and disc instantons of Lagrangian torus fibers in Fano toric manifolds*, Asian J. Math. **10** (2006), no. 4, 773–814.

[10] I. Ciocan-Fontanine, B. Kim, *Wall-crossing in genus zero quasimap theory and mirror maps*, arXiv:1304.7056.

[11] D. Cox, J. Little, H. Schenck, *Toric Varieties*, Graduate Studies in Mathematics, **124**, American Mathematical Society, Providence, RI, 2011.

[12] K. Fukaya, Y.-G. Oh, H. Ohta, K. Ono, *Lagrangian Floer theory on compact toric manifolds, I*, Duke Math. J. **151** (2010), no. 1, 23–174.

[13] K. Fukaya, Y.-G. Oh, H. Ohta, K. Ono, *Lagrangian Floer theory and mirror symmetry on compact toric manifolds*, arXiv:1009.1648.

[14] W. Fulton, *Introduction to Toric Varieties*, Annals of Mathematics Studies, **131**, The William H. Roever Lectures in Geometry. Princeton University Press, Princeton, NJ, 1993.

[15] A. Givental, *Equivariant Gromov-Witten invariants*, Internat. Math. Res. Notices (1996), no. 13, 613–663.

[16] A. Givental, *A mirror theorem for toric complete intersections*, in: "Topological Field Theory, Primitive Forms and Related Topics (Kyoto, 1996)", 141–175, Progr. Math., **160**, Birkhäuser Boston, Boston, MA, 1998.

[17] A. Givental, *Gromov-Witten invariants and quantization of quadratic Hamiltonians*, Mosc. Math. J. **1** (2001), no. 4, 551–568.

[18] E. González, H. Iritani, *Seidel elements and mirror transformations*, Selecta Math. (N.S.) **18** (2012), no. 3, 557–590.

[19] E. González, H. Iritani, *Seidel elements and potential functions of holomorphic disc counting*, arXiv:1301.5454.

[20] E. Ionel, T. Parker, *Relative Gromov-Witten Invariants*, Ann. Math. (2), **157** (2003), no. 1, 45–96.

[21] E. Ionel, T. Parker, *The Symplectic sum formula for Gromov-Witten invariants*, Ann. Math. (2), **159** (2004), no. 3, 935–1025.

[22] H. Iritani, *Convergence of quantum cohomology by quantum Lefschetz*, J. Reine Angew. Math. **610** (2007), 29–69.

[23] S.-C. Lau, N. C. Leung, B. Wu, *A relation for Gromov-Witten invariants of local Calabi-Yau threefolds*, Math. Res. Lett. **18** (2011), no. 5, 943–956.

[24] S.-C. Lau, N. C. Leung, B. Wu, *Mirror maps equal SYZ maps for toric Calabi-Yau surfaces*, Bull. London Math. Soc. **44** (2012), 255–270.

[25] A.-M. Li, Y. Ruan, *Symplectic surgery and Gromov-Witten invariants of Calabi-Yau 3-folds*, Invent. Math. **145** (2001), no. 1, 151–218.

[26] J. Li, *Stable morphisms to singular schemes and relative stable morphisms*, J. Differential Geom. **57** (2001), no. 3, 509–578.

[27] J. Li, *A degeneration formula of GW-invariants*, J. Differential Geom. **60** (2002), no. 2, 199–293.

[28] B. Lian, K. Liu, S.-T. Yau, *Mirror principle, III*, Asian J. Math. **3** (1999), no. 4, 771–800.

[29] D. McDuff, *Quantum homology of fibrations over* S^2, Internat. J. Math. **11** (2000), no. 5, 665–721.

[30] D. McDuff, S. Tolman, *Topological properties of Hamiltonian circle actions*, IMRP Int. Math. Res. Pap. (2006), 72826, 1–77.

[31] P. Seidel, π_1 *of symplectic automorphism groups and invertibles in quantum homology rings*, Geom. Funct. Anal. **7** (1997), no. 6, 1046–1095.

[32] H.-H. Tseng, D. Wang, *Seidel representation for symplectic orbifolds*, arXiv:1207.4246.

[33] H.-H. Tseng, D. Wang, *Seidel representations and quantum cohomology of toric orbifolds*, arXiv:1211.3204.

[34] C. Woodward, *Quantum Kirwan morphism and Gromov-Witten invariants of quotients*, arXiv:1204.1765.

Department of Mathematics, Ohio State University
100 Math Tower, 231 West 18th Ave., Columbus, OH 43210, USA
E-mail address: hhtseng@math.ohio-state.edu

Proceedings of the 6th International Congress of Chinese Mathematicians
ALM 36, pp. 505–518

Strongly Liftable Schemes and the Kawamata-Viehweg Vanishing in Positive Characteristic

Qihong Xie

Abstract

This paper is a survey on the theory of strongly liftable schemes, which is based on the author's talk at the Sixth International Congress of Chinese Mathematicians (ICCM, 2013) in Taipei. In this paper, we first recall some results on vanishing theorems in algebraic geometry. Secondly, we give a systematic introduction to the theory of strongly liftable schemes and its applications to vanishing theorems in positive characteristic. Finally, some remarks and further problems are put forward.

Mathematics Subject Classification: 14F17, 14E30.

Keywords and Phrases: Strongly liftable scheme, Kawamata-Viehweg vanishing, positive.

1 Vanishing theorems in algebraic geometry

As for vanishing theorems in algebraic geometry, the most famous one is the so-called Kodaira vanishing theorem.

Theorem 1.1. (Kodaira vanishing) *Let X be a smooth projective complex variety, ω_X the canonical invertible sheaf on X, and $\mathcal{L}$ an ample invertible sheaf on X. Then $H^i(X, \omega_X \otimes \mathcal{L}) = 0$ holds for any $i > 0$.*

For convenience of citation, we also state the Kodaira-Akizuki-Nakano vanishing theorem in the logarithmic form as follows.

This paper was partially supported by the National Natural Science Foundation of China (Grant No. 11231003 and 11271070), and the Shanghai Center for Mathematical Sciences, Fudan University.

Theorem 1.2. (Kodaira-Akizuki-Nakano vanishing) *Let X be a smooth projective complex variety, D a simple normal crossing divisor on X, and $\mathcal{L}$ an ample invertible sheaf on X. Then $H^j(X, \Omega_X^i(\log D) \otimes \mathcal{L}^{-1}) = 0$ holds for any $i + j < \dim X$.*

There are many generalizations of the Kodaira vanishing theorem. One of the most important generalizations is the Kawamata-Viehweg vanishing theorem. As is well known, it is inevitable to run the higher dimensional minimal model program in the categories of varieties with suitable singularities, hence we have to consider $\mathbb{Q}$-divisors instead of integral divisors. It turns out that the Kawamata-Viehweg vanishing theorem is indispensable and plays a crucial role in birational geometry of higher dimensional complex varieties.

The Kawamata-Viehweg vanishing theorem is of several forms. The one dealing with ample $\mathbb{Q}$-divisors follows directly from the Kodaira vanishing theorem via the Kummer covering trick [Ka82, Vi82].

Theorem 1.3. (Kawamata-Viehweg vanishing, smooth case) *Let X be a smooth projective complex variety, and H an ample $\mathbb{Q}$-divisor on X such that the fractional part $\langle H \rangle$ has simple normal crossing support. Then $H^i(X, K_X + \ulcorner H \urcorner) = 0$ holds for any $i > 0$.*

The most general form is stated for log pairs which have only Kawamata log terminal singularities [KMM87, Theorem 1-2-5].

Theorem 1.4. (Kawamata-Viehweg vanishing, general case) *Let X be a normal projective complex variety, $B = \sum b_i B_i$ an effective $\mathbb{Q}$-divisor on X, and D a $\mathbb{Q}$-Cartier Weil divisor on X. Assume that (X, B) is Kawamata log terminal (KLT for short), and $D - (K_X + B)$ is ample. Then $H^i(X, D) = 0$ holds for any $i > 0$.*

In what follows, we always work over *a perfect field or an algebraically closed field k of characteristic $p > 0$* unless otherwise stated.

When we study algebraic geometry in positive characteristic, many pathologies naturally appear as counterexamples to some well-known theorems in complex algebraic geometry. For instance, Raynaud [Ra78] has first found counterexamples to the Kodaira vanishing theorem on certain smooth projective surfaces.

Theorem 1.5. (Raynaud's counterexample) *There exist counterexamples to the Kodaira vanishing theorem on a smooth projective surface X, where X is either a quasi-elliptic surface of Kodaira dimension 1 or a surface of general type, and X admits a fibration $f : X \to C$ such that each fiber is a singular rational curve.*

On the other hand, the author [Xie11a] gave a characterization of counterexamples to the Kodaira vanishing theorem.

Theorem 1.6. *If there exists a counterexample to the Kodaira vanishing theorem on a smooth projective surface X, then X must be either a quasi-elliptic surface of Kodaira dimension 1 or a surface of general type, and X admits a fibration $f : X \to C$ such that each fiber is a singular curve.*

Theorems 1.5 and 1.6 show that counterexamples to the Kodaira vanishing theorem are really restrictive from the viewpoint of the classification theory of algebraic surfaces in positive characteristic. Whereas considering the Kawamata-Viehweg vanishing theorem, we found that there exist counterexamples to it even on geometrically ruled surfaces, which shows that pathologies do exist widely in algebraic geometry of positive characteristic.

We recall the following definition from [Ta72b, Definitions 9 and 11].

Definition 1.7. Let $f \in K(C)$ be a rational function on a smooth projective curve C.

$$n(f) := \deg\left[\frac{(df)}{p}\right],$$

where $(df) = \sum_{x\in C} v_x(df)x$ is the divisor associated to the rational differential 1-form df. The Tango invariant $n(C)$ of the curve C is defined by

$$n(C) := \max\{n(f) \mid f \in K(C), f \notin K^p(C)\}.$$

If $n(C) > 0$, then C is called a Tango curve.

If $f \notin K^p(C)$, then (df) is a canonical divisor on C with degree $2(g(C)-1)$. It is easy to see that $n(C) \leq [2(g(C)-1)/p]$. Therefore, if $n(C) > 0$ then $g(C) \geq 2$ holds. On the other hand, it follows from [Ta72b, Proposition 14] that $n(C) \geq 0$ holds if $g(C) \geq 1$. By an elementary calculation, we can show that $n(\mathbb{P}^1) = -1$ and $n(C) = 0$ holds for any elliptic curve C.

Example 1.8. There do exist Tango curves C for each characteristic $p > 0$.

(i) Assume $p \geq 3$. Let $h \geq 3$ be an odd integer, and let C be the projective completion at infinity of the affine curve defined by $y^2 = x^{ph} + x^{p+1} + 1$. It is easy to verify that C is a smooth hyperelliptic curve and that $(d(y/x^p)) = (ph-3)z_\infty$, where z_∞ is the infinity point of C (cf. [Sh94, Chapter III, §6.5]). Hence $n(C) = n(y/x^p) = h - 1 > 0$.

(ii) Let $h > 2$ be an integer, and let C be the projective completion at infinity of the Artin-Schreier cover of the affine line defined by $y^{hp-1} = x^p - x$. It is easy to verify that C is a smooth curve and that $(dy) = p(h(p-1)-2)z_\infty$, where z_∞ is the infinity point of C. Hence $n(C) = n(y) = h(p-1) - 2 > 0$ (cf. [Ra78]).

(iii) Assume $p \geq 3$. Let $C \subset \mathbb{P}^2$ be the projective curve defined by $x_0^{p+1} = x_1x_2(x_0^{p-1} + x_1^{p-1} - x_2^{p-1})$. Then we can show that C is a smooth curve with $n(C) = n(x_0/x_1) = p - 2 > 0$ (cf. [Ta72b, Proposition 28]).

The author [Xie10a] proved that we can construct counterexamples to the Kawamata-Viehweg vanishing theorem on geometrically ruled surfaces over Tango curves.

Theorem 1.9. (Xie's counterexample) *If C is a Tango curve, then there is a $\mathbb{P}^1$-bundle $f : X \to C$, and an ample $\mathbb{Q}$-divisor H on X such that $C' = \mathrm{Supp}\langle H\rangle$ is a smooth curve and $f|_{C'} : C' \to C$ is the Frobenius morphism. However, $H^1(X, K_X + \ulcorner H \urcorner) \neq 0$.*

We saw that, in general, vanishing theorems fail definitively in positive characteristic. In order to obtain some positive results on vanishing theorems, usually algebraic varieties should be put on certain lifting conditions.

Definition 1.10. Let $W_2(k)$ be the ring of Witt vectors of length two of k. Then $W_2(k)$ is flat over $\mathbb{Z}/p^2\mathbb{Z}$, and $W_2(k) \otimes_{\mathbb{Z}/p^2\mathbb{Z}} \mathbb{F}_p = k$. For the explicit construction and further properties of $W_2(k)$, we refer the reader to [Se62, II.6]. The following definition [EV92, Definition 8.11] generalizes the definition [DI87, §1.6] of liftings of k-schemes over $W_2(k)$.

Let X be a noetherian scheme over k, and $D = \sum D_i$ a reduced Cartier divisor on X. A lifting of (X, D) over $W_2(k)$ consists of a scheme $\widetilde{X}$ and closed subschemes $\widetilde{D}_i \subset \widetilde{X}$, all defined and flat over $W_2(k)$ such that $X = \widetilde{X} \times_{\operatorname{Spec} W_2(k)} \operatorname{Spec} k$ and $D_i = \widetilde{D}_i \times_{\operatorname{Spec} W_2(k)} \operatorname{Spec} k$. We write $\widetilde{D} = \sum \widetilde{D}_i$ and say that $(\widetilde{X}, \widetilde{D})$ is a lifting of (X, D) over $W_2(k)$, if no confusion is likely.

Let $\mathcal{L}$ be an invertible sheaf on X. A lifting of $(X, \mathcal{L})$ consists of a lifting $\widetilde{X}$ of X over $W_2(k)$ and an invertible sheaf $\widetilde{\mathcal{L}}$ on $\widetilde{X}$ such that $\widetilde{\mathcal{L}}|_X = \mathcal{L}$. For simplicity, we say that $\widetilde{\mathcal{L}}$ is a lifting of $\mathcal{L}$ on $\widetilde{X}$, if no confusion is likely.

Deligne and Illusie [DI87] have used the method—"decomposition of de Rham complex" to prove the Kodaira-Akizuki-Nakano vanishing theorem (absolute form) on a smooth projective variety X provided that X is liftable over $W_2(k)$, and used the reduction modulo p method to give a purely algebraic proof of the Kodaira-Akizuki-Nakano vanishing theorem (absolute form) in characteristic zero. Furthermore, they have also claimed the corresponding results [DI87, §4.2] in the logarithmic form for a log pair (X, D), where D is a simple normal crossing divisor on X. The explicit statements and proofs of those results have been given by Esnault and Viehweg [EV92, §8–§11]. In particular, if (X, D) is liftable over $W_2(k)$, then the Kodaira-Akizuki-Nakano vanishing theorem (logarithmic form) holds for (X, D).

Theorem 1.11. (Kodaira-Akizuki-Nakano vanishing in char. $p > 0$) *Let X be a smooth projective variety with $\dim X = d$, D a simple normal crossing divisor on X, and $\mathcal{L}$ an ample invertible sheaf on X. Assume that (X, D) is liftable over $W_2(k)$. Then $H^j(X, \Omega^i_X(\log D) \otimes \mathcal{L}^{-1}) = 0$ holds for any $i + j < \inf(d, p)$.*

Later, Hara [Ha98] and Matsuki-Olsson [MO05] have independently proved the Kawamata-Viehweg vanishing theorem in positive characteristic under certain lifting conditions of log pairs over $W_2(k)$. The method of Hara is to use a quasi-isomorphism between the logarithmic de Rham complex and its variant by adding certain modulo p fractional parts, while Matsuki and Olsson replaced the Kummer covering trick with the stack technique, which behaves well in arbitrary characteristic, and interpreted the Kawamata-Viehweg vanishing on varieties as the Kodaira vanishing on stacks.

Theorem 1.12. (Kawamata-Viehweg vanishing in char. $p > 0$) *Let X be a smooth projective variety with $\dim X = d$, H an ample $\mathbb{Q}$-divisor on X, and D a simple normal crossing divisor containing $\operatorname{Supp}(\langle H \rangle)$. Assume that (X, D) is liftable over $W_2(k)$. Then $H^j(X, \Omega^i_X(\log D)(-\ulcorner H \urcorner)) = 0$ holds for any $i + j < \inf(d, p)$. In particular, $H^i(X, K_X + \ulcorner H \urcorner) = 0$ holds for any $i > d - \inf(d, p)$.*

In addition, Illusie [Il90] has generalized the result in [DI87] to the relative case to prove the decomposition of de Rham complexes with coefficients in the

Gauss-Manin systems and the Kollár vanishing theorem [Ko86, Theorem 2.1] in positive characteristic for semistable reductions. Furthermore, the author [Xie11c] generalized Illusie's result to prove the decomposition of de Rham complexes with smooth horizontal coefficients and the corresponding Kollár vanishing theorem for semistable reductions.

2 The theory of strongly liftable schemes

In order to obtain a practical version of the Kawamata-Viehweg vanishing theorem in positive characteristic instead of Theorem 1.12, we have to take the lifting property of all log pairs into account. For this purpose, we shall introduce a new notion— "strongly liftable scheme", which was first defined by the author [Xie10c, Definition 2.3].

Let X be a smooth scheme over k, and $\widetilde{X}$ a lifting of X over $W_2(k)$. Then $\mathcal{O}_{\widetilde{X}}$ is flat over $W_2(k)$, hence flat over $\mathbb{Z}/p^2\mathbb{Z}$. Note that there is an exact sequence of $\mathbb{Z}/p^2\mathbb{Z}$-modules:

$$0 \to p \cdot \mathbb{Z}/p^2\mathbb{Z} \to \mathbb{Z}/p^2\mathbb{Z} \xrightarrow{r} \mathbb{Z}/p\mathbb{Z} \to 0$$

and a $\mathbb{Z}/p^2\mathbb{Z}$-module isomorphism $p : \mathbb{Z}/p\mathbb{Z} \to p \cdot \mathbb{Z}/p^2\mathbb{Z}$. Tensoring the above by $\mathcal{O}_{\widetilde{X}}$, we obtain an exact sequence of $\mathcal{O}_{\widetilde{X}}$-modules:

$$0 \to p \cdot \mathcal{O}_{\widetilde{X}} \to \mathcal{O}_{\widetilde{X}} \xrightarrow{r} \mathcal{O}_X \to 0, \tag{1}$$

and an $\mathcal{O}_{\widetilde{X}}$-module isomorphism

$$p : \mathcal{O}_X \to p \cdot \mathcal{O}_{\widetilde{X}}, \tag{2}$$

where r is the reduction modulo p satisfying $p(x) = p\widetilde{x}$, $r(\widetilde{x}) = x$ for $x \in \mathcal{O}_X$, $\widetilde{x} \in \mathcal{O}_{\widetilde{X}}$. Combining the exact sequence (1) and the isomorphism (2), we obtain an exact sequence of $\mathcal{O}_{\widetilde{X}}$-modules:

$$0 \to \mathcal{O}_X \xrightarrow{p} \mathcal{O}_{\widetilde{X}} \xrightarrow{r} \mathcal{O}_X \to 0. \tag{3}$$

Definition 2.1. Let X be a smooth scheme over k. X is said to be strongly liftable over $W_2(k)$, if there is a lifting $\widetilde{X}$ of X over $W_2(k)$, such that for any prime divisor D on X, (X, D) has a lifting $(\widetilde{X}, \widetilde{D})$ over $W_2(k)$ as in Definition 1.10, where $\widetilde{X}$ is fixed for all liftings $\widetilde{D}$. In other words, X is strongly liftable over $W_2(k)$, if X and all prime divisors on X can be lifted simultaneously over $W_2(k)$.

Let X be a smooth scheme over k, $\widetilde{X}$ a lifting of X over $W_2(k)$, D a prime divisor on X and $\mathcal{L}_D = \mathcal{O}_X(D)$ the associated invertible sheaf on X. Then there is an exact sequence of abelian sheaves:

$$0 \to \mathcal{O}_X \xrightarrow{q} \mathcal{O}^*_{\widetilde{X}} \xrightarrow{r} \mathcal{O}^*_X \to 1, \tag{4}$$

where $q(x) = p(x) + 1$ for $x \in \mathcal{O}_X$, $p : \mathcal{O}_X \to p \cdot \mathcal{O}_{\widetilde{X}}$ is the isomorphism (2) and r is the reduction modulo p. The exact sequence (4) gives rise to an exact sequence of cohomology groups:

$$H^1(\widetilde{X}, \mathcal{O}^*_{\widetilde{X}}) \xrightarrow{r} H^1(X, \mathcal{O}^*_X) \to H^2(X, \mathcal{O}_X). \tag{5}$$

If $r : H^1(\widetilde{X}, \mathcal{O}^*_{\widetilde{X}}) \to H^1(X, \mathcal{O}^*_X)$ is surjective, then $\mathcal{L}_D$ has a lifting $\widetilde{\mathcal{L}}_D$, i.e. $\widetilde{\mathcal{L}}_D$ is an invertible sheaf on $\widetilde{X}$ with $\widetilde{\mathcal{L}}_D|_X = \mathcal{L}_D$. Tensoring the exact sequence (3) by $\widetilde{\mathcal{L}}_D$, we have an exact sequence of $\mathcal{O}_{\widetilde{X}}$-modules:

$$0 \to \mathcal{L}_D \xrightarrow{p} \widetilde{\mathcal{L}}_D \xrightarrow{r} \mathcal{L}_D \to 0,$$

which gives rise to an exact sequence of cohomology groups:

$$H^0(\widetilde{X}, \widetilde{\mathcal{L}}_D) \xrightarrow{r} H^0(X, \mathcal{L}_D) \to H^1(X, \mathcal{L}_D). \tag{6}$$

There is a criterion for strong liftability [Xie11b, Proposition 2.5].

Proposition 2.2. *Let X be a smooth scheme over k, and $\widetilde{X}$ a lifting of X over $W_2(k)$. Then X is strongly liftable if the following two conditions hold:*

(i) $r : H^1(\widetilde{X}, \mathcal{O}^*_{\widetilde{X}}) \to H^1(X, \mathcal{O}^*_X)$ *is surjective;*

(ii) *For any prime divisor D on X, there is a lifting $\widetilde{\mathcal{L}}_D$ of $\mathcal{L}_D = \mathcal{O}_X(D)$ such that $r : H^0(\widetilde{X}, \widetilde{\mathcal{L}}_D) \to H^0(X, \mathcal{L}_D)$ is surjective.*

Remark 2.3. It follows from [Xie11b, Lemma 2.6] that if X is strongly liftable, then the condition (i) always holds. However, an example given in [Xie11b, Remark 2.7] showed that even if X is strongly liftable, the condition (ii) does not hold in general. Thus Proposition 2.2 is a sufficient condition but not a necessary condition.

By means of Proposition 2.2, we can give many concrete examples of strongly liftable schemes.

Theorem 2.4. *The following schemes are strongly liftable over $W_2(k)$:*

(i) $\mathbb{A}^n_k$, $\mathbb{P}^n_k$ *and a smooth projective curve;*

(ii) *A smooth projective variety of Picard number 1 which is a complete intersection in $\mathbb{P}^n_k$;*

(iii) *A smooth projective rational surface;*

(iv) *A smooth projective toric variety;*

(v) *A cyclic cover over $\mathbb{P}^n_k$ $(n \geq 3)$ with the reduced part of the branch locus being smooth.*

Proof. (i) see [Xie10c, Lemmas 3.1 and 3.2]; (ii) see [Xie10c, Lemma 3.3]; (iii) see [Xie11b, Proposition 5.2]; (iv) see [Xie11b, Theorem 3.2]; (v) see [Xie13b, Corollary 4.5]. □

Remark 2.5. (1) By Deligne-Illusie's result [DI87], Raynaud's counterexamples to the Kodaira vanishing theorem are not liftable over $W_2(k)$. Thus in general, quasi-elliptic surfaces and surfaces of general type are not liftable over $W_2(k)$, hence not strongly liftable over $W_2(k)$ by definition.

(2) By Theorem 1.12, the log pair (X, C') constructed in Theorem 1.9 is not liftable over $W_2(k)$. Thus ruled surfaces are liftable over $W_2(k)$ [Xie10b, Corollary 4.2], but in general, not strongly liftable over $W_2(k)$.

(3) Oort has proved that abelian varieties are liftable over $W_2(k)$, but not strongly liftable over $W_2(k)$ (see [Xie10c, Remark 3.11]).

We have the cyclic cover trick over $W_2(k)$ [Xie11b, Theorem 4.1], which can be used to study the behavior of cyclic covers over strongly liftable schemes.

Theorem 2.6. *Let X be a smooth scheme, and $\mathcal{L}$ an invertible sheaf on X. Let N be a positive integer prime to p, and D an effective divisor on X with $\mathcal{L}^N = \mathcal{O}_X(D)$.*

(i) *Let $\mathcal{A} = \bigoplus_{i=0}^{N-1} \mathcal{L}^{-i}([\frac{iD}{N}])$. Then $\mathcal{A}$ is an $\mathcal{O}_X$-algebra. Let $Y = \mathbf{Spec}\,\mathcal{A}$. Then Y is normal and the natural projection $\pi : Y \to X$ is a finite surjective morphism of degree N, which is called the cyclic cover obtained by taking the N-th root out of D. Furthermore, if $\mathrm{Sing}(D_{\mathrm{red}}) = \emptyset$, then Y is smooth.*

(ii) *If X has a lifting $\widetilde{X}$ over $W_2(k)$, $\mathcal{L}$ has a lifting $\widetilde{\mathcal{L}}$ on $\widetilde{X}$, and D has a lifting $\widetilde{D}$ on $\widetilde{X}$ with $\widetilde{\mathcal{L}}^N = \mathcal{O}_{\widetilde{X}}(\widetilde{D})$, then Y is liftable over $W_2(k)$.*

Remark 2.7. In Theorem 2.6(ii), if D cannot be lifted over $W_2(k)$, then Y is not necessarily liftable over $W_2(k)$. Such an example has been given in [Ra78] and was generalized in [Xie10a, Theorem 3.6] (see [Xie11b, Remark 4.2]).

There are two corollaries to Theorem 2.6 (see [Xie11b, Corollaries 4.3 and 4.4]).

Corollary 2.8. *Let X be a smooth projective variety, and $\mathcal{L}$ an invertible sheaf on X. Let N be a positive integer prime to p, and D an effective divisor on X with $\mathcal{L}^N = \mathcal{O}_X(D)$. Let $\pi : Y \to X$ be the cyclic cover obtained by taking the N-th root out of D. If X is strongly liftable over $W_2(k)$, $H^1(X, \mathcal{L}^N) = 0$ and $\mathrm{Sing}(D_{\mathrm{red}}) = \emptyset$, then Y is a smooth projective scheme which is liftable over $W_2(k)$.*

Corollary 2.9. *Let X be a smooth projective toric variety, and $\mathcal{L}$ an invertible sheaf on X. Let N be a positive integer prime to p, and D an effective divisor on X with $\mathcal{L}^N = \mathcal{O}_X(D)$ and $\mathrm{Sing}(D_{\mathrm{red}}) = \emptyset$. Let $\pi : Y \to X$ be the cyclic cover obtained by taking the N-th root out of D. Then Y is a smooth projective scheme which is liftable over $W_2(k)$.*

By using Corollary 2.9, we can construct a large class of liftable smooth projective varieties obtained by taking cyclic covers over toric varieties. In particular, there do exist many liftable smooth projective varieties of general type.

We have also a criterion [Xie13b, Theorem 4.1] for that cyclic covers over strongly liftable schemes are still strongly liftable.

Theorem 2.10. *With notation and assumptions as in Theorem 2.6, fix such liftings $\widetilde{X}, \widetilde{\mathcal{L}}$ and $\widetilde{s}$ corresponding to $\widetilde{D}$ as in Theorem 2.6(ii). Assume further that for any prime divisor E on X which is not contained in $\mathrm{Supp}(D)$, there exists a lifting $\widetilde{E} \subset \widetilde{X}$ of $E \subset X$ such that $\widetilde{s}|_{\widetilde{E}} \in H^0(\widetilde{E}, \widetilde{\mathcal{L}}^N|_{\widetilde{E}})$ is a divisible lifting of $s|_E \in H^0(E, \mathcal{L}^N|_E)$. Then Y is strongly liftable over $W_2(k)$.*

Definition 2.11. A noetherian scheme X is said to satisfy the H^i-vanishing condition, if $H^i(X, \mathcal{L}) = 0$ holds for any invertible sheaf $\mathcal{L}$ on X. For example, the projective space $\mathbb{P}^n_k$ satisfies the H^i-vanishing condition for any $1 \leq i \leq n-1$.

There is a corollary to Theorem 2.10 (see [Xie13b, Corollary 4.4]).

Corollary 2.12. *Let X be a smooth projective variety satisfying the H^i-vanishing condition for $i = 1, 2$. Then X is strongly liftable over $W_2(k)$, and for any cyclic cover $\pi : Y \to X$ constructed as in Theorem* 2.6(i)*, Y is also strongly liftable over $W_2(k)$.*

We have the Kummer covering trick over $W_2(k)$ [XW13, Theorem 3.1], which was proved by means of the logarithmic techniques developed by K. Fujiwara, K. Kato and C. Nakayama.

Theorem 2.13. *Let X be a smooth projective variety strongly liftable over $W_2(k)$, D a $\mathbb{Q}$-divisor on X such that the fractional part $\langle D \rangle = \sum_{i \in I} \frac{a_i}{b_i} D_i$ satisfies that $0 < a_i < b_i$, $(a_i, b_i) = 1$ and $p \nmid b_i$ for all $i \in I$, and $\sum_{i \in I} D_i$ is simple normal crossing. Then there exists a finite Galois morphism $\tau : Y \to X$ from a smooth projective variety Y with Galois group $G = \mathrm{Gal}(K(Y)/K(X))$ which satisfies the following conditions:*

(i) *$\tau^* D$ becomes an integral divisor on Y;*

(ii) *$\mathcal{O}_X([D]) \cong (\tau_* \mathcal{O}_Y(\tau^* D))^G$, $\mathcal{O}_X(K_X + \ulcorner D \urcorner) \cong (\tau_* \mathcal{O}_Y(K_Y + \tau^* D))^G$, where G acts naturally on $\tau_* \mathcal{O}_Y(\tau^* D)$ and on $\tau_* \mathcal{O}_Y(K_Y + \tau^* D)$. Via these isomorphisms, $\mathcal{O}_X([D])$ (resp. $\mathcal{O}_X(K_X + \ulcorner D \urcorner)$) turns out to be a direct summand of $\tau_* \mathcal{O}_Y(\tau^* D)$ (resp. $\tau_* \mathcal{O}_Y(K_Y + \tau^* D)$); and*

(iii) *Y is liftable over $W_2(k)$.*

By means of Theorem 2.13, we can construct a large class of liftable smooth projective varieties from strongly liftable schemes. Furthermore, the Kawamata-Viehweg vanishing theorem on strongly liftable schemes can also be deduced from Theorem 2.13 (see [XW13, Corollary 3.9]).

3 Applications to vanishing theorems

If X is a smooth projective variety strongly liftable over $W_2(k)$, then the Kawamata-Viehweg vanishing theorem (smooth case) holds on X, which is a direct consequence of Definition 2.1 and Theorem 1.12.

When considering algebraic surfaces, we can weaken the strong liftability condition to obtain a more practical result on the Kawamata-Viehweg vanishing theorem (general case) [Xie10c, Theorem 4.2].

Theorem 3.1. *Let X be a normal projective surface, D a $\mathbb{Q}$-Cartier Weil divisor on X, and B an effective $\mathbb{Q}$-divisor such that (X, B) is KLT and that $D - (K_X + B)$ is ample. If X is birational to a strongly liftable smooth projective surface S, then $H^1(X, D) = 0$ holds.*

Definition 3.2. Let X be a normal projective surface, and B an effective $\mathbb{Q}$-divisor. The log pair (X, B) is called a log del Pezzo surface if (X, B) is KLT and $-(K_X + B)$ is ample. X is called a log del Pezzo surface (resp. weak log del Pezzo surface) if $(X, 0)$ is KLT and $-K_X$ is ample (resp. nef and big).

We can prove that for a log del Pezzo surface (X, B), X is a rational surface, i.e. X is birational to the projective plane $\mathbb{P}_k^2$, which is strongly liftable by Theorem 2.4. By Theorem 3.1, we have the following result [Xie10b, Theorem 1.4 and Corollary 1.6].

Theorem 3.3. *The Kawamata-Viehweg vanishing theorem (general case) holds for rational surfaces, hence holds for log del Pezzo surfaces.*

Theorem 3.3 immediately implies the following corollary [Xie10b, Corollary 1.7].

Corollary 3.4. *Let X be a (weak) log del Pezzo surface. Then $H^1(X, \mathcal{O}_X) = 0$ holds.*

Remark 3.5. (1) Corollary 3.4 gives an affirmative answer to an open problem put forward by Shepherd-Barron in [SB91, Remark after Proposition 21].

(2) A Fano variety, by definition, is a projective variety X with the anticanonical divisor $-K_X$ ample. Fano surface is conventionally called del Pezzo surface. As is well known, Fano variety has appeared as a kind of outcome of running the minimal model program, so the study of Fano varieties is of certain interest in birational geometry of algebraic varieties. Let us recall some known vanishing or non-vanishing results concerning Fano varieties in positive characteristic, which show that Corollary 3.4 is just a result as expected.

(i) Tango [Ta72a] has proved that the Kodaira vanishing theorem holds on smooth projective ruled surfaces, hence holds on *smooth* del Pezzo surface.

(ii) Reid [Re94] has found *nonnormal* del Pezzo surfaces X with $H^1(X, \mathcal{O}_X) \neq 0$.

(iii) Schröer [Sc07] proved that over any *nonperfect* field k of characteristic $p = 2$, there is a normal del Pezzo surface X with $H^1(X, \mathcal{O}_X) \neq 0$.

(iv) Shepherd-Barron [SB97] established that $H^1(X, \mathcal{O}_X) = H^2(X, \mathcal{O}_X) = 0$ holds for smooth Fano *threefolds*.

(v) Lauritzen and Rao [LR97] has constructed counterexamples to the Kodaira vanishing theorem on certain smooth Fano varieties of *dimension at least* 6.

In fact, we have a generalization of Theorem 3.1, where no singularity assumption is made [Xie11b, Theorem 3.7].

Theorem 3.6. *Let X be a normal projective surface and H a nef and big $\mathbb{Q}$-divisor on X. If X is birational to a strongly liftable smooth projective surface S, then $H^1(X, K_X + \ulcorner H \urcorner) = 0$ holds.*

By Theorem 2.4(iv), smooth projective toric varieties are strongly liftable, hence by Theorem 1.12, the Kawamata-Viehweg vanishing theorem (smooth case) holds for smooth projective toric varieties. In fact, we can use the liftability of the relative Frobenius morphism of toric varieties and the strong liftability of toric varieties to prove the Bott vanishing theorem, the degeneration of the Hodge to de Rham spectral sequence and the Kawamata-Viehweg vanishing theorem for projective toric varieties.

Let X be a normal scheme over k, and D a reduced Weil divisor on X. Then there exists an open subset U of X such that $\mathrm{codim}_X(X-U)\geq 2$, U is smooth over k, and $D|_U$ is simple normal crossing on U. Such an U is called a required open subset for the log pair (X,D).

Example 3.7. (1) Let X be a smooth scheme over k, and D a simple normal crossing divisor on X. Then we can take the largest required open subset $U=X$.

(2) Let $X=X(\Delta,k)$ be a toric variety, and D a torus invariant reduced Weil divisor on X. Take U to be $X(\Delta_1,k)$, where Δ_1 is the fan consisting of all 1-dimensional cones in Δ. Then it is easy to show that U is an open subset of X with $\mathrm{codim}_X(X-U)\geq 2$, U is smooth over k, and $D|_U$ is simple normal crossing on U. Hence U is a required open subset for (X,D).

Definition 3.8. Let X be a normal scheme over k, and D a reduced Weil divisor on X. Take a required open subset U for (X,D), and denote $\iota:U\hookrightarrow X$ to be the open immersion. For any $i\geq 0$, define the Zariski sheaf of differential i-forms of X with logarithmic poles along D by

$$\widetilde{\Omega}^i_X(\log D)=\iota_*\Omega^i_U(\log D|_U).$$

Since $\mathrm{codim}_X(X-U)\geq 2$ and $\Omega^i_U(\log D|_U)$ is a locally free sheaf on U, $\widetilde{\Omega}^i_X(\log D)$ is a reflexive sheaf on X, and the definition of $\widetilde{\Omega}^i_X(\log D)$ is independent of the choice of the open subset U. Furthermore we have the Zariski-de Rham complex $(\widetilde{\Omega}^\bullet_X(\log D),d)$.

The following results [Xie13a, Theorems 1.1–1.3] generalize those results of Danilov [Da78], Buch-Thomsen-Lauritzen-Mehta[BTLM97], Mustaţă [Mu02] and Fujino [Fu07] to the case where concerned Weil divisors are not necessarily torus invariant.

Theorem 3.9. (Bott vanishing) *Let X be a projective toric variety over k, D a reduced Weil divisor on X, and $\mathcal{L}$ an ample invertible sheaf on X. Then $H^j(X,\widetilde{\Omega}^i_X(\log D)\otimes\mathcal{L})=0$ holds for any $j>0$ and any $i\geq 0$.*

Theorem 3.10. (Hodge to de Rham spectral sequence) *Let X be a projective toric variety over k, and D a reduced Weil divisor on X. Then the Hodge to de Rham spectral sequence degenerates in E_1:*

$$E_1^{ij}=H^j(X,\widetilde{\Omega}^i_X(\log D))\Longrightarrow\mathbf{H}^{i+j}(X,\widetilde{\Omega}^\bullet_X(\log D)).$$

Theorem 3.11. (Kawamata-Viehweg vanishing) *Let X be a projective toric variety over k, and H a nef and big $\mathbb{Q}$-divisor on X. Then $H^i(X,K_X+\ulcorner H\urcorner)=0$ holds for any $i>0$.*

Remark 3.12. In fact, the conclusions of Theorems 3.9–3.11 hold for projective toric varieties of arbitrary characteristic, which is proved in [Xie13c, Theorems 1.1–1.3] by means of the reduction modulo p method.

4 Some remarks and further problems

In fact, we can generalize almost all of the results in Section 2 to the case where everything is considered over $W(k)$, the ring of Witt vectors of k. A smooth scheme X is said to be strongly liftable over $W(k)$, if X and all prime divisors on X can be lifted simultaneously over $W(k)$. The following result [XW13, Theorem 1.3] is a generalization of Theorem 2.4.

Theorem 4.1. *The following schemes are strongly liftable over $W(k)$:*

(i) $\mathbb{A}^n_k$, $\mathbb{P}^n_k$ *and a smooth projective curve;*

(ii) *A smooth projective variety of Picard number* 1 *which is a complete intersection in* $\mathbb{P}^n_k$*;*

(iii) *A smooth projective rational surface;*

(iv) *A smooth projective toric variety.*

By means of cyclic covers over toric varieties, we can obtain a large class of smooth projective varieties which are liftable over $W(k)$ [XW13, Corollary 1.4].

Corollary 4.2. *Let X be a smooth projective toric variety, and $\mathcal{L}$ an invertible sheaf on X. Let N be a positive integer prime to p, and D an effective divisor on X with $\mathcal{L}^N = \mathcal{O}_X(D)$ and $\mathrm{Sing}(D_{\mathrm{red}}) = \emptyset$. Let $\pi : Y \to X$ be the cyclic cover obtained by taking the N-th root out of D. Then Y is a smooth projective scheme liftable over $W(k)$.*

Finally, we put forward some related problems to the theory of strongly liftable schemes. As shown in Remark 2.5, we would like to ask:

Problem 4.3. *As for $K3$ surfaces, Kummer surfaces, Enriques surfaces, elliptic surfaces and so on, are they strongly liftable over $W_2(k)$? If possible, find counterexamples to the Kawamata-Viehweg vanishing theorem on them.*

We would also like to put forward the following conjecture, i.e. cyclic covers over toric varieties should be strongly liftable over $W_2(k)$, where the liftability was proved in Theorem 2.6 and an easy case was settled in Corollary 2.12.

Conjecture 4.4. *Let X be a smooth projective toric variety, and $\mathcal{L}$ an invertible sheaf on X. Let N be a positive integer prime to p, and D an effective divisor on X with $\mathcal{L}^N = \mathcal{O}_X(D)$ and $\mathrm{Sing}(D_{\mathrm{red}}) = \emptyset$. Let $\pi : Y \to X$ be the cyclic cover obtained by taking the N-th root out of D. Then Y is a smooth projective scheme which is strongly liftable over $W_2(k)$.*

Remark 3.5(v) implies that there exist smooth Fano varieties of dimension at least 6 which are not liftable over $W_2(k)$, whereas Remark 3.5(iv) gives us a hint that smooth Fano threefolds would be strongly liftable over $W_2(k)$. Therefore, it is of great interest to ask:

Problem 4.5. *Are smooth Fano threefolds strongly liftable over $W_2(k)$?*

References

[BTLM97] A. Buch, J. F. Thomsen, N. Lauritzen, V. Mehta, The Frobenius morphism on a toric variety, *Tohoku Math. J.*, **49** (1997), 355–366.

[Da78] V. I. Danilov, The geometry of toric varieties, *Russ. Math. Surv.*, **33** (1978), 97–154.

[DI87] P. Deligne, L. Illusie, Relèvements modulo p^2 et décomposition du complexe de Rham, *Invent. Math.*, **89** (1987), 247–270.

[EV92] H. Esnault, E. Viehweg, *Lectures on vanishing theorems*, DMV Seminar, **20**, Birkhäuser, 1992.

[Fu07] O. Fujino, Multiplication maps and vanishing theorems for toric varieties, *Math. Zeit.*, **257** (2007), 631–641.

[Ha98] N. Hara, A characterization of rational singularities in terms of injectivity of Frobenius maps, *Amer. J. Math.*, **120** (1998), 981–996.

[Il90] L. Illusie, Réduction semi-stable et décomposition de complexes de Rham á coefficients, *Duke Math. J.*, **60** (1990), 139–185.

[Ka82] Y. Kawamata, A generalization of Kodaira-Ramanujam's vanishing theorem, *Math. Ann.*, **261** (1982), 43–46.

[KMM87] Y. Kawamata, K. Matsuda, K. Matsuki, Introduction to the minimal model problem, Alg. Geom. Sendai 1985, *Adv. Stud. Pure Math.*, **10** (1987), 283–360.

[Ko86] J. Kollár, Higher direct images of dualizing sheaves, I, *Ann. Math.*, **123** (1986), 11–42.

[LR97] N. Lauritzen, A. Rao, Elementary counterexamples to Kodaira vanishing in prime characteristic, *Proc. Indian Acad. Sci. Math. Sci.*, **107** (1997), 21–25.

[MO05] K. Matsuki, M. Olsson, Kawamata-Viehweg vanishing as Kodaira vanishing for stacks, *Math. Res. Lett.*, **12** (2005), 207–217.

[Mu02] M. Mustaţă, Vanishing theorems on toric varieties, *Tohoku Math. J.*, **54** (2002), 451–470.

[Ra78] M. Raynaud, Contre-exemple au "vanishing theorem" en caractéristique $p > 0$, C. P. Ramanujam — A tribute, *Studies in Math.*, **8** (1978), 273–278.

[Re94] M. Reid, Nonnormal del Pezzo surfaces, *Publ. Res. Inst. Math. Sci.*, **30** (1994), 695–727.

[Sc07] S. Schröer, Weak del Pezzo surfaces with irregularity, *Tôhoku Math. J.*, **59** (2007), 293–322.

[Se62] J.-P. Serre, *Corps Locaux*, Hermann, 1962.

[Sh94] I. R. Shafarevich, *Basic Algebraic Geometry I*, second, revised and expanded edition, Springer-Verlag, 1994.

[SB91] N. I. Shepherd-Barron, Unstable vector bundles and linear systems on surfaces in characteristic p, *Invent. Math.*, **106** (1991), 243–262.

[SB97] N. Shepherd-Barron, Fano threefolds in positive characteristic, *Compositio Math.*, **105** (1997), 237–265.

[Ta72a] H. Tango, On the behavior of cohomology classes of vector bundles under the Frobenius map (in Japanese), *Res. Inst. Math. Sci., Kôkyûroku*, **144** (1972), 93–102.

[Ta72b] H. Tango, On the behavior of extensions of vector bundles under the Frobenius map, *Nagoya Math. J.*, **48** (1972), 73–89.

[Vi82] E. Viehweg, Vanishing theorems, *J. Reine Angew. Math.*, **335** (1982), 1–8.

[Xie10a] Q. Xie, Counterexamples to the Kawamata-Viehweg vanishing on ruled surfaces in positive characteristic, *J. Algebra*, **324** (2010), 3494–3506.

[Xie10b] Q. Xie, Kawamata-Viehweg vanishing on rational surfaces in positive characteristic, *Math. Zeit.*, **266** (2010), 561–570.

[Xie10c] Q. Xie, Strongly liftable schemes and the Kawamata-Viehweg vanishing in positive characteristic, *Math. Res. Lett.*, **17** (2010), 563–572.

[Xie11a] Q. Xie, A characterization of counterexamples to the Kodaira-Ramanujam vanishing theorem on surfaces in positive characteristic, *Chin. Ann. Math., Series B*, **32** (2011), 741–748.

[Xie11b] Q. Xie, Strongly liftable schemes and the Kawamata-Viehweg vanishing in positive characteristic, II, *Math. Res. Lett.*, **18** (2011), 315–328.

[Xie11c] Q. Xie, Decomposition of de Rham complexes with smooth horizontal coefficients for semistable reductions, *Pacific J. Math.*, **254** (2011), 465–497.

[XW13] Q. Xie, J. Wu, Strongly liftable schemes and the Kawamata-Viehweg vanishing in positive characteristic, III, *J. Algebra*, **395** (2013), 12–23.

[Xie13a] Q. Xie, Vanishing theorems on toric varieties in positive characteristic, to appear in *Math. Zeit.*, DOI: 10.1007/s00209-013-1193-2.

[Xie13b] Q. Xie, Cyclic covers over strongly liftable schemes, preprint.

[Xie13c] Q. Xie, Vanishing theorems on toric varieties, preprint.

School of Mathematical Sciences, Fudan University
Shanghai 200433, China
E-mail address: qhxie@fudan.edu.cn

Proceedings of the 6th International Congress of Chinese Mathematicians
ALM 36, pp. 519–539

Birational Geometry in the Study of Dynamics of Automorphisms and Brody/Mori/Lang Hyperbolicity

De-Qi Zhang

Abstract

We survey our recent papers (some being joint ones) about the relation between the geometry of a compact Kähler manifold and the existence of automorphisms of positive entropy on it. We also use the language of log minimal model program (LMMP) in birational geometry, but not its more sophisticated technical part. We give applications of LMMP to positivity of log canonical divisor of a Mori / Brody / Lang hyperbolic (quasi-) projective (singular) variety.

Mathematics Subject Classification: 14J50, 32M05, 32H50, 37B40, 32Q45, 14E30.

Keywords and Phrases: Automorphism groups of complex varieties, positive entropy, Tits alternative for automorphism groups, Brody/Kobayashi/Lang/Mori/ hyberbolicity, ample canonical divisor.

1 Introduction

We work over the field $\mathbb{C}$ of complex numbers. In §2, we consider automorphisms or automorphism groups of positive (topological) entropy on compact Kähler manifolds or normal projective varieties. This part is the continuation of our earlier survey paper [29]. In §3, we consider $\mathrm{Aut}(X)$ for an almost homogeneous compact Kähler manifold X. In §4, we consider positivity of log canonical divisor of Mori hyperbolic or Brody hyperbolic quasi-projective varieties. In §5, we consider positivity of canonical divisor of algebraic Lang hyperbolic singular varieties.

We use the convention in [8, 20], and the book of Hartshorne.

This work is supported by an ARF of NUS.

2 Automorphisms of positive entropy

This section is mainly based on [33, 32] and the joint paper [4]. We refer to Dinh [7] for another survey on results of Tits alternative type.

For a linear transformation L on a finite-dimensional vector space V over $\mathbb{C}$ or its subfields, its *spectral radius* is defined as

$$\rho(L) := \max\{|\lambda| \, : \, \lambda \in \mathbb{C} \text{ is an eigenvalue of } L\}.$$

Let X be a compact complex Kähler manifold and Y a normal projective variety, and let $g \in \mathrm{Aut}(X)$ and $f \in \mathrm{Aut}(Y)$. Define the (topological) *entropy* $h(*)$ and *first dynamical degrees* $d_1(*)$ as:

$$\begin{aligned} h(g) :&= \log \rho(g^* \mid \oplus_{i \ge 0} H^i(X, \mathbb{C})), \\ d_1(g) :&= \rho(g^* \mid H^2(X, \mathbb{C})) \ (= \rho(g^* \mid H^{1,1}(X))), \\ d_1(f) :&= \rho(f^* \mid \mathrm{NS}_{\mathbb{C}}(Y)) \end{aligned}$$

where $\mathrm{NS}_{\mathbb{C}}(Y) := \mathrm{NS}(Y) \otimes_{\mathbb{Z}} \mathbb{C}$ is the *complexified Neron-Severi group*. By the fundamental work of Gromov and Yomdin, the above definition of entropy is equivalent to its original definition (cf. [8, §2.2] and the references therein). Further, when Y is smooth, the above two definitions of $d_1(*)$ coincide; for $\mathbb{Q}$-factorial Y (cf. [20, 0.4(1)]), we have $d_1(f) = d_1(\widetilde{f})$ where $\widetilde{f}$ is the lifting of f to the one on an $\mathrm{Aut}(Y)$-equivariant resolution of Y. We call $\tau := g$ or f, of *positive entropy* (resp. *null entropy*) if $d_1(\tau) > 1$ (resp. $d_1(\tau) = 1$), or equivalently $h(\tau) > 0$ (resp. $h(\tau) = 0$) in the case of compact Kähler manifold.

We say that the induced action $G \mid H^{1,1}(X)$ is *Z-connected* if its Zariski-closure in $\mathrm{GL}(H^{1,1}(X))$ is connected with respect to the Zariski topology; in this case, the *null set*

$$N(G) := \{g \in G \mid g \text{ is of null entropy}\}$$

is a (necessarily normal) subgroup of G (cf. [30, Theorem 1.2]). In [30], we have proved:

Theorem 2.1. (cf. [30]) *Let X be an n-dimensional ($n \ge 2$) compact complex Kähler manifold and G a subgroup of* $\mathrm{Aut}(X)$. *Then one of the following two assertions holds:*

(1) *G contains a subgroup isomorphic to the non-abelian free group $\mathbb{Z} * \mathbb{Z}$, and hence G contains subgroups isomorphic to non-abelian free groups of all countable ranks.*

(2) *There is a finite-index subgroup G_1 of G such that the induced action*

$$G_1 \mid H^{1,1}(X)$$

is solvable and Z-connected. Further, the subset

$$N(G_1) := \{g \in G_1 \mid g \text{ is of null entropy}\}$$

of G_1 is a normal subgroup of G_1 and the quotient group $G_1/N(G_1)$ is a free abelian group of rank $r \le n-1$. We call this r **the rank of G_1 and denote it as** $r = r(G_1)$.

A compact Kähler manifold X is *ruled* if it is bimeromorphic to a manifold with a $\mathbb{P}^1$-fibration. By a result of Matsumura, X is ruled if $\mathrm{Aut}_0(X)$ is not a compact torus (cf. [10, Proposition 5.10]). When X is a compact complex Kähler manifold (or a normal projective variety), set $L := H^2(X, \mathbb{Z})/(\text{torsion})$ (resp. $L := \mathrm{NS}(X)/(\text{torsion})$), $L_{\mathbb{R}} := L \otimes_{\mathbb{Z}} \mathbb{R}$, and $L_{\mathbb{C}} := L \otimes_{\mathbb{Z}} \mathbb{C}$.

In [4], we have strengthened Theorem 2.1 to the following:

Theorem 2.2. (cf. [4, Theorem 1.5]) *Let X be a compact Kähler (resp. projective) manifold of dimension n and $G \le \mathrm{Aut}(X)$ a subgroup. Then one of the following properties holds.*

(1) *$G|L_{\mathbb{C}} \ge \mathbb{Z} * \mathbb{Z}$ (the non-abelian free group of rank two), and hence $G \ge \mathbb{Z} * \mathbb{Z}$.*

(2) *$G|L_{\mathbb{C}}$ is virtually solvable and*

$$G \ge K \cap L(\mathrm{Aut}_0(X)) \ge \mathbb{Z} * \mathbb{Z}$$

where $L(\mathrm{Aut}_0(X))$ is the linear part of $\mathrm{Aut}_0(X)$ (cf. [10, Definition 3.1, p. 240]) *and*

$$K = \mathrm{Ker}(G \to \mathrm{GL}(L_{\mathbb{C}})),$$

so X is ruled (cf. [10, Proposition 5.10]).

(3) *There is a finite-index solvable subgroup G_1 of G such that the null subset $N(G_1)$ of G_1 is a normal subgroup of G_1 and*

$$G_1/N(G_1) \cong \mathbb{Z}^{\oplus r}$$

for some $r \le n - 1$.

In particular, either $G \ge \mathbb{Z} * \mathbb{Z}$ or G is virtually solvable. In Cases (2) and (3) above, $G|L_{\mathbb{C}}$ is finitely generated.

In view of Theorem 2.1, we are more interested in the group $G \le \mathrm{Aut}(X)$ where $G \mid H^{1,1}(X)$ is solvable and Z-connected and that the rank $r(G) = \dim X - 1$ (maximal value).

In the following, denote by $\mathrm{Aut}_0(X)$ the *identity connected component* of $\mathrm{Aut}(X)$. A group *virtually has a property* (P) if a finite-index subgroup of it has the property (P). Let X be a normal projective variety with at worst canonical singularities (cf. [20, Definition 2.34]) and $G \le \mathrm{Aut}(X)$ a subgroup.

The pair (X, G) is *non-minimal* if: there are a finite-index subgroup G_1 of G and a *non-isomorphic* G_1-equivariant birational morphism $X \to Y$ onto a normal projective variety Y with at worst *isolated* canonical singularities. The pair (X, G) is *minimal* if it is not non-minimal.

A complex torus has lots of symmetries. Conversely, our Theorem 2.3 in [33] (see also Theorem 2.4 for non-algebraic manifolds) says that the maximality $r(G) = \dim X - 1$ occurs only when X is a quotient of a complex torus T and G is mostly descended from the symmetries on the torus T.

Consider the following:

Hypothesis (C): The discrete image (a lattice) of every quasi-nef sequence induced injective homomorphism $\overline{\varphi} : G/N(G) \to (\mathbb{R}^{\oplus n-1}, +)$ is, up to finite-index, the standard lattice. Namely, $G/N(G)$ is freely generated by cosets $g_i N(G)$ so that $\varphi(g_i)$ equals the i-th coordinate $(0, \cdots, 0, \log \chi_i(g_i), 0, \cdots, 0)$.

Theorem 2.3. (cf. [33, Theorem 1.1]) *Let X be an n-dimensional ($n \ge 3$) normal projective variety and $G \le \mathrm{Aut}(X)$ a subgroup such that the induced action $G \,|\, \mathrm{NS}_{\mathbb{C}}(X)$ is solvable and Z-connected and that the rank $r(G) = n - 1$ (i.e., $G/N(G) = \mathbb{Z}^{\oplus n-1}$). Assume the three conditions:*

(i) [1] *X has at worst canonical, quotient singularities* (cf. [20, Definition 2.34]).

(ii) *X is a minimal variety, i.e., the canonical divisor K_X is nef* (cf. [20, § 0.4(3)]).

(iii) *The pair (X, G) is minimal in the sense above.*

(iv) *Hypothesis* (C).

Then the following four assertions hold.

(1) *The induced action $N(G) \,|\, \mathrm{NS}_{\mathbb{C}}(X)$ is a finite group.*

(2) *$G \,|\, \mathrm{NS}_{\mathbb{C}}(X)$ is a virtually free abelian group of rank $n - 1$.*

(3) *Either $N(G)$ is a finite subgroup of G and hence G is a virtually free abelian group of rank $n-1$, or X is an abelian variety and the group $N(G) \cap \mathrm{Aut}_0(X)$ has finite-index in $N(G)$ and is Zariski-dense in $\mathrm{Aut}_0(X)$ ($\cong X$).*

(4) *We have $X \cong T/F$ for a finite group F acting freely outside a finite set of an abelian variety T. Further, for some finite-index subgroup G_1 of G, the action of G_1 on X lifts to an action of G_1 on T.*

When the X below is non-algebraic, we don't require the minimality of the pair (X, G).

Theorem 2.4. (cf. [32, Theorem 2.1]) *Let X be an n-dimensional ($n \ge 3$) compact complex Kähler manifold which is not algebraic. Let $G \le \mathrm{Aut}(X)$ be a subgroup such that the induced action $G \,|\, H^{1,1}(X) \le \mathrm{Aut}(H^{1,1}(X))$ is solvable and Z-connected and that the rank $r(G) = n - 1$ (i.e., $G/N(G) = \mathbb{Z}^{\oplus n-1}$). Assume that X is minimal, i.e., the canonical divisor K_X is contained in the closure of the Kähler cone of X. Assume also Hypothesis* (C). *Then the following four assertions hold.*

(1) *The induced action $N(G) \,|\, H^{1,1}(X)$ is a finite group.*

(2) *$G \,|\, H^{1,1}(X)$ is a virtually free abelian group of rank $n - 1$.*

[1] In Theorem 1.2, the condition (i) can be weakened to: (i)' X has at worst canonical singularities and is smooth in codimension two. This is by the same proof of ours and due to Theorem 1.16 in the recent paper of D. Greb, S. Kebekus and T. Peternell: "Étale fundamental groups of Kawamata log terminal spaces, flat sheaves, and quotients of Abelian varieties", arXiv:1307.5718.

(3) *Either $N(G)$ is a finite subgroup of G and hence G is a virtually free abelian group of rank $n-1$, or X is a complex torus and the group $N(G) \cap \mathrm{Aut}_0(X)$ has finite-index in $N(G)$ and is Zariski-dense in* $\mathrm{Aut}_0(X)$ $(\cong X)$.

(4) *We have $X \cong T/F$ for a finite group F acting freely outside a finite set of a complex torus T. Further, for some finite-index subgroup G_1 of G, the action of G_1 on X lifts to an action of G_1 on T.*

In Theorem 2.6 below, we will assume that (i) G is abelian and (ii) the absence of point-wise G-fixed subvarieties of positive dimension or G-periodic rational curves or Q-tori. In Theorem 2.3, these two restrictions are replaced by the natural minimality condition on X and the pair (X, G), and that $G \mid \mathrm{NS}_{\mathbb{C}}(X)$ is solvable, the latter of which is natural in view of Theorem 2.1. The quotient singularities assumption in Theorem 2.3 is necessary because an effective characterization of torus quotient is only available in dimension three by [28] where the bulk of the argument is to show that the variety has only quotient singularities.

The lack of the abelianness assumption on G makes our argument much harder, for instance we cannot simultaneously diagonalize $G \mid \mathrm{NS}_{\mathbb{C}}(X)$ or find enough number of linearly independent common nef eigenvectors of G as required in [8] for abelian groups.

Theorems 2.3 and 2.4 answer [30, Question 2.17], assuming the conditions here. When G is abelian, the finiteness of $N(G)$ is proved in the inspiring paper of Dinh-Sibony [8, Theorem 1] (cf. also [31]), assuming only $r(G) = n-1$. For non-abelian G, the finiteness of $N(G)$ is not true and we can at best expect that $N(G)$ is virtually included in $\mathrm{Aut}_0(X)$ (as done in Theorems 2.3 and 2.4), since a larger group $\widetilde{G} := \mathrm{Aut}_0(X)\, G$ satisfies

$$\widetilde{G} \mid \mathrm{NS}_{\mathbb{C}}(X) = G \mid \mathrm{NS}_{\mathbb{C}}(X), \; N(\widetilde{G}) = \mathrm{Aut}_0(X) \cdot N(G) \geq \mathrm{Aut}_0(X),$$
$$\widetilde{G}/N(\widetilde{G}) \cong G/N(G).$$

There are examples (X, G) with rank $r(G) = \dim X - 1$ and X complex tori or their quotients (cf. [8, Example 4.5], [31, Example 1.7]).

The proof of Theorem 2.3 is much harder than that of Theorem 2.4 because of the presence of singularities on X.

The conditions (i)–(iii) in Theorem 2.3 are quite necessary in deducing $X \cong T/F$ as in Theorem 2.3(4). Indeed, if $X \cong T/F$ as in Theorem 2.3(4), then X has only quotient singularities and $dK_X \sim 0$ (linear equivalence) with $d = |F|$, and we may even assume that X has only canonical singularities if we replace X by its global index-1 cover; thus X is a minimal variety. If the pair (X, G) is not minimal so that there is a non-isomorphic G_1-equivariant birational morphism $X \to Y$ for some finite-index subgroup G_1 of G, then the exceptional locus of this morphism is G_1- and hence G-periodic, contradicting the fact that the rank $r(G) = n - 1$.

The *first key step* in proving Theorem 2.3 is the analysis of our quasi-nef sequence $L_1, \ldots, L_k$ $(0 < k < n)$ (cf. [30, §2.2]) and we are able to show that L_i can actually be taken to be nef, when the rank $r(G) = n - 1$. The *second key step* is to split G as $N(G)\, H$ such that $H \mid H^{1,1}(X)$ is free abelian; thus we have a nef and big class A as the sum of nef common eigenvectors of H, leading to the vanishing of $A^{n-i}.c_i(X)$, where $c_i(X)$ $(i = 1, 2)$ are Chern classes (cf. [28, pp. 265–267]). The *third key step* is to use the minimality of (X, G) and Kawamata's base point freeness result for $\mathbb{R}$-divisor (cf. [3, Theorem 3.9.1], available only for projective variety at the moment) to deduce the vanishing of $c_2(X)$ as a linear form; this vanishing does not directly follow from the vanishing of $A^{n-2} \cdot c_2(X)$ because A may not be ample. Now Theorem 2.3(4) follows from the vanishing of $c_i(X)$ $(i = 1, 2)$ and the characterization of torus quotient originally deduced from Yau's deep result (cf. [2]).

Remark 2.5. (1) When $\dim X = 3$, the (i) in Theorem 2.3 can be replaced by: (i)' X has at worst canonical singularities. See [28, Corollary at p. 266].

(2) Theorems 2.3 and 2.4 are not true when $n := \dim X = 2$; see Remark 2.9 below. We used $n \geq 3$ to deduce the vanishing of $c_2(X) \cdot A^{n-2}$ as commented above.

For the definitions of *Kodaira dimension* $\kappa(X)$ and singularities of *terminal*, *canonical* or *klt type*, we refer to [20, Definitions 2.34 and 7.73]. A subvariety $Y \subset X$ is called *G-periodic* if Y is stabilized (set theoretically) by a finite-index subgroup of G. Denote by $q(X) := h^1(X, \mathcal{O}_X)$ the *irregularity* of X. A projective manifold Y is a *Q-torus* if $Y = A/F$ for a finite group F acting freely on an abelian variety A.

Without assuming the minimality of the pair (X, G) we have proved in [32]:

Theorem 2.6. (cf. [32, Theorem 1.2]) *Let X be an n-dimensional normal projective variety and let $G := \mathbb{Z}^{\oplus n-1}$ act on X faithfully such that every element of $G \setminus \{\mathrm{id}\}$ is of positive entropy. Then the following hold:*

(1) *Suppose that $\tau : A \to X$ is a G-equivariant finite surjective morphism from an abelian variety A. Then τ is étale outside a finite set (hence X has only quotient singularities and is klt); $K_X \sim_{\mathbb{Q}} 0$ ($\mathbb{Q}$-linear equivalence); no positive-dimensional proper subvariety $Y \subset X$ is G-periodic.*

(2) *Conversely, suppose that no positive-dimensional G-periodic subvariety $Y \subset X$ is either fixed (point wise) by a finite-index subgroup of G, or is a Q-torus with $q(Y) > 0$, or has $\kappa(Y) = -\infty$. Suppose also one of the following two conditions.*

(2a) *$n = 3$, and X is klt.*

(2b) *$n \geq 3$, and X has only quotient singularities.*

Then $X \cong A/F$ for a finite group F acting freely outside a finite set of an abelian variety A. Further, for some finite-index subgroup G_1 of G, the action of G_1 on X lifts to an action of G_1 on A.

As a consequence of Theorems 2.6 and [8, Theorem 1] (or Theorem 2.1), we have:

Corollary 2.7. (cf. [32]) *Let X be a normal projective variety of dimension $n \geq 3$ with only quotient singularities, and let $G := \mathbb{Z}^{\oplus r}$ act on X faithfully for some $r \geq n-1$ such that every element of $G \setminus \{\mathrm{id}\}$ is of positive entropy. Then $r = n-1$. Further, $X \cong A/F$ for a finite group F acting freely outside a finite set of an abelian variety A, if and only if X has no positive-dimensional G-periodic proper subvariety $Y \subset X$.*

The proof of [32, Theorem 2.2] gives the following, essentially already proved in [31].

Corollary 2.8. (cf. [32]) *Let X be a normal projective variety of dimension $n \geq 3$ and let $G := \mathbb{Z}^{\oplus r}$ act on X faithfully for some $r \geq n-1$ such that every element of $G \setminus \{\mathrm{id}\}$ is of positive entropy. Then $r = n-1$. Suppose further that both X and (X, G) are minimal in the sense of* [32, *Definition* 2.1] *(which is slightly different from the one given before Theorem* 2.3) *and either X has only quotient singularities, or X is a klt threefold. Then $X \cong A/F$ for a finite group F acting freely outside a finite set of an abelian variety A.*

Remark 2.9. (1) In Theorem 2.6 (2) we need to assume that $\dim X \geq 3$ which is used at the last step of the proof to show the vanishing of the second Chern class $c_2(X)$. In fact, inspired by the comment of the referee of [32], one notices that a complete intersection $X \subset \mathbb{P}^2 \times \mathbb{P}^2$ of two very general hypersurfaces of type $(1, 1)$ and $(2, 2)$ is a $K3$ surface (called Wehler's surface) of Picard number two, X has an automorphism g of entropy $2\log(2+\sqrt{3}) > 0$, and X contains no (-2)-curve, so there is no g-periodic curve on X. Thus, both X and the pair $(X, \langle g \rangle)$ are minimal in the sense of [32, Definition 2.1]. However, X is not birational to the quotient of a complex torus, because a (smooth) projective $K3$ surface X birational to the quotient of a complex torus has the transcendental lattice of rank $\leq$ (that of a complex 2-torus), i.e., ≤ 5, and hence has Picard number $\geq (h^2(X, \mathbb{C}) - 5)$ which is 17.

(2) In Theorem 2.6 (2) we can weaken the assumption on G as a condition on $G^* := G|_{\mathrm{NS}_{\mathbb{R}}(X)}$ (cf. [8, Theorem 4.7]):

"$G^* \cong \mathbb{Z}^{\oplus n-1}$ and every element of $G^* \setminus \{\mathrm{id}\}$ is of positive entropy."

But we need also to replace the last sentence

"Further, ... of G_1 on A." in Theorems 2.6 (2) as:

"Further, the action of G on X lifts to an action of a group $\widetilde{G}$ on A with $\widetilde{G}/\operatorname{Gal}(A/X) \cong G$."

Our bimeromorphic point of view, in terms of the minimality assumption in 2.8 towards the dynamics study seems natural, since one may blow up some Zariski-closed and G-stable proper subset of X (if such subset exists) to get another pair (X', G) which is essentially the same as the original pair (X, G).

The very starting point of our proof is the existence of enough nef eigenvectors L_i of G, due to the fundamental work of Dinh-Sibony [8]. The minimal model program (cf. [20]) is used with references provided for non-experts. Our main contribution lies in showing that the pair (X, G) can be replaced with an equivariant one so that $H := \sum_{i=1}^{n} L_i$ is an ample divisor. To conclude, we prove a result of Hodge-Riemann type for singular varieties to show the vanishing of Chern classes $c_i(X)$ $(i = 1, 2)$, utilizing $H^{n-i} \cdot c_i(X) = 0 = H^{n-2} \cdot c_1(X)^2$. Then we use the characterization of étale quotient of a complex torus as the compact Kähler manifold X with vanishing Chern classes $c_i(X)$ $(i = 1, 2)$ (cf. [2, §1]) and its generalization to singular varieties (cf. [28]).

For a possible generalization of the proof to a Kähler n-fold X, we remark that the restriction "rank $r(G) = n - 1$" implies that X is either Moishezon and hence projective, or has algebraic dimension $a(X) = 0$ (cf. [30, Theorem 1.2]). Thus the case $a(X) = 0$ remains to be treated. See a related remark in [6, §3.6].

3 Automorphism groups of almost homogeneous varieties

This section is based on the joint paper [9]. Let X be a compact Kähler manifold. Denote by $\mathrm{Aut}(X)$ the automorphism group of X and by $\mathrm{Aut}_0(X)$ the identity connected component of $\mathrm{Aut}(X)$. By [10], $\mathrm{Aut}_0(X)$ has a natural meromorphic group structure. Further there exists a unique meromorphic subgroup, say $L(X)$, of $\mathrm{Aut}_0(X)$, which is meromorphically isomorphic to a linear algebraic group and such that the quotient $\mathrm{Aut}_0(X)/L(X)$ is a complex torus. In the following, by a subgroup of $\mathrm{Aut}_0(X)$ we always mean a meromorphic subgroup and by a linear algebraic subgroup of $\mathrm{Aut}_0(X)$ we mean a Zariski closed meromorphic subgroup contained in $L(X)$.

For a subgroup $G \le \mathrm{Aut}(X)$, the pair (X, G) is called *strongly primitive* if for every finite-index subgroup G_1 of G, X is not bimeromorphic to a non-trivial G_1-equivariant fibration, i.e., there does not exist any compact Kähler manifold X' bimeromorphic to X, such that X' admits a G_1-equivariant holomorphic map $X' \to Y$ with $0 < \dim Y < \dim X$ and $G_1 \le \mathrm{Aut}(X')$. See Oguiso-Truong [26] for the very first examples of strongly primitive pairs $(X, \langle g \rangle)$ with X a Calabi-Yau threefold or rational threefold. From the dynamical point of view, these manifolds are essential. Our Theorem 3.1 of [9] says that for these manifolds, unless it is a complex torus, there is no interesting dynamics if its automorphism group has positive dimension.

Theorem 3.1. (cf. [9, Theorem 1.1]) *Let X be a compact Kähler manifold and $G \le \mathrm{Aut}(X)$ a subgroup of automorphisms. Assume the following three conditions:*

(1) $G_0 := G \cap \mathrm{Aut}_0(X)$ *is infinite.*

(2) $|G : G_0| = \infty$.

(3) *The pair (X, G) is strongly primitive.*

Then X is a complex torus.

As a key step towards Theorem 3.1, we prove the following result in [9]. A proof for Theorem 3.2(2) is long overdue (and we do it geometrically via 3.2(1)), but the authors could not find it in any literature, even after consulting many experts across the continents.

Theorem 3.2. (cf. [9, Theorem 1.2]) *Let X be a compact Kähler manifold and $G_0 \subset \mathrm{Aut}_0(X)$ a linear algebraic subgroup. Assume that G_0 acts on X with a Zariski open dense orbit. Then we have:*

(1) *X is projective; the anti canonical divisor $-K_X$ is big, i.e. $\kappa(X, -K_X) = \dim X$.*

(2) *$\mathrm{Aut}(X)/\mathrm{Aut}_0(X)$ is finite.*

Remark 3.3. (i) The condition (2) in Theorem 3.1 is satisfied if G acts on $H^2(X, \mathbb{C})$ as an infinite group, or if G has an element of positive entropy.

(ii) Theorem 3.2 implies that when $\dim X \geq 3$ the case (4) in [30, Theorem 1.2] does not occur, hence it can be removed from the statement.

(iii) Theorem 3.1 generalizes [4, Theorem 1.2], where it is proven for $\dim X = 3$ and under an additional assumption.

(iv) A similar problem for endomorphisms of homogeneous varieties has been studied by Cantat in [5].

Two applications are given. The first one generalizes the following result due to Harbourne [13, Corollary 1.4] to higher dimension: if X is a smooth projective rational surface with $\mathrm{Aut}_0(X) \neq (1)$, then $\mathrm{Aut}(X)/\mathrm{Aut}_0(X)$ is finite. To state it, recall ([10, Theorem 4.1]) that for any connected subgroup $H \leq \mathrm{Aut}(X)$, there exist a quotient space X/H and an H-equivariant dominant meromorphic map $X \dashrightarrow X/H$, which satisfies certain universal property.

Corollary 3.4. (cf. [9, §1.4]) *Let X be a compact Kähler manifold with irregularity $q(X) = 0$. Suppose that the quotient space $X/\mathrm{Aut}_0(X)$ has dimension ≤ 1. Then X is projective and $\mathrm{Aut}(X)/\mathrm{Aut}_0(X)$ is finite.*

The second application essentially says that when we study dynamics of a compact Kähler manifold X, we may assume that $\mathrm{Aut}_0(X)_{\mathrm{lin}} = (1)$, where $\mathrm{Aut}_0(X)_{\mathrm{lin}}$ is the largest connected linear algebraic subgroup of $\mathrm{Aut}_0(X)$.

Corollary 3.5. (cf. [9, §1.5]) *Let X be a smooth projective variety and $G_0 \lhd G \leq \mathrm{Aut}(X)$. Suppose that G_0 is a connected linear closed subgroup of $\mathrm{Aut}_0(X)$. Let Y be a G-equivariant resolution of the quotient space X/G_0 and replace X by a G-equivariant resolution so that the natural map $\pi : X \to Y$ is holomorphic. Then for any $g \in G$, we have the equality of the first dynamical degrees:*

$$d_1(g_{|X}) = d_1(g_{|Y}),$$

where $d_1(g_{|X}) := \max\{|\lambda| \,;\, \lambda$ is an eigenvalue of $g^* \mid H^{1,1}(X)\}$.

In particular, $G_{|X}$ is of null entropy if and only if so is $G_{|Y}$.

Remark 3.6. If $q(X) = 0$, then $\mathrm{Aut}_0(X)$ (hence G_0) is always a linear algebraic group. On the other hand, if G_0 is not linear, then Corollary 3.5 does not hold. For example, let $X = T_1 \times T_2$ be the product of two complex tori and $G_0 = \mathrm{Aut}_0(T_1)$. The quotient space $Y := X/G_0$ is T_2. Suppose that $g \in \mathrm{Aut}(T_1)$ is an element of positive entropy which acts trivially on T_2. Then we have $d_1(g_{|X}) > d_1(g_{|Y}) = 1$.

4 Positivity of log canonical divisors and Mori/ Brody hyperbolicity

This section is based on the results in the joint paper [23]. An algebraic variety X is called *Brody hyperbolic* = BH (resp. Mori hyperbolic = MH) if the following hypothesis (BH) (resp. (MH)) is satisfied:

(BH) Every holomorphic map from the complex line $\mathbb{C}$ to X is a constant map.

(MH) Every algebraic morphism from the complex line $\mathbb{C}$ to X is a constant map. (Equivalently, no algebraic curve in X has normalization equal to $\mathbb{P}^1$ or $\mathbb{C}$.)

Clearly, (BH) $\Rightarrow$ (MH).

Consider a pair (X, D) of an (irreducible) algebraic variety X and a (not necessarily irreducible or equi-dimensional) Zariski-closed subset D of X. When X is an irreducible curve and D a finite subset of X, the pair (X, D) is *Brody hyperbolic* (resp. MH) if $X \setminus D$ is Brody hyperbolic (resp. MH). Inductively, for an algebraic variety X and a Zariski-closed subset D of X, with $D = \cup_i D_i$ the irreducible decomposition, the pair (X, D) is *Brody hyperbolic* (resp. MH) if $X \setminus D$ and the pairs $(D_k, D_k \cap (\cup_{i \ne k} D_i))$ for all k, are so. Note that the Zariski-closed subset $D_k \cap (\cup_{i \ne k} D_i)$ may not be equi-dimensional in general.

Here, it is crucial from the perspective of hyperbolic geometry that D is a divisor and all the D_k's are at least $\mathbb{Q}$-Cartier (so as to have equidimensionality above for example). But we will work with a bit less from the outset as demanded by our general setup.

The pair (X, D) is called *projective* if X is a projective variety and *log smooth* if further X is smooth and D is a reduced *divisor* with simple normal crossings.

From now on, we always assume that D is a reduced Weil divisor for a pair (X, D).

Recall that a divisor F on a smooth projective variety X is called *numerically effective* (nef), if $F \cdot C \ge 0$ for all curves C on X. We consider the following conjecture:

Conjecture 4.1. *Let (X, D) be a log smooth projective Mori hyperbolic (resp. Brody hyperbolic) pair. Then the log canonical divisor $K_X + D$ is nef (resp. ample).*

Our main theorems of [23] below give an affirmative answer to Conjecture 4.1 but with further assumptions for the ampleness of $K_X + D$, such as the ampleness of at least $n-2$ irreducible Cartier components of D for $n = \dim(X, D) := \dim X$. To precise them, we will need to refer to the following two standard conjectures on the structure of algebraic varieties.

Conjecture 4.2. Abundance(l): *Let (X, D) be an l-dimensional log smooth projective pair whose log canonical divisor $K_X + D$ is nef. Then some positive multiple of $K_X + D$ is base point free, i.e. $K_X + D$ is semi-ample.*

Conjecture 4.3. CY(m): *Let X be an m-dimensional simply connected non-singular projective variety with trivial canonical line bundle, i.e. a Calabi-Yau manifold. Then X contains a rational curve.*

We remark that Abundance(l) is known to hold for $l \leq 3$ (even for dlt or lc pairs) and CY(m) is known to hold for $m \leq 2$ (see [20, §3.13], [25]).

Let D be a reduced divisor on a projective variety X and $\mathcal{D}$ the collection of terms in its irreducible decomposition $D = \sum_{i=1}^{s} D_i$. A stratum of D is a set of the form $\cap_{i \in I} D_i$ ($= X$ when $I = \emptyset$) for some partition $\{1, \dots, s\} = I \coprod J$. We call a *$\mathcal{D}$-rational curve* a rational curve ℓ in a stratum $\cap_{i \in I} D_i$ of D such that the normalization of $\ell \setminus (\cup_{j \in J} D_j)$ contains the complex line $\mathbb{C}$. We call a *$\mathcal{D}$-algebraic 1-torus* a rational curve ℓ in a stratum $\cap_{i \in I} D_i$ of D such that $\ell \setminus (\cup_{j \in J} D_j)$ has the 1-dimensional algebraic torus $\mathbb{C}^* = \mathbb{C} \setminus \{0\}$ as its normalization. We call a closed subvariety $W \subseteq X$ a *$\mathcal{D}$-compact variety* if W lies in some stratum $\cap_{i \in I} D_i$ of D and $W \cap \cup_{j \in J} D_j = \emptyset$. A $\mathcal{D}$-compact variety $W \subseteq X$ is called a $\mathcal{D}$-compact rational curve, a *$\mathcal{D}$-torus*, a *$\mathcal{D}$-$\mathbb{Q}$-torus*, or a *$\mathcal{D}$-$\mathbb{Q}$-CY* variety if W is a rational curve, an abelian variety, a $\mathbb{Q}$-torus, or a $\mathbb{Q}$-CY variety, respectively.

Here a projective variety is called a *$\mathbb{Q}$-torus* if it has an abelian variety as its finite étale (Galois) cover. A projective variety X with only klt singularities is called a *$\mathbb{Q}$-CY variety* if some positive multiple of its canonical divisor K_X is linearly equivalent to the trivial divisor. A simply connected $\mathbb{Q}$-CY surface with only Du Val singularities is called a *normal K3 surface.* A $\mathbb{Q}$-torus is an example of a smooth $\mathbb{Q}$-CY.

Theorem 4.4. (cf. [23, Theorem 1.4]) *Let X be a smooth projective variety of dimension n and D a reduced divisor on X with simple normal crossings. Then the closure $\overline{\text{NE}}(X)$ of effective 1-cycles on X is generated by the $(K_X + D)$-non negative part and at most a countable collection of extremal $\mathcal{D}$-rational curves $\{\ell_i\}_{i \in N}$, $N \subseteq \mathbb{Z}$:*

$$\overline{\text{NE}}(X) = \overline{\text{NE}}(X)_{K_X + D \geq 0} + \sum_{i \in N} \mathbb{R}_{>0}[\ell_i].$$

Further, $-\ell_i \cdot (K_X + D)$ is in $\{1, 2, \dots, 2n\}$, and is equal to 1 when ℓ_i is not a $\mathcal{D}$-compact *rational curve. In particular, if the pair (X, D) is Mori hyperbolic then $K_X + D$ is nef.*

Theorem 4.5. (cf. [23, Theorem 1.5]) *Let X be a smooth projective variety of dimension n and D a reduced divisor on X with simple normal crossings. Fix an r in $\{1, 2, \dots, n\}$. Assume that* Abundance(l) *holds for $l \leq r$ and that D has at least $n - r + 1$ irreducible components amongst which at least $n - r$ are ample. If $K_X + D$ is not ample, then either it has non-positive degree on a $\mathcal{D}$-rational curve or on a $\mathcal{D}$-algebraic 1-torus, or some positive multiple of it is linearly equivalent* *to the trivial divisor on a smooth $\mathcal{D}$-$\mathbb{Q}$-CY variety T of dimension $< r$. We can take T to be a $\mathbb{Q}$-torus if further* CY(m) *holds for $m < r$.*

In particular, if Abundance(l) *and* CY(m) *hold for $l \leq r$ and $m < r$ and the pair (X, D) is Brody hyperbolic then $K_X + D$ is ample.*

Since the two conjectures are known for $r = 3$ (see the remark above), an immediate corollary is the following generalization, to arbitrary dimensions for pairs, of results that were only known up to dimension two in the case $D = 0$ concerning hyperbolic varieties.

Theorem 4.6. (cf. [23, Theorem 1.6]) *Let X be a smooth projective variety of dimension n and D a reduced divisor on X with simple normal crossings such that D has at least $n-2$ irreducible components amongst which at least $n-3$ are ample. If K_X+D is not ample, then either it has non-positive degree on a $\mathcal{D}$-rational curve or on a $\mathcal{D}$-algebraic 1-torus, or some positive multiple of it is linearly equivalent to the trivial divisor on a $\mathcal{D}$-torus T with* $\dim T \leq 2$. *In particular, if (X, D) is Brody hyperbolic, then $K_X + D$ is ample.*

We remark that Theorems 4.4, 4.6 and 4.5 are special cases of Theorems 4.8 and 4.9 (see also Theorem 4.10) where we allow the pair (X, D) to be singular[2] and D to be augmented with a fractional divisor Γ in a setting that is natural to our approach. Theorem 4.4 is also obtained in McQuillan-Pacienza [24] by a different method, but not our more general result in the singular case, Theorem 4.8. That article obtained the results in the setting of complete intersection "stacks" by a direct analysis of Mori's bend and break procedure and includes for the most part our refined geometric version of the cone theorem for smooth pairs, Theorem 4.4, which we obtain however in our more general singular (dlt) setting, Theorem 4.8, and by a different and independent method.

By running the minimal model program, an easy consequence of this cone theorem is the following criterion for $K_X + D$ to be peudo-effective (the weakest form of positivity in birational geometry) under a mild condition on D. It is a special case of Theorem 4.11.

Theorem 4.7. (cf. [23, Theorem 1.7]) *Let X be a smooth projective variety and D a nonzero simple normal crossing reduced divisor no component of which is uniruled. Then $K_X + D$ is not pseudo-effective if and only if D has exactly one irreducible component and X is dominated by $\mathcal{D}$-rational curves ℓ with*

$$-\ell \cdot (K_X + D) = 1, \quad \ell \cong \mathbb{P}^1, \quad \ell \setminus \ell \cap D \cong \mathbb{C}$$

for a general ℓ.

In particular, $K_X + D$ is pseudo-effective for a log smooth pair (X, D) such that D contains two or more non-uniruled components.

[2]Here by a singular pair, we will assume that it is a dlt pair. The assumption is natural (and in many respects the most general) as our proof is by induction on dimension from running the LMMP for singular pairs. It implies an explicit adjunction formula and that k-fold intersections of components of D are of pure codimension-k in X, crucial in our inductive procedure.

Key to our proof in the presence of Cartier boundary divisors is Kawamata's result on the length of extremal rays, a fundamental result in the subject. Our proofs of Theorems 4.9 and 4.10 are inductive in nature and reduce the problem to questions on adjoint divisors in lower dimensions by adjunction. The log minimal model program (LMMP) is run formally without going into its technical details. Our inductive procedure is naturally adapted to answer some fundamental questions concerning adjoint divisors.

Theorems 4.8, 4.9 and 4.11 below include Theorems 4.4, 4.5 and 4.7 as special cases.

Let X be a normal projective variety and D a reduced Cartier divisor on X. The pair (X, D) is called *BH, or MH with respect to a (Cartier) decomposition* of $D = \sum_{i=1}^{s} D_i$, if each D_i is a reduced (Cartier) divisor on X, not necessarily irreducible, and both $X \setminus D$ and $(\cap_{i \in I} D_i) \setminus (\cup_{j \in J} D_j)$ are respectively BH, or MH for every partition of $\{1, \ldots, s\} = I \coprod J$. Note that, for a log smooth pair (X, D), the respective definitions of hyperbolicity given early on are equivalent to the definitions of hyperbolicity here with respect to the irreducible decomposition of D.

Theorem 4.8. (cf. [23, Theorem 3.1]) *Let X be a projective variety of dimension n, D a reduced divisor that is the sum of a collection $\mathcal{D} = \{D_i\}_{i=1}^{s}$ of reduced Cartier divisors and Γ an effective Weil $\mathbb{Q}$-divisor on X such that the pair $(X, D + \Gamma)$ is divisorial log terminal. Then the closure $\overline{\mathrm{NE}}(X)$ of effective 1-cycles on X is generated by the $(K_X + D + \Gamma)$-non negative part and at most a countable collection of extremal $\mathcal{D}$-rational curves $\{\ell_i\}_{i \in N}$, $N \subseteq \mathbb{Z}$:*

$$\overline{\mathrm{NE}}(X) = \overline{\mathrm{NE}}(X)_{K_X + D + \Gamma \geq 0} + \sum_{i \in N} \mathbb{R}_{>0}[\ell_i].$$

*Furthermore, $-\ell_i \cdot (K_X + D + \Gamma)$ is in $(0, 2n]$, and is even in $(0, 2)$ if ℓ_i is not $\mathcal{D}$-*compact. *In particular, if the pair (X, D) is Mori hyperbolic with respect to the Cartier decomposition $D = \sum_{i=1}^{s} D_i$ then $K_X + D + \Gamma$ is nef.*

One may take $\Gamma = 0$ in Theorems 4.8 – 4.10. Such a Γ naturally appears in our inductive procedure as the "different" in the adjunction formula.

We remark that Abundance(l) (for dlt pairs) and CY(m) always hold for $l \leq 3$ and $m \leq 2$ ([20, §3.13], [25]) and that a normal K3 surface has infinitely many elliptic curves ([25]).

Theorem 4.9. (cf. [23, Theorem 3.2]) *Let X be a projective variety of dimension n, D a reduced Weil divisor and Γ an effective Weil $\mathbb{Q}$-divisor on X such that the pair $(X, D + \Gamma)$ is divisorial log terminal (dlt). Assume one of the following three conditions:*

(1) *$n \leq 2$.*

(2) *$n = 3$, D is nonzero and Cartier.*

(3) *$n \geq 4$, the pair $(X, D + \lfloor \Gamma \rfloor)$ is log smooth, there is an $r \in \{1, 2, \ldots, n\}$ such that D has at least $n - r + 1$ irreducible components amongst which at least $n - r$ are ample and* Abundance(l) *(for dlt pairs) holds for $l \leq r$.*

Then $K_X+D+\Gamma$ is ample, unless either $(K_X+D+\Gamma)$ has non-positive degree on a $\mathcal{D}$-rational curve or on a $\mathcal{D}$-algebraic 1-torus, or the restriction $(K_X+D+\Gamma)_{|T} \sim_{\mathbb{Q}} 0$ for a $\mathcal{D}$-$\mathbb{Q}$-CY variety T. When T is singular, it is a normal K3 surface and $n \in \{2,3\}$. Otherwise, T can be taken to be an abelian surface or an elliptic curve in Cases (1), (2) and, if further CY(m) holds for $m < r$, taken to be a $\mathbb{Q}$-torus with $\dim T < r$ in Case (3).

In particular, if the pair (X, D) is Brody hyperbolic then $K_X+D+\Gamma$ is ample, provided that either $n \le 3$ or CY(m) holds for $m < r$.

Theorem 4.10. (cf. [23, Theorem 3.3]) *Let X be a projective variety of dimension n, D a reduced divisor that is the sum of a collection $\mathcal{D} = \{D_i\}_{i=1}^{s}$ of reduced Cartier divisors for some $s \ge 1$ and Γ an effective Weil $\mathbb{Q}$-divisor on X such that the pair $(X, D+\Gamma)$ is divisorial log terminal. Assume one of the following two conditions:*

(1) *$n = 4$, $s \ge 2$ and D_1 is irreducible and ample.*

(2) *$n \ge 4$, $s \ge n-r+1$ for some $r \in \{1, 2, \dots, n\}$, D_j is ample for all $j \le n-r+1$ and* Abundance(l) *(for dlt pairs) holds for $l \le r$.*

Then $K_X+D+\Gamma$ is ample, unless either $(K_X+D+\Gamma)$ has non-positive degree on a $\mathcal{D}$-rational curve or on a $\mathcal{D}$-algebraic 1-torus, or the restriction $(K_X+D+\Gamma)_{|T} \sim_{\mathbb{Q}} 0$ for some $\mathcal{D}$-$\mathbb{Q}$-CY variety T ($\dim T < r$ in Case (2)). T is smooth when the pair $(X, D+\lfloor \Gamma \rfloor)$ is log smooth and can be taken to be a $\mathbb{Q}$-torus when further CY(m) *holds for $m < r$.*

Theorem 4.11. (cf. [23, Theorem 3.4]) *Let X be a $\mathbb{Q}$-factorial projective variety, and D a nonzero reduced Weil divisor with $D = \sum_{i=1}^{s} D_i$ the irreducible decomposition such that the pair (X, D) is divisorial log terminal.*

(1) *Assume that no component of D is uniruled and $K_X + D$ is not pseudo-effective. Then D has exactly one irreducible component and X is dominated by $\mathcal{D}$-rational curves ℓ with*

$$-\ell \cdot (K_X + D) = 1, \quad \ell \cong \mathbb{P}^1, \quad \ell \setminus \ell \cap D \cong \mathbb{C}$$

for a general ℓ.

(2) *Conversely, assume that X has at worst canonical singularities and X is dominated by $\mathcal{D}$-rational curves. Then $K_X + D$ is not pseudo-effective.*

Of interest here is that the above geometric criterion for the pseudo-effectivity of $K_X + D$ in Theorem 4.11 is obtained naturally from Theorem 4.8 by running the log minimal model program for the pair (X, D). In fact, by running the log minimal model program for a normal surface we get the following slightly stronger consequence. It is noteworthy that Keel-McKernan [19] has obtained the same conclusion in dimension two without the assumption of the absence of $\mathcal{D}$-rational curves in D.

Proposition 4.12. (cf. [23, Proposition 3.5]) *Let X be a projective surface and D a nonzero reduced Weil divisor with $D = \sum_{i=1}^{s} D_i$ the irreducible decomposition such that the pair (X, D) is divisorial log terminal and there are no $\mathcal{D}$-rational curves in D. Assume that $K_X + D$ is not pseudo-effective. Then D has exactly one irreducible component, and X is dominated by $\mathcal{D}$-rational curves ℓ with*

$$-\ell \cdot (K_X + D) = 1, \quad \ell \cong \mathbb{P}^1, \quad \ell \setminus \ell \cap D \cong \mathbb{C}$$

for a general ℓ.

N.B. In Theorem 4.10 (1), D_1 needs to be irreducible so that Theorem 4.9 (2) can be applied: if $D_1 = D_{11} + D_{12}$ is reducible then the restriction $D_{2|D_{11}}$ might be zero. *The ampleness assumption on some D_i's in various theorems is due to the same reason.*

5 Positivity of canonical divisors of singular varieties and Lang hyperbolicity

This section is based on the paper [15]. Let X be a variety. X is *Brody hyperbolic* (resp. *algebraic Lang hyperbolic*) if every holomorphic map $V \to X$ where V is the complex line $\mathbb{C}$ (resp. V is an abelian variety) is a constant map. Since an abelian variety is a complex torus, Brody hyperbolicity implies algebraic Lang hyperbolicity. When X is compact complex variety, Brody hyperbolicity is equivalent to the usual Kobayashi hyperbolicity (cf. [21]).

In the first part (Theorem 5.3) of this section, we let X be a normal projective variety and aim to show the ampleness of the *canonical divisor K_X of X*, assuming that X is algebraic Lang hyperbolic. We allow X to have arbitrary singularities and assume only that X is $\mathbb{Q}$-*Gorenstein* (so that the ampleness of K_X is well-defined), i.e., K_X is $\mathbb{Q}$-*Cartier:* mK_X is a Cariter divisor for some positive integer m.

For related work, it was proved in [27] that a 3-dimensional hyperbolic smooth projective variety X has ample K_X unless X is a Calabi-Yau manifold where every nonzero effective divisor is ample. The authors of [14] prove the ampleness of K_X when X is a smooth projective threefold having a Kähler metric of negative holomorphic sectional curvature; they also generalize the results to higher dimensions with some additional conditions.

In the second part of this section (Theorems 5.4) aiming to prove Lang's conjecture in Corollary 5.5, even the normality of X is not assumed. Our approach is to take a projective resolution of X and run the relative Minimal Model Program (MMP) over X. We use only the frame work of MMP, but not its detailed technical part. Certain mild singularities occur naturally along the way. See [20, Definition 2.34] for definitions of *canonical, Kawamata log terminal* (klt), and *divisorial log terminal* (dlt) *singularities.*

In the last part (Proposition 5.7), we try to avoid assuming conjectures.

Now let X be a normal projective variety. X is a *Calabi-Yau variety* if X has at worst canonical singularities, its canonical divisor is $\mathbb{Q}$-linearly equivalent to zero: $K_X \sim_{\mathbb{Q}} 0$ (and hence X has Kodaira dimension $\kappa(X) = 0$) and the *irregularity* $q(X) := h^1(X, \mathcal{O}_X) = 0$. X is *of general type* if some (equivalently every) projective resolution X' of X has maximal Kodaira dimension: $\kappa(X') = \dim X'$. A $\mathbb{Q}$-Gorenstein variety X is *minimal* if the canonical divisor K_X is *numerically effective* (= *nef*).

Conjecture 5.1 below is long standing. When $\dim X \leq 2$, it is true by the classification of complex surfaces and the *fact* $(*)$: a (smooth) $K3$ surface has infinitely many (singular) elliptic curves (cf. [25, Theorem in Appendix]).

In Conjecture 5.1, the conclusion means the existence of at least one non-constant holomorphic map $f : V \to X$ from an abelian variety V, but does *not* require the image $f(V)$ (or the union of such images) to be Zariski-dense in X. This does not seem sufficient for our purpose to show the non-existence of subvariety X' of Kodaira dimension zero in an algebraic Lang hyperbolic variety W as in Corollary 5.5 below (see 5.9, and think about a proof of the non-hyperbolicity of every normal $K3$ surface using the fact $(*)$ above). Fortunately, we are able to show in Corollary 5.5 (or Theorems 5.4) that the normalization X of $X' \subseteq W$ is a Calabi-Yau variety and hence f composed with the finite morphism $X \to X' \subseteq W$ produces a non-constant holomorphic map from the abelian variety V, thus deducing a contradiction to the hyperbolicity of W.

Conjecture 5.1. *Suppose that X is an absolutely minimal* (in the sense of 5.8) *Calabi-Yau variety. Suppose further that every birational morphism $X \to Y$ onto a normal projective variety is an isomorphism. Then X is not algebraic Lang hyperbolic.*

We recall the definition of the *nef reduction* $f_L : X \dashrightarrow Y$ of a nef $\mathbb{Q}$-Cartier divisor L on a normal projective variety (cf. [1]). f_L is almost holomorphic with F a compact general fibre; the restriction $L_{|F}$ is numerically trivial; a general curve C on X is contracted by f_L if and only if $L \cdot C = 0$; this f_L is unique up to birational equivalence of Y; $\dim Y$ is called the *nef dimension* of L and denoted as $n(L)$. We have $n(L) = \dim X$ (maximal case) if and only if we have the positivity: $L \cdot C > 0$ for every curve C which is not contained in a fixed countable union of Zariski-closed proper subsets of X; also $n(L) = 0$ holds if and only if $L \equiv 0$ (numerically zero).

The abundance conjecture below is slightly weaker than the usual one where the extra hypothesis $(*)$ is usually not assumed. When K_X is nef and big or when $\dim X \leq 3$, both Conjectures 5.2(1) and 5.2(2) (and their log version, even without the hypothesis $(*)$) are true; see [20, Theorem 3.3, §3.13].

Conjecture 5.2. *Let X be an n-dimensional minimal normal projective variety, i.e., the canonical divisor K_X is a nef $\mathbb{Q}$-Cartier divisor. Assume the hypothesis $(*)$: the nef dimension $n(K_X)$* (cf. [1]) *satisfies $n(K_X) = n$.*

(1) *If X has at worst klt singularities, then K_X is semi-ample, i.e., the linear system $|mK_X|$ is base-point free for some $m > 0$.*

(2) *If X has at worst canonical singularities and $K_X \not\equiv 0$ (not numerically zero), then the Kodaira dimension $\kappa(X) > 0$.*

In this paper, by hyperbolic we mean algebraic Lang hyperbolic.

When X has at worst klt singularities, Theorem 5.3 below follows from the MMP and has been generalized to the quasi-projective case in [23].

In Theorem 5.3, we don't impose any condition on the singularities of X, except the $\mathbb{Q}$-Cartierness of K_X for the concluding ampleness of K_X to make sense.

Without assuming Conjecture 5.1 or 5.2, we can at least say that K_X is movable or nef in codimension-one (cf. Remark 5.8).

Theorem 5.3. (cf. [15]) *Let X be a $\mathbb{Q}$-Gorenstein normal projective variety which is algebraic Lang hyperbolic. Assume that Conjectures* 5.1 *and* 5.2(1) *hold. Then K_X is ample at smooth points and klt points of X. To be precise, there is a birational morphism $f_c : X_c \to X$ such that X_c has at worst klt singularities, K_{X_c} is ample, and*

$$E_c := f_c^* K_X - K_{X_c}$$

is effective and f_c-exceptional with $f_c(E_c) \subseteq \mathrm{Nklt}(X)$, the non-klt locus of X (cf. [11, §4.4]). *In particular, $f_c = \mathrm{id}$ and K_X is ample, if X has at worst klt singularities* (cf. 5.8).

Even the normality of X is not assumed here in Theorem 5.4.

Theorem 5.4. (cf. [15]) *Let W be an algebraic Lang hyperbolic projective variety, and $X \subseteq W$ a projective subvariety (with $X = W$ allowed). Assume either* $\dim X \le 3$ *or Conjecture* 5.2(2) *(resp. either* $\dim X \le 3$ *or Conjecture* 5.2(2) *with the hypothesis* $(*)$ *removed). Then there is a birational surjective morphism $g_m : X_m \to X$ such that X_m is a minimal variety with at worst canonical singularities and one of the following is true.*

(1) *K_{X_m} is ample. Hence both X_m and X are of general type.*

(2) *$g_m : X_m \to X$ is the normalization map; X_m is an absolutely minimal* (in the sense of 5.8) *Calabi-Yau variety with* $\dim X_m \ge 3$.

(3) *There is an almost holomorphic map $\tau : X_m \dashrightarrow Y$ (resp. a holomorphic map $\tau : X_m \to Y$) such that its general fibre F is an absolutely minimal Calabi-Yau variety with $3 \le \dim F < \dim X_m$, and $(g_m)_{|F} : F \to g_m(F) \subset X$ is the normalization map.*

Given a projective variety W, let $\widetilde{W} \xrightarrow{\sigma} W$ be a projective resolution. We define the *albanese variety of* W as $\mathrm{Alb}(W) := \mathrm{Alb}(\widetilde{W})$, which is independent of the choice of the resolution $\widetilde{W}$, since every two resolutions of W are dominated by a third one and the albanese variety, being an abelian variety, contains no rational curves. We define the *albanese (rational) map* $\mathrm{alb}_W : W \dashrightarrow \mathrm{Alb}(W)$ as the composition

$$W \overset{\sigma^{-1}}{\dashrightarrow} \widetilde{W} \xrightarrow{\mathrm{alb}_{\widetilde{W}}} \mathrm{Alb}(\widetilde{W}).$$

Lang's [21, Conjecture 5.6] (one direction only; see 5.8 for the other direction) follows from Conjectures 5.1 and 5.2(2):

Corollary 5.5. (cf. [15]) *Let W be an algebraic Lang hyperbolic projective variety. Assume either* $\dim W \leq 3$ *or Conjecture* 5.2(2). *Then we have:*

(1) *If Conjecture* 5.1 *holds, then W and all its subvarieties are of general type.*

(2) *If the albanese map* $\mathrm{alb}_W : W \dashrightarrow \mathrm{Alb}(W)$ *has general fibres of dimension* ≤ 2, *then W is of general type.*

If we don't assume Conjecture 5.1 in Theorem 5.3, then the proof shows:

Corollary 5.6. (cf. [15]) *Let X be a $\mathbb{Q}$-Gorenstein normal projective variety of dimension n, which is algebraic Lang hyperbolic. Assume either $n \leq 3$ or Conjecture* 5.2(1). *Then either K_X is ample at smooth points and klt points of X as detailed in Theorem* 5.3; *or X is a Calabi-Yau variety of dimension $n \geq 3$; or X is covered by subvarieties $\{F'_t\}$ whose normalizations are Calabi-Yau varieties of the same dimension k with* $3 \leq k < n$.

Without assuming Conjectures 5.2 (and 5.1), we have:

Proposition 5.7. (cf. [15]) *Let W be a projective variety which is algebraic Lang hyperbolic. Then W is of general type provided that either the albanese map* $\mathrm{alb}_W : W \dashrightarrow \mathrm{Alb}(W)$ *is generically finite (but not necessarily surjective); or the Kodaira dimension* $\kappa(W) \geq \dim W - 2$; *or* $\kappa(W) \geq \dim W - 3$ *and Conjecture* 5.1 *holds in dimension three.*

Remark 5.8. (1) In Theorem 5.3, by the equality $f_c^* K_X = K_{X_c} + E_c$ and the ampleness of K_{X_c}, the *exceptional locus* $\mathrm{Exc}(f_c)$ (the subset of X_c along which f_c is not isomorphic) is contained in $\mathrm{Supp}\, E_c$. This and the effectivity of E_c justifies the phrasal: K_{X_c} is ample outside $f_c(E_c)$.

(2) Without assuming Conjecture 5.1 or 5.2, the proof of Theorem 5.3 shows that $f^* K_X = K_{X'} + E'$ with $K_{X'}$ nef and E' f-exceptional. Hence $K_X = f_* K_{X'}$ is movable, or nef in codimension-one.

(3) Let $X_2 \to X_1$ be a finite morphism (but not necessarily surjective). If X_1 is Brody hyperbolic or algebraic Lang hyperbolic then so is X_2. The converse is not true.

(4) Every algebraic Lang hyperbolic projective variety X_1 is *absolutely minimal* in the sense: if X_2 is a normal projective variety with at worst klt singularities then every birational map $h : X_2 \dashrightarrow X_1$ is a well defined morphism. (This result was proved by S. Kobayashi when X_2 is nonsingular). Indeed, let Z be a resolution of the graph of h such that we have birational surjective morphisms $p_i : Z \to X_i$ satisfying $h \circ p_2 = p_1$. Then every fibre $p_2^{-1}(x_2)$ is rationally chain connected by [12, Corollary 1.5] and hence $p_1(p_2^{-1}(x_2))$ is a point since hyperbolic X_1 contains no rational curve. Thus h is extended to a well defined morphism by [16, Proof of Lemma 14], noting that X_2 is normal and p_2 is surjective, and using the Stein factorization.

(5) If Y is an algebraic Lang hyperbolic Calabi-Yau variety (like X_m and F in 5.4 (2) and (3), respectively), then every birational morphism $h : Y \to Z$ onto a normal projective variety is an isomorphism. Indeed, by [18, Corollary 1.5], Z has only canonical singularities. Thus the exceptional locus $\mathrm{Exc}(h)$ is covered

by rational curves by [12, Corollary 1.5]. Since Y is hyperbolic and hence has no rational curve, we have $\mathrm{Exc}(h) = \emptyset$ and hence h is an isomorphism, Z being normal and by the Stein factorization.

(6) About the converse of Corollary 5.5, i.e., another direction of Lang [21, Conjecture 5.6]: Assume that every non-uniruled projective variety has a minimal model with at worst canonical singularities and that abundance Conjecture 5.2(2) holds. If a projective variety W and all its subvarieties are of general type, then W is algebraic Lang hyperbolic. Indeed, let $f : V \to W$ be a morphism from an abelian variety V and let $V \to X \to f(V)$ be its Stein factorization, where $V \to X$ has connected general fibre F and $X \to f(V)$ is a finite morphism. Since V is non-uniruled, so is F. Hence $\kappa(F) \geq 0$ by the assumption. The assumption and Iitaka's $C_{n,m}$ also imply that

$$0 = \kappa(V) \geq \kappa(F) + \kappa(X) \geq \kappa(X) \geq \kappa(f(V)) = \dim f(V)$$

(cf. [17, Corollary 1.2]). Hence f is a constant map.

5.9. *Comments about the proofs:* In our proofs, neither the existence of minimal model nor the termination of MMP is assumed. To show that every subvariety X of a hyperbolic W is of general type, one key observation is that a relative minimal model of X birational over X which exists by the main Theorem 1.2 in [3], *is indeed a minimal model of* X since W is hyperbolic. One natural approach is to take a general fibre F (which may not even be normal) of an Iitaka (rational) fibration of X (assuming $\kappa(X) \geq 0$) and prove that F has a minimal model F_m. Next, one tries to show that $q(F_m) = 0$ and F_m is a Calabi-Yau variety and then tries to use Conjecture 5.1 to produce a non-hyperbolic subvariety S of F_m, but this does not guarantee the same on $F \subset X$ (to contradict the hyperbolicity of X) because such $S \subseteq F_m$ might be contracted to a point on F. In our approach, we are able to show that the *normalization of F is a Calabi-Yau variety, which is the key of the proofs.* It wouldn't help even one assumes the smoothness of the ambient space W since its subvariety X may not be smooth, or at least normal or Cohen-Macaulay to define the canonical divisor K_X meaningfully to pull back or push forward.

References

[1] T. Bauer, F. Campana, T. Eckl, S. Kebekus, T. Peternell, S. Rams, T. Szemberg and L. Wotzlaw, A reduction map for nef line bundles, Complex Geometry (Göttingen, 2000), 27–36, Springer, 2002.

[2] A. Beauville, Some remarks on Kähler manifolds with $c_1 = 0$, Classification of Algebraic and Analytic Manifolds (Katata, 1982, ed. K. Ueno), Progr. Math., **39**, Birkhäuser, 1983, pp. 1–26.

[3] C. Birkar, P. Cascini, C. D. Hacon and J. McKernan, Existence of minimal models for varieties of log general type, J. Amer. Math. Soc. **23** (2010), 405–468.

[4] F. Campana, F. Wang and D. -Q. Zhang, Automorphism groups of positive entropy on projective threefolds, Trans. Amer. Math. Soc. **366** (2014), no. 3, 1621–1638. arXiv: 1203.5665.

[5] S. Cantat, Endomorphismes des variétes homogènes, Enseign. Math. **49** (2004), 237–262.

[6] S. Cantat, Quelques aspects des systèmes dynamiques polynomiaux: existence, exemples et rigidité (version préliminaire); Panorama et Synthèse, **30**, à paraitre.

[7] T. -C. Dinh, Tits alternative for automorphism groups of compact Kähler manifolds, arXiv:1210.3681.

[8] T. -C. Dinh and N. Sibony, Groupes commutatifs d'automorphismes d'une variété kählerienne compacte, Duke Math. J. **123** (2004), 311–328.

[9] B. Fu and D. -Q. Zhang, A characterization of compact complex tori via automorphism groups, Math. Ann. **357** (2013), no. 3, 961–968.

[10] A. Fujiki, On automorphism groups of compact Kähler manifolds, Invent. Math. **44** (1978), 225–258.

[11] O. Fujino, Fundamental theorems for the log minimal model program, Publ. Res. Inst. Math. Sci. **47** (2011), no. 3, 727–789.

[12] C. D. Hacon and J. McKernan, On Shokurov's rational connectedness conjecture, Duke Math. J. **138** (2007), no. 1, 119–136.

[13] B. Harbourne, Rational surfaces with infinite automorphism group and no antipluricanonical curve, Proc. Amer. Math. Soc. **99** (1987), no. 3, 409–414.

[14] G. Heier, S. S. Y. Lu and B. Wong, On the canonical line bundle and negative holomorphic sectional curvature, Math. Res. Lett. **17** (2010), no. 6, 1101–1110.

[15] F. Hu, S. Meng and D. -Q. Zhang, Ampleness of canonical divisors of hyperbolic normal projective varieties, Math. Z. **278** (2014), no. 3-4, 1179–1193.

[16] Y. Kawamata, Characterization of abelian varieties, Compositio Math. **43** (1981), no. 2, 253–276.

[17] Y. Kawamata, Minimal models and the Kodaira dimension of algebraic fiber spaces, J. Reine Angew. Math. **363** (1985), 1–46.

[18] Y. Kawamata, Crepant blowing-ups of 3-dimensional canonical singularities and its application to degeneration of surfaces, Ann. of Math. **127** (1988), 93–163.

[19] S. Keel and J. McKernan, Rational curves on quasi-projective surfaces, Mem. Amer. Math. Soc. **140** (1999), no. 669, 81–83.

[20] J. Kollár and S. Mori, Birational geometry of algebraic varieties, Cambridge Tracts in Math., **134**, Cambridge University Press, 1998.

[21] S. Lang, Hyperbolic and Diophantine analysis, Bull. Amer. Math. Soc. (N.S.) **14** (1986), 159–205.

[22] D. I. Lieberman, Compactness of the Chow scheme: applications to automorphisms and deformations of Kähler manifolds, Fonctions de plusieurs variables complexes, III (Sém. François Norguet, 1975–1977), pp. 140–186, Lecture Notes in Math., **670**, Springer, 1978.

[23] S. S. Y. Lu and D. -Q. Zhang, Positivity criteria for log canonical divisors and hyperbolicity, J. Reine Angew. Math. (to appear), 2015, DOI:10.1515/crelle-2015-0013, arXiv:1207.7346.

[24] M. McQuillan and G. Pacienza, Remarks about bubbles, arXiv:1211.0203.

[25] S. Mori and S. Mukai, The uniruledness of the moduli space of curves of genus 11, Algebraic Geometry (Tokyo/Kyoto, 1982), 334–353, Lecture Notes in Math., **1016**, Springer, Berlin, 1983.

[26] K. Oguiso and T. T. Truong, Explicit examples of rational and Calabi-Yau threefolds with primitive automorphisms of positive entropy, arXiv:1306.1590.

[27] T. Peternell, Calabi-Yau manifolds and a conjecture of Kobayashi, Math. Z. **207** (1991), no. 2, 305–318.

[28] N. I. Shepherd-Barron and P. M. H. Wilson, Singular threefolds with numerically trivial first and second Chern classes, J. Alg. Geom. **3** (1994), 265–281.

[29] D. -Q. Zhang, Dynamics of automorphisms of compact complex manifolds, Proceedings of the Fourth International Congress of Chinese Mathematicians (ICCM2007), Section: Algebraic Geometry, 17–22 December 2007, HangZhou, China, ICCM 2007, Vol II, pp. 678–689, arXiv:0801.0843.

[30] D. -Q. Zhang, A theorem of Tits type for compact Kähler manifolds, Invent. Math. **176** (2009), 449–459.

[31] D. -Q. Zhang, Automorphism groups of positive entropy on minimal projective varieties, Adv. Math. **225** (2010), no. 5, 2332–2340.

[32] D. -Q. Zhang, Algebraic varieties with automorphism groups of maximal rank, Math. Ann. **355** (2013), no. 1, 131–146.

[33] D. -Q. Zhang, Compact Kähler manifolds with automorphism groups of maximal rank, Trans. Amer. Math. Soc. **366** (2014), no. 7, 3675–3692.

Department of Mathematics, National University of Singapore
10 Lower Kent Ridge Road, Singapore 119076
E-mail address: `matzdq@nus.edu.sg`

Proceedings of the 6th International Congress of Chinese Mathematicians
ALM 36, pp. 541–555

Mumford Compactification of Siegel Variety from the Viewpoint of Kähler-Einstein Metric

Yi Zhang

Abstract

This article is a survey on recent progress of the author's joint program with Professor S.-T. Yau. We study Mumford smooth toroidal compactifications of smooth Siegel varieties thoroughly from the viewpoint of Kähler-Einstein metric. We obtain a global volume form formula for an arbitrary Siegel variety $\mathcal{A}_{g,\Gamma}(g > 1)$ with smooth toroidal compactification $\overline{\mathcal{A}}_{g,\Gamma}$ having normal crossing boundary divisor $D_\infty := \overline{\mathcal{A}}_{g,\Gamma} \setminus \mathcal{A}_{g,\Gamma}$, we use this volume form formula to show that the logarithmical canonical bundle degenerate sharply even though it is big and numerically effective.

Mathematics Subject Classification: 14J15.

Keywords and Phrases: Moduli of Abelian varieties, Siegel variety, smooth toroidal compactification, volume form.

When we begin to talk about Kähler-Einstein metrics on noncompact manifolds, we must go back to Yau's ICCM talk in Helsinki in 1978(cf.[Yau80]). After the calabi conjecture was solved by Yau, Cheng and Yau developed the Kähler Einstein manifolds with negative scalar curvature(cf.[CY80]). There were two types of noncompact Kähler Einstein manifolds that they worked with: One class is those with infinite volume and in most cases, they are Stein manifolds. They proved, in particular, that there is a complete Kähler Einstein metric on any Stein manifold which supports a volume with strongly negative Ricci tensor. This includes any bounded pseudoconvex domain. The second class of manifolds are those quasi-projective manifolds. Hence this is a Zariski open set M of some algebraic manifold $\overline{M}$, the complement of some divisor D on $\overline{M}$.

The motivation of our study is to find conditions on the canonical line bundle $K_{\overline{M}}$ and D to make sure that one can construct such a metric with reasonable detail informations on the metric. If $K_{\overline{M}} + D$ is positive and D is normal crossing, this is rather straight forward, and was described in Yau's congress talk in Helsinki[Yau80] and some interesting results were obtained in [TY86] and [Yau93]. However, in most cases of geometry, $K_{\overline{M}} + D$ is big and nef, the most interesting class of quasi-projective manifold is the quotient of Hermitian symmetric domain whose Bergman metric is Kähler-Einstein. There are Mumford toroidal compactifications of the manifold which gives rise to normal crossing. Therefore, we focus on the model of Siegel variety, because it is very important and interesting in both algebraic geometry and number theory, also because it arises as a certain moduli space of Abelian varieties, the purpose of our research is to study smooth toroidal compactifications of Siegel varieties and their applications. We try to understand the Kähler-Einstein metrics on Siegel varieties through the compactifications, and we finally show that $K_{\overline{M}} + D$ is never positive on $\overline{M}$ in case of Mumford compactification of Siegel variety.

On the other hand, by private communication Yau has told the author that he believes that the intrinsic Kähler-Einstein metric on a quasi-projective manifold M should be helpful for finding a nice compactification $\overline{M}$ of M and he conjectures that there should be some correspondence between nice compactifications and Kähler-Einstein metric of a quasi-projective manifold, and he has thought this problem for a long time. We realize this thought of Yau's partially for Siegel variety, actually we show that any Mumford compactification is totally determined by the Bergman metric on Siegel variety.

Let g be a fixed positive integer more than two. The Siegel space of genus g

$$\mathfrak{H}_g := \{\tau = X + \sqrt{-1}Y \in \mathrm{Sym}_{g\times g}(\mathbb{C}) \mid Y > 0\}$$

is a homogenous space($\cong \mathrm{Sp}(g,\mathbb{R})/\mathrm{U}(g)$), also can be realized as a bounded domain

$$\begin{array}{ccl} \mathfrak{H}_g & \xrightarrow{\cong} & \{F^1 \in \mathrm{Grass}(g, \mathbb{C}^{2g}) \mid \psi(F^1, F^1) = 0, \sqrt{-1}\psi(F^1, \overline{F^1}) > 0\}, \\ \tau & \longmapsto & F^1_\tau \end{array}$$

where F^1_τ is the subspace of $\mathbb{C}^{2g}$ spanned by the column vectors of the matrix $\begin{pmatrix} \tau \\ I_g \end{pmatrix}$ and $\psi = \begin{pmatrix} 0 & -I_g \\ I_g & 0 \end{pmatrix}$ is a symplectic bilinear form on $\mathbb{C}^{2g}$. Thus, the compact dual $\mathfrak{H}_g$ is

$$\check{\mathfrak{H}}_g := \{F^1 \in \mathrm{Grass}(g, \mathbb{C}^{2g}) \mid \psi(F^1, F^1) = 0\}.$$

We note that $F^\bullet_\tau := (0 \subset F^1_\tau \subset \mathbb{C}^{2g})$ is a Hodge filtration for any τ in $\check{\mathfrak{H}}_g$.

1 Cusps of the Siegel space $\mathfrak{H}_g$ from the viewpoint of mixed Hodge theory

Beside the compact dual $\check{\mathfrak{H}}_g$, the $\mathfrak{H}_g$ has a nature small closure

$$\overline{\mathfrak{H}_g} = \{F^1 \in \mathrm{Grass}(g, \mathbb{C}^{2g}) \mid \psi(F^1, F^1) = 0, \sqrt{-1}\psi(F^1, \overline{F^1}) \geq 0\}.$$

The boundary $\partial\mathfrak{H}_g := \overline{\mathfrak{H}_g}\backslash\mathfrak{H}_g$ is in fact a disjoint union of all boundary components with the following form of $\{F^1 \in \overline{\mathfrak{H}_g} \mid F^1 \cap \overline{F^1} = W_{\mathbb{R}} \otimes \mathbb{C}\}$ such that $W_{\mathbb{R}}$ is an isotopic rational subspace of $V_{\mathbb{R}} := V_{\mathbb{Q}} \otimes \mathbb{R}$ and $V_{\mathbb{Q}} := \mathrm{span}_{\mathbb{Q}}\{e_1, \dots, e_g, e_{g+1}, \dots, e_{2g}\}$ is a rational vector space spanned by the standard symplectic base $\{e_i\}_{i=1}^{2g}$ of $\mathbb{C}^{2g}$.

As the set of rational numbers is dense in $\mathbb{R}$, the most important boundary components are rational. A cusp (or rational boundary component) of the Siegel space $\mathfrak{H}_g$ is a subset in $\partial\mathfrak{H}_g$ defined by $\mathfrak{F}(W_{\mathbb{Q}}) := \{F^1 \in \overline{\mathfrak{H}_g} \mid F^1 \cap \overline{F^1} = W_{\mathbb{Q}} \otimes \mathbb{C}\}$ where $W_{\mathbb{Q}}$ is an isotopic rational subspace of $V_{\mathbb{Q}}$. Any cusp can be transformed into a standard one. Let $\mathfrak{F}(W_{\mathbb{Q}})$ be any cusp with $\dim_{\mathbb{Q}} W_{\mathbb{Q}} = g - k (0 \le k \le g)$. There always exists a $M \in \mathrm{Sp}(g, \mathbb{Z})$ such that $M(W_{\mathbb{Q}}) = V_{\mathbb{Q}}^{(k)} := \mathrm{span}_{\mathbb{Q}}\{e_{k+1}, \dots, e_g\}$, and that cusp isomorphic to a Siegel space of low genus:

$$\begin{aligned} M(\mathfrak{F}(W_{\mathbb{Q}})) &:= \mathfrak{F}(M(W_{\mathbb{Q}})) \\ &= \mathfrak{F}(V_{\mathbb{Q}}^{(k)}) \\ &\cong \text{genus } k \text{ Siegel space } \mathfrak{H}_k. \end{aligned}$$

There are some algebraic groups over $\mathbb{Q}$ given by any cusp $\mathfrak{F}$

$$\begin{aligned} \mathcal{N}^{\mathfrak{F}} &:= \{g \in \mathrm{Sp}(g, \mathbb{R}) \mid g\mathfrak{F} = \mathfrak{F}\}, \\ \mathcal{W}^{\mathfrak{F}} &:= \text{the unipotent radical of } N^{\mathfrak{F}}, \\ U^{\mathfrak{F}} &:= \text{the center of } W^{\mathfrak{F}} \end{aligned}$$

and for each $N \in u^{\mathfrak{F}}$ has $N^2 = 0$ we have an identification

$$\exp : u^{\mathfrak{F}} \xrightarrow{\cong} U^{\mathfrak{F}}, \ N \mapsto I + N$$

where $u^{\mathfrak{F}} = \mathrm{Lie}(U^{\mathfrak{F}})$ is the commutative Lie algebra of $U^{\mathfrak{F}}$, therefore the $U^{\mathfrak{F}}$ can be regarded as a rational linear space; there is a positive cone in $U^{\mathfrak{F}}$:

$$C(\mathfrak{F}) := \left\{ N \in u^{\mathfrak{F}(W_{\mathbb{Q}})} \;\middle|\; \psi(\cdot, N(\cdot)) > 0 \text{ on } \frac{V_{\mathbb{Q}}}{W_{\mathbb{Q}}^{\perp}} \right\}$$

where $W_{\mathbb{Q}}^{\perp} := \{v \in V \mid \psi(W_{\mathbb{Q}}, v) = 0\}$ is the perpendicular and $\psi(\cdot, N(\cdot))$ is the induced bilinear symmetric real form on the quotient space $\frac{V_{\mathbb{Q}}}{W_{\mathbb{Q}}^{\perp}}$.

Example 1.1. Define $\mathfrak{F}_k := \mathfrak{F}(V^{(k)})$ for $V_{\mathbb{Q}}^{(k)} := \mathrm{span}_{\mathbb{Q}}\{e_{k+1}, \dots, e_g\}$.

$$u^{\mathfrak{F}_k} = \left\{ \begin{pmatrix} 0_k & 0 & 0 & 0 \\ 0 & 0_{g-k} & 0 & u \\ 0 & 0 & 0_k & 0 \\ 0 & 0 & 0 & 0_{g-k} \end{pmatrix} \;\middle|\; u \in \mathrm{Sym}_{g-k}(\mathbb{R}) \right\} \cong \mathrm{Sym}_{g-k}(\mathbb{R}),$$

$$C(\mathfrak{F}_k) = \left\{ \begin{pmatrix} 0_k & 0 & 0 & 0 \\ 0 & 0_{g-k} & 0 & u \\ 0 & 0 & 0_k & 0 \\ 0 & 0 & 0 & 0_{g-k} \end{pmatrix} \in u^{\mathfrak{F}_k} \;\middle|\; u > 0 \right\} \cong \mathrm{Sym}^{+}_{g-k}(\mathbb{R}).$$

Denote the dual cusp of a cusp $\mathfrak{F}(W_{\mathbb{Q}})$ by

$$\mathfrak{F}^{\vee} := \mathfrak{F}(W_{\mathbb{Q}}^{\vee})$$

where $W_{\mathbb{Q}}^{\vee}$ is the dual space of $W_{\mathbb{Q}}$ under $(V_{\mathbb{Q}}, \psi)$ so that $W_{\mathbb{Q}}^{\vee}$ is also an isotropic rational subspace of $V_{\mathbb{Q}}$ with same dimension of $W_{\mathbb{Q}}$. For example, if $W_{\mathbb{Q}} = V_{\mathbb{Q}}^{(k)}$ then $W_{\mathbb{Q}}^{\vee} = \mathrm{span}_{\mathbb{Q}}\{e_{g+k+1}, \ldots, e_{2g}\}$.

For a cusp $\mathfrak{F}(W_{\mathbb{Q}})$ and an arbitrary N in the positive cone $C(\mathfrak{F})$, there are two weight filtrations: A $\mathfrak{F}$-cusp weight filtration on $V_{\mathbb{Q}}$ given by $W_{\bullet}^{\mathfrak{F}} := (0 \subset W_0^{\mathfrak{F}}(:= W_{\mathbb{Q}}) \subset W_1^{\mathfrak{F}}(:= W_{\mathbb{Q}}^{\perp}) \subset W_2^{\mathfrak{F}}(:= V_{\mathbb{Q}}))$ and a weight filtration $W_{\bullet}(N) = (0 \subset W_{-1}(N) \subset W_0(N) \subset W_1(N)$ on $V_{\mathbb{R}} := V_{\mathbb{Q}} \otimes \mathbb{R}$ by setting $W_1(N) := V_{\mathbb{R}}, W_{-1}(N) := \mathrm{Im}(N : V_{\mathbb{R}} \to V_{\mathbb{R}})$ and $W_0(N) := \mathrm{Ker}(N : V_{\mathbb{R}} \to V_{\mathbb{R}})$. Actually these two filtrations are matched by shifting some weight. Moreover, any cusp of Siegel space $\mathfrak{H}_g$ can be identified with a set of certain weight one polarized mixed Hodge structures, this topic is related to the degenerate limits of Abelian varieties.

Theorem 1.2. ([YZ12]) *Let $\mathfrak{F}$ be a cusp of Siegel space $\mathfrak{H}_g$. Then, we have:*

1. *For any Hodge filtration $F^{\bullet} \in \mathfrak{F}^{\vee}$ and any $N \in C(\mathfrak{F})$, the pair $(F^{\bullet}, W_{\bullet}^{\mathfrak{F}})$, together with N and the standard symplectic form ψ on $V_{\mathbb{R}}$, determines a polarized mixed Hodge structure of weight one on $V_{\mathbb{R}}$.*

2. *Any pair $(F_{\tau}^{\bullet}, W_{\bullet}^{\mathfrak{F}})$ with $F_{\tau}^{\bullet} \in \mathfrak{H}_g$ is a mixed Hodge structure of weight one on $V_{\mathbb{Q}}$.*

3. *Furthermore,*

$$\mathfrak{F}^{\vee} = \{F^{\bullet} \in \check{\mathfrak{H}}_g \mid (F^{\bullet}, W_{\bullet}^{\mathfrak{F}}, N, \psi) \text{ is a polarized mixed Hodge structure of weight one for all } N \in C(\mathfrak{F})\}.$$

For mixed Hodge theory and polarized mixed Hodge structure, one can refer to [Del71] and [CKS86]. This theorem has some interesting applications in the theory of Siegel modular forms, and also help us to get the unique group-invariant Kähler-Einstein metric on the Siegel space $\mathfrak{H}_g$ is a Hodge metric(cf.[YZ12-2]).

2 Infinity boundary divisors on Mumford compactifications

From now on, we fix an arithmetic neat subgroup $\Gamma \subset \mathrm{Sp}(g, \mathbb{Z})$. For example,we take Γ as

$$\Gamma_g(N) := \{\gamma \in \mathrm{Sp}(g, \mathbb{Z}) \mid \gamma \equiv I_{2g} \mod N\},\ N \geq 3.$$

We obtain a smooth quasi-projective **Siegel variety** $\mathcal{A}_{g,\Gamma} := \Gamma \backslash \mathfrak{H}_g$ of dimension $g(g+1)/2$ and the algebraic group $U^{\mathfrak{F}}$ has a nature integral lattice $U^{\mathfrak{F}}(\mathbb{Q}) \cap \Gamma$.

We now sketch the construction of smooth toroidal compactifications of $\mathcal{A}_{g,\Gamma}$(cf. [AMRT] and [FC]). Let $\mathfrak{F}$ be any cusp of $\mathfrak{H}_g$.

It was shown in [BB66] and [AMRT] that the $N^{\mathfrak{F}}$ is a maximal parabolic subgroup, the Levi-decomposition of $N^{\mathfrak{F}}$ becomes the following semi-product

$$N^{\mathfrak{F}} = \underbrace{(G_h^{\mathfrak{F}} \times G_l^{\mathfrak{F}})}_{\text{direct product}} \cdot \mathcal{W}^{\mathfrak{F}}$$

since $\mathrm{Sp}(g,\mathbb{R})$ is a simple group, and the $G_h^{\mathfrak{F}} \cdot \mathcal{W}^{\mathfrak{F}}$ centralizes $U^{\mathfrak{F}}$. Let $\Gamma_{\mathfrak{F}} := \Gamma \cap N^{\mathfrak{F}}$. We have a group homomorphism $\varphi : \Gamma_{\mathfrak{F}} \to \mathrm{Aut}^0(U^{\mathfrak{F}})$ by sending γ to

$$\overline{\gamma} : U^{\mathfrak{F}} \longrightarrow U^{\mathfrak{F}}, \quad u \longmapsto \gamma u \gamma^{-1}.$$

Then we get an automorphism of $C(\mathfrak{F})$ preserving the lattice $U^{\mathfrak{F}}(\mathbb{Z})$,

$$\overline{\Gamma_{\mathfrak{F}}} := \varphi(\Gamma_{\mathfrak{F}}) = \varphi(\Gamma \cap G_l^{\mathfrak{F}}) \subset \mathrm{Aut}(C(\mathfrak{F})).$$

Example 2.1. For cusp $\mathfrak{F}_k := \mathfrak{F}(V^{(k)})$,

$$G_l^{\mathfrak{F}_k} = \left\{ \left(\begin{array}{cccc} 1_k & 0 & 0 & 0 \\ 0 & g & 0 & 0 \\ 0 & 0 & 1_k & 0 \\ 0 & 0 & 0 & {}^t g^{-1} \end{array} \right) \middle| \; g \in \mathrm{GL}(g-k,\mathbb{R}) \right\}$$
$$\cong \mathrm{GL}(g-k,\mathbb{R}).$$

Thus, the action

$$G_l^{\mathfrak{F}_k}(\mathbb{R}) \times C(\mathfrak{F}_k) \longrightarrow C(\mathfrak{F}_k), \; (M,A) \longmapsto MAM^{-1}$$

is equivalent to the action

$$\mathrm{GL}(g-k,\mathbb{R}) \times \mathrm{Sym}^+_{g-k}(\mathbb{R}) \longrightarrow \mathrm{Sym}^+_{g-k}(\mathbb{R}), \; (g,u) \longmapsto g \cdot u \cdot^t g.$$

Let $\mathfrak{F}_1 = \mathfrak{F}(W_1), \mathfrak{F}_2 = \mathfrak{F}(W_2)$ be two cusps such that $\mathfrak{F}(W_2) \prec \mathfrak{F}(W_1)$(i.e., by definition $W_1 \subset W_2$). Then $U^{\mathfrak{F}_1}$ is a rationally defined subspace of $U^{\mathfrak{F}_2}$ and

$$\overline{C(\mathfrak{F}_1)} = U^{\mathfrak{F}_1} \cap \overline{C(\mathfrak{F}_2)}, \quad U^{\mathfrak{F}_1} \cap C(\mathfrak{F}_2) = \emptyset.$$

We define rational closure $\overline{C(\mathfrak{F})}^{\mathrm{rc}} := \bigcup_{\mathfrak{F} \preceq \mathfrak{F}'} C(\mathfrak{F}')$. There is $\overline{C(\mathfrak{F}_1)}^{\mathrm{rc}} = U^{\mathfrak{F}_1} \cap \overline{C(\mathfrak{F}_2)}^{\mathrm{rc}}$. In [AMRT], there always exists a $\overline{\Gamma_{\mathfrak{F}}}$-admissible polyhedral decomposition of $C(\mathfrak{F})$, i.e., a collection of convex rational polyhedral cones (infinitely many in general) $\Sigma_{\mathfrak{F}} = \{\sigma_\alpha^{\mathfrak{F}}\}_\alpha \subset \overline{C(\mathfrak{F})}$ such that

i. $\overline{C(\mathfrak{F})}^{\mathrm{rc}} = \bigcup_\alpha \sigma_\alpha^{\mathfrak{F}}$,

ii. $\Sigma_{\mathfrak{F}}$ is a fan(i.e., if two polyhedral cones $\delta_1, \delta_2 \in \Sigma_{\mathfrak{F}}$ then $\delta_1 \cap \delta_2 \in \Sigma_{\mathfrak{F}}$),

iii. the group $\overline{\Gamma_{\mathfrak{F}}}$ has an action on $\Sigma_{\mathfrak{F}}$ with finitely many orbits.

One can use the technique of toroidal geometry to construct a partial compactification of $Z_{\mathfrak{F}}$ of $\mathcal{A}_{g,\Gamma}$ in the direction of $\mathfrak{F}$ by the $\overline{\Gamma_{\mathfrak{F}}}$-admissible polyhedral decomposition. The key points in toric geometry are:

- Let M be the dual lattice of $N := U^{\mathfrak{F}}(\mathbb{Q}) \cap \Gamma$. For any σ in the AMRT decomposition Σ, $X_\sigma := \operatorname{Spec}\mathbb{C}[\sigma^\vee \cap M]$ is an affine toric variety where $\sigma^\vee = \{m \in M_{\mathbb{R}} \mid < m, u >\geq 0, \forall u \in \sigma\}$ (In particular, $\mathbb{T}_{\mathfrak{F}} := \operatorname{Spec}[\{0\}^\vee \cap M] = \operatorname{Spec}[M]$ is a torus).

- If τ is a face of σ then X_τ is an open subvariety of X_σ. So we patch together all affine toric varieties and get a toric scheme X_Σ(it is not of finite type in general) and the torus $\mathbb{T}_{\mathfrak{F}}$ acts on X_Σ.

- There is a bijection between the set of cones in Σ and the set of orbits in X_Σ, and there holds $\dim_{\mathbb{R}} \sigma + \dim_{\mathbb{C}} \mathcal{O}^\sigma = \dim_{\mathbb{C}} \mathbb{T}_{\mathfrak{F}}$. Moreover, $\tau \subseteq \sigma$ if and only if $\mathcal{O}^\sigma \subset \overline{\mathcal{O}^\tau}^{\mathrm{cl}}$.

- In particular, each edge $\rho \in \Sigma$ gives a codimension one closed subvariety $D_\rho := \overline{\mathcal{O}^\rho}^{\mathrm{cl}}$ in X_Σ, which actually is a $\mathbb{T}_{\mathfrak{F}}$-invariant prime divisor of X_Σ.

- If σ is a regular cone of top-dimensional (i.e., $\dim_{\mathbb{R}} \sigma = \operatorname{rank} M$ and the set of edges $\{\rho_i\}$ of σ is a base of the lattice N) then $X_\sigma \setminus \bigcup_i D_\rho = \mathbb{T}_{\mathfrak{F}}$.

Assume that $\mathfrak{F} = \mathfrak{F}(W_{\mathbb{Q}})$ has $\dim_{\mathbb{Q}} W_{\mathbb{Q}} = k$. There is an analytic isomorphism(Realization of $\mathfrak{H}_g$ as Siegel domain of the third kind)

$$\begin{aligned}\Phi : u_{\mathbb{C}}^{\mathfrak{F}} \times M_{g-k,k}(\mathbb{C}) \times \mathfrak{F}(W_{\mathbb{Q}}) &\xrightarrow{\cong} D(\mathfrak{F}),\\ (a+\sqrt{-1}b, C, F) &\longmapsto \exp(a+\sqrt{-1}b)\cdot C\cdot \check{F}\end{aligned}$$

with

$$\Phi_{\mathfrak{F}}^{-1}(\mathfrak{H}_g) = (u^{\mathfrak{F}} + \sqrt{-1}C(\mathfrak{F})) \times M_{g-k,k}(\mathbb{C}) \times \mathfrak{F}(W_{\mathbb{Q}})$$

where $D(\mathfrak{F}) := \bigcup_{C\in u^{\mathfrak{F}}} \exp(\sqrt{-1}C)\mathfrak{H}_g \subset \check{\mathfrak{H}}_g$ and $\check{F} \in \mathfrak{F}(W_{\mathbb{Q}}^\vee)$ is the dual of F. The torus $\mathrm{T}_{\mathfrak{F}} := \operatorname{Spec}\mathbb{C}[M]$ is isomorphic analytically to $\frac{U^{\mathfrak{F}}(\mathbb{C})}{\Gamma\cap U^{\mathfrak{F}}(\mathbb{Q})} \cong (\mathbb{C}^\times)^{\dim u^{\mathfrak{F}}}$. There is an analytic isomorphism

$$\frac{D(\mathfrak{F})}{\Gamma\cap U^{\mathfrak{F}}(\mathbb{Q})} \cong \frac{u_{\mathbb{C}}^{\mathfrak{F}}}{\Gamma\cap U^{\mathfrak{F}}(\mathbb{Q})} \times M_{g-k,k}(\mathbb{C}) \times \mathfrak{F} \cong \mathbb{T}_{\mathfrak{F}} \times M_{g-k,k}(\mathbb{C}) \times \mathfrak{F},$$

so that $D(\mathfrak{F})/\Gamma\cap U^{\mathfrak{F}}(\mathbb{Q}) \longrightarrow D(\mathfrak{F})/u_{\mathbb{C}}^{\mathfrak{F}}$ is a principal $\mathbb{T}_{\mathfrak{F}}$-bundle. Replace the fiber $\mathbb{T}_{\mathfrak{F}}$ with X_Σ and do more as in [AMRT] and [YZ12], we finally have the partial compactification of $Z_{\mathfrak{F}}$.

The description with coordinates gives us a geometric picture of the normal crossing boundary divisor: Let σ be a regular cone in $C(\mathfrak{F})$ of top-dimension $k(k+1)/2(= \dim_{\mathbb{R}} u^{\mathfrak{F}})$. Take a $\mathbb{Z}$-basis $\{\zeta_\alpha\}_1^{k(k+1)/2}$ of $\Gamma\cap U^{\mathfrak{F}}(\mathbb{Q})$ such that they generate all edges of σ. Let $\{\delta_\alpha\} \subset \mathrm{Hom}(U^{\mathfrak{F}}(\mathbb{Z}), \mathbb{Z})$ be the dual basis of $\{\zeta_\alpha\}$. Then, we have

$$\begin{array}{ccc}
\mathfrak{H}_g & \xrightarrow{\subset} & \mathbb{C}^{k(k+1)/2} \times M_{k,g-k}(\mathbb{C}) \times \mathfrak{H}_{g-k} \text{ with } (u_\alpha,\, s_i,\, t_j) \\
\downarrow & & \Big\downarrow {\scriptstyle (z_\alpha := \exp((2\pi\sqrt{-1})\delta_\alpha(u)),\, s_i,\, t_j)} \\
\dfrac{\mathfrak{H}_g}{\Gamma \cap U^{\mathfrak{F}}} & \xrightarrow{\subset} & \underbrace{(\mathbb{C}^\times)^{k(k+1)/2}}_{\text{key point for compactification}} \times M_{k,g-k}(\mathbb{C}) \times \mathfrak{H}_{g-k}
\end{array}$$

as $\mathbb{C}^\times + \{z = 0\} = \mathbb{C}$, and the $\frac{\mathfrak{H}_g}{\Gamma \cap U^{\mathfrak{F}}}$ is an analytically étale(local isomorphism) cover of $\mathcal{A}_{g,\Gamma}$.

After partial compactifications along all directions of all cusps, we consider the global compactification given by an admissible decomposition $\Sigma_{\mathfrak{F}_0}$ for the standard minimal cusp $\mathfrak{F}_0 := \mathfrak{F}(\mathrm{span}_{\mathbb{Q}}\{e_1, \ldots, e_g\})$. From a $\overline{\Gamma_{\mathfrak{F}_0}}$-admissible polyhedral decomposition $\Sigma_{\mathfrak{F}_0} := \{\sigma_\alpha^{\mathfrak{F}_0}\}$ of $C(\mathfrak{F}_0)$, we can get(cf. [AMRT] and [FC]) a family $\{\Sigma_{\mathfrak{F}}\}_{\mathfrak{F}}$ such that each $\Sigma_{\mathfrak{F}}$ is a $\overline{\Gamma_{\mathfrak{F}}}$-admissible polyhedral decomposition of $C(\mathfrak{F})$ by the following two steps:

- (Transformation)For a arbitrary minimal cusp $\mathfrak{F}_{\min} = M(\mathfrak{F}_0)$ where $M \in \mathrm{Sp}(g, \mathbb{Z})$, we define

$$\Sigma_{\mathfrak{F}_{\min}} := M(\Sigma_{\mathfrak{F}_0}) = \{M\sigma_\alpha^{\mathfrak{F}_0}M^{-1} \mid \sigma_\alpha^{\mathfrak{F}_0} \in \Sigma_{\mathfrak{F}_0}\}.$$

- (Restriction)For any cusp $\mathfrak{F}$, if $\mathfrak{F}_{\min}$ is a minimal cusp with $\mathfrak{F}_{\min} \prec \mathfrak{F}$ then we define

$$\Sigma_{\mathfrak{F}} := \Sigma_{\mathfrak{F}_{\min}}|_{\overline{C(\mathfrak{F})}} = \{\sigma_\alpha^{\mathfrak{F}_{\min}} \cap \overline{C(\mathfrak{F})} \mid \sigma_\alpha^{\mathfrak{F}_{\min}} \in \Sigma_{\mathfrak{F}_{\min}}\}.$$

By [AMRT], we have following relations among partial compactfications:

a) For any cusp $\mathfrak{F}$ and any element $\gamma \in \Gamma$, there is an analytic isomorphism $\Pi_{\mathfrak{F},\gamma\mathfrak{F}} : Z_{\mathfrak{F}} \xrightarrow{\cong} Z_{\gamma\mathfrak{F}}$.

b) For any two cusp $\mathfrak{F}_1 \prec \mathfrak{F}_2$, there is an local isomorphism $\Pi_{\mathfrak{F}_2,\mathfrak{F}_1} : Z_{\mathfrak{F}_2} \longrightarrow Z_{\mathfrak{F}_1}$.

Define an equivalent relation R on $\widetilde{\mathfrak{H}_g/\Gamma} := \overset{\circ}{\bigsqcup_{\mathfrak{F}}} Z_{\mathfrak{F}}$(disjoint union) as follows: $x \sim^R y$ for $x \in Z_{\mathfrak{F}_1}, y \in Z_{\mathfrak{F}_2}$ if and only if there are a cusp $\mathfrak{F}_3$ and a $\gamma \in \Gamma$ such that

- $\mathfrak{F}_1 \preceq \mathfrak{F}_3$ and $\gamma\mathfrak{F}_2 \preceq \mathfrak{F}_3$,
- there is a point $z \in Z_{\mathfrak{F}_3}$ with

$$\begin{array}{ccc}
 & (Z_{\mathfrak{F}_3}, z) & \\
{\scriptstyle \Pi_{\mathfrak{F}_3,\mathfrak{F}_1}} \swarrow & & \searrow {\scriptstyle \Pi_{\mathfrak{F}_3,\gamma\mathfrak{F}_2}} \\
(Z_{\mathfrak{F}_1}, x) & & (Z_{\gamma\mathfrak{F}_2}, \Pi_{\mathfrak{F}_2,\gamma\mathfrak{F}_2}(y))
\end{array}$$

Then we glue all partial compactifications $Z_{\mathfrak{F}}$ to get a smooth global compactification

$$\overline{\mathcal{A}}_{g,\Gamma} := \frac{\widetilde{\mathfrak{H}_g/\Gamma}}{\sim_R}$$

of $\mathcal{A}_{g,\Gamma}$, which is a compact Hausdroff analytic variety. A theorem of Tai in [AMRT] says that one can refine the decomposition $\Sigma_{\mathfrak{F}_0}$ to get a projective toroidal compactification $\overline{\mathcal{A}}_{g,\Gamma}$. One can take a sufficiently fine decomposition of $C(\mathfrak{F}_0)$ so that $\overline{\mathcal{A}}_{g,\Gamma}$ is projective and the boundary divisor $D_\infty := \overline{\mathcal{A}}_{g,\Gamma} \setminus \mathcal{A}_{g,\Gamma}$ is normal crossing(cf. [AMRT] and [Mum77]).

But, there a question left:

Question 2.2. How to get the infinity divisor D_∞ normal crossing?

Now we assume AMRT's decomposition for $\mathfrak{F}_0$ is regular (any cone in $\Sigma_{\mathfrak{F}_0}$ is regular with respect the lattice). For any cusp $\mathfrak{F}$, the induced decomposition $\Sigma_{\mathfrak{F}}$ is then regular and we have:

- the projection $\pi_{\mathfrak{F}} : Z_{\mathfrak{F}} \longrightarrow \overline{\mathcal{A}}_{g,\Gamma}$ is analytic étale.
- $Z_{\mathfrak{F}}$ is smooth and $B_{\mathfrak{F}} := Z_{\mathfrak{F}} \setminus \frac{\mathfrak{H}_g}{\Gamma \cap U^{\mathfrak{F}}}$ is a normal crossing divisor.
- But some irreducible components of $\pi_{\mathfrak{F}}(B_{\mathfrak{F}})$ may have self-intersections in $\overline{\mathcal{A}}_{g,\Gamma}$ in general.

We introduce a separable condition: A $\overline{\Gamma_{\mathfrak{F}}}$-admissible polyhedral regular decomposition $\Sigma_{\mathfrak{F}}$ of $C(\mathfrak{F})$ is **Γ-separable** if a $\gamma \in \overline{\Gamma_{\mathfrak{F}}}$ satisfies $\gamma(\sigma) \cap \sigma \neq \{0\}$ for some cone $\sigma \in \Sigma_{\mathfrak{F}}$ then γ acts as the identity on the cone σ.

Theorem 2.3. ([YZ12]) *The boundary divisor $D_\infty := \overline{\mathcal{A}}_{g,\Gamma} \setminus \mathcal{A}_{g,\Gamma}$ is normal crossing if and only if the AMRT's regular decomposition $\Sigma_{\mathfrak{F}_0}$ is Γ-separable.*

Here is a remark: For a Mumford toroidal compactification $\overline{\mathcal{A}}_{g,\Gamma}$ with normal crossing infinity divisor $D_\infty := \overline{\mathcal{A}}_{g,\Gamma} \setminus \mathcal{A}_{g,\Gamma}$, we find that if $D_1, \ldots, D_N (N = \dim_{\mathbb{C}} \mathcal{A}_{g,\Gamma})$ is some irreducible components of the normal crossing infinity divisor $D_\infty :=$ such that $\bigcap_{i=1}^{N} D_i \neq \emptyset$ then the divisors D_i's correspond to edges of a certain top-dimensional polyhedral cone σ in $\Sigma_{\mathfrak{F}_0}$.

3 Volume form on Siegel variety

We take coordinate system $\tau = (\tau_{ij}) \in \mathfrak{H}_g$ of the Siegel space $\mathfrak{H}_g = \{\tau = \mathrm{Re}(\tau) + \sqrt{-1}\mathrm{Im}(\tau) \in \mathrm{Sym}_g(\mathbb{C}) \mid \tau =^t \tau, \mathrm{Im}(\tau) > 0\}$. The unique group-invariant Kähler Einstein metric on $\mathfrak{H}_g$ is

$$ds^2 = \mathrm{Tr}(d\tau \mathrm{Im}(\tau)^{-1} d\overline{\tau} \mathrm{Im}(\tau)^{-1})$$

with Kähler form ω_{KE}, and the group-invariant volume form is

$$\left(\frac{\sqrt{-1}}{2}\right)^{\frac{g(g+1)}{2}} \frac{2^{\frac{g(g-1)}{2}} dV_g}{(\det \mathrm{Im}(\tau))^{g+1}},$$

where $dV_g(\tau) := (\bigwedge_{1\leq i\leq j\leq g} d\tau_{ij}) \wedge (\bigwedge_{1\leq i\leq j\leq g} d\overline{\tau_{ij}})$ is the standard Euclidian volume form. The Kähler-Einstein equation is

$$\sqrt{-1}\partial\overline{\partial}\log(\text{Volume form}) = \omega_{\text{KE}}.$$

Let $N = g(g+1)/2$. Let $\Sigma_{\mathfrak{F}_0}$ be any decomposition of $C(\mathfrak{F}_0)$ such that the boundary divisor $D_\infty = \overline{\mathcal{A}}_{g,\Gamma} \setminus \mathcal{A}_{g,\Gamma}$ is normal crossing. By the construction, the most essential object in toroidal compactification is $Z_{\mathfrak{F}_0}$, the partial compactification in the direct of the minimal cusp $\mathfrak{F}_0$. We focus on the model associate to a top-dimensional polyhedral cone $\sigma_{\max}$ (i.e., $\dim_{\mathbb{R}} \sigma_{\max} = N$) in the decomposition $\Sigma_{\mathfrak{F}_0}$, and use it to give a global coordinate chart of $\mathcal{A}_{g,\Gamma}$. Take a $\mathbb{Z}$-basis $\{\zeta_\alpha\}_1^n$ of $\Gamma \cap U^{\mathfrak{F}_0}$ such that $\zeta_1, \ldots, \zeta_n$ generate all edges of the cone $\sigma_{\max}$. Let $\{\delta_\alpha\}$ be the dual basis of $\{\zeta_\alpha\}$. Then, we have an open embedding $\lambda : \mathfrak{H}_g \xrightarrow{\hookrightarrow} U^{\mathfrak{F}_0}(\mathbb{C}) \xrightarrow{\cong} \mathbb{C}^N$. Consider the exponential

$$\begin{array}{ccccc} \mathfrak{H}_g & \xrightarrow{\subset} & \mathbb{C}^N & & \\ \downarrow & & \downarrow{\scriptstyle (w_\alpha := \exp((2\pi\sqrt{-1})\delta_\alpha(u)))} & & \\ \mathfrak{H}_g/\Gamma\cap U^{\mathfrak{F}} & \xrightarrow{\subset} & (\mathbb{C}^\times)^N & \xrightarrow{\subset} & \mathbb{C}^N \end{array}$$

$$\widetilde{\Delta}_{\sigma_{\max}} := \left\{ x = (w_1, \ldots, w_N) \in \mathbb{C}^N \;\middle|\; \begin{array}{l}\text{there exists a neighborhood} \\ \Delta_x \text{ of } x \text{ such that } \Delta_x \cap (\mathbb{C}^\times)^n \subset \dfrac{\mathfrak{H}_g}{\Gamma \cap U^{\mathfrak{F}_0}}\end{array} \right\}.$$

We have

$$\bigcup_{\beta=1}^{n} \{(w_1, \ldots, w_N) \in \widetilde{\Delta}_{\sigma_{\max}} \mid w_\beta = 0\} \xrightarrow{\subset} (\widetilde{\Delta}_{\sigma_{\max}}, (w_1, \ldots, w_N)).$$

The $\widetilde{\Delta}_{\sigma_{\max}}$ is open dense in $Z_{\mathfrak{F}_0}$ and there is a commutative diagram of morphisms

$$\begin{array}{ccc} \frac{\mathfrak{H}_g}{\Gamma\cap U^{\mathfrak{F}_0}} & \xrightarrow{\subset} & \widetilde{\Delta}_{\sigma_{\max}} \\ {\scriptstyle\pi}\downarrow & & {\scriptstyle\pi_{\mathfrak{F}_0}}\downarrow \text{étale} \\ \mathcal{A}_{g,\Gamma} & \xrightarrow{\subset} & \overline{\mathcal{A}}_{g,\Gamma} \end{array}$$

Define $W_{\mathfrak{F}_{\sigma_{\max}}} := \pi_{\mathfrak{F}_0}(\widetilde{\Delta}_{\sigma_{\max}})$. Since D_∞ is normal crossing, The $(W_{\sigma_{\max}}, (w_1, \ldots, w_n))$ becomes a coordinate chart of $\overline{\mathcal{A}}_{g,\Gamma}$. We have

$$W^*_{\sigma_{\max}} := W_{\sigma_{\max}} \setminus \bigcup_{i=1}^{n} \{w_i = 0\} = W_{\sigma_{\max}} \setminus D_\infty = \mathcal{A}_{g,\Gamma}.$$

By this coordinate chart, the Φ has a new representation associated to one arbitrary top-dimensional polyhedral lattice cone in the positive convex cone $C(\mathfrak{F}_0)$:

Theorem 3.1. ([YZ12]) *Let $\Sigma_{\mathfrak{F}_0}$ be any decomposition of $C(\mathfrak{F}_0)$ such that the corresponding Mumford toroidal compactification $\overline{\mathcal{A}}_{g,\Gamma}$ of $\mathcal{A}_{g,\Gamma}$ has normal crossing boundary divisor $D_\infty = \overline{\mathcal{A}}_{g,\Gamma} \setminus \mathcal{A}_{g,\Gamma}$. Let σ be an arbitrary top-dimensional polyhedral cone in $\Sigma_{\mathfrak{F}_0}$ and Let $D_1, \ldots, D_N (N = \dim_{\mathbb{C}} \mathcal{A}_{g,\Gamma})$ be some different irreducible components of D_∞ corresponding to edges of σ. Then the volume $\Phi_{g,\Gamma}$ on $\mathcal{A}_{g,\Gamma}$ can be represented by*

$$\Phi_{g,\Gamma} = \frac{d\mathcal{V}_g}{(\prod_{j=1}^{N} ||s_i||_i^2) F_\sigma^{g+1}(\log ||s_1||_1, \ldots, \log ||s_N||_N)},$$

where

- *$d\mathcal{V}_g$ is a continuous volume form on a partial compactification $\mathcal{U}_{\sigma_{\max}}$ of $\mathcal{A}_{g,\Gamma}$ with $\mathcal{A}_{g,\Gamma} \subset \mathcal{U}_{\sigma_{\max}} \subset \overline{\mathcal{A}}_{g,\Gamma}$,*
- *the $|| \cdot ||_i$ is a suitable Hermitian metric of the line bundle $[D_i]$ on $\overline{\mathcal{A}}_{g,n}$ for every integer $i \in [1, N]$,*
- *the s_i is global section of $\mathcal{O}_{\overline{\mathcal{A}}_{g,\Gamma}}(D_i)$ such that $D_i = \{s_i = 0\}$,*
- *the $F_\sigma \in \mathbb{Z}[x_1, \ldots, x_N]$ is a homogenous polynomial of degree g and the coefficients of F_σ are integers dependent only on Γ and σ together with marking order of edges.*

4 Applications of the new formulation of volume form in Algebraic Geometry

We have a further observation: Whether $D_\infty := \overline{\mathcal{A}}_{g,\Gamma} \setminus \mathcal{A}_{g,\Gamma}$ is normal crossing or not, we always have a partial compactification given by the diagram

$$\begin{array}{ccc} \mathfrak{H}_g & \xrightarrow{\subset} & U^{\mathfrak{F}_0}(\mathbb{C}) \cong \mathbb{C}^n \\ \downarrow & & \downarrow {\scriptstyle w_i := \exp(2\pi\sqrt{-1} l_i(z))} \\ \mathfrak{H}_g/\Gamma \cap U^{\mathfrak{F}}(\mathbb{Q}) & \xrightarrow{\subset} & (\mathbb{C}^*)^n \end{array}$$

where $\{l_i\}_{i=1}^n$ is the dual base of edges of a regular polyhedral cone σ. Then, the $(w_1, \ldots, w_n)$ is always a local coordinate system of the partial compactification even though it can not be a local coordinate system of $\overline{\mathcal{A}}_{g,\Gamma}$. The quotient manifold $\Gamma \cap U^{\mathfrak{F}}(\mathbb{Q}) \backslash \mathfrak{H}_g$ also has an induced Kähler-Einstein metric with volume form

$$\Phi_\sigma = \frac{(\frac{\sqrt{-1}}{2})^n 2^{\frac{g(g-1)}{2}} \mathrm{vol}_\Gamma(\sigma)^2 \bigwedge_{1 \le i \le n} dw_i \wedge d\overline{w_i}}{(\prod_{1 \le i \le n} |w_i|^2)(F_\sigma(\log |w_1|, \ldots, \log |w_n|))^{g+1}}.$$

Due to the Kähler-Einstein metric on $\Gamma \cap U^{\mathfrak{F}}(\mathbb{Q}) \backslash \mathfrak{H}_g$, the function $H := -\log F_\sigma$ must satisfy the following elliptic Monge-Ampére equation

$$\det \left(\frac{\partial^2 H}{\partial x_i \partial x_j} \right)_{i,j} = 2^{\frac{g(g-1)}{2}} \mathrm{vol}_\Gamma(\sigma)^2 \exp((g+1)H) \qquad (4.0.1)$$

on the domain $\{(x_1, \ldots, x_n) \in \mathbb{R}^n \mid x_i \leq -C < 0, \forall i\}$.

Example 4.1. ([Wang93]) Let $g = 2$ and $\Gamma = \Gamma(n) := \{\gamma \in \mathrm{Sp}(2, \mathbb{Z}) \mid \gamma \equiv I_4 \mod n\}$. Any minimal cusp of the Baily-Boral compactification $\overline{\mathcal{A}}_{2,\Gamma}$ is a point. Fix a minimal cusp $\mathfrak{F}_0$, the positive convex cone $C(\mathfrak{F}_0)$ is the following set

$$\{u \mid u \text{ is a } (2 \times 2) \text{ positive real matrix }\}.$$

In the rational closure of the positive convex cone, we choose a top-dimensional polyhedral lattice cone(**principal cone** by J. Igusa)

$$\sigma = \left\{ \sum_{1 \leq i \leq j \leq 2} \lambda_{ij} \zeta_{ij}^n | \lambda_{ij} \geq 0 \right\}$$

where

$$(\star) \qquad \begin{cases} \zeta_{i,i}^n = I_4 + nE_{i,i}, & \text{for } 1 \leq i \leq 2; \\ \zeta_{i,j}^n = I_4 + n(-E_{i,j} - E_{j,i} + E_{i,i} + E_{j,j}), & \text{for } 1 \leq i < j \leq 2. \end{cases}$$

Here e.g. $E_{1,1} = \begin{pmatrix} 1 & 0 \\ 0 & 0 \end{pmatrix}$ and $E_{1,2} = \begin{pmatrix} 0 & 1 \\ 0 & 0 \end{pmatrix}$.

The global compactification $\overline{\mathcal{A}}_{2,\Gamma}$ by Igusa's decomposition does not have normal crossing D_∞ on global compactification, but the local coordinate chart of the corresponding partial compactification associated to σ has a normal crossing boundary divisor $\bigcup_{i=1}^{3} \{w_i = 0\}$, the volume form in this partial compactification can be represented as

$$\Phi_{2,\Gamma} = \left(\frac{\sqrt{-1}}{2}\right)^3 \frac{2 \bigwedge_{1 \leq i \leq 3} dw_i \wedge d\overline{w_i}}{(\prod_{1 \leq i \leq 3} |w_i|^2) F(\log|w_1|, \log|w_2|, \log|w_3|)^3},$$

where $F(X, Y, Z) = XY + XZ + YZ$.

Certainly, the function F in Wang's example satisfies the Monge-Ampére equation.

This volume form formula implies that any intrinsic Kähler-Einstein metric(or Hodge metric) on an quasi-projective manifold M should be helpful for finding a nice compactification $\overline{M}$ of M. This is a mysterious and interesting phenomenon.

There is an interesting application of our volume form.

For any projective smooth toroidal compactification $\overline{\mathcal{A}}_{g,\Gamma}$ with normal crossing boundary divisor D_∞, the following results are well known:

a) $\overline{\mathcal{A}}_{g,\Gamma}$ is a resolution of the Satake-Baily-Borel compactification $\mathcal{A}^*_{g,\Gamma}$ ([AMRT]).

b) The logarithmical canonical line bundle $\omega_{\overline{\mathcal{A}}_{g,\Gamma}}(D_\infty)$ is the pull back of the ample line bundle $\omega_{\mathcal{A}^*_{g,\Gamma}}$ on $\mathcal{A}^*_{g,\Gamma}$(cf. [Mum77]). So that the line bundle $\omega_{\overline{\mathcal{A}}_{g,\Gamma}}(D_\infty)$ is big and nef on $\overline{\mathcal{A}}_{g,\Gamma}$.

To understand compactifications more, we try to study properties of $\omega_{\overline{\mathcal{A}}_{g,\Gamma}}(D_\infty)$ and the degeneration of $\omega_{\mathcal{A}_{g,\Gamma}}$. We obtain that the divisor $K_{\overline{\mathcal{A}}_{g,\Gamma}} + D_\infty$ is never ample on $\overline{\mathcal{A}}_{g,\Gamma}$ by studying the intersection numbers

$$(K_{\overline{\mathcal{A}}_{g,\Gamma}} + D_\infty)^m \cdot D_1 \dots D_{n-m}$$

for any irreducible components $D_1, \dots, D_{n-m}$ of D_∞.

Consider the stratification of the boundary $D_\infty = \bigcup_{j=1}^{m} D_j$, define

$$Y_\infty := \bigcup_{D_j \neq Y} (D_j \cap Y) \text{ and } Y^* := Y \setminus Y_\infty$$

for every irreducible component Y of D_∞. Let $\Psi := (\partial\overline{\partial} \log \Phi_{g,\Gamma})^p$. Here are interesting observations:

- In sense of limit, the
$$\mathrm{Res}_Y((\partial\overline{\partial} \log \Phi_{g,\Gamma})^p) := \Psi|_{Y^*}$$
is well defined for any irreducible component Y of D_∞. The form
$$\mathrm{Res}_Y((\partial\overline{\partial} \log \Phi_{g,\Gamma})^p)$$
is a closed smooth form on Y^*.
- For each irreducible component Y of D_∞, the form $\mathrm{Res}_Y((\partial\overline{\partial} \log \Phi_{g,\Gamma})^p)$ has Poincaré growth on the simple normal crossing divisor Y_∞ of Y so that
$$\left(\frac{\sqrt{-1}}{2\pi}\right)^p \mathrm{Res}_Y((\partial\overline{\partial} \log \Phi_{g,\Gamma})^p)$$
becomes a positive closed current on Y.

Example 4.2. Take a local coordinate $\{(U, (w_1, \dots, w_N))\}$ of $\overline{\mathcal{A}}_{g,\Gamma}$ such that

$$U \cap D_\infty = \bigcup_{i=1}^{N} \{w_i = 0\} \text{ and } D_1 \cap U = \{w_1 = 0\}.$$

On $U \setminus D_\infty$, the volume form $\Phi_{g,\Gamma}$ becomes

$$\Phi_{g,\Gamma} = \frac{(\frac{\sqrt{-1}}{2})^N \bigwedge_{1\le i\le N} dw_i \wedge d\overline{w_i}}{(\prod_{1\le i\le N} |w_i|^2)(F(\log|w_1|, \dots, \log|w_N|))^{g+1}},$$

where $F \in \mathbb{R}[x_1, \dots, x_N]$ is a homogenous polynomial in of degree g. On $U \setminus D_\infty$, there is

$$\partial\overline{\partial} \log \Phi_{g,\Gamma} = \frac{-(g+1)}{4} \sum_{1\le i,j\le N} \frac{T_{i,j}(\log|w_1|, \dots, \log|w_N|)}{F^2(\log|w_1|, \dots, \log|w_N|)} \frac{dw_i \wedge d\overline{w_j}}{w_i \overline{w_j}},$$

where $T_{i,j} := F\frac{\partial^2 F}{\partial x_i \partial x_j} - \frac{\partial F}{\partial x_i}\frac{\partial F}{\partial x_j}$. Write

$$F = S(x_2, \ldots, x_N)x_1^m + \text{terms of lower degree of } x_1,$$

where $m := \deg_{x_1} F(x_1, \ldots, x_N)$. On the set $V := \{w_1 = 0\} \setminus \bigcup_{i \neq 1} \{w_i = 0\}$,we have

$$\begin{aligned}\mathrm{Res}_{D_1}(\partial\overline{\partial}\log\Phi_{g,\Gamma}) &= \frac{-(g+1)}{4}\sum_{2\leq i,j\leq N}\frac{T_{i,j}(\log|w_1|, \ldots, \log|w_N|)}{F^2(\log|w_1|, \ldots, \log|w_N|)}|_V \frac{dw_i \wedge d\overline{w_j}}{w_i\overline{w_j}}\\ &= \frac{-(g+1)}{4}\sum_{2\leq i,j\leq N}\frac{T^{(1)}_{i,j}(\log|w_2|, \ldots, \log|w_N|)}{S^2(\log|w_2|, \ldots, \log|w_N|)}\frac{dw_i \wedge d\overline{w_j}}{w_i\overline{w_j}}\end{aligned}$$

where $T^{(1)}_{i,j} := S\frac{\partial^2 S}{\partial x_i \partial x_j} - \frac{\partial S}{\partial x_i}\frac{\partial S}{\partial x_j}$ for $2 \leq i, j \leq N$. Moreover, if $g = 2$ then

$$\mathrm{Res}_{D_1}((\partial\overline{\partial}\log\Phi_{g,\Gamma})^2) = (\mathrm{Res}_{D_1}(\partial\overline{\partial}\log\Phi_{g,\Gamma}))^2 = 0$$

since $\deg S \leq 1$ and the matrix $(T^{(1)}_{i,j})$ is at most of rank one.

Let $D_1, \ldots, D_d$ ($d < n$) be any different irreducible components of D_∞. By Mumford's goodness along D_∞([Mum77]) and some arguments of Kollár in [Kol87], we have

$$\begin{aligned}&(K_{\overline{\mathcal{A}}_{g,\Gamma}} + D_\infty)^{n-d} \cdot D_1 \ldots D_d\\ &= \int_{\mathcal{A}_{g,\Gamma}} c_1([K_{\mathcal{A}_{g,\Gamma}}], h_{KE})^{n-d}\Big(\bigwedge_{1\leq j\leq d} c_1([D_j], ||\cdot||_j)\Big)\\ &= \int_{\mathcal{A}_{g,\Gamma}} \left(\frac{\sqrt{-1}}{2\pi}\partial\overline{\partial}\log\Phi_{g,\Gamma}\right)^{n-d}\Big(\bigwedge_{1\leq j\leq d} c_1([D_j], ||\cdot||_j)\Big)\\ &= \int_{D_i} \mathrm{Res}_{D_i}\left(\left(\frac{\sqrt{-1}}{2\pi}\partial\overline{\partial}\log\Phi_{g,\Gamma}\right)^{n-d}\right) \wedge \Big(\bigwedge_{1\leq j\leq d, j\neq i} c_1([D_j], ||\cdot||_j)\Big)\end{aligned}$$

for any i, where $||\cdot||_j$ is an arbitrary Hermitian metric on the line bundle $[D_j]$. By recursion, we obtain the following result.

Theorem 4.3. ([YZ12]) *Let $\overline{\mathcal{A}}_{g,\Gamma}$ be any Mumford toroidal compactification of the smooth Siegel variety $\mathcal{A}_{g,\Gamma}$ such that boundary divisor $D_\infty = \overline{\mathcal{A}}_{g,\Gamma} \setminus \mathcal{A}_{g,\Gamma}$ is normal crossing. Then, we have:*

$$\begin{aligned}&(K_{\overline{\mathcal{A}}_{g,\Gamma}} + D_\infty)^{\dim_{\mathbb{C}}\mathcal{A}_{g,\Gamma}-d} \cdot D_1 \cdots D_d\\ &= \int_{\bigcap_{k=1}^{d} D_k} \mathrm{Res}_{D^{(d)}}\left(\mathrm{Res}_{D^{(d-1)}} \cdots \left(\mathrm{Res}_{D^{(1)}}\left(\left(\frac{\sqrt{-1}}{2\pi}\partial\overline{\partial}\log\Phi_{g,\Gamma}\right)^{\dim_{\mathbb{C}}\mathcal{A}_{g,\Gamma}-d}\right)\right)\cdots\right),\end{aligned}$$

where $D^{(1)} := D_1, D^{(2)} := D^{(1)} \cap D_2, \ldots, D^{(d)} := D^{(d-1)} \cap D_d$. In particular, if $d \geq g - 1$ then the intersection number

$$(K_{\overline{\mathcal{A}}_{g,\Gamma}} + D_\infty)^{N-d} \cdot D_1 \cdots D_d = 0.$$

There are many interesting applications of our volume form in Algebraic Geometry, for example, the applications to intersection cohomology and L^2 cohomology of Siegel variety and new proof of Kazdan's theorem on arithmetic variety, they will appear in further.

References

[AMRT] A. Ash, D. Mumford, M. Rapoport and Y. Tai: *Smooth compactifications of locally symmetric varieties. Second edition. With the collaboration of Peter Scholze.* Cambridge Mathematical Library. Cambridge University Press, Cambridge, 2010.

[BB66] W.L. Baily and A. Borel: *Compactification of arithmetic quotients of bounded symmetric domains.* Ann. of Math. **84** (1966), 442–528.

[CKS86] E. Cattani, A. Kaplan and W. Schmid: *Degeneration of Hodge structures.* Ann. Math. **38** (1986), 457–535.

[CY80] S.-Y. Cheng and S.-T. Yau: *On the existence of a complete Kähler metric on noncompact complex manifolds and the regularity of Fefferman's equation.* Comm. Pure Appl. Math. **33** (1980), no. 4, 507–544.

[Del71] P. Deligne: *Théorie de Hodge II.* I.H.É.S. Publ. Math. **40** (1971), 5–57.

[FC] G. Faltings and C.-L. Chai: *Degeneration of abelian varieties. With an appendix by David Mumford.* Ergebnisse der Mathematik und ihrer Grenzgebiete(3) [Results in Mathematics and Related Areas(3)], **22**, Springer-Verlag, Berlin, 1990.

[Kol87] J. Kollár: *Subadditivity of the Kodaira dimension: Fibres of general type.* Algebraic Geometry, Sendai, 1985. Advanced Studies in Pure Mathematics **10** (1987), 361–398.

[Mum77] D. Mumford: *Hirzebruch's proportionality theorem in the non-compact case.* Inventiones math. **42** (1977), 239–272.

[TY86] G. Tian and S.-T. Yau: *Existence of Kähler-Einstein metrics on complete Kähler manifolds and their applications to algebraic geometry.* Mathematical aspects of string theory(San Diego, Calif., 1986), 574–628, Adv. Ser. Math. Phys., **1**, World Sci. Publishing, Singapore, 1987.

[Wang93] W. Wang: *On the smooth compactification of Siegel spaces.* J. Differential Geom. **38** (1993), no. 2, 351–386.

[Yau78] S.-T. Yau: *On the Ricci curvature of a compact Kähler manifold and the complex Monge-Ampère equation*, I, Comm. Pure Appl. Math. **31** (1978), 339–411.

[Yau80] S.-T. Yau: *The role of partial differential equations in differential geometry.* Proceedings of the International Congress of Mathematicians (Helsinki, 1978). pp. 237–250, Acad. Sci. Fennica, Helsinki, 1980.

[Yau93] S.-T. Yau: *A splitting theorem and an algebraic geometric characterization of locally Hermitian symmetric spaces.* Comm. Anal. Geom. **1** (1993), no. 3–4, 473–486.

[YZ12] S.-T. Yau and Y. Zhang: *The geometry on smooth toroidal compactifications of Siegel varieties.* The American Journal of Mathematics **136** (2014), no.4, 859–942.

[YZ12-2] S.-T. Yau and Y. Zhang: *Hodge bundles on smooth compactifications of Siegel varieties and applications.* arXiv:1201.3784.

School of Mathematical Sciences, Fudan University
Shanghai 200433, China
E-mail address: zhangyi_math@fudan.edu.cn

Made in the USA
Monee, IL
16 March 2026

46289701R00326